Carpenter
210 Stratford Drive
Chapel Hill, NC 27516

CLINICAL TRIALS

A Methodologic Perspective

Second Edition

STEVEN PIANTADOSI
Johns Hopkins School of Medicine, Baltimore, MD

WILEY-INTERSCIENCE

A JOHN WILEY & SONS, INC., PUBLICATION

Published by John Wiley & Sons, Inc., Hoboken, New Jersey.
Published simultaneously in Canada.

For general information on our other products and services please contact our Customer Care Department
within the U.S. at 877-762-2974, outside the U.S. at 317-572-3993 or fax 317-572-4002.

Wiley also publishes its books in a variety of electronic formats. Some content that appears in print,
however, may not be available in electronic format.

Library of Congress Cataloging-in-Publication Data:

Piantadosi, Steven.
   Clinical trials : a methodologic perspective / Steven Piantadosi—2nd ed.
      p. ; cm. – (Wiley series in probability and statistics)
   Includes bibliographical references and index.
   ISBN-13: 978-0-471-72781-1 (cloth : alk. paper)
   ISBN-10: 0-471-72781-4 (cloth : alk. paper)
   1. Clinical trials—Statistical methods. I. Title. II. Series.
   [DNLM: 1. Biomedical Research—methods. 2. Research—methods. 3. Clinical
   Trials—methods. 4. Statistics—methods. W 20.5 P581c 2005]
   R853.C55P53 2005
   610'.72—dc22
                                                              2005040833

Printed in the United States of America.

10 9 8 7 6 5 4 3 2 1

CONTENTS

PREFACE

A respectable period of time has passed since the first edition of this book, during which several pressures have necessitated a second edition. Most important, there were the inadvertent errors needing correction. The field has also had some significant changes, not so much in methodology perhaps as in context, regulation, and the like. I can say some things more clearly now than previously because many issues are better defined or I care more (or less) about how the discussion will be received.

I have added much new material covering some gaps (and therefore some new mistakes), but always hoping to make learning easier. The result is too much to cover in the period that I usually teach—one academic quarter. It may be appropriate for a semester course. Many students tell me that they consult this book as a reference, so the extra material should ultimately be useful.

Many colleagues have been kind enough to give their valuable time to review drafts of chapters, despite apparently violating Nabokov's advice: "Only ambitious nonentities and hearty mediocrities exhibit their rough drafts. It is like passing around samples of one's sputum." Nevertheless, I am grateful to Elizabeth Garrett-Mayer, Ph.D., Bonnie Piantadosi, M.S.W, M.H.S., Anne Piantadosi, Irene Roach, Pamela Scott, Ph.D., Gail Weinmann, M.D., and Xiaobu Ye, M.D., M.S. for such help. Chris Szekely, Ph.D. reviewed many chapters and references in detail. Sean Roach, M.L.S. and Alisa Moore provided valuable assistance with references. Alla Guseynova, M.S. reviewed and corrected the computational code for the book.

As usual, students in my class *Design and Analysis of Clinical Trials* offered through the Hopkins Department of Biostatistics and the Graduate Training Program in Clinical Investigation provide the best motivation for writing this book by virtue of their lively discussion and questions. I would also like to thank the faculties, students, and sponsors from the AACR/ASCO Vail Workshop, as well as the sister workshops, FECS in Flims, Switzerland, and ACORD in Cairns, Australia, who have provided many practical questions and examples for interesting clinical trial designs over the years. Such teaching venues require conciseness and precision from a clinical trialist, and illustrate the very heart of collaborative research.

Finally there are two institutional venues that have helped motivate and shape my writing regarding clinical trials. One is the Protocol Review and Monitoring Committee in the Cancer Center, a scientific review forum on which I have served for many years. The second is the Institutional Review Board, on which I have more recently begun to serve. My colleagues in both of these settings encourage and take a careful, constructive, and detailed view of a multitude of diverse trials, and have taught me a great deal.

STEVEN PIANTADOSI
Baltimore, Maryland, 2005

Books are fatal: they are the curse of the human race. Nine-tenths of existing books are nonsense, and the clever books are the refutation of that nonsense. The greatest misfortune that ever befell man was the invention of printing. [Benjamin Disraeli]

Writing is an adventure. To begin with, it is a toy and an amusement. Then it becomes a mistress, then it becomes a master, then it becomes a tyrant. The last phase is that just as you are about to be reconciled to your servitude, you kill the monster and fling him to the public. [Winston Churchill]

Writing is easy; all you do is sit staring at a blank sheet of paper until the drops of blood form on your forehead. [Gene Fowler]

PREFACE TO THE FIRST EDITION

In recent years a great deal has been written about clinical trials and closely related areas of biostatistics, biomathematics, biometry, epidemiology, and clinical epidemiology. The motive for writing this book is that there still seems to be a need among both physicians and biostatisticians for direct, relevant accounts of basic statistical methods in clinical trials. The need for both trialists and clinical investigators to learn about good methodology is particularly acute in oncology, where investigators search for treatment advances of great clinical importance, but modest size relative to the variability and bias which characterize studies of human disease. A similar need with the same motivation exists in many other diseases.

On the medical side of clinical trials, the last few years have seen a sharpened focus on training of clinical investigators in research methods. Training efforts have ranged from short intensive courses to research fellowships lasting years and culminating in a postgraduate degree. The evolution of teaching appears to be toward defining a specialty in clinical research. The material in this book should be of interest to those who take this path. The technical subjects may seem difficult at first, but the clinician should soon become comfortable with them.

On the biostatistical side of clinical trials, there has been a near explosion of methods in recent years. However, this is not a book on statistical theory. Readers with a good foundation in biostatistics should find the technical subjects practical and quite accessible. It is my hope that such students will see some cohesiveness to the field, fill in gaps in their knowledge, and be able to explore areas such as ethics and misconduct that are important to clinical trials.

There are some popular perceptions about clinical trials to which this book does not subscribe. For example, some widely used terminology regarding trials is unhelpful and I have attempted to counteract it by proposing alternatives. Also noncomparative trial designs (e.g., early developmental studies) are often inappropriately excluded from discussions of methods. I have tried to present concepts that unify all designed studies rather than ideas that artificially distinguish them. Dealing with pharmacokinetic-based designs tends to complicate some of the mathematics, but the concepts are essential for understanding these trials.

The book is intended to provide at least enough material for the core of a half-semester course on clinical trials. In research settings where trials are used, the audience for such a course will likely have varying skills and directions. However, with a background of basic biostatistics, an introductory course in clinical trials or research methods, and appropriate didactic discussion, the material presented here should be useful to a heterogeneous group of students.

Many individuals have contributed to this book in indirect, but significant ways. I am reminded of two colleagues, now deceased, who helped to shape my thinking about clinical trials. David P. Byar, M.D. nurtured my early academic and quantitative interest in clinical trials in the early 1980s at the National Institutes of Health. After I joined the Johns Hopkins School of Medicine, Brigid G. Leventhal, M.D. showed me a mature, compassionate, and rigorous view of clinical trials from a practitioner's perspective. I hope the thoughts in this book reflect some of the good attitudes and values of these fine scholars.

Other colleagues have taught me many lessons regarding clinical trials through their writings, lectures, conversations, and willingness to answer many questions. I would particularly like to thank Mitchell H. Gail, M.D., Ph.D. and Curtis L. Meinert, Ph.D. for much valuable advice and good example over the years. One of the most worthwhile experiences that a trial methodologist can have is to review and influence the designs of clinical trials while they are being developed. My colleagues at Johns Hopkins have cooperated in this regard through the Oncology Center's Clinical Research Committee, especially Hayden Braine, M.D., who has chaired the Committee wisely for many years.

Through long-standing collaborations with the Lung Cancer Study Group, I met many clinical scholars with much to say and teach about trials. I would like to thank them, especially E. Carmack Holmes, M.D., John C. Ruckdeschel, M.D., and Robert Ginzberg, M.D. for showing me a model of interdisciplinary collaboration and friendship, which continued to outlive the financial arrangements. Recent collaborations with colleagues in the New Approaches to Brain Tumor Therapy Consortium have also enhanced my appreciation and understanding of early developmental trials.

In recent months many colleagues have assisted me by reading and offering comments on drafts of the chapters that follow. For this help, I would like to thank Lina Asmar, Ph.D., Tatiana Barkova, Ph.D., Jeanne DeJoseph, Ph.D., C.N.M., R.N., Suzanne Dibble, D.N.Sc., R.N., James Grizzle, Ph.D., Curt Meinert, Ph.D., Mitch Gail, M.D., Ph.D., Barbara Hawkins, Ph.D., Steven Goodman, M.D., Ph.D., Cheryl Enger, Ph.D., Guanghan Liu, Ph.D., J. Jack Lee, Ph.D., Claudia Moy, Ph.D., John O'Quigley, Ph.D., Thomas F. Pajak, Ph.D., Charles Rohde, Ph.D., Barbara Starklauf, M.A.S., Manel C. Wijesinha, Ph.D., and Marianna Zahurak, M.S. Many of the good points belong to them—the errors are mine.

Students in my classes on the *Design and Analysis of Clinical Trials* and *Design of Experiments* at the Johns Hopkins School of Hygiene and Public Health have contributed to this book by working with drafts, making helpful comments, working problems, or just discussing particular points. I would especially like to thank Maria Deloria, Kathleen Weeks, Ling-Yu Ruan, and Jeffrey Blume for their input. Helen Cromwell and Patty Hubbard have provided a great deal of technical assistance with the preparation of the manuscript. Gary D. Knott, Ph.D., Barry J. Bunow, Ph.D., and the staff at Civilized Software, Bethesda, Maryland (www.civilized.com) furnished me with MLAB software, without which many tasks in the following pages would be difficult.

This book was produced in LaTeX using *Scientific Workplace* version 2.5. I am indebted to Kathy Watt of TCI Software Research in Las Cruces, New Mexico, (www.tcisoft.com) for assistance in preparing the style. A special thanks goes to Irene Roach for editing an early draft of the manuscript and to Sean Roach for collecting and translating hard-to-find references.

It is not possible to write a book without stealing a large amount of time from one's family. Bonnie, Anne L., and Steven T. not only permitted this to happen but also were supportive, helpful, and understood why I felt it was necessary. I am most grateful to them for their patience and understanding throughout this project.

STEVEN PIANTADOSI
Baltimore, Maryland,
March 1997

1

PRELIMINARIES

1.1 INTRODUCTION

The best time to contemplate the quality of evidence from a clinical trial is before it begins. High-quality evidence about the effects of a new treatment is a consequence of good study design and execution, which themselves are the results of careful planning. This book attempts to acquaint investigators with ideas of design methodology that are helpful in planning, conducting, analyzing, and assessing clinical trials. The discussion covers a number of subjects relevant to early planning and design; some that find general agreement among methodologists and others that are contentious. It is unlikely that a reader experienced with clinical trials will agree with all that I say or what I choose to emphasize, but my perspective should be mainstream, internally consistent, and useful for learning.

This book is not intended to be an introduction to clinical trials. It should be part of a one- or two-quarter second structured postgraduate course for an audience with quantitative skills and a biological focus. The first edition evolved over a dozen years from the merging of two courses: one in experimental design and one in clinical trials. This second edition is the result of seven additional years of teaching and concomitant changes in the field. The book assumes a working knowledge of basic biostatistics and some familiarity with clinical trials, either didactic or practical. It is also helpful if the reader understands some more advanced statistical concepts, especially lifetables, survival models, and likelihoods. I recognize that clinicians often lack this knowledge. However, many contemporary medical researchers are seeking the required quantitative background through formal training in clinical investigation methods or experimental therapeutics. No clinical knowledge is needed to understand the concepts in this book, although it will be helpful throughout.

Many readers of this book will find the discussion uneven, ranging from basic to technically complex. This is partly a consequence of the very nature of clinical trials and

Clinical Trials: A Methodologic Perspective, 2E, by S. Piantadosi
Copyright © 2005 John Wiley & Sons, Inc.

partly the result of trying to address a heterogeneous population of students. My classes typically contain an equal mixture of biostatistics graduate students, medical doctors in specialty or subspecialty training (especially working toward a degree in clinical investigation), and other health professionals training to be sophisticated managers or consumers of clinical trials. For such an audience the goal is to provide breadth and to write so as not to be misunderstood.

This book should be supplemented with lecture and discussion, and possibly a computer lab. The reader who does not have an opportunity for formal classroom dialogue will need to explore the references more extensively. Exercises and discussion questions are provided at the end of each chapter. Most are intentionally made open-ended, with a suggestion that the student answer them in the form of a one- or two-page memorandum, as though providing an expert opinion to less-experienced investigators.

1.2 AUDIENCE AND SCOPE

The audience for this book is clinical trialists. It is not a simple matter to define a clinical trialist, but operationally it is someone who is immersed in the science of trials. Being a truly interdisciplinary field, trialists can be derived from a number of sources: (1) quantitative or biostatistical, (2) administrative or managerial, (3) clinical, or (4) ethical. Therefore students can approach the subject primarily from any of these perspectives.

It is common today for rigorous trialists to be strongly statistical. This is because of the fairly rapid recent pace of methods for clinical trials coming from that field, and also because statistics pertains to all of the disciplines in which trials are conducted. However, the discussion in this book does not neglect the other viewpoints that are also essential to understanding trials. Many examples will relate to cancer because that is the primary field in which I work, but the concepts will generalize to other areas.

Scientists who specialize in clinical trials are frequently dubbed "statisticians." I will sometimes use that term with the following warning regarding rigor: statistics is an old and broad profession. There is not a one-to-one correspondence between statisticians or biostatisticians and knowledge of clinical trials. However, trial methodologists, whether statisticians or not, are likely to know a lot about biostatistics and will be accustomed to working with statistical experts. Many trial methodologists are not statisticians at all, but evolve from epidemiologists or clinicians with a strongly quantitative orientation, as indicated above.

I have made an effort to delineate and emphasize principles common to all types of trials: translational, developmental, safety, comparative, and large-scale studies. This follows from a belief that it is more helpful to learn about the similarities among trials rather than differences. However, it is unavoidable that distinctions must be made and the discussion tailored to specific types of studies. I have tried to keep such distinctions, which are often artificial, to a minimum. Various clinical contexts also treat trials differently, a topic discussed briefly in Chapter 4.

There are many important aspects of clinical trials not covered here in any detail. These include administration, funding, conduct, quality control, and the considerable infrastructure necessary to conduct trials. These topics might be described as the technology of trials, whereas my intent is to focus on the science of trials. Technology is vitally important, but falls outside of the scope of this book. Fortunately there are excellent sources for this material.

No book can be a substitute for regular interaction with a trial methodologist during both the planning stages of a clinical investigation and its analysis. I do not suggest passive reliance on such consultations, but intend to facilitate disseminating knowledge from which true collaborations between clinicians and trialists will result. Although many clinicians think of bringing their final data to a statistician, a collaboration will be most valuable during the design phase of a study when an experienced trialist may prevent serious methodologic errors, help streamline a study, or suggest ways to avoid costly mistakes.

The wide availability of computers is a strong benefit for clinical researchers, but presents some dangers. Although computers facilitate efficient, accurate, and timely keeping of data, modern software also permits or encourages researchers to produce "statistical" reports without much attention to study design and without fully understanding assumptions, methods, limitations, and pitfalls of the procedures being employed. Sometimes a person who knows how to run procedure-oriented packages on computerized data is called the "statistician," even though he or she might be a novice at the basic theory underlying the analyses. It then becomes possible to produce a final report of a trial without the clinical investigator understanding the limitations of analysis and without the analyst being conversant with the data. What a weak chain this is.

The ideas in this book are intended to counteract these tendencies, not by being old-fashioned but by being rigorous. Good design inhibits errors by involving a statistical expert in the study as a collaborator from the beginning. Most aspects of the study will improve as a result, including reliability, resource utilization, quality assurance, precision, and the scope of inference. Good design can also simplify analyses by reducing bias and variability and removing the influence of complicating factors. In this way number crunching becomes less important than sound statistical reasoning.

The student of clinical trials should also understand that the field is growing and changing in response to both biological and statistical developments. A picture of good methodology today may be inadequate in the near future. This is probably more true of analytic methods than design, where the fundamentals will change more slowly. Analysis methods often will depend on new statistical developments or theory. These in turn depend on (1) computing hardware, (2) reliable and accessible software, (3) training and re-training of trialists in the use of new methods, (4) acceptance of the procedure by the statistical and biological communities, and (5) sufficient time for the innovations to diffuse into practice.

It is equally important to understand what changes or new concepts do not improve methodology but are put forward in response to non-science issues or because of creeping regulation. The best recent example of this is the increasing sacrifice of expertise in favor of objectivity in the structure and function of clinical trial monitoring (discussed in Chapter 14). Such practices are sometimes as ill considered as they are well meaning, and may be promulgated by sponsors without peer review or national consensus.

Good trial design requires a willingness to examine many alternatives within the confines of reliably answering the basic biological question. The most common errors related to trial design are devoting insufficient resources or time to the study, rigidly using standard types of designs when better (e.g., more efficient) designs are available, or undoing the benefits of a good design with a poorly planned (or executed) analysis. I hope that the reader of this book will come to understand where there is much flexibility in the design and analysis of trials and where there is not.

1.3   OTHER SOURCES OF KNOWLEDGE

The periodical literature related to clinical trials is large. I have attempted to provide current useful references for accessing it in this book. Aside from individual study reports in many clinical journals, there are some periodicals strongly related to trials. One is *Controlled Clinical Trials*, which has been the official journal of the Society for Clinical Trials (SCT) (mostly a U.S. organization). The journal was begun in 1980 and is devoted to trial methodology. The SCT was founded in 1981 and its 1500 members meet yearly. In 2003 the SCT changed its official journal to *Clinical Trials*, the first issue of which appeared in January 2004. This reincarnated journal should be an excellent resource for trialists. A second helpful periodical source is *Statistics in Medicine*, which frequently has articles of interest to the trialist. It began publication in 1982 and is the official publication of the International Society for Clinical Biostatistics (mostly a European organization). These two societies have begun joint meetings every few years.

Many papers of importance to clinical trials and related statistical methods appear in various other applied statistical and clinical journals. Reviews of many methods are published in *Statistical Methods in Medical Research*. One journal of particular interest to drug development researchers is the *Journal of Biopharmaceutical Statistics*. A useful general reference source is the journal *Biostatistica*, which contains abstracts from diverse periodicals. Statistical methodology for clinical trials appears in several journals. The topic was reviewed with an extensive bibliography by Simon (1991). A more extensive bibliography covering trials broadly has been given by Hawkins (1991).

In addition to journals there are a number of books and monographs dealing with clinical trials. The text by Meinert (1986) is a practical view of the infrastructure and administrative supports necessary to perform quality trials, especially randomized controlled trials. Freidman, Furberg, and DeMets (1982) and Pocock (1996) also discuss many conceptual and practical issues in their excellent books, which do not require extensive statistical background. A nice encyclopedic reference regarding statistical methods in clinical trials is provided by Redmond and Colton (2001). There is a relatively short and highly readable methodology book by Silverman (1985), and a second more issue oriented one (Silverman, 1998) with many examples. Every trialist should read the extensive work on placebos by Shapiro and Shapiro (1997).

In recent years many Web-based sources of information regarding clinical trials have been developed. The quality, content, and usefulness are highly variable and the user must consider the source when browsing. Resources of generally high quality that I personally find useful are listed in Table 1.1. My list is probably not complete with regard to specialized needs, but it provides a good starting point.

In the field of cancer trials, Buyse, Staquet, and Sylvester (1984) is an excellent source, although now becoming slightly dated. A contemporary view of cancer trials is given by Girling et al. (2003). The book by Leventhal and Wittes (1988) is useful for its discussion of issues from a strong clinical orientation. A very readable book with many good examples is that by Green, Benedetti, and Crowley (2002). In the field of AIDS, a useful source is Finkelstein and Schoenfeld (1995). The serious student should also be familiar with the classic papers by Peto et al. (1977a, b).

Spilker (e.g., 1993) has written a large volume of material about clinical trials, much of it oriented toward pharmaceutical research and not statistical methods. Another useful reference with an industry perspective is Wooding (1994). Data management

TABLE 1.1   Some Web Resources for Clinical Trials Information

Link	Description
assert-statement.org	A standard for the scientific and ethical review of trials: a structured approach for ethics committees reviewing randomized controlled clinical trials.
cochrane.org	The Cochrane Collaboration provides up-to-date information about the effects of health care.
clinicaltrials.gov	Provides information about federally and privately supported clinical research.
consort-statement.org	CONSORT statement: an evidence-based tool to improve the quality of reports of randomized trials.
jameslindlibrary.org	Evolution of fair tests of medical treatments; examples from books and journal articles including key passages of text.
icmje.org	Uniform requirements for manuscripts submitted to biomedical journals
gpp-guidelines.org	Encourages responsible and ethical publication of clinical trials sponsored by pharmaceutical companies
ncbi.nlm.nih.gov/entrez/query.fcgi	PubMed: includes over 14 million citations for biomedical articles back to the 1950's
mcclurenet.com/ICHefficacy.html	ICH efficacy guidelines
controlled-trials.com	Current controlled trials: provides access to peer reviewed biomedical research

is an important subject for investigators, but falls outside the scope of this book. The subject is probably made more complex by the fact that vastly more data are routinely collected during developmental trials than are needed to meet the objectives. Good sources of knowledge concerning data management include the books by McFadden (1998) and that edited by Rondel, Varley, and Webb (1993), and an issue of *Controlled Clinical Trials* (April 1995) devoted to data management. Books on other relevant topics will be mentioned in context later.

Even in a very active program of clinical research, a relatively short exposure to the practical side of clinical trials cannot illustrate all the important lessons. This is because it may take years for any single clinical trial, and many such studies, to yield all of their information useful for learning about methodology. Even so, the student of clinical trials will learn some lessons more quickly by being involved in an actual study, compared with simply studying theory. In this book, I illustrate many concepts with published trials. In this way the reader can have the benefit of observing studies from a long-term perspective, which would otherwise be difficult to acquire.

1.3.1   Terminology

The terminology of clinical trials is not without its ambiguities. A recent firm effort has been made to standardize definitions in a dictionary devoted to clinical trial terminology (Meinert, 1996). Most of the terms within this book are used in a way consistent with such definitions. A notable exception is that I propose and employ explanatory alternatives to the widely used, uninformative, inconsistent, and difficult-to-generalize "phase I, II, III, or IV" designations for clinical trials. This topic is discussed in Chapter 6.

Much of the terminology of clinical trials has been derived directly from drug development. Because of the heavy use of clinical trials in the development of cytotoxic drugs for the treatment of cancer between the 1960s and the 1990s, the terminology for this setting has found its way inappropriately into other contexts. Drug development terminology is often ambiguous and inappropriate for clinical trials performed in many other areas, and is even unsuitable for many new cancer therapies that do not act through a direct cytotoxic mechanism. For this reason I have not employed this outdated terminology in this book and have used descriptive alternatives (Section 6.3.2).

1.3.2   Review of Notation and Terminology Is Helpful

There is no escaping the need for mathematical formalism in the study of clinical trials. It would be unreasonable, a priori, to expect mathematics to be as useful as it is in describing nature (Wigner, 1959). Nevertheless, it is, and the mathematics of probability is the particular area most helpful for clinical trials. Galileo said:

> The book of the universe is written in mathematical language, without which one wanders in vain through a dark labyrinth.

Statisticians light their dark labyrinth using abstract symbols (e.g., Greek letters) as a shorthand for important mathematical quantities and concepts. I will also use these symbols when appropriate in this book, because many ideas are troublesome to explain without good notation.

However, because this book is not oriented primarily toward statistical theory, the use of symbols will be minimal and tolerable, even to nonmathematical readers. A review and explanation of common usage of symbols consistent with the clinical trials literature is given in Appendix B. Because some statistical terms may be unfamiliar to some readers, definitions and examples are also listed in that chapter. Abbreviations used in the book are also explained there.

This book does not and cannot provide the technical statistical background that is needed to understand clinical trial design and analysis thoroughly. As stated above, much of this knowledge is assumed to be present. Help is available in the form of practical and readable references. Examples are the books by Armitage and Berry (1994) and Marubini and Valsecchi (1995). More concise summaries are given by Campbell and Machin (1990) and Everitt (1989). A comprehensive reference with good entries to the literature is the *Encyclopedia of Biostatistics* (Armitage and Colton, 1998). More specialized references will be mentioned later.

1.4   EXAMPLES, DATA, AND PROGRAMS

It is not possible to learn all the important lessons about clinical trials from classroom instruction or reading, nor is it possible for every student to be involved with actual trials as part of a structured course. This problem is most correctable for topics related to the analysis of trial results, where real data can be provided. For some examples and problems used in this book, the data are provided through the author's Web site (www.cancerbiostats.onc.jhmi.edu/). Throughout the book, I have made a concerted effort to provide examples of trials that are instructive but small, so as to be digestible by the student. Computerized data files and programs to read and analyze them are provided on the Web site. It also contains some sample size and related programs that are helpful for design calculations. More powerful sample size (and other) design software that is available commercially is discussed in Chapter 11.

Many, but not all, tables, figures, and equations in the text have been programmed in Mathematica, Version 5 (Wolfram, 2003). The computer code related to the book is available from the author's Web site. Mathematica, which is required to use the relevant programs, is commercially available. The stand-alone programs mentioned above and Mathematica code are made available for instructional purposes without warranty of any kind—the user assumes responsibility for all results.

1.5 SUMMARY

The premise of this book is that well-designed experimental research is a necessary basis for therapeutic development and clinical care decisions. The purpose of this book is to address issues in the methodology of clinical trials in a format accessible to interested statistical and clinical scientists. The audience is intended to be practicing clinicians, statisticians, trialists, and others with a need for understanding good clinical research methodology. The reader familiar with clinical trials will notice a few substantive differences from usual discussions, including the use of descriptive terms for types of trials and an even-handed treatment of different statistical perspectives. Examples from the clinical trials literature are used, and data and computer programs for some topics are available. A review of essential notation and terminology is also provided.

2

CLINICAL TRIALS AS RESEARCH

2.1 INTRODUCTION

In the late nineteenth and early twentieth century, therapeutics was in a state of nihilism. Nineteenth-century science had discovered that many diseases improved without therapy, and that many popular treatments, among them certain natural products and bloodletting, were ineffective. The nihilism was likely justifiable because it could be claimed that the entire history of therapeutics up to that point was essentially only the history of the placebo effect (Shapiro and Shapiro, 1997). However, when scientists showed that diseases like pellagra and diabetes could have their effects relieved with medicinals, belief in treatment began to return. Following the discovery of penicillin and sulfanilamide in the twentieth century, the period of nihilism ended (Thomas, 1977; Coleman, 1987).

More recently discovery of effective drugs for the treatment of cancer, cardiovascular disease, infections, and mental illness as well as the crafting of vaccines and other preventive measures have demonstrated the value of therapeutics. There is economic evidence and opinion to support the idea that the strong overall economic status of the United States is substantially due to improved health of the population (Funding First, 2000), itself dependent on effective public health and therapeutic interventions. Clinical investigation methods are important in the search for effective prevention agents and treatments, sorting out the benefits of competing therapies, and establishing optimum treatment combinations and schedules.

Experimental design and analysis have become essential because of the greater detail in modern biological theories and the complexities in treatments of disease. The clinician is usually interested in small, but biologically important, treatment effects that can be obscured by uncontrolled natural variation and bias in nonrigorous studies.

This places well-performed clinical trials at the very center of clinical research today, although the interest in small effect sizes also creates problems for other aspects of clinical investigation (Ahrens, 1992).

Other contemporary pressures also encourage the application of rigorous clinical trials. Societal expectations to relieve suffering through medical progress, governmental regulation of prescription drugs and devices, and the economics of pharmaceutical development all encourage or demand efficient and valid study design. Nearly all good clinical trials have basic biological, public health, and commercial value, encouraging investigators to design studies that yield timely and reliable results.

In the early twenty-first century the pendulum of nihilism has swung strongly in the opposite direction. Today there is belief in the therapeutic efficacy of many treatments. Traditional medicine, complimentary, alternative, fringe, and other methods abound, with their own advocates and practitioners. Many patients put their confidence in untested or inadequately tested treatments. Even in disease areas where therapies are evaluated rigorously, many patients assume treatments are effective, or at least worth the risk, or they would not be under investigation. Other patients are simply willing to take a chance that a new treatment will work, especially when the side effects appear to be minimal.

To a rigorous modern clinical investigator, these comprise opportunities to serve the needs of patients and practitioners by providing the most reliable evidence about treatment effects and risks. These circumstances often provide pressure to use clinical trials. However, the same forces can create incentives to bypass rigorous evaluation methods because strong beliefs of efficacy can arise from unreliable data, as has been the case historically. Whether contemporary circumstances encourage or discourage clinical trials depends largely on mindset and values.

A trialist must understand two different modes of thinking that support the science—clinical and statistical. They both underlie the re-emergence of therapeutics as a modern science. Each method of reasoning arose independently and must be combined skillfully if they are to serve therapeutic questions effectively.

2.1.1   Clinical Reasoning Is Based on the Case History

The word *clinical* is derived from the Greek *kline,* which means bed. In modern usage, *clinical* not only refers to the bedside but pertains more generally to the care of human patients. The quantum unit of clinical reasoning is the case history, and the primary focus of clinical inference is the individual patient. Before the widespread use of experimental trials, clinical methods of generalizing from the individual to the population were informal. The concepts of person-to-person variability and its sources were also described informally. Medical experience and judgment was not, and probably cannot be, captured in a set of rules. Instead, it is a form of "tacit knowledge" (Polanyi, 1958), and is very concrete.

New and potentially useful clinical observations are made against this background of reliable experience. Following such observation, many advances have been made by incremental improvement of existing ideas. This process explains much of the progress made in medicine and biology up to the twentieth century. Incremental improvement is a reliable but slow method that can optimize many complex processes. For example, the writing of this book proceeded largely by slightly improving earlier drafts, especially true for the second edition. However, there was a foundation of design that greatly facilitated the entire process. Clinical trials can provide a similar foundation of design for clinical inference, greatly amplifying the benefits of careful observation.

There often remains discomfort in clinical settings over the extent to which population-based estimates (i.e., those from a clinical trial) pertain to any individual, especially a new patient outside the study. This is not so much a problem interpreting the results of a clinical trial as a difficulty trying to use results to select the best treatment for a new individual. There is no formal way to accomplish this generalization in a purely clinical framework. It depends on judgment, which itself depends on experience. However, clinical experience historically has been summarized in nonstatistical ways.

A stylized example of clinical reasoning, and a rich microcosm of issues, can be seen in the following case history, transmitted by Francis Galton (1899):

> The season of strawberries is at hand, but doctors are full of fads, and for the most part forbid them to the gouty. Let me put heart to those unfortunate persons to withstand a cruel medical tyranny by quoting the experience of the great Linnæus. It will be found in the biographical notes, written by himself in excellent dog-latin, and published in the life of him by Dr. H. Stoever, translated from German into English by Joseph Trapp, 1794. Linnæus describes the goutiness of his constitution in p. 416 (*cf.* p. 415) and says that in 1750 he was attacked so severely by siatica that he could hardly make his way home. The pain kept him awake during a whole week. He asked for opium, but a friend dissuaded it. Then his wife suggested "Won't you eat strawberries?" It was the season for them. Linnæus, in the spirit of experimental philosopher, replied, "*tentabo*—I will make the trial." He did so, and quickly fell into a sweet sleep that lasted two hours, and when he awoke the pain had sensibly diminished. He asked whether any strawberries were left: there were some, and he ate them all. Then he slept right away till morning. On the next day, he devoured as many strawberries as he could, and on the subsequent morning the pain was wholly gone, and he was able to leave his bed. Gouty pains returned at the same date in the next year, but were again wholly driven off by the delicious fruit; similarly in the third year. Linnæus died soon after, so the experiment ceased.

> What lucrative schemes are suggested by this narrative. Why should gouty persons drink nasty waters, at stuffy foreign spas, when strawberry gardens abound in England? Let enthusiastic young doctors throw heart and soul into the new system. Let a company be run to build a curhaus in Kent, and let them offer me board and lodging gratis in return for my valuable hints.

The pedigree of the story may have been more influential than the evidence it provides. It has been viewed both as quackery and as legitimate (Porter and Rousseau, 1998), but as a trialist and occasional sufferer of gout, I find the story both quaint and enlightening with regard to a clinical mindset. Note especially how the terms "trial" and "experiment" were used, and the tone of determinism.

Despite its successes clinical reasoning by itself has no way to deal formally with a fundamental problem regarding treatment inefficacy. Simply stated, that problem is "why do ineffective treatments frequently appear to be effective?" The answer to this question may include some of the following reasons:

- The disease has finished its natural course
- There is a natural exacerbation–remission cycle
- Spontaneous cure has occurred
- The placebo effect

- There is a psychosomatic cause and, hence, a cure by suggestion
- The diagnosis is incorrect
- Relief of symptoms has been confused with cure
- Distortions of fact by the practitioner or patient
- Chance

However, the attribution of effect to one or more of these causes cannot be made reliably using clinical reasoning alone.

Of course, the same types of factors can make an effective treatment appear ineffective, a circumstance for which pure clinical reasoning also cannot offer reliable remedies. This is especially a problem when seeking small magnitude, but clinically important, benefits. The solution offered by statistical reasoning is to control the signal-to-noise ratio by design.

2.1.2   Statistical Reasoning Emphasizes Inference Based on Designed Data Production

The word *statistics* is derived from the Greek *statis* and *statista,* which mean *state.* The exact origin of the modern usage of the term *statistics* is obscured by the fact that the word was used mostly in a political context to describe territory, populations, trade, industry, and related characteristics of countries from the 1500s until about 1850. A brief review of this history was given by Kendall, who stated that scholars began using data in a reasoned way around 1660 (Kendall, 1960). The word *statistik* was used in 1748 to describe a particular body of analytic knowledge by the German scholar Gottfried Achenwall (1719–1772) in *Vorbereitung zur Staatswissenschaft* (Achenwall, 1748; Hankins, 1930). The context seems to indicate that the word was already used in the way we mean it now, but some later writers suggest that he originated it (Fang, 1972; Liliencron, 1967). Porter (1986) gives the date for the use of the German term *statistik* as 1749.

Statistics is a highly developed information science. It encompasses the formal study of the inferential process, especially the planning and analysis of experiments, surveys, or observational studies. It became a distinct field of study only in the twentieth century (Stigler, 1986). Although based largely on probability theory, statistics is not, strictly speaking, a branch of mathematics. Even though the same methods of axioms, formal deductive reasoning, and logical proof are used in both statistics and mathematics, the fields are distinct in origin, theory, practice, and application. Barnett (1982) discusses various views of statistics, eventually defining it as:

> the study of how information should be employed to reflect on, and give guidance for action in, a practical situation involving uncertainty.

Making reasonable, accurate, and reliable inferences from data in the presence of uncertainty is an important and far-reaching intellectual skill. It is not merely a collection of ad hoc tricks and techniques, an unfortunate view occasionally held by clinicians and some grant reviewers. Statistics is a way of thinking or an approach to everyday problems that relies heavily on designed data production. An essential impact of statistical thought is that it minimizes the chance of drawing incorrect conclusions from either good or bad data.

Modern statistical theory is the product of extensive intellectual development in the early to middle twentieth century and has found application in most areas of science. It is not obvious in advance that such a theory should be applicable across a large number of disciplines. That is one of the most remarkable aspects of statistical theory. Despite the wide applicability of statistical reasoning, it remains an area of substantial ignorance for many scientists.

Statistical reasoning is characterized by the following general methods, in roughly this order:

1. Establish an objective framework for conducting an investigation.
2. Place data and theory on an equal scientific footing.
3. Employ designed data production through experimentation.
4. Quantify the influence of chance on outcomes.
5. Estimate systematic and random effects.
6. Combine theory and data using formal methods to make inferences.

Reasoning using these tools enhances validity and permits efficient use of information, time, and resources.

Perhaps because it embodies a sufficient degree of abstraction but remains grounded by practical questions, statistics has been very broadly successful. Through its mathematical connections, statistical reasoning permits or encourages abstraction that is useful for solving the problem at hand and other similar ones. This universality is a great advantage of abstraction. In addition abstraction can often clarify outcomes, measurements, or analyses that might otherwise be poorly defined. Finally abstraction is a vehicle for creativity.

2.1.3 Clinical and Statistical Reasoning Converge in Research

Because of their different origins and purposes, clinical and statistical reasoning could be viewed as fundamentally incompatible. But the force that combines these different types of reasoning is research. A clinical researcher is someone who investigates formal hypotheses arising from work in the clinic (Frei, 1982; Frei and Freireich, 1993). This requires two interdependent tasks that statistics does well: generalizing observations from few to many, and combining empirical and theory-based knowledge.

In the science of clinical research, empirical knowledge comes from experience, observation, and data. Theory-based knowledge arises from either established biology or hypothesis. In statistics, the empirical knowledge comes from data or observations, while the theory-based knowledge is that of probability and determinism, formalized in mathematical models. Models specifically, and statistics in general, are the most efficient and useful way to combine theory and observation.

This mixture of reasoning explains both the successful application of statistics broadly and the difficulty that some clinicians have in understanding and applying statistical modes of thought. In most purely clinical tasks, as indicated above, there is relatively little need for statistical modes of reasoning. The best use and interpretation of diagnostic tests is one interesting exception. Clinical research, in contrast, demands critical and quantitative views of research designs and data. The mixture of modes of reasoning provides a solution to the inefficacy problem outlined in Section 2.1.1.

To perform, report, and interpret research studies reliably, clinical modes of reasoning must be reformed by statistical ideas. Carter, Scheaffer, and Marks (1986) focused on this point appropriately when they said:

> Statistics is unique among academic disciplines in that statistical thought is needed at every stage of virtually all research investigations, including planning the study, selecting the sample, managing the data, and interpreting the results.

Failure to master statistical concepts can lead to numerous and important errors and biases in medical research, a compendium of which is given by Andersen (1990). Coincident with this need for statistical knowledge in the clinic, it is necessary for the clinical trials statistician to master fundamental biological and clinical concepts relevant to the disease under study. Failure to accomplish this can also lead to serious methodological and inferential errors. A clinical researcher must consult the statistical expert early enough in the conceptual development of the experiment to improve the study. The clinical researcher who involves a statistician only in the "analysis" of data from a trial can expect a substantially inferior product overall.

2.2 DEFINING CLINICAL TRIALS FORMALLY

2.2.1 Mixing of Clinical and Statistical Reasoning Is Recent

The historical development of clinical trials has depended mostly on biological and medical advances, as opposed to applied mathematical or statistical developments. A broad survey of mathematical advances in the biological and medical sciences supports this interpretation (Lancaster, 1994). For example, the experimental method was known to the Greeks, especially Strato of Lampsacus (c. 250 BCE) (Magner, 2002). The Greek anatomists, Herophilus and Erasistratis in the third century BCE, demonstrated by vivisection of prisoners that loss of movement or sensation occurred when nerves were severed. Such studies were not perpetuated, but it would be two millennia before adequate explanations for the observations would be formulated (F.R. Wilson, 1998; Staden, 1992).

There was considerable opposition to the application of statistics in medicine, especially in the late eighteenth century and early nineteenth century when methods were first being developed. The numerical method, as it was called, was proposed and developed in the early nineteenth century and has become most frequently associated with Pierre Charles Alexander Louis. His best-known and most controversial study was published in 1828, and examined the effects of bloodletting as treatment for pneumonia (see Louis, 1835). Although the results did not clearly favor or disfavor bloodletting, his work became controversial because it appeared to challenge conventional practice on the basis of numerical results. The study was criticized, in part, because the individual cases were heterogeneous and the number of patients was relatively small. There was even a claim in 1836 by d'Amador that the use of probability in therapeutics was antiscientific (d'Amador, 1836).

Opposition to the application of mathematical methods in biology also came from Claude Bernard (1865). His argument was also based partly on individual heterogeneity. Averages, he felt, were as obscuring as they were illuminating. Gavarret gave a formal specification of the principles of medical statistics in 1840 (Gavarret, 1840).

He proposed that at least two hundred cases, and possibly up to five hundred cases were necessary for reliable conclusions. In the 1920s R.A. Fisher demonstrated and advocated the use of true experimental designs, especially randomization, in studying biological problems (Fisher, 1935; Box, 1980).

Yet it was the mid-1900s before the methodology of clinical trials began to be applied earnestly. The delay in applying existing quantitative methods to clinical problem solving was probably a consequence of many factors, including inaccurate models of disease, lack of development of drugs and other therapeutic options, physician resistance, an authoritarian medical system that relied heavily on expert opinion, and the absence of the infrastructure needed to support clinical trials. In fact the introduction of numerical comparisons and statistical methods into assessments of therapeutic efficacy has been resisted by medical practitioners at nearly every opportunity over the last 200 years (Matthews, 1995).

Even today in biomedical research institutions and clinical trial collaborations, there is a firm tendency toward the marginalization of statistical thinking. The contrast with the recognized and overwhelming utility of mathematics in the physical sciences is striking. Among others, Eugene Wigner (1959) pointed this out, stating that:

> ... the enormous usefulness of mathematics in the natural sciences is something bordering on the mysterious and that there is no rational explanation for it.

It is noteworthy that he referred broadly to natural science.

Part of the difficulty fitting statistics into clinical and biomedical science lies with the training of health professionals, particularly physicians who are often the ones responsible for managing clinical trials. Most medical school applicants have minimal mathematical skills, and the coverage of statistical concepts in medical school curricula is usually brief, if at all. Statistical reasoning is often not presented correctly or effectively in postgraduate training programs. The result is ignorance, discomfort, and the tendency to treat statistics as a post hoc service function. For example, consider the important synergistic role of biostatistics and clinical trials in cancer therapeutics over the last 40 years. Biostatistical resources in cancer centers sponsored by the National Institutes of Health are evaluated for funding using essentially the same guidelines and procedures as for resources such as glass washing, animal care, and electron microscopes (NCI, 2003).

An excellent historical review of statistical developments behind clinical trials is given by Gehan and Lemak (1994). Broader discussions of the history of trials are given by Bull (1959), Meinert (1986), and Pocock (1996). It is interesting to read early discussions of trial methods (e.g., Herdan, 1955) to see issues of concern that persist today.

Since the 1940s clinical trials have seen a widening scope of applicability. This increase is a consequence of many factors, including the questioning of medical dogma, notable success in applying experimental designs in both the clinical and basic science fields, governmental funding priorities, regulatory oversight of drugs and medical devices with its more stringent demands, and development of applied statistical methods. In addition the public, governmental, industrial, and academic response to important diseases like cardiovascular disease, cancer, and AIDS has increased the willingness of, and necessity for, clinicians to engage in structured experiments to answer important questions reliably (Gehan and Schneiderman, 1990; Greenhouse,

1990; Halperin, DeMets, and Ware, 1990). Finally the correct perception that well-done clinical trials are robust (i.e., insensitive to deviations from many underlying assumptions) has led to their wider use. Health professionals who wish to conduct and evaluate clinical research need to become competent in this field.

Discomfort with, and arguments against, the use of rigorous experimental methods in clinical medicine persist. No single issue is more of a focal point for such objections than the use of randomization because of the central role it plays in comparative trials. Typical complaints about randomization are illustrated by Abel and Koch (1997, 1999), who explicitly reject randomization as a (1) means to validate certain statistical tests, (2) basis for (causal) inference, (3) facilitation of masking, and (4) method to balance comparison groups. Similar arguments are given by Urbach (1993). These criticisms of randomization are extreme and I believe them to be incorrect. I will return to this discussion in Chapter 13.

2.2.2   Clinical Trials Are Rigorously Defined

An *experiment* is a series of observations made under conditions controlled by the scientist. To define an investigation as an experiment, the scientist must control the application of treatment (or intervention). In other words, the essential characteristic that distinguishes experimental from nonexperimental studies is whether or not the scientist controls or manipulates the treatment (factors) under investigation. In nonexperimental studies, treatments are applied to the subjects for reasons beyond the investigator's control. The reasons why treatments were applied may be unknown, or possibly even known to be confounded with prognosis.

Frequently in experiments there is a second locus of control over extraneous influences. The play of chance is an influence that the scientist usually intends to reduce, for example. This additional control isolates the effect of treatment on outcome and makes the experiment efficient. However, efficiency is a relative concept and is not a required characteristic of a true experiment. *Design* is the process or structure that controls treatment administration and isolates the factors of interest.

A *clinical trial* is an experiment testing a medical treatment on human subjects. Based on the reasoning above, a control group internal to the study is not required to satisfy the definition. In particular, nonrandomized studies can be clinical trials. The clinical trialist also attempts to control extraneous factors that may affect inference about the treatment. Which factors to control depends on context, resources, and the type of inference planned. Generically, the investigator will control (minimize) factors that contribute to outcome variability, selection bias, inconsistent application of the treatment, and incomplete or biased ascertainment of outcomes. We might think of these extraneous factors as noise with random and nonrandom components.

Use of the term "observational" to describe medical studies that are not clinical trials (e.g., epidemiologic studies) is common but inaccurate. All scientific studies depend on observation. The best terminology to distinguish clinical trials from other studies is *experimental* versus *nonexperimental*. Conceptual plans for observation, data capture, follow-up of study participants, and analysis are similar for many types of medical studies.

Use of an explicit comparison does not distinguish a clinical trial from a nonexperimental study. All medical studies are at least implicitly comparative—what do we observe compared to that which is expected under different conditions. In an important class of clinical trials the comparison is explicit in the design of the experiment, for example, randomized studies. But it is not true that trials are the only type of comparative studies.

2.2.3 Experiments Can Be Misunderstood

The denotations outlined in the previous section are not always appreciated by researchers. Even so, most medical practitioners and clinical researchers have a positive disposition towards clinical trials based on their experience. This is not always the case for the public, where use of the word "experiment" is often pejorative and makes some individuals uncomfortable. Terminology can convey values, especially to patients considering participation in a clinical trial.

An informal use of the word "experiment" often implies an unacceptable situation in which the experimenter lacks respect for the study participants. For example, consider this article from a national daily newspaper:

> Patients are not always fully informed that they are guinea pigs in medical research studies authorized by the Food and Drug Administration, *The* (Cleveland) *Plain Dealer* said. The newspaper analysis of FDA files found that in 4,154 inspections of researchers testing new drugs on humans, the FDA cited more than 53% for failing to fully disclose the experimental nature of the work. (*USA Today*, 1996)

This was the entire article, and it tends to equate being "guinea pigs," with "experiment," and deception, some of it on the part of the FDA. This article is little more than an advertisement favoring misconceptions. It is not the shortness of discussion that contributes to such corruptions, as evidenced in a lengthy piece by Lemonick and Goldstein (2002). This article, discussed in more detail in Chapter 3, is similarly misinformed and suffers principally from its own double standards.

Despite corruptions such as this, the use of the word "experiment" in a scientific context to describe clinical trials is appropriate and has a tradition (Beecher, 1970; Fleiss, 1986a; Freund, 1970; Fried, 1974; Hill, 1963; Katz, 1972; McNeill, 1993; Silverman, 1985). The usage of this term in this book will always be according to the definition above, as in *conducting an experiment,* rather than indicating disrespect for, or *experimenting on,* study participants.

Aside from terminology, the newspaper blurb illustrates another issue about populist perceptions regarding clinical trials, particularly in the United States. To most nonscientists and the lay press, clinical trials are viewed as a component of the medical establishment. Stories about wasting resources, abuses, and difficulties with clinical trials fit well with popular anti-establishment thinking. Similarly there is broad sympathy with the following stereotypes: the lone practitioner idealized as representing the best interests of the patient; one who has discovered a cure; new, less costly, or alternative approaches; oppression by rigor, unthinking critics, or tradition. Alternative medicine appeals to some of these stereotypes, which can make good newspaper copy. The history and current status of the failed anticancer agent, hydrazine sulfate (discussed in Section 16.7.4), fits this paradigm almost perfectly.

Unfortunately, such populist thinking about clinical trials and the mass media perspective that encourages it are exactly backward. Historically as well as to a large extent today, clinical trials have been an anti-establishment tool, challenging authoritarian views and those aligned with special or conflicted interests. Trials may never receive a great deal of public sympathy or the perspective they deserve because they are expensive and require government or corporate funding, and don't appear to advocate directly for the individual. Nevertheless, clinical trials are our most powerful tool to counteract dogma.

2.2.4   Clinical Trials as Science

The use of clinical trials is consistent with, if not at the heart of, the character of the scientific process in medicine. Clinical trials usually do not require that science be defined more rigorously than a statement due to Thomas Henry Huxley (1880), who said:

> science is simply common sense at its best; that is, rigidly accurate in observation and merciless to fallacy in logic.

Accordingly we could suppose that medicine is not science because of its historical origins, occasional lack of sense, inaccurate observation, illogic, and considerable other nonscience content and activity. But there is at least a strong consilience (E.O. Wilson, 1998) between the methods of medicine and those of science, much as there is between statistics and mathematics.

Differentiating between science and nonscience in a formal way turns out to be a difficult philosophical problem, often approached by debates over "demarcation" criteria (e.g., Curd and Cover, 1998). One influential demarcation criterion put forward as a hallmark of the scientific method is falsification of theory (Popper, 1959). Used alone, it is probably inadequate broadly as well as for clinical trials. I cannot review this complex issue here, but it is safe to say that clinical trials incorporate definitive characteristics of the scientific method as outlined below.

Science can be differentiated readily from other fields (e.g., see Mayr, 1997, for one perspective) partly on the basis of experimentation. Scientific medicine advances through its experimental construct, the clinical trial. Sir Austin Bradford Hill (1963) stated the case quite well:

> In the assessment of a treatment, medicine always has proceeded, and always must proceed, by way of experiment. The experiment may merely consist in giving the treatment to a particular patient or series of patients, and of observing and recording what follows—with all the difficulty of interpretation, of distinguishing the *propter hoc* from the *post hoc*. Nevertheless, even in these circumstances and in face of the unknown a question has been asked of Nature, and it has been asked by means of trial in the human being. There can be no possible escape from that. This is *human* experimentation—of one kind at least. *Somebody* must be the first to exhibit a new treatment in man. *Some patient*, whether for good or ill, must be the first to be exposed to it.

Clinical trials represent the empirical arm of scientific medicine. However, experimentation alone is insufficient to make scientific progress. The rational arm of scientific medicine, represented by biological theory, is equally necessary. In this context, *theory* means an organized system of accepted knowledge with wide applicability, as opposed to simply a speculation. Clinical trials have become so well recognized as necessary to the method of scientific medicine that some researchers ignore or discount the necessary role of biological theory. Scientific inference is not a *structure*, as represented by an experiment, but a *process* of reconciliation between experiments and theory. This point is vital to understanding why a clinical trial alone does not represent a scientific test of a therapy in the absence of a plausible mechanism of action for that therapy. Experiments or trials cause us to modify theory—more specifically to replace one theory with another. Science requires both the theory and the experiment but never requires us to replace the theory with the experiment. Furthermore in a mature science the right experiment is determined by questions that arise from prevailing theory (Friedlander, 1995). This issue will surface again in the discussion of trials for complementary and alternative medicine (Section 4.5).

In addition to being experiments, clinical trials reflect important general characteristics of the scientific method. These include *instrumentalizing perception* or *measurement,* which enhances repeatability and quantification; *externalizing plans and memory* in a written record, to facilitate reference and defense; *control of extraneous factors* as part of the study design (e.g., using internal controls and methods to control bias); and submitting completed work to *external recognition, verification, or disproof.* All these fundamental and general characteristics of scientific inquiry are integrated in the modern practice of clinical trials.

Sometimes investigations that superficially appear to be clinical trials are not. Examples are so-called seeding trials, occasionally conducted by pharmaceutical companies as marketing tools because they encourage physicians to prescribe a new drug (Kessler et al., 1994). The distinction between such efforts and true clinical trials can be made by examining the purposes and design of the study. Characteristics of seeding trials include (1) a design that cannot support the research goals, (2) investigators recruited because of their prescribing habits rather than scientific expertise, (3) the sponsor that provides unrealistically high reimbursements for participation, (4) minimal data collected or those of little scientific interest, (5) the study conducted through the sponsor's marketing program rather than a research division, and 6) the agent tested being similar to numerous therapeutic alternatives.

2.2.5   Trials and Statistical Methods Fit within a Spectrum of Clinical Research

Clinical experiments must be taken in context, both for the evidence they yield and for the methodology they employ. Trial results coexist with clinical and preclinical research of all types, some of which is supportive of their findings and some of which is not. The following categorization of clinical research, adapted from Ahrens (1992), shows how designed experiments fit within the spectrum of clinical research methods. Based on scientific objectives, sponsor funding, and technical skills and training of investigators, clinical research can be divided into seven areas:

1. Studies of disease mechanisms. The studies may be either descriptive or analytic but require laboratory methods and clinical observations under controlled conditions. Examples are metabolic and pharmacologic studies.
2. Studies of disease management. These studies involve evaluations of developing treatments, modalities, or preventive measures and have outcomes with direct clinical relevance. Often internal controls and true experimental designs will be used. Most clinical trials fit within this category.
3. In vitro studies on materials of human origin. These are studies using observational designs and blood, tissue, or other samples that attempt to show associations thought to be clinically important. Series of surgical or pathologic studies are often this type.

4. Models of human health and disease processes. These studies include animal and theoretical (e.g., mathematical) models of disease, often used to guide laboratory and clinical investigations.

5. Field surveys. These are descriptive or analytic studies of risk factors or disease correlates in populations. Examples include epidemiologic and genetic studies.

6. Technology development. This type of research includes development and mechanistic testing of new diagnostic and therapeutic methods. Examples of technology developed in these types of studies include imaging devices and methods, diagnostics, vaccines, and applied analytic methods such as biostatistical methods.

7. Health care delivery. These studies focus on economic and social effects of health care practice and delivery. Examples include studies of education and training, cost effectiveness, health care access, and financing.

Any one or several of these types of clinical research studies can relate directly to clinical trials. For example, they may be precursor, follow-up, or parallel studies essential to the design and interpretation of a clinical trial. In any case, experimental designs have a central role in the spectrum of clinical research studies.

2.3 PRACTICALITIES OF USAGE

2.3.1 Predicates for a Trial

A clinical trial cannot take place unless key circumstances support it. The list of requirements includes (1) having a good question, (2) uncertainty/equipoise in the scientific community, (3) an appropriate level of risk–benefit for the interventions, (4) receptivity in the context for the trial (Chapter 4), (5) a design appropriate to the scientific question, and (6) the necessary resources and technology to perform the study. A deficiency in any of these components will stop the trial from either getting underway or being successful.

The extent to which these prerequisites are met is a matter of judgment at the outset but will be obvious by the time a study is well underway. If we view trials through these principles, we can get a good perspective on their likely success, quality, and perhaps even impact. Furthermore these underpinnings provide a way to see the considerable commonality of methodology across the wide spectrum to which trials are applied.

2.3.2 Trials Can Provide Confirmatory Evidence

Aside from testing novel ideas, clinical trials are frequently used to test or reinforce weak evidence from earlier studies. There are a variety of reasons why more than one trial might be performed for essentially the same research question (usually by different sets of investigators). In some circumstances this may happen inadvertently. For example, in pharmaceutical studies the onset or planning of a trial may be confidential and could be undertaken simultaneously by different companies using slightly different treatments. In retrospect these trials may be seen as confirmatory. The same situation may arise in different countries for scientific or political reasons.

Confirmatory trials may also be planned. When the results of a new trial seem to contradict prevailing biological theory, many researchers or practitioners may be unconvinced by the findings, particularly if there are methodological problems with the study design, as there frequently are. This could be true, for example, when there is relatively little supporting evidence from preclinical experiments or epidemiologic studies. Confirmatory trials may be needed to establish the new findings as being correct and convince researchers to modify their theories. When the magnitude of estimated treatment effects is disproportionate to what one might expect biologically, a similar need for confirmatory studies may arise. This use of trials is difficult, and can be controversial, but is approachable by structured reasoning (Parmar, Ungerleider, and Simon, 1996).

There are numerous types of design flaws, methodologic errors, problems with study conduct, or analysis and reporting mistakes that can render clinical trials less than convincing. We would expect the analysis and reporting errors to be correctable if the design of the study is good, but they are frequently accompanied by more serious shortcomings. In any case, trials with obvious flaws are open to criticism, which may limit their widespread acceptance. Such skepticism may motivate a confirmatory study.

Finally there is a tendency for studies with positive results to find their way preferentially into the literature (publication bias), which can distort the impression that practitioners get (Chapter 18). This implies that even a published positive result may leave some doubt as to true efficacy. Also some positive studies are the result of chance, data-driven hypotheses, or subset analyses that are more error prone than a p-value would suggest. Readers of clinical trials tend to protect themselves by reserving final judgment until findings have been verified independently or assimilated with other knowledge. The way in which practitioners use the information from clinical trials seems to be very complex. An interesting perspective on this is given by Henderson (1995) in the context of breast cancer clinical trials. Medicine is a conservative science, and behavior usually does not change on the basis of one study. Thus confirmatory trials of some type are often necessary to provide a firm basis for changing clinical practice.

2.3.3 Clinical Trials Are Unwieldy, Messy, and Reliable

Like many scientific endeavors, clinical trials are constrained by fiscal and human resources, ethical concern for the participants, and the current scientific and political milieu. Trials are frequently unwieldy and expensive to conduct. They require collaborations among patients, physicians, nurses, data managers, and methodologists. They are subject to extensive review and oversight at the institutional level, by funding agencies, and by regulators. Trials consume significant time and resources and, like all experiments, can yield errors. Multi-institutional trials may cost from millions up to hundreds of millions of dollars and often take many years to complete. Clinical trials studying treatments for prevention of disease are among the most important but cumbersome, expensive, and lengthy to conduct.

Certain studies may be feasible at one time, given all these constraints, but infeasible a few months or years later (or earlier). In other words, many studies have a "window of opportunity" during which they can be accomplished and during which they are more likely to have an impact on clinical practice. For comparative trials, this window often exists relatively early after the process of development of a therapy. Later clinicians' opinions about the treatment solidify, even without good data, and economic incentives can dominate the application of a new treatment. Then many practitioners may be reluctant to undertake or participate in a comparative clinical trial. The issues surrounding a trial of extracorporeal membrane oxygenation (ECMO) in infants with respiratory failure illustrate, in part, the window of opportunity. This trial is discussed in Sections 3.5.3, 13.4.3, and 13.7.

Clinical trials are frequently messier than we would like. A highly experienced and respected basic science researcher tells the story of an experiment he instructed a lab technician to perform by growing tumors in laboratory rats:

> Once we decided to set up an immunologic experiment to study host defenses. I was out of town, and I had a new technician on board, and I asked him to order 40 rats. When I returned, I discovered we had a mixture of Sprague-Dawley, Wistar, and Copenhagen rats. Furthermore, their ages ranged from 3 months to 2 years. I had asked him to inject some tumor cells and was horrified to find out that he had used three tumor lines of prostate cancer cells and administered 10^3 to 10^8 cells in various parts of the animals and on different days during a period of 2 weeks. Unfortunately, 20% of the animals escaped during the experiment, so we could present data only on age-matched controls. (Coffey, 1978)

At this point we might begin to lose confidence in the findings of this experiment as a controlled laboratory trial. However, these heterogeneous conditions might well describe a clinical trial under normal circumstances, and it could be a useful one because of the amount of structure used successfully. These types of complications are common in clinical trials, not because of the investigator's laxity but because of the nature of working with sick humans. Alternatives to clinical trials are usually messier than trials themselves.

Trials are more common now than ever before because they are the most reliable means of making correct inferences about treatments in many commonly encountered clinical circumstances. When treatment differences are about the same size as patient-to-patient variability and/or the bias from flawed research designs, a rigorous clinical trial (especially a randomized design) is the only reliable way to separate the treatment effect from the noise. Physicians should be genuinely interested in treatments that improve standard therapy only moderately. We can be sure that patients are always interested in even small to moderate treatment benefits. The more common or widespread a disease is, the more important even modest improvements in outcome will be.

The bias and variability that obscures small but important treatment effects has larger consequences if we expect benefits to accrue mostly in the future. This is the case for prevention and treatment of many chronic diseases. Without rigorous design and disciplined outcome ascertainment, significant advances in treatment aggregated from several smaller improvements at earlier times would be missed. This lesson is quite evident from the treatment of breast cancer since about 1985, for example.

A paradoxical consequence of the reliability of rigorous trials to demonstrate modest sized treatment effects, and/or those evident only after years of observation, is that the methodology might sometimes be viewed as not contributing to substantial advancement. It is easy to undervalue trials when they each produce only a modest effect. Hence it is important to ask the right research questions and identify the optimal research designs for them.

Well-performed clinical trials offer other advantages over most uncontrolled studies. These include a complete and accurate specification of the study population at base line, rigorous definitions of treatment, bias control, and active ascertainment of endpoints. These features will be discussed later.

2.3.4   Other Methods Are Valid for Making Some Clinical Inferences

Historically many medical advances have been made without the formal methods of comparison found in controlled clinical trials, in particular, without randomization, statistical design, and analysis. For example, vitamins, insulin, many drugs and antimicrobials (e.g., penicillin), vaccines, and tobacco smoke have had their effects convincingly demonstrated without using controlled clinical trials. Studies that demonstrated these effects were epidemiological, historically controlled, and, in some cases, uncontrolled. Thus true experimental designs are not the only path to advancing medical knowledge.

For example, ganciclovir was approved by the U.S. Food and Drug Administration for the treatment of retinitis caused by cytomegalovirus in patients with human immunodeficiency virus (HIV) infection. The evidence in favor of ganciclovir efficacy consisted of a relatively small but well-performed study using historical controls (Jabs, Enger, and Bartlett, 1989). The treatment benefit was large, supported by other data, and the toxicity profile of the treatment was acceptable. (However, it took a clinical trial to establish that another drug, foscarnet, was associated with longer survival and similar visual outcome as ganciclovir.)

In contrast, consider the difficulty in inferring the benefit of zidovudine (AZT) treatment in prolonging the survival of HIV positive patients from retrospective studies (e.g., Moore et al., 1991). Although AZT appeared to prolong survival in this study, the control group had a shorter survival than untreated patients earlier in the epidemic. Sicker patients did not live long enough to receive AZT, making the treatment appear beneficial. This illustrates the difficulty in defining treatment in retrospect. Some issues related to using observational data for assessing AIDS treatments are discussed by Gail (1996).

Studies based on nonexperimental comparative designs can sometimes provide valid and convincing evidence of treatment efficacy. Using nonexperimental designs to make reliable treatment inferences requires five things: (1) the treatment of interest must occur naturally, (2) the study subjects have to provide valid observations for the biological question, (3) the natural history of the disease with standard therapy, or in the absence of the intervention, must be well-known, (4) the effect of the treatment or intervention must be large enough to overshadow random error and bias, and (5) evidence of efficacy must be consistent with other biological knowledge. These criteria are often difficult to satisfy convincingly because they rely on circumstances beyond the investigators' control. A potentially helpful method for some alternative therapies is to evaluate "best case series" as a way to assess any therapeutic promise (Nahin, 2002).

A humorous angle on this point was recently provided by Smith and Pell (2003), who examined parachute use to prevent death or trauma and poked fun at overzealous evidence-based requirements that treatment efficacy be established only by systematic reviews of randomized trials. Several of my clinical colleagues were more than a little sympathetic to the implications of this humor, indicating that an appropriate balance is needed when assessing the types of studies that support a proposed treatment and the magnitude of the effect. Although Smith and Pell appeal to "common sense," there is ample evidence in medical history that such a basis alone is inadequate. The five points mentioned above can be used to supplement common sense.

Nonexperimental Designs

There are at least three types of nonexperimental designs that can yield convincing evidence of treatment differences when the assumptions above are met: epidemiologic studies, historically controlled trials, and databases. Epidemiologic designs are an efficient way to study rare events and can control some important sources of random error and bias. They represent a lower standard of evidence for treatment studies because, without randomization, the investigator cannot control unknown confounders. Nevertheless, they are a valuable source of evidence. These strengths and weaknesses are typified by case-control designs, which compare exposures (determined retrospectively) in subjects with a condition versus "controls" without the condition. The design is not as useful for studying the effects of treatment because of the reasons cited and because it is difficult to define "treatment" retrospectively.

Jenks and Volkers (1992) outlined a number of factors possibly associated with increased cancer risk. Some were biologically plausible while others seem to be based only on statistical associations. Apparent increases in cancer risks for factors such as electric razors and height could be due to bias (or chance). In any case, small increases in risk are at the limit of resolution of observational study designs. Many provocative associations between human disease and environmental exposures or other factors are not amenable to study by clinical trial designs because the exposures or factors are not under the experimenter's control. A good example of this is the question of right-versus left-handedness and its relationship to longevity (Halpern and Coren, 1991).

Historically controlled studies can also provide convincing evidence. Although improvements in supportive care, temporal trends, or different methods of evaluation for the same disease can render these studies uninterpretable, there are circumstances in which they are valid and convincing. An example might be the study of promising new treatments in advanced diseases with a uniformly poor prognosis. A specific example of this is genetically engineered tumor vaccines, tested in patients with advanced cancer. Large beneficial treatment effects could show themselves by prolonging survival far beyond that which is known to be possible using conventional therapy. However, consider how unreliable treatment inferences based on historical controls would be in a disease like AIDS, where supportive care and treatment of complications have improved so much in the last 10 to 15 years. Silverman (1985) gives some striking examples of time trends in infectious disease mortality. Stroke mortality has also shown clinically significant secular trends (Bonita and Beaglehole, 1996; Brown et al., 1996). Similar problems can be found with historical controls in many diseases (Dupont, 1985).

Database studies are sometimes suggested as a valid way of making treatment inferences. Such studies are convenient when the data are available, inexpensive to conduct, and may be based on a very large number of patients. For example, patients treated at a single institution may be followed with the help of a database that facilitates both clinical management and asking research questions. In such situations treatment selection is partly an outcome and partly a predictor of how the patient is doing. Therefore treatment comparisons are suggestive, but not definitive tests of relative efficacy. The limitations of database analyses are well known (Byar, 1991) but often overlooked.

Database analyses often carry fancier names today than they did some years ago. For example, we speak of "outcomes research" and "data mining." When initiating such studies, it is important to ask why the database exists in the first place, to determine whether or not it can provide reliable evidence for the therapeutic question at hand. The presence of treatment assignments and outcomes in a large happenstance database is not sufficient to guarantee that an analysis of relative treatment effects will yield a correct or reliable answer. Simply stated, the reason is lack of control over observer bias, selection bias, and outcome ascertainment. An excellent example of the inadequacy of such studies is provided by Medicare data and the background to lung volume reduction surgery prior to the National Emphysema Treatment Trial (Section 4.6.6).

2.3.5   Trials Are Difficult to Apply in Some Circumstances

Not all therapeutic areas can tolerate an infusion of rigor, however needed it may be. Comparative trials may be difficult to apply in some settings because of human factors, such as ethical considerations, or strongly held beliefs of practitioners or participants. Logistical difficulties prevent many trials that otherwise could be appropriate. Early on, the AIDS epidemic motivated trialists to examine many basic assumptions about how studies were designed and conducted (Byar et al., 1990) and relax some of the developmental rigor.

In oncology there are many conditions requiring multiple therapies or modalities for treatment. Although numerous factors are theoretically manageable by study design, practical limitations can thwart systematic evaluation of treatment combinations and interactions. Similarly, when minor changes in a complex therapy are needed, a trial may be impractical or inappropriate. In very early developmental studies when the technique or treatment is changing significantly or rapidly, formal trials may be unnecessary and could slow the developmental process.

Trials are probably more difficult to apply when studying treatments that depend on proficiency than when studying drugs. Proficiency may require specific ancillary treatments or complex diagnostic or therapeutic procedures (e.g., surgery). Well-designed trials can in principle isolate the therapeutic component of interest, but this may be difficult. When treatments are difficult to study because they depend strongly on proficiency, they may also be of lesser utility because they are hard to generalize to clinical practitioners at large.

Time and money may prevent the implementation of some trials. Trials are also not the right tool for studying rare outcomes in large cohorts or some diagnostic modalities. However, in all these cases the methods used to gather data or otherwise to design methods to answer the biological question can benefit from quantification, structure, and statistical modes of thought. Bailar et al. (1984) suggest a framework for evaluating studies without internal controls. This provides additional support for the value of studying trial methodology.

Biological knowledge may suggest that trials would be futile unless carefully targeted. For example, suppose a disease results from a defect in any one of several metabolic or genetic pathways, and that the population is heterogeneous with respect to such factors. Treatments that target only one of the pathways are not likely to be effective in everyone. Therefore clinical trials would be informative and feasible only if done in a restricted population.

2.3.6   Randomized Studies Can Be Initiated Early

While comparative clinical trials may not be the method of choice for making treatment inferences in some circumstances, there are many situations where it is advantageous to implement trials early in the process of developing a new therapy. Some clinical trials methodologists have called for "randomization from the first patient" to reflect these circumstances (Chalmers, 1975a, b; Spodick, 1983). The window of opportunity for performing a randomized trial may only exist early and can close as investigators gather more knowledge of the treatment. Many disincentives to experimentation can arise later, such as practitioner bias, economic factors, and ethical constraints.

Reasons why very early initiation of randomized trials are desirable include the following: (1) the ethics climate is conducive to, or even requires it, (2) high-quality scientific evidence may be most useful early, (3) early trials might delay or prevent widespread adoption of ineffective therapies, and (4) nonrandomized designs used after a treatment is widespread may not yield useful evidence. These points may be particularly important when no or few treatments are available for a serious disease (consider the history of AIDS) or when the new therapy is known to be safe (e.g., some disease-prevention trials). Many times randomized trials yield important information about ancillary or secondary endpoints that would be difficult to obtain in other ways.

There are also reasons why randomization from the first patient may be difficult or impossible to apply: (1) the best study design and protocol may depend on information that is not available early (e.g., the correct dose of a drug), (2) adequate resources may not be available, (3) investigators may be unduly influenced by a few selected case reports, (4) patients, physicians, and sponsors may be biased, (5) the trial may not accrue well enough to be feasible, and (6) sample size and other design parameters may be impossible to judge without preliminary trials. When safe and effective treatments are already available, there may not be much incentive to conduct randomized trials early. For examples of randomized trials initiated early, see Chalmers (1975b), Garceau et al. (1964), and Resnick et al. (1968).

2.4 SUMMARY

Clinical and statistical thinking are not incompatible, but complementary for the purposes of research. The goal of both types of reasoning is generalization. The clinical approach generalizes primarily on a biological basis, while statistical modes of reasoning generalize primarily on the basis of data. Research broadly and clinical trials specifically require a carefully planned combination of clinical and statistical reasoning. This leads to a formal definition of a clinical trial as a designed experiment.

Although trials can be unwieldy because of their size, complexity, duration, and cost, they are applicable to many areas of medicine and public health. While not the only method for making valid inferences, clinical trials are essential in situations where it is important to learn about treatment effects or differences that are about the same magnitude as the random error and bias that invariably plague medical studies. Investigators should be alert to circumstances where the window of opportunity to conduct randomized trials occurs early.

2.5 QUESTIONS FOR DISCUSSION

1. Discuss some specific situations in which it would be difficult or inefficient to perform clinical trials.

2. Give some circumstances or examples of clinical trials that do not have all the characteristics of the scientific method.

3. Historically the development of statistical methods has been stimulated by gambling, astronomy, agriculture, manufacturing, economics, and medicine. Briefly sketch the importance of these for developing statistical methods. Within medicine, what fields have stimulated clinical trials methods the most? Why?

4. Discuss reasons why and why not randomized prevention trials, particularly those employing treatments like diet, trace elements, and vitamins, should be conducted without extensive "developmental" trials.

3

WHY CLINICAL TRIALS ARE ETHICAL

3.1 INTRODUCTION

Clinical trials are ethically appropriate and necessary in many circumstances of medical uncertainty. However, ethics considerations preclude using trials to address some important questions. The ethics landscape can change within the time frame of a single trial, and principles of ethics can conflict with one another. In many respects, ethics are as determining as the science, which explains why trials are examined so critically from this perspective.

Patients place a great deal of trust in clinical investigators and often participate in trials when the chance of personal benefit is low. They accept or even endorse the dual role of physicians as practitioners and researchers. Clinical investigators must continually earn the privilege to perform research by protecting the interests of those who grant it.

For these reasons I will discuss ethics early in this book so that its impact can be appreciated. There are many thoughtful and detailed discussions of ethics related to medicine broadly (Reich, 1995), medical research (Woodward, 1999), and clinical trials specifically (Beauchamp and Childress, 2001; Fried, 1974; Freund, 1970; Katz, 1972; Levine, 1986; Ramsey, 1975). Special problems related to cancer treatments are discussed in depth and from a modern perspective in Williams (1992). An international view is given by McNeill (1993). A very thoughtful perspective is given by Ashcroft et al. (1997). The discussion here is focused on the design of clinical trials but should provide a perspective on more general concerns.

I will partition the ethics landscape into five areas: the nature and consequences of duality in the obligation of physicians, historically derived principles of medical ethics, contemporary foundational principles for clinical trials, concerns about specific experimental methods like randomization or the use of placebos, and professional conduct. All these areas have a significant relationship to the design and conduct of clinical trials.

Some discussions about ethics concerns in clinical trials are mainly relevant to randomized designs. However, the most fundamental issues do not depend on the method of treatment allocation but on the fact that the practitioner is also an experimentalist. The discussion here is intended to include trials of all types. The term "physician" can be interpreted to indicate health care practitioners broadly.

Clinical trials routinely raise legitimate ethics concerns (Hellman and Hellman, 1991), but advocates for these studies sometimes respond to challenges largely by reiterating important discoveries facilitated by clinical experiments (Passamani, 1991). Such discussions are usually held under the premise that practitioners can behave ethically without performing designed experiments—a questionable assumption in my opinion. It is not necessary to choose between rigorous trials and ethical behavior. The importance of science or the knowledge gained is not a counterweight to concerns about ethics. But a large number of the clinical questions we face require a true experimental approach balanced by concern for ethics.

3.1.1 Science and Ethics Share Objectives

Circumstances guarantee that some patients will receive inferior treatments. Suboptimal or ineffective treatments are given, either by choice or dogma, because the condition is poorly understood, or when the best therapy has not yet been determined and widely accepted. Our task is to learn efficiently and humanely from this unavoidable circumstance in a way that balances individual rights with the collective good. Clinical trials are often the best way to do this. Furthermore no historical tradition nor modern ethics principle proscribes either the advocacy of clinical trials by physicians or the role of physicians as experimentalists. When properly timed, designed, presented to subjects, and conducted, a clinical trial is an ethically appropriate way to acquire new clinical knowledge. Although much knowledge can be acquired outside the scope of definitive clinical trials, ignorance can be unethical. For example, clinical decisions based only on weak evidence, opinion, or dogma, without the benefit of reasonable scientific certainty, raise their own ethics questions.

The practitioner has an obligation to acknowledge uncertainty honestly and take steps to avoid and resolve ignorance. Well-designed and conducted clinical trials are ethically more appropriate than some alternatives, such as acquiring knowledge ineffectively, by tragic accident, or failing to learn from seemingly unavoidable but serious clinical errors. Thus there are imperatives to conduct medical research (Eisenberg, 1977). The objectives of science and ethics are convergent.

Medical therapies, diagnostics, interventions, or research designs of any type, including clinical trials, can be carelessly applied in specific instances or for certain patients. When discussing ethics, we must distinguish between the *methodology* of trials as a science and the *characteristics, implementation, or application of a particular trial*. If problems regarding ethics arise, they will nearly always be a consequence of the context, circumstance, or details of the scientific question, design, or conduct of a specific trial. When sponsors of research and investigators follow the basic principles of ethics and informed consent for human subjects experiments that have evolved over the last 50 years, the resulting studies are likely to be appropriate.

3.1.2   Equipoise and Uncertainty

Equipoise is the concept that a clinical trial (especially a randomized trial) is motivated by collective uncertainty about the superiority of one treatment versus its alternative. The existence of equipoise helps to satisfy the requirement that study participants not be disadvantaged. It is a practical counterweight to the idealization of the personal care principle (discussed later), and supports a comparative trial as the optimal course of action to resolve scientific uncertainty, rather than merely a tolerable alternative. From the original description:

> ... at the start of the trial, there must be a state of clinical equipoise regarding the merits of the regimens to be tested, and the trial must be designed in such a way as to make it reasonable to expect that, if it is successfully conducted, clinical equipoise will be disturbed. (Freedman, 1987)

Although often discussed from the perspective of the individual practitioner, equipoise is a collective concept (in contrast to the personal care principle). It means that uncertainty exists among experts in the medical community about the treatments being compared. It is not necessary for each investigator in a trial to be in a state of personal equipoise. Equipoise is often a consequence of individual doubt of certitude and respect for contrary expert opinion. It can exist even when individual investigators have weakly held preferences for one treatment or another. An investigator might hold strong beliefs about certain treatments and consequently not feel comfortable participating in a trial. This often happens but does not dismantle the notion of equipoise.

There are some problems with the idea of equipoise. It may be lost with preliminary data from a trial or be unattainable in the presence of weak or biased evidence. It also ignores the uncertainty inherent in the common practice of holding a hunch. As a practical matter, equipoise is often difficult to establish.

A second concept may correct some of these difficulties and can supplement or displace equipoise as a foundational requirement for a trial. This is the uncertainty principle, which is defined by the individual practitioner and his or her comfort in recommending that a patient participate in a clinical trial. If the practitioner has genuine uncertainty about the appropriate treatment, then the patient can be encouraged to participate in a trial:

> A patient can be entered if, and only if, the responsible clinician is substantially uncertain which of the trial treatments would be most appropriate for that particular patient. A patient should not be entered if the responsible clinician or the patient for any medical or non-medical reasons [is] reasonably certain that one of the treatments that might be allocated would be inappropriate for this particular individual (in comparison with either no treatment or some other treatment that could be offered to the patient in or outside the trial). (Peto and Baigent, 1998)

The uncertainty principle has more appeal at the level of the individual physician where it allows them to be unsure if their hunch is correct (e.g., Sackett, 2000). It does not correspond perfectly with equipoise because an investigator could be uncertain but represent a distinct minority opinion. A sizable number of practitioners who are uncertain will define a state of equipoise.

3.2   DUALITY

Health care provides benefits both to the individual and to society. Physician practice is weighted by ideals toward the individual, but societal benefit is at least indirect. Clinical trials also provide benefits to the individual and society, balanced less strongly in favor of the individual. They highlight the potential conflict of obligations that physicians and other health care practitioners have between their current patients and those yet to be encountered. I will refer to the roles of the physician on behalf of the individual and society as duality.

3.2.1   Clinical Trials Sharpen, but Do Not Create, the Issue

Discussions of the dual obligations of physicians have been taking place in the literature for many years (Gilbert, McPeek, and Mosteller, 1977; Herbert, 1977; Hill, 1963; Tukey, 1977). Guttentag (1953) acknowledged this duality and suggested that research and practice for the same patient might need to be conducted by different physicians. Schafer (1982) points out the conflict of obligations, stating:

> In his traditional role of healer, the physician's commitment is exclusively to his patient. By contrast, in his modern role of scientific investigator, the physician engaged in medical research or experimentation has a commitment to promote the acquisition of scientific knowledge.

The tension between these two roles is acute in clinical trials but is also evident in other areas of medicine. For example, epidemiologists recognize the need for protection of their research subjects and have proposed ethical guidelines for studies (Coughlin, 1995; Fayerweather, Higginson, and Beauchamp, 1991). Quantitative medical decision-making methods can also highlight ethical dilemmas (Brett, 1981).

Research and practice are not the only categories of physician activity. Levine (1986) also distinguishes "nonvalidated practices" and "practice for the benefit of others." Nonvalidated or investigational practices are those that have not been convincingly shown to be effective. The reasons for this might be that the therapy is new or that its efficacy was never rigorously shown in the first place. Practice for the benefit of others includes organ donation, vaccination, and quarantine. Ethical issues related to these are complex and cannot be covered here.

Clinical trials are a lightning rod for ethics concerns because they bring into sharp focus two seemingly opposed roles of the physician in modern medicine (Levine, 1992). But these competing demands on the physician have been present throughout history and are not created uniquely by clinical trials. Before the wide use of clinical trials, Ivy (1948) suggested that the patient is always an experimental subject. Shimkin (1953) also took this view. Even the ancient Greek physicians acknowledged this (see Section 3.2.4). Today the roles of patient advocate and researcher are not the only two points of conflict for the physician. For example, the desires for academic and financial success compete with both patient care and research. Thus the practitioner is faced with numerous potential ethics conflicts.

3.2.2   A Gene Therapy Tragedy Illustrates Duality

One of the most poignant cases highlighting the dual nature of medical care and the ethics concerns surrounding it occurred at the University of Pennsylvania in 1999. An eighteen-year-old man with ornithine transcarbamylase (OTC) deficiency died after participating in a clinical trial testing a gene-based therapy for the condition. OTC deficiency is an inherited disorder in liver cells that prevents them from properly metabolizing nitrogen. In its common form, it causes death in affected newborn males. The clinical trial in question was a dose-ranging study, in which a modified human adenovirus was being used to introduce a normal OTC gene into liver cells. The participant who died was the last of 18 originally planned for the trial and received a relatively high dose of the viral vector. He died from multiple organ system failure due to an immunologically mediated reaction.

The patient's clinical condition prior to participating in the trial was not urgent. Altruism played a strong role in his decision to volunteer, as did the expectation of benefit. However, the investigators were (prematurely) optimistic of producing a lasting therapeutic benefit. In retrospect, they probably had too favorable a view of the safety and efficacy of the treatment. If so, this was an error of judgment on their part but not a violation of the standards of performing research in a practice setting. Their behavior reflected the principles of dual obligation in the same way as that of clinical researchers at all major medical centers. In the aftermath of this incident, reviewers found deficiencies in the oversight and monitoring of the program of gene therapy clinical trials. The mixture of a fatal toxicity with ethics issues, regulation, and gene therapy puts a high profile on this case.

3.2.3 Research and Practice Are Convergent

One of the principal discomforts about the ethics of clinical trials arises from the widely held, but artificial, distinction between research and practice in medicine. Some activities of the physician seem to be done primarily for the good of the individual patient, and we place these in the domain of practice. Other actions seem to be performed for the collective good or to acquire generalizable knowledge, and we label these as research. This distinction is often convenient, but artificial, because expert physician behavior is nearly always both research and practice.

A more accurate distinction could be based on the physician's degree of certainty. When the physician is certain of the outcome from a specific therapy for a particular patient, applying the treatment might be described as *practice*. When the physician is unsure of the outcome, applying the treatment could be considered *research*, at least in part. The continuum in degrees of certainty illustrates the artificial nature of the research-practice dichotomy.

There are very few actions that the physician can carry out for the benefit of the individual patient and not yield some knowledge applicable to the general good. Likewise, nearly all knowledge gained from research can be of some benefit to individual patients. It is possible for patients to be misinformed about, or harmed by, either inappropriate practice actions or improper research activities. Both activities can have large or small risk–benefit ratios.

Practice versus research seems to be partly a result of the setting in which the action takes place, which may be a consequence of the motives of the physician. Physician behavior can be "practice" in one setting, but when part of a different structure, the same behavior becomes "research." Comparison of two standard therapies in a randomized trial might illustrate this. For example, there are two widely used treatments for early stage prostate cancer: surgery and radiotherapy. Because of history, practice and referral patterns, specialty training and belief, economics, and stubbornness, no randomized comparison has been done. If one were conducted, it would require more careful explanation of risks, benefits, and alternatives than either practice now demands.

In modern academic medical centers the convergence of research and practice is clearly evident. In these settings physicians are trained not only to engage in good practice but also to maintain a mind set that encourages learning both from clinical practice and from formal experimentation. Even with strong financial incentives to separate them, it is clear that research and clinical practice are one and the same in many of these institutions. Not surprisingly, these academic settings are the places most closely associated with performing clinical trials. However, academic medicine is a development of recent history. In the past many medical advances were made in mostly traditional practice settings, precisely because physicians both took advantage of, and created, research opportunities.

Double Standards

The artificial distinction between research and practice can create double standards regarding ethical conduct. The *practitioner* is free to represent his or her treatment preferences for the patient with relatively informal requirements for explaining alternatives, rationale, risks, and benefits. Similarly a second practitioner who prefers a different treatment may have minimal requirements to offer alternatives. However, the *investigator* studying a randomized comparison of the two standard treatments to determine which one is superior will likely incur ethical obligations well beyond either practitioner. A perspective on this point is given by Lantos (1993, 1994).

A second, and more far-reaching double standard can be seen in cases where individuals have been injured as a result of treatments that were not thoroughly studied during development. Thalidomide for sedation during pregnancy is one example. Physicians cannot expect to provide any individual with health care that is in his or her best interest without knowing the detailed properties of that therapy, including risks and benefits. This knowledge often comes most efficiently from clinical trials. A discussion of the sources and potential problems arising from this double standard is given by Chalmers and Silverman (1987).

A third source of double standards arises in considering different cultural expectations regarding medical care, patient information, and research. Some cultures place fewer restrictions on research practices in humans than does the United States, or perhaps the individuals in the population have a greater expectation that these types of studies will be done. One cannot make value judgments from this fact. For example, there are considerable differences in the expectations regarding informed consent for research studies among the United States, Europe, Africa, and Asia. These differences are a consequence of many factors, including the structure and economics of health care, physician attitudes, patient expectations, attitudes toward litigation, the nature of threats to the public health, and cultural norms, especially regarding risk acceptance and consent.

Even our own cultural perspective is inconsistent in this regard. I have seen no better illustration of this than the article by Lemonick and Goldstein (2002) and the accompanying cover of *Time* magazine. This article should be read by everyone working in clinical trials, not for its factual content but because it is a perfect example of cultural double standards regarding medical research broadly and clinical trials specifically. As a result it strongly reinforces the negative light in which such studies are frequently

TABLE 3.1   Rational Motivations for Risk Acceptance in Clinical Trials

Societal:	Direct and indirect benefits to the current population from past studies
	Cost savings from new treatments for serious diseases
	Use of appropriate risk–benefit considerations to develop new therapies
	Proven therapies replace ineffective but seemingly low-risk ones, and not vice versa
Individual:	Altruism
	Amelioration of disease or improved quality of life
	Improved supportive care
	Hope

presented to the public. The sensational costs of clinical trials in human terms are easy to display—deaths of two research volunteers who should not have been harmed. In short, we desire risk-free medical studies. However, acceptance of risk at both the societal and individual level is not only inevitable, but rational behavior. Some reasons are listed in Table 3.1.

The wish for risk-free medical trials is as vain as the desire for risk-free automobiles or air travel, for example, both of which predictably kill sizable numbers of healthy volunteers. The societal and scientific necessities are to manage risk–benefit, informing participants and removing coercion, and learning from accidents and mistakes. Clinical trials have a very respectable track record in these regards, probably much better than biomedical research in general. Even so, the reaction typified by Lemonick and Goldstein is to recapitulate mistakes of medical research, attribute them to clinical trials, and ignore the improvements and safeguards implemented after learning from mistakes and accidents.

To add perspective, there has been much attention in recent years to deaths attributable to medical mistakes—largely medication errors. There may be tens of thousands of such deaths in the United States each year. It seems likely that the risk of serious harm on clinical trials is proportionately much lower because of the safeguards surrounding such studies. Factors that contribute to the safety of clinical trials include, but are not limited to, peer review of research protocols, Institutional Review Board oversight, reporting of adverse events, independent trial monitoring committees, attentive clinical care at centers of excellence, and openness of research results. Routine clinical care will probably never implement such an array of safeguards.

Personal Care

The "personal care principle" (PCP) has been raised by some as an important ethical guideline for practitioner conduct. The PCP appears to have roots in the Hippocratic Oath and states that physicians have an obligation to act always in the best interests of the individual patient (Fried, 1974). There are echoes of this in the Helsinki Declaration (see below). In a sense, physicians are taught to believe in, and endorse, the personal care ideal. Many accept the notion that they should always make the best recommendation based on the knowledge available and the patient's wishes. This does not mean that such recommendations are always based on scientific fact or well-formed opinion. Sometimes physician preferences are based on important nonrational factors. Other times preferences are based on unimportant factors. Discussions of issues surrounding the PCP is given by Markman (1992), Freedman (1992), and Royall (1992).

The PCP does not support the lack of knowledge or opinion that often motivates making treatment choices as part of a structured experiment, particularly by random assignment. If the PCP were in fact a minimal standard of conduct or ethics principle, it could create serious problems for conducting randomized clinical trials (e.g., Royall et al., 1991)—except that patients could always agree to participate in trials despite their practitioners' opinions. For this reason I will spend a little effort arguing against it.

In its simplest form the PCP makes the physician–patient relationship analogous to the attorney–client relationship. As such it appears to be a product of, and contribute to, an adversarial relationship between society and the patient, making the sick person in need of physician protection. Although physicians and patients are partners, the therapeutic environment is not usually adversarial. Even research and clinical trials are not adversarial circumstances for participants, where one compromises to have an overall best outcome. Current political demands for representative research studies, although misguided in other ways, illustrate this.

There are fundamental problems with the PCP. For example, the PCP does not encourage patients and physicians to take full advantage of genuine and widespread uncertainty or differences of opinion that motivate randomization. It does not acknowledge the lack of evidence when treatments are being developed. It raises moral and ethical questions of its own, by discouraging or removing an objective evaluation tool that can correct for economic and other biases that enter decisions about treatment selection. All these issues are illustrated by the history of Laetrile, discussed briefly on page 138. The ascendancy of a personal care ideal may have been an artifact of a time when medical resources were very abundant.

An insightful physician will probably recognize that many of his/her treatment "preferences" are artificial and carry no ethical legitimacy. More to the point, *the PCP is not a defining ethical principle* but an idealization. The PCP is violated in many instances that have nothing to do with randomization or clinical trials, as discussed below.

3.2.4   The Hippocratic Tradition Does Not Proscribe Clinical Trials

The Hippocratic Oath has been a frequently cited modern-day code of conduct for the practice of medicine. It represents an ideal that has been extensively modified by modern medical practice, scientific discovery, cultural standards, and patient expectations. Many physicians no longer repeat the full oath upon graduation from medical school and some have rarely (or never) seen it. Most clinical trialists have not studied it, and occasionally someone will imply that the oath is inconsistent with experimentation.

To assess its actual implications for ethics and clinical trials, it is useful to examine the full text. This somewhat literal translation comes from Miles (2004, pp. xiii–xiv), who also discusses in great detail the ethical implications of the oath:

> I swear by Apollo the physician and by Asclepius and by Health [the god Hygeia] and Panacea and by all the gods as well as goddesses, making them judges [witnesses], to

bring the following oath and written covenant to fulfillment, in accordance with my power and judgment:

to regard him who has taught me this techné [art and science] as equal to my parents, and to share, in partnership, my livelihood with him and to give him a share when he is in need of necessities, and to judge the offspring [coming] from him equal to [my] male siblings, and to teach them this techné, should they desire to learn [it], without fee and written covenant;

and to give a share both of rules and of lectures, and all the rest of learning, to my sons and to the [sons] of him who has taught me and to the pupils who have both made a written contract and sworn by a medical convention but by no other.

And I will use regimens for the benefit of the ill in accordance with my ability and my judgment, but from [what is] to their harm or injustice I will keep [them].

And I will not give a drug that is deadly to anyone if asked [for it], nor will I suggest the way to such a counsel.

And likewise I will not give a woman a destructive pessary.

And in a pure and holy way I will guard my life and my techné.

I will not cut, and certainly not those suffering from stone, but I will cede [this] to men [who are] practitioners of this activity.

Into as many houses as I may enter, I will go for the benefit of the ill, while being far from all voluntary and destructive injustice, especially from sexual acts both upon women's bodies and upon men's, both of the free and of the slaves.

And about whatever I may see or hear in treatment, or even without treatment, in the life of human beings—things that should not ever be blurted out outside—I will remain silent, holding such things to be unutterable [sacred, not to be divulged].

If I render this oath fulfilled, and if I do not blur and confound it [making it to no effect] may it be [granted] to me to enjoy the benefits of both life and of techné, being held in good repute among all human beings for time eternal. If, however, I transgress and perjure myself, the opposite of these.

Although it underscores the morality needed in the practice of medicine, and therefore in medical research, it is easy to see why the oath is of marginal value to modern physicians. It contains components that are polytheistic, sexist, proscriptive, celibate, and superstitious. Specific clinical proscriptions (i.e., abortion, surgery for urinary stones, and the use of "deadly" drugs) have long since fallen, as has the pledge for free instruction. Principles that have found widespread acceptance or are reflected in the ethics of today's medical practice are "doing no harm," prescribing for "the benefit of the ill," confidentiality, and an obligation to teach. Even confidentiality in modern medicine is open to some question (Siegler, 1982), particularly in the situations described below surrounding reportable diseases.

One could not expect the Hippocratic Oath to mention clinical trials, even implicitly, because the concept as we know it did not exist when it was written. However, acting for the benefit of the ill (implied plural) is noteworthy because it is an obligation to society as well as to the individual. The idea of experiment was known to ancient Greek physicians, but it was taken to mean a change in treatment for a patient on recommended therapy (Miles, 2004). Many medical treatises were simply the compilation of such cases.

Modern Ethic

The Hippocratic Oath is primarily an idealization filtered through modern views, and not a required code of conduct for all circumstances. These limitations also apply to our contemporary explicit ethical principles. The American Medical Association (AMA) does not endorse the Hippocratic Oath. The AMA's own principles of ethics contain similar sentiments but are modernized and oriented toward practitioners (AMA, 2001):

> The medical profession has long subscribed to a body of ethical statements developed primarily for the benefit of the patient. As a member of this profession, a physician must recognize responsibility to patients first and foremost, as well as to society, to other health professionals, and to self. The following Principles adopted by the American Medical Association are not laws, but standards of conduct which define the essentials of honorable behavior for the physician.
>
> 1. A physician shall be dedicated to providing competent medical care, with compassion and respect for human dignity and rights.
>
> 2. A physician shall uphold the standards of professionalism, be honest in all professional interactions, and strive to report physicians deficient in character or competence, or engaging in fraud or deception, to appropriate entities.
>
> 3. A physician shall respect the law and also recognize a responsibility to seek changes in those requirements which are contrary to the best interests of the patient.
>
> 4. A physician shall respect the rights of patients, colleagues, and other health professionals, and shall safeguard patient confidences and privacy within the constraints of the law.
>
> 5. A physician shall continue to study, apply, and advance scientific knowledge, maintain a commitment to medical education, make relevant information available to patients, colleagues, and the public, obtain consultation, and use the talents of other health professionals when indicated.
>
> 6. A physician shall, in the provision of appropriate patient care, except in emergencies, be free to choose whom to serve, with whom to associate, and the environment in which to provide medical care.
>
> 7. A physician shall recognize a responsibility to participate in activities contributing to the improvement of the community and the betterment of public health.
>
> 8. A physician shall, while caring for a patient, regard responsibility to the patient as paramount.
>
> 9. A physician shall support access to medical care for all people.

Principles 5 and 7 can be interpreted explicitly to recognize the obligation to conduct research, and therefore to perform clinical trials.

3.2.5 Physicians Always Have Multiple Roles

There are many circumstances besides clinical trials in which the physician's duty extends beyond the individual patient or is otherwise conflicting with the societal obligations stated in the AMA principles. In some circumstances significant obligations extend to other members of society or to patients yet to be encountered. In other circumstances the physician may have conflicting obligations to the same patient (e.g., care of the terminally ill patient). Dual roles for the physician are acknowledged by the codes of conduct cited in this chapter.

The examples listed below are circumstances consistent with both tradition and current ideas about "good" health care. They illustrate the somewhat mythical nature of acting only in the best interests of the individual patient. Although clinical trials highlight dual obligations, medical practitioners encounter and manage similar conflicts of commitment in other places. There have been occasional historical experimental studies that illustrate this point (e.g., Barber et al., 1970).

Examples

The first example is the *teaching and training* of new physicians and health care professionals. This is necessary for the well-being of future patients, and is explicitly required by the Hippocratic tradition, but is not always in the best interests of the current patient. It is true that with adequate supervision of trainees, the risks to patients from teaching are small. In some ways the patient derives benefit from being on a teaching service. However, the incremental risks attributable to inexperienced practitioners are not zero, and there may be few benefits for the patient. Even if safe, teaching programs often result in additional inconvenience to the patient. Aside from outright errors, teaching programs may perform more physical exams, diagnostic tests, and procedures than are strictly necessary. Thus the safest, most comfortable, and most convenient strategy for the individual patient would be to have an experienced physician and to permit no teaching around his or her bedside.

Vaccination is another example for which the physician advocates, and with the individual patient, accepts risks for the benefit of other members of society. With a communicable disease like polio, diphtheria, or pertussis, and a reasonably safe vaccine, the most *practical* strategy for the individual is to be vaccinated. However, the *optimal* strategy for the individual patient is to have everyone *else* vaccinated so that he or she can derive the benefit without any risk. For example, almost all cases of polio between 1980 and 1994 were caused by the oral polio vaccine (Centers for Disease Control, 1997). A strategy avoiding vaccination is impractical for the patient because the behavior of others cannot be guaranteed. This is not a workable policy for the physician to promote either, because it can be applied only to a few individuals, demonstrating the obligation of the physician to all patients.

Triage, whether in the emergency room, battlefield hospital, or domestic disaster, is a classic example of the physician placing the interests of some patients above those of others and acting for the collective good. It is a form of *rationing of health care*, a broader circumstance in which the interests of the individual are not paramount. Although in the United States we are not accustomed to rationing (at least in such stark terms), constrained health care resources produce it, and may increasingly do so in the future. For example, physicians working directly for profit-making companies and managed health plans, rather than being hired by the individual patient, could create competing priorities and highlight the dual obligations of the physician.

Abortion is another circumstance in which the physician has obligations other than to the individual patient. No stand on the moral issues surrounding abortion is free of this dilemma. Whether the fetus or the mother is considered the patient, potential conflicts arise. This is true also of other special circumstances such as the separation of conjoined twins or in utero therapy for fatal diseases in which there is no obvious line of obligation. For a discussion of the many problems surrounding fetal research, see Ramsey (1975).

Organ donation is almost never in the best medical interests of the donor, especially when the donor gives a kidney, bone marrow, blood, or other organ while still alive and healthy. It is true that donors have little risk in many of these situations, especially ones from which they recover quickly, such as blood donation, but the risk is not zero. Even in the case of the donor being kept alive only by artificial means, it seems that no medical benefit can be derived by donating organs, except to people other than the donor. However, it is clear that physicians endorse and facilitate altruistic organ donation, again illustrating their dual roles.

Issues surrounding *blood banking* apart from viewing blood products as organ donation also highlight the dual obligations of the physician. Many practices surrounding blood banking are regulated (dictated) by the government to assure the safety and availability of the resource. The physician overseeing the preparation and storage of blood products will consider needs and resources from a broad perspective rather than as an ideal for each patient. The practitioner may not be free to choose ideally for the patient with regard to such issues as single donor versus pooled platelets, age of the transfused product, and reduction of immunoactive components such as leukocytes.

The dual obligations of the physician can be found in requirements for *quarantine, reporting,* or *contact tracing* of certain diseases or conditions. Circumstances requiring this behavior include infectious diseases, sexually transmitted diseases, suspected child abuse, gunshot wounds, and some circumstances in forensic psychiatry. Although these requirements relate, in part, to nonmedical interests, reporting and contact tracing highlight dual obligations and seem to stretch even the principle of physician confidentiality. The fact that required reporting of infectious diseases compromises some rights of the patient is evidenced by government and public unwillingness to require it for HIV positive and AIDS diagnoses in response to concerns from those affected. There is no better contemporary example of this than the isolations and restrictions placed on patients with severe acute respiratory syndrome (SARS).

Treatments to control epidemics (aside from quarantine) can present ethical conflicts between those affected and those at risk. For example, in the summer of 1996 there was an epidemic of food poisoning centered in Sakai, Japan, caused by the O157 strain of *E. coli.* At one point in the epidemic almost 10,000 people were affected, and the number of cases was growing by 100 per day. Serious complications were due to hemolytic uremic syndrome. At least seven people died, and nearly all those infected were schoolchildren. As a means to stop or slow the epidemic, Sakai health officials considered treating with antibiotics 400 individuals who were apparently in their week-long incubation period. The treatment was controversial because, while it kills O157 *E. coli,* endotoxin would be released that could make the individuals sick. The obligations of health providers in such a situation cannot be resolved without ethical conflicts.

A final example of dual obligations can be found in *market-driven health care financing.* In this system employers contract with insurance companies for coverage and reimbursement. Both the patient and physician are effectively removed from negotiating important aspects of clinical care. If the physician is employed by a health care provider rather than the patient, competing obligations are possible (Levinsky, 1996). The U.S. Supreme Court recently ruled that health maintenance organizations (HMOs) cannot be sued for denial of services, which is simply their form of rationing health care for economic reasons.

These examples of the competing obligations of the physician are no more or less troublesome than clinical trials. Like clinical trials each has reasons why, and circumstances in which, they are appropriate and ethical. We are familiar with many of them and have seen their place in medical practice more clearly than the relatively recent arrival of clinical trials. However, the dual obligations of the physician are present and have always been with us. Obligations to individuals as well as to "society" can be compassionately and sensibly managed in most circumstances. This inevitable duality of roles that confronts the physician investigator is not a persuasive argument against performing clinical trials.

3.3 HISTORICALLY DERIVED PRINCIPLES OF ETHICS

Biomedical experimentation has taken a difficult path to reach its current state. A brief discussion of history is offered by Reiser (2002). The landmarks that have shaped current ethics are relatively few, but highly visible. The major landmarks tend to be crises that result from improper treatment of research subjects. A concise review is given by McNeill (1993, ch. 1). It can be difficult to extract the required principles, but it is essential to understand the troubled history.

3.3.1 Nuremberg Contributed an Awareness of the Worst Problems

The term "experimentation," especially related to human beings, was given a dreadful connotation by the events of World War II. The evidence of criminal and unscientific behavior of physicians in the concentration camps of Nazi Germany became evident worldwide during the 1946–1947 Nuremberg trials. There were numerous incidents of torture, murder, and experimentation atrocities committed by Nazi physicians. In fact 20 physicians and 3 others were tried for these crimes at Nuremberg (Annas and Grodin, 1992). Sixteen individuals were convicted and given sentences ranging from imprisonment to death. Four of the 7 individuals executed for their crimes were physicians.

At the time of the trial there were no existing international standards for the ethics of experimentation with human subjects. The judges presiding at Nuremberg outlined 10 principles that are required to satisfy ethical conduct for human experimentation. This was the Nuremberg Code, adopted in 1947 (U.S. Government, 1949). Full text is given in Appendix D. The Code established principles for the following points:

- Study participants must give voluntary consent.
- There must be no reasonable alternative to conducting the experiment.
- The anticipated results must have a basis in biological knowledge and animal experimentation.
- The procedures should avoid unnecessary suffering and injury.
- There is no expectation for death or disability as a result of the study.
- The degree of risk for the patient is consistent with the humanitarian importance of the study.
- Subjects must be protected against even a remote possibility of death or injury.
- The study must be conducted by qualified scientists.
- The subject can stop participation at will.

- The investigator has an obligation to terminate the experiment if injury seems likely.

The Nuremberg Code has been influential in United States and in international law to provide the groundwork for standards of ethical conduct and protection of research subjects.

3.3.2 High-Profile Mistakes Were Made in the United States

Despite the events at Nuremberg a persistent ethical complacency in the United States followed. In the late 1940s and early 1950s the American Medical Association (AMA) was keenly aware of Nuremberg and felt that its own principles were sufficient for protecting research subjects (Ivy, 1948; Judicial Council of the AMA, 1946). The principles advocated by the AMA at that time were (1) patient consent, (2) safety as demonstrated by animal experiments, and (3) investigator competence. A few examples of ethically inappropriate studies are sketched here. Some others are given by Beecher (1966).

In 1936 the U.S. Public Heath Service had started a study of the effects of untreated syphilis in Tuskegee, Alabama. Three-hundred-ninety-nine men with advanced disease were studied along with 201 controls. The study continued long after effective treatment for the disease was known, coming to public attention in 1972 (Brandt, 1978; Department of Health, Education, and Welfare, 1973; Jones, 1981). In fact there had been numerous medical publications relating the findings of the study, some after the development of penicillin (e.g., Schuman et al., 1955). The study had no written protocol.

Another study at the Jewish Chronic Diseases Hospital in Brooklyn in 1963 saw cancer cells injected into 22 debilitated elderly patients without their knowledge to see if they would immunologically reject the cells (Katz, 1972). Consent was said to have been obtained orally, but records of it were not kept. The hospital's Board of Trustees was informed of the experiment by several physicians who were concerned that the patients did not give consent. The Board of Regents of the State University of New York reviewed the case and concluded that the investigators were acting in an experimental rather than therapeutic relationship, requiring patient consent.

At Willowbrook State Hospital in New York, retarded children were deliberately infected with viral hepatitis as part of a study of its natural history (Katz, 1972). Some subjects were fed extracts of stool from those with the disease. Investigators defended the study because nearly all residents of the facility could be expected to become infected with the virus anyway. However, even the recruitment was ethically suspect because overcrowding prevented some patients from being admitted to the facility unless their parents agreed to the study.

3.3.3 The Helsinki Declaration Was Widely Adopted

In 1964 the eighteenth World Medical Association (WMA) meeting in Helsinki, Finland, adopted a formal code of ethics for physicians engaged in clinical research (World Medical Association, 1964). This became known as the Helsinki Declaration, which has been revised by the WMA in 1975 (Tokyo), 1983 (Venice), 1989 (Hong Kong), 1996 (South Africa), and 2000 (Scotland). This declaration is intended to be reviewed and updated periodically. The current version is provided in Appendix E.

The WMA was founded in 1947, and is an amalgamation of medical associations from the different countries of the world. The U.S. representatives come from the

AMA. The WMA has a service objective. In recent years the Association has been concerned about the potential to exploit developing countries for the conduct of human experimentation. Among clinical researchers in the United States, both the AMA and the WMA have low visibility.

Considerable controversy surrounds the 2000 version of the Helsinki Declaration. Before the exact language was adopted, the document was criticized as having a utilitarian ethic and serving to weaken the researchers' responsibility to protect study subjects. An exchange of points on these and related ideas is offered by Brennan (1999) and Levine (1999). The most substantive and controversial issues arise from discussions surrounding 16 trials investigating the vertical transmission of AIDS, conducted by U.S. investigators in Africa (Lurie and Wolfe, 1997). Some studies were randomized placebo-controlled trials testing the ability of drug therapy to reduce mother–infant transmission of HIV. The studies were planned by academic investigators, including some at Johns Hopkins, and sponsored by NIH and the CDC. All research protocols underwent extensive ethics reviewed by IRBs and the host country's health ministries, researchers, and practitioners.

The basic criticisms of these trials focus on the use of a placebo treatment when drug therapy is known to be effective at reducing HIV transmission, and placement of the trials in a developing country setting where no treatment was the norm, and hence the placebo would be seen as appropriate. The studies would not be considered appropriate in the United States, for example, and were consequently judged by some to be globally unethical.

The facts of the studies, criticisms, and rejoinders can be seen from the literature (Angell, 2000a; Varmus and Satcher, 1997). The issues are complex and relate to cultural differences, scientific needs, humanitarian efforts, medical and scientific paternalism, and ethics. I cannot resolve the issues here. But I do think many of the complaints about ethics were not well founded. For a mature, reasoned perspective on this, see the discussion by Brody (2002).

The essential issue for this discussion is that the Helsinki Declaration was revised in 2000 to make such studies more difficult or impossible. The relevant text is paragraph 29:

> The benefits, risks, burdens and effectiveness of the method should be tested against those of the best current prophylactic, diagnostic, and therapeutic methods. This does not exclude the use of placebo, or no treatment, in studies where no proven prophylactic, diagnostic or therapeutic method exists.

Although overtly ambiguous, the language is explicit when one considers the history behind why it was revised. This provision is poised to hinder needed research in developing countries, be ignored on occasion, or be modified. The FDA has historically pointed to the Helsinki Declaration in its guidelines but does not permit others to write its regulations. I suspect the FDA would not be able to embrace this version of the Helsinki Declaration unreservedly.

In October 2001 the WMA Council took the unusual step of issuing a note of clarification regarding paragraph 29. It was adopted by the WMA General Assembly, 2002. It says:

> The WMA is concerned that paragraph 29 of the revised Declaration of Helsinki (October 2000) has led to diverse interpretations and possible confusion. It hereby reaffirms its

position that extreme care must be taken in making use of a placebo-controlled trial and that in general this methodology should only be used in the absence of existing proven therapy. However, a placebo-controlled trial may be ethically acceptable, even if proven therapy is available, under the following circumstances:

Where for compelling and scientifically sound methodological reasons its use is necessary to determine the efficacy or safety of a prophylactic, diagnostic or therapeutic method; or

Where a prophylactic, diagnostic or therapeutic method is being investigated for a minor condition and the patients who receive placebo will not be subject to any additional risk of serious or irreversible harm.

All other provisions of the Declaration of Helsinki must be adhered to, especially the need for appropriate ethical and scientific review.

The second paragraph of the clarification is odd indeed because it seems to disregard situations where a placebo control is scientifically useful but ethically inappropriate. But the major difficulty with paragraph 29 is its dogmatism and the supposition that either equivalence or superiority trials will always be achievable.

Other provisions of the Helsinki Declaration contain elements that illustrate the dual role of the physician. For example, it begins by stating "It is the mission of the physician to safeguard the health of the people." However, in the next paragraph it endorses the Declaration of Geneva of the World Medical Association, which states regarding the obligations of physicians, "the health of my patient will be my first consideration," and the International Code of Medical Ethics, which states "a physician shall act only in the patient's interest when providing medical care which might have the effect of weakening the physical and mental condition of the patient." Later it states "Medical progress is based on research which ultimately must rest in part on experimentation involving human subjects."

It is noteworthy that the principle of acting only in the individual patient's interest seems to be qualified and that the declaration presupposes the ethical legitimacy of biomedical and clinical research. Among other ideas it outlines principles stating that research involving human subjects must conform to generally accepted scientific principles, be formulated in a written protocol, be conducted only by qualified individuals, and include written informed consent from the participants.

3.3.4   Other International Guidelines Have Been Proposed

The United Nations General Assembly adopted the International Covenant on Civil and Political Rights in 1976, which states (Article 7): "No one shall be subjected to torture or to cruel, inhuman or degrading treatment or punishment. In particular, no one shall be subjected without his free consent to medical or scientific experimentation." In 1982 the World Health Organization (WHO) and the Council for International Organizations of Medical Sciences (CIOMS) issued a document, *Proposed International Guidelines for Biomedical Research Involving Human Subjects,* to help developing countries apply the principles in the Helsinki Declaration and the Nuremberg Code. The guidelines were extended in a second document in 1991 dealing with epidemiologic studies, in part, in response to needs arising from field trials testing AIDS vaccines and drugs. The second document was called *International Guidelines for Ethical Review of Epidemiologic Studies.* In 1992 the Guidelines were revised at a meeting in Geneva, resulting in the *International Ethical Guidelines for Biomedical Research Involving Human Subjects.*

Some areas of medical research are not mentioned in the guidelines, including human genetic research, embryo and fetal research, and research using fetal tissue (Council for International Organizations of Medical Sciences, 1993).

Guideline 11 of these regulations states: "As a general rule, pregnant or nursing women should not be subjects of any clinical trials except such trials as are designed to protect or advance the health of pregnant or nursing women or fetuses or nursing infants, and for which women who are not pregnant or nursing would not be suitable subjects." This Guideline is directly opposite current thinking by many women's advocacy groups in the United States, the FDA, and NIH, which have relaxed or removed such restrictions in favor of allowing pregnant women more self-determination regarding participation in clinical research of all types.

3.3.5 Institutional Review Boards Provide Ethical Oversight

In response to mistakes and abuses in the United States, Congress established the National Commission for the Protection of Human Subjects of Biomedical and Behavioral Research through the 1974 National Research Act. Interestingly it was the first national commission to function under the new Freedom of Information Act of 1974, so all its deliberations were public and fully recorded. The Act required the establishment of Institutional Review Boards (IRB) for all research funded in whole or in part by the federal government. In the form of the 1978 Belmont Report, this Commission provided a set of recommendations and guidelines for the conduct of research with human subjects and articulated the principles for actions of IRBs (National Commission for Protection of Human Subjects of Biomedical and Behavioral Research, 1978).

In 1981 the federal regulations were modified to require IRB approval for all drugs or products regulated by the FDA. This requirement does not depend on the funding source, the research volunteers, or the location of the study. Regulations permitting compassionate use of experimental drugs were disseminated in 1987 and 1991. IRBs must have at least five members with expertise relevant to safeguarding the rights and welfare of patients participating in biomedical research. At least one member of the IRB should be a scientist, one a nonscientist, and at least one should be unaffiliated with the institution. The IRB should be made up of individuals with diverse racial, gender, and cultural backgrounds. Individuals with a conflict of interest may not participate in deliberations. The scope of the IRB includes, but is not limited to, consent procedures and research design.

The Belmont Report outlined ethical principles and guidelines for the protection of human subjects. A major component of this report was the nature and definition of informed consent in various research settings. In the Belmont Report three basic ethical principles relevant to research involving human subjects were identified. The report recognized that "these principles cannot always be applied so as to resolve beyond dispute particular ethical problems." These principles are discussed in the next section.

IRBs approve human research studies that meet specific prerequisites. The criteria are (1) the risks to the study participants are minimized, (2) the risks are reasonable in relation to the anticipated benefits, (3) the selection of study participants is equitable, (4) informed consent is obtained and appropriately documented for each participant, (5) there are adequate provisions for monitoring data collected to ensure the safety of the study participants, and (6) the privacy of the participants and confidentiality of the data are protected.

Informed consent is a particularly important aspect of these requirements. The consent procedures and documents must indicate that the study involves research, describes reasonable foreseeable risks and discomfort, and describes potential benefits and alternatives. In addition the consent document must describe the extent to which privacy of data will be maintained, treatment for injuries incurred, and who to contact for questions. Finally the consent indicates that participation is voluntary and no loss of benefits will occur if the patient does not enter the study.

3.3.6 Ethical Principles Relevant to Clinical Trials

The principles of ethics to which physicians aspire probably cannot be applied universally and simultaneously. Furthermore one cannot *deduce* an ethical course of action in all circumstances, even after accepting a set of principles. However, there are three principles of ethics outlined in the Belmont Report that are widely accepted in modern medical practice: respect for persons (individual autonomy), beneficence, and justice (Levine, 1986; National Commission for Protection of Human Subjects of Biomedical and Behavioral Research, 1978).

Taken together, they provide guidance for appropriate behavior when conducting human experimentation. The National Commission addressed the ethics of human research specifically in outlining the principles and acknowledged the conflicts that can occur in specific circumstances and even between the principles themselves.

Respect for Persons (Autonomy)

Autonomy is the right of self-governance and means that patients have the right to decide what should be done for them during their illness. Because autonomy implies decision, it requires information for the basis of a decision. Autonomous patients need to be informed of alternatives, including no treatment when appropriate, and the risks and benefits associated with each.

The principle of autonomy is not restricted to clinical trial settings but is broadly applicable in medical care. Practitioners who prefer a particular treatment usually recognize that realistic alternatives are possible and that the individual needs to make informed selections. In situations where the patient is incapacitated or otherwise unable to make informed decisions, the principle of autonomy extends to those closest to the patient. This may mean that patients without autonomy (e.g., children or incapacitated patients) should not be allowed to participate in research.

Clinical trials often ask that patients surrender some degree of autonomy. For example, in a trial the patient may not be able to choose between two or more appropriate treatments (randomization may do it), or the patient may be asked to undergo inconvenient or extensive evaluations to comply with the study protocol. In a masked clinical trial the patient may be made aware of risks and benefits of each treatment but be unable to apply that information personally with certainty. However, to some extent the patient's autonomy is retrievable. When clinical circumstances require it or when reliable information becomes available, the patient can be more fully informed or trial participation can be ended. The "re-consent" process is an example of this.

Temporarily giving up autonomy is not unique to clinical trials; it is a feature of many medical circumstances. Consider the patient who undergoes surgery using a general anesthetic. Although informed beforehand of the risks and benefits, the patient is not autonomous during the procedure, particularly if the surgeon encounters something unexpected.

Respect for persons is an idea that incorporates two ethical convictions. The first is autonomy, as discussed above, and second, that persons with diminished autonomy need protection from potential abuses. Some individuals, especially those who have illnesses, mental disability, or circumstances that restrict their personal freedom, have diminished autonomy. People in these categories may need protection, or even exclusion, from certain research activities. Other individuals may need only to acknowledge that they undertake activities freely and are aware of potential risks.

A circumstance in which application of this principle is problematic occurs in using prisoners for research purposes. One could presume that prisoners should have the opportunity to volunteer for research. However, prison conditions could be coercive on individuals who appear to volunteer for research activities. This is especially true if there are tangible benefits or privileges to be gained from participation in this research. Consequently, as the Belmont Report states, it is not clear whether prisoners should be allowed to volunteer or should be protected in such circumstances.

Beneficence and Nonmaleficence

Beneficence is a principle that reflects the patient's right to receive advantageous or favorable consideration, namely derive benefit. Nonmaleficence is the physician's duty to avoid harm (*primum non nocere*) and to minimize the risk of harm. We can refer to these principles jointly as beneficence. Because physicians also have a duty to benefit others when possible, the principle of beneficence has the potential to conflict with itself. For example, knowledge of what provides benefit and causes harm comes from research. Therefore investigators are obliged to make practical and useful assessments of the risks and benefits involved in research. This necessitates resolving the potential conflict between risk to participants and benefit to future patients.

Research can create more than minimal risk without immediate direct benefit to the research subject. In some cases such research will not be permitted by oversight committees. However, in other cases it may be justified. For example, many arguments have been made by patients infected with HIV that unproven but potentially beneficial treatments should be available to them. Some of these treatments carry the possibility of harm with a low potential for benefit. The use of baboon bone marrow transplantation in AIDS is an example.

The assessment of risks and benefits requires that research studies be scientifically valid and therefore properly designed. However, valid studies do not automatically have value or significance for science, the participants, or future patients. In addition to validity, the study must investigate an important question and have an appropriate risk−benefit ratio for the participants. Investigators should probably establish the scientific validity of a proposed study prior to considering the ethical question of whether or not it has value or significance. The principle of beneficence can be satisfied only when both components are favorable.

Justice

The principle of justice addresses the question of fairly distributing the benefits and burdens of research. Compensation for injury during research is a direct application of this principle. Injustice occurs when benefits are denied without good reason or when the burdens are unduly imposed on particular individuals. In the early part of this century, burdens of research fell largely upon poor patients admitted to the public wards of the hospital. In contrast, benefits of improvements learned at their expense often accrued to private patients. The injustice of denying treatment to men in the Tuskegee Syphilis Study has already been discussed.

There are some circumstances where distinctions based on experience, competence, age, and other criteria justify differential treatment. For reasons already mentioned, research should be conducted preferentially on adults rather than children. Institutionalized patients and prisoners should be involved in research only if it relates directly to their conditions and there are not alternative subjects with full autonomy.

3.4   CONTEMPORARY FOUNDATIONAL PRINCIPLES

Principles of biomedical ethics imply several contemporary requirements for the ethical conduct of research (Wells, 1992). These include informed consent of the participants, assessment and disclosure of risks and benefits, and appropriate selection of research subjects. In today's practice, application of these principles requires other components such as optimal study design, investigator competence, a balance of risk and benefit for study participants, patient privacy, and impartial oversight of consent procedures (Sieber, 1993).

The best synthesis of these ideas is given by Emanuel et al. (2000), who provided requirements for evaluating the ethics of clinical research studies. (Table 3.2). These were augmented later to add the requirement for a collaborative partnership (Emanuel, 2003). The individual requirements are discussed below. Today, at least in academic medical centers, the appropriateness of these components is assured by a combination of investigator training, infrastructure, institutional review, and peer review. However, there is no formula to guarantee these, and as with all scientific endeavors, much rests on judgment and trust.

3.4.1   Collaborative Partnership

A collaborative partnership implies that the research involves the community in which it takes place. Members of the community participate in planning, oversight, and use of research results. Typical mechanisms for accomplishing this include forming community advisory boards, placing patient advocates on review or monitoring committees, and using the community to advocate for research support. This type of partnership may be a critical foundation for ethically conducting some types of research. An example

TABLE 3.2   Principles for Ethical Clinical Trials

Collaborative partnership
Scientific value
Scientific validity
Fairness of subject selection
Favorable risk–benefit
Independent review
Informed consent
Respect for enrolled subjects

Source: Emanuel et al. (2000).

might be studies in emergency situations (e.g., cardiopulmonary resuscitation) where informed consent is not possible. Prospective community awareness of such research would therefore be essential.

3.4.2 Scientific Value

If the study has scientific value, useful knowledge will be derived from the research. This means not only that the question of importance but also that the results will be made available to society at large whether "positive" or "negative." Value is also relevant to the use of scarce resources. Studies of high value will be allowed to consume resources preferentially.

A study that has value should produce a product (result) that is visible to the scientific community through publication and presentation. An extension of this idea is that results should not be kept confidential because of proprietary concerns. More to the point, the investigator should not participate in a clinical trial that is likely to be valueless if the sponsor disapproves of the findings.

3.4.3 Scientific Validity

Scientific validity is a consequence of good study design, and means that patients on a trial are contributing to answering a question that is important and that has a high chance of being answered by the experiment being undertaken. Research designs that are grossly flawed or those that cannot answer the biological question are not ethical. Similarly those that ask unimportant questions are unethical, even if they pose minimal risk. As Rutstein (1969) said:

> It may be accepted as a maxim that a poorly or improperly designed study involving human subjects ... is by definition unethical. Moreover, when a study is in itself scientifically invalid, all other ethical considerations become irrelevant.

A second component of validity derives from investigator competence: technical, research, and humanistic. Technical competence is assessed by education, knowledge, certification, and experience. In addition to technical competence, the investigator must have research competence. This may be based on both training and experience. One tangible aspect of research competence might be that the investigator has performed a competent systematic review of current knowledge and previous trials to be certain that the planned study is justified. When the clinical trial is completed, a valid, accurate, and complete description of the results should be published to ensure dissemination of the knowledge (Herxheimer, 1993). A component of a valid finished trial is an appropriate statistical analysis.

Humanistic competence requires compassion and empathy. These cannot be taught in the same way that technical and research competence can, but the proper clinical and research environment and good research mentoring facilitate it.

3.4.4 Fair Subject Selection

Historically participation in research has been viewed as a risk. At the level of the individual subject it is both a risk and a benefit with acceptable balance as indicated below. When things go awry, such as the extreme case of death attributable to the treatment, participation will be viewed retrospectively purely as a risk. Distributing the risks of research fairly and avoiding exploitation of vulnerable people continues to be a principle of ethical behavior. Voluntary selection is a foundation of fairness. It follows that removal of obstacles to participation is also within the principle of fairness.

In recent times participation in research has been viewed increasingly as a benefit or a right. This view is not so much at the level of the individual research subject but pertains to sociodemographic, advocacy, or political groups claiming that their health needs have been neglected. This view makes a direct connection between representation in a study cohort and (disproportionate) derived benefits for similar members of society. It has political origins rather than scientific ones.

The view of research participation as a right embraces several fallacies. First, it does not fully acknowledge the view based on risk. Second, it assumes that sociodemographic groups derive benefits when their members enter research studies, which is not always true. Third, it assumes that knowledge about groups is gained by direct participation alone, namely that studies have primarily empirical external validity.

3.4.5 Favorable Risk–Benefit

Patients must be excluded from the study if they are at undue risk or are otherwise vulnerable (Weijer and Fuks, 1994). Having the study reviewed by the IRB or other ethics board, satisfying eligibility criteria, and using informed consent do not eliminate this duty. Patients could be at high risk as a result of errors in judgment about their risk, atypical reactions or side effects from the treatment, or for unknown reasons. The individuals affected and others likely to be affected should be excluded from further participation.

The assessments of risk and benefits implies that the research is properly designed and has had competent objective review. If alternative ways of providing the anticipated benefits to the patient without involving research are known, they must be chosen. Investigators must distinguish between the probability of harm and the severity of the effect. These distinctions can be obscured when terms like "high risk" or "low risk" are used. For example, if a life-threatening or fatal side effect is encountered with low frequency, is this high risk? Similarly benefits have magnitudes and probabilities associated with them. Furthermore risks or benefits may not accrue only to the research subject. In some cases the risks or benefits may affect patient families or society at large.

A favorable risk–benefit setting does not require that the subjects be free of risk or that they be guaranteed benefit. The tolerance for risk increases as the underlying severity of the disease increases. In the setting of a fatal disease, patients and research subjects may be willing to accept a modest chance for minimal symptomatic improvement with even a significant risk of death. A good example of the reasonableness of this is lung volume reduction surgery for patients with advanced emphysema. Even in properly selected patients, surgery carries a risk of death far in excess of most elective procedures, and there may be only a 30% chance of modest (but symptomatically important) benefit.

3.4.6 Independent Review

Review of proposed and ongoing research studies is performed by institutions through at least two mechanisms. In the United States the first is the Institutional Review Board

(IRB), which is responsible for the ethical oversight of all Public Health Service sponsored investigation. In other countries this role is covered by an Independent Ethics Committee (IEC) (Australia), a Local Research Ethics Committee (LREC) (England), or a Research Ethics Board (Canada). For convenience, I will refer to all such committees as IRBs. IRB committees are typically composed of medical practitioners, bioethicists, lawyers, and community representatives. They review planned trials from an ethical perspective, including consent documents and procedures. Increasingly IRBs are also being asked to review the scientific components of research studies. This can be helpful when the expertise is available and a hindrance when it is not.

A second method by which institutions or collaborative clinical trial groups review studies is by using a Treatment Effects Monitoring Committee (TEMC). These committees oversee ongoing clinical trials with regard to treatment efficacy and safety. If convincing evidence about efficacy is provided by the trial before its planned conclusion, the DSMC will recommend early termination. Similarly, if serious unforeseen toxicities or side effects are discovered, the trial might be halted. Designing and executing this aspect of a trial can be quite complex and is discussed in Chapter 14.

Most large multicenter clinical trials have additional layers of concept development and/or review, although perhaps not focused only on ethics. Reviews might be performed by program staff at sponsors, the FDA, or collaborators. Such reviews tend to be focused on scientific issues but always include a perspective on ethics. At NCI designated Comprehensive Cancer Centers, an extra formal structured scientific and ethics review is required prior to the submission of a research project to the IRB. The mechanism of study review is itself peer reviewed, and the process provides a layer of quality assurance for cancer trials that usually makes IRB scrutiny more efficient and productive.

3.4.7 Informed Consent

Informed consent is a complex but important aspect of the practice and regulation of clinical trials. The requirement for consent is grounded in moral and legal theory and clinical practice. A perspective on this and historical developments in informed consent is given by Faden and Beauchamp (1986). Broad reviews of issues surrounding informed consent are given by Appelbaum, Lidz, and Meisel (1987) and Rozovsky (1990). In the context of AIDS, a useful review has been written by Gray, Lyons, and Melton (1995).

Common errors in consent documents include excessive length and too high a reading level for adequate comprehension. Both of these are easily correctable. Also investigators often wrongly view consent as a document rather than a process. The goal is to transmit culturally valid information regarding risks and benefits of participation in research. Because of the liability climate in the United States, there is sometimes a perspective here to view the consent document and process largely as a protection for the investigator.

Elements of Informed Consent

Elements of informed consent include information provided to the patient, comprehension of that information by the patient and his or her family, and an assessment of the voluntary nature of the consent. The information required in a consent document generally includes the nature of the research procedure, its scientific purpose, and alternatives to participation in the study. Patient comprehension is facilitated by careful attention to the organization, style, and reading level of consent forms. When children are subjects of research, it is frequently necessary to obtain informed consent from the legal guardian and obtain assent from the child. To verify that consent is given voluntarily, the conditions under which it is obtained must be free of coercion and undue influence. Conditions should not permit overt threats or undue influence to affect the individual's choice.

A problem area for consent is in conducting research on emergency medical treatments. It is instructive to review the problems with consent in the emergency room setting as a microcosm of consent issues (Biros et al., 1995). The principal difficulty is in being able to obtain valid consent from critically ill patients to test promising new treatments with good research designs. Some patients may be unconscious, relatives may be unavailable, or there may not be sufficient time to meet the same standards of informativeness and consent, as in ordinary hospital or clinic settings.

In 1996 the FDA and NIH proposed new measures for the protection of research subjects in emergency settings. The new FDA rules and NIH policies on emergency consent waiver make it easier to study drugs and devices in patients with life-threatening conditions who are unable to give informed consent. The new regulations permit enrolling patients in research studies without their consent provided the following criteria are met:

1. An independent physician and an IRB agree to the research and that it addresses a life-threatening situation.
2. The patient is in a life-threatening situation.
3. Conventional treatments are unproven or unsatisfactory.
4. The research is necessary to determine the safety and efficacy of the treatment and it cannot be carried out otherwise.
5. Informed consent cannot feasibly be obtained from the patient or legal representative.
6. The risks and potential benefits of the experimental procedure are reasonable compared with those for the underlying medical condition of the patient and standard treatments.
7. Additional protections are in place such as consultations with the community, advance public disclosure of the study design and risks, public disclosure of the study results, and FDA review of the study protocol.

The merits and weaknesses of these specialized rules will be evident in the next few years as they are applied in specific research projects.

3.4.8 Respect for Subjects

Respect for subjects pertains to the way that potential participants are approached and given options, as well as the ongoing engagement of those who agree to participate in the study. One feature of respect is privacy, discussed below. Second is allowing participants to change their mind and withdraw from the trial without incurring penalties. Third, the new information gathered from the study must be made available to participants, perhaps during the course of the trial. Results should be available to the participants at the end of the study. Finally, the interests of the participants should be continually monitored while the study is taking place. Treatment effects monitoring (discussed in Chapter 14) provides a mechanism to assure certain of these rights.

Patient rights to privacy has a long tradition and has been underscored recently by the Health Insurance Portability and Accountability Act of 1996 (HIPAA) regulations (DHHS, 2000). The purpose of HIPAA was to guarantee security and privacy of health information. The burden on researchers is likely to be considerable and, as of this writing, not fully evident.

The privacy right has been made ascendant even in circumstances such as the AIDS epidemic, where certain benefits to society could be gained by restricting the right to privacy (e.g., contact tracing and AIDS screening). It is maintained by appropriate precautions regarding written records and physician conduct. In the information age, extra care is required to maintain privacy on computerized records or other electronic media that can be easily shared with colleagues and widely disseminated.

Maintaining privacy requires steps such as patient consent, restricting the collection of personal information to appropriate settings and items, ensuring security of records, and preventing disclosure. These and other privacy principles have been crafted into a comprehensive set of guidelines in Australia (Cooper, 1991) and are relevant broadly.

3.5 METHODOLOGIC REFLECTIONS

The principles of ethics outlined above are directly represented in the conduct of trials, especially with respect to formal ethics review, investigator competence, informed consent, and disclosure of risks. There are further connections between principles of ethics and important study design features discussed later in this section. Some of these design characteristics are occasionally contentious.

To begin with, it is worthwhile to examine the ethics imperative to conduct sound research—that is, to generate reliable evidence regarding therapeutic efficacy.

3.5.1 Practice Based on Unproven Treatments Is Not Ethical

Despite knowledge, skill, and empathy for the patient the practitioner sometimes selects treatments without being absolutely certain about relative effectiveness. Many accepted therapies have not been developed or tested with the rigor of evidence based medicine. Treatment preference based on weak evidence does not provide an ethics imperative as strong as that for therapies established on a rigorous scientific foundation, if at all. Astute practitioners are alert to both the ethical mandate to use the best treatment, and the imperative to understand the basis of a practice and challenge it with rigorous experiments when appropriate. These mandates are not in opposition. To the contrary, the obligation to use the best treatment implies a requirement to test existing ones and develop new therapies.

Some untested practices are sensible and are not likely to cause harm, such as meditation, exercise, relaxation, and visualization. These and related activities may be lifestyles as much as they are therapies. They may improve a patient's quality of life, especially when applied to self-limited conditions, and normally do not replace treatments of established benefit. The discussion here does not pertain to such adjunctive treatments.

Ethics concerns arise when unproven or untested therapies replace proven ones, particularly for chronic or fatal illnesses (Kennedy et al., 2002). Ignorance about the relative merits of any treatment cannot carry ethical legitimacy for the physician, even in a purely "practice" setting. To summarize:

> A physician's moral obligation to offer each patient the best available treatment cannot be separated from the twin clinical and ethical imperatives to base that choice of treatment on the best available and obtainable evidence. The tension between the interdependent responsibilities of giving personal and compassionate care, as well as scientifically sound and validated treatment, is intrinsic to the practice of medicine today . . . . Controlled clinical trials—randomized when randomization is feasible, ethically achievable, and scientifically appropriate—are an integral part of the ethical imperative that physicians should know what they are doing when they intervene into the bodies, psyches, and biographies of vulnerable, suffering human beings. (Roy, 1986)

Lack of knowledge can persist purely as a consequence of the newness of a therapy. In this circumstance clinical trials may be the most appropriate way to gain experience and acquire new knowledge about the treatment. However, ignorance can remain even in the light of considerable clinical experience with a therapy, if the experience is outside of a suitably structured environment or setting that permits learning. This can happen when proponents of a treatment have not made sufficient efforts to evaluate it objectively, even though they may be clinically qualified and have good intentions.

There are widely used unproven therapies for many conditions including AIDS, cancer, arthritis, diabetes, musculoskeletal conditions, skin diseases, and lupus. Such therapies are particularly common in cancer and AIDS (Astin, 1998; Sparber, Bauer, Curt et al., 2000; Sparber, Wootton, Bauer et al., 2000; Fairfield et al., 1998; Paltiel et al., 2001). The National Center for Complementary and Alternative Medicine at NIH defines "alternative" therapies to be those that are unproven and has attempted to investigate some of them.

There is a wide spectrum of such therapies, even if one restricts focus to a single disease like cancer (Aulas, 1996; Brigden, 1995; Cassileth and Chapman, 1996; Office of Technology Assessment, 1990). The large number of unproven therapies suggests that weak or biased methods of evaluation are the norm. Many fringe treatments for various diseases are discussed on an Internet site that monitors questionable treatments (www.quackwatch.org). A broad view of ethics related to the complementary and alternative medicine context is provided in the book by Humber and Almeder (1998).

Example: Unproven Cancer Therapies
The American Cancer Society (ACS) has investigated and summarized scientific evidence concerning some unproven cancer treatments through its Committee on Questionable Methods of Cancer Management (e.g., ACS, 1990, 1993). Over 100 unproven or untested treatments for cancer are listed on the above-named Internet site for fringe therapies. A few examples are provided in Table 3.3. Although many or all of these treatments have professional advocates and some have been associated with anecdotes of benefit, they are most often used in an environment that inhibits rigorous evaluation. See Barrett (1993) for a review of this and related subjects.

Unconventional methods like those in Table 3.3 often sound reasonable to patients who have few options and lack knowledge about cancer treatment. Few, if any, have been shown to be safe and effective for any purpose using rigorous clinical trials. Therefore evidence supporting them remains unconvincing despite the "clinical experience" that has accumulated. Some treatments listed in Table 3.3 have been studied

TABLE 3.3   Some Questionable Cancer Therapies Listed on the American Cancer Society Web Site

Antineoplastons	Immuno-augmentative therapy
Anvirzel	Induced remission therapy
Bio-ionic system	Induced hypoglycemic treatment
BioResonance Tumor Therapy	Insulin potentiation therapy
CanCell (also called Cantron, Entelev, and	Intra-cellular hyperthermia therapy
Protocel)	Krebiozen
Canova method	Laetrile
Cansema system	Livingston-Wheeler regimen
Chaparral	Macrobiotic diet
Controlled amino acid therapy	Metabolic therapy
Coral calcium	Mistletoe/Iscador
Di Bella therapy	Moerman diet
Dimethyl sulfoxide	Multi-wave oscillator
Electron replacement therapy	Oncolyn
Elixir Vitae	PC-SPES
Escharotic salves	Polyatomic oxygen therapy
Essiac	Resan antitumor vaccine
Galavit	Revici cancer control
Galvanotherapy	Shark cartilage
Gerson method	Stockholm Protocol
Gonzalez (Kelley) metabolic therapy	Sundance Nachez mineral water
Grape cure	Ultraviolet blood irradiation
Greek cancer cure	"Vitamin B-17" tablets
Hoxsey treatment	Vitamin C
Hydrazine sulfate	Wheat grass
Hyperthermia, whole body	Zoetron therapy

and found to be ineffective (e.g., Laetrile, hydrazine). Others have proven effective uses, but not when employed unconventionally (e.g., vitamin C, dimethyl sulfoxide).

As many as 5% of cancer patients abandon traditional therapies in favor of alternative methods and up to 70% of cancer patients employ some alternative therapy (McGinnis, 1990). Unconventional therapy is used by a sizeable fraction of people in the United States (Eisenberg et al., 1993). Issues related to the evaluation of complementary therapy are discussed in Section 4.5 and an example of the evaluation of a particular unconventional treatment is given in Section 16.7.4. For a perspective on the potentially false basis for, and consequences of, some alternative therapies, see the discussion of shark cartilage by Ostrander et al. (2004).

Unproven treatments for other diseases have proponents, but can be found in similar settings that discourage objective evaluation. In any case, strongly held opinion cannot ethically substitute for experimental testing. Efforts by NIH to study some unconventional treatments have been subject to controversy (e.g., Marshall, 1994). Critical views are offered by Atwood (2003), Green (2001), Frazier (2002), and Stevens (2001).

3.5.2   Ethics Considerations Are Important Determinants of Design

Clinical trials are designed to accommodate ethical concerns on at least two levels. The first is when the practitioner assesses the risks and benefits of treatment for the individual patient on an ongoing basis. The physician always has an obligation to terminate experimental therapy when it is no longer in the best interests of an individual patient. Second, the physician must be constantly aware of information from the study and decide if the evidence is strong enough to require a change in clinical practice, either with regard to the participants in the trial or to new patients. In many trials this awareness comes from a Treatment Effects Monitoring Committee, discussed in Chapter 14. Both of these perspectives require substantive design features.

Consider early pharmacologic oriented studies in humans as an example. The initial dose of drug employed, while based on evidence from animal experiments, is conservatively chosen. It is unlikely that low doses will be of benefit to the patient, particularly with drugs used in life-threatening diseases like cancer or AIDS. Ethical concern over risk of harm requires these designs to test low doses ahead of high doses. Such studies often permit the patient to receive additional doses of drug later, perhaps changing their risk–benefit ratio.

In the next stage of drug development, when a therapeutic dosage has been established, staged designs, or those with "stopping rules," are frequently used to minimize exposing patients to ineffective drugs, and to detect large improvements quickly. Similar designs are used in comparative trials. These designs, developed largely in response to ethical concerns, are discussed in more depth in Chapter 14. In these and other circumstances ethical concerns drive modifications to the clinically, statistically, or biologically optimal designs.

The ethics needed in research design have been illustrated exceptionally well in recent circumstances surrounding trials in acute respiratory distress syndrome (ARDS) conducted by the ARDS Network. This multicenter group was initiated in 1994 by the National Heart Lung and Blood Institute (NHLBI). A more detailed history of the controversy sketched here is provided by Steinbrook (2003a). The ARDS Network randomized study of tidal volumes took place between 1996 and 1999, and was published in 2000 (ARDS Network, 2000) indicating the superiority of low tidal volume management. A randomized study of the use of pulmonary artery catheter versus central venous catheter was begun in July 2000.

Both studies were criticized by two physicians in complaints to the Office for Human Research Protections (OHRP) because of their concerns regarding the nature of the control groups. The tidal volume study was published at the time concerns were voiced, but the fluid and catheter trial was less than half completed. Because of the OHRP attention, the sponsor voluntarily suspended the trial in 2002 and convened an expert review panel. Although the panel firmly supported the study design and conduct, the trial remained closed while OHRP commissioned a separate review. In July 2003 after nearly a year, the OHRP allowed the trial to continue without modification (Steinbrook, 2003b).

This case illustrates the attention given to the ethics of research design and the extraordinary—perhaps inappropriate—power that individuals can have when such questions are raised. I say inappropriate for two reasons. First, the suspension ignored the considerable process leading to the study implementation, including multidisciplinary scientific review and oversight and redundant independent reviews by Institutional Review Boards (Drazen, 2003). The IRBs are explicitly approved by OHRP. Second, the suspension process did not itself provide due process for disagreements (e.g., the OHRP did not make public the names of its eight consultants). The study may well have been damaged, yet the OHRP did not require any design changes, asking only for additional IRB reviews and changes to the consent documents.

3.5.3   Specific Methods Have Justification

Rigorous scientific evaluation of therapies fits well with imperatives of ethics. Tensions that arise can be lessened through scrupulous attention to principles but tend to recur with some specific elements of design. However, most elements of good study design do not trouble our sense of ethics, and they may be required as part of good research practice. Examples of design elements that are always appropriate include a written protocol, prospective analysis plan, defined eligibility, adequate sample size, and safety monitoring.

Other components of design may raise ethics questions, at least in some settings. Examples include use of a control group, method of treatment allocation (e.g., randomization), placebos, and the degree of uncertainty required to conduct a trial. Concern over these points can be illustrated by recent randomized trials of extracorporeal membrane oxygenation (ECMO) for neonates with respiratory failure. This series of studies has been discussed at length in the pediatric and clinical trials literature. Some issues related to treatment allocation and the ECMO trials are discussed in Chapter 13. For a recent review with useful references from an ethical perspective, see Miké, Krauss, and Ross (1993).

Randomization

The justification for randomization follows from two considerations: (1) a state of relative ignorance—equipoise (Freedman, 1987) or uncertainty—and (2) the ethical and scientific requirement for reliable design. If sufficient uncertainty or equipoise does not exist, randomization is not justified. Because convincing evidence can develop during the conduct of a trial, the justification for continuing randomization can be assessed by a Treatment Effects Monitoring Committee (Data Monitoring Committee) (Baum, Houghton, and Abrams, 1994). This is discussed briefly below and in Chapter 14.

Randomization is a component of reliable design because it is the main tool to reduce or eliminate selection bias. It may be the only reliable method for doing so. Randomization controls effects due to known predictors (confounders), but importantly it also controls bias due to unobserved or unknown factors. This latter benefit is the major, and often overlooked, strength of randomization. Methods of treatment allocation, including randomization, are discussed in Chapter 13. It should be noted that the merits of randomization are not universally accepted either on ethical or methodological grounds (e.g., Urbach, 1993), but mainstream opinion agrees on its usefulness and appropriateness.

Treatment Preference

A physician with a preference for a particular treatment should not be an investigator in a clinical trial. This assumes that the preference is strong enough so that the physician does not recognize a state of equipoise. A potential participant that strongly prefers one treatment in a comparative trial should not participate in the study. Even if individuals have preferences based on weak evidence or nonrational factors, they are not eligible to participate in the trial. A similar proscription applies to patients who prefer some treatment not under study, for example, in a single-arm trial. Subjects with firm preferences or reservations are more likely to withdraw from the study or develop strong dissatisfaction with their results. Exclusions because of strong preferences are required by the principle of autonomy. Investigators should not exert too much effort to overcome the reservations of potential participants because it might be misinterpreted as coercion. Such concerns work against over enthusiastic "inclusiveness"; for example, see Section 12.4.4.

Informed Consent

Informed consent is a vital element of clinical trials. There are numerous examples of studies where patients have been exposed to potentially or definitively harmful treatments without being fully apprised of the risk. Some possible examples from tests of radioactive substances from 1945 to 1975 recently came to light (U.S. Congress, 1980). Presumably patients would not voluntarily accept some of the risks inherent in those studies if properly informed. Standards for informed consent were different then than they are now, making definitive interpretation of those events somewhat difficult.

When problems with adequate informed consent occur, it is frequently the result of investigator error. Patients and their families are vulnerable, and technical information about new treatments is hard to understand, especially when it is presented to them quickly. This can also be a problem when using treatments already proven to be effective. Informed consent procedures, originally developed to protect research subjects from exploitation, are now viewed by some researchers as a means to protect them from litigation.

Monitoring

During the conduct of a trial, convincing evidence regarding the outcome could become available. Trial designs routinely incorporate interim analyses and guidelines to deal with this possibility without disrupting the statistical properties of the study. If investigators find early evidence convincing, they should recommend stopping the study to minimize the number of subjects receiving inferior therapy. Interim analysis is especially of concern in comparative experiments.

Safety monitoring of data collected during a clinical trial is also a required oversight. Adverse events attributable to therapy must be reported to sponsors and regulators promptly. Both safety and efficacy concerns may be addressed by a Treatment Effects Monitoring Committee (TEMC). This aspect of trial design and conduct is discussed in Chapter 14.

Use of Placebo Controls

It has been argued that the history of medical therapeutics up to the recent age of experimental development has been the history of the placebo effect (Shapiro and Shapiro, 1997). Regardless of the accuracy of the claim, the power of suggestion needs to be considered when evaluating any treatment. The effect can be controlled by using a placebo (or sham) in a comparative experiment, and it is often necessary and practical to do so. Thus the use of a placebo, like randomization, is justified by the imperative to employ valid design.

It is not always ethical to employ a placebo—for example, if withholding treatment would place subjects at risk. Some clinical trials continue to be performed using placebos in questionable situations. Rothman and Michels (1994) discussed this problem and offered some examples. The choice of an appropriate control depends, in part, on the nature and strength of physicians' belief about the treatment in question and alternatives to it. Except for trivial situations, it is not appropriate to replace effective therapy with a placebo, even with "informed consent." It is better to evaluate the new treatment against standard therapy without compromising the benefit to the patients in the control group.

CLINICAL TRIALS

Placebos and "no treatment" are not exactly equivalent. In a comparative design, subjects might be randomized to a treatment, B, or its placebo, P. If these were the only treatments, then P would be a "no treatment" arm. However, if there is an active treatment, A, that ethically should be administered to everyone, then B might still be rigorously tested by randomizing between $A + B$ and $A + P$. Thus tests of the incremental effects of a new treatment can often be placebo controlled. If B is a potential substitute for A, then a direct comparison, with the need for a placebo, might be warranted.

There are circumstances in which investigators and patients might be comfortable not using standard therapy, and comparing a new treatment to a placebo. This might be the case, for example, if the standard treatment is weakly effective and associated with high morbidity and the new treatment being tested has low morbidity. Such a situation is not unique to trials because patients and physicians could sensibly refuse any treatment where the risk–benefit ratio is unfavorable. Also placebo controls might be scientifically and ethically justified when new evidence arises suggesting that standard therapy is ineffective.

The placebo question is made more worrisome when considering surgical treatments, for which a placebo would be a sham procedure. Sham procedures would seem to always place subjects at unacceptable risk and therefore be unacceptable. Such a conclusion is not sufficiently refined and is discussed further in Chapter 4.

Demonstration Trials

When ample evidence already exists or investigators are already convinced that a particular treatment is best, it may be unethical to conduct a trial to convince colleagues of its superiority. Such studies are sometimes termed "demonstration trials" and are subject to comparator bias (see Mann and Djulbegovic, 2004). There are situations where not all practitioners are convinced by the same evidence or same standard of evidence. This happens often in different countries and/or depending on strength of prior belief and local standards of practice. Ethical standards are also local, and some studies are acceptable to physicians in one culture but viewed as inappropriate in another. In any case, most practitioners would discourage comparative trials performed only to strengthen evidence.

Sometimes the standard of care contains procedures or treatments that are unnecessary. Electronic fetal monitoring during labor may be an example. A demonstration trial may be appropriate in such circumstances. More generally, demonstration trials to show equivalence are less problematic than those intended to show a difference.

3.6 PROFESSIONAL CONDUCT

3.6.1 Conflict of Interest

Objectivity is an unattainable ideal for all scientists, especially those working in most clinical research studies. Elsewhere I have discussed the need for using methods, procedures, and outcomes that encourage objectivity. It is equally important to be objective regarding behavioral factors that affect investigators themselves. Breakdown in objectivity often arises from conflicts of interest, a descriptive term that many organizations also define explicitly. There is much evidence that objectivity in clinical research can be degraded by commercial interests (Angell, 2000b).

The generic approach to conflict of interest is to define it in financial terms. For most scientists and research settings, this can be a serious error. Most of us are far more likely to encounter influential intellectual conflicts of interest than financial ones. The actual currency of clinical research is ideas, in which it is all too easy to become invested and lose balance. Even when financial conflicts are managed according to policy, as most are nowadays, there can remain a residual intellectual conflict. This might help explain the observation (e.g., Bekelman, Li, and Gross, 2003; Lexchin et al., 2003) that industry sponsored research is more likely to result in positive findings or those favorable to the sponsor. Concerns over this issue are heightened by the fact that many formal collaborations between medical schools and industry sponsors do not follow guidelines for data access and publication set up by the International Committee of Medical Journal Editors (Schulman et al., 2002).

The Code of Federal Regulations (42 CFR 50) specifies rules to ensure objectivity in research sponsored by the Public Health Service. Investigators must disclose significant financial interests defined as equity exceeding $10,000 or 5% or more ownership that would reasonably appear to affect the design, conduct, or reporting of research supported by the PHS. Personal financial interests as well as those of a spouse and dependent children are included.

Significant financial interest is defined as anything of monetary value including salary, consulting fees, honoraria, equity such as stocks or other types of ownership, and intellectual property rights such as patents, copyrights, and royalties. Significant financial interest does not include income from seminars, lectures, or teaching sponsored by public or not-for-profit organizations below the thresholds cited above.

Management of financial interests that could create bias may require investigators to reduce or eliminate the holdings, recognition, and oversight by the research institution, public disclosure, or modification of the research plan. Investigators can anticipate such problems and disqualify themselves from participation in studies that present the opportunity for bias. Many experienced investigators have potential conflicts, often of a minor or indirect nature. An example is when their university or institution benefits directly from a particular study, although they do not. Such cases can often be managed simply by public disclosure.

Research misconduct or fraud is an extreme example of loss of objectivity. This important subject is discussed in Chapter 22. Misconduct can have many contributing factors, but a lack of objectivity about the role of the investigator is often one cause.

Although often unrecognized, competing research interests are a potential source of lack of objectivity. Consider the investigator who has multiple research opportunities, or a mix of research and "practice" options, available to patients. For example, there might be competing clinical trials with the same or similar eligibility criteria. This can lead to the physician acting as an investigator for some patients, a different investigator for others, and as a practitioner for still others. If treatment or research preferences are driven partly by financial concerns, reimbursement issues, ego, research prowess, or other pressures, the potential for conflicting interests is high.

Although this situation is common in academic medical centers where many studies are available, investigators often do not recognize it as a source of conflict. A solution to such problems is facilitated by nonoverlapping investigations, a clear priority for any competing studies, and not concurrently conducting research and practice in the same patient population. This may mean, for example, that when conducting a clinical trial, the physician might best refer ineligible patients or those declining participation because of a treatment preference to a trustworthy colleague not participating in the study.

There are two principal methods for dealing with conflicts of interest. The first is to decline to participate in any activity where one's conflict of interest would be seen as detrimental to the process. The relevant perceptions are not necessarily our own, but the view likely to be held by others, particularly the public. The second method for dealing with conflicts of interest is disclosure to others involved in the process. This allows them to compensate or replace the affected individual if they deem it necessary. Disclosure may need to be an ongoing process as one's personal or professional circumstances change, or as the issues and climate of the collaboration evolve.

3.6.2   Professional Statistical Ethics

The ethical conduct of statistical science does not have a high profile among either the public or other scientific disciplines that draw heavily from its methods. However, professional societies have examined ethical issues surrounding the practice of statistical methods and have had long-standing guidelines for the profession. The American Statistical Association (ASA) (www.amstat.org/index.cfm), the Royal Statistical Society (RSS) (www.rss.org.uk), and the International Statistical Institute (ISI) (www.cbs.nl/isi/nutshell.htm) have published guidelines for the conduct of their members. The guidelines are similar in content and spirit (American Statistical Association, 1995; Royal Statistical Society, 1993).

The ASA is the oldest professional society in the United States. Its Committee on Professional Ethics perceives the potential benefit and harm from the use of statistical methods in science and public policy. Circumstances where statistical thinking can be misused include not only clinical trials and other areas of medicine but also statistical findings presented as evidence in courts of law, and some political issues. The ASA's position on ethical matters follows closely those of W. E. Deming, one of its most prominent members (Deming, 1986). Although the Society for Clinical Trials is more directly concerned and involved with clinical trials and has many statisticians as members, they have not dealt directly with professional ethics of statisticians in this context.

The ASA and RSS guidelines do not articulate fundamental professional ethical principles directly in the way that has been done for patients' rights in medical research, nor do they deal directly with clinical trials. However, the message implicit in the guidelines has several components similar in spirit to those outlined earlier in this chapter and is relevant to statistical practice surrounding clinical trials. These include (1) investigator competence, (2) disclosure of potential conflicts, (3) confidentiality, (4) documentation of methods, and (5) openness. Guidelines from the ASA are summarized here. Guidelines from the RSS are given in Appendix H. The ISI guidelines are quite thoughtful and include a bibliography. They are too extensive to reproduce in this book but are available on the ISI Web site.

General

Statisticians have a duty to maintain integrity in their professional work, especially where private interests may inappropriately affect the development or application of statistical knowledge (potential conflicts of interest). Statisticians should

- honestly and objectively present their findings and interpretations;
- avoid untrue, deceptive, or undocumented statements;
- disclose conflicts of interest.

Data Collection

Collecting data for a statistical investigation may impose a burden on respondents, may be viewed as an invasion of privacy, and involves legitimate confidentiality considerations. Statisticians should

- collect only the data needed for the purpose of the investigation;
- inform each potential respondent about the general nature and sponsorship of the investigation and the intended use of the data;
- establish their intentions to protect the confidentiality of information collected from respondents, strive to ensure that these intentions realistically reflect their ability to do so, and clearly state pledges of confidentiality and their limitations to the respondents;
- ensure that the means are adequate to protect confidentiality to the extent pledged or intended, that processing and use of data conform with the pledges made, that appropriate care is taken with directly identifying information (using such steps as destroying this type of information or removing it from the file when it is no longer needed for the inquiry), and that appropriate techniques are applied to control disclosures;
- ensure that whenever data are transferred to other persons or organizations, this transfer conforms with the established confidentiality pledges, and require written assurance from the recipients of the data that the measures employed to protect confidentiality will be at least equal to those originally pledged.

Openness

Statistical work must be open to assessment of quality and appropriateness, and such assessments may involve an explanation of the assumptions, methodology, and data processing used in a study. Statisticians should

- delineate the boundaries of the investigation as well as the boundaries of the statistical inferences that can be derived from it;
- emphasize that statistical analysis may be an essential component of a study and should be acknowledged in the same manner as other essential components;
- be prepared to document data sources used in a study, known inaccuracies in the data, and steps taken to correct or refine the data and statistical procedures applied to the data, and the assumptions required for their application;
- make the data available for analysis by other responsible parties with appropriate safeguards for privacy concerns;
- recognize that the selection of a statistical procedure may to some extent be a matter of judgment and that other statisticians may select alternative procedures;
- direct any criticism of a statistical study to the study methods and not to the individuals conducting it.

Client Relationships

A client or employer may be unfamiliar with statistical practice and be dependent upon the statistician for expert advice. Statisticians should

- make clear their qualifications to undertake that statistical inquiry at hand;

- inform a client or employer of all factors that may affect or conflict with their impartiality;
- accept no contingency fee arrangements;
- fulfill all commitments in any inquiry undertaken;
- apply statistical procedures without concern for a favorable outcome;
- state clearly, accurately, and completely to a client the characteristics of alternate statistical procedures along with the recommended methodology and the usefulness and implications of all possible approaches;
- disclose no private information about or belonging to any present or former client without the client's approval.

3.7   SUMMARY

Clinical trials highlight some of the competing obligations that physicians and patients face in health care today. There is potential for clinical trials to be poorly planned or conducted, compromising the ethical treatment of research subjects. This potential exists in many areas of medical practice and must be counterbalanced by safeguards and standards of ethical conduct. Features of some clinical trials such as randomization and the use of placebos illustrate clearly competing obligations. However, trials do not necessarily generate unique ethical concerns nor make them unsolvable.

The history of medical experimentation in the twentieth century illustrates the potential for infringing on the rights of patients and an evolving standard for conducting medical research of all types. Protections for patients are based on international agreements and guidelines, governmental regulations, institutional standards and review, and the ethical principles of autonomy, beneficence, and justice. Careful practical implementation of these protections usually yields clinical trials that are ethically acceptable. Ethical norms, especially those for clinical trials, appear to be culturally dependent.

There are circumstances where the most ethical course of action for the physician is to recommend participation in a clinical trial. This may be the case when the physician is genuinely uncertain about the benefit of a treatment or the difference between alternative treatments. Opinions about treatments based on evidence of poor quality are not ethically legitimate, even if they are firmly held. Opinions based on scientifically weak evidence are subject to influence by financial, academic, or personal pressures.

Clinical trial statisticians are also held to standards of professional conduct. These standards require competence, confidentiality, impartiality, and openness. Like other investigators, the statistician must demonstrate an absence of conflicts of interest.

3.8   QUESTIONS FOR DISCUSSION

1. Financial and academic pressures appear not to be the primary ethical conflict in clinical trials. Is this accurate or not? Discuss how these pressures can compromise the rights of the research subject.

2. Some clinical trials aren't conducted in the United States because of either ethical concerns or lack of patient acceptance. Trials are sometimes more feasible in other countries even when conducted by U.S. sponsors. An example is the pertussis vaccine trial (Greco et al., 1994; Leary, 1994; Marwick, 1994). Comment on the ethics of this practice.

3. Are risks and benefits evenly distributed between trial participants and future patients? Discuss your point of view.

4. In recent years there has been heightened concern over potential differences in treatment effects between men and women, especially if a trial enrolled patients of only one sex. In some cases trials have been essentially repeated by sex, such as the Physicians Health Study (Steering Committee of the Physicians' Health Study Research Group, 1989) and the Women Physicians' Health Study (Frank, 1995). When is this practice ethical and when is it not? Are there ethical concerns about resource utilization?

5. There is a need for physicians to conduct research in some settings where patients (and their families) can neither be informed nor give consent in the conventional manner. An example might be in the emergency department for cardiopulmonary resuscitation or head injury. Discuss whether or not clinical trials can be conducted ethically in such circumstances.

6. Developing interventions and conducting clinical trials in specific disease areas like medical devices, biological agents, surgery, AIDS, cytotoxic drugs, and disease prevention can be quite different. Discuss how clinical trials in these various areas raise different ethical issues.

4

CONTEXTS FOR CLINICAL TRIALS

4.1 INTRODUCTION

The purpose of this chapter is to discuss the way clinical trials are used in different medical contexts. Context is defined primarily by treatment modality rather than medical specialty because it is the nature of the therapy that relates most directly to clinical trial design. The specific contexts discussed are drugs, devices, surgery, complementary and alternative medicine (CAM), and prevention. The intent is not to fragment the principles that underlie trials but to illustrate common themes in these diverse areas. It is helpful to understand how the setting has an impact on the design, ethics, interpretation, and frequency of studies performed.

Statistical principles of experimental design are very reliable, flexible, and broad, having been of service to agriculture, industrial quality control, reliability testing, and medicine. Trials have been used for decades to develop and test interventions in nearly all areas of medicine and public health (Office of Technology Assessment, 1983; Starr and Chalmers, 2004). Trials seem to be most actively employed in cancer, cardiovascular disease, and AIDS. These areas share a need to evaluate new drugs, drug combinations, devices, surgery, other treatment modalities, and diagnostics. However, trial methods can also be used to assess treatment algorithms as compared with specific agents, making them applicable to a wide array of therapeutic and prevention questions.

Despite extensive use, clinical trials have not diffused evenly into all medical disciplines or contexts. The heterogeneous application of trial methodology can be partly explained by two general factors: demands by practitioners and patients, and key external pressures. Demands by practitioners and patients are different in various specialities and diseases, affecting the perceived necessity and acceptance of clinical trials. For example, the training of practitioners in some disciplines may not cultivate a reliance or insistence on clinical trials as rigorous evaluation tools. Also patients may place inconsistent demands on different therapies or those administered at different points in the progression of disease. A good example is treatment for cancer. Soon after diagnosis most patients demand cutting-edge and safe therapies. If the disease progresses, the same patients may turn to alternative therapies that appear safe but are marginally effective or completely unproven, without the same demands for demonstrated efficacy. This illustrates that patients' risk–benefit assessments change as a disease advances.

Different contexts and areas of medicine also experience distinct key external pressures. These influences arise from regulation, the pace of therapeutic development—which itself may depend on scientific developments in other fields, chance, and the economics of care. Regulation of drugs is a major determinant of the use of clinical trials in that context (discussed extensively below). Innovation and rigor are frequently at odds with the external pressure of cost control. Clinical trials are often inhibited when health insurance companies are reluctant to pay for routine costs surrounding experimental therapy. For insurers the definition of experimental therapy can be quite broad. For example, if one compares two standard therapies for a disease in a randomized clinical trial, an insurance company might refuse to pay for either therapy or the associated costs, under the claim that the treatments are "experimental." In contrast, the same insurer might pay for either treatment outside of the trial.

Fortunately this double standard is lessening. In 1998 the state of Maryland was among the first to pass legislation requiring insurers to pay for nonexperimental patient care costs incurred during clinical trials. This covers parallel costs or care that would have been reimbursed as standard of care. The National Emphysema Treatment Trial (NETT Research Group, 1999) (discussed in more detail in Section 4.6.6) is a study in which the Center for Medicare and Medicaid Services (CMS, formerly the Health Care Financing Administration) agreed to reimburse for lung volume reduction surgery (only in the context of the randomized trial). At the federal level, in June 2000, there was an executive memorandum from the U.S. president to CMS to cover routine costs of qualifying clinical trials. Advocacy groups are pursuing federal legislation that will obligate all third-party payers to do the same.

4.1.1 Some Ways to Learn about Trials in a Given Context

Learning about the full scope and number of clinical trials in any field is probably impossible. There are three targets for such information: patients, practitioners, and researchers. In the field of cancer, practitioners and researchers can obtain details regarding many clinical trials through the Physician Data Query (PDQ) system. Some of the information therein often is transmitted to the patient. PDQ was begun in 1984 and reflects an effort to make physicians treating cancer patients aware of the studies being performed, expert opinion about the trials, and sketches of the treatment protocol (Hubbard, Martin, and Thurn, 1995). Currently there are over 2000 active and 13,000 closed cancer clinical trials listed in PDQ, and these are updated bimonthly. On-line computer access is at www.cancer.gov/cancer_information/pdq.

A similar resource is available for AIDS trials at the AIDSInfo Internet site, www. aidsinfo.nih.gov/clinical_trials. The CenterWatch Clinical Trials Listing Service maintains an international listing of more than 41,000 clinical trials that are currently active. It is available at www.centerwatch.com.

The U.S. government has a patient resource registry of clinical trials that is not disease specific. The registry is maintained by the National Library of Medicine and can be found at clinicaltrials.gov. It contains over 10,600 studies being performed at thousands of clinical sites nationwide and in 90 countries.

The Cochrane Collaboration is a registry of all randomized clinical trials performed in specified disease areas. It is named after the epidemiologist Archie Cochrane (1909–1988; see Cochrane, 1972), who first suggested the idea in the 1980s (Dickersin and Manheimer, 1998). Since 1993 numerous investigators around the world have been exhaustively compiling the studies and data that will facilitate meta-analyses (Chapter 21) and high-quality evidence-driven treatment decisions. This database will be of great use to researchers and practitioners. Additional details can be found at the Cochrane Collaboration Web site (www.cochrane.org).

The recently initiated Project ImpACT (Important Achievements of Clinical Trials, www.projectimpact.info) is an attempt to identify and characterize clinical trials that have had a major impact on therapeutic decisions in various medical disciplines and modalities. This will take several years to complete but will provide a resource for studying influential trials.

All these resources depend on voluntary submission of information by researchers or sponsors of trials and are incomplete. Lack of public awareness of certain trials and their results can have serious consequences. For example, regulatory decisions regarding safety, efficacy, or labeling could be affected by incomplete information about study results (Steinbrook, 2004a). The same might be said of more routine therapeutic decision in circumstances where rigorous evidence is scarce. In 2004 the International Committee of Medical Journal Editors (ICMJE) announced a new requirement (effective July 2005) for registration of clinical trials in a public registry as a prerequisite for publishing results in their 11 member journals (De Angelis et al., 2004). The 1997 FDA Modernization Act required registration of trials in the National Library of Medicine database, but only about 49% of industry sponsored trials and 91% of government sponsored trials have been registered (Steinbrook, 2004b). The new requirements by the ICMJE should help but may not completely fix the problem.

4.1.2 Issues of Context

The most consequential contextual issue for clinical trials is, in my opinion, the scientific view versus public perception of trials. This dichotomy is seldom discussed but is important for several reasons. First, the two perspectives are often starkly different because of priorities and understanding. Second, it is the public who eventually participate in trials (or not), providing definitive evidence of how they are perceived. Third, the difference in perspective is often overlooked by investigators who have a difficult time with perspectives other than science, even when a study encounters important nonscience issues like fraud, conflict of interest, funding, or ethics.

Public perception is more important than science whenever a trial is stressed, controversial, or held up as a bad example. Because this book is about science, I will not pursue this issue further here. However, many examples of this dichotomy can be found by reading accounts of trials in the lay press and following them back into the scientific literature and back in time (e.g., Shalala 2000).

Within the science domain the relationship between the frequency, quality, and nature of clinical trials and the context in which they are performed can be assessed using a framework based on the following characteristics:

- Role of regulation for the therapy
- Ease with which observer bias can be controlled (e.g., placebos)

TABLE 4.1   Context Characteristics with Implications for Clinical Trials

Characteristic	Context				
	Drugs	Devices	Prevention	CAM[a]	Surgery
Strength of regulation	Strong	Moderate	Strong	Minimal	None
Ease of bias control	Easy	Difficult	Easy	Easy	Difficult
Treatment uniformity	High	High	High	Low	Low
Likely effect size	Small	Large	Moderate	Small	Large
Use of incremental improvement	Common	Common	Minimal	None	Common
Short-term risk–benefit	Favorable	Varied	Favorable	Unknown	Varied
Long-term risk–benefit	Varied	Varied	Favorable	Unknown	Favorable
Tradition for rigorous trials	Strong	Weak	Strong	None	Varied

[a]Complementary and alternative medicine.

- Uniformity, or lack thereof, of the treatment
- Expected magnitude of treatment effects
- Relevance of incremental improvement
- General assessment of risk–benefit
- Tradition and training of practitioners, especially with respect to the scientific method.

Table 4.1 provides an overview of the characteristics of some contexts discussed in this chapter with regard to this framework. For specific trials there may be additional factors explaining their acceptance or lack thereof, such as patient acceptance or window of opportunity. However, the perspective in the following sections regarding details of contexts and characteristics is more universal.

4.2 DRUGS

Most of the discussion in this book pertains directly to the development and evaluation of drugs. Drugs are the most common therapeutic modality employed today, so there is ample reason for them to be the subject of many clinical trials. This context is very heterogeneous, both because of the vast universe of drug products and the different entities performing drug trials worldwide. The latter includes such diverse groups as pharmaceutical companies, single academic investigators, collaborative academic and government cooperative groups (themselves fairly heterogeneous), groups of practitioners, managed care, and various permutations of these groups in different countries.

In the United States there is a dichotomy between the pharmaceutical and academic models for performing trials. The pharmaceutical perspective tends to emphasize discovery, systematic early development, market forces, the regulatory overlay, product formulation, toxicity, and liabilities. The academic view tends to emphasize development and comparison with less regard for regulatory considerations or other market forces.

Pharmaceutical companies that do most of the drug development in the United States are motivated by a mixture of profit and idealized or regulatory mandated concern for public health and safety. Often the statistical tools and study designs applied by both developers and regulators are not completely adequate. The resulting mix of forces applies stress to most clinical trials, rarely making them ideal for the regulatory decisions that they support.

One of the strongest motivations for performing clinical trials when developing and evaluating drugs is the regulatory apparatus in the United States. To appreciate fully the impact that drug regulation has on the design and conduct of clinical trials, the student should become acquainted with the regulatory history of the Food and Drug Administration (FDA). A reasonable introductory source for this information is Young (1983) or the FDA Internet sites. The essential FDA perspective is that randomized controlled trials are excellent and irreplaceable tools for clarifying safety and efficacy concerns about new treatments prior to marketing approval (Leber, 1991).

The regulations that govern FDA control over drug products have largely been reactions to specific crises in public health and safety that result in legislation from Congress. This has limited thoughtful regulatory design. The FDA has the difficult job of balancing the conflicting needs of the public within statutory limitations. The public requires safe and effective medications but also requires that they be developed and made available in a timely fashion. RCTs are not strictly required by the FDA's statutory "adequate and well-controlled" clause, but they are the most reliable way to satisfy it. An FDA perspective on statistical issues in drug development is given by Anello (1999), and specifically on oncology drugs by Hirschfeld and Pazdur (2002).

An interesting perspective on drug regulation arose from the withdrawal of rofecoxib from the market in 2004. This nonsteroidal anti-inflammatory agent had been on the market since 1999 and was designed to be safer than earlier drugs in its class, although cardiovascular safety was questioned early in its history. After five years of use and the premature stopping of a polyp prevention trial by a monitoring committee, data were convincing that the drug was associated with an unacceptable risk of myocardial infarction or stroke (FitzGerald, 2004; Topol, 2004). The clinical trials on which approval was based did not adequately address cardiovascular safety, and definitive studies of the question were never undertaken. Thus deficiencies in evidence and an unexpected finding led to the largest withdrawal of a prescription drug in history, highlighting both effective and ineffective aspects of the regulatory process.

Some additional insight into drugs as a context can be gained by considering dietary supplements or botanicals, with their bewildering and faddish history. These agents, mostly natural products, are taken like drugs but are explicitly outside FDA regulation. A perspective on this is given by Marcus and Grollman (2002). The rationale for this is unclear now because of the way dietary supplements are taken, their availability and composition, and implicit or explicit health claims by manufacturers. They represent a microcosm of the drug context in the absence of regulation. The premise of safety is more a reflection of their traditional use rather than evidence from systematic trials.

It is essentially impossible for patients and practitioners to know the risk–benefit of these substances because of the absence of rigorous testing.

4.2.1   Are Drugs Special?

The drug context is very heterogeneous and defies generalizations. It can be as different from itself as it is from other settings discussed below. There is little about drugs that makes them uniquely suitable or unsuitable for clinical trials. Often drugs can be relatively easily administered, but there are a large number of exceptions. Drugs are often used to gain small-sized or short-term benefits, especially when their risk is low. Nonprescription analgesics (and the differences among them) are an example of this.

One characteristic that makes drugs amenable for comparative clinical trials is the readiness with which placebos can be used to control observer bias, at least in principle. A placebo should have the same appearance as the active drug, sometimes requiring the same color or similar taste. These are usually not significant restrictions. The feasibility of making a placebo encourages the use of rigorous clinical trials to evaluate drugs. Other modalities, such as therapeutic devices, are not as placebo-friendly. The reduction in bias that placebo controls afford enhances reliable estimation of small treatment differences.

There are characteristics of drugs that can significantly complicate the design and conduct of clinical trials. These include the proper scheduling or relative timing of treatments and titration of the therapeutic ratio. The best schedule of drug administration to satisfy simple optima (e.g., constant serum level) can often be determined from pharmacokinetic experiments. However, if efficacy depends on some other measure of exposure, such as peak concentration or time above a threshold, the schedule problem can be more complex. The situation can be intractable to theory when drug combinations are used because there may be interactions that cannot be predicted from simple time–concentration data.

Another relevant characteristic of drugs is that they typically have uniform formulation and manufacture. This is usually the case even in developmental clinical trials. This uniformity simplifies many trial design considerations and provides reassurance for inference and external validity, especially for single agent studies. It also means that the only feasible customization of many drugs for the individual is through dosing. For example, the therapeutic ratio is almost never under direct investigator control. Sometimes a change in formulation may improve the therapeutic ratio. This contrasts with radiotherapy, discussed below, where the therapeutic ratio can be controlled.

One interesting exception is regional perfusion of drug to maximize exposure to a limb or organ while minimizing toxicity to the remainder of the body. This technique is used for the treatment of some cancers. Even this complex procedure affords only crude control over the therapeutic ratio because the target is imperfectly isolated from the normal tissues and some drug escapes systemically.

Doses for many drugs are established early in development based on relatively small experiments. This is usually optimal from a developmental perspective. However, the dose question sometimes needs to be addressed in comparative studies because of changes in formulations, better ancillary care, side effects, or other practical reasons. As outlined in Chapter 10, investigators must distinguish sharply between the dose-safety and the dose-efficacy relationship.

4.2.2   Why Trials Are Used Extensively for Drugs

Regulation

Government regulation of drug treatments requiring them to be safe and effective has been constructed in response to various unfortunate events (e.g., thalidomide and birth defects). Nearly all drugs expose the entire body to their actions, creating the potential to produce side effects in unexpected ways. Regulation is the force that balances this risk against evidence of health benefit and ignorance, meaning lack of knowledge about harm. The need for regulation is created, in part, because such treatments are advertised and marketed on a large scale. It remains probably the most dominant influence for the rigorous testing of drugs.

Because the universe of drugs is so large, regulatory considerations for them are not uniform and allow for the basis of drug approval to depend on clinical circumstances. For example, Subpart H of the regulation allows approval on the basis of a surrogate endpoint or on a clinical endpoint other than survival or irreversible morbidity (21 CFR 314.510). Approval can also be restricted to assure safe use (21 CFR 314.520). Subpart E is intended to make drugs for life-threatening illnesses available more rapidly by encouraging an early collaboration between sponsors and the FDA to develop efficient preclinical studies and human trials that could lead to approval without large randomized studies (21 CFR 312.80-312.88).

Early in the AIDS epidemic, alterations in the drug approval process hastened getting promising agents to the public to fill unmet medical needs. Another example is drugs to treat cancer, where some unique characteristics of the disease and its treatment also change the regulatory perspective. The oncology setting is characterized by a very heterogeneous life-threatening disease, multiple modes of therapy, a unique perspective on risk–benefit and serious adverse events, specialists trained to use dangerous drugs, investigative nature of the discipline, wide variety of products used, and relatively high risk of drug development. These features make oncology drug regulation subjectively somewhat different than for other settings, with frequent interest in pathways for rapid approval.

Tradition

There is a strong tradition of experimental trials for drugs, especially in disease areas such as cardiovascular disease and cancer. Because of the training of practitioners, the culture is respectful of the need for trials and good study design. There is also considerable public expectation that prescription drugs will be safe and effective. Perhaps because of perceived risk, the expectation for drugs is much higher than for, say, alternative medicine therapies.

Confounding

Uncontrolled drug studies are subject to the same selection bias and observer bias as other contexts. For drugs that do not produce strong side effects or stress the recipient, selection bias might be a slightly reduced concern. In any case, there is still a strong potential for observer bias. In short, drugs do not appear to be special with regard to the potential for confounded treatment effects.

Incremental Improvement

Drugs are often designed by incremental improvement of existing agents. Modifications to the compound might reduce unwanted side effects, increase absorption, or prolong half-life, for example. Usually even small changes in a drug require rigorous testing. Minor chemical modifications to a molecule can substantially alter its behavior and efficacy, as has been amply illustrated by drug development. Thus incremental improvement does not provide a vehicle for minimizing the need for rigorous testing of new agents.

Masking and Placebos

Drugs are almost universally suited to the use of masking and placebos. The clinical setting or features of a drug may prevent placebo use, but the principle holds. Only occasionally do logistical difficulties prevent the use of a placebo.

Economics

The economics of drug therapy is tightly bound to regulation. Despite the high cost of getting a new drug to the commercial market and the high cost to the patient of some newer drugs, they often remain very cost effective therapies. This encourages their use. Individual practitioners usually do not profit directly from the use of a particular drug, making the economic incentives somewhat different than for procedure-oriented therapies. These factors tend to make the economic considerations supportive of rigorous testing.

Psychology

The psychological acceptance of most drugs by patients is relatively high, particularly in the short term. They often represent a convenient, cost effective, safe, and effective solution to symptoms and underlying causes of disease. Analgesics and antibiotics might be typical examples of this. For longer term use even mild side effects can diminish acceptance or decrease compliance. Knowing that a drug has had fairly rigorous testing for safety and efficacy is a substantial benefit to the psychology of their acceptance. This firmly supports clinical trial testing during development.

4.3 DEVICES

The Food, Drug, and Cosmetic Act of 1938 defines a medical device as

> an instrument, apparatus, implement, machine, contrivance, implant, in vitro reagent, or other similar or related article, including any component, part, or accessory that is:
>
> a) recognized in the official National Formulary, or the United States Pharmacopoeia, or any supplement to them,
>
> b) intended for use in the diagnosis of disease or other conditions, or in the cure, mitigation, treatment, or prevention of disease in man or other animals, or
>
> c) intended to affect the structure or any function of the body of man or other animals, and which does not achieve its primary intended purposes through chemical action within or on the body of man or other animals and which is not dependent upon being metabolized for the achievement of any of its principal intended purposes. (U.S. Congress, 1938)

I am tempted to define a medical device simply as any object used in/on the body about which a health claim is made; however, the exact definition is not critical to this discussion. Medical devices, loosely characterized, function through physical or electromechanical actions. This definition encompasses diverse entities including rubber gloves, surgical instruments, cardiac pacemakers, tongue depressors, medical software, and diagnostic test kits.

4.3.1 Use of Trials for Medical Devices

Regulation

There are currently over 1700 medical devices in the United States with a market of $70 billion. Many medical devices have not been subject to the same development or comparative testing as drugs for reasons discussed below. The regulatory apparatus surrounding the development and evaluation of medical devices is important but not as familiar to many clinical trialists as that for drugs. See Hutt (1989) for a brief history of device regulation and Murfitt (1990) for a discussion of software regulation. It may not seem obvious that medical software should be classified as a device, but in most instances it is no less "mechanical." Software is certainly more similar to a device than to a drug.

The regulatory climate surrounding devices is substantially different than that for drugs. Before 1976 devices could be marketed and used without demonstration of safety and effectiveness. The FDA began regulating devices for evidence of safety and effectiveness based on the 1976 Medical Device Amendment to the Food, Drug, and Cosmetic Act of 1938. An interesting perspective on device regulation from a surgical perspective was given by Brantigan (1995). The summary statistics regarding device regulation presented here are derived from the work of Scott (2004).

Some medical devices are explicitly exempt from FDA clearance based on established safety, although registration, labeling, and good manufacturing practices are required (21 CFR Parts 862-892). For medical devices that are not exempt, there are two mechanisms allowed by law under which the FDA can approve marketing. The first is premarket notification (510(k)), which is based on a device being "substantially equivalent" to an existing device before 1976. Additionally the Safe Medical Devices Act of 1990 allows substantial equivalence claims to be based on 510(k) approvals granted after 1976. Clinical data proving safety and effectiveness are not required using this mechanism, and only about 10% of 501(k) devices have supporting clinical data. New or high-risk medical devices must be approved under a premarket approval application (PMA), which are based on data demonstrating clinical safety and effectiveness. These account for only about 10% of all devices (Scott, 2004).

Because of differences in statute, regulation is not nearly as effective at maintaining rigor in device development as it is for drugs. Between the 1976 Medical Device Amendment, and 2002, 102,533 devices were cleared for marketing in the United States through the 510(k) mechanism, as compared with 883 approved by the more demanding PMA process (Scott, 2004). "Substantial equivalence," which has no statutory definition, was rejected by the FDA only 2% of the time for the 510(k) pathway, leaving the majority of medical devices to enter the market on the basis of being equivalent to something available prior to 1976 or its successor. Given the technological advancements since 1976, this seems incredible. In any case, the level of evidence supporting device approval is much lower than that for drugs, the supporting information is difficult to obtain, and there is no comprehensive documentation of safety and effectiveness for devices (Scott, 2004).

Tradition

The testing of medical devices should not be, in principle, different from evaluating drugs. Some special exceptions are mentioned below. The practicalities and traditions are different, however. Device studies have a dominant concern over functional and mechanistic features as compared to clinical outcomes. In contrast, drug studies have a more even balance of formulation, bioavailability (and related aspects analogous to device function), and clinical benefit.

Like surgical treatments, medical devices can be developed inexpensively by small groups of investigators. At some point in development, investigators must distinguish between device function and clinical outcome. Case series with selected patients tend to confuse the difference between device function and clinical outcome, which should remain separate. A properly designed and functioning device does not automatically produce a beneficial effect in the patient. Trials that test differences in patient outcome attributable to a device need to be as large and rigorous as those for drugs.

Because of the informality of the early developmental process and selection of favorable case reports, a device can sometimes gain widespread use without rigorous testing of clinical outcome. We may require more extensive follow-up to observe device failures than the amount of time required to see side effects from drugs or other systemically acting agents. Thus the risk–benefit picture may change substantially over time or take an extended period to become clear. Examples where this may have been a problem because complications became evident only after a long time, include prosthetic heart valves found years later to be catastrophically failing, some intrauterine birth control devices, and silicone breast implants, which have been implicated in a variety of ailments. Rigorous developmental trials, long-term animal studies, and postmarketing surveillance studies might have ameliorated each of these problems.

4.3.2 Are Devices Different from Drugs?

There can be reasons why in vivo developmental studies of some devices may not need to be as rigorous as many trials for drugs and biological agents. First, the action of most devices is physiologically localized rather than systemic. This leads to a greater determinism in how they work, how they might fail, and the possible manifestations of failure. Examples are catheters or implantable electrodes, which have a relatively small set of possible side effects or complications compared with drugs and biologicals. Second, devices are often constructed of materials that have been tested in similar biological contexts and whose properties are well known. Again, this effectively rules out, or reduces the probability of, certain complications. Examples of this are biologically inert materials, such as some metals and synthetics.

Third, devices typically operate on physical, chemical, or electronic principles that are known to be reliable because of previous testing or evaluation outside human subjects. Based on these characteristics, investigators may know much more about a device at a given point in its development than is known about a drug or biological at a comparable stage. These characteristics may contribute to less of a need for extensive developmental testing of devices compared with drugs in humans, but not less rigorous testing overall.

The literature surrounding medical devices contains many justifications as to why randomized controlled trials are not needed or are actually the wrong approach. There are some reasons for the frequent use of designs other than randomized trials to evaluate medical devices. However, the basic perspective of the clinical trialist should be that rigorous, unbiased evaluation methods are appropriate for devices, and that there are important therapeutic questions that are best answered by such methods. In some instances devices might need more extensive testing than other treatments. Historically, devices have had their share of trouble: a slightly humorous angle on this fact is provided by the Museum of Questionable Medical Devices in Minneapolis (www.mtn.org/quack/welcome.htm).

If devices were inherently safer than drugs, it might explain the difference in attitude toward these contexts. A device usually does not have a direct analogy to "dose," which is one way that drugs produce safety failures or adverse effects. An important safety concern for device relates to their initial use, such as, implantation. This procedure, often a complex invasive one, may carry significant risks, which although are not literally attributable to the device, are inseparable from it. This might be associated with a relatively high short-term risk, namely that of surgery, and a lower long-term risk. The only analogous concepts for drugs would be allergic or idiosyncratic reactions, or the risks of a large loading dose.

Drugs and devices both carry risks associated with duration of exposure. Apart from initial use, duration of exposure is the principle dimension by which safety fails for devices, whereas for drugs it is common but secondary to dose, allergy, intolerance, and the like. Said another way, for drugs the probability of an adverse effect increases strongly with dose and usually less strongly with exposure time. For devices the probability of an adverse event might increase with exposure time.

A rationale for small simple designs for device trials is their potentially large effect size. The natural history is well known for the patient with an irregular heart rhythm, a failed heart valve, degenerated joint, unstable fracture, uncorrected anatomic defect, or many other conditions amenable to treatment by a device. Furthermore small treatment effects attributable to the use of a particular device are usually not important. Many devices are designed and used only in situations where they can provide a large and specific treatment effect, for example, due to their mechanical properties. These situations in no way obviate the need for study rigor or infrastructure, but they may limit the size, duration, or architecture of a trial.

In other circumstances a new device represents an incremental improvement of an item known by experience to be safe and effective. Perhaps a new material is employed that is more durable, less reactive, or less prone to infection than one previously in use. The device may offer improved electronics or other features that can be convincingly evaluated ex vivo. The need for trials in these circumstances is not eliminated, but their design might be relaxed compared to drugs, where analogous minor changes in formulation could affect bioavailability or action. We might say that the actions of devices are less dependent than drugs on their respective "initial conditions," meaning drug effects are potentially more chaotic.

Although the considerations above may apply frequently to devices, there are numerous situations in which devices need to be evaluated for moderate to small treatment effects. These might arise when a device is a new treatment for a condition where standard therapy exists. Questions about efficacy and risk benefit are important, and are answerable using rigorous study designs. In addition we cannot equate mechanical function and reliability with clinical benefit, as discussed above. The former is partly knowable from preclinical studies, whereas the latter requires definitive trials.

4.3.3   Case Study

Many of the issues surrounding the development, regulation, and clinical use of medical devices are illustrated by deep-brain stimulation for control of manifestations of advanced Parkinson's disease. Based on animal models and exploratory human studies, electrical stimulation of certain regions of the brain has been shown to reduce tremors and similar symptoms in some patients. The devices utilized, their anatomical targets, and their method of implantation evolved through relatively small developmental trials. Initial regulatory approval was given for a device for unilateral use to control Parkinsonian tremor, indicating basic comfort with safety and efficacy.

In 1996 small feasibility trials of bilateral electrical stimulation of the globus pallidus or subthalamic nucleus were begun in the United States and Europe, in patients with advanced Parkinson's disease, who had been responsive to L-Dopa but were progressing symptomatically. These trials were designed with sample sizes in the range of two dozen patients and with subjective clinical endpoints evaluated at three months—essentially immediately relative to the course of the disease.

Promising early results caused these trials to be expanded to about 150 patients (absurdly large sizes for feasibility) and longer term follow-up. The data, when presented in a regulatory forum in 2000, raised relatively little concern over device function or safety but raised significant questions about clinical efficacy, endpoint selection, study design, statistical analysis, and practitioner expertise. If the treatment effect had not been so large in magnitude (strong improvement in motor subscore of the Unified Parkinson's Disease Rating Scale), the regulatory approval would likely have failed because of poor clinical trial methodology. Some of the methodological issues can be seen in a reported clinical trial from the Deep-Brain Stimulation for Parkinson's Disease Study Group (2001). This illustrates that while devices may proceed through early development taking advantage of some of the features discussed above, questions about clinical efficacy are no different than for drugs.

4.4   PREVENTION

Clinical trials that assess methods for prevention of disease are among the most important, complex, and expensive types of studies that can be done. Aside from this complexity, prevention trials also highlight ethics concerns and risk–benefit judgments because the participants are often at lower risk than individuals with a disease. Even risks at "epidemic" levels (a few percent) in the population may be much lower than risks encountered by individuals who actually have the disease.

Prevention trials can be loosely categorized as primary, secondary, or tertiary (Bertram, Kolonel, and Meyskens, 1987). Here again, terminology seems to have arisen in oncology and then crept into broader contexts. Primary prevention trials assess an intervention in individuals who are initially free of disease. In some cases such individuals will be at very low absolute risk for the outcome of interest. If a substantial number of cases are required to assess the effects of the treatment, as is often the case, then these trials will need to be quite large and have extensive follow-up to generate the information (events) to meet the scientific objectives. An example is diet and life-style change to prevent cancer or cardiovascular disease.

Secondary prevention attempts treatments or interventions in individuals with characteristics or precursors of the disease, or with early stage disease, to prevent progression or sequelae. An example would be lowering of blood pressure to prevent stroke or myocardial infarction. A precursor condition is something like keratotic skin lesions, which often precede the occurrence of basal cell carcinoma. Individuals with such lesions might be an appropriate population in which to test agents that could reduce the frequency of this skin cancer. Secondary prevention trials tend to be smaller than primary prevention trials, because the population is at higher risk and will yield more events.

Tertiary prevention attempts to prevent recurrences of the disease in individuals who have already had one episode. An example is the use of vitamins to prevent or delay second malignancies in patients who have already had a primary cancer. Such a population is usually at the highest risk of having an event, and consequently these might be the most efficient of all prevention trials with respect to size. However, there may be fundamental biological differences between preventing a disease in a basically healthy population and attempting to prevent a disease when it has already become manifest on one occasion. It is quite possible for the treatment to work under one condition but not under the other.

4.4.1   The Prevention versus Therapy Dichotomy Is Overworked

Therapeutics is too broad to be considered a context in the sense of this chapter. It includes much of drugs, devices, and surgery, for example. Prevention is also a wide field, although I believe it can be sensibly discussed as a context. As prevention widens the nature of its interventions, it may become less cohesive and therefore less of a context for trials. Already we see prevention incorporating diverse interventions such as vaccines, diet, lifestyle, herbals, vitamins, trace elements, and drugs.

Contrasts between prevention trials and therapeutic trials are often made on the basis of stereotypes. Consider the difference between trials that test diet for prevention of coronary heart disease versus those that evaluate coronary bypass grafting for its treatment. Prevention and therapy trials are usually said to be different based on the origins of ideas to be tested, nature of specific interventions, absolute risk of the study cohort, scientific expertise required, study size, duration, and cost.

When we compare prevention and therapeutics broadly, there are few differences in methodologic principles such as those outlined in Section 2.3.1. The nature of prevention and therapeutic questions is not fundamentally different. We often speak of prevention when we merely delay or exchange causes of death (e.g., cancer versus cardiovascular disease, and vice versa). Most of the differences between studies of prevention and therapy relate to technology—size, cost, and duration. But these are a consequence almost entirely of the different absolute risk of the study cohort, rather than a reflection of principles. Thus the dichotomy is overemphasized, especially for comparative trials.

Some areas where therapeutic and prevention trials have substantive differences, though not necessarily of principle, include the following. Preventive agents are usually not developed with the same structured early phases as drugs. The suggested efficacy of prevention agents often comes from epidemiologic or other nonexperimental studies. Because many prevention agents are virtually nontoxic, early safety and dose-finding studies are less critical and may not need to be oriented toward pharmacokinetics. Because of the long study duration that is typical of prevention trials, poor adherence and delayed onset of treatment effectiveness may diminish or distort the treatment effect. Special trial designs and analyses that are insensitive to these problems are needed. Finally, because many prevention treatments have a low incidence of overlapping side effects, they are suitable for simultaneous or combined administration in factorial designs (discussed in Chapter 19).

4.4.2 Vaccines and Biologicals

Vaccines and biologicals are similar to the extent that both typically utilize natural substances to stimulate the body's own beneficial responses. Examples of biologics are cells or humoral agents that stimulate responses in the recipient. The components of some vaccines (proteins from bacteria or viruses, live, or attenuated cells) are not necessarily "naturally present" in health, but the response is. This characterization is crude because some drugs are natural products also (e.g., steroids, plant products), and some vaccines are not. Through triggering host mechanisms, vaccines and biologics also seem to have the potential for great amplification of effect, substantially changing our perspective about dose. Therapeutic benefit may rely on a different mechanism than drug metabolism or direct effect on receptors. In any case, the distinction is imperfect, and this creates some difficulties for sketching the role of clinical trials in the development and evaluation of these substances. In this discussion I will refer mostly to "vaccines," although similar remarks often apply to other biological agents.

Many vaccines are among the most effective and cost effective interventions for public health. They are discussed here separately from other prevention agents only to reflect their uniqueness and as a practical distinction that persists today. Vaccines have been produced mostly in response to infectious diseases, but now some other illnesses may be amenable to vaccine treatment or prevention. Cancer is one possible example. A good deal of effort is also being devoted to both treatment and preventive vaccines for HIV. Gilbert (2000) discusses some of the statistical design issues for HIV vaccine trials. Much of the public and professional attitude toward vaccines is predicated on their high degree of effectiveness and safety.

Although somewhat forgotten today, the American public has had a favorable experience with an important vaccine trial, perhaps the largest clinical trial ever done. I refer to the 1954 Salk polio vaccine trial (Meldrum, 1998; Dawson, 2004), in which 1.8 million children were involved. Conclusive and impressive efficacy caused the vaccine to be licensed by the PHS the same day the results were announced. Not only did this study address one of the most important public health questions of its time, it was financed entirely without government money. In the early days of applying the new vaccine, several lots were contaminated with live virus, resulting in 59 cases of paralytic polio and 5 deaths. Despite this public confidence in the program remained high. The contrast between this history and contemporary attitudes and practice is striking. The polio vaccine trials are a classic example of prevention trials (Lambert and Markel, 2000).

A traditional view of contemporary vaccine trials is given by Farrington and Miller (2001). Vaccine development and use is typified by the following features. They are often intended to treat individuals at risk but without the disease, they often produce a long-lasting or permanent effect, and the diseases they prevent may be at seemingly low frequency in the population. Therapeutic vaccines may not fit this paradigm. These characteristics sharpen the focus on safety. Some vaccine opponents emphasize uncommon or hypothetical side effects and question the benefit. For example, there was some concern that polio vaccines from 50 years ago could have been contaminated by viruses that might be causing or contributing to AIDS or cancer now (Horowitz,

2001). Pertussis (whole-cell) vaccination seems to be perpetually criticized, and the anti-vaccine movement had demonstrable negative effects worldwide on the control of this serious disease (Gangarosa et al., 1998). A more acrid view of the pertussis case is given by Hoyt (2004).

The demand for a strong risk–benefit ratio and concern over uncommon but serious side effects has motivated the application of rigorous large clinical trials for vaccine development. These trials examine short-term outcomes and adverse events as carefully as prevention endpoints. The measure of efficacy might be immunological response, or more definitively, reduction in morbidity, mortality, or the incidence of specific infection. Some studies might also show cross immunity manifest by reduction in other infections.

There are several incentives to perform vaccine trials in developing countries. Delivery of the agent is often technologically simple. The frequency of the target disease may be considerably higher in a developing country. For this reason or because of cultural values, the acceptance of an experimental study by participants can be great. Companies may prefer such a setting because the liability climate is also usually more favorable than in the United States. In any case, trials in developing countries have raised serious and incompletely resolved ethical questions (Shapiro and Meslin, 2001).

Performing such trials in developing countries always raises concerns about "exploitation." This term is often used but not justified in my opinion. Exploitation implies that by design, the subject is worse off afterward than before entering some agreement. This is almost never the case in a clinical trial. The resolution of such questions depends on examining the risk–benefit for the study participants, the benefit to the culture in which the study is conducted, the peer review and ethics review of the study in both cultures, and addressing concerns raised by trials in similar contexts. None of our cherished concepts translate perfectly across cultures.

4.4.3 A Perspective on Risk–Benefit

It is often said that (primary) prevention trials are performed in "healthy" subjects. If this were true literally, it would present serious difficulties for the usual risk–benefit calculus. I will argue here that health is not the absence of disease, and that either individual or collective risk is a state intermediate between health and disease. This is certainly how we all behave. Low risk (or perceived low risk) induces behavior and judgments close to those taken by healthy individuals. High risk induces behavior similar to individuals with disease. Attributing the idea to Hamilton, Pickering et al. (1954), Rose (1992) stated

> ... the idea of a sharp distinction between health and disease is a medical artefact for which nature, if consulted, provides no support.

Consequently, to view the prevention context as black and white creates a trap. Even relatively safe interventions such as vaccines lose their attractiveness if risk of disease is assumed away. Would the polio field trials have ever been done if the perception of risk was low? Dramatic prevention issues, such as smallpox vaccination in the United States in the wake of terrorist activities, center as much on risk perception (i.e., the unhealth of the healthy population) as they do on the properties of the preventive. The continuum that exists between health and disease has been articulated previously.

An illustration of this trap was the concern over the use of Tamoxifen for the prevention of breast cancer in high risk women. The effectiveness of this therapy is established now, being close to 50% reduction of risk of breast cancer, and it may seem strange that serious objections were raised about investigating the question. However, strong concerns about the Tamoxifen prevention trial were raised by some in the breast cancer advocacy community during the design phase of the study (e.g., Breast Cancer Action Board of Directors, 1996). The notion that the target population was healthy was central to the objections.

A congressional hearing was held on this question, rationalized by Rep. Donald Payne (D-NJ):

> It is crucial that the Federal Government conduct research to prevent the epidemic of breast cancer that is frightening women throughout our country. However, exposing healthy women to a potentially fatal drug may not be the best way to prevent breast cancer. It is important to make sure that federally funded research protects the patients who participate from unnecessary risks and makes sure they are accurately informed of the likely risks and benefits. (Bruning, 1992)

Superficially this sounds quite reasonable, but the logical inconsistency of an "epidemic of breast cancer that is frightening women throughout our country" and "healthy women" is the error to which I referred above. Labeling the drug as "potentially fatal" was an overstatement. The sentiment in the last sentence of the quote was an argument against a problem that the proposed study did not have. These mistakes regarding the context of the trial were not unique to politicians. See Fugh-Berman (1992) for a further example.

There was also significant opposition to the first RCT of zidovudine for prevention of maternal–infant HIV transmission. Oddly the opposition came from AIDS advocacy groups who perceived the trial as disfavoring the rights of women relative to their offspring. It should not come as a surprise that ideology sometimes opposes science, but finding such a prevention question within the scope of conflict is remarkable. It was soon discovered that prevention worked when the trial was stopped at its first interim analysis (Connor et al., 1994). The debate eventually shifted to criticism of the ethics of subsequent HIV prevention trials, under the assumption that efficacy had been established by this first study (e.g., Lurie and Wolfe, 1997).

Recent examples of botched concern over prevention trials can also be found. The Alzheimer's Disease Anti-inflammatory Prevention Trial (ADAPT) was a randomized placebo-controlled trial testing the effect of naproxen and celecoxib to reduce the incidence of AD (Martin et al., 2002). Naproxen is an over-the-counter drug thought to be a relatively safe at the appropriate dose. Celecoxib is designed to have fewer side effects than other nonsteroidal anti-inflammatory agents, presently the most widely used class of drugs. ADAPT was publicly criticized and labeled "unethical" for two reasons: unacceptable risk (because it is "known" that the treatments are "incorrect") and lack of adequate information in its consent form (Barbehenn, Lurie, and Wolfe, 2002). The criticisms were derived from an unsophisticated interpretation of scientific evidence combined with an overestimate of the risk of gastrointestinal bleeding from the study interventions. The critics of ADAPT built their case on demonstrably false scientific conclusions (Breitner and Meinert, 2002). They also chose a nonscientific and inflammatory venue for their criticism, a technique that is often the signature of an ulterior motive.

Interestingly, in late 2004, ADAPT enrollment was suspended based on complex safety concerns unrelated to the above criticisms. Rofecoxib, a drug in the same class as celecoxib, was removed from the market earlier in the year because of its association with elevated risk of thromboembolic events (Fitzgerald, 2004). Celecoxib was then being studied for a wide variety of indications, and its safety subsequently came under intense scrutiny. A collaborative NCI trial of celecoxib for prevention of cancer was halted because of increased risk of major cardiovascular events. Based on this and practical concerns, but not due to an internal safety signal, ADAPT enrollment was suspended. At the time of this writing both the plans for ADAPT and the fate of celecoxib are uncertain.

As a last illustration of risk–benefit issues, consider the role of surgery as a preventive intervention. It is uncommon for surgery to be used in this role for obvious reasons. But it might be appropriate for women at exceptionally high risk of developing breast cancer to have prophylactic mastectomy, for example. Joint replacement is another circumstance in which surgery is arguably used for prevention. This is especially true considering the high mortality surrounding hip fracture in the elderly. Thus risk represents the gradation between the absence and presence of a condition, and the appropriate behavior follows accordingly.

4.4.4 Methodology and Framework for Prevention Trials

Prevention trials do not have many fundamental methodologic differences from therapeutic trials. Pharmaceutical agents are widely used for prevention, although they are often derived from substances that are considered safer than many drugs. In this sense prevention agents bear a resemblance to complementary and alternative medicine, except that they are derived directly from mainstream scientific concepts. Additionally they are invariably supported by extensive preclinical evidence. Examples are vitamins, their analogues, and trace elements. An interesting exception is Tamoxifen for prevention of breast cancer, where prevention efficacy was suggested by its performance in treatment trials. A strong safety profile is essential because we expect them to be used in individuals without the targeted disease and administered for prolonged periods of time. It is the nature of the questions being asked, size, and expense of prevention trials that tend to distinguish them from treatment trials. Some prevention agents are literally drugs. These and vaccines are regulated in much the same way as drugs.

Because of these same characteristics, prevention trials are often amenable to testing multiple treatments at the same time. For example, in a study designed to prevent atherosclerosis, it might be possible to test one or more additional treatments, provided they do not have overlapping side effects or toxicities. One could combine dietary changes, exercise, and medication in a single clinical trial. Treatments could be studied jointly or independently.

This reasoning leads to the frequent use of factorial designs (discussed in Chapter 19), where more than one treatment (or factor) administered simultaneously is studied in a single trial. If the treatments do not interact with one another, then independent estimates of individual treatment effects can be obtained efficiently from such a design. If treatments do interact with one another, the same factorial designs can be used (with a larger sample size) to assess the direction and strength of interaction.

Many interventions proposed for disease prevention studies are not amenable to placebo control or masking. For example, dietary changes, exercise, and the like do not have corresponding placebo controls. This can lead to the same type of observer bias that arises in other unmasked studies. It may also produce "drop-ins," where participants on the standard treatment or no-treatment arm adopt the experimental therapy after learning about the trial. Drop-ins are more likely when the intervention is safe and easy, like dietary changes or exercise. They can also occur in drug treatment trials if the risk–benefit appears favorable to those with the condition under study. This happened extensively in the 1980s and 1990s AIDS trials.

Another concern that arises frequently in prevention trials is compliance or treatment adherence. Because these studies frequently require participants to take medication or maintain their treatment for several years, compliance may be considerably worse than on short-term treatment trials. How to measure and improve compliance is often an important, if not critical, issue. Fortunately many prevention strategies have demonstrated their efficacy despite imperfect compliance.

The early development of prevention agents is often replaced by evidence from epidemiologic studies. Unlike therapeutic trials, where the intermediate step in development is to ascertain clinical activity and compare it to an external standard, the middle developmental step in prevention studies is often a randomized trial using a surrogate outcome (Chapter 8). Surrogate outcomes make such trials shorter, less expensive, and more efficient. Promising findings in such a study could lead to a large-scale prevention trial employing a definitive outcome.

When designing definitive comparative prevention trials, some extra consideration should be given to the type I and II error rates. These rates should be chosen to reflect the consequences of making the respective error. When the treatment is thought to be safe and it is vital to demonstrate it, the type II rate should probably be set quite low, whereas the type I rate might be relaxed compared to the typical therapeutic trial.

4.5 COMPLEMENTARY AND ALTERNATIVE MEDICINE

It is not easy to define "alternative medicine," now often called "complementary and alternative medicine" (CAM). Two modern textbooks on the subject (Yuan and Bieber, 2004; Freeman 2004) do not explicitly define CAM but list a range of topics thought to be within its scope. Yuan and Bieber (2004) state that CAM consists of those practices "not currently considered an integral part of conventional therapies."

Another definition was offered by Raso (1998):

> Alternative healthcare (alt-care, alternative care, alternative healing, alternative healing therapies, alternative health, alternative medicine, alternative therapeutics, alternative therapies, alt-med, CAM, complementary health care, complementary medicine, extended therapeutics, Fringe Medicine, holistic healing, holistic health, holistic medicine, innovative medicine, mind-body medicine, natural healing, natural health, natural medicine, New Age medicine, New Medicine, planet medicine, unconventional medicine, unconventional therapies, unconventional therapy, unorthodox healing, unorthodox therapies, wholistic medicine): A limitless hodgepodge of health-related methods distinguished from biomedical healthcare partly, perhaps chiefly, by its almost unambiguous acceptance of "spiritual health" as a medical concern. One of its general principles is that a practitioner is a teacher who can "empower" one. Its purported goal is not to cure, but to effect "healing": an experience of physical, mental, and spiritual "wholeness."

This illustrates how difficult it can be to craft a suitable definition for this diverse context (see also Angell and Kassirer, 1998).

The National Center for Complementary and Alternative Medicine (NCCAM), the NIH Center that is responsible for investigating these therapies, classifies a number of common practices as shown in Table 4.2. This listing is fine for some purposes,

TABLE 4.2   Common CAM Practices

Practice	Examples
Biologically based systems	Diet; herbals
Energy	Reiki; magnets; qi qong
Alternative systems	Homeopathy; naturopathy
Mind–body	Yoga; prayer; meditation
Manipulative	Massage; chiropractic

but it hardly constitutes a definition. Many traditional medical practices also fall into those or similar domains. For example, diverse areas such as botanicals, biologicals, devices, psychotherapy, and radiotherapy all have accepted roles in traditional medicine. Many CAM treatments are characterized by only regional use. Others are little more than attitude, dietary, or lifestyle changes and may be widely used. Some proponents make extraordinary claims in light of the natural history of the disease, while others attribute to the treatment improvements that are consistent with typical variations in disease course.

It is not sensible to define CAM on the basis of content that is constantly in flux. For the purposes of discussing methodology and context, I define CAM as *a treatment whose mechanism of action is poorly defined or incompatible with established biology.* Poorly defined or incompatible is partly subjective, but this escapes the pitfall of defining CAM by the nature of the therapy. It also takes advantage of the somewhat easier task of directly characterizing scientific medicine, which is mechanistically founded, and ultimately justified by both biological theory and empirical efficacy. The huge contemporary push in genomic sciences illustrates the emphasis on basic biological mechanisms in medicine. I reject the occasional claim that the mechanism of action of a CAM therapy is unknowable or inaccessible to scientific means.

Some CAM therapies have elaborate justifications inconsistent with established science (e.g., homeopathy). Others invoke ill-defined or incomplete mechanisms of action (e.g., therapeutic touch). Some therapies often classified as CAM suggest mechanisms that are not well understood but are very plausible or evident in scientific medicine (e.g., mind–body interactions). Thus the dividing line between CAM and scientific medicine is imperfect. A few treatments used in scientific medicine work via unknown mechanisms of action (e.g., lithium for manic depressive illness), which is far from saying that they will be found to be inconsistent with established biology. When a therapy of any origin is based on well-defined, established biological mechanisms, and is also shown to be effective, it becomes incorporated into scientific medicine.

The power of mechanistic theory and integration with existing biological knowledge can be seen in recent investigations into the salutary effects of cinnamon on blood sugar, cholesterol, and triglycerides. Based on the origins and contemporary medicinal use of the spice, it falls within the CAM domain. However, there is rigorous science that reveals the molecule responsible for producing cinnamon's insulin-like effects and its cellular mechanisms of action (Jarvill-Taylor, Anderson, and Graves, 2001). Some CAM proponents are endorsing the use of the spice on the basis of this single study. It remains to be seen if additional research will support therapeutic uses for cinnamon, or produce derivatives or new compounds, but the pathway for potential incorporation into scientific medicine is visible. Adoption of natural products or derivatives into medicine in this way is a historical norm. Many other substances with conceptually similar origins and therapeutic claims in the CAM context have not been able to cross the divide.

4.5.1   The Essential Paradox of CAM and Clinical Trials

The problem and paradox regarding the relationship between CAM (by whatever definition) and clinical trials is the following. To test any treatment rigorously, it must at least temporarily be brought into the domain of orthodox science, with the attendant requirements regarding mechanisms of action, consistency with other accepted findings, and evidentiary implications such as proof of failure. CAM proponents often desire the imprimatur of science but frequently reject the methods necessary to gain it. In some cases findings not in accord with prior belief are simply discarded (e.g., Section 16.7.4). Others contend that CAM by its very nature cannot be evidence based (Tonelli and Callahan, 2001). See also commentary on this idea by Bloom (2001).

The greatest difficulty for evaluating CAM therapies using clinical trials arises when the therapeutic question is only partially brought into the scientific method. This occurs when the evaluation structurally proceeds by way of a rigorous experimental design, but the hypothetical means of action relies on unfounded or demonstrably false principles. This issue was introduced in Section 2.2.4. Mechanism (and support from other evidence) is absolutely essential to the experimental method because it is the only basis for resolving the difference between type I errors and true findings. Experiments without a foundation in established biology are little more than elaborate speculations.

An excellent illustration of this problem is a randomized clinical trial, funded by the Uniformed Services University of the Health Sciences (USUHS), studying the effect of therapeutic touch (TT) on pain and infection in 99 burn victims. In the words of the lead investigator:

> The idea behind the practice of Therapeutic Touch is that the human energy field is abundant and flows in balanced patterns in health but is depleted and/or unbalanced in illness or injury. The Therapeutic Touch practitioner assesses the patient's energy field patterns with his/her hands to identify areas of depleted, congested, blocked, or unbalanced energy. Then, the Therapeutic Touch treatment consists of a series of techniques implemented to replenish, clear, modulate and rebalance the patient's energy field patterns. (Turner, 1994)

The practice of TT is loosely based on the theory of unitary human beings (Rogers, 1970), which essentially equates human beings with energy fields. Beyond this, other humans can both manipulate and assess these energy fields. Although a staple of "nursing theory" and usually couched in scientific terms, these energy-based explanations appear supernatural.

In this randomized trial one group of patients was treated by a nurse trained in TT, while the other was treated by a nurse who was instructed to mimic the treatment. Because the patients had first-, second-, or third-degree burns on 5% to 70% of their bodies, the practitioners did not actually touch the study subjects but only moved their hands over them. Care was taken to train the mimics so their hand movements appeared real.

This study design was severely criticized by Selby and Scheiber (1996) in advance of the results because of the unconventional (unscientific) model of disease and treatment, the fact that patients were not actually touched, and the nature of the mimic or sham treatment control group. This study illustrates exactly the problem of omitting strong biological rationale for mechanism of effect, thereby bringing the therapeutic paradigm only partially into the domain of science.

Ultimately the investigators reported that TT significantly reduced pain and anxiety compared to sham TT (Turner et al., 1998). They interpreted the finding as support for the proposed mechanism, ignoring obvious ways of testing it directly. For example, energy fields can be measured, and no new anatomical energy receptors have been found or proposed in humans. Furthermore the investigators did not discuss the substantial likelihood that the result arises from methodologic flaws or random error under the assumption that the proposed mechanism is false. A scientific perspective on therapeutic touch is offered by Scheiber and Selby (2000).

Poorly defined mechanisms such as "energy fields" are often invoked as their own explanation. Frequently the names chosen to represent such treatments are value laden (e.g., *therapeutic* touch), in contrast with traditional medical nomenclature. Therapies based on such ideas may be demonstrably safe (provided they do not displace a needed treatment) but not so easily shown to be effective. Contrast this with radiotherapy, for example, which also might be said to work through an "energy field" but has exquisite mechanistic explanations and demonstrated efficacy and indications.

4.5.2   Why Trials Have Not Been Used Extensively in CAM

Formal clinical trials have not been widely applied either inside CAM practice or using CAM treatments in traditional settings. Reasons for this include the following: Many CAM therapies are used in settings where impartial evaluation is inhibited or discouraged by practitioners and patients alike. There may be strong financial incentives for practitioners to maintain unchallenged belief in the efficacy of such treatments. Also rigorous methods developed in conventional medicine are sometimes thought (incorrectly) to be inapplicable to CAM (e.g., Carter, 2003).

Many medical scientists and patients are willing to apply lower standards of acceptance to some CAM treatments because the absolute risk appears to be low. This seems to obviate the need for rigorous testing. This might be appropriate when there is a reasonable risk–benefit ratio or even when there is only an expectation of such, but it is not sensible in the absence of efficacy. It must be said that nursing and supportive care professionals often provide the gateway for this phenomenon, a good example being therapeutic touch as just discussed.

Perhaps the principal reason why CAM treatments have not been extensively and rigorously studied is foundational. Many alternative systems of therapeutic effect do not rely on structural and functional relationships such as those in scientific based medicine. This eliminates the requirement to propose and study anatomical, physiological, or biochemical mechanisms. The alternative systems in Table 4.2 are examples. Nuland (2000) said it well:

> Scientific medicine is the only tradition in which knowledge of the body's actual structure and functioning is acknowledged not only as the basis of understanding disease but also as something unattainable without dissection of the dead and study of the living, whether at surgery or by means of various biochemical or imaging techniques. Since other systems do not rely on directing therapy at specific well-identified abnormalities within individual organs, detailed knowledge is superfluous and useless to them. Like humoral medicine, such schemes depend for their efficacy on generalized readjustments of entire conditions of constitutional abnormality.

Whether or not it is theoretically possible to test every CAM question experimentally is an open question. There is one level at which this seems to be true, given only a suitable outcome measure and the superficial structure of an experiment. However, an experiment consists of both an empirical component and a mechanistic one, which is sometimes implicit. Absence of a mechanistic component disqualifies an empirical test as science. It is nearly impossible to generate true scientific evidence unless the investigator tests outcome and mechanism simultaneously.

Regulation

There is no direct government regulation of CAM treatments requiring them to be safe or effective. Many such treatments employ dietary supplements or other similar substances that fall explicitly outside FDA oversight. Even so, the claims made on behalf of such therapies, the attitudes of practitioners and patients, and the potential for CAM therapy to displace treatments of proven benefit and to cause their own adverse effects, suggest that some regulation might be appropriate (Stein, 2002; Marcus and Grollman, 2002).

Tradition

There is not a strong tradition of experimental trials among CAM practitioners, for reasons outlined above. However, the outcomes of CAM therapies are subject to the same selection bias and observer bias as any other treatment, perhaps more so because of the setting in which they are often used. It is possible that many CAM treatments appear more effective than they really are, further complicating our ability to do rigorous trials. There is also not a strong tradition of testing CAM therapies within scientific medicine. This situation is beginning to change with the application of NIH money. Because of the essential paradox discussed above, it remains to be seen how successful this will be.

Incremental Improvement

Most CAM treatments evolve by incremental improvement, although it is not clear that they are titrated to efficacy as much as to safety and cultural acceptability. There does not appear to be any "science of" or formalized developmental process for CAM therapies, thus alleviating the need to provide mechanistic rationale. However, many CAM treatments are well suited to masking, randomization, and placebo control.

Economics

Many CAM treatments are supported by relatively few practitioners, even a single one. This reduces the need for resources and collaboration, meaning peer review and funding. Economic incentives can be quite favorable for the CAM proponent and correspondingly unfavorable for rigorous voluntary testing. Because many CAM therapies use natural products or existing modalities, the opportunity for patents is limited. This restricts the resources that practitioners and proponents are willing to apply.

Psychology

Although difficult to generalize, patients can be psychologically very accepting of CAM therapies, provided they appear safe and are accompanied by an appropriate veneer of favorable performance and rationale. Decisions to take a CAM therapy are often based on trust, just as with conventional therapy. I can provide no definitive explanation for why patients appear to accept CAM therapies less critically than traditional ones.

Perhaps this is actually a misperception, and patients are broadly very accepting of any treatment provided they trust the practitioner and the risk–benefit appears appropriate to their understanding of their disease. Again, we must be mindful that the apparent safety of some therapies may be seemingly a product of how they are used rather than evidence from rigorous trials.

In the absence of rigorous evaluation, there is little to prevent ineffective therapies from remaining in the CAM domain. We pay a double price for this. Patients can become preoccupied with useless treatments, and science misses a much needed opportunity to broaden its view. Even a relatively small amount of rigor injected into the CAM setting would probably provide a large benefit to society.

4.5.3   Some Principles for Rigorous Evaluation

Applying clinical trials to the study of CAM requires several principles. First, we must accept the idea that all natural phenomena are accessible to scientific knowledge. Books could be written about this, but I don't consider it controversial. (We might even *define* science as the study of natural phenomena, but that's not essential.) Some advocates claim that scientific methods must be modified for, or are inapplicable to, the study of CAM. This claim is internally inconsistent.

Second, disease processes, administration of a treatment, the treatment itself, recovery from disease, and death are all natural phenomena. CAM proponents often support this idea by the frequent claim that its treatments are more natural than traditional healing, apart from the fact that the observables are natural. Although early in investigation, a CAM therapy may be stereotyped as involving an unexplained phenomenon, this is not a final explanation. Nor is it proof that such observations and phenomena are inaccessible to science.

Third, benefit is assessed by repeated structured observation. Strong evidence is gained by reproducibility. This is explicit in CAM by the use of the case history, case series, and occasional study report to influence opinion. Strong evidence (near proof or high reliability) based on repeatability is claimed by CAM advocates when and because they urge the application of treatments to new patients.

Traditional science also uses the method of structured observation and reliability, albeit usually quite rigorously. The essential differences between CAM as it has been traditionally supported and rigorous scientific clinical trials are (1) integration with biological knowledge about mechanism and (2) degree of structure applied on a purely empirical basis to generate data. CAM evaluations have been weaker on both counts.

Thus there are several components of structure that can be introduced to evaluate CAM claims more rigorously:

- Establish the biological framework for acquiring new knowledge. Provide background for a reasonable biological mechanism of action.
- Document promising cases/results. Even a handful of such cases could be *evidence of benefit* and might motivate more reliable and extensive study. Failure to provide it, when large numbers of patients have been treated, can be taken as evidence of no benefit.
- Prospectively write a study plan.
- Adhere to objective, definitive outcomes.
- Account for all patients treated and all time at risk following therapy. Accounting is based simply on active follow-up of patients and active ascertainment of outcomes.

The results, when compared to the natural history of the disease, provide informal quantitative evidence as to *degree of benefit*. This component of structure can be attained easily using a noncomparative (safety and activity) clinical trial design.

- Employ a comparative study design. This design should be enhanced by masking, randomization, placebo control, or other design maneuvers to strengthen the reliability of its findings.
- Solicit peer review of plans, process, and results.
- Obtain independent external verification or validation.

The application of any one of these components to many CAM claims would help considerably with evaluation, and using some or all of them systematically would be a service to patients everywhere. There is nothing in this list that works systematically against CAM, but all the steps work against ineffective treatments, deception, and quackery. In the case of Laetrile, for example (page 138), the application of fairly minimal structure to evaluate efficacy claims was enormously helpful. In the case of hydrazine (page 443), rigorous structure and strong evidence was not accepted as disproof by proponents.

4.6   SURGERY AND SKILL-DEPENDENT THERAPIES

The frequency and rigor of clinical trials in surgery and skill-dependent therapies is less than in many other medical disciplines (Solomon and McLeod, 1998; Anyanwu and Treasure, 2003). The problem has been evident from the surgical literature for some time but seems persistent (McCulloch et al., 2002). Surgical and other skill-dependent therapies, like certain medical devices, can have substantive differences from those that test drugs or biologicals. These include the potential for strong bias, the tradition of practitioners, and an expectation of large effect sizes. None of these eliminates the need for rigorous evaluation methods based on clinical trials.

Uncontrolled studies, traditionally the most common design, are vulnerable to confounding by selection and observer bias. Reliance on case series in particular for therapeutic comparisons aggravates the potential for bias. For a practical discussion related to this point, see Horton (1996) and the studies he discusses (Casey et al., 1996; Cuschieri et al., 1996; Majeed et al., 1996; Marubini et al., 1996). The insular attitude of some surgeons regarding rigorous trials is superficially similar to CAM practitioners, in that it is not unusual for surgeons to distinguish themselves explicitly from "physicians." However, surgery and CAM are worlds apart with regard to biologically based justifications for their respective treatments.

Because of the absence of rigorous evaluation, many widely used surgical procedures have probably been unsafe, expensive, ineffective, or otherwise suboptimal. Grimes (1993) provides some examples. Compared to drugs, surgical procedures may have a higher frequency of being ineffective or unsafe, based on the same reasoning. Beecher (1961) provides some examples of historical interest where previously accepted surgical treatments were later rejected, such as prefrontal lobotomies for schizophrenia and colectomies for epilepsy.

In the middle developmental stage of therapeutic development, investigators must distinguish between the feasibility of the procedure or technique and clinical benefit. Usually the failure of a surgical procedure is associated with an unfavorable outcome, but this does not mean that success of the procedure will yield a clinical benefit. From a methodologic point of view, we cannot substitute trials demonstrating successful completion of the procedure for those demonstrating improved patient outcome. Technical improvements that seem like a good idea may not easily demonstrate improved patient outcomes in rigorous studies. One example is laparoscopic assisted colectomy (Weeks et al., 2002; Petrelli, 2002).

Historically many surgeons have been reluctant to apply randomization to studies of patient outcomes, tending to make comparisons of surgical methods more difficult. Nevertheless, clinical trials of all types have been applied to surgical questions and have produced important information in areas like vascular, thoracic, and oncologic surgery (e.g., Law and Wong 1997). For a review of the role of statistical thinking in (thoracic) surgery, see Piantadosi and Gail (1995). McPeek, Mosteller, and McKneally (1989) discuss general issues regarding randomized trials in surgery, as does McCulloch et al. (2002). A broad discussion of statistical methods in surgical research is given by Murray (1991a, b). When both treatments under study are surgical, a randomized trial might sometimes be easier to motivate (e.g., Prinssen et al., 2004).

Surgery and skill-dependent therapies have other substantive differences from drugs, biologicals, and the like. We need to understand these because they can have a strong impact on the development and evaluation of treatments. Unlike drugs, surgical procedures are often developed by single or small groups of investigators at relatively little cost. Surgical procedures have a high degree of determinism associated with them and often work on physical principles, unlike pharmacologic agents. Surgical treatments are only occasionally amenable to evaluation versus placebo. Apart from the "placebo effect" and selection factors favoring those patients well enough to undergo surgery, a true surgical placebo, such as sham surgery, is nearly always ethically problematic. However, there have been notable trials in which sham surgery was justified. This is discussed in Section 4.6.5.

Among surgeons there is not a uniformity of perspective regarding controlled trials. In multidisciplinary collaborations for single therapeutic questions or in disease-oriented groups, surgical collaborators tend to be enthusiastic about studies. This reflects both enlightened attitudes by individual investigators and the fact that surgery is integral to the treatment of some diseases. Cancer is a case in point. For example, there is a collaborative oncology group sponsored by the American College of Surgeons. At the other ends of the spectrum of therapeutics, where surgery is the primary mode of therapy, or it has been little used, attitudes among surgeons toward controlled trials are highly variable and can be quite negative.

Examples of successful rigorous surgical trials include studies of adjuvant therapy in lung cancer, breast sparing surgery for cancer, and studies of the benefits of coronary artery bypass grafting. Coronary artery bypass surgery probably would have had its benefits demonstrated earlier if rigorous trials had been a part of its development. The paradigm of how new surgical therapies are developed is different than that for drugs. Developing surgical techniques does not necessarily follow the developmental phases discussed previously for drugs. Techniques tend to evolve slowly and are often amenable to improvement, even while being studied, to the point of becoming individualized to certain surgeons or institutions. In principle, new surgical techniques could be readily and quickly compared with standard methods in formal clinical trials soon after being shown to be feasible. However, despite the relatively relaxed need for developmental testing prior to a CTE trial, new surgical techniques seem to be infrequently compared formally with old techniques.

4.6.1   Why Trials Have Not Been Used Extensively in Surgery

For many practitioners the prevailing attitude regarding developmental studies of surgery appears to be that such treatments are intuitively justified, have large treatment effects, and have favorable risk–benefit when applied in the right patients. Under these assumptions relatively informal evaluation methods, such as case series or small prospective studies, are all that is needed to generate evidence. There are specific circumstances in which this confidence is justified, but generally these assumptions are over optimistic. Surgical interventions are not inherently justified or guaranteed to have either a large treatment effect or favorable risk–benefit.

A critical view of why formal trials are not more widely used in surgery is given below. It follows the framework described above for drugs, and reflects a mix of foundational and practical issues.

Regulation

There is no government regulation of surgical treatments requiring them to be safe or effective. Regulation is absent, in part, because such treatments are not advertised and marketed on a commercial scale the way that drugs are. Although there are no corporate entities that market surgical treatments on a large scale, there is compelling reason to have such treatments be safe and effective because a given procedure is likely to be applied extensively by many practitioners. For example, consider the marketing status of vision correction using laser surgery. This technique is currently marketed in a fashion similar to drugs.

Tradition

There is not a strong tradition of experimental trials in the surgical community, although it is improving in some disease areas such as cancer. Because of teaching and training, surgical culture is more respectful of experience and opinion than study design. The same traditions also limit formal training in research methods. I have indicated elsewhere that medicine historically has always resisted the introduction of rigorous evaluation methods. This tendency is aggravated in surgery, especially when surgical leaders set a poor example by their own investigational methods, and/or actively oppose reasonable but rigorous approaches. These factors can mistakenly reduce a surgeon's uncertainty, and correspondingly the equipoise concerning a trial that might actually be necessary.

Confounding

The outcomes of skill-dependent therapies confound three important effects: (1) the selection of patients i.e., prognosis, (2) the skill and supportive care of the practitioner, and (3) the efficacy of the procedure. Early in development this confounding makes some treatments look promising and may therefore inhibit rigorous independent evaluation of the therapy. It is hard to overstate the potential for this confounding to obscure the real treatment effects.

Incremental Improvement

Incremental improvement of accepted procedures is the historical norm for development of skill-dependent treatments. Accepted and familiar procedures can appear to carry ethical mandates before they are proven to be safe and effective. Incrementally improved procedures may therefore inherit a similar mandate. There is a myth in surgery that some procedures work when applied by skilled hands but not when used by inexperienced surgeons, or even that some complex procedures cannot be "taught." This is nonsense—nearly every operative procedure being applied routinely today was at some time in the past a complex and specialized procedure.

What is demonstrably true is that experience with a surgical procedure reduces risk to the patient, whether through the appropriate application of selection criteria, enhanced technical skill of the surgeon, or ancillary care. Demonstrations of this effect in the literature are old and convincing (Luft et al., 1979; Birkmeyer et al., 2002; Begg et al., 2002; Halm et al., 2002). Here again, this fact cannot be extrapolated to mean that experience also increases efficacy.

Masking and Placebos

Surgery is not very amenable to masking or the use of placebos, diminishing the expected rigor with which some such trials can be done. Many investigators would say that placebo or sham surgery is categorically unethical. This is a reaction to the pure risk and no benefit one would expect on the sham surgery treatment. Although this perspective is reasonable most of the time, it is important to recognize circumstances in which a sham procedure conveys little or no risk, making a placebo trial possible. Also there are many important therapeutic questions that do not require the use of a placebo control, such as comparison of surgery versus another modality.

Selection

Surgical treatments are often developed in, and applied to, patients with a good prognosis. Although they often increase the short-term risks to the patient, the subsequent risk may be lower than that in a nonsurgical group, simply by selection.

Economics

Surgical treatments are often developed at low cost by a single practitioner or a small group. This cost feasibility reduces the need for resources and collaboration. As a consequence there is less funding for (and less pressure to develop) the considerable infrastructure needed to support good clinical trials. The economics of health care favors the adoption and continued use of a surgical procedure that shows promise. A technique or procedure may become widespread early based largely on economic factors, making it very difficult to perform the needed comparative trials.

Psychology

Patients are psychologically very accepting of surgical therapies because they imply rapid and substantial benefit by correcting defects or removing disease. In reality surgical treatments are a very heterogeneous lot—some representing obvious benefits with large treatment effects, others more questionable. An example of a large treatment effect is surgical therapy for trauma, such as control of bleeding or correction of fractures. An example of a potentially small or adverse treatment effect is surgery for Parkinson's disease, as mentioned above.

It is sometimes claimed that a new surgical procedure must be used by a trained practitioner, implying that the standard of evidence in support of safety and efficacy can be relaxed. This second level of application reduces both risk and the need for rigor.

However, prescription drugs also anticipate the intervention of a skilled practitioner. The skill of the practitioner may reduce risk, but neither skill nor a favorable attitude can substitute for efficacy.

4.6.2   Reasons Why Some Surgical Therapies Require Less Rigorous Study Designs

A clinical trial with an internal control is a method for estimating a treatment effect that is clinically important but relatively small (compared to natural variation). Treatments that yield sufficiently large effects (e.g., penicillin for pneumococcal pneumonia) can have their benefits convincingly demonstrated by study designs that do not control all sources of error. Some surgical interventions are likely to be of this type. Examples are trauma care, correction of fatal congenital defects, or control of bleeding. Any therapy that counteracts short-term rapid demise is likely to be judged effective based on minimalist study designs.

Large surgical treatment effects might also be seen in some chronic disease settings. An example of this is joint replacement for degenerative arthritis. Confidence that the clinical and biological setting and preclinical data are consistent with large treatment effects could cause investigators to reduce the rigor of planned trials. Of course, such thinking is not restricted to surgical therapy, and it is often wishful thinking rather than reality. But it is common in surgery, perhaps because only large treatment effects truly offset short-term operative morbidities.

The theme seems to me to be whether a proposed treatment is *corrective* as opposed to *mitigating*. We might expect corrective therapies to have large effects, whereas mitigating treatments would have smaller effects. Cures, so to speak, are dramatic and easy to detect. However, the investigator who plans a study only to detect a large corrective effect reliably is likely to miss valuable mitigating effects. Studies should be designed based on the magnitude of effect that is clinically important rather than what one hopes to see.

Another potential rationale for a small or informal study is when the treatment is substantially the same as one about which a great deal is known. This notion is used often with regard to devices and seems relevant to many operative procedures that are widely used and then modified. Instinct tells us that incremental improvements do not require extensive evaluation, except that we can occasionally be wrong about the "improvement" of an increment. A sequence of increments constitutes drift, with the same implication. Also we do not always know as much about the benefits–risks of the standard procedure as we would like, rigorous studies often being lacking there as well.

4.6.3   Sources of Variation

The comparison of two surgical procedures in a single CTE trial may be less prone to bias because the problem of differential stresses on the patients is removed. However, it may be unlikely that two different surgical procedures for the same condition carry precisely the same risks. In fact this may be one of the primary reasons for comparing them. In this situation we can expect to encounter the same difficulties utilizing rigorous methods of bias control (masking, placebos, objective endpoints) as for trials with a single surgical treatment arm.

Operator skill is another source of variation that may also create difficulties for surgical trials. Some surgeons, particularly those who develop a treatment or technique, may have a higher success rate or a lower morbidity rate than practitioners who are just learning to use the procedure. In most cases the highest success rates and lowest complication rates appeared to occur in series where the surgeon performs a large number of similar procedures each year. Whatever the source, different practitioners may produce substantially different outcomes from ostensibly the same surgical procedure. This can aggravate the variation or heterogeneity with which clinical trial designs must cope, and may also affect the generalizability of the results. A frequent criticism of clinical trials is that they are beyond the scope of ordinary practice and that the patient cohorts are highly self-selected, rendering the results of questionable utility to routine practice. I do not find this criticism compelling, for reasons outlined elsewhere in this book. However, if a treatment, surgical or otherwise, cannot be applied as intended outside the scope of clinical trial, the concerns regarding external validity are justified.

4.6.4 Difficulties of Inference

The principal difficulty for inference arising from the absence of rigorous evaluation methods for skill-dependent therapies is the triple confounding of the patient's prognosis (selection), the practitioner's expectation (observer bias), and the true efficacy of the treatment. This confluence of selection and observer bias confuses the assessment of outcomes and makes reliable comparisons essentially impossible, except within the context of a well-designed and conducted RCT. In addition to this confounding of effects, surgeons may be seduced into equating the feasibility of the operative procedure (technique) with clinical efficacy when in fact the two are distinct.

The influence of patient selection is often not fully appreciated. Selection is widely used by surgical practitioners who speak of "surgical judgment," defined to be not only the proper timing and application of a particular procedure but also the selection of appropriate surgical candidates. Patients who are not robust enough to survive the surgical procedure will not receive the operation. Consequently the results of a surgical case series cannot be compared to patients who did not undergo the procedure, even if the cohorts appear superficially or statistically similar. Somewhat paradoxically, the better surgical judgment is, the stronger the selection effect. Even so, it is surprising how often inappropriate comparisons are made.

A classic example of selection effects was seen in a randomized surgical trial of portacaval shunt for the treatment of esophageal varices, using survival as the primary outcome (Garceau et al., 1964). The investigators followed both randomized groups, as well as a group of 288 patients not selected for the trial. Although the randomized comparison showed no beneficial effect of shunt, both enrolled groups did strikingly better than the unselected cohort. This trial is interesting both for its methodologic implications about selection bias and for its historical value. Although it appeared in a prominent journal 40 years ago, it is surprising how resistant surgical research design has been to the possible effects of selection bias.

Comparisons of case series, done so frequently in surgery, raise concerns regarding selection bias. This is similar to a well-known bias in epidemiology, where one version is termed the "healthy worker effect." Individuals in the workplace are an unsuitable comparison group (controls) for patients hospitalized with a particular illness (cases). Despite common exposures workers are healthier in aggregate than hospitalized patients or else would not be able to remain at work. It is plausible that individuals hospitalized at different institutions might also have been subject to substantial selection effects.

Comparison of a surgical treatment to nonsurgical therapy in an RCT presents challenges absent from most drug trials. A problem arises if surgery induces a stress on one group, that is, a differential stress following randomization. Higher mortality in the surgical group, for example, can make the survivors appear to perform better or live longer than subjects in the comparison group even in the absence of a true treatment benefit. In other words, the weaker patients may fare worse under surgery, making the survivors appear to have improved. Comparisons of mortality may be valid, but comparisons of survivors may not be.

Exactly this situation arose in the National Emphysema Treatment Trial (NETT), a randomized comparison of lung volume reduction surgery versus medical management for patients with emphysema (discussed in Section 4.6.6). Short-term mortality in the surgical group might have biased functional outcomes in favor of surgical survivors. Any group comparison based only on survivors, such as averages of functional test measures, is subject to this effect. To correct for it, a dichotomous outcome can be defined in all patients: improved versus unimproved, where the unimproved category includes deaths or those too ill to perform the test. Thus such a trial can provide an unbiased estimate of mortality or certain derived outcomes, but it is not guaranteed to yield an unbiased estimate of the magnitude of functional improvement. Fortunately NETT was designed to give precise estimates of the mortality risk ratio.

4.6.5   Control of Observer Bias Is Possible

An often-cited problem with performing surgical comparative trials is the difficulty in using masking and placebos, the goal being control of observer bias. It is well-known that unmasked randomized trials tend to favor the expectations of the investigators. This is even more of a problem when the methods of evaluation are (partly) subjective, as is often the case in surgical studies. Symptom relief is a common outcome of many trials and is subject to this bias.

Surgical placebos necessarily take the form of sham procedures, which are widely claimed to be unethical. The reason is because of the unfavorable risk–benefit ratio that results from a sham procedure. This is distinctly different than the use of a drug placebo, which carries only the risk associated with denial or delay of treatment. A classic example of a sham procedure in a surgical trial came from studying internal mammary artery ligation for treatment of angina. The procedure was used in several uncontrolled studies, observed to be "effective," and was being widely accepted by surgeons (Kitchell et al., 1958). Two studies were performed in which a control group was treated by a sham procedure (skin incision only) and compared to mammary artery ligation. Cobb et al. (1959) reported similar improvements in both groups, and Dimond et al. (1960) reported no benefit in either group. These trials convinced practitioners that the procedure was ineffective but also that sham operations were inappropriate.

However, there are circumstances in which the risk of a sham surgical procedure is near zero, and its use may be essential to help resolve an important therapeutic question. Investigators should probably keep an open mind about sham procedures. An example of this occurred in two small NIH sponsored randomized trials of fetal tissue implantation for treatment of advanced Parkinson's disease (Freed et al., 2001; Olanow et al., 2003). Parkinson's disease is one where it may be vital to control for a placebo effect (Shetty et al., 1999). The treatment required a stereotactic surgical procedure to implant dopamine-producing fetal cells in the substantia nigra of patients with the disease. One anatomical approach to accomplish this is through burr holes in the frontal bone. Small bilateral skin incisions were made in the forehead, burr holes drilled through the bone, and the needle was passed through the dura and advanced until the tip was in the correct location. This procedure is performed under sedation and local anesthesia, meaning the patients are "awake."

These trials were done using a sham procedure in the control group. The ideal placebo would have been transplantation of an inactive cell suspension. However, the investigators could not justify the risk attributable to actually passing needles into the brain of placebo patients. The sham procedure consisted of the steps up to, but not including, the penetration of the dura. There was broad agreement among neurosurgeons that risk was negligible up to that point. Additional time and maneuvers were conducted in the operating room to be certain that the patient could not tell if the procedure was a sham. The study design also called for placebo-treated patients to return after an evaluation period to undergo the real procedure, at which time the previously drilled holes would be used. Postsurgical evaluations of the disease were performed by neurologists who were unaware of the actual treatment assignment. The results of these trials do not support the initial enthusiasm for fetal tissue implantation.

These trial designs were reviewed extensively prior to implementation. The reviews included peer review for funding, sponsor review, local IRB review, and examination by an independent performance and safety monitoring board that also carefully followed the conduct of the studies. Even so, ethical questions were raised (Macklin, 1999; Freeman et al., 1999; Albin, 2002). Each reader should evaluate these trials for his or her level of comfort. My sense is that they were appropriately designed (although small), well timed in the window of opportunity, and would likely have yielded biased results without the unusual sham procedures employed. A strict prohibition against sham surgery is probably not sensible (Horng and Miller, 2002; Albin, 2002).

We might expect a sham surgical procedure to be more appropriate under a few fairly rigorous circumstances. These are when the sham is convincing but incomplete enough to produce negligible risk, necessary use of a subjective outcome, and when no other design can truly meet the needs of the investigation. A sham procedure was used by Ruffin et al. (1969) in their trial of gastric freezing for duodenal ulcers and recently by Moseley et al. (2002) in a study of arthroscopic knee surgery. Other procedural (but perhaps not technically surgical) shams have been reported for ulcers (Feldman et al., 1980a, b), hemodialysis (Nissenson et al., 1979; Schulz et al., 1981; Carpenter et al., 1983), peritoneal dialysis (Whittier et al., 1983), and laser photocoagulation (Robertson and Ilstrup, 1983).

Investigators must thoroughly examine the clinical contexts and risk of the sham procedure. We cannot pretend that all widely applied surgical procedures are effective or carry favorable risk–benefit ratios. This represents a risk of its own that might be offset by the use of sham controls in appropriate circumstances. Investigators must also be alert to surgical questions in which masking and placebos can be used without requiring a sham procedure. Examples of this are the biodegradable polymer studies in patients with brain tumors (Brem et al., 1995; Westphal et al., 2003).

4.6.6 Illustrations from an Emphysema Surgery Trial

Many of the contextual issues regarding surgery are illustrated by the National Emphysema Treatment Trial (NETT), which was a randomized comparison of lung volume reduction surgery (LVRS) plus medical management versus medical management alone for patients with advanced emphysema (NETT Research Group, 1999). From its inception in 1997, NETT was surrounded by controversy rooted deeply in the context of surgery. The principal source of difficulty was that many thoracic surgeons believed that the benefits of lung volume reduction had already been established based on uncontrolled studies of postoperative lung function and the resulting interpretation by experts. Some believed a randomized trial was therefore unnecessary and unethical. Other surgeons were supportive of a randomized trial, especially when discussed in private, but group dynamics made it difficult for that view to become dominant. The poor reception of NETT by the academic surgical community was evident again when the trial was completed (discussed below).

Many surgeons failed to recognize the weaknesses of uncontrolled study designs, and therefore the poor standard of evidence upon which LVRS was based. This reflects both undue respect for experience and opinion, and lack of training in methodology.

Because of deficiencies in the literature regarding LVRS, there was equipoise in the multidisciplinary collaboration needed to perform a definitive randomized trial. Uncontrolled studies could not correct for the confounding of selection, experience, and treatment effect mentioned above. Follow-up data in case series were short and often incomplete. No long-term studies of survival had been done. Optimal selection criteria for surgery were uncertain. Medicare claims showed a very broad application of LVRS and a high mortality rate. The operative mortality ranged between 4% and 15%. Medicare data indicated a six-month mortality of 17% and suggested a similarly high one-year mortality following surgery. Thus the information necessary for physicians and patients to assess accurately the risk–benefit profile of this major surgery was felt by many to be inadequate.

The Center for Medicare and Medicaid Services (CMS) was particularly interested in the LVRS question because of the high cost, wide application, and high mortality. Even so, arguments were being made to have the procedure routinely covered by Medicare—an implicit endorsement of an accepted standard therapy. CMS initiated a collaboration with the National Heart Lung and Blood Institute (NHLBI) to test the value of LVRS rigorously and base a coverage decision on the results of the scientific study (Carino et al., 2004).

This setting was nearly ideal for an RCT. Several factors were less than optimal, however: some strong opinions had already been formed on the basis of weak evidence, the window of opportunity was rapidly closing, and there was a lack of support or even hostility among some members of the medical and lay community for a randomized trial. Two assessments of the LVRS question had been undertaken by the NHLBI and the Center for Health Care Technology Assessment to review what was known and identify research gaps. Both assessments endorsed the advisability of an ethically feasible randomized clinical trial (Weinmann and Hyatt, 1996). CMS developed new policies based on conditional coverage to allow a controlled trial to take place.

Design

Drafting the NETT protocol to eliminate deficiencies in earlier studies required much attention to ethics, clinical practice, management, design methodology, and trial conduct. The design objectives of NETT were to

- employ definitive clinical outcomes defined for every subject to reduce observer bias,
- require a lengthy active follow-up,

- use a large sample size and inclusive cohort,
- randomize treatments to reduce selection bias,
- analyze risk factors and definitive outcomes to determine selection criteria,
- base the trial on biological characteristics and assessments of the disease to enhance external validity.

Meeting these objectives would yield needed data on the risks and benefits of LVRS and help delineate optimal patient selection.

Randomization in each of multiple centers occurred only after patients had been evaluated and assessed to be good surgical candidates, removing the effects of selection bias. Even so, operative mortality would create differential selection after randomization that could confound the treatment effect. For example, if weaker patients in the surgical group had a higher short-term mortality than stronger patients, the average pulmonary function postoperatively in the surgical group will be superior to that in the medical group even when the treatments were equally efficacious. The concern was appropriate, as illustrated by the early findings from NETT, where a small subgroup of patients were found to be at high risk of mortality following surgery (NETT Research Group, 2001). Randomization of this subset of patients was discontinued early.

This survivor's bias was counteracted in two ways: one using design and a second using analysis. First, NETT used overall survival as its primary outcome measure. There was controversy about this because some investigators felt that short-term pulmonary function, exercise, or quality of life were more appropriate. But considerations regarding statistical precision and clinical utility led to a larger lengthier trial with survival as the primary outcome. Second, functional outcomes in the treatment groups were not summarized as mean values defined only in survivors, a pervasive flaw in the historical observational studies. Instead, each subject was classified as either improved or unimproved with respect to a given outcome, where the unimproved category included those deceased or unable to complete the evaluation. Thus a functional outcome was defined for every subject.

Results

NETT was executed smoothly and effectively at 17 collaborating centers. Recruitment was substantially lower than originally anticipated, reflecting the high rate of comorbidities in patients with advanced emphysema and overestimates of the applicability of LVRS. Treatment adherence and completeness of follow-up were outstanding—95% and 99% respectively. This completeness contrasts with many of the nonexperimental studies of LVRS, illustrating one of the values of designed data production. The overall survival findings from NETT showed no difference (Figure 4.1). However, based on NETT's prospective evaluation of selection criteria proposed by investigators, a qualitative treatment–covariate interaction was observed (Figure 4.2). This survival result identifies a subset whose survival is improved by surgery (upper lobe disease, low exercise capacity; Figure 4.2, upper left panel), and one in which surgery should be avoided (non–upper lobe disease, high exercise capacity; Figure 4.2, lower right panel).

The remaining two subsets have modest but clinically significant improvements in parameters such as exercise, pulmonary function, and quality of life (NETT Research Group, 2003a). An example is shown in Figure 4.3, which demonstrates the increased exercise capacity at two years in non–high-risk patients assigned to LVRS. This data

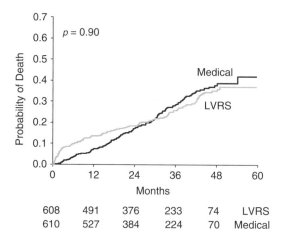

Figure 4.1   Overall survival following randomization in NETT. Data from NETT Research Group (2003a).

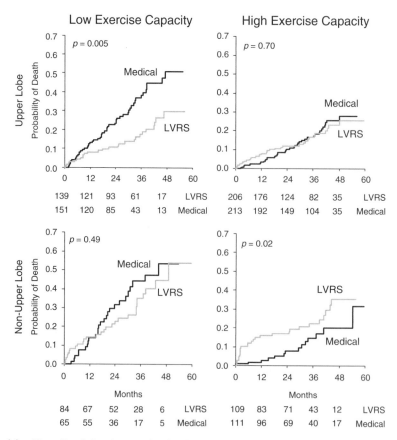

Figure 4.2   Mortality following randomization in NETT non-high risk subgroups. Data from NETT Research Group (2003a).

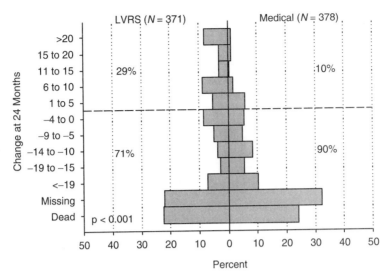

Figure 4.3 Distribution of maximum exercise changes at two years in non–high-risk NETT patients. Data from NETT Research Group (2003a).

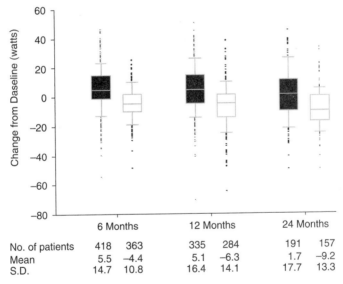

No. of patients	418	363	335	284	191	157
Mean	5.5	−4.4	5.1	−6.3	1.7	−9.2
S.D.	14.7	10.8	16.4	14.1	17.7	13.3

Figure 4.4 Change in maximum exercise measured by cycle ergometry in NETT survivors. Data from NETT Research Group (2003a).

display accounts for every patient, including those who could not complete the intended evaluation due to death or illness, who are classified as unimproved. This contrasts with summaries such as Figure 4.4 for maximum work measured by cycle ergometry, which demonstrates improvements subject to the survivor bias. Although typical of pre-NETT reasoning and flawed, Figure 4.4 does demonstrate the remarkable heterogeneity in the outcome and the absence of functional determinism that results from LVRS.

The clinical findings from NETT were that (1) a high-risk subgroup could be defined by easily determined disease characteristics and parameters, (2) no overall improvement in survival is attributable to LVRS, (3) selection criteria could be improved for patients undergoing surgery, (4) survival is improved in optimally selected patients, (5) safety is improved by optimally excluding patients, (6) modest but clinically significant improvements were made in exercise capacity, pulmonary function, and quality of life in optimally selected patients, and (7) these findings presented a strong degree of statistical certainty and external validity. Also NETT yielded a cost effectiveness analysis that demonstrated costs per quality-adjusted life-year were substantially in excess of $50,000 in the near term but were likely to drop below this threshold after 5 to 10 years of follow-up (NETT Research Group, 2003b). NETT findings yielded a quantitative basis for decision making at all levels. CMS implemented a restricted coverage decision based on the trial.

Methodological implications of NETT reinforce (1) the feasibility of well- executed trials of complex surgical questions, (2) the need for rigorous design (e.g., control of bias, long-term follow-up), and (3) the deleterious effects of lost data. The trial also provides some insights into the larger context of surgery, especially when examining events before the study was initiated and after it was completed.

Surgical Context after the Trial

The single most telling characteristic regarding the surgical view of NETT was a nonevent; specifically the failure of the American Association for Thoracic Surgery (AATS) to accommodate the initial presentation of primary results from the RCT at its 2003 annual meeting. The reasons for this appeared superficially complex—scheduling, journal selection, and society procedures. The actual reasons, in my opinion, are a reflection of weakened scientific values with respect to clinical experimentation in the surgical discipline. Sadly, the AATS suggested that the NETT primary results be presented at a continuing education program prior to its meeting rather than at a regular scientific session. This proposal was extremely unusual for an original scientific study and unacceptable to the NETT Steering Committee, who felt that the trial was too important for such a venue. Even if NETT had been afforded a regular AATS slot, the program allows only an impossibly short 8 minutes for presentation. The first presentation of NETT results took place just days later in a 3 hour scientific session at the 2003 American Thoracic Society, a multidisciplinary medical meeting.

Despite overcoming the obstacles to applying a rigorous randomized study in this context and the value of the clinical findings, NETT continued to attract criticism (e.g., Berger et al., 2001). Paradoxically, much of this criticism came from within the thoracic surgical community after completion of the trial, despite NETT yielding a conceptual concordance with prior belief and strong evidence supporting an LVRS indication in properly selected patients. As long as a year after completion of NETT, thoracic surgeons were still "debating" the trial (Cooper, 2004). Although lively scientific debate is healthy, complaints about NETT were a potpourri of concerns about study motivation and administration that have no effect on the results, minor methodological points, misunderstandings regarding analytic methods, grievances regarding restrictions on investigators, and even criticism of the way the results were peer reviewed. It also became clear that some surgeons did not fully understand the concept of equipoise.

Most of the concerns voiced about NETT a year after its publication had long since received due process. Thoughtful debate took place at many levels typical of a large multidisciplinary investigation: sponsor preparation, Steering Committee discussion about protocol design and implementation, independent ethics panel review, multiple IRB reviews, quarterly DSMB oversight, investigator's meetings, manuscript preparation, peer review, and public debate. The fact that some of the issues covered continued to resonate almost exclusively within the surgical community illustrates the contextual problem.

On all of these points—presentation of NETT results, methodologic criticisms, due process, and interpretation of results—surgeons remained privately divided. This unfortunately placed added strain on surgical investigators who saw the merits of what had been accomplished. The NETT experience overall emphasizes the undue respect that surgeons traditionally have for uncontrolled studies and expert opinion. For example, critical examinations of the type aimed at NETT were never directed at the studies that formed a basis for support of LVRS in the first place. It also demonstrates that large, rigorous, complex clinical trials are outside the experience of many surgeons, even those in academic based practice. Until teaching, training, and the culture changes, such trials will continue to be feared and misunderstood in this context.

4.7 A BRIEF VIEW OF SOME OTHER CONTEXTS

4.7.1 Screening Trials

Screening is a particular form of diagnostic testing, applied to a population at risk of developing a disease, in an attempt to diagnose the condition earlier than the natural history would manifest it. A premise of screening is that early diagnosis makes the condition more treatable or curable. If there are no good treatments, then screening and early diagnosis will not be helpful. Screening tests are often repeatedly applied to diagnose newly incident cases. Screening for cancer in older persons is a standard example.

The purpose of a screening trial is to assess the utility of the intervention compared to usual care in preventing deaths or complications from the index disease, and sometimes to compare different methods of diagnosis. Unlike therapeutic studies, screening trials nearly always carry a positive image, perhaps because there is a presupposition that early detection of disease is a helpful thing to do. Like therapeutic trials, a number of different study designs have been applied to screening questions, but the most reliable is the RCT. A general discussion of this context is given by Prorok (2001).

Most design issues in screening trials are similar to those for preventive and therapeutic agents, including the best choice for an outcome measure. Survival is often seen as the definitive outcome measure, but others have been proposed including the population incidence of advanced disease and stage shifts. Early detection induces a bias in the comparison of survival times that artificially makes screen-detected cases appear to live longer. This lead time bias must be corrected to estimate the true benefit of screening.

The number and interval of screenings is a unique design consideration, as is the interplay among sample size, trial duration, and the (temporally delayed) onset of the screening effect. A large relatively short duration trial might not provide sufficient time for the benefits of screen-detected cases relative to other cases to be seen. Thus definitive screening trials may need to be as long as prevention studies, even if the disease under study has a relatively short natural history.

Screening methods and studies are similar to prevention trials with regard to risk–benefit and many other methodologic considerations. For example, the role of developmental studies is similar in that early feasibility trials are not formalized and may be replaced by observational evidence. Some methodologic considerations for such trials, including the use of reciprocal controls for screening trials have been given by Byar (1990) and Freedman and Green, (1990). Because it employs diagnostic tests, procedures, or algorithms, and may use standard clinical evaluations, the framework for evaluating screening is different than that for prevention.

Prorok (2001) distinguishes screening trials that address a single question from those that investigate more than one question. The former is typified by the Health Insurance Plan (HIP) trial of breast cancer screening (Shapiro et al., 1988). There are numerous examples of screening trials that examine multiple questions, also depending on how screening is applied (e.g., continuous or split). For some prototypical designs, see the Prostate, Lung, Colorectal and Ovarian (PLCO) Cancer Screening Trial (Gohagan et al., 1994, 2000) and the Stockholm Breast Cancer Screening Trial (Frisell et al., 1989).

Role of Regulation

Screening methods are not explicitly regulated, but their components may be (e.g., radiological procedures, diagnostic tests). Screening is complex and expensive, and usually comes with sponsorship or the recommendation of governmental authorities. Therefore it is a virtually regulated context in that studies are carefully designed, reviewed, conducted, analyzed, and debated before conclusions are made.

Observer Bias

Observer bias can be controlled in screening studies in much the same way as therapeutic or prevention trials by using masking and objectively defined outcomes. Lead time bias presents some difficulties that cannot be corrected in a straightforward way by design.

Treatment Uniformity

Easy to define for the purposes of the trial. Uniformity is more difficult to achieve in the way screening is actually implemented on a population basis.

Magnitude of Effect

Screening is expected to have a relatively large effect both in gaining early diagnosis and reducing mortality and morbidity. This partly explains the positive view that many have regarding it. However, the real benefits of screening can only be gained through treatments whose effects are greatly amplified by early detection. Such treatments are difficult to develop in diseases like cancer. In contrast, for heart disease secondary to high blood pressure, screening and treatment may yield very large benefits.

Incremental Improvement

The role of incremental improvement can be substantial for screening interventions, particularly when they employ standard methods that are constantly being improved. Examples of this are radiographic procedures, especially computer-based ones, and genetic tests. The ability to improve a screening algorithm may not require extensive testing if convincing technical improvements are made in it components.

Risk–Benefit

Screening is generally very low risk, making any benefits that accrue favorably balanced from this perspective. However, cost effectiveness may be unfavorable for some screening methods or when applied to low-risk populations. The true positive rate is a strong function of the background probability of disease, so the same algorithm can have very different properties when applied in different settings.

Tradition and Training of Practitioners

The tradition and training of practitioners greatly supports rigorous evaluation of screening interventions.

4.7.2 Diagnostic Trials

Diagnostic trials are those that assess the relative effectiveness of two or more diagnostic tests in detecting a disease or condition. Often one test will be a diagnostic "gold standard" and the other will be a new, less expensive, more sensitive, or less invasive alternative. Although not strictly required, the diagnostic procedures are often both applied to the same experimental subjects, leading to a pairing of the responses. In that case it is the discordancies of the diagnostic procedures that carry information about their relative effectiveness.

Diagnostic tests are often not developed in as structured a fashion as drugs and other therapeutic modalities. This is the case because the risks of diagnostic procedures are often low. They may be applied in apparently healthy populations, as in screening programs, and may therefore require exceptional sensitivity and specificity to be useful. Developing better diagnostic modalities presupposes that effective treatment for the disease is available. Otherwise, there is little benefit to learning about it as early as possible.

Diagnostic trials can also be affected by biases that do not trouble treatment trials. Suppose that a disease is potentially detectable by screening during a window of time between its onset (or sometime thereafter) and the time when it would ordinarily become clinically manifest. The "lead time" is the interval from the screening detection to the usual clinical onset. Because of person-to-person variability, those with a longer window are more likely to be screened in the window. Consequently they will be overrepresented in the population detected with the disease. This produces length-biased sampling. See Walter and Day (1983) or Day and Walter (1984) for a discussion of this problem.

4.7.3 Radiotherapy

The characteristics that distinguish radiotherapy (RT) from other treatments, especially drugs, include the following. RT uses a dose precision that is higher than that for drugs. The RT dose precision is not based on weight but on the amount of the treatment (e.g., rads) delivered to the target tissue. Practitioners of RT manufacture or produce their own medicine at the time of treatment—it is not produced remotely. In principle, RT can control the therapeutic ratio. The dose delivered to the target tissue and that delivered to the adjacent tissue can be determined with considerable accuracy. The ratio can be titrated using source, dose rate, geometry, and other factors to produce the desired effects and minimize toxicities on surrounding tissue.

For the highest standard of validity and reliability, clinical trials in RT must account for these and other characteristics. See Perez et al. (1984) for a discussion of some quality control topics. Some of the factors related to dosing and treatment administration include the following:

- Dose and dose rate
- Source (e.g., photons or electrons)
- Equipment calibration
- Dose schedule (fractionation)
- Distance from source
- Target volume
- Dose delivered and tolerance of adjacent tissue
- Heterogeneity of dose delivered
- Intervals between fractions or treatments
- Use of concomitant treatment (e.g., drugs, oxygen, or radiosensitizers)

4.8   SUMMARY

The differential use of clinical trials in various medical contexts is a consequence of factors such as the role of regulation, ease with which observer bias can be controlled, treatment uniformity, expected magnitude of treatment effects, relevance of incremental improvement, and the tradition and training of practitioners. In a context such as drug therapy (a useful reference point because of its wide applicability), these factors combine to encourage the routine use of rigorous clinical trials. In contrast, the characteristics of a context such as complementary and alternative medicine generally discourage the use of clinical trials.

Recognizing the forces at work and the characteristics of various contexts helps to understand the strengths and weaknesses of clinical trials as a broad tool in medicine and public health. There are legitimate reasons why the stereotypical paradigm for drug development clinical trials does not generalize into some other contexts. Devices and surgical therapies sometimes share a characteristic (or expectation) that the treatment effect will be large, and therefore evident using relatively informal evaluation methods. They also often proceed using incremental improvement of existing therapies. However, it is critical to recognize the many circumstances where this is not the case, which would then require rigorous trials. The early development of preventive agents also often proceeds without the same types of clinical trials as drugs, using laboratory and epidemiologic evidence. The later development of preventive agents invariably uses the same type of rigorous trials as drugs.

Use of clinical trials in complementary and alternative medicine represents somewhat of a paradox. There is an obvious empirical method that seems relevant to this context, and some trials in CAM have proceeded accordingly. Trials are not purely empirical, however, and require biological underpinnings of mechanism to be truly scientific. CAM does not routinely supply rationale based on biological mechanism, undermining the utility of trials in this context.

4.9 QUESTIONS FOR DISCUSSION

1. Discuss other specific contexts in which it would be unnecessary or inefficient to perform clinical trials.

2. Give specific examples of clinical trials that do not have all the characteristics of the scientific method.

3. What medical disciplines have stimulated clinical trials methods the most? Why?

4. Discuss reasons why and why not randomized prevention trials, particularly those employing treatments like diet, trace elements, and vitamins, should be conducted without "developmental" trials.

5. In the therapeutic touch example from Section 4.5.1, note the difference between the grant title and the report title (Turner, 1994; Turner et al., 1998). Discuss the implications of this.

5

STATISTICAL PERSPECTIVES

5.1 INTRODUCTION

Clinical trials are a hybrid of both clinical and statistical reasoning and are applied in a variety of clinical contexts (Chapter 4). What comes as a surprise, disappointment, and fascination to many investigators is that clinical trials do not embody a uniform statistical method for inference. This is actually a statement about statistics generally because no single perspective is accepted uniformly in the field. The purpose of this chapter is to sketch the principal statistical perspectives through which trials are viewed and discuss their respective strengths and weaknesses.

Although this issue is more interesting to statisticians than clinicians, it is important for all to understand it because it has an impact on the tools used for design, analysis, and inference, and can occasionally lead to differing interpretations of the same data. It is certainly possible to ignore differences in statistical perspectives or "schools of thought" and still achieve a solid understanding of clinical trials. However, ignoring differences means that a single perspective must be adopted. Any single perspective cannot be applied ideally and may not deal effectively with all of the practical difficulties that clinical trials present.

Statistics is used as a descriptive tool, an analytic science, and an aid to making decisions. Differences in statistical perspectives reflect critical dualities in the use of the science and, to a certain extent, in the notion of probability. Some methods may be tailored primarily to one purpose but not the others. Probability can be sensibly defined as either a constant of nature, evidenced after repeated experimentation, or as a subjective degree of belief. These ideas have been discussed by many authors, and a brief summary is given by Barnett (1982) in the context of formal statistical inference.

5.2 DIFFERENCES IN STATISTICAL PERSPECTIVES

5.2.1 Models and Parameters

As discussed in Chapter 2, statistics is both a descriptive and an analytic science. Methods for data description do not strongly highlight different statistical perspectives. However, if we attempt to uncover relationships in the data, analytic tools become necessary, and these lead to foundational differences in perspective or philosophy of inference. This is a reflection of the fact that there is no single best way to connect the empirical findings of data with the truth of nature.

Analytic statistical methods use models and their attendant parameters. A model is a mathematical or probability statement that describes the properties of a single observable or the relationship among two or more observables. It is therefore a statement *in* mathematics, but also a statement *about* nature. A model is never observed directly—it is a theoretical construct. Because it is partly a description of data, a model can be viewed as meta-data.

A classic simple model relating two quantities, x and y, is linear,

$$y = a + bx,$$

where a is the intercept and b the slope. Models have a deterministic part, shown here for the linear model, and a random part, omitted here for clarity. One reason that models are so useful is that they separate three essential components: the structure of relationships among observables, numerical properties that are captured in the parameters (a and b in the linear model above), and the random component.

Data yield quantitative information about model parameters, which in turn provide insight into nature. How to characterize and quantify the connection between data and model parameters is one of the sources of disagreement among statisticians. For example, in the linear model above, we could view a and b as constants of nature. Any imprecision in their values would then be attributed to sampling of the data. Alternatively, we could view a and b as random variables, in which case uncertainty about the parameters would be described in terms of their probability distribution(s). Additional consequences of how parameters are viewed and used are discussed below.

5.2.2 Philosophy of Inference Divides Statisticians

At least three major philosophies of statistical thought coexist today. My discussion here is not intended as an endorsement of any perspective. I do intend to emphasize points of agreement and disagreement and point out strengths and weaknesses of each. An important point of agreement among all approaches is the value of designed data production and the principles of experimental design discussed earlier in this book. Differences arise mostly with regard to technical features of analysis and proper interpretation of estimates. Thus, despite differences that are consequential to many statisticians, the different philosophies do not carry strong implications for design-oriented researchers such as clinical trialists.

The view and practice held by a majority of statisticians is "frequentist" (or "classical"), a name derived from the definition of probability in terms of frequencies of outcomes or long-run behavior. This method of inference makes use of tools familiar to many clinicians, such as point estimation, hypothesis and significance tests, confidence intervals, and unbiased estimators. The basic framework for frequentist inference hypothesizes repetitions of the experiment under identical conditions. This leads to somewhat unintuitive probability statements about the data conditional on specific hypotheses about the parameters.

A second school of thought, Bayesian, uses probability models for the data and the unknown parameters that give rise to it. Bayesian inference uses data to make probability statements about the parameters. It is also appropriate, or even necessary, to make probability statements about the parameters prior to the experiment, depending on subjective probability, degree of belief, or other evidence. The subjectivity allowed to initiate Bayesian inference routinely draws criticism. The basic Bayesian machinery is theoretically sound, perhaps even more so than frequentist methods, but subjective probability remains controversial.

The distinction between frequentist and Bayesian views is partly the result of viewing the quantities to be estimated (parameters) as random variables (Bayesian) or fixed constants of nature (frequentist). If the unknown parameters are taken to be random, then it is sensible to use the observed data to make probability statements about them. In contrast, if the parameters are viewed as fixed constants, then we cannot make probabilistic inferences about them, but can make probability statements about *data* conditional on hypothetical values of the parameters. This distinction is essential and will be discussed in more detail below.

A third philosophy is termed "likelihood" because of its emphasis and use of the likelihood principle and the statistical likelihood function (discussed below) as the bases of inference. The likelihood has a natural interpretation as a quantification of relative evidence, and in some form it is used in all methods of statistical inference. Whether because of its newness or other incompletely resolved issues, pure likelihood-based methods have not become as widely used in statistics as frequentist or Bayesian.

5.2.3 Resolution

Any approach to the statistical aspects of clinical trials must depend partly on practical grounds. Because of computational difficulties, Bayesian methods have not been widely accessible or used by statisticians until recently. Reflecting this, the approaches to clinical trial design and analysis outlined in this book are predominantly frequentist in origin. This is not so much an endorsement of that philosophical perspective as an acquiescence to simplicity and the preponderance of techniques likely to be encountered in the clinical trials literature. Frequentist techniques have been presented essentially without alternatives to clinicians for 50 years.

In some places in this book, a Bayesian and likelihood perspective is sketched alongside the frequentist one. In other circumstances, Bayesian methods exist but have not been discussed. Whenever possible, the student of clinical trials should investigate related Bayesian and/or likelihood methods to see if they might be better suited than traditional methods to the problem at hand. One area where the approaches yield substantially different results is in monitoring and interim analyses of trials, discussed in Chapter 14. It is my sense that many clinicians, given the chance, might naturally prefer inference based on Bayesian thinking, except that the techniques are often mathematically more complex than frequentist procedures and have not had as many advocates in the statistical profession.

5.2.4   Points of Agreement

Although differences between statistical schools of thought are consequential, there are more important methodological concerns in clinical trials on which there is wide agreement. For example, general principles of experimental design, careful data production, and using methods to increase precision and eliminate bias when designing studies are universally regarded. The proper counting of events and emphasis on estimation is essential when conducting analyses. When reporting results, objectivity and thoroughness are the keys. Each of these issues is probably of more practical importance than the philosophical differences between schools of inference.

Aside from agreement regarding points of experimental design, frequentist, Bayesian, and likelihood schools of thought agree on some fundamental technical aspects of how to use data. The likelihood function (discussed below) represents a basic probability model for data and is fundamental to all methods of analysis and inference. This provides considerable consistency among various statistical perspectives—the role and principles of experimental design and the fundamental use of probability models and data. This serves to emphasize that differences important to statisticians may not always be consequential for the design and interpretation of clinical trials.

Binomial Analysis Example

To compare and contrast, it is helpful to have the same practical example for each of the sections below. I will take a simple case as illustration—a one-sample binomial experiment. The example will highlight some important differences regarding analysis and interpretation. Some perspectives on design, particularly with regard to sample size, are given in Chapter 11.

Assume that we have 20 independent samples of a Bernoulli random variable (success or failure). Suppose that we observe in our experiment 8 successes and 12 failures. The chance of success is θ and the chance of failure is $1 - \theta$. The principal statistical issue is to make inferences about the unknown probability of success, θ. The first step of analysis is to take a plausible probability model for the data. In this case we can assume as usual that the number of successes and failures follows a binomial probability distribution with unknown parameter θ. All three of the statistical approaches agree on this.

Likelihood Function

Next we might address the question: How likely were we to observe the data under the assumed probability model? The answer to this question is the *likelihood* of the data, which according to our binomial probability model is

$$L(\theta) = \binom{n}{r}\theta^r(1-\theta)^{n-r} = \binom{20}{8}\theta^8(1-\theta)^{12},$$

where n ($= 20$) is the number of samples or trials, r ($= 8$) is the number of successes, and θ is the probability of success. The observed data enter the likelihood function as n and r.

The notation $L(\theta)$ suggests that the likelihood is a function only of θ, which is not necessarily correct for all purposes. It depends explicitly on both θ and the data, $\mathbf{x}$, which could be expressed as $L(\theta, \mathbf{x})$. Bayesian procedures would view the likelihood as $L(\mathbf{x}|\theta)$, the likelihood of the data, $\mathbf{x}$, conditional on θ. This will be made more explicit below.

In this simple case the likelihood is the same as the binomial probability of observing r successes out of n trials. Not all probability models will yield such a simple likelihood. The likelihood is appropriately viewed as a function of the unknown parameter θ.

We have a very powerful tool in the likelihood function. It is a model whose structure informs us about the data and whose parameters hold information from the data. Although not obvious, it contains all the information we have regarding θ. It captures both the probability model and the observed data, and yields a quantitative measure for any hypothetical value of θ. Figure 5.1 illustrates this, where the value of the likelihood function for our data is plotted against hypothetical choices for θ. In Figure 5.1 the vertical axis is logarithmic to compress the scale and facilitate comparing ratios.

Algebraically, likelihoods are often constructed from the product of independent probability terms. Such functions are usually more tractable on a log scale where the logarithm of products is converted to the sum of logarithms. Functional maxima, discussed below, are not affected by this re-scaling. The value scale chosen for the abscissa is not critical. In Figure 5.1 the scale includes the binomial coefficient.

Several things are evident from the likelihood plot. The likelihood has a highest value or maximum for a particular value of θ, which therefore seems most consistent with the data. This means that the data are most likely to have been observed under a certain alternative, $\widehat{\theta}$, termed the "maximum likelihood estimate" (MLE). Values for θ distant from the MLE are much less likely to be responsible for the data.

The maximum of $L(\theta)$, or equivalently the maximum of $\log[L(\theta)]$, can be found using straightforward methods of calculus. Define

$$\mathcal{L}(\theta) = \log[L(\theta)] = r \log(\theta) + (n-r) \log(1-\theta).$$

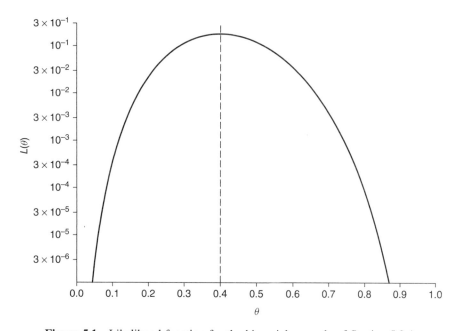

Figure 5.1   Likelihood function for the binomial example of Section 5.2.4.

Note that the maximum occurs at the value $\widehat{\theta}$ where the derivative with respect to θ (slope) is zero. Therefore

$$\frac{d\mathcal{L}(\theta)}{d\theta} = \frac{r - n\theta}{\theta - \theta^2},$$

so that

$$0 = \frac{r - n\widehat{\theta}}{\widehat{\theta} - \widehat{\theta}^2},$$

and

$$\widehat{\theta} = \frac{r}{n}.$$

Thus our instinctive estimate $\widehat{\theta} = r/n = 0.4$ has an analytic pedigree—it is the MLE for θ, the true probability of success. For general estimation problems, the parameter(s) associated with the maximum of the likelihood function yield a broad class of useful estimates with good statistical properties.

5.3   FREQUENTIST

Frequentist methods view unknown parameters, the objects of statistical inference, as fixed constants of nature. In the binomial example above the true probability of success, θ, is taken to be a constant that must be estimated from the data. Using the ideas of likelihood from Section 5.2.4, the MLE of θ is $\widehat{\theta}$, the value of the parameter that seems most likely to have given rise to the observed data. If it is unlikely that the data arose from a hypothetical value of θ, this is taken as evidence against that being the correct value.

We can construct a probability distribution for the observed data as a function of θ. The frequentist frame of reference for this distribution is replication of the experiment under a hypothetical value of the parameter. Although the value of θ is assumed to be fixed, sampling variation gives rise to the probability distribution. Probability statements can be made about data, under specific hypotheses about the fixed value of the parameter. This procedure is a hypothesis test: a specified hypothetical value for θ is rejected if it is relatively unlikely that the observed data arose from it. One value of particular interest is θ_0, the null hypothesis value.

A closely related frequentist procedure is the "significance test," which is also based on the probability distribution under θ_0. Rather than a dichotomous rejection or no rejection of θ_0, the significance test determines a tail area of the distribution for values more extreme than $\widehat{\theta}$. This tail area, or p-value, is the chance that the observed result or one more extreme will arise by chance under the null hypothesis in repetitions of the experiment. Small tail areas (low p-values) suggest that $\widehat{\theta}$ is inconsistent with θ_0, meaning significantly different.

A confidence interval is a potentially superior idea to either the hypothesis test or the significance test. It defines an interval around $\widehat{\theta}$ in which the p-value (or equivalently, the hypothesis test) for any θ would be nonsignificant at a specified level. Unlike the hypothesis test and the significance test, confidence intervals are determined under $\widehat{\theta}$ rather than θ_0.

For a given experiment either the true value of θ lies within our confidence interval or it does not. There is no probability associated with that event. If the experiment could be repeated a large number of times, identical constructions of the confidence interval would capture the true θ with the specified frequency. This idea is somewhat more useful than significance testing because the confidence interval is centered around the observed value and its width reflects the precision of the experiment. Even so, many people misinterpret the confidence interval as a probability statement about the true parameter value.

Frequentist inference is a somewhat odd but familiar process. Many clinicians find it awkward or inappropriate to use frequentist procedures literally. A 95% confidence interval would include the true parameter 95% of the time, if the experiment were repeated a large number of times. This repetition never occurs. However, we tend to deliberately misunderstand this and behave as if there is a 95% chance that the true parameter lies within the interval from our single experiment.

5.3.1   Binomial Case Study

In the binomial example from Section 5.2.4, we might have hypothesized a priori that the true probability of success was $\theta_0 = 0.5$ (null hypothesis). Under such a hypothesis, would it be unusual to observe the data at hand? The answer can be seen in Figure 5.2 where the sampling distribution of $\widehat{\theta}$ under the null hypothesis is drawn. This curve is a binomial distribution with $\theta = 0.5$. Obtaining a value of 0.4 from such a distribution does not appear to be unlikely. In fact the chance of a value more extreme than 0.4 compared to the null using the tail areas of this distribution is 0.49.

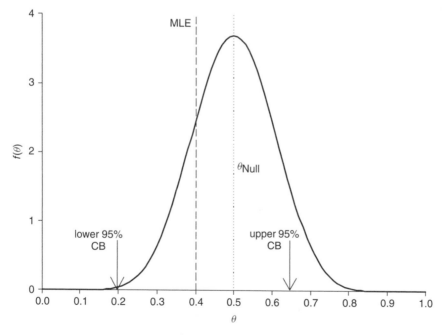

Figure 5.2   Sampling distribution of $\widehat{\theta}$ under the null hypothesis that $\theta = 0.5$. CB represents the 95% confidence bound.

That is, nearly 50% of the distribution contains values as "unusual" as 0.4. Thus there is insufficient evidence to reject the possibility that $\theta = 0.5$ as an explanation for the observed outcome. In this example we could report $p = 0.49$ as the two-sided p-value, indicating that 0.4 is not statistically significantly different from the null value 0.5. Also shown in Figure 5.2 is a confidence interval based on the binomial distribution centered at the MLE. The confidence interval includes the null value 0.5. In a given circumstance, it either does or does not contain θ—there is no probability associated with it.

This type of "hypothesis test" can be performed using an approximate but usually simpler procedure. A normalizing transformation (subtract the mean and divide by the standard deviation) yields

$$Z = \frac{\widehat{\theta} - \theta_0}{\sqrt{\theta_0(1 - \theta_0)/n}} = \frac{0.4 - 0.5}{0.5/\sqrt{20}} = -0.894,$$

which is an approximate normal deviate. The two-sided tail area under the normal distribution corresponding to $Z = -0.894$ is 0.57, consistent with the findings above. Another broadly useful class of hypothesis tests can be based entirely on the ratio of the likelihood function evaluated at two hypothetical values of θ. This will be discussed below in the section on likelihood inference.

5.3.2   Other Issues

There are potential practical problems with frequentist inferences. The properties of p-values are ill-suited for assessing strength of evidence, yet they are often used in this capacity. There is more discussion of this topic in Chapter 16. Confidence intervals, having a direct correspondence to a hypothesis test, are often used as rejection regions. Use of a confidence interval in this way essentially equates all the values of θ between the bounds, when in fact, the points at the ends of the interval have much less evidence in their favor than points near the middle.

This use of confidence intervals is a problem, for example, in some comparative trials intended to show equivalence. Suppose that Δ is a relative treatment effect. It is common practice to define equivalence as when the lower 95% confidence bound on Δ exceeds a prespecified tolerance, say 0.8. What should be the conclusion when the lower bound is 0.79? Failing to meet the specified criterion, one might conclude that the treatments are not equivalent. However, the data might strongly support a relative effect of 1.0 (point estimate), but our final conclusion would be based on hypothetical parameter values that have little support. There is more discussion of this topic in Section 11.5.9.

When designing comparative experiments, the primary frequentist approach is to postulate a hypothesis test and then to control the false positive (type I) and false negative (type II) error rates (Chapter 11). The resulting experiments are usually analyzed in terms of significance testing, a different procedure, despite the hypothesis test design. Even many noncomparative trial designs make use of an implicit comparison or hypothesis test to quantitate the experiment in terms of the type I and type II error rates. The fact that the dominant method of sample size determination is derived from hypothesis testing should be viewed primarily as a matter of convenience.

5.4   BAYESIAN

An alternative view or method of inference is called "Bayesian," named for the English clergyman Thomas Bayes. The Bayesian school uses ideas loosely associated with Bayes' posthumous publication from 1763 (Bayes, 1763; 1958). Bayes' theorem regarding probabilities is not controversial and represents the way to invert conditional probabilities. If B_1, B_2, ..., B_n are disjoint events, $\Pr\{\cup B_i\} = 1$, and A is any other event,

$$\Pr\{B_i \mid A\} = \frac{\Pr\{A \mid B_i\}\Pr\{B_i\}}{\sum_{j=1}^{n}\Pr\{A \mid B_j\}\Pr\{B_j\}} \ . \tag{5.1}$$

A similar form can be obtained for continuous distributions and will be illustrated below. In the present context, if A represents the data and B a set of possible parameter values in the probability model, Bayes theorem allows us to make probability statements about B_i after observing the likelihood, $\Pr\{A \mid B_i\}$.

There is nothing controversial about this formula. For example, it is used in Section 7.2.7 to calculate the chance that a trial result is correct as a function of the type II error rate and the prior probability of treatment success. The issues at play in Bayesian inference are foundational, as discussed later.

Bayesian inference represents knowledge (and ignorance) in the form of probability distributions and yields posterior probability distributions and credible intervals (Spiegelhalter, Freedman, and Parmar, 1994) that may be unfamiliar to many clinical trialists. Bayesian inference treats unknown parameters as random objects about which probability statements can be made. It formally incorporates information from prior knowledge or belief about these parameters, or from outside the experiment, as part of the inferential process. This necessitates accepting subjective notions of probability, meaning representing degree of belief as a probability distribution. An example of the use of subjective information is given in the discussion of the continual reassessment method for dose-finding trials in Section 10.4.2.

The use of subjective probabilities is one of the major sources of controversy about Bayesian methods. This is not to say that frequentist methods are more objective because they do not permit degree of belief to enter the inferential process as a probability distribution. Frequentists methods may appear to treat choices for important inferential parameters as objective, when in fact they are subjective. Simple examples are choices for the type I and type II error levels for hypothesis tests.

Bayesian methods hold some potential advantages for the clinical investigator. Once uncertainty is characterized in the form of a probability distribution, the Bayesian method offers a formal and consistent process for solving inferential problems. The method is coherent, which means that it does not violate logical axioms of uncertainty. Non-Bayesian methods violate such axioms in some circumstances. A large body of knowledge from outside the study may be important in designing, analyzing, and interpreting it. Bayesian methods can account for this information in a natural, formal way.

In many cases Bayesian approaches lead to the same or similar inferences as those derived from frequentist methods. In other cases the procedures and results are different, fueling a debate as to how the approaches can be justified. Discussion of Bayesian inference and approaches can be found in Cornfield (1969), O'Hagan (1994), Lindley

(1965), Barnett (1982), and Spiegelhalter, Freedman, and Parmar (1994). The contrast between Bayesian and frequentist views of clinical trials can be seen succinctly in papers (and the accompanying discussion) by Berry (1993) and Whitehead (1993). For an in-depth discussion of a Bayesian perspective in a real-world clinical trial, see Kadane (1996). A concise and practical discussion of some issues related to confidence intervals is given by Burton (1994).

5.4.1   Choice of a Prior Distribution Is a Source of Contention

Before the experiment begins, the Bayesian summarizes knowledge about the unknown statistical parameter (e.g., treatment effect) in the form of a probability distribution, called the "prior distribution." When data exist that bear on this question, this procedure is not controversial. In the absence of data, other knowledge or opinion must substitute as the basis for the prior.

The principal source of disagreement regarding Bayesian methods is the use of subjective probability. A strongly optimistic investigator will produce a different conclusion via his/her prior than a skeptical one. Although this type of inconsistency is the way the world works, it is an obvious problem if we assume that scientific inference is an objective process where the same data should produce essentially the same inference. Perhaps inference is not completely objective. Of course, the *relative* effect of the data in modifying the belief of the optimist and the skeptic is the same, but the actual outcomes might be quite different.

This issue cannot be covered comprehensively in this book. Background and discussions are given by Martz and Waller (1982), O'Bryan and Walter (1979), Lindley (1965), and Savage (1954). Aside from the notion of subjective probability, it is not obvious that representing knowledge (or ignorance) in the form of a probability distribution is the correct approach, although in the case of Bayesian inference, it is a powerful convenience. When prior information is available in the form of a probability distribution, Bayesian methods are not controversial. For example, this can happen when making inferences about samples from production runs.

There are some practical difficulties implementing Bayesian methods. Because clinical researchers do not automatically represent their knowledge in the form of a probability distribution, getting them to put information in this form can be a problem. This process, called "eliciting a prior," requires skill and planning on the part of the trial methodologist. Different clinical investigators are likely to provide different prior distributions for the same parameter. Groups of investigators may not agree on a single prior, even if they agree on the need for a clinical trial.

The evidence for prior distributions can be based on previous randomized trials, nonrandomized studies, preclinical data, or investigator opinion. Bayesian statisticians discuss various types of prior distributions corresponding to different opinions about the external data. *Reference* priors are intended to represent a minimal amount of prior information. Some of them are "improper" probability distributions that essentially spread belief over all mathematically possible values. Although inference using these types of priors can yield conclusions equivalent to some frequentist inferences, these prior distributions may be unrealistic because they give equal weight to any possible value of the parameter.

Clinical priors are intended to represent the opinions and beliefs of particularly well-informed scientists about the treatment effect. Because clinicians are usually not

Bayesian statisticians, eliciting this information in a quantitative form can be difficult. *Skeptical* prior distributions are those that attempt to quantify the belief that large treatment effects are unlikely. They might be appropriate for use by regulatory officials. A skeptical prior might be equivalent to assuming that a certain fraction of the trial has been completed without seeing a treatment effect. *Enthusiastic* priors are those that attempt to quantify the urge to continue a clinical trial, even when the results support the null hypothesis. These might be used in circumstances where adopting a particular promising treatment, even if it is not as effective as hoped, has few drawbacks.

5.4.2   Binomial Case Study

Consider the binomial example from Section 5.2.4 and that some evidence or opinion regarding the true probability of success, θ, is available before the experiment. Suppose that 0.6 is thought to be the most likely value for the success probability, and that this value is not strongly favored over 0.5 or 0.7 but is strongly favored over lower values such as 0.4. In principle, any value between 0 and 1 could be correct.

Investigators could summarize this evidence as a probability distribution $f(\theta)$. Bayesian probability calculations will be simplified if a suitable mathematical form for $f(\theta)$ is chosen. In this case suppose that we use a beta distribution

$$f(\theta) = \frac{\theta^{a-1}(1-\theta)^{b-1}}{B(a,b)},$$

where a and b are constants that characterize the shape and location of the distribution and $B(a,b)$ is the beta function. The benefits of choosing this functional form will be evident below.

Within this family of distributions, a and b capture the evidence or opinion regarding θ. Figure 5.3 shows some possibilities based on the numerical values given above. Curve A ($a = 13$, $b = 9$) shows fairly firm support for 0.6 being the correct value, as though it were based on a sizable number of actual observations. Curve B ($a = 4$, $b = 3$) shows fairly weak support for 0.6. Curve C ($a = 1, b = 1$) shows equal support for any value of θ, meaning it is "noninformative."

The next step in Bayesian inference is to form the likelihood function, $L(\mathbf{x}|\theta)$, which is viewed as the probability of the data, $\mathbf{x}$, conditional on the parameter θ. The likelihood given in Section 5.2.4 can be used directly.

Finally the prior distribution is combined with the likelihood to yield a "posterior distribution" $g(\theta|\mathbf{x})$, the probability of θ conditional on $\mathbf{x}$, from which probability statements about the parameter can be made. The posterior distribution is calculated from the continuous form of Bayes theorem as

$$g(\theta|\mathbf{x}) = \frac{L(\mathbf{x}|\theta)\,f(\theta)}{\int_{-\infty}^{+\infty} L(\mathbf{x}|\theta)\,f(\theta)\,d\theta}, \tag{5.2}$$

which is a continuous version of equation (5.1). The integral in the denominator is taken over an appropriate range for θ.

For mathematically complex priors and/or likelihoods, calculating the posterior distribution can be difficult. However, we have purposely chosen a form for $f(\theta)$ that

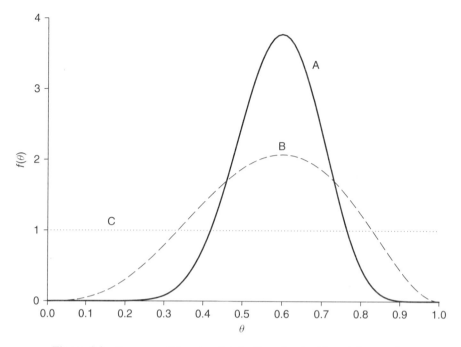

Figure 5.3   Some possible prior distributions for the binomial example.

will facilitate the calculation. Observe that

$$L(\mathbf{x}|\theta)\, f(\theta) = \binom{n}{r}\theta^r(1-\theta)^{n-r}\frac{\theta^{a-1}(1-\theta)^{b-1}}{B(a,b)}$$

$$= \frac{\theta^{r+a-1}(1-\theta)^{n-r+b-1}}{B(r+1, n-r+1)B(a,b)}.$$

Also

$$\int_{-\infty}^{+\infty} L(\mathbf{x}|\theta)\, f(\theta)\, d\theta = \frac{\int_0^1 \theta^{r+a-1}(1-\theta)^{n-r+b-1}d\theta}{B(r+1, n-r+1)B(a,b)}$$

$$= \frac{B(r+a, n-r+b)}{B(r+1, n-r+1)B(a,b)},$$

so

$$g(\theta|\mathbf{x}) = \frac{\theta^{r+a-1}(1-\theta)^{n-r+b-1}}{\int_0^1 \theta^{r+a-1}(1-\theta)^{n-r+b-1}d\theta}$$

$$= \frac{\theta^{r+a-1}(1-\theta)^{n-r+b-1}}{B(r+a, n-r+b)}, \tag{5.3}$$

which is itself a beta distribution. This makes clear why a beta distribution was chosen for the prior. Doing so makes the posterior distribution have the same form, which is an algebraic and computational convenience.

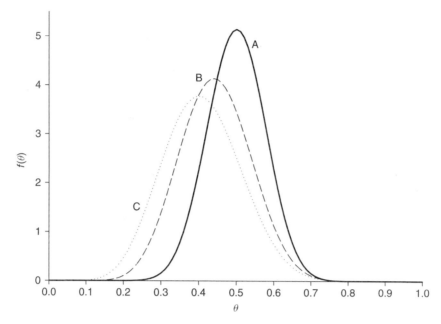

Figure 5.4   Posterior distributions for the binomial example using equation (5.3).

Posterior Distributions Yield Inferences

We can now see the corresponding posterior distributions that result from each of the three prior distributions in Figure 5.3. The posterior distributions calculated from equation (5.3) are shown in Figure 5.4. The experiment has sharpened the prior distributions and pulled them in the direction of the new data. The extent to which this happens depends substantially on the degree of certainty described by the prior. From curve C derived from the noninformative prior, numerical probability statements can be made about θ as illustrated in Table 5.1. The true probability of response lies between 0.245 and 0.583 with 90% certainty, for example. Constraints can be applied either to the tail areas of the credible regions (e.g., making them equal) or to the width of the interval (e.g., making it symmetric).

5.4.3   Bayesian Inference Is Different

Even this brief sketch should highlight some of the important differences between the Bayesian and frequentist approaches to designing clinical trials. A major component of Bayesian design is deciding on the prior distribution that represents knowledge (or belief) about the treatment effect before the start of the trial. This can be controversial when it requires subjective probability, that is, summarizing degree of belief in the form of a probability distribution. Also an implicit Bayesian assumption is that a probability distribution is the way to represent investigator knowledge. Despite these problems, for every frequentist procedure there is an equivalent Bayesian one. In other words, the Bayesian can view frequentist procedures as having implicitly chosen a prior distribution that, if used as described above, yields essentially the same inference.

Apart from questions about prior distributions, the Bayesian paradigm provides a consistent and understandable mechanism for solving inferential problems. This is a

TABLE 5.1   Credible Regions for Bayesian Posterior Distribution from curve C in Figure 5.4

Coverage	Bounds		Property
%	Lower	Upper	
50%	0.337	0.479	Equal tails
50%	0.338	0.480	Symmetric
90%	0.245	0.583	Equal tails
90%	0.240	0.578	Symmetric
95%	0.218	0.616	Equal tails
95%	0.211	0.608	Symmetric
99%	0.171	0.677	Equal tails
99%	0.155	0.663	Symmetric

Note: Two types of intervals are presented: symmetric about the mean of curve C (which may not have equal upper and lower tail areas) and equal tails (which may be asymmetric).

considerable strength. Furthermore Bayesian inferences are made in the form of probability statements about the unknown parameter or treatment effect. This is intuitively appealing and broadly useful for solving clinical problems. Bayesian approaches can be crafted for all the trial types discussed in the next section.

The design considerations discussed above are also important to the Bayesian method. For example, if preliminary data about a treatment effect arise from biased studies, one cannot use them uncritically to form a prior distribution for a subsequent comparative trial. Similarly the Bayesian seeks the same control over random error and bias in comparative trials as the frequentist does. However, randomization does not play the same role for Bayesian inference as it can for certain frequentist procedures (see Chapter 13).

5.5   LIKELIHOOD

There is a third school of statistical inference that offers many strengths. It is termed "likelihood," based on the central "likelihood principle," and main tool, the statistical likelihood function. As seen from the examples above, the likelihood is at the heart of all methods of statistical inference. Frequentists use it for tests of hypotheses, and Bayesians use the likelihood to modify their prior probability distribution, yielding a posterior distribution. Likelihood statisticians use it alone to represent differences in evidence regarding the unknown parameter. See Edwards (1972) and the more recent important work by Royall (1997) for a discussion of likelihood methods. Blume (2002) presents an overview and some applications to clinical trials. An example of a likelihood-based design for safety and efficacy trials is given in Chapter 11.

The likelihood principle states that two experiments or observations that generate identical likelihoods represent equivalent evidence. In particular, the sample space under which the data are collected is irrelevant with regard to the evidence generated. This principle has met with some resistance among statisticians, although it appears to be derivable from other fundamental, widely accepted principles.

In the binomial example from Section 5.2.4, 20 observations were made and the number of successes was the random variable outcome. Suppose, instead, that the study design was to take observations until 8 successes had been observed. The random variable would then be the number of trials required to get 8 successes. Suppose further that this happened after 20 observations. The likelihood for this circumstance is the same as before, apart from a scale change due to the binomial coefficient. The likelihood principle states that the evidence with regard to θ is the same in both cases, whereas frequentist inferences such as estimators (bias), confidence intervals, and p-values would be different.

The likelihood principle applies even if the meaning of θ is different in the two contexts, but the likelihood functions are the same (by virtue of the same probability model). For example, one experiment might be the success rate of a new drug treatment and another the frequency of heads from a possibly biased coin. If 8 events out of 20 trials are observed, the evidence supporting each θ is the same.

The likelihood function cannot be viewed as measuring absolute evidence, but does represent relative evidence regarding different values of θ. For such purposes, it is helpful to scale the likelihood function as $L(\theta)/L(\widehat{\theta})$, where $\widehat{\theta}$ is the MLE, so that the numerical values represent evidence relative to that supporting $\widehat{\theta}$. This dimensionless scale seems unfamiliar at first but is completely general and useful for summarizing relative evidence.

5.5.1   Binomial Case Study

For our binomial example, Figure 5.5 shows the likelihood function with the vertical axis scaled by the MLE. The vertical scale remains logarithmic. The horizontal axis has also been slightly truncated at each end. This figure is otherwise the same as Figure 5.1. Two reference points for $L(\theta)/L(\widehat{\theta})$ are indicated: likelihood

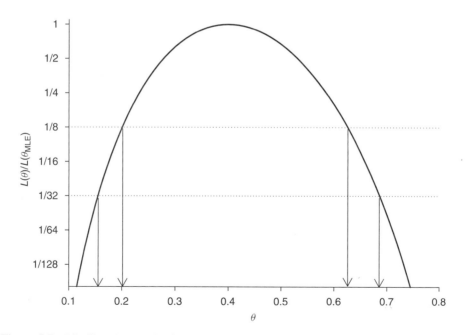

Figure 5.5   Likelihood ratio plot for the binomial example showing intervals of moderate and strong support.

ratios of 1/8 and 1/32, often suggested as representing moderate and strong evidence respectively. There is moderate evidence from the data that θ lies in the interval 0.2 to 0.63, and strong evidence that θ lies in the interval 0.16 to 0.68. Such statements define support intervals that are analogous to confidence intervals and credible intervals.

As appealing as likelihood inference is, there are some practical difficulties with it. One is the unfamiliar scale of quantification of likelihood ratios. We have little experience with evidence summarized on a ratio scale and reference values such as 8 (moderate) and 32 (strong) are difficult to interpret at first. Experience can correct this, but it does not appear to be strongly intuitive. Also likelihood methods are not as amenable to testing composite hypotheses as are point comparisons. For example, how do we quantify the relative evidence that θ exceeds 0.25? Some type of weighted average of $L(\theta)/L(\widehat{\theta})$ over the region in question might be appropriate. However, such a procedure begins to have a Bayesian-like feel.

5.5.2 Likelihood-Based Design

The likelihood function by itself is also a suitable device for addressing questions about sample size and precision. Consider the shape of $L(\theta)/L(\widehat{\theta})$ as the sample size increases, as shown in Figure 5.6. There is no sampling variation in Figure 5.6 because $\widehat{\theta}$ was held to the constant value 0.4. The 1/32 support intervals corresponding to Figure 5.6 are shown in Table 5.2. A sample size might be selected such that the width of the strongly supported values for θ is within a prespecified tolerance. A more sophisticated approach based on evidentiary criteria is discussed by Royall (1997). Some other straightforward approaches are used in Chapter 11.

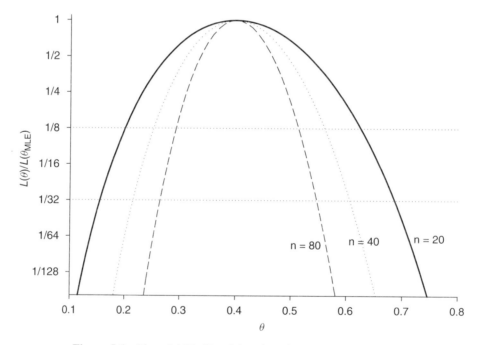

Figure 5.6 Binomial likelihood functions for various sample sizes.

TABLE 5.2   Binomial Likelihood-Based 1/32 Support Intervals for Various Sample Sizes

Sample Size	$\frac{1}{32}$ Support Interval		
	Lower	Upper	Width
20	0.155	0.686	0.531
40	0.215	0.607	0.391
60	0.246	0.569	0.324
80	0.264	0.547	0.282

5.6   AFTERTHOUGHTS

5.6.1   Statistical Procedures Are Not Standardized

All statistical views of clinical trials reflect features of true experimental designs. Statistical reasoning is essential in designing the architecture of the trial, choosing a sample size, deciding to stop an ongoing study early, and analyzing and interpreting trial results. Even with this common backdrop, there are many issues and approaches to clinical trials that are not standardized. Some lack of standardization is a consequence of making study designs and analyses responsive to the needs of a specific biological question. Nuances in the scientific question may drive substantial differences in design. An example is the use of placebos in comparative trials. It may not be ethical or scientifically appropriate to use placebo controls when standard treatments are known to provide benefit for a serious illness. In self-limited diseases, or those with no known treatments, a trial with a placebo group could be the most appropriate and informative.

Lack of standardization also reflects the fact that study designs and statistical methods of analysis are not fixed consequences of the research question or data collected. To understand this, it is important to distinguish between *standards* and *standardization*. Having standards implies a minimal acceptable quality or depth of sophistication in design and analysis of trials. This is usually considered a good idea. Standardization implies uniformity, which is almost never the best idea for independent clinical investigations.

Different researchers could perform different analyses, possibly yielding different inferences. When this occurs, it is often the result of selecting one statistical approach instead of another, although either may be perfectly appropriate for a given research question. Also the same data may be relevant to different clinical, biological, or analytic issues. However, the same situation arises because of fundamental philosophical differences of opinion about the correct quantitative approach to the biological question. In these and other ways, clinical trials are as subject to differences of opinion as other areas of scientific inquiry.

5.6.2   Practical Controversies Related to Statistics Exist

Philosophical differences aside, there are many practical issues about clinical trials that generate disagreement. For a condensed discussion, see Byar, Simon, Friedewald, et al. (1976). Lack of uniformity in a developing discipline can be good because it stimulates creativity and questioning of assumptions. But nonuniformity can also imply that incorrect or suboptimal procedures are being applied in some clinical trials. This could be a problem if such errors are not evident from the published reports or if other investigators are not knowledgeable enough to detect the mistakes. In a number of areas of trial design and conduct, there is no general agreement about the best standard to recommend. Some of these, listed below, are actively evolving, and others are likely to remain contentious.

Although on solid statistical ground *randomization* can be controversial when applied in some clinical circumstances. Its use appears to highlight dual obligations, since physicians do not select treatments on the basis of the personal care ideal (Chapter 3). Randomization may be inconvenient or unnecessary in some circumstances. However, the advantages of randomization are considerable, and it is not very controversial when appropriately employed. Details are discussed in Chapter 13.

Intention to treat is the idea that subjects in a randomized clinical trial should be analyzed according to the treatment group to which they were assigned, even if they did not receive the intended treatment or received only a portion of it. This perspective is usually supported by statisticians because it yields desirable properties for testing the null hypothesis of no-treatment difference. Many clinicians and other investigators often favor analyses based on treatment actually received, an approach that is advertised as more accurately estimating the true biological effect of the treatment. Their claim may not be true. This is an important topic and will be covered in depth in Chapter 15.

Even in randomized trials, imbalances between the treatment groups can occur by chance. Randomization theory permits us to ignore such imbalances when testing treatment differences, although pragmatic concerns often lead to the use of *adjustment* procedures through statistical modeling. These methods essentially view the trial as a database (although perhaps an exceptionally high-quality one) and yield estimates of treatment effect "controlled" for covariate imbalances. Trialists do not agree on the need to perform these types of analyses or what factors should be used for adjustment. This topic is discussed in Chapter 17.

An approach to eliminating the need for adjusted analyses is to ensure that the treatment groups are balanced with respect to influential prognostic variables. One method for inducing balance is to use *stratification* with blocked treatment assignments, which forces exact balance of the stratification variables frequently during the trial. Although theoretically unnecessary as suggested above, debates about whether or not to use stratification routinely are a minor issue in trial design. Additional details are discussed in Chapter 13.

The expense and complexities of clinical trials limit their wide applicability. However, there is a broad need for reliable and unbiased inferences like those that usually result from well-performed randomized comparisons. This tension has led to *large-scale trials* (LST), discussed in Chapter 6, which test the worth of interventions in large populations without the extensive data collection and other infrastructure needed for many trials. The idea of conducting LSTs is not especially controversial, but establishing the best methods for doing so and defining relaxed notions of treatment and eligibility in such studies can be a problem.

Despite high-profile instances of enhanced inferences arising from their use, *overviews* or *meta-analyses* remain a somewhat controversial tool for assessing evidence from multiple clinical trials. Sources of difficulty that can limit the utility of overviews include the fact that each study included may have a slightly different design, and the possibility that not all relevant studies or the underlying data are accessible to the persons performing the overview. This could be a problem if only the published literature is used as a source of information about trials performed. Meta-analyses are discussed in Chapter 21.

There are other areas that create controversy among trialists but are not primarily statistical in nature. These include aspects of reporting, quality control of trial data, and political intrusions into and regulation of trials. Although each of these topics is important, they can not be covered here.

5.7 SUMMARY

Statistics is not a perfectly unified field, particularly with regard to the best method for making inferences from data. There is general agreement among statisticians on the role of study design and the representation of data and probability models in the form of a likelihood function. Most commonly encountered statistical analysis procedures are frequentist, an approach that views unknown parameters as constants of nature and defines probability as long-term frequency of outcome. This method uses familiar constructs such as p-values, confidence intervals, and hypothesis tests for inference.

The Bayesian view defines probability as a subjective degree of belief, and views unknown parameters as random variables. Prior belief is combined with data from the experiment to yield (posterior) probability statements about the parameters. The representation of knowledge, ignorance, or belief about parameters in the form of a probability distribution before data are collected is the source of most of the controversy regarding Bayesian methods.

Likelihood methods are based entirely on the likelihood function, which combines the observed data with an assumed probability model. Relative evidence regarding parameter values can be assessed from the likelihood function. Likelihood ratios (relative evidence) and support intervals (analogous to confidence intervals) are the primary tools of inference using this approach.

It is relatively easy to construct examples that highlight the differences in statistical philosophy behind each of these approaches. It is more difficult to find practical examples, especially in clinical trials, where different approaches yield qualitatively different inferences about treatments. The trialist should be aware of differences because one approach or another may be better suited to a particular problem. However, all have their limitations, and we must be flexible enough to choose the best approach to a given situation.

5.8 QUESTIONS FOR DISCUSSION

1. Can belief about a parameter be converted into (pseudo-) data? Suppose that you are about to observe the behavior of a possibly biased coin. How might your prior belief be represented as data?

2. A frequentist confidence interval for the binomial example was given in Section 5.3.1. If that interval were taken as a Bayesian posterior credible interval, what would it say about the "prior" distribution that the frequentist implicitly used?

6

CLINICAL TRIALS AS EXPERIMENTAL DESIGNS

6.1 INTRODUCTION

Formal definitions for *experiment*, *clinical trial*, and *design* were offered in Section 2.2.2. We can view an experiment or a clinical trial as a series of carefully collected observations made under conditions controlled or arranged by the investigator (Stebbing, 1961). True experimental designs incorporate features to control systematic error, reduce random variation, increase precision, and sometime assess treatment interactions or the effects of covariates. By removing systematic error and breaking apart natural correlations, in designed experiments we can isolate the effects of interest and estimate them with a high degree of precision. Proper design and analysis allows the investigator to attribute observed differences reliably either to the factors under control or to random error.

The field of experimental design began with works by R.A. Fisher (1925; 1935). It originated around agricultural trials, and evolved through industrial and laboratory experiments, and animal studies (Cochran and Cox, 1957; Cox, 1958; Haaland, 1989; Heiberger, 1989; Hinkelmann and Kempthorne, 1994; Lorenzen and Anderson, 1993; Winer, 1962), followed by clinical trials. In nonclinical settings, investigators have the greatest degree of control over the structure and size of the study and the experimental units. The clinical trial is a relatively new science to draw on this body of knowledge (Fleiss, 1986). In fact the field of clinical trial design has not borrowed as much strength as it could from classic experimental design until recently. An interesting area of overlap is factorial experiments, discussed in Chapter 19.

The classically described principles of experimental design in field trials are *local control* to reduce experimental error, *replication* to estimate variability, and *randomization* for validity. The idea of local control pertained to field conditions literally but is also valid today, for example, when blocking and stratifying by institution in multicenter trials. Replication of experimental units may seem an obvious principle today, but historically it certainly was not widely acknowledged by physicians.

Clinical Trials: A Methodologic Perspective, 2E, by S. Piantadosi
Copyright © 2005 John Wiley & Sons, Inc.

Although many clinical trials are not randomized, using randomization to reduce bias and increase validity is widely accepted. Thus the classic principles remain operative today. They will be elaborated below, where much of the discussion pertains implicitly to randomized studies.

6.1.1   Trials Are Relatively Simple Experimental Designs

Although they have the essential characteristics of true experimental designs, clinical trials often lack some of the more intricate features of studies on inanimate objects, plants, or animals. First, human responses to medical treatments tend to be more variable than those from genetically identical plants or animals or from tightly controlled physical and chemical processes. The clinical investigator is not usually able to control as many sources of variability through design as can the laboratory or industrial experimenter. This leads clinical experiments to have a relatively large number of subjects assigned to each treatment to control random variation.

Second, experiments on nonhuman subjects require relatively few constraints compared with clinical trials. The study can often be made as large as scientifically necessary apart from economic considerations. Ethical issues, if they exist, are not usually internal to the experiment. In contrast, clinical trials are constrained by cost and respect for the rights of research subjects. Also some important biological questions cannot be answered because patients with the needed characteristics simply do not exist.

Third, the clinical investigator usually cannot test all subjects simultaneously, as might be done in an agricultural or industrial trial. This leads to long study durations while patients are accrued and followed. Some outcomes in clinical trials require exceptionally long periods of follow-up to manifest themselves (e.g., survival or disease incidence/progression in a prevention trial). This also prolongs study duration and creates opportunities for data to be lost. Lengthy clinical trials allow additional sources of variability to be introduced into the treatment assessment. Changes in the health care system, new methods of supportive or ancillary care, adjuvant therapies, and other factors not anticipated in the initial design can affect the treatment estimate or comparison.

Fourth, in therapeutic trials it is often impossible to test multiple treatments in the same patient because of overlapping side effects. Agricultural, industrial, and animal experiments can frequently study more than one treatment effect. Two possible exceptions to this general maxim are trials of diagnostic tests and disease prevention studies, where the interventions are usually very low risk for the study participants.

Fifth, the use of uncertainty to motivate clinical trials is more formal and fundamental than in nonhuman studies. There is always an epistemological importance to uncertainty in medicine and it can be an important tool in experimentation. The degree of uncertainty that motivates a clinical trial is likely to be higher than that underlying many experiments in agriculture, industry, or the laboratory. A high degree of certainty in the outcome before the study is conducted is a hallmark of "demonstration trials," which may be appropriate (and common) in fields other than clinical trials.

Despite being relatively simple in design, clinical trials are quite complex undertakings. Nothing said above is meant to dispute that fact. However, their complexity arises much more from ethics, biology, logistics, and execution than from the structure of the experimental design. The distinction is important because the consequences of deficiencies in a clinical trial depend largely on the source. Problems in trials tend to arise more readily from the complex logistics and execution.

6.1.2   Study Design Is Critical for Inference

Students of clinical trials tend to approach the subject with a focus on performing correct analyses. Although there are many pitfalls for the inexperienced or uncritical data analyst, good trial design and conduct is considerably more important than analysis. Skillful analyses cannot be relied upon to correct design faults. For example, the two principal types of bias, selection bias and observer bias, cannot be reliably fixed by analysis.

Furthermore good designs are usually simple to analyze correctly, often requiring only technically simple methods such as proper counting of events. When a trial is well designed and executed, analyses can be corrected or repeated until they reveal the evidence needed from the study. Investigators hope that all this will take place prior to publication, but the data can be analyzed again later if new methods are required.

Two serious shortcomings of poorly designed and implemented trials are not amenable to correction with analysis. These are systematic error, such as selection bias, and imprecision in the estimate of the treatment effect. Bias reduction results from design features, such as using objective endpoints, active ascertainment of outcomes, no post hoc exclusions based on outcome, and impartial treatment assignment (e.g., randomization in comparative studies). High precision is mainly a consequence of having employed an adequate sample size. If investigators fail in these or other important aspects of the design, there are no analysis methods that will allow them to recover reliably from the error.

Careful attention to design yields several benefits that cannot be guaranteed by any analysis. Proper design:

- allows investigators to satisfy ethical constraints,
- permits efficient use of scarce resources,
- isolates the treatment effect of interest from confounders,
- controls precision,
- reduces selection bias and observer bias,
- minimizes and quantifies random error or uncertainty,
- simplifies and validates the analysis,
- increases the external validity of the trial.

These advantages should provide sufficient reason for investigators to place an emphasis on study design.

6.2   GOALS OF EXPERIMENTAL DESIGN

6.2.1   Control of Random Error and Bias Is the Goal

The simplest statement of purpose for experimental design, and therefore clinical trials, is to control the effects of random error and bias (nonrandom error). The appropriate degree of such control is not necessarily to minimize error but to make such influences small with respect to the magnitude of the treatment effect of interest. For example, random error can be made arbitrarily small (minimized), but only at great expense in terms of sample size. The art of designing trials is to provide a degree of error reduction that is efficient and sufficient for the purpose of the clinical objectives.

The biases of most concern are selection bias and observer bias. I would add the bias due to uncontrolled covariates (prognostic factors) as a fourth major concern, although it is not completely independent of the idea of selection bias. These four effects are ubiquitous and strong and are discussed more in Chapter 7. The design and analytic maneuvers that we apply are directed at one or more of them. There are many different ways to produce each type of bias depending on the circumstances of a given trial, and sometimes more than a single way to control each. This yields the seeming complexity of design.

6.2.2   Conceptual Simplicity Is Also a Goal

The most critical and difficult prerequisite for a good study is to select an important feasible question to answer. Accomplishing this is a consequence primarily of biological knowledge:

> The best of science doesn't consist of mathematical models and experiments, as textbooks make it seem. Those come later. It springs fresh from a more primitive mode of thought, wherein the hunter's mind weaves ideas from old facts and fresh metaphors and the scrambled crazy images of things recently seen. To move forward is to concoct new patterns of thought, which in turn dictate the design of the models and experiments. (Wilson, 1992)

Apart from such insights, good experimental design and avoiding methodological errors will facilitate answering the right question.

The purpose of most trials is to estimate the magnitude of effects (or differences) in treatments that have been carefully crafted on biological grounds. Trials are most necessary and useful when the treatment effects are likely to be masked by person to person variation. To accomplish its goals, a trial should

- reduce errors resulting from chance,
- reduce or eliminate bias and other extraneous effects,
- yield relevant estimates of clinical effects,
- quantify precision,
- be simple in design and analysis,
- provide a high degree of credibility, reproducibility, and external validity,
- influence the design and conduct of subsequent studies and/or future clinical practice.

Conceptual simplicity in design and analysis is a very important feature of good trials. Some investigators worry that the infrastructure and the statistical machinery needed to design and analyze clinical trials are too complex to be reliable. It is true that clinical trials fall outside the range of experience of many medical practitioners and that they are by no means the only way of making valid clinical inferences. However, simple ideas often lead to simple solutions only after encountering complexities along the way. Fortunately the complexities in clinical trials are usually logistical rather than conceptual. This is analogous to the way in which surgical treatments are often conceptually straightforward but technically complex.

When trials cannot be made simple, they should be made as parsimonious as possible. This means that they should be designed to be as simple as feasible, given the complexities of the research question, treatment algorithm, assessment method, and analysis. Even in complex contexts, the effect of interest can usually be isolated by the study design.

Another principal goal of a clinical trial is to influence in a beneficial way the design of subsequent experiments. I have tried to illustrate this idea most clearly for translational trials (Chapter 9). Dose-finding establishes the optimal dose for much of the developmental process, for example. Safety and activity (SA) trials can abort development, at least within certain disease classes. Even comparative trials may alter the design of later studies, or suggest new hypotheses and experiments.

It is interesting that "good" clinical trials do not always influence clinical practice, at least not immediately. This may not be a shortcoming of the study but a consequence of practitioners' habit or dogma, and/or seeming inconsistencies with prior knowledge. In the long run, well-performed trials may be verified (apart from outright errors of chance) and improve clinical practice.

6.2.3 Encapsulation of Subjectivity

Most investigators who invest intellectually in the experimental method are reluctant to accept that any part of it has a significant subjective component. In reality, all clinical trials require subjective judgments on the part of the investigator at any of several points in design, conduct, or analysis. This is especially true of early developmental trials (e.g., see the discussion in Section 10.2.5) or trials without internal controls. It is a mistake not to recognize the subjectivity that is present, or to design and interpret studies in formulaic ways. We could more appropriately view experimental designs as devices that encapsulate both objective plans along with unavoidable subjectivity.

A common way that subjectivity enters design is when choosing the type I or type II error rate. Typical choices of 5% and 10% for these rates are often not given much thought or are assumed to be objective, when in fact they are subjective. It is more appropriate to set these error rates to reflect the consequences of making the respective mistake, but this also involves subjective assessments.

Subjective judgments are required in many other places in a clinical trial. Some are listed in Table 6.1. Rigorous study design minimizes, but does not eliminate, the need for subjectivity. For example, assessing outcomes could be made more objective by masking and placebo control, which are not always possible. Also, no assumptions regarding statistical philosophy are implied by Table 6.1. In other words, both Bayesian and frequentist methods require this type of subjective input.

6.2.4 Leech Case Study

A stimulating example of design goals can be found in a study of the appetite of leeches for human subjects (Baerheim and Sandvik, 1994). Medicinal leeches have been used in recent times in microsurgery to reduce swelling. Different treatments were applied to the leeches to observe the effect on the time to initiation of feeding. This trial is

TABLE 6.1   Decisions in Clinical Trials Where Judgments Are Necessarily Subjective

Selecting the treatments to study
Specifying error rates
Specifying eligibility criteria
Evaluating and attributing side effects of treatment
Assessing outcomes
Choosing between analytic methods
Data analysis methods and pathway
Interpreting and describing results

made more interesting by the fact that one of the treatments was Guinness stout—the brewery in question having played a role in the early development of statistical methods through the works of Student (William S. Gosset). The other treatments applied to the leeches included bock, water, garlic, soured cream, and no intervention.

The study illustrates many of the goals of experimental design discussed above. There were maneuvers to control both random and systematic error, including random ordering of treatments. The trial was designed, executed, and analyzed in a simple fashion, and yielded quantitative effect measures and estimates of precision. It also allowed a transparent view of subjectivity both within the study and regarding the history behind the question being asked. There were imperfections in the trial: small size, no accounting for treatment crossovers in the design or analysis, and missing outcomes. But overall it is a worthwhile and entertaining illustration of design goals.

6.3   TRIAL TERMINOLOGY

It is not possible to define or classify clinical trials universally. Medical disciplines tend to view their own research issues as being unique, which often creates terminology within each specialty. But to facilitate explaining and discussing important design concepts, it is helpful to have terminology that is independent of the application. In this section I review basic types of trials and attempt to provide a broad descriptive terminology for them. After descriptions and definitions, the discussion will pertain largely to trials of pharmacological therapy.

6.3.1   Drug Development Traditionally Recognizes Four Trial Designs

In therapeutic drug (especially cytotoxic drug) development, clinical trials are often classified simply as phases I, II, III, and IV. This terminology is inadequate for universal application, but I will review it here because it is so widely used. *Phase I studies are pharmacologically oriented* and usually attempt to find the best dose of drug to employ. *Phase II trials look for evidence of activity, efficacy*, and *safety* at a fixed dose. Phases I and II are not hypothesis driven, meaning that formal comparisons to other treatments are not usually internal to the experimental design.

In phase III new treatments are compared with alternatives, no therapy, or placebo. The comparison group is internal to the design. Investigators would not undertake the greater expense and effort of phase III comparative testing unless there was preliminary evidence from phase I and II that a new treatment was safe and active or effective. *Phase IV is postmarketing surveillance* and may occur after regulatory approval of a new treatment or drug to look for uncommon side effects. This type of study design is also used for purposes other than safety and efficacy, such as for marketing.

Although this terminology was developed in, and is applied widely to, cancer therapeutics, some of the definitions are different in cancer prevention. In cancer prevention trials, phase II is divided into IIa and IIb. Phase IIa trials are small-scale feasibility studies using intermediate endpoints, such as cancer precursor lesions or biomarkers. (Intermediate endpoints are defined and discussed in Chapter 8.) Phase IIb trials are randomized comparative studies using intermediate endpoints. Phase III cancer prevention trials are comparative designs (like IIb) using definitive clinical endpoints such as cancer incidence (Kelloff, Johnson, Crowell, et al., 1995). Some cancer prevention investigators have used the term "phase IV" to mean a defined population study (Greenwald, 1985; Greenwald and Cullen, 1985). The same authors define phase V to be demonstration and implementation trials.

6.3.2 Descriptive Terminology Is Broader and Recognizes More Types of Trials

Differences in terminology, even within the field of cancer, illustrate the problem with nondescriptive labels. Most nonpharmacologic therapies are developed in steps that do not fit drug development terminology. New targeted agents and biologicals in oncology do not always fit the cytotoxic paradigm. The "phase I, II, III" labels have become too strong a metaphor, and frequently interfere with our need to think creatively about study purposes and designs. It is common to see protocols with "phase I/II" titles, and I have even seen one or two titled "phase I/II/III," indicating how investigators labor under the restrictive labels.

A more general and useful description of studies should take into account the purposes of the design, independently of the treatment being studied. Descriptive terminology and its relationship to old fashioned drug development labels is provided in Table 6.2. Although versatile and somewhat obvious, the descriptions are unconventional, so I will use them in parallel with traditional labels when appropriate. In this book the terms phase I, II, III, and IV, if used at all, refer narrowly to drug trials.

TABLE 6.2   Descriptive Terminology for Clinical Trials

Developmental Stage	Terminology	
	Old	Descriptive
Early	None	Translational
↓	Phase I	Treatment mechanism, TM
	↓	Dose-finding, DF
		Dose-ranging
Middle	Phase II	Safety and activity, SA
↓	Phase IIa	↓
Late	Phase IIb	Comparative, CTE
↓	Phase III	↓
	Phase IV	Expanded safety, ES
	Large simple	Large scale

The descriptive terminology also accounts for translational trials, an important class discussed in Chapter 9. It distinguishes several types of early developmental designs, particularly those aimed at establishing a safe dose of a new drug or agent to study. Comparative treatment efficacy (CTE) is an important class of trial designs that embodies many of the rigorous design fundamentals discussed throughout this book. More details about these design types are given in Section 6.5.

6.4 DESIGN CONCEPTS

Experimental design is a method to advance knowledge efficiently and reliably. There are ways to increase knowledge that do not rely as much on designed observation. These include taking advantage of chance occurrences and using incremental improvements based on experience. Using chance or "natural experiments" is a powerful but inefficient method for advancing knowledge. Incremental improvements are reliable but slow, and may reach a local rather than a global optimum.

Both of these ways of acquiring new knowledge can incorporate theory, but they subordinate it to observation. In contrast, the results of experiments carry relatively more information because they put theory on an equal footing with data. Consider this from Osler (1910):

> Man can do a great deal by observation and thinking, but with them alone he cannot unravel the mysteries of Nature. Had it been possible the Greeks would have done it; and could Plato and Aristotle have grasped the value of experiment in the progress of human knowledge, the course of European history might have been very different.

6.4.1 The Foundations of Design Are Observation and Theory

Medical investigations, like all science, are based on careful observation and recording. However, there is more to science than observation and compilation. Theory is required to help structure, interpret, and guide observation. Without theory, we could not separate passive experience from generalizable inference (active experience). Medical investigations can be classified and distinguished from one another on the basis of the quantity and quality of both observation and theory. In particular, true experiments test theory by controlling treatment assignments, endpoint ascertainment, and analysis, whereas nonexperimental designs lack one or more of these characteristics.

Several types of clinical observations have traditionally been used, and might in some circumstances provide convincing evidence that can advance treatment. On the other hand, it is an often overlooked fact that uncontrolled observations can inhibit therapeutic advancement because they are error prone. If uncontrolled observations are wrong, and given too much credence, the effect will be to retard or divert therapeutic development. Historically practitioners may have been frequently misled in this way.

Case reports usually constitute the weakest clinical evidence because they demonstrate only that an event of some interest is possible. A single such observation, if sufficiently far outside of common experience, can yield important evidence. But most often case reports can only generate hypotheses rather than test them because they are roughly consistent with everyday experience. The investigator does not control treatment assignment, endpoint ascertainment, or confounders in case reports, and no formal statistical analysis is possible.

As an example of the information in case reports, consider that even a life-threatening disease like cancer is not uniformly fatal. There are numerous individual case reports and case series of spontaneous remission of cancer in the medical literature (Cole, 1976; Challis and Stam, 1990; O'Regan and Hirshberg, 1993). These are sufficient to prove that spontaneous remission happens, and to generate interesting clinical hypotheses, but not convincing evidence of the efficacy of any particular treatment associated with the cases.

Case series are more helpful than case reports because they carry the weight of some experience in proportion to their size and quality. However, they also only generate a hypothesis rather than test it. Case series are a type of accurate compilation and are arguably not science. They are typical of much historical medical writing and afford only modest opportunity for therapeutic advancement. Although some minimal statistical analysis may be possible for extensive case series, the investigator lacks control over treatment assignment, selection bias, confounders, or endpoint ascertainment as in true experiments.

Consider the inferential importance of the small series of patients who received Jarvik-7 artificial hearts (Fox and Swazey, 1992). Beginning with Barney Clark at the University of Utah Medical Center, five patients were given these artificial hearts by two different surgeons at three different institutions between December 2, 1982 and April 14, 1985. Patients lived between 10 and 620 days. Despite the heterogeneity, limited number of patients, and duration of follow-up, convincing clinical information was obtained from the case series suggesting that the device was not ready for wider use. Even so, that experience offered only modest information to advance the development of the artificial heart.

Database analyses are similar to case series but usually larger in scope and often include attempts to evaluate treatment differences. However, database analyses often have shortcomings that prevent convincing comparison of treatments. Problems include selection bias, the possibility that endpoints have been ascertained passively or in a biased way, and unidentifiable confounders secondary to unknown rationale for treatment selection (Byar, 1991; Green and Byar, 1984). Consequently, rather than being used for treatment comparisons, databases are best used for (1) storing and retrieving information about an individual patient to help clinical management or case reporting, (2) to find or describe a group of similar patients (a cohort), and (3) to study patterns in the data with descriptive models or other statistical methods (e.g., prognostic factor studies).

Observational (e.g., epidemiologic) studies yield evidence about treatment (exposure) differences, and usually control endpoint ascertainment and analysis. These studies lack key design strengths of experiments, such as control over treatment assignment and control over unknown confounders. In actual epidemiological settings, the exposure information may be determined retrospectively (i.e., remembered by the subjects) and may therefore be affected by recall bias. This might not be a problem for treatments that are documented in a medical record. However, such studies are subject to other biases because of factors confounded with treatment selection (Green and Byar, 1984).

Comparative clinical trials control all the key components of true experimental designs. When properly designed and conducted, they yield a strong expectation that bias has been eliminated or minimized. Also the precision of a trial can be increased by

TABLE 6.3   Strength of Evidence Provided by Types of Medical Studies

Study Type	Evidence Obtained
Case report	Demonstration only that some event of clinical interest is possible.
Case series	Demonstration of certain possibly related clinical events but subject to large selection biases.
Database analysis	Treatment not determined by experimental design but by factors such as physician or patient preference. The data are unlikely to have been collected specifically to evaluate efficacy.
Observational study	Investigator takes advantage of "natural" exposures or treatment selection and chooses a comparison group by design.
Controlled clinical trial	Treatment assigned by design. Endpoint ascertainment is actively performed and analyses are planned in advance.
Replicated clinical trials	Independent verification of treatment efficacy estimates.

Source: Byar (1978).

studying more subjects, a conceptually simple maneuver. The credibility of comparative trials is enhanced with *independent replication* or other verification.

Byar (1978) called this progression from case reports to confirmed randomized clinical trials a "hierarchy of strength of evidence" (Table 6.3). Replication of rigorous clinical trial findings is rarely if ever done, but the formal assessment of consistency in the results of multiple trials is often attempted through meta-analyses. Olkin (1995b) has suggested that meta-analyses with original data be taken as the highest level of credibility. Meta-analyses are discussed in Chapter 21.

6.4.2   A Lesson from the Women's Health Initiative

The idea that comparative clinical trials alone convey the strongest quality of evidence about treatment effects is occasionally challenged. Examples are reports by Benson and Hartz (2000) and Concato et al. (2000), in which the authors show similar risk estimates for some randomized trials and nonexperimental designs addressing the same questions. Pocock and Elbourne (2000) pointed out some methodological concerns that provide cautions about the conclusions.

An important practical lesson can be seen in findings from the Women's Health Initiative (WHI) randomized placebo-controlled trial of 16,608 healthy postmenopausal women assessing the effects of estrogen and progestin (WHI Investigators Writing Group, 2002). Many researchers assumed that the WHI trial would find that hormone replacement therapy reduces coronary heart disease and provides an overall benefit to recipients (e.g., Stampfer and Colditz 1991; Grady et al., 1992).

After 5.2 years of average follow-up, increases in breast cancer, coronary heart disease (CHD), stroke, and pulmonary embolism due to hormone therapy more than offset reductions in colorectal cancer and hip fracture. The study ended early and surprisingly because of these findings. The difference between what was expected and what was observed is qualitative and important because up to 38% of postmenopausal women have been treated with hormone replacement therapy.

The RCT will most likely convince investigators, practitioners, and patients, and it will be seen as more reliable than contradictory opinion or earlier findings from uncontrolled studies. Only the RCT can reliably control unknown or unobserved prognostic factors. Prior to the WHI, studies were probably biased—women who received hormone replacement therapy may have been healthier than those who did not. A research design that prevents such a bias is more reliable then one that does not.

Knowing that experimental and nonexperimental designs yield similar estimates in a portfolio of studies extracted from the medical literature in no way reduces the potential bias in nonexperimental results. The hierarchy of evidence is based on the quality of the research *design*, which gives us reasonable expectations regarding reliability. The central issue (regarding randomization) is control of unobserved confounders. Even if we accept the Concato et al. (2000) analyses at face value, a few such findings cannot reassure us that uncontrolled prognostic factors are usually inconsequential.

6.4.3 Experiments Use Three Components of Design

In the statistical field of experimental design, there are three components that support analysis and inference: treatment design, error control design, and sampling and observation design (Hinkelmann and Kempthorne, 1994). These components are present in clinical trials, although the exact architecture of the study depends on its specific purpose. Table 6.4 characterizes the complexity of design components in some clinical trials relative to each other.

The simplest treatment design is usually found in the single cohort study, meaning all patients receive the same treatment. However, the treatment algorithm itself could be clinically quite complex. Even comparative trials usually employ relatively simple treatment designs: A versus B or A + B + C versus A + B. In contrast, factorial designs (discussed in Chapter 19) have more complex treatment designs. Dose-finding studies also have fairly sophisticated treatment designs because of the escalation/de-escalation rules. I have characterized large-scale trials as having intermediate complexity in treatment design because they might very loosely define treatment, leading to heterogeneity in the therapies actually administered.

The error control design for most clinical trials is relatively simple. It is not unusual to employ a completely randomized design, for example, in a comparative trial. Stratifying and blocking adds some complexity and better control over variation, but trials seldom require or tolerate the more sophisticated error control of elaborate experimental designs.

Sampling and observation design is the place where clinical trials reveal more complexity than many other biological experiments. The sampling design begins with the

TABLE 6.4   Relative Sophistication of Design Components in Various Clinical Trials

Trial Type	Treatment Design	Error Design	Sampling Design
Translational	Simple	Simple	Simple
Dose-finding	Complex	Simple	Simple
SA	Simple	Simple	Moderate
CTE	Moderate	Complex	Moderate
Factorial	Very complex	Complex	Complex
Large scale	Moderate	Moderate	Simple
Crossover	Complex	Moderate	Moderate

crafting of the study cohort. Many trials require fairly lengthy periods of observation. The observations themselves might require complex procedures or assessments. Often multiple longitudinal observations are taken, and there may be several key outcome measures needed to answer the scientific questions. This complexity adds time, effort, logistical difficulties, and cost to the study. There is no general relationship between the size of an investigation and the complexity of the sampling and observation design. Relatively small dose-finding trials may have complex pharmacokinetic studies attached.

Understanding the three components of design in a true experiment allows insight into the strengths and weaknesses of many medical studies. The essential locus of control of a clinical trial is the assignment of subjects to the treatment groups. I say this because the error control design and the sampling and observation design implicitly follow this. Admittedly a near fatal flaw of many clinical trials is an inadequate error control design caused by small sample size. However, nonexperimental studies may have no treatment design. Some studies such as case series may have deficient sampling and observation design.

Example 1   The consequences of a lack of true experimental design can be seen in the history of Laetrile from the 1970s and 1980s. Laetrile gained widespread acceptance by the public as an effective cancer treatment despite a lack of controlled evaluations and considerable skepticism by the medical establishment. Although as many as 70,000 cancer patients received the drug, the environments in which it was typically used discouraged one or more components of design and impartial assessments. In a review of the drug's possible efficacy, investigators from the NCI were able to evaluate only 93 cases with sufficient data, of which 6 may have had objective responses (Ellison et al., 1978). Moertel et al. (1983) studied 178 patients who received the drug in a trial without an internal control group and concluded that there was no evidence of efficacy.

Replication

The most basic feature of experimental design is independent replication. It is such a natural idea that it hardly seems worthy of explication. However, there are two components of replication that need some thought: independence and quantification. Usually individual study subjects provide independent replication. However, when individuals are not the unit of observation as explained below, replication requires more planning. Replication is the only way to reduce random error, leading immediately to considering the quantitative relationship between the number of replicates (e.g., sample size) and the magnitude of error.

Selection effects can directly undo the benefit of replication. Case selection, such as dropping some patients from a case series or retrospectively removing subjects from a single cohort trial, can create a bias that is uncorrectable by replication.

Finally replication can exist at more than one level. Sometimes blocks or other experimental units are replicated within an experiment. The study center is an example. Replication of the trial itself is a powerful method for increasing reliability and precision. Meta-analyses (Chapter 21) attempt to draw on this potential source of replication. The question or concept under investigation can also be subject to replication. For example, is it a good idea to control high blood pressure? Verification from studies of blood pressure control by different classes of drugs provides irrefutable support for the answer.

Experimental and Observational Units

Treatment is applied to the "experimental units," and measurements are made on the "observational units." In most clinical trials both the experimental unit and the observational unit are almost always the individual patient.

There are circumstances in which groups of individuals are the units of randomization. Groups might be defined by families, schools, or communities, for example. This design might be chosen to reduce costs, for ethical reasons, to reduce the complexity of a trial, or to control the contamination of treatment effects from members of the same group. Also political and social constraints might make individual randomization impractical, and some interventions naturally affect or target all members of a group (Cornfield, 1978). Cluster designs have a lower efficiency relative to the same number of subjects randomized independently. Some details are discussed in Chapter 11.

Community intervention trials usually apply treatment to more than one person as part of the experimental unit and consequently often use such designs. Similarly, they also often take measurements from aggregates of individuals. These designs are applied in studies of disease prevention and therapies in developing countries (Chowdhury et al., 1991; Kassaye, Larson, and Carlson, 1994; Mahfouz et al., 1995; Meyer et al., 1991; Sommer et al., 1986; Whitworth et al., 1992). The design of such studies is discussed by Gail et al. (1996), Donner and Klar (2000), and Murray (1998).

Example 2 Suppose that we study the transmission of a communicable disease where family members of each case are at highest risk. For practical reasons the different interventions tested might have to be given to families as the experimental unit. The observational unit might still be the individual patient. Treatments could be compared based on the overall proportion of new cases in the treatment groups. Similarly suppose that we were to test strategies to reduce cigarette smoking. One intervention might be tried as a campaign in several cities and another in a different group of cities. Here the experimental unit is the city and the observational unit might also be the city with smoking quit rates as the outcome. A prevention clinical trial, where large groups of individuals are the experimental units, is a smoking cessation trial (COMMIT, 1995a, 1995b).

In some clinical trials the experimental and observational unit is neither the individual nor a group of individuals. An example of such a study might be an ophthalmological trial. One or both eyes could be entered in the trial. This presumes that the treatment can be isolated to one eye, as might be the case for surgery.

Treatments and Factors

In the classical experimental design literature, treatment combinations are often called "factors." An industrial or agricultural experiment may involve questions about several control variables, each at more than one level. For example, temperature and pressure may be factors determining the yield from a chemical reaction. Combinations could be tested to determine how to maximize the overall yield of product. The need for testing multiple factors and levels is common because many agricultural or manufacturing and other industrial processes typically involve complex processes.

Clinical trials seldom employ more than two or three treatments (factors). When they do, the treatments are often at a single dose (level) each. Even when trials use three or more arms, the treatment groups usually consist of either different levels of the same treatment (e.g., dose levels) or single levels of different treatments (e.g., different drugs at fixed dose levels). Thus the simplest sorts of one-factor analyses can often be applied routinely. Were it not for special endpoints in clinical trials, basic methods of analysis, such as analyses of variance (ANOVA), could be used as widely as they are in other experiments.

Some clinical trials investigate treatment interactions, requiring factorial designs. This is an important class of designs, especially for disease prevention studies. Factorial trials have important subtleties and complexities that also bear on single factor designs. Factorial clinical trials are discussed in depth in Chapter 19.

Nesting

Nesting occurs when a factor or effect lies entirely within another factor or effect in a hierarchical arrangement. For example, in the usual two-group (or parallel or independent groups) clinical trial, patients are nested within treatments. In contrast, in cross-over trials, each patient eventually receives both treatments and so patients are not nested within treatment. Such non-nested designs are discussed in Chapters 19 and 20. Some non-nested designs allow greater efficiency in estimating treatment effects. However, they usually require stronger assumptions to support the applicability of the design.

Nesting is also relevant to blocking and stratification. RCTs are often blocked and stratified by study center, for example, as part of the error control design (see blocking discussed below). If centers are nested within larger units (e.g., states or countries), then such a study is also blocked and stratified by that larger unit, albeit with a larger block size. To maximize efficiency, the error control design must be taken into account when analyzing the results of an experiment. Classically the ANOVA model is written with appropriate effects for blocks, for example. For a clinical trial a stratified analysis would be used. I have seen exactly this situation misunderstood by trialists who should have known better.

Randomization

Randomization is used to remove systematic error and to justify type I error probabilities. Applied first to agricultural experiments by R.A. Fisher, it has also become an important tool in clinical trials after the pioneering work by Hill and Doll in Great Britain. Randomization has been used so successfully in comparative studies that its use seems to require a formal hypothesis test. However, randomization does not always motivate an explicit hypothesis test, and it may be used merely to remove selection bias. Formal treatment comparisons in clinical trials are sometimes based on quasi-random allocation methods such as "minimization," where patient assignments are made to induce prognostic factor balance between the groups. These and other ideas related to treatment allocation are discussed in Chapter 13.

Blocking

Blocking is a technique where subjects are grouped within more homogeneous larger units to control variation. It originated in agricultural field trials where treatment comparisons within smaller geographic areas could help overcome local effects such as soil quality or sunlight. The extraneous source of variation would affect each treatment equally and cancel out.

In comparative clinical trials, the classical idea of blocking exists in a slightly modified form. The term "blocking" usually implies a forced balance of treatment assignments within a larger experimental unit. For example, all the treatment assignments might be blocked in a fixed sample size trial so that the number in each treatment group is exactly equal at the end of the trial. The order is random but the totals are fixed.

More typically blocking is performed within strata, thereby removing the stratum to stratum effect. Somewhat homogeneous strata could be defined based on risk levels or other important covariates. A prototypical example in multicenter clinical trials is blocking by clinic or investigator, so that each study center yields an estimate of the treatment difference. Variation from clinic to clinic is considered uninteresting in this scenario. Blocking in clinical trials is usually based on one of a few criteria: (1) proximity such as clinic, (2) important covariates that convey risk, (3) time period, or (4) resources such as drug lot. Blocking and stratification are discussed in more detail in Chapters 7 and 13.

Example 3   Suppose that we are comparing two treatments, A and B, over a three-day period and we can switch easily between them within a day. We could sequence the process in the following way: day 1, $ABAB$; day 2, $BAAB$; day 3, $BABA$. Each day is a block and within each day, we can obtain a valid comparison of A and B and eliminate the day-to-day variability by averaging the $A - B$ difference over three days. Formally, a block is a set of experimental units (the number of units equals a multiple of the number of treatments) chosen so that if the treatments are equivalent, the responses will be nearly equivalent. The order of the treatments is randomly assigned within each block. This arrangement eliminates the effect of variation between blocks.

The idea of blocking can be extended to Latin square designs. The utility of these designs is best understood by considering the need for two-way elimination of heterogeneity or two factors that must be controlled when studying several treatments. If we have k treatments, a $k \times k$ Latin square has each treatment represented exactly once in each row and column. In agricultural experiments, the rows and columns might be plots of land, but need not have any particular shape. In industrial experiments, the plots might be different runs of a process with different control variables, where the rows and columns represent different locations in time (e.g., week and day within week).

Suppose that we have four antibiotics to test in different hospitals or clinics with various types of wards, and that the effects of both clinic and ward represent extraneous sources of variation. The treatments could be laid out in one of the arrangements shown in Figure 6.1, with a treatment occurring exactly once in each row and each column. There are $4! = 24$ possible arrangements that satisfy this constraint. More complete control over the extraneous effects could be taken using replicates of this layout that permuted the treatment assignments. Such designs are infrequently used in clinical trials.

Stratification

The idea of stratification is used in two ways in clinical trials: as a design technique to improve the comparability of treatment groups, and as an analytic technique to increase precision or account for risk factors. These two uses of stratification have in common the notion of controlling extraneous variation. In either circumstance stratification implies a second maneuver to render it useful. When using stratified randomization,

		Ward 1	Ward 2	Ward 3	Ward 4
Clinic	I	A	B	C	D
	II	B	C	D	A
	III	C	D	A	B
	IV	D	A	B	C

		Ward 1	Ward 2	Ward 3	Ward 4
Clinic	I	A	C	B	D
	II	B	D	C	A
	III	C	A	D	B
	IV	D	B	A	C

		Ward 1	Ward 2	Ward 3	Ward 4
Clinic	I	D	B	C	A
	II	B	C	A	D
	III	C	A	D	B
	IV	A	D	B	C

		Ward 1	Ward 2	Ward 3	Ward 4
Clinic	I	A	D	C	B
	II	D	C	B	A
	III	C	B	A	D
	IV	B	A	D	C

Figure 6.1 Latin square layouts with four treatments or factors. For factors A through D there are $4! = 24$ such layouts.

blocking is also required to accomplish the balance. Treatments will then be balanced within the stratum, which is the same as balancing the stratum variable within treatment. This is discussed in Chapter 13.

When analyzing treatment effects, stratification also requires pooling of estimates. Strata are usually defined by absolute risk. Within each stratum the relative treatment effect is estimated and pooled across strata to yield an overall treatment effect. (Assume there is in fact a common treatment effect, i.e., no treatment by stratum interactions.) This yields a treatment effect estimate uninfluenced by the extraneous variation of the stratum variable.

Stratified randomization requires the use of a stratified analysis, a point occasionally ignored by data analysts. The best way to understand this is by observing that stratified randomization explicitly acknowledges the stratum factor as an unwanted source of variation. This source of variation can be *removed* by balancing the treatment assignments within stratum, provided the analysis is conducted as indicated above. Otherwise, the between-stratum variation will be combined with the within-stratum variation, decreasing precision and defeating the purpose of stratifying.

Finally with regard to stratification I note that the term is occasionally used to imply subsetting of the data and estimating separate effects in each stratum that are not pooled, presumably because there is not a common effect. Strictly speaking, this situation is not "stratification." Such a case implies interaction between treatment and stratum (treatment by covariate interaction) and must be handled using other analytic methods.

Treatment Masking

Treatment masking, or blinding as it has been called historically, is an effective way to increase the objectivity of the person(s) observing experimental outcomes. When the treatments are masked, the bias or expectations of the observer are not likely to influence the measurements taken. For example, animal testing of a cytotoxic drug might consist of longitudinal measurements of tumor diameter. If the observer knows which animals have received drug and which were given placebo, the measurements could partially reflect his or her bias about the treatment. In contrast, if the observer is masked, the measurements are less likely to be affected by prejudice. Masking is particularly important in clinical trials that require self-reporting or self-assessments by the patients. Masking sounds like a modern idea, but in fact it has been used in medical and related evaluations for two hundred years (Kaptchuk, 1998).

Investigators often underestimate the value of treatment and assessment masking. There is a tendency to believe that biases are small in relation to the magnitude of treatment effects (when in fact the converse is usually true), or that practitioners can compensate for their prejudice and subjectivity. A classic counterexample is provided by Noseworthy et al. (1994) in their randomized trial of cyclophosphamide and plasma exchange for multiple sclerosis. In this trial all clinical outcome assessments were made by both masked and unmasked neurologists. The results demonstrated a significant treatment effect, as measured by unmasked neurologists, but not by masked ones. The investigators concluded that masking prevented a type I error, with clear implications for other trials.

6.5 SURVEY OF DEVELOPMENTAL TRIAL DESIGNS

6.5.1 Early Development

Translation between Laboratory and Clinic
A vitally important, widely used, but poorly described type of clinical trial is termed *translational*. This term is intended to capture the connection between the laboratory and the clinic, where developmental reasoning is circular, meaning moves back and forth between the lab and clinic depending on the results of early clinical trials. Many ideas tested in translational trials will be discarded, but a few will spawn the widely understood linear development sequence described below. Translational trials are explicitly defined and discussed in Chapter 9.

Mechanistic Trials
Nearly all early developmental studies have a similar goal. The common purpose is testing *treatment mechanism*. To the biomedical engineer, the treatment mechanism has a conventional interpretation in terms of device function. To the clinical pharmacologist, the treatment mechanism means bioavailability of the drug; to the surgeon, it means the operative procedure or technique; to the gene therapist, it means the function of the cells or engineered gene; and so on. Even diagnostic and screening trials have *treatment mechanisms* in this expanded sense of the word. I refer to these early developmental studies generally as treatment mechanism (TM) trials. Mechanistic success is not equivalent with clinical outcome, which must be assessed in later and larger clinical trials.

Early drug studies are often concerned with more than the mechanism of drug availability. They also frequently investigate the relationship between dose and safety using *dose-finding* strategies. For example, for many new drugs or preventive agents, we might be interested in the minimum effective dose. In cytotoxic drugs developed for oncology, the focus is usually the highest tolerated dose because of the belief that it will provide the best therapeutic effect. All types of dose-finding studies can be called DF trials. Thus phase I (drug) studies are TM and DF trials, but the converse is not true.

Another characteristic of TM and DF studies is their close connection to biological models. For treatments based on drug action, the biological models are usually pharmacokinetic or dose–response models. Other treatments may be based on more complex models of normal function or disease. In any case, the outcome of TM and DF studies is usually best described in terms of such a model. In later developmental trials the model of the treatment and its interaction with the disease process or clinical outcome is less important. Even if a plausible model has not been formulated, later developmental studies can provide convincing empirical evidence of efficacy.

Can Randomization Be Used in Early Developmental Trials?

Early developmental trials are not explicitly comparative. Selection bias is not a major concern when using primarily treatment mechanism and pharmacologic endpoints. Therefore randomization seems to have no role in such studies.

However, consider the development of new biological agents that are expensive to manufacture and administer and may have large beneficial treatment effects. An example of such a treatment might be genetically engineered tumor vaccines that stimulate immunologic reactions to tumor cells and can immunize laboratory animals against their own tumors. DF study objectives for these agents are to determine the largest number of cells (dose) that can be administered and to test for side effects of the immunization.

Side effects can arise from two sources: the new genes in the transfected tumor cells or the cells themselves. One way to establish the safety of the transfected cells is to compare their toxicity with that from untransfected cells. This can be done during DF trials by using a randomized design. Such considerations led one group of investigators to perform a randomized dose escalation trial in the gene therapy of renal cell carcinoma (Simons, Jaffee, Weber, et al., 1997). An advantage of this design was that the therapy had the potential for large beneficial effects that might have been demonstrated convincingly from the trial.

6.5.2 Middle Development

SA Trials Determine Feasibility and Estimate Treatment Effects

Middle developmental studies address questions related to clinical outcome and treatment "tolerability." Tolerability has three components: feasibility, safety, and activity/efficacy. Because treatment feasibility is often verified earlier in development, *safety and activity* determination is a primary purpose of such studies regardless of the treatment being tested. I refer to these studies as safety and activity (SA) trials.

The cancer drug paradigm for middle development presupposes that many therapies are in the pipeline for testing, and that most will fail to be of clinical benefit. Such a perspective is realistic for oncology but may not hold as well in other contexts. If potential therapeutic advances are encountered less frequently (e.g., treatment of degenerative neurologic disease), alternative middle developmental strategies might be better. In such cases, it might make sense to skip this step of development entirely (see below). In other circumstances, middle developmental studies with rigorous design features such as a randomized comparison group would be appropriate. The general characteristics of SA trials presented here are broadly serviceable.

In this stage of development the critical information gained is an estimate of the probability that patients will benefit from the therapy (or have serious side effects from it). These risk–benefit estimates will be compared with knowledge about conventional treatment to determine if additional (larger) trials are justified. Unsafe, inactive, or ineffective treatments will be discarded. This is in essence the overriding purpose of middle development—to discard inactive treatments without investing heavily in them.

There are several classic design features of SA trials: (1) focused eligibility, (2) fixed dose or treatment algorithm, (3) single cohort, results of which are compared to an external reference standard, (4) use of a surrogate clinical outcome to shorten trial time and lower cost, (5) modest number of patients, and (6) explicit decision parameters regarding continued development. Such studies are subject to selection bias, which can be minimized (but not eliminated) by using explicit eligibility criteria, adequate sample size, multicenter designs, standard outcome criteria, and disciplined counting of subjects and events.

Oncology phase II studies are SA trials. Safety is assessed by organ system toxicity measured according to defined criteria. Evidence of activity is usually assessed by the surrogate outcome "response," which is a measure of tumor shrinkage. In some phase II oncology trials, efficacy can be estimated from more definitive clinical outcomes such as disease recurrence, progression, or duration of survival. In cancer prevention, phase IIa studies are SA trials that employ surrogate or intermediate outcomes.

The classic phase II surrogate outcome is tumor shrinkage (response rate). This outcome appears to be in the causal pathway for definitive outcomes like survival. With discipline it can be measured reliably and is known fairly soon after the completion of treatment. In reality, tumor shrinkage is not an excellent surrogate, but it makes these trials reasonably efficient and reliable. SA trials can easily be adapted for definitive outcomes like survival. Designs for this are discussed in Chapter 11.

The questions addressed by SA trials are relevant in most developmental areas, but different designs may be necessary depending on the context. For example, a good standard therapy may already exist. This means we would favor a replacement only if it appears to be substantially more effective than standard, or if it has fewer side effects. This has quantitative design implications. For certain types of biological agents like vaccines or gene therapies, we might be willing to pursue only large treatment effects. Designs in high-risk populations or less rigorous comparisons might be appropriate.

SA studies usually involve between 25 and 100 patients. Resources are seldom available to study more patients, and the gain in precision that would be obtained by doing so is usually not worth the cost. The design can be fixed sample size, staged, or fully sequential. Staged designs are those conducted in groups with an option to terminate the trial after each group of patients is assessed. Fully sequential designs monitor the treatment effect after each study subject (Chapter 14). Staging allows a reliable decision to be made at the earliest possible time. An excellent review of phase II (SA) design types is given by Herson (1984); sample size is discussed in Chapter 11.

Skipping the SA Step

Most discussions of the developmental paradigm implicitly assume that middle development (phase II or SA) trials are always performed. However, the middle developmental step was designed to answer specific questions in a specialized context. Not all therapeutic development fits the need for such trials.

Middle development was intended to avoid doing expensive, lengthy, comparative trials on drugs that were unlikely to be beneficial. This was an issue in drug development in cancer therapy throughout much of the last 30 years of the twentieth century because there were many potential therapeutics, most were unlikely to work, and resources would not support testing all of them rigorously. Middle developmental designs with the efficiency of a single cohort and short-term surrogate outcomes can reject ineffective treatments as early as possible. By spending a relatively small amount of calendar time and money on a study with reasonable operating characteristics, the overall development process can be made faster and more efficient.

Sometimes the development pipeline does not require SA trials. It may be the case that resources for performing a trial are relatively abundant, the number of potential therapeutics is small, and those in development are very promising. Also the market-place may offer high profit and/or patient needs may be pressing. This could be the case for the development of some very specifically targeted drugs, for example. If so, investigators should move as quickly as possible to definitive CTE designs (versus standard therapy), shortening the pipeline, reducing the total cost of development, and maximizing benefits and profits, assuming efficacy will be demonstrated.

The risks in skipping middle developmental trials are considerable, and derive mostly from the scenario where the new therapy is shown to be ineffective after an expensive comparative trial. The following points should be considered before abandoning the SA step:

1. The sponsor must be willing to accept the financial risk of a negative CTE trial.
2. The intellectual risk of a long comparative study that yields a null result must be acceptable. This may not be the case for young investigators who need academic credit from activities early in their careers.
3. There should be enough promising information regarding safety and likelihood of efficacy in advance of the trial. This may come from dose-finding or other early developmental trials.
4. The treatment must have a very compelling biological rationale. Examples are drugs or other agents that are targeted to receptors or pathways that are known to be essential in the disease pathway.
5. Calendar time must be critical. The potential for lost opportunity or profit can make this so.
6. The chance of unforeseen events, both within and outside the study, must be low based on the development history up to that point.
7. The cost of middle developmental trials must be high relative to the information gained.

In cases that appear to fit these points, we always intend to follow the principle of least chagrin, sometimes evident only in retrospect.

Middle development in oncology drugs classically depends on tumor response as a surrogate outcome. For some drugs short-term shrinkage of the tumor does not appear to be biologically possible (e.g., agents that retard or stop growth without killing cancer cells). This would seem to provide an additional rationale for eliminating the middle developmental step and proceeding directly to a comparative trial. Doing so on this basis is risky. The history of the development of matrix metalloproteinase inhibitors (MMPI) reflects this risk. Many drugs in this class were pushed forward into comparative studies without the basis of SA trials. Over 20 comparative trials were performed without a single positive result (Canetta, 2004).

Randomization in SA Trials

Use of randomization in safety and activity trials seems to be a widely misunderstood topic. The question arises most naturally when there are two or more treatments in development for the same population (Simon, Wittes, and Ellenberg, 1985). We could perform the needed studies either serially or in parallel. Both strategies would use the same amount of calendar time and patients. However, using a parallel design permits subjects to be randomized to the treatment groups, removing selection and temporal effects that might influence studies done serially. Parallel groups provide a basis for selection of the best-performing treatment without regard to the magnitude of differences. Determining the ranking of an outcome measure is usually easier (more efficient) than estimating relative effect size.

Randomization would be used in this setting to remove selection bias and to control temporal trends. Its purpose is not to motivate a formal hypothesis test of the type routinely used in comparative studies. In other words, a randomized SA trial is not a cheap way to obtain a definitive answer to comparative questions. Randomization assures that selection of the winning treatment is free of selection bias. I compare this situation to a race (horse, human, etc.) where the starting positions are chosen randomly and the winner is selected regardless of formal statistical comparisons to the loser(s).

A second point of frequent confusion regarding randomized SA trials is whether or not to include a standard therapy arm. As a matter of routine, none of the treatments in a randomized SA trial should be standard therapy. The principal reason to exclude them is resource utilization. Although including a standard treatment reference arm would improve the quality of decisions, we generally do not want to trade maximum efficiency for it. We expect that most new treatments tested in SA trials will fail to be real advances, placing a premium on efficient evaluation and elimination of them. This perspective originated with the cancer drug development model but is valid more generally.

Thus the prototypical randomized SA trial will have two or more new treatments, no standard arm, and a primary objective of choosing for subsequent CTE testing the best-performing therapy (e.g., highest success rate) without regard to magnitude of differences. Having made these generalizations, there are occasional circumstances in which randomization versus standard therapy would be appropriate. For example, if we have few middle developmental questions, a high degree of confidence in one, and substantial resources, then a randomized design versus standard therapy is easier to justify. However, in such circumstances, it may make sense to skip the SA trial entirely and proceed directly to a true comparative design as discussed above. It is also important to design the error properties of such trials thoughtfully so that a good treatment is not discarded.

Some special practical circumstances are required to implement randomization in SA trials. In particular, the accrual rate must be sufficient to complete a parallel-groups design while the question is still timely. Also the decision to conduct studies of some of the treatments cannot be a consequence of the results in other treatments. If so, the parallel design may be unethical, wasteful, or not feasible. Randomized SA studies may be particularly well suited for selecting the best of several treatment schedules for drugs or drug combinations because these questions imply that the same population of patients will be under study. Similarly, when addressing questions about sequencing of different modes of treatment, such as radiotherapy with chemotherapy for cancer, randomized SA studies may be good designs for treatment development.

Other Design Issues in SA Trials

There are important design considerations for SA trials that are more of a clinical concern than statistical. For example, if standard therapy is available, investigators may be reluctant to displace it with an unproven therapy. It might be possible to administer the new treatment early, then proceed to standard therapy, if needed, after the delay. This design is sometimes called the "up front window" design—hideously bad terminology in my opinion. Some discussions of this type of design are given by Groninger et al. (2000), Balis (1997), Frei (1998), and Zoubek et al. (1995).

The principal concern in such designs is that delay of standard therapy will be a detriment to the patient. It is also possible that some residual effect of the first treatment will interfere with the benefit of the second. Because this is more of a clinical question than a study design issue, it is not possible to provide uniformly sound advice about it. It might be possible to establish that delay of a standard therapy for a given disease is not harmful, but one cannot generalize from that circumstance to other diseases or even to other treatments for the same condition.

6.5.3   Late Development

Comparative Studies

Later developmental trials usually have definitive clinical endpoints and address questions of *comparative treatment efficacy*. They employ a concurrent control group that receives standard therapy, a placebo, or an appropriate alternative treatment. Such studies are designed to provide valid and fairly precise estimates of differences in clinical outcome attributable to the treatments or interventions. Studies that are designed to show the equivalence of a new treatment to standard therapy also fit this description. These studies are comparative treatment efficacy (CTE) trials and correspond to phase III.

In cancer prevention, both phase IIb and III studies are CTE trials. Phase IIb studies are comparative trials that employ intermediate endpoints. Thus they are not definitive unless the intermediate endpoint has been validated. Phase III cancer prevention studies are CTE trials directly analogous to treatment trials because they use definitive endpoints such as disease incidence or progression.

Comparative trials are sometimes performed on an exceptionally *large scale*. A typical CTE trial conducted in oncology by a multicenter collaborative group might randomize a few hundred patients. In comparison, some trials in oncology, cardiovascular disease, or prevention randomize thousands of patients. The purpose of such studies might be to assess a small treatment difference or to provide empirical evidence of external validity by employing a large heterogeneous study cohort similar to the disease population. Some investigators have advocated simple methods of treatment definition and data collection to minimize the cost of these trials, giving rise to the term "large simple trials" (Yusuf, Collins, and Peto, 1984). I will refer to such studies as large-scale (LS) trials, acknowledging the fact that they are frequently not simple. See also Freedman (1989) and Souhami (1992) for a discussion of these types of studies.

CTE Trial Features

CTE trials are definitive steps in the evaluation of new treatments. In a regulatory context such well-controlled trials are called *pivotal* because the evidence they provide is central to critical decision making regarding approval. (In a treatment development context SA trials are pivotal because they afford an opportunity to terminate development.) The reliability of these trials is a consequence of design. Most often we think of comparative clinical trials as those employing a concurrent (perhaps randomized) control group. However, some CTE trials are designed to compare new therapy versus historical controls. The credibility of historically controlled trials may depend heavily on the magnitude of the estimated treatment effect. Results may be convincing only when there are large benefits that cannot be explained by chance or bias.

The sampling design for most comparative trials uses a single subject as the unit of study. This is straightforward. The treatment design might be more complex, however. Single modality trials tend to be the most simple and the results of such trials are likely to be easily interpretable. However, in many chronic diseases combined modality treatments such as drugs, surgery, radiotherapy, immunotherapy, endocrine therapy, and gene therapy might be employed in various combinations. Testing combined modality treatments requires that the investigators consider the sequencing and timing of the various treatments. The design must permit isolating the effect of the treatment(s) under investigation without confounding by other modalities. More complex treatment designs are needed to study treatment interactions.

Adjuvant trials are combined modality studies in which one of the treatments is given before or after the primary treatment in an attempt to enhance its effect. An example is the use of systemic chemotherapy following (or preceding) surgery for resection of tumors. Selecting the right combination of drugs and their timing relative to another therapy is a challenge for CTE clinical trials.

Error control for CTE trials nearly always consists of either a completely randomized design or blocking and stratification. In either case the precision is increased primarily by increasing the sample size. There are some circumstances where further reductions in random variation can be accomplished using within subject differences (e.g., crossover trials). Matching, which might be likened to an extreme form of stratification, is seldom used.

Like SA studies CTE trials can employ fixed sample sizes, staged designs, or fully sequential methods. Because of the size and complexity of the studies, most designs are either fixed sample size or staged. There are other important considerations in comparative trials that lead to the use of crossover designs or factorial designs. These are discussed in Chapter 20.

Another distinguishing feature of comparative trials is that they are frequently performed using the resources of several institutions or hospitals simultaneously. This is necessary because most single institutions do not accrue a sufficient number of patients to complete comparative testing of therapies in a reasonable length of time. Thus the logistics and organization of these studies can be complex. Furthermore the potential heterogeneity that arises from using patients in various institutions may be a consideration in the design, analysis, and interpretation of these studies.

For many diseases fairly effective treatments are already available and widely used. Developing new treatments in this setting presents special problems for comparative trial design. For example, suppose that we wish to demonstrate that a new anti-inflammatory agent is as effective for chronic arthritis as an agent already in widespread use. Such a clinical trial would not be designed necessarily to show the superiority of the new treatment. Rather, it might be designed to demonstrate the equivalence of the new treatment with standard therapy, at least as far as pain relief and mobility are concerned. The new agent might also have a lower incidence of side effects. These designs, often called "equivalence trials," are important, but sometimes difficult to implement.

ES (Phase IV) Studies Look for Uncommon Effects of Treatment

When treatments are beginning to be widely applied, as is often the case after regulatory approval, there is still an opportunity to learn about uncommon side effects, interactions with other therapies, or unusual complications. Uncommon events may affect the use or indication of the treatment if they are serious enough. Some treatments may become widely used after being administered to relatively few patients in CTE trials, as in the case of a few AIDS drugs or orphan drugs. Regulatory approval may be gained after studying only a few hundred individuals. An *expanded safety* study, or ES trial, could provide important information that was not gathered earlier in development.

Some phase IV or postmarketing surveillance studies are ES trials. However, most phase IV studies capture only serious side effects and may not precisely ascertain the number of patients who have received the treatment. Furthermore some phase IV studies are intended to be marketing research, to uncover new product indications that may protect patents or yield other financial benefits. A true ES trial would be designed to provide a reliable estimate of the incidence of serious side effects.

There have been several circumstances in which these types of studies have resulted in removal of new drugs from the market because of side effects that were not thought to be a problem during development. Since 1974 at least 10 drugs have been removed from the market in the United States because of safety concerns. The rate is higher in some other countries (Bakke et al., 1995). In other circumstances, performing such studies would have been of benefit to both the patients and the manufacturers. An example of this is the recent problems attributed to silicone breast implants.

Because investigators are potentially interested in very uncommon side effects, especially those that are life threatening or irreversible, ES clinical trials generally employ large sample sizes. However, these studies need not be conducted only in a setting of postmarketing surveillance. For example, similar questions arise in the application of a "standard" treatment in a common disease. Even single institutions can perform studies of this type to estimate long-term complication rates and the frequency of uncommon side effects.

Other late development trials may recapitulate designs discussed above but be applied to subsets or defined populations of patients. Such studies may investigate or target patients thought to be the most likely to benefit from the treatment and may be driven by hypotheses generated during earlier development. These types of late developmental trials should be governed by methodologic and design concerns similar to those discussed elsewhere in this book.

One of the difficulties in interpreting these large-scale postmarketing safety studies is being able to attribute complications or uncommon side effects reliably to the treatment in question, rather than to other factors associated with the underlying disease process or to unrelated factors. This is even more of a problem when events occur at long intervals after the initial treatment. A second problem arises because many ES studies do not determine how many individuals have received the treatment, only the number who experience adverse events. An incidence rate for adverse events cannot be calculated unless the number of individuals "at risk" can be estimated. In some other countries closer tracking of drug use after marketing permits true incidence rates to be estimated.

6.6 SPECIAL DESIGN ISSUES

6.6.1 Placebos

Ineffective treatments can appear to produce convincing improvements in diseases. There are at least three mechanisms by which this happens: (1) spontaneous improvement in the disease, (2) regression to the mean, and (3) the placebo effect. Many disease processes have natural remissions and exacerbations. If a patient begins treatment during an exacerbation, which might be likely, then a natural remission may be interpreted as an effect of treatment.

Regression to the mean is a similar phenomenon, but in a purely statistical context. Disease severity is always measured imperfectly and with random variation. If treatment is undertaken when the disease appears most severe, subsequent measures are likely to be closer to the mean (i.e., less severe), even if the underlying condition is unchanged. The change could be mistakenly attributed to treatment. This phenomenon can affect perfectly objective measurements.

The placebo effect is a term used to describe the apparent efficacy of an inert substance, particularly in the context of positive patient expectations (Beecher, 1955; Roberts et al., 1993). A rigorous definition of a placebo is somewhat difficult to craft (Grunbaum, 1986), but the central idea is that a placebo is used for its psychological or psychophysical effect when in fact it has no activity for the condition in question (Shapiro, 1960; Sullivan, 1993). Placebo effects have been seen in many different diseases including cancer (Moertel, 1976), cardiovascular disease (Archer and Leier, 1992; Amsterdam, Wolfson and Gorlin, 1969), psychiatric disorders (Laporte and Figueras, 1994; Berman, Sapers, Chang et al., 1995), pain (Sullivan, 2004), and neurologic disease (Shetty et al., 1999).

In principle, such effects would be measurable if a placebo-treated group were compared to a no-treatment control. Historically placebos have had a definite role in therapy—perhaps as much to relieve the psychological stress of the practitioner as that of the patient. In any case, these *therapeutic placebos* are known to the practitioner to be biologically inert. The patient is therefore deceived but may receive tangible benefits anyway, for reasons discussed below.

The use of an *experimental placebo* in a clinical trial is slightly different. In a trial the placebo imitates the treatment applied to the comparison group. Assuming the patients and researchers are masked, the placebo neutralizes effects of the therapy that presumably the investigators do not care about. Such effects might be a consequence only of patient expectations and/or might be mediated through real neurologic mechanisms. Effects observed beyond this are more purely biologically based.

Said another way, any therapeutic effect might be viewed as the sum of at least two components: a pure biological response plus a psychosocial response (perhaps itself partly mediated by some biological mechanism). A placebo-controlled design offers the chance of isolating the first component. This is not to say that the psychosocial component is necessarily small, useless, or uninteresting, only that it is sometimes

**TABLE 6.5   Possible Mechanisms
Underlying the Placebo Response**

Regression to the mean
Disease remission
Conditioned response
Expectancy
Health beliefs model (social learning)
Subordination/politeness
"Meaning" response
Subsiding negative effects of previous treatment
Supportive care
Relief of anxiety

viewed as a nuisance with regard to the biological component. Possible mechanisms by which the second component is produced are listed in Table 6.5.

There are two dangers to ignoring the placebo effect. First, the apparent benefit of an inert treatment may be offset by less obvious risks. This means the biological effects could be seriously negative while the psychosocial effects are positive. This is unacceptable because we could at least choose a safe placebo.

Second, placebo effects can be misleading. The patient may turn away from more effective therapies, especially those that have some risk but are offset by even greater biological gain. The investigator could incorrectly attribute placebo effects to biology, confounding knowledge and progress.

An experimental placebo affects the observer (investigator) as well as the study subject. The investigator may be biased in favor of a particular treatment, or in favor of some treatment compared to no treatment. A placebo group with masking of investigators and subjects removes this bias. Control of observer bias may be particularly important when the trial uses subjective outcome assessments. Masking is the principal design maneuver that controls observer bias, but it is virtually inseparable from the placebo when comparing therapy to "no treatment."

An in-depth discussion of the placebo from the perspective of history, therapeutics, ethics, and study design is given by Shapiro and Shapiro (1997). Experimental maneuvers to control the placebo effect are usually necessary and important. Shapiro and Shapiro provide an excellent explanation:

> An important conclusion to be drawn from review of the placebo's role in medical history is that the overwhelming majority of clinicians, including the gifted ones, were astonishingly and abysmally wrong. Intelligence, learnedness, professional status, and elaborate theorizing are no guarantee against the placebo effect and invariably have produced the most elaborately ineffective remedies—often those with monstrous potential for harming or killing patients.

Recently it has been argued that the placebo effect does not exist. Hrobjartsson and Götzsche (2001) reported an analysis of 114 clinical trials that, among other things, compared placebo with no treatment. Trials were classified into one of four categories according to the nature of their outcome measurement: subjective versus objective and binary (relative risk) versus difference of means. The results are abstracted in Table 6.6, where it is seen that three of the four confidence intervals on the effect include the null.

TABLE 6.6   Data on Question of Whether Placebo Effect Exists

Outcome		Number of		Overall	95% CI	
Type	Quality	Patients	Studies	Effect	Lower	Upper
Binary	Subjective	1928	23	0.95	0.86	1.05
	Objective	1867	9	0.91	0.80	1.04
	Overall	3795	32	0.95	0.88	1.02
Mean	Subjective	3081	53	−0.36	−0.47	−0.25
	Objective	1649	29	−0.12	−0.27	0.03
	Overall	4730	82	−0.28	−0.38	−0.19

Source: Hrobjartsson and Götzsche (2001).

This study was widely interpreted as disproving the existence of the placebo effect—a conclusion not strongly supported by the data (Bailar 2001; Spiegel, Kraemer, Carlson et al., 2001). Weak evidence of an effect should not be interpreted as evidence of no effect. The number of studies in each category is fairly small, but the effect estimates consistently support the presence of a placebo effect. In the largest category, subjective assessments of mean differences, the 95% confidence interval on the effect size excludes the null (Table 6.6). One might argue from these results that the placebo effect is quite small if it exists at all. This could be true, but it is not possible to generalize from an analysis restricted to clinical circumstances that permit both a placebo and a no-treatment control. For example, the authors found a significant placebo effect in pain studies. It is not clear why it should exist there but not elsewhere.

In many circumstances placebos are not ethically appropriate, for example, when an effective treatment already exists for a serious illness. A placebo group in a comparative trial would expose some patients to the risks of an ineffective treatment. If the disease is mild and self-limited or no treatment is the standard of practice, then a placebo might be appropriate. It is widely stated that when evaluating surgical treatments, a placebo (i.e., sham surgery) is not appropriate. Some surgeries place the patient at risk, and doing so without the potential for benefit is unacceptable. Examining circumstances from the essential perspective of risk will reveal occasional opportunities for appropriate use of sham surgery.

Sometimes *accepted* treatments exist but have not been proven more effective than placebo. Recognizing this situation requires experience, skill, and some bravery. Convincing one's colleagues to participate in a rigorous test of such a therapy can be nearly impossible. For that reason it is important to use rigorous designs as early as possible in therapeutic development, before the wide acceptance of a treatment is mistaken for effectiveness.

Wise use of a placebo can facilitate comparisons that might be hard to interpret otherwise. One example is when comparing treatments that require different routes of administration. Suppose that we wish to compare an oral medication versus a different drug given intravenously for severe nausea. One treatment group might consist of oral medication followed by an intravenous placebo. The second group might receive a placebo pill followed by the intravenous drug. This so-called double dummy design would make the treatments more directly comparable than if no placebos were used.

6.6.2   Equivalence or Noninferiority

Some trials are designed to demonstrate the equivalence of two treatments, or perhaps better stated that a new treatment is not inferior to a standard one. These studies are sometimes called active control trials. However, the purpose of a comparative trial employing an active control could be either to demonstrate the superiority of a new treatment over the standard (the most familiar circumstance) or to demonstrate the equivalence (noninferiority) of the new treatment. Therefore I will use the term "noninferiority" to be as specific as possible. Superiority trials are concerned essentially only with the relative effect of treatment. Noninferiority trials must address both the relative and absolute effects of treatments, making them more complex.

Conventional comparative study designs based on hypothesis tests are not adequate to deal with these questions because low power or low precision favors a conclusion of equivalence. Blackwelder (1982) pointed this out and also suggested a sample size approach (Blackwelder and Chang, 1982). The literature on equivalence designs and sample size has grown. A readable review of methods is given by Bristol (1999). Some ethics perspectives are provided by Djulbegovic and Clarke (2001). An interesting discussion of problems of equivalence interpretation in the context of coronary heart disease trials is given by Ware and Antman (1997) and also discussed later by Fleming (2000) in a useful review.

If the control treatment has an established absolute effect (e.g., versus placebo), the question of equivalence is direct. But if standard therapy is only presumed to be better than placebo, there is little sense in showing equivalence unless we can also demonstrate an appropriate degree of efficacy for the standard. Noninferiority questions arise often in chronic disease settings like oncology where many treatments have some activity but it is difficult to make large therapeutic gains. Some quantitative aspects of these trials are discussed in Section 11.5.9.

A new treatment could be slightly less effective (or not clinically significantly superior) than standard therapy on a definitive outcome such as survival, but have fewer side effects, lower cost, greater convenience, or higher quality of life. A small decrease in efficacy for survival may be well worth other benefits. Even in the absence of outright superiority for secondary outcomes, a new treatment that provides essentially the same therapeutic benefit as standard therapy is a potentially appropriate alternative. This principle is certainly operative in regulatory circumstances. The main issues involve the clinical and quantitative definitions of what it means for two therapies to be equivalent. With a definitive outcome such as survival, what hazard ratio (HR) *disfavoring* the new treatment should we be willing to accept?

It is not sufficient simply to examine the estimated HR for new versus standard. We might want firm evidence that the HR is equal to or greater than 0.8, for example. This value could be taken as the lower tolerance for a 95% confidence interval for the HR. Any point estimate for the HR would be considered noninferior as long as the lower bound exceeds 0.8. To allow the possibility that the HR < 1.0 and the lower confidence bound is greater than 0.8, the sample size for a trial might need to be quite large.

6.6.3   Nonuniformity of Effect

A recurring issue with treatments that do not appear to fulfill their biological promise is the belief that the therapy must help subjects with certain characteristics but not benefit others. This concern also arises by design, meaning some treatments target specific aspects of the disease or its cure that not all patients share. If we account for these characteristics in the design of our trial, we could see clearly a biological principle and a benefit for a subset of patients. Typical trials that admit a heterogeneous study cohort are likely to miss the effect because they yield an overall treatment effect that is diluted by subjects who cannot possibly benefit.

A straightforward example of this might be antibiotic therapy. If a new antimicrobial is targeted to a particular organism or to a biological characteristic that not all bacteria have, then it may appear to be ineffective in a trial that does not control the type of microbe, even though the clinical presentation of study subjects is fairly uniform. Other examples arise in cancer treatment that target particular receptor, enzyme, or genetic pathways that only a fraction of tumors have. Even a clinically well-defined disease could be produced by different pathways, not all of which are amenable to the same therapy. The more specific treatments become, the more likely we are to encounter this possibility.

In principle, this problem can always be fixed by sample size. An overall treatment effect that is small by virtue of its being diluted in the study population can be discerned by increasing precision (higher sample size). However, this fix may be unsatisfactory in several ways. The larger size and expense needed to overcome the problem is inefficient and strongly discouraging for developmental trials. Also an important biological finding may be obscured. Finally reliance on subset analyses post hoc is error prone. The most satisfactory approach is to incorporate biological knowledge into the design of the study.

This problem touches on two important areas of study design and analysis. One is treatment–covariate interaction, which provides a general framework for investigating questions of nonuniformity of effect. This is mostly an issue for comparative trials, hence the difficulty of coping with this problem in developmental studies. Relevant discussion of interaction is in Section 12.4.4 and Chapter 19. A second area of concern is subset analysis and multiplicity, discussed in Section 16.7.

6.6.4 Randomized Discontinuation

Sometimes we need to investigate the duration that therapy should be given—perhaps after patients appear to stabilize or benefit from treatment. Patients in a treated cohort could be randomly assigned to have therapy discontinued (or switched), then followed for disease relapse or progression. This is a randomized discontinuation (RD) design. It might also be applicable in circumstances where investigators are uncomfortable withholding all initial treatment.

Such a design might be fine from an inferential perspective, but it does present some ethical difficulties. In the usual clinical trial, a negative effect on the disease process is (would be) an unintended consequence. However, in the RD design, a negative effect is intended. This difference can be consequential and can limit the applicability of such designs.

Hypothetically, the study cohort for a RD trial could be selected so that patients most likely to withstand discontinuation (not be harmed) are preferentially selected. But this would seem to be as much a premonition as a workable strategy. Even if it could be accomplished, the treatment effect would likely be pushed toward the null, which is an anticonservative bias. Based on these considerations, we might expect RD designs to be infrequently applicable.

6.6.5   Hybrid Designs May Be Needed for Resolving Special Questions

Although many trial designs fit into the types discussed above, there are occasional circumstances in which other designs are more appropriate. For example, a drug study might combine dose-finding with estimation of response and toxicity rates, particularly if the basic pharmacology has already been worked out. Similarly, if one or more SA trials have already been finished, additional studies might focus more on efficacy rather than safety. These are sometimes called phase IIa and phase IIb trials. If several SA agents are being tested in the same population, a randomized design might be employed, where the least efficacious appearing treatments are dropped in favor of an expanded comparison of the remaining ones.

While these and many other hybrid designs are possible and have found some use, investigators should not sacrifice the goals and objectives of the drug or treatment development to a seemingly faster but more error-prone design. Although some savings in sample size or development time might be achieved with these and other hybrids, the costliest errors are those that cause us to pursue ineffective treatments, wasting time and resources, or those that cause us to miss improvements. These errors can only be controlled by having adequate information on which to base decisions regarding further development. Good study design, appropriate to the level of development, is the best way to provide the information needed.

6.6.6   Clinical Trials Cannot Meet Some Objectives

All clinical trials are performed in the presence of significance constraints. Some of these constraints are based on ethical concerns, others are a consequence of practical issues such as available resources, and still others are the result of scientific limitations. Because clinical trials, especially comparative ones, are reliable tools for evaluating treatment effects, it is understandable that the method and approach might be employed to answer other types of questions. However, inappropriately broad use of this methodology can lead to problems of the type that are described next.

Demonstration Trials

One inappropriate use of clinical trials is to attempt to demonstrate treatment effects that are already known to exist. This presents ethical problems, particularly if patients assigned to receive an inferior treatment could suffer permanent harm. Investigators cannot endorse accruing patients to a clinical trial if they believe that convincing evidence is already available concerning the superiority of one of the treatments.

Economic Evaluations

Another potentially problematic use of clinical trials is for economic evaluations. In many settings it is worthwhile and important to include cost considerations in the evaluation of new treatments. However, two problems can arise when such evaluations are performed early in treatment development. First, treatments may not be produced or applied in their most economical fashion when first tested. A seemingly high-cost therapy may prove to be less expensive with more refinement. Similarly it could become more cost effective if refinements improve its efficacy.

Second, there is often little uncertainty associated with the cost of particular interventions, making them unsuitable as outcome measures. The cost might be essentially fixed and could be calculated directly from a protocol or regimen specification and compared directly with an alternative. In other words, the variability associated with cost can be relatively small and may not require an experimental structure to study it. Cost effectiveness does depend on the magnitude of the treatment benefit and is subject to variation. While such questions are important, they are not always best answered as part of an efficacy evaluation.

Cost efficacy evaluations were planned and conducted as part of the National Emphysema Treatment Trial (NETT) (NETT Research Group, 2003b). These provided a useful perspective on the efficacy evaluations, but the time horizon was probably too short during the clinical trial proper for them to be definitive. Attempted cost projections depended heavily on survival, a major unknown. After the trial it was no longer possible to follow the study cohort for costs associated with treatment in an unperturbed way. The longer term cost outcomes were disrupted by the efficacy findings themselves.

Overviews

In recent years formal methods have been proposed and adopted for combining evidence from many clinical trials, each of which may have been too small to convincingly demonstrate a subtle treatment effect. These "overviews" or "meta-analyses" are popular ways of looking for consistency in trial results and can be very helpful for making clinically relevant inferences. They are subject to other kinds of error and biases, as discussed in Chapter 21. In any case, the primary purpose of clinical trials is not to serve as building blocks for overviews. Investigators should not routinely conduct clinical trials with low power, anticipating that a later overview will yield the correct answer about treatment efficacy.

Nonspecific Therapy

Another purpose for which clinical trials have been suggested is to evaluate treatments defined in a nonspecific way. For example, one might ask a question like; Is chemotherapy for advanced stage lung cancer beneficial? A clinical trial could be proposed to answer such a question. In this hypothetical trial supportive care might be compared with chemotherapy of *any type*. The reasons for not specifying the treatment more explicitly are that individual practitioners may have differences of opinion about the appropriate therapy to use and might be unwilling to participate in a clinical trial which *requires* the use of specific agents. We might expect to overcome this problem with a trial that permits a nonspecific treatment definition and employs a sufficiently large sample size. Trials such as this may not yield useful information because any outcome will immediately raise questions about heterogeneity in the treatment group. What if certain regimens appear to be helpful and others appear to be harmful? It is unlikely that the nonspecific question will continue to be of interest after the data are collected.

6.7 IMPORTANCE OF THE PROTOCOL DOCUMENT

A protocol is the document that specifies the research plan for a clinical trial. As seen below, the protocol should address every aspect of the investigation: background, clinical, administrative, managerial, statistical, and informed consent, to name a few. It is the keystone of the planning process and deserves much of the investigators' attention before a study is submitted for ethics review.

The most difficult parts of writing a protocol are (1) formulating and developing an important, feasible question, and (2) being certain that resources (funds, patients, time) are available to meet the objectives. These aspects of clinical investigation cannot be dealt with here. However, given an important question and the resources to answer it, developing the research protocol is the crucial next step in planning a clinical trial.

A useful first step in preparing a complete research protocol is developing a "concept sheet," a one- or two-page document describing the essential scientific and clinical features of the proposed study. Aside from helping to structure the final document, the concept sheet can be used to communicate basic scientific aspects of the trial and gather feedback. This is particularly helpful in a collaborative group setting and for study sponsors. Changes in concepts can be made before the investigator invests in a complete protocol.

"Protocol" also has a second meaning—it is the *logical plan* or prescription for the study apart from the written document. It is important to have a high degree of fidelity between the written document and the intended logical plan. If the two do not agree closely, circumstances with individual patients will almost certainly arise that result in differences of interpretation, inconsistency, and nonadherence to the plan.

There are many documents besides the research protocol that are essential for conducting a clinical trial. Examples include an Investigator's Brochure, which describes the investigational product or device, documentation of ethical review, case report or data forms, and numerous other regulatory documents. For studies that will be part of regulatory approval applications, the necessary list of documents is extensive and must be obtained from the appropriate regulatory agency. Examples of such agencies are the FDA and the Health Protection Branch (HPB) in some other countries. The International Conference on Harmonization of Technical Requirements for Registration of Pharmaceuticals for Human Use (ICH) has outlined in detail required documents and guidelines for regulatory approval in cooperating countries (ICH, 1994).

6.7.1   Protocols Have Many Functions

The protocol is the single most important quality control tool for all aspects of a clinical trial because it contains a complete specification for both the research plan and treatment for the individual patient. The written protocol is also the only effective method for communicating research ideas and plans in detail to other investigators. The protocol is a peer review document because, in the planning stages, a scientific study cannot be reviewed and critiqued without the protocol. For collaborative groups and the NIH, the protocol is the quantum unit of research. For the FDA and other regulatory bodies, the protocol is a legal document in addition to its other important functions. The protocol also specifies all aspects of the statistical design of a clinical trial and, therefore, the quantitative conclusions of the study are conditional upon it. The protocol and associated consent forms also have medical-legal and ethical implications and will be examined in this light by institutional review committees (e.g., IRBs). Investigators who need more motivation for careful preparation of research protocols should probably not be conducting clinical research!

Some investigators have suggested publication or registration of clinical trial protocols in periodicals to make the fact of a planned experiment known to the scientific community at large. For some high-profile or controversial trials the protocol rationale is published (e.g., NSABP Breast Cancer Prevention Trial; NSABP, 1992). Although protocol documents are not published, placing such information in the public domain would emphasize the importance of the research plan and provide a way to minimize publication bias.

6.7.2   Deviations from Protocol Specifications Are Common

Despite the general importance of the study protocol, investigators must be mindful that, to varying degrees, the details of the trial plans are not followed precisely in all patients. This can happen, for example, because of differences in interpretation. More frequently protocol deviations are a consequence of unanticipated events and corresponding clinical judgment about what good medical practice requires for a particular patient. Some deviations from the protocol are inconsequential, while others are substantive and could affect the validity of the trial.

The importance of deviations depends on how they affect the inferences from the trial, not on the magnitude of the errors. For example, many studies require that patients have laboratory evidence of normal major organ system function at study entry. Permitting patients with small differences in these "normal values" to enter the study will rarely affect the validity of the trial. In this situation these deviations are minor and might be ignored. However, if we intend to study treatment effects in patients with a particular diagnosis or condition, permitting those with the wrong diagnosis to enter the trial could affect the validity of the conclusions. In this situation entering subjects with the wrong diagnosis might be considered major deviations. Even so, in randomized trials, this problem would have to occur frequently or affect the treatment groups differentially to have consequences for the study validity.

Protocol deviations, even major ones, are not necessarily a sign of a poor quality study, even though some regulatory agencies take a dim view of them. Some regulators view the protocol rigidly, as a virtual *contract* to perform a study in a particular fashion in every patient. To a strong regulatory mind set, deviations are evidence of sloppiness or, if frequent enough, scientific misconduct. In my opinion, this view is inappropriate and unfortunate. The protocol represents a plan or intent, not a guarantee that every clinical circumstance will fit the mold. Furthermore it is not obvious to what extent the protocol can or should refer to the individual patient as opposed to an idealized research subject. Human subjects are autonomous and many medical decisions supersede the scientific objectives of the trial.

This is an argument in favor of a balanced view of protocol deviations. On the one hand, the protocol is a design specification and the investigators should adhere to the plans as closely as possible. On the other hand, the design exists in an environment that changes from patient to patient, and some deviations are likely. The more detailed and rigid the specifications are, the more likely it is that deviations will occur. Therefore the protocol should require only those procedures and data that are essential to the scientific and ethical integrity of the trial.

In any case, the perfect protocol will never be written. An important part of planning for a clinical trial is to establish a mechanism for modifying and correcting ambiguities in the protocol that experience with it will uncover. Protocol amendments and ancillary documents that accomplish this (e.g., policy and procedural memoranda) carry the full force of protocol. Methods for communicating such changes to the research group are an important part of study planning and conduct.

Design Validity versus Biological Validity

When assessing a trial with a protocol that has not been followed perfectly, investigators may need to distinguish between *design validity* and *biological validity*. A technically well-designed and conducted trial can address a weak, meaningless, or inappropriate biological question. In contrast, a flawed design or imperfectly executed trial could provide valid and convincing biological evidence. Efforts to enhance design validity by coping with protocol deviations may not be needed if the study already provides biological validity.

The best recent example of this situation is the National Surgical Adjuvant Breast and Bowel Project (NSABP) misconduct case (see Chapter 22), in which eligibility criteria were fabricated by one investigator for a small proportion of study participants. The resulting trial, even including the ineligible patients, almost certainly had biological validity. However, many people agonized excessively over design validity and sought to support or increase it by discarding various fractions of the data. This raised questions about the entire trial, which prompted worries about the collaborative group, the treatment itself, the principal investigator, and the sponsor.

Consider that nonexperimental (retrospective) study designs can sometimes yield convincing evidence of treatment efficacy, even though patients were treated without a research protocol. The principal problems in making reliable efficacy assessments from these designs arise because (1) it is often difficult to define retrospectively who received treatment, (2) investigators cannot control selection bias and confounders convincingly, and (3) outcomes may not be assessed in a systematic fashion. Problems with inferences from nonexperimental designs do not occur primarily because individuals did not all follow the same protocol exactly. Prospective clinical trials correct these deficiencies by defining the treatment group rigorously and by controlling bias and random error. Thus protocol deviations in a trial that do not affect these features of the design will probably not strongly degrade the validity of the study.

6.7.3 Protocols Are Structured, Logical, and Complete

There is no way to describe a universal trial protocol, except to say that it contains detailed accurate information from which knowledgeable investigators can review or conduct a study. Usually protocols omit details about the considerable infrastructure required to perform a trial, such as administrative procedures, unless they are integral to the design. Even without such details, protocols are important and useful documents. This section contains a sketch of the depth and breadth of a typical trial protocol. Additional details, at least with respect to oncology protocols, can be found in Leventhal and Wittes (1988).

This outline and brief discussion is intended to assist investigators in drafting, reviewing, and improving the quality of their own clinical trials. The emphasis of this outline is on comparative trials performed in a collaborative setting. However, almost all studies would benefit from having their protocols address most of the topics below. The headings presuppose a pharmacologic therapy, but there should be little difficulty in adapting it to trials using other treatment modalities. The total length of a protocol will often be 50 typed pages or more, written in a structured formal style with page numbers and references, and containing most of the following sections.

1. **Title Page** (essential for all protocols). The title page should include (a) study title, (b) date of current revision, (c) principal investigator and phone number, (d) other study collaborators, and (e) sponsor or administrative office and phone number. Adding the revision date prevents confusion when protocols are developed in a collaborative setting. Some sponsors, such as pharmaceutical companies, omit naming a principal investigator, but responsibility for the study and the protocol document must be clear.

2. **Contents or Index** (optional). Short protocols may omit it.

3. **Protocol Synopsis** (optional). A one- or two-page description of complicated or lengthy protocols can be a great help to reviewers.

4. **Schema** (essential for complex SA and CTE protocols). Patient flow with major relevant landmarks such as study entry, evaluations and treatments, and randomization are presented in diagrammatic form. The basic architecture of the study design will be evident from the schema alone. Doses, schedules, and other details regarding therapy should be deleted from the schema to prevent using it for treatment without consulting the full protocol.

5. **Objectives of the Study** (essential for all protocols). The objectives pose important scientific questions addressed by the trial. Usually there is a single primary objective with several secondary ones. The objectives should be stated clearly, concisely, and quantitatively, and correspond with the hypotheses to be tested by the study. It is also helpful to indicate the relevant endpoints. This section is usually quite brief and is formatted as an outline with details.

6. **Introduction and Background** (essential for all protocols). This section provides the introduction and scientific background/rationale for the study and should be an adequate introduction to the subject for knowledgeable reviewers. Summaries of similar or earlier developmental studies will provide justification for the stated objectives. If a new treatment is being tested in SA or CTE trials, pilot data may be required to justify the larger study. Keep in mind that some readers and potential critics of the protocol will not be as expert in the science as the authors. References must be included, and this section should be written in a narrative style. Supporting or recent historical data that justify the design of the study need to be included. Relevant preclinical data should also be provided.

7. **Study Design** (essential). This includes details regarding the treatment plan and regimen, doses and modifications, expected side effects and actions to take.

8. **Drug Information** (essential for studies employing drugs). All drugs used in the study need to be described in alphabetical order with respect to human toxicity, pharmaceutical data, administration, storage and stability, and supplier. The investigator should list the National Service Center (NSC) number, Investigational New Drug (IND) number, if relevant, and any other standard names for the drug.

9. **Staging Criteria** (essential for all studies). This section lists the criteria or system by which extent or severity of disease will be established. In cancer studies, extent of disease may be determined by stage, the assessment of which is often standardized. For some other diseases, extent of disease may be less standardized, or even controversial. Therefore the criteria used to establish extent may need to be listed explicitly.

10. **Patient Eligibility and Exclusionary Criteria** (essential for all protocols). This section is a detailed, specific, and quantitative description of eligibility requirements and exclusions. For example, eligibility criteria may depend on subcategory of disease, prior therapy, measurability of disease extent or response, performance status or disease severity, and required organ function. Exclusionary criteria should also be outlined in the same way. Criteria for exclusions should be named on the basis of knowledge obtained prior to the

assessment of response or study outcome. Exclusions should not be permitted on the basis of information that only becomes known after the initiation of treatment. Any competing studies or factors that will interfere with accrual should be listed here or in a separate section.

11. **Registration or Randomization Procedures and Stratification** (essential for all protocols). Phone numbers, contacts, and procedures for placing a patient on the study are provided in this section. This should include information required for central eligibility checks. In many cases the individuals responsible for the mechanistic parts of placing a patient on study are not physicians or persons otherwise knowledgeable about the subtleties of the study. They may require detailed procedural specifications here to avoid confusion or entry of ineligible patients later. Stratification factors may affect the randomized assignment or treatment and need to be available at the time a treatment assignment or registration is requested.

12. **Treatment Program** (essential for all protocols). This includes details of required therapy such as chemotherapy dosage, biopsies, and masking procedures. For TM studies, the dose escalation will be detailed. For studies employing more than one treatment modality, details of each need to be provided in separate sections. Within the treatment specification, secondary registrations or randomizations should be indicated. This section should also list criteria for discontinuing protocol treatment, such as completion of therapy, disease progression, unacceptable side effects, or patient preference.

13. **Dosage Modification/Side Effects** (essential for all protocols). This includes details of required dose modifications and reductions when toxicities or side effects are encountered. The side effects or toxicities to be monitored should be listed. Dose modifications should address the following concerns: baseline conditions that necessitate changes in the initial dose, criteria for multiple courses of treatment, dose decrements or increments to be employed, criteria for substituting treatment delays for dose reductions, an investigator to contact concerning dose modification questions, and reporting requirements for unexpected side effects.

14. **Agent Information** (essential for all agents). This section explains details of drug preparation, storage, dilution, and stability. For experimental agents, availability and drug-specific background data should be given. The drug brochure for investigational agents should accompany the protocol.

15. **Treatment Evaluation** (essential for all protocols). This section provides the details for the methods of evaluation and for determining endpoints and toxicities. In studies like SA trials of cytotoxic agents in oncology, the methods for assessing tumor response are critical to the evaluation and generalizability of the trial. Scoring, intensity, or other evaluation criteria should be listed.

16. **Adverse Events and Toxicity Management** (optional as a separate section). The definitions of adverse events and methods for reporting them should be provided. It should also be clear what is not an adverse event. A definition for a serious adverse event is also needed. This section may also describe how to handle laboratory abnormalities attributable to the therapy.

17. **Serial Measurements/Study Calendar** (essential for SA and CTE studies). This lists the study parameters and milestones and the time at which they are to

be determined while the patient is on protocol treatment. This should include a listing of baseline measurements. In addition this section specifies the actions to be taken in response to specific study complications, such as steps for removing a patient from the treatment protocol because of side effects or toxicity. This section should cross-reference the data forms and their submission schedule.

18. **Statistical Considerations** (essential for DF, SA, and CTE studies, optional for TM studies). This section addresses the following relevant points: recap of study objectives and endpoint definitions; required sample size, power, and the way it was estimated; precision of the design for determining major study endpoints; projected accrual rate and duration of the study; and analysis methods. Also plans for interim analyses and guidelines for early stopping must be included. The statistical section should outline the quantitative properties of the study with respect to secondary objectives. A sketch of the final analysis is also useful for most studies but may be critical for those used in a regulatory setting. In those cases the analysis plan may need to be specified in considerable detail, and a separate document describing the analysis may be needed.

19. **External Collaborations or Reviews** (essential for all multi-institutional studies). This section describes arrangements and procedures with investigators outside of the parent institution. For example, the details of how to obtain pathological specimen or imaging review should be given. Names, addresses, and phone numbers should be listed for outside investigators if not given on the title page. In the event that extensive or critical portions of the study depend on collaborations outside the parent institution, supplementary protocols may need to be submitted (e.g., extensive pathologic review).

20. **Data Recording, Management, and Monitoring** (essential for all protocols). This section provides details for the collection and review of data while the study is in progress. It includes the contact person with names and addresses for submission of data forms or related collection instruments. This will include addresses and phone numbers for questions, emergencies, and the reporting of fatal, life-threatening, or unexpected reactions. The submission schedule for data forms must be outlined. In addition, the membership of a safety monitoring committee should be outlined. Requirements for retention of study records should also be specified. Other topics to address include IRB and informed consent review, indemnification, and confidentiality.

21. **Special Instructions**. This section contains instructions to study managers for obtaining and shipping specimens for special laboratory tests or analyses. Any special data or data forms required by the study sponsor should be listed with instructions on how they should be handled. Pharmacokinetic assessments are the type of special specimens that might be described in this section.

22. **Communication and Publication of Data** (essential for all studies which involve collaborations, especially data management, outside of the institution, particularly pharmaceutical company sponsored studies). This section outlines agreements regarding the communication and publication of study data. Any limitation or restrictions of investigator access to the study data must be detailed. This section should also specify prior approvals, if required, for publishing the data. The format of this section is narrative.

23. **Peer Review** (this section is optional). This section outlines the peer review procedures that the protocol may have already passed. For example, for cooperative group protocols, peer review may include the group mechanism and NIH.

24. **Patient Consent** (essential for all protocols). This section may be submitted as an appendix with patient consent forms. The reading level of the patient consent document should be eighth grade or lower. An appropriate language should be used if there is a plan to enroll patients whose primary language is not English, and "back-translation" or other method should be used to be certain that important concepts are translated properly.

25. **References** (essential for all protocols). Include publications based on the trial (if any) and copies of important references from the protocol.

26. **Data (Case Report) Forms** (optional). Investigators are encouraged to submit proposed data forms for the study. Ideally the forms should be self-coding for computer entry. The forms will be reviewed for capturing of items required to meet study objectives. Data forms may be submitted as an appendix.

27. **Protocol Amendments** (essential for all protocols if any amendments exist). This section includes a summary of all protocol modifications that affect patient treatment arranged in chronological order. A brief explanation of each amendment is also useful.

28. **Other Appendices** (optional). These may include ethics review documentation, agent information, toxicity criteria, flow sheets, and so forth.

29. **Glossary** (optional). Include definitions of abbreviations, special, or unfamiliar terms. This section might be useful for psychosocial research terminology, and new drugs. For complex studies, consider putting the glossary up front.

6.8 SUMMARY

Clinical trials are true experimental designs and can be distinguished from nonexperimental studies by the way the investigator controls the administration of treatment and unwanted sources of variation that can obscure the scientific objectives. Controlled trials and verified comparative trials provide the strongest evidence about treatment effects, whereas nonexperimental designs provide the weakest. Effort spent by investigators achieving an optimal design is more worthwhile and productive than trying to correct deficiencies after the trial is finished and the data are analyzed. A good trial design will be simple, measure clinically important outcomes, be precise (reduce errors of chance), eliminate selection and observer bias, and provide externally valid results.

Most clinical trial designs require patients to be nested within treatments. Parallel groups designs that use this nested structure can be improved by using randomization, blocked assignments for balance, and placebo treatments when feasible to control observer bias.

Many clinical trials fit into one of five types. Translational trials are used to refine therapeutic ideas, inform subsequent experiments, and plan additional preclinical studies. They require a circularity between the clinic and the laboratory. Treatment mechanism studies are early developmental trials with human subjects. Their objectives are usually dose-finding (the relationship between dose and safety) and pharmacokinetics.

SA studies employ a fixed treatment dose or schedule and look for evidence of treatment feasibility, efficacy, and side effects. CTE (phase III) studies compare seemingly active treatments to standard therapy or other treatments. Expanded safety (phase IV) studies are undertaken after a new treatment comes into wide use and are intended to observe uncommon side effects of treatment.

The research protocol is a written scientific plan for a clinical trial. While it does not usually describe the considerable infrastructure necessary to carry out a trial, the protocol provides enough information about the study to serve developmental and peer review purposes. Protocols specify eligibility and exclusion criteria, treatment plans and modifications, endpoint assessments and other evaluations, statistical plans, and informed consent procedures and documents. Minor deviations from protocol specifications are common and usually do not undermine the integrity of the trial. Frequently occurring major deviations may affect the validity of the trial or be a sign of other problems with the investigation.

6.9   QUESTIONS FOR DISCUSSION

1. Many methodologic problems in trials can be avoided reliably by proper design features but not by analyses. List and discuss as many as you can think of.

2. What are circumstances in which epidemiologic designs would be preferred over comparative clinical trials for judging treatment efficacy?

3. Interventions to prevent disease often do not undergo TM and SA testing. What type of evidence can replace the information from such developmental trials?

4. In device, surgical, and biologic trials, what is meant by "device function"? Compare and contrast questions about "device function" with questions of clinical outcome.

5. Any of the following may be the experimental unit in a comparative trial: treatment course, patient, family, physician practice, village, or city. Sketch examples of where each might be used.

6. Suppose that investigators conduct an SA trial evaluating response to treatment as the endpoint. The study is stopped due to loss of funding. Months later the same study is re-instituted by the same investigators after additional support is found. Is this one or two trials? Can the data be combined and, if so, how?

7. Suppose that drug A is an established treatment administered at a fixed dose. Investigators wish to study the combination of drugs A and B but are unsure of the best dose of either one in the presence of the other. What type of design would permit studying this question? Compare your suggestion with a typical dose-finding trial. What if the correct doses are known but the order of A and B is in question?

8. A randomized CTE trial has five treatment arms: A, B, C, $A + B$, and $B + C$. What issues would you expect to be discussed in the analysis plan of the statistics section in the protocol? What if only the first three groups were involved?

7

RANDOM ERROR AND BIAS

7.1 INTRODUCTION

Error has two components, a purely random one and systematic one called bias. Understanding the differences between randomness and bias, and the sources of each, is the first step in being able to control them using experimental design. The terms "random error," "random variability," or just "variability" are often used to describe the play of chance, particularly when the effects of explanatory or predictive factors have already been taken into account. An operational definition of randomness might be *unexplainable fluctuations*, these are fluctuations that remain beyond our ability to attribute them to specific causes.

"Bias" means systematic error—deviations that are not a consequence of chance alone. Bias can arise from numerous sources, such as selection effects, uncontrolled prognostic factors, procedural flaws, and statistical methods, and also from perceptual errors, attitudes, and beliefs. These last three sources are often collectively called observer bias. The exact consequences of bias may be difficult to know in advance, but it is usually simple to understand factors that can contribute to it. Strong sources of bias can often be anticipated well enough to be controlled. In clinical trials we tend to focus on bias as much as random variation as a source of error because many biases are strong relative to the size of treatment effects (e.g., selection effects).

Errors in experiments can be discussed in either a purely statistical context or in a larger clinical one. Although mathematical formalism is required to discuss the statistical context, random error and bias are conceptually the same in a clinical context. Most statistical discussions of random error (and bias) assume that a primary focus of the investigation is hypothesis testing. This framework is convenient and will be used here. However, adopting the usual perspective is not intended to be a general endorsement of hypothesis tests as the preferred method for making inferences.

Clinical Trials: A Methodologic Perspective, 2E, by S. Piantadosi
Copyright © 2005 John Wiley & Sons, Inc.

Estimates of treatment effect and confidence intervals can be similarly affected by random error and bias.

Pure random error has no preferred direction. Statistically, we expect its net or average effect to be zero. Clinically, we can always make its relative effect inconsequentially small by averaging over a large number of observations or a long enough period. This does not mean that chance will not affect a particular observation, but that randomness averages out to have a small relative effect in the long run. Replication (increasing the number of observations or repeating the experiment) reduces the magnitude of random error. Because of sampling variability, subject-to-subject differences, measurement error, or other sources of noise, we can never completely eliminate random error. However, in most experiments it can be controlled and reduced to acceptably low levels by careful attention to the experimental design, particularly sample size.

Bias, unlike random fluctuation, is a component of error that has a net direction and magnitude. In a purely statistical context, bias can sometimes be quantified and may be quite small. In a clinical context, bias has four deadly characteristics: it can arise in numerous, diverse, but common ways; its direction or magnitude usually cannot be predicted; it can easily be as large as the treatment effect of interest; and it cannot be corrected by replication. As a result bias can dominate the guesses of practitioners making unstructured treatment evaluations, or invalidate poorly planned or conducted treatment comparisons. The factors that produce bias may not be amenable to quantification but can usually be removed or reduced by good design and conduct of the experiment.

The difference and importance of distinguishing between random variation and bias is analogous to sound or signal reproduction. In this analogy the treatment effect is the sound level or signal strength, variation is like background noise, and bias is distortion. Background noise (variation) is important when the sound or signal level (treatment effect) is low. Reducing variation is like increasing the signal-to-noise ratio. In contrast, even when the noise is low, the sound produced may be distorted unless the system is designed and performing properly. *Even when the variation is low, the treatment effect may be biased unless the trial is designed and conducted properly.* Having a strong signal is not sufficient if there is too much distortion.

Example

Suppose that we have two different estimators (or *methods of estimation*) of a treatment effect, Δ, one unbiased and the other biased (Figure 7.1). Because of sampling variation, repeated use of either estimate would produce a distribution of values. One estimator, $\hat{\Delta}$, has random error but no bias, and the distribution of values we would obtain is centered around the true value, $\Delta = 0$. The second estimator, $\tilde{\Delta}$, has both random error and bias, and its distribution is not centered around the true value. Many times $\tilde{\Delta}$ yields answers similar to $\hat{\Delta}$, but on average, $\tilde{\Delta}$ does not give the true value.

In practice, the investigator does not see a full distribution of values for any estimate because the experiment is performed only once. Thus the actual estimate obtained for Δ is a single sample from a distribution. Because both $\hat{\Delta}$ and $\tilde{\Delta}$ in this example are subject to randomness, either can yield a value that is far enough away from the true value and lead us to conclude $\Delta \neq 0$ (e.g., after a hypothesis test). In this example, $\tilde{\Delta}$ is closer, on average, to the true treatment effect than $\hat{\Delta}$. Although not generally true, sometimes the overall performance (bias + random error) of a biased estimator can be better than that of an unbiased one.

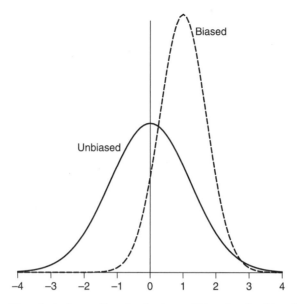

Figure 7.1 Hypothetical sampling distributions of biased and unbiased estimators.

7.2 RANDOM ERROR

In this section, I assume that the estimators being employed are free of bias. I will return to examine the consequences of bias later in the chapter.

7.2.1 Hypothesis Tests versus Significance Tests

Hypothesis testing has had a prominent role in developing modern ideas about the design and analysis of clinical trials and still provides a useful perspective on errors of inference. Following ideas put forth by Neyman and Pearson (e.g., 1933; Lehmann, 1986), hypothesis testing is an approach for choosing between two competing statistical hypotheses. Let the competing hypotheses be labeled H_0 and H_a, and denote a summary of the data (or *statistic*), by T. The hypothesis testing approach requires that we define a *critical region* in advance of taking data values, and then choose H_0 or H_a, depending on whether or not T falls within the critical region.

In practice, investigators seldom employ hypothesis testing in exactly this way. A different procedure, called *significance testing* (Cox and Hinkley, 1974) is used more commonly. A nonmathematical comparison of the two procedures is given by Salsburg (1990). Assume that the probability distribution of the test statistic, T, is known or can be approximated when H_0 is true. The more extreme the actual value of T is, relative to this distribution, the less likely H_0 is to be true. The significance level is

$$p = \Pr\{T^* \geq T \mid H_0\},$$

where T is the value of the statistic based on the observed data. Thus this "test" yields a significance level (or p-value) instead of a decision. The p-value is intended to help the investigator assess the strength of evidence for or against H_0. In reality, the *p*-value does not have this nice interpretation. The deficiencies of this approach will be discussed in Chapter 16.

The basic structures of hypothesis and significance tests are the same as are their properties with respect to random error. This is discussed in the next section.

7.2.2   Hypothesis Tests Are Subject to Two Types of Random Error

The two types of random error that can result from a formal hypothesis test are shown in Table 7.1. The type I error is a "false positive" result and occurs if there is no treatment effect or difference but the investigator wrongly concludes that there is. The type II error is a "false negative" and occurs when investigators fail to detect a treatment effect or difference that is actually present. The *power* of the test is the chance of declaring a treatment effect or difference of a given size to be statistically significantly different from the null hypothesis value when the alternative hypothesis is true (i.e., the probability of *not* making a type II error).

These ideas are illustrated in Figure 7.2, which shows the distributions (assume they are normal) of a treatment effect estimator, $\widehat{\Delta}$, under both the null hypothesis ($\Delta = 0$) and under an alternative hypothesis ($\Delta \neq 0$). The short vertical lines represent a "critical value" chosen from the type I error to reject the null hypothesis. For example, the critical value could be ± 1.96 standard deviations from the null hypothesis mean,

TABLE 7.1   Random Errors from Hypothesis Tests and Other Types of Dichotomized Inferences

	Truth and Consequences	
Result of Test	H_0 True	H_0 False
Reject H_0	Type I error	No error
Don't reject H_0	No error	Type II error

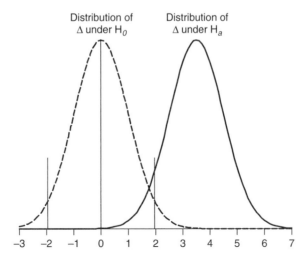

Figure 7.2   Distributions of a treatment effect estimate under null and alternative hypotheses.

which would yield a (two-sided) type I error of 5%. The following discussion will focus on only the upper critical value, although both have to be considered for a two-sided test.

If the null hypothesis is true, $\widehat{\Delta}$ will come from the distribution of values centered at 0 ($\Delta = 0$). If $\widehat{\Delta}$ exceeds the critical value, the experimenter would find it unlikely that it came from the distribution centered at 0 and would reject the null hypothesis. This would constitute a type I error if the null hypothesis were true. If $\widehat{\Delta}$ does not exceed the critical value, the experimenter would not reject the null hypothesis.

If the alternative hypothesis is true, $\widehat{\Delta}$ comes from the distribution centered away from 0 ($\Delta \neq 0$). As before, the experimenter will reject the null hypothesis if $\widehat{\Delta}$ is greater than the critical value. However, if $\widehat{\Delta}$ is less than the critical value, the experimenter will not reject the null hypothesis, resulting in a type II error. When the alternative hypothesis is true, one would like to reject the null most of the time, that is, have a test with high power.

The properties of the hypothesis test are determined by choosing the alternative hypothesis and the type I error level. If the alternative hypothesis is taken to be far away from the null, the test will have a high power. However, alternatives far away from the null may be unrealistic or uninteresting. The consequences of type I and type II errors are different. Aside from the fact that they are described and quantified under different assumptions about the true state of nature, maneuvers to control them need to be different. These are discussed in the following sections.

7.2.3 Type I Errors Are Relatively Easy to Control

Usually there is only one factor that governs the chance of making a type I error, that is, the critical value of the test. The experimentalist is free to set the critical value for the type I error at any desired level, controlling it even during the analysis of an experiment. The type I error rate does not depend on the size of the experiment. For example, in Figure 7.2, the type I error can be reduced by moving the critical value (short vertical line) to the right.

There are some circumstances in which the investigator must consider more than just the critical value to control the type I error. This error can become inflated when multiple tests are performed. This is likely to happen in three situations. The first is when investigators examine accumulating data and repeatedly perform statistical tests, as is done in sequential or group sequential monitoring of clinical trials. A second situation in which the type I error can become inflated occurs when many outcomes or treatment groups are examined using multiple hypothesis tests. A third circumstance arises when multiple subsets, interactions, or other exploratory analyses are performed. Although the error of each test can be controlled in the manner outlined above, the overall (experimentwide) probability of an error will increase. Some corrections during analysis are possible, but the overall error rate should be carefully considered during the design of the trial. This point will arise again in the discussion of sequential methods in Chapter 14. Inflation of the type I error is also discussed in Section 16.7.2.

7.2.4 The Properties of Confidence Intervals Are Similar

In many circumstances, summarizing the observed data using point estimates and confidence intervals gives more information than using only hypothesis tests. Confidence

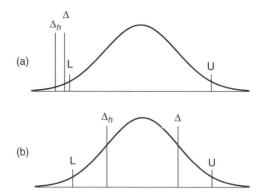

Figure 7.3   Confidence intervals. Δ_h and Δ denote a hypothetical and true parameter value, respectively. U and L are the confidence bounds around the estimate. (The true parameter value remains unknown.)

intervals are always based (centered) on the observed or estimated effect and convey useful information regarding the precision of the estimate. This makes them more descriptive and useful than hypothesis tests or p-values, which are often inadequate for summarizing data (see Chapter 18). However, hypothesis tests and confidence intervals share some common properties. For example, we can usually reconstruct hypothesis tests from the confidence intervals. More specifically, a confidence interval is a collection of hypotheses that cannot be rejected at a specified α-level.

Suppose that Δ is the true parameter value and that investigators hypothesize Δ_h to be correct. A situation analogous to a type I error can occur if the confidence interval around the estimate of Δ excludes Δ_h (Figure 7.3a). Then the sample on which the confidence interval has been based was an atypical one, and we would wrongly conclude that the data were inconsistent with Δ_h, even though $\Delta_h \approx \Delta$. An error analogous to the type II error can occur when the confidence interval includes both Δ and Δ_h (Figure 7.3b). This can happen even when Δ_h is substantially different from Δ.

7.2.5   Using a One- or Two-Sided Hypothesis Test Is Not the Right Question

In comparative experiments investigators frequently ask if a one-sided hypothesis test is appropriate. When there is biological knowledge that the treatment difference can have only one direction, a one-sided test may be appropriate. Firm knowledge concerning the direction of a treatment effect is not easy to acquire. As one possibility we might imagine testing the effects of a nontoxic adjuvant, such as A versus A + B, where B can only augment the effect of A. Given sufficient knowledge about B, a one-sided test may be sensible.

This is not the same as "only being interested in" differences in one direction, which happens frequently. Not all hypothesis tests are inferentially symmetric. If a treatment is being compared with a placebo, investigators will not be interested in demonstrating conclusively that the new treatment is worse than the placebo. Similarly, if a new treatment is compared to standard therapy, learning that the new treatment is not superior to the standard is sufficient. Some questions have only one interesting, useful, and appropriate side.

However, the directionality of the biological question in no way settles the appropriate probability level to use in a significance or hypothesis test. Should investigators accept a lower standard of significance just because there is a known or preferred direction for the treatment difference compared to when it is not known? I think not, but this is precisely what happens if a one-sided critical value is chosen for $\alpha = 0.05$ when otherwise a two-sided $\alpha = 0.05$ test would be used. In other words, using the same type I error probability for one- and two-sided questions amounts to using a lower critical value for the one-sided case. A more consistent procedure would be to employ a one-sided $\alpha = 0.025$ type I error, which would yield the same standard of significance as the conventional two-sided test. This discussion is not meant to define evidence in terms of the p-value.

Thus the right question is the standard of evidence (critical value) and not the direction of the hypothesis test. It is not appropriate to employ a one-sided test as a falsely clever way to reduce the sample size. It is always necessary to adjust the type I error rate to suit the question at hand.

7.2.6   *P*-Values Quantify the Type I Error

P-values are probability statements made under the null hypothesis (usually the hypothesis of no clinical effect or difference). Suppose that the null hypothesis is correct. Because of random variability the estimated effect or difference will be different from zero. We could ask, If the null hypothesis is correct, how likely were we to obtain the observed result, or one more extreme? If the observed result is unlikely, we would take this as evidence that the null hypothesis is false. If we reject the null hypothesis when it is correct, it is a type I error. The significance level, or p-value, is the probability of obtaining the observed result (or one further away from the null) when the null hypothesis is in fact true. If the observed result (or one more extreme) is relatively likely, we would not reject the null hypothesis.

If we could repeat our experiment many times, the estimates of clinical effect obtained would average out close to the true value (assuming no bias). Then the probability distribution of the estimates would be evident, and there would be little trouble in deciding if the assumption of no effect (or no difference) was correct. However, we usually perform only a single study, meaning that we cannot conclusively judge our hypothesis. We judge instead the estimate obtained, under an assumption of no effect. In this way of thinking the probability distribution or uncertainty refers to the estimate obtained and not to the true treatment effect. Thus the p-value is not a statement about the true treatment effect but about estimates that might be obtained if the null hypothesis were true.

7.2.7   Type II Errors Depend on the Clinical Difference of Interest

There are three factors that influence the chance of making a type II error. These are the critical value for the rejection of the null hypothesis, the width of the distribution (variance) of the estimator under the alternative hypothesis, and the distance between the centers of the null and alternative distributions, meaning the alternative treatment effect or difference (Figure 7.2). In principle, investigators have control over the rejection region (discussed above) and the width of the distribution, which is a direct consequence of sample size.

The magnitude of the alternative hypothesis is not truly under investigator control: it is a consequence of the minimally important difference (MID) that is specified from clinical considerations. We can calculate power for different hypothetical MIDs. Larger assumed MIDs decrease the type II error from a given sample size. Thus all trials have a high power to detect MIDs that are hypothetically large enough. Unfortunately, large treatment differences are usually not plausible clinically, so the high power to detect them is not useful. The utility of a large study is that it can reliably detect realistic but important small differences.

These ideas can be made more quantitative by considering the accompanying power curve (Figure 7.4) for a hypothetical clinical trial comparing survival in two treatment groups. The power of the study is plotted against the assumed treatment difference measured as a hazard ratio (ratio of median event times). A study with 100 events (solid line) has only 53% power to detect a hazard ratio of 1.5, but 90% power to detect a ratio of 2 or higher. Smaller studies have lower power for all effect sizes, but they eventually reach 90% power to detect hazard ratios greater than 2. In the presence of censoring there is not a perfect correspondence between events and sample size, but this is unimportant for the purposes here. A more detailed discussion of power appears in Chapter 11.

It is useful to examine the consequences of type I and type II error rates beyond the context of a single trial. If these error rates are habitually relaxed when conducting trials, as is sometimes the case with the type II error, there is a direct consequence on the reliability of any finding. Specifically, the chance that the treatment is actually ineffective even when a trial indicates a positive result increases as the designed error rates increase. If $T^{\pm}$ represents the truth about treatment efficacy and $S^{\pm}$ is the study

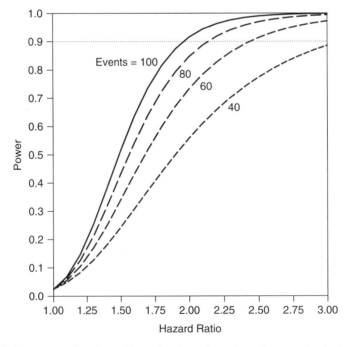

Figure 7.4 Power as a function of hazard ratio and number of events for the logrank test.

TABLE 7.2   Chance of a False Positive Study Result as a Function of Power and Prior Probability of Success

p	β	$P[T^-|S^+]$	p	β	$P[T^-|S^+]$	p	β	$P[T^-|S^+]$
0.01	0.1	0.73	0.10	0.1	0.20	0.60	0.1	0.02
	0.2	0.76		0.2	0.22		0.2	0.02
	0.5	0.83		0.5	0.31		0.5	0.03
0.02	0.1	0.58	0.20	0.1	0.10	0.70	0.1	0.01
	0.2	0.60		0.2	0.11		0.2	0.01
	0.5	0.71		0.5	0.17		0.5	0.02
0.03	0.1	0.47	0.30	0.1	0.06	0.80	0.1	0.01
	0.2	0.50		0.2	0.07		0.2	0.01
	0.5	0.62		0.5	0.10		0.5	0.01
0.04	0.1	0.40	0.40	0.1	0.04	0.90	0.1	<0.01
	0.2	0.43		0.2	0.04		0.2	<0.01
	0.5	0.55		0.5	0.07		0.5	<0.01
0.05	0.1	0.35	0.50	0.1	0.03			
	0.2	0.37		0.2	0.03			
	0.5	0.49		0.5	0.05			

Note: In all cases, $\alpha = 0.05$.

result, Bayes's theorem shows that

$$P[T^-|S^+] = \frac{P[S^+|T^-]P[T^-]}{P[S^+|T^-]P[T^-] + P[S^+|T^+]P[T^+]} = \frac{\alpha(1-p)/2}{\alpha(1-p)/2 + (1-\beta)p},$$

$$(7.1)$$

where p is the unconditional (prior) probability of a true treatment advance.

The form of equation (7.1) demonstrates that $(1-\beta)p$ must be high or most positive study findings are likely to be incorrect. The joint effects of p and β can be seen in the examples in Table 7.2. In many chronic diseases the prior chance of a treatment advance is quite small, increasing the need for high power. Safety and activity trials, despite their limitations, essentially increase p, improving the reliability of subsequent comparative trials.

7.2.8   Post hoc Power Calculations Are not Helpful

At the conclusion of a comparative clinical trial we know the estimated treatment effect and its variability. The power of the experiment is usually no longer a meaningful concern. The power characterizes uncertainty before the trial is conducted, in particular, the chance of rejecting the null if a specified alternative hypothesis is true. However, the experiment removes this uncertainty and provides useful information about the treatment effect.

The power of a completed trial against the *observed* difference is likely to be low if that difference is smaller than the alternative hypothesis used to plan the study—a common enough occurrence. However, the alternative treatment effect used to design the trial is not supported by the data. In other words, a post hoc power calculation will indicate only that the trial had lower power to detect effects smaller than that for which it was designed, which should be obvious. For these reasons power calculations after a trial is completed are essentially useless.

A potential exception to the futility of post hoc power calculations is if the study results were expressed only as a dichotomy, that is reject the null hypothesis or not. If we don't know the magnitude of the treatment effect, the power of the trial against hypothetical effect sizes might continue to be of some interest. However, it is not appropriate to summarize the results of a clinical trial merely as rejecting the null or not.

Example 4   Suppose that a trial is designed to detect a hazard ratio of 1.9 with 90% power between the treatment groups using a sample size of 100 events (Figure 7.4). Suppose also that when the study is over, the estimated hazard ratio is 1.25. The power of the study to detect a hazard ratio of 1.25 as being statistically significant is only about 15% (Figure 7.4). Because $\Delta = 1.25$ is closer to the null hypothesis than $\Delta = 2.0$, it is clear that the power against the observed difference is lower than the power against the original alternative hypothesis. Knowing this does not help us to interpret the evidence provided by the trial. Calculating the power of the study against other alternative hypotheses that are not supported by the data will not be especially informative. The question of sample size that could detect a hazard ratio of 1.25 with some specified power could arise when planning a new trial.

7.3   CLINICAL BIASES

In this section, I consider clinical assessments or estimates that are subject to both bias and random error. Two major categories of such biases are selection bias and observer bias. Either of these in turn can be the result of other specific effects. For example, either a selection bias or an observer bias could arise from certain attitudes and beliefs. The history of clinical trials shows a distinct evolution of methods to control these biases along with random error. Excellent brief discussions of selection bias and observer bias are given by Chalmers (2003) and Kaptchuk (2002) respectively.

7.3.1   Relative Size of Random Error and Bias Is Important

When bias is present, its relative magnitude compared with random error is important. If the bias component is relatively small, it will not be a large component of the total error. However, bias could be the principal component of the total error (Figure 7.5). A strong bias can yield an estimate very far from the true value, or even drive the estimate in the wrong direction. In other words, a truly beneficial treatment might be seen as harmful, and vice versa.

If we only knew the direction of a bias, we might be able to design informative studies or interpret results more clearly. For example, if we consistently overestimate the effect of treatment A, but a randomized trial shows it to be inferior to B, the study result will remain convincing. Unfortunately, investigators seldom know either the relative magnitude or direction of bias, so corrections are not generally possible. The only reliable strategy is to prevent bias by using good design.

7.3.2   Bias Arises from Numerous Sources

There are many potential sources and types of bias in medical studies. For a very detailed listing, see Sackett (1979). Here I discuss only a few common types that can seriously affect inferences from clinical trials. I will refer to biases that are a

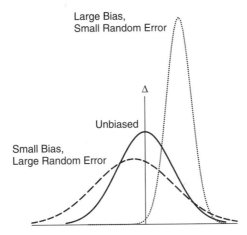

Figure 7.5   Relative size of bias and random error.

TABLE 7.3   Some Bias Types and Their Potential Effects on the Results of Clinical Trials

Type of Bias	Potential Consequences
Selection bias	May affect external validity, especially if the study has no internal controls.
Procedure selection	Healthier patients selected for a particular treatment may cause it to look good.
Postentry exclusion	Over optimistic view of the treatment, especially side effects.
Selective loss of data	Variety of effects depending on the mechanism of data loss.
Ascertainment bias	Treatment effect is brought more into alignment with prior expectation.
Retroactive definitions	Strong anticonservative bias is possible.
Uncontrolled covariate	Confounding of treatment effect with covariate effect.

consequence of inattention to study design and assessment criteria as structural biases. Some of these are listed in Table 7.3. Statistical biases are those that arise from the selection of certain procedures, tests, or data manipulations, and are correctable by using alternative methods. Publication or reporting bias results from preferential selection of positive study results (i.e., those showing significant differences) over negative results (i.e., those failing to show a significant difference) when clinical trials are published (see Chapter 18).

Bias in the selection of the study cohort is a potentially limiting factor in the external validity of trial findings, especially SA studies. For randomized trials, selection bias may not be as serious a problem because it affects all groups equally. Therefore it does not influence the estimated treatment difference unless the selected characteristics modulate the relative effects of treatment (treatment–covariate interactions). In contrast, procedure selection (e.g., surgery versus medical treatment) can cause the treatment groups to be composed or analyzed differently in a comparative trial and can therefore have an impact on the estimated treatment difference.

Galen (c. 130–210 CE) created a system of medicine from the works of Hippocrates, Aristotle, and Plato, dominating Western and Islamic medicine for hundreds of years. His teachings were so highly respected that they were not disputed, despite credible observations contradicting some of them. Such authoritarianism can foster a serious source of bias in clinical studies, for example, by retroactive definitions. This is illustrated by the following quote, attributed to Galen:

> All who drink of this remedy recover in a short time, except those whom it does not help, who all die. Therefore, it is obvious that it fails only in incurable cases. (Strauss, 1968)

This reasoning is circular and defines prognosis in terms of outcome. The bias that results can be very strong.

Inappropriate exclusion of eligible study subjects from the analysis is another way to create potentially strong biases. This can be a difficult source of bias to eliminate when there are seemingly excellent clinical reasons for making exclusions. However, whenever eligible patients are excluded from analysis, the experimental paradigm is broken, possibly leading to unwanted or unanticipated effects on estimation. This idea is developed further below and in Chapter 15, where handling of data imperfections is discussed.

Selective loss of data can result from choosing unworkable or suboptimal outcomes or from errors in study conduct. This is one of the most serious problems with nonexperimental data because of the potential to produce strong spurious effects and because it may not be obvious what data are missing. In a clinical trial, prior specification of required follow-ups and the data collection schedule tend to minimize selective loss of data. Poor choice for an outcome measure can also create the problem. For example, attrition in a seriously ill cohort will prevent outcome measurements, such as patient reported assessments, from being taken, resulting in a substantial survivors bias.

Assessment bias is another familiar potential difficulty. Patients' self-assessments and clinicians' judgments often lack objectivity when they have expectations about the treatment under study. Similar circumstances can arise from study groups, collaborators, or companies under external pressures such as time, money, or ego. Because trial designs that use subjective assessments are common, necessary, and important, controlling this bias is especially important.

Uncontrolled covariates present a potential source of strong confounding or bias. Failure to account for the effects of covariates will reduce precision at a minimum, and can mask or amplify true treatment effects. Influential covariates can be responsible for, or mediate, other biases such as selection bias. Age, severity, extent of disease, or comorbidities are examples of covariates that can appear as selection effects.

TM Trials

Treatment mechanism trials generally employ objective methods of assessing the function, feasibility, or action of the treatment. Examples include pharmacologic outcomes, and physical or anatomic measurements. Because of this, properly performed TM studies can be subject to less bias than some later developmental trials. However, if subjective evaluation criteria are used or the investigator does not guard against his or her own assessment bias, the results of a TM trial can mislead. An example is if the demonstration of a successful mechanism such as drug delivery or device function is equated with clinical efficacy.

DF Trials

Dose-finding trials can be subject to biases that are not adequately appreciated. For example, dose escalations in oncology using traditional decision and stopping rules with prespecified dose levels are essentially sequential trials. When they terminate because of toxicity, there is a tendency to underestimate the dose associated with the target level of toxicity. For example, if a dose escalation terminates because dose limiting toxicity is seen in 1 out of 3 patients, the true probability of toxicity is usually less than $1/3$. Operationally this is not a problem, but it can leave investigators with additional dose-finding necessary later in development. When properly implemented, the continual reassessment method of dose-finding is not as subject to this bias.

Recently there have been suggestions that patients can select their own dose of a new drug in DF trials (within limits and after being properly informed about the risks and benefits). This strategy responds to ethical concerns that many patients on such trials receive ineffective doses and cannot benefit because the drug levels are conservative early in the study. Using this strategy may confound dose with performance status or other important prognostic factors, causing investigators to misestimate the target dose or to overestimate efficacy. Determining whether or not such compromises are worthwhile will require a substantial amount of investigation.

SA Trials

Safety and activity trials may be the most subject to bias of all types of trials because (1) they employ partly subjective or surrogate clinical endpoints, (2) they do not have internal controls, (3) they are relatively small with highly selected patient populations, and (4) some investigators tend to perform and emphasize subset and/or post hoc analyses. None of these characteristics necessarily induce bias, but they do nothing to discourage it. Patient selection effects can be strong in such trials, especially when they are performed within a single institution. Removal of patients from the analysis for reasons that are both outcomes and predictors, such as failing to complete therapy, can also bias the results.

CTE Trials

Comparative treatment efficacy trials, particularly those with methodologic flaws, are also potentially subject to significant biases. Problems can result from using subjective endpoints, invalid surrogate endpoints, inappropriate comparison groups, and from emphasizing analyses that are prone to error. However, well-designed and conducted CTE trials will minimize or eliminate such biases. Methods for doing so are sketched in the next section.

7.3.3   Controlling Structural Bias Is Conceptually Simple

Relatively few methods are available, or needed, to eliminate or reduce structural bias. None of the methods and procedures discussed here is complicated, although each of them, at times, can conflict with other constraints imposed on a study. When large treatment effects are present, these methods of reducing bias may be unnecessary. However, when small to moderate treatment effects are present, these methods increase the strength of evidence from the trial.

Randomization

Randomization is the principal method available to the investigator designing a comparative study for reducing selection bias. It is also the only method that can reliably control the effects of unknown covariates. This benefit is often overlooked, but it cannot be achieved by any other design maneuver. Randomization effectively guarantees that both observed and unobserved baseline differences between the treatment groups are attributable to chance, the effects of which can be quantified by the statistician. After accounting for chance, the remaining differences can be attributed to the treatment(s) reliably, if other sources of bias have been eliminated.

The benefits and necessity of randomization for reducing bias and controlling extraneous effects are well established. But occasional complaints about it are raised by methodologists, and randomization is not viewed favorably by all potential trial participants. We do know that inadequately performed randomization or its lack is associated with large treatment effects that are likely to be biased (Schulz, 1995). Furthermore there is good evidence that adequately concealed randomization results in relative treatment effects less subject to bias than trials in which the randomization has not been well concealed (Kunz and Oxman, 1998). Randomization is discussed in more detail in Chapter 13.

Blocking and Stratification

Blocking and stratification used together is a simple and effective means for controlling the effects of known covariates. It will have no effect on unknown covariates, unlike randomization. Operationally, blocking and stratifying balances treatment groups within covariate levels (strata), thereby balancing covariates within treatment groups. The resulting induced balance is reassuring (but not essential) with regard to removing the influence of the covariates. Blocking and stratification are discussed in more detail in Chapter 13.

Masking

Masking (blinding) reduces assessment bias. Single masking means that the patients in the study are unaware of which treatment they receive. Masking can be accomplished by ensuring that both treatments look, feel, or taste the same and that the investigators do not know or reveal the treatment assignment to the patients. A masked placebo control is different from a no-treatment control. In the latter case the treatment difference will consist of the effects of the treatment plus the "placebo effect," which might be sizable. For example, in studies comparing analgesics, patients who know they are receiving a new drug or procedure may have a bias in favor of its efficacy, causing them to overstate its effect. Thus masking can improve the objectivity of partially subjective outcomes.

In many drug trials, masking can be accomplished effectively by using a placebo tablet or formulation that appears the same as the agent being tested. Double masking (double blind) implies that both patient and investigator responsible for assessing the outcome are unaware of which treatment is being administered. This type of masking further increases the usefulness of subjective endpoints because investigators can also be influenced by their expectations. This is especially true if the investigator has been exposed to seemingly favorable preclinical data, believes strongly in the biological basis on which the therapy was developed, and/or has professional or financial interest in the success of the study. Effective treatment masking is essential in such cases.

Some trialists have made a case for triple masking, which is a situation where the Data Safety Monitoring Committee (DSMC) for a clinical trial is also unaware of the treatment assignment. Triple masking is intended to increase the objectivity of their decision to stop or continue a study. I believe the DSMC should not be masked. The DSMC is charged with making important decisions about the ongoing ethical and efficacy evidence from the study (Chapter 14). This role is difficult enough when treatment assignments are known to the DSMC members and may be impossible if they are masked. For example, suppose a major difference in non–life-threatening side effects is seen during a trial comparing standard therapy with a new treatment, but no difference in efficacy is evident. If standard therapy has a higher incidence of side effects, it might well be wise to allow the study to remain open to gather more knowledge about, and clinical experience with, the new treatment. Conversely, if the new treatment is the more toxic one, the trial could be stopped in favor of standard therapy. Masking of the DSMC in this situation would prevent them from doing an effective job.

Concurrent Controls

A concurrent control group is a relatively resource intensive but effective method for reducing bias. It eliminates the confounding of treatment with calendar time and facilitates the use of randomization, which is the most reliable method for reducing bias due to treatment selection. There are many examples of strong trends in outcomes such as disease mortality over time (Silverman, 1985). Such trends could render a historically controlled study uninterpretable, as discussed in Chapter 6. One does not need to examine a long interval of time to see time trends. For example, they are evident over short time intervals in areas such as supportive care of cancer and AIDS patients. Similar problems can affect inferences based on databases (Byar, 1980, 1991).

Objective Assessments

In most situations investigators have a choice about the methods that will be used to make clinical assessments of the major study endpoints. Whenever possible, the methods employed should be objective ones, to reduce assessment bias and increase the reproducibility of the findings. Objective assessments are those on which independent reviewers would agree. Examples ranging from most objective to least objective are quantitative laboratory measurements, vital status or survival time, time to disease progression or recurrence, outcomes based on predefined criteria such as toxicity or side effects, physician-based judgments like functional indexes, quality of life measurements, psychosocial assessments, and patient self-reports like pain intensity.

In a few circumstances the most objective assessment may not be the best method of evaluating treatment. For example, when making assessments of pain intensity, the physician or nurse assessment may be poorly correlated with that from the patient (Grossman et al., 1992). Health professionals probably tend to underestimate pain and overestimate functional status. A better method for a clinical trial would be to use the patient's own assessment and employ masking to improve the reliability of the study.

Active Follow-up and Endpoint Ascertainment

Even though endpoints such as disease recurrence and death are objective, the methods used to ascertain them may not be objective if investigators do not plan properly for conducting follow-up examinations. If a trial relies only on passive reporting of such events, the chance for bias is increased. For example, we cannot assume that patients who do not return to clinic for scheduled follow-up visits are alive and well.

In contrast, if follow-up status is determined actively, such as with scheduled clinic visits or follow-up and phone contacts or home visits, the chance of ascertainment bias is reduced.

No Post hoc Exclusions

After study entry, especially after treatment has begun, many study-related events that occur are likely to be correlated with one another. Some events will be causally related, while others are only weakly correlated, being connected through known or unknown third factors. In any case, one cannot realistically expect to select a subset of eligible study participants on the basis of one event and have them be comparable with regard to all other factors. For example, if we select a subset based on good treatment compliance, they are likely to differ from the subset of patients with poor compliance.

The consequence of exclusions is that analyses of subsets of trial participants are subject to bias. One bias that can occur is a "selection bias," where patients predisposed to a particular outcome may be selected preferentially by the subset factor. The strength of this bias depends mostly on unobservable correlations between the selection factor and the outcome. Also the results in small subsets can be heavily influenced by the outcomes in a few patients. Because little is typically known about the interplay of important factors defining and influencing the subset, it is better to prevent bias by avoiding these analyses.

In SA trials, these forces operate in addition to the selection effects of the eligibility criteria, which may be considerable. Thus our usual inability to compare results, even informally, among institutions is worsened by subset analyses. In CTE randomized trials, the effect on treatment differences is not as great, unless the selection effect acts differently in the treatment groups. Although subset analyses do not yield statistical tests with desirable properties, they can help explain the overall results of a trial or be useful for planning new studies.

7.4 STATISTICAL BIAS

Statisticians invariably try to avoid or reduce bias. However, there are circumstances where bias cannot be avoided. Even if it could be, a study might require too many resources to exclude all possible sources of bias. Unfortunately, even qualitative information about bias is difficult to obtain. In a few circumstances investigators can quantify and correct a statistical bias.

Statistical bias can arise from some methods of analysis or estimation. Fortunately such biases are usually not large and often are amenable to quantification and correction by other statistical means. One example of this is the effect of missing covariates in some nonlinear regression models commonly used to analyze clinical trials. The relative treatment effects obtained from some models can be biased if necessary prognostic factors are omitted from the regression (Gail, Wieand, and Piantadosi, 1984; Gail, Tan, and Piantadosi, 1988). This is in contrast to omitting terms from linear regressions where the remaining effects are estimated without bias (although with higher variances). Fortunately the bias is not large in situations likely to be encountered commonly in clinical trials (Chastang, Byar, and Piantadosi, 1993).

Another situation that can produce a statistical bias occurs when a clinical trial is stopped early according to any of several guidelines commonly used for this purpose (discussed in Chapter 14). When a trial is terminated early because evidence favors rejecting the null hypothesis, the estimated treatment effect is biased in the direction of the alternative hypothesis. The earlier a trial is stopped, the larger is the potential for bias. Unfortunately, in this case there is not a good correction for the bias because its magnitude depends on the size of the true treatment effect, which is unknown. Thus designs that permit early stopping can be unfavorable if investigators absolutely require an unbiased estimate of the treatment effect, even when an extreme alternative hypothesis is true.

7.4.1 Some Statistical Bias Can Be Corrected

Perhaps the most commonly encountered statistical bias arises when estimating the variance of a random variable. Suppose that s^2 is the estimate of a true population variance, σ^2, in which the sum of squared deviations (SSD) is customarily divided by $n - 1$:

$$s^2 = \frac{\sum_{i=1}^{n}(x_i - \overline{x})^2}{n - 1}.$$

Students often ask intuitively why the SSD is not divided by n, and are told that $n - 1$ is the "degrees of freedom." In fact

$$v^2 = \frac{\sum_{i=1}^{n}(x_i - \overline{x})^2}{n}$$

is the maximum likelihood estimate (MLE) of σ^2, but it is biased by the factor $(n - 1)/n$. Specifically, the expected value of v^2 is $\sigma^2(n - 1)/n$. Thus it is not very biased, except in small samples. This bias can be corrected by multiplying v^2 by the factor $n/(n - 1)$, which yields s^2. A common feature of maximum likelihood estimates is that they are biased in small samples. Their widespread use is a consequence of the fact that they are usually unbiased in large samples (e.g., consider the case just sketched), are efficient, and have asymptotic normal distributions.

Furthermore, when the sample comes from a normal distribution, the square root of the unbiased estimate of the variance is *not* an unbiased estimate of the standard deviation; that is s is not unbiased for the standard deviation, σ (Holtzman, 1950; Cureton, 1968). This fact is often ignored. An unbiased estimator for σ is

$$\widehat{\sigma} = \sqrt{\frac{\sum_{i=1}^{n}(x_i - \overline{x})^2}{k}},$$

where

$$k = 2\left(\frac{\Gamma(n/2)}{\Gamma((n - 1)/2)}\right)^2.$$

7.4.2 Unbiasedness Is Not the Only Desirable Attribute of an Estimator

Although unbiasedness is a desirable characteristic of statistical estimators, there are other important and useful attributes, such as a small average overall error or mean squared error (MSE). (Consider the previous discussion regarding Figure 7.1.) Suppose that η is an estimator of a parameter. The MSE is defined as

$$\text{MSE} = \int_{-\infty}^{\infty} (\eta - \theta)^2 f(\eta) \, d\eta, \tag{7.2}$$

where θ is the true value of the parameter and $f(\eta)$ is the probability density distribution of η. Because the MSE measures the average error, it can be smaller for a biased estimator compared with an unbiased one (Figure 7.5). Suppose that $\bar{\eta}$ is the average or expected value of η. Equation (7.2) yields

$$
\begin{aligned}
\text{MSE} &= \int_{-\infty}^{\infty} (\eta - \bar{\eta} + \bar{\eta} - \theta)^2 f(\eta) \, d\eta \\
&= \int_{-\infty}^{\infty} \left((\eta - \bar{\eta})^2 + (\bar{\eta} - \theta)^2 + 2(\eta - \bar{\eta})(\bar{\eta} - \theta) \right) f(\eta) \, d\eta \\
&= \int_{-\infty}^{\infty} (\eta - \bar{\eta})^2 f(\eta) \, d\eta + (\bar{\eta} - \theta)^2 + 0.
\end{aligned}
$$

The first term is the variance of η, the second term is b^2, the square of the bias of η for θ, and the third term is zero. Thus

$$\text{MSE} = \text{var}(\eta) + b^2.$$

This demonstrates that the relative sizes of bias and variance are important in determining the overall error, at least in a statistical sense. Unfortunately, these quantitative statistical relationships have no direct analogues in the domain of clinical bias where the magnitudes of possible systematic errors are usually unknown.

To follow the example in the previous section regarding the unbiased estimates of σ^2 and σ from the normal distribution, we write the minimum MSE estimate of the variance of a normal distribution (Stuart and Ord, 1987) as

$$\tilde{\sigma}^2 = \frac{\sum_{i=1}^{n} (x_i - \bar{x})^2}{n+1}.$$

Similarly a minimum MSE estimate of σ is

$$\tilde{\sigma} = \sqrt{\frac{\sum_{i=1}^{n} (x_i - \bar{x})^2}{k}},$$

where

$$k = 2 \left(\frac{\Gamma((n+1)/2)}{\Gamma(n/2)} \right)^2.$$

An example of where it is useful to consider (and minimize) MSE instead of bias arises in problems commonly encountered when estimating parameters in regression models with highly correlated predictor variables. In this circumstance the ordinary maximum likelihood estimates are unstable and can yield values that have the wrong algebraic sign or incorrect significance levels. The problem can be corrected by using "ridge regression," in which a small amount of bias is permitted in the estimates because it greatly reduces the total variability. Although the procedure has drawbacks, there are circumstances where it is appropriate. See Draper and Smith (1981) for a discussion of this topic in the setting of linear models. The method is essentially equivalent to penalized likelihood estimation, applied to proportional hazards regression by Verweij and Van Houwelingen (1994).

7.5 SUMMARY

Random error and bias are qualitatively different types of errors that can affect inferences. Random error is a consequence only of chance and is most relevant to clinical trials in a statistical context. Bias is systematic error and is relevant primarily in a clinical context, although it can also arise in statistical estimation. A good experimental design will control both types of error.

When making inferences from hypothesis tests, there are two random errors possible. If the null hypothesis is true but the test rejects, this is a type I error. P-values quantify the type I error assuming the null hypothesis is true. If the alternative hypothesis is true (null hypothesis false) but the test fails to reject, this is a type II error. The power of a trial for a specific alternative is 1 minus the probability of a type II error.

Type I errors can be controlled because they depend on the critical value chosen to reject the null hypothesis. The data analyst has control over this critical value. Type II errors are more difficult to control because they depend on the precision of the estimate or the sample size of the experiment. Therefore the type II error can only be controlled by the design of the study rather than the analysis.

Like type II errors, bias can only be controlled by design rather than analyses. There are many sources of bias in clinical trials including patient selection, postentry exclusion of subjects for reasons related to prognosis, selective loss of data, assessment bias, and improperly defined eligibility, response, or evaluability. Relatively simple methods can reduce or eliminate these and other biases. Bias-reducing methods include randomization, treatment masking, objectively defined endpoints, active rather than passive ascertainment of outcomes, and not permitting post hoc exclusions.

Statistical biases, especially in estimation, occur frequently. They can usually be corrected mathematically or are of small magnitude compared with errors in design and conduct of trials. In a few circumstances biased estimation methods are deliberately utilized in an attempt to reduce the total (mean squared) error.

7.6 QUESTIONS FOR DISCUSSION

1. Which provides stronger evidence against the null hypothesis: a small experiment with $p = 0.05$ or a large experiment with $p = 0.01$?

2. Discuss the difference between what is unknown and what is random in a statistical analysis. Are they the same? Are they modeled the same way?

3. Type I and II error rates are often chosen entirely by convention. Discuss the most appropriate ways that you can think of for setting the probabilities of type I and

II error rates. Apply your reasoning to trials studying disease prevention with a relatively safe intervention compared with disease treatment using a risky therapy.

4. When testing the equivalence of two treatments, what is the null hypothesis? In these so-called equivalence or noninferiority trials, what happens to the type I and II errors? Should the test be one or two sided?

5. One can reconstruct a hypothesis test from confidence intervals. Do confidence intervals suffer from the same strengths and weaknesses as hypothesis tests? Discuss why or why not.

6. Suppose that the treatment difference estimated in a particular randomized trial is exactly equal to the true treatment difference, and that the p-value that results is $p = 0.025$. If the trial is repeated in the same way, what is the chance that the p-value from the new study will be less than 0.025?

7. SA trials frequently show promise of treatment benefit, even when informally compared with standard therapy. However, when the promising new treatments are studied in CTE trials, they often fail to show improvement over conventional therapy. Discuss reasons why this might happen.

8

OBJECTIVES AND OUTCOMES

8.1 INTRODUCTION

Objectives of clinical trials can be described on several levels. Clinical trials are typically used as components of a larger process whose goal is to improve the treatment of a disease. As such, the objective of a trial may be qualitative rather than quantitative. Its contribution or utility to the process is a matter of subjective interpretation and must take into account information external to the experiment. Achieving broad objectives of process depends on the specific findings of the study. For example, suppose that a new treatment is found to be superior to conventional therapy in a trial. The management of disease might not improve unless the new treatment is feasible to apply widely, is cost effective, and produces trial results that are judged to be generalizable. The regulatory process is another example that depends on both actual findings and their subjective interpretation.

The internal objectives of a clinical trial are the research questions or goals of the experiment. It is essential to separate this level from the overall process. An internal objective for a comparative trial might be to determine which therapy has superior efficacy and/or fewer side effects. Achieving internal objectives does not depend on which therapy is better, only on obtaining a valid result. Additional examples of internal objectives include evaluating safety or pharmacokinetics, or establishing equivalence. For example, the objectives of a safety and activity trial can be met whether or not the treatment is judged worthy of continued development.

At a third level, the objectives of a trial are specifically linked to outcomes. Outcomes are quantitative measurements implied or required by the objectives. An *outcome* is determined in each study subject, whereas the specific *objectives* are met by analyzing the aggregate of outcomes. For example, a specific objective might be to estimate the relative hazard of all-cause mortality in two treatment groups based on survival time outcomes. The word "endpoint" is often used synonymously with outcome. I prefer the latter term because the occurrence of a particular outcome may not imply the "end" of follow-up or anything else for the subject. The best outcome to measure depends on the specific objective and the way in which it is quantified.

Suppose that the objective of a trial is to determine if a new surgical procedure "reduces peri-operative morbidity" relative to some standard. The assessment of operative morbidity can be partly subjective, which may be a substantial issue if more than one surgeon participates in the study. At least three aspects of morbidity might need to be defined. The first is a window of time, during which morbid events could be plausibly attributed to the operative procedure. The second is a list of events, diagnoses, or complications to be included. The third is specification of procedures or tests required to establish each event or diagnosis definitively. Using these criteria, "morbid events" can be well-defined and a definitive outcome established for each study subject.

A clinical objective may have more than one way of being quantified or may be described by more than one outcome. For example, "improved survival" or "survival difference" might mean prolonged median survival, higher five-year survival, or a lower death rate in the first year. Each definition might require different methods of assessment, and need not yield the same sample size, overall design, or analysis plan for the trial. In the peri-operative morbidity example, the outcome could be defined in terms of any of several events or a composite. Knowing which outcome and method of quantification to use in particular clinical circumstances is an essential skill for the statistician.

Trials invariably have a single primary objective (and consequent outcome) and possibly many secondary ones. This is largely a bookkeeping device because more than one objective is usually vital. Other times it may not be feasible, or there may not be sufficient resources to answer more than one primary question reliably. We can only actively tailor control over the type II error for one objective. Secondary objectives have type II error properties determined passively by the sample size for the primary objective. Trials with many objectives require a multiplicity of statistical analyses, some of which may be based on subsets of the study cohort. This can increase the possibility of error as discussed in Section 16.7.2.

Each trial setting is unique with respect to the efficiency and practicality of various outcome measurements. Investigators may also need to consider the cost efficiency of various outcomes and choose a feasible one or allocate resources appropriately. A discussion from this perspective is given by Terrin (1990).

8.2 OBJECTIVES

8.2.1 Estimation Is the Most Common Objective

Most clinical objectives translate into a need to *estimate* an important quantity. For example, the objective of the surgery study mentioned above might be to estimate the rate (risk or probability) of peri-operative morbidity. In oncology dose-finding trials, the primary purpose is usually to estimate an optimal dose, such as the maximum tolerated dose (MTD). More generally, we might view these trials as estimating the dose–toxicity function in a clinically relevant region (although the conventional designs employed are not very good at this). The objective of SA studies is usually to estimate response and toxicity probabilities using a fixed dose of drug or a specific treatment.

CTE trials typically estimate treatment differences or risk ratios, and ES trials estimate rates of complications or side effects.

The importance of estimation goes beyond characterizing clinical trial objectives. Estimation also carries strong implications with respect to how trials should be analyzed, described, and interpreted. A key to each is to emphasize the relevant effect estimates.

8.2.2   Selection Can Also Be an Objective

In some circumstances trials are intended primarily to *select* a treatment that satisfies a set of important criteria, as opposed to simply estimating the magnitude of an effect. For example, the treatment with the highest response rate could be selected from among several alternatives in a multi-armed randomized SA trial. The size of each treatment group in such a study might not be sufficient to permit pairwise comparisons between the treatment groups with high power. Such a design is not intended to test hypotheses about pairwise differences in the treatment effects but rather to rank the estimated response rates.

Frequentist sequential trial designs (see Chapter 14) use significance tests to select the treatment that yields the most favorable outcome. When sequential designs terminate early, the overall type I error is controlled, but the treatment effect or difference may be overestimated. These designs select the best treatment but do not provide an unbiased estimate of differences. There are many situations where this is useful. For example, when choosing from among several developing treatments, it may be important to know which one performs best without certainty about the magnitude of the difference.

8.2.3   Objectives Require Various Scales of Measurement

Clinical trial objectives use measurements that fall into one of four numeric categories. The first is *classification*, or determining into which of several categories an outcome fits. In these cases the numerical character of the outcome is nominal, namely it is used for convenience only. There is no meaning ascribed to differences or other mathematical operations between the numbers that label outcomes. In fact the outcomes need not be described numerically. An example of such a measurement is classifying a test result as normal, abnormal, or indeterminate. The results could be labeled 1, 2, or 3, but the numbers carry no intrinsic meaning.

A second type of objective requires *ordering* of outcomes that are measured on a degree, severity, or ordinal scale. An example of an ordinal measurement is severity of side effects or toxicity. In this case we know that category 2 is worse (or better) than category 1, but not how much worse (or better) it is. Similarly category 3 is more extreme, but the difference between 3 and 2 is not necessarily the same as the difference between 2 and 1. Thus the rankings are important, but the differences between values on the ordinal scale have no meaning.

Still other objectives are to *estimate differences* and lead to interval scales of measurement. On interval scales, ordering and differences between values is meaningful. However, there is an arbitrary zero point, so ratios or products of measures have no meaning. An example of an interval scale is pain severity, which can be measured with an arbitrary zero point. A difference in pain on a particular scale is meaningful, but the ratio of two scores is not.

TABLE 8.1   Examples of Different Data Types Frequently Used in Clinical Trials

ID Number	Age	Sex	Toxicity Grade	Age Rank	Sex Code
1	41	M	3	1	0
2	53	F	2	5	1
3	47	F	4	2	1
4	51	M	1	4	0
5	60	F	1	9	1
6	49	M	2	3	0
7	57	M	1	7	0
8	55	M	3	6	0
9	59	F	1	8	1
⋮	⋮	⋮	⋮	⋮	⋮
Scale:	Ratio	Category	Ordinal	Interval	Nominal

Finally the objectives of some studies are to *estimate ratios*. These lead to ratio scales on which sums, differences, ratios, and products are all meaningful. Elapsed time from a clinical landmark, such as diagnosis (time at risk), is an example of a ratio scale. Both differences and ratios of time from diagnosis are meaningful.

Examples of various scales of measurements for hypothetical data concerning age, sex, and treatment toxicity are shown in Table 8.1. Toxicity grade is an ordinal scale, age rank is an interval scale, age is a ratio scale, and sex code is nominal. The outcomes that are appropriate for each type of objective are discussed in the next section.

8.3   OUTCOMES

8.3.1   Mixed Outcomes and Predictors

Measurements taken during a clinical trial can potentially serve one of several purposes. The first purpose is to capture the effect of treatment, that is, as an outcome. A second use is to predict subsequent events or outcomes. There is not necessarily a problem with this duality, unless the roles are confused by design or analysis. Measurements can represent independent domains in which to evaluate treatment effects, such as assessments of side effects (risk) versus efficacy (benefits). They can also serve as markers of disease, triggers for changing therapy, and surrogate outcomes. The investigator must be thoughtful and explicit about how a given measurement will be used. Both the study design and analysis depend on this clarity.

Measurements that are fixed at the time of study entry can only be used as predictors. Examples are patient attributes such as sex and other demographic characteristics. Any measurement that can change during follow-up could, in principle, serve any role mentioned above. An exception is survival, which is a pure outcome. The most appropriate use of a given measurement depends on biology, the setting of the study, the nature of the treatment, and the purposes of the trial. For example, a subject's functional abilities could be the most relevant outcome for a palliative therapy, but may also predict very strongly later function or other outcomes. Thus measures of function could have both roles within the same study.

8.3.2   Criteria for Evaluating Outcomes

There are a handful of general considerations for evaluating or selecting an outcome for a clinical trial (Table 8.2). Most of these characteristics are intuitive and require no additional explanation. The degree to which a proposed outcome satisfies ideal characteristics could be the basis for choosing it, although other considerations such as tradition often play a major role. The outcome employed should be methodologically well-established so that the investigators can expect the results of the trial to be widely understood and accepted.

The role of the outcome(s) employed in a clinical trial is analogous to that of diagnostic criteria for a disease. Proper assessment and classification depend on each, as does clinical relevance. For translational, early, and middle developmental trials, evidence of biological marker activity is usually sufficient for clinical relevance. Biological marker activity is often evidenced by appropriate laboratory measurements on serum or tissue, or imaging studies, but the trial itself may contain marker validation as a secondary objective. For definitive comparative trials, clinical relevance can only be established by suitably defined measures of efficacy. If a relevant clinical efficacy parameter is not used in such trials, the role of the treatment in practice will not be well defined regardless of the outcome.

8.3.3   Prefer "Hard" or Objective Outcomes

"Hard" outcomes are clinical landmarks that rank high with respect to the above-stated characteristics—they are well defined biologically (and in the study protocol), are definitive with respect to the disease process, and require no subjectivity. Thus they are not prone to observer bias. Examples include death, disease relapse or progression, and multiple laboratory measurements. "Soft" outcomes are those that do not rank as

TABLE 8.2   Considerations for Evaluating and Selecting Outcomes

Characteristic	Meaning
Relevant	Clinically important/useful
Quantifiable	Measured or scored on an appropriate scale
Valid	Measures the intended construct or effect
Objective	Interpreted the same by all observers
Reliable	Same effect yields consistent measurements
Sensitive	Responds to small changes in the effect
Specific	Unaffected by extraneous influences
Precise	Has small variability

high with regard to the ideal characteristics. This distinction is not the same as that between definitive and "surrogate" outcomes, discussed below.

An example of why subjectivity is undesirable in clinical trial outcomes is the so-called *Hawthorne effect*, named after experiments at the Hawthorne Plant of the General Electric Company in the 1920s and 1930s (Homans, 1965). These studies tested the effects of working conditions on productivity and demonstrated that even adverse changes could improve productivity. However, it turned out that the research subjects were affected by the knowledge that they were being tested, illustrating that study participants can respond in unexpected ways to support the research hypothesis if they are aware of it. While such effects may be more likely when subjective outcomes are used, they can also influence hard outcomes that depend on changes in behavior.

Some useful and reliable outcome measures fall between the extremes of "hard" and "soft." An example is pathologic classification, which is usually based on expert, experienced, and objective judgment. Such outcomes are likely to be useful in clinical trials and prognostic factor studies because they are valid and reliable. The important underlying issue with outcome measures is how error prone they are. Even a good outcome, such as vital status, can be made unreliable if investigators use poor methods of ascertainment.

8.3.4 Outcomes Can Be Quantitative or Qualitative

Defined outcomes and prospective methods of assessing them are among the characteristics that distinguish rigorous experimental designs from other types of studies. Thus the strength of evidence from a trial depends greatly on these aspects of design. The most important beneficial characteristics of the outcome used in a study are that it must correspond to the scientific objective of the trial, and the method of outcome assessment must be accurate and free of bias. These characteristics are important, not only for subjective outcomes like functional status or symptom severity but also for more objective measures such as survival and recurrence times. These outcomes usually become evident long after the treatment is begun, providing a chance that incomplete follow-up can affect the results.

There are several classes of outcomes that are used in many types of trials. These include continuously varying measurements, dichotomous outcomes, event times, counts, ordered categories, unordered categories, and repeated measures. Measurements with established reliability and validity might be called *measures*. In the next section each of these will be described with specific examples.

8.3.5 Measures Are Useful and Efficient Outcomes

Measurements that can theoretically vary continuously over some range are common and useful types of assessments in clinical trials. Examples include many laboratory values, blood or tissue levels, functional disability scores, or physical dimensions. In a study population these measurements have a distribution, often characterized by a mean or other location parameter, and variance or other dispersion parameter. Consequently these outcomes will be most useful when the primary effect of a treatment is to raise or lower the average measure in a population. Typical statistical tests that can detect differences such as these include the t-test or a nonparametric analogue and analyses of variance (for more than two groups). To control the effect of confounders or prognostic factors on these outcomes, linear regression models might be used, as is often done in analyses of (co-)variance.

8.3.6 Some Outcomes Are Summarized as Counts

Count data also arise frequently in clinical trials. Count data are most common when the unit of observation is a geographic area or an interval of time. Geographic area is only infrequently a source of data in treatment trials, although we might encounter it in disease prevention studies. Counting events during time intervals is quite common in trials in chronic diseases, but the outcomes counted (e.g., survival) usually number 0 or 1. These special types of counts are discussed below.

Some assessments naturally yield counts that can take on values higher than 0 or 1. For example, we might be interested in the cancer prevention effects of an agent on the production of colonic polyps. Between each of several examination periods, investigators might count the number of polyps seen during an endoscopic exam of the colon, and then remove them. During the study period each patient may yield several correlated counts. In the study population counts might be summarized as averages, or as an average intensity (or density) per unit of observation time.

8.3.7 Ordered Categories Are Commonly Used for Severity or Toxicity

Assessments of disease severity or the toxicity of treatments are most naturally summarized as ordered categories. For example, the functional severity of illness might be described as mild, moderate, or severe. The severity of a disease can sometimes be described as an anatomic extent, as in staging systems used widely in cancer. In the case of toxicities from cytotoxic anticancer drugs, five grades are generally acknowledged. Classifying individuals into a specific grade, however, depends on the organ system affected. Cardiac toxicity, for example, ranges from normal rate and rhythm to atrial arrhythmias to ventricular tachycardia. Neurologic toxicity ranges from normal function, to somnolence, to coma. There is no overall scale of toxicity independent of organ system.

When used as a primary outcome or effect of treatment, ordered categories can capture much of the information in, but may be easier and more convenient to apply, than quantitative scores. For example, ordered categories might be useful in assessing outcomes for degree of impairment due to chronic arthritis, functional ability and muscular strength in multiple sclerosis, or functional ability in chronic heart disease. When measures are categorized for simplicity or convenience, investigators should retain the continuous measurements, whenever possible, to facilitate different analyses or definitions in the future.

Summarizing outcomes from ordered categories requires some special methods of analysis. One would not automatically look for linear trends across the categories because the ordering itself may not be linear. For example, the difference between a grade IV and grade V side effect may not imply the same relative change as between grade I and grade II. Consequently care must be exercised in the choice of analytic methods.

8.3.8 Unordered Categories Are Sometimes Used

Outcomes described as unordered categories are uncommon in clinical trials but are occasionally necessary. For example, following bone marrow transplantation for acute leukemia, some patients might develop acute graft versus host disease, others might develop chronic graft versus host disease, and others might remain free of either outcome. These outcomes are not graded versions of one another. In most circumstances unordered categorical outcomes can be reclassified into a series of dichotomies to simplify analyses.

8.3.9 Dichotomies Are Simple Summaries

Some assessments have only two possible values, for example, present or absent. Examples include some imprecise measurements such as shrinkage of a tumor, which might be described as responding or not, and outcomes like infection, which is either present or not. Inaccuracy or difficulty in accurate grading can make a measured value or ordinal assessment into a dichotomous one. In the study population these outcomes can often be summarized as a proportion of "successes" or "failures." Comparing proportions leads to tests such as the chi-square or exact conditional tests. Another useful population summary for proportions is the odds, log-odds, or odds ratio for the outcome. The effect of one or more prognostic factors (or confounders) on this outcome can be modeled using logistic regression.

When summarizing proportions, one should be certain of the correct denominator to use. Problems can arise in two areas. First, there is a tendency to exclude patients from the denominator for clinical reasons, such as failure to complete an assigned course of treatment. The dangers of this are discussed in Chapter 15.

Second, the units of measurement of the denominator must be appropriate. For a proportion, the units of the denominator must be persons, as compared with person-years for a hazard. For example, if we compare the proportions of subjects with myocardial infarction in two cohorts, we might calculate r_1/n_1 and r_2/n_2, where r_1 and r_2 are the number of patients with the event. However, if the follow-up time in the two cohorts is very different, this summary could be misleading. It might be better to use the total follow-up time in each cohort as the denominator. This treats the outcome as a risk rate rather than a proportion.

8.3.10 Event Times May Be Censored

Measurements of the time interval from treatment, diagnosis, or other baseline landmarks to important clinical events such as death (event times) are common and useful outcomes in chronic disease clinical trials. Survival time is an often cited example of a "definitive" outcome because it is usually determined with minimal error. Many other intervals might be of clinical importance, such as time to hospital discharge or time spent on a ventilator. The distinguishing complication of event time measurements, like many longitudinal assessments, is the possibility of censoring. This means that some subjects being followed on the trial may not experience the event of interest by the end of the observation period.

The nature of accruals and event times in a clinical trial with staggered patient entry is shown in Figure 8.1. In this hypothetical study patients are accrued until calendar time T. After T there is a period of additional follow-up lasting to $T + \tau$. Some subjects are observed to have events (denoted by x's) during the study period (e.g., #3 – #6). Others are lost to follow-up during the study, as denoted by the circles (e.g., #1), and may have events after the study period. Still others remain event free

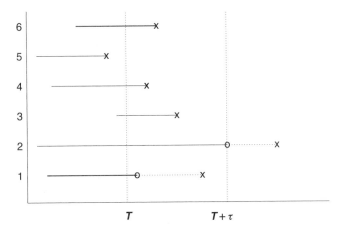

Figure 8.1   Accrual, follow-up, and censoring on an event time study.

at the end of the study, also denoted by circles (e.g., #2). Thus subjects #1 and #2 are censored.

Censoring of event times most often occurs when an individual is followed for a period of time but is not observed to have the event of interest. Thus we know only that the event time was greater than some amount but do not know its exact value. This is often called *right censoring*. Censoring can also occur if we observe the presence of a state or condition but do not know when it began. For example, suppose that we are estimating the distribution of times from seropositivity to clinical AIDS in patients at high risk of HIV. Some members of the cohort will already be seropositive at the start of the observation period. The time to AIDS is censored for these observations because we do not know the precise point of seroconversion. This is often called *left censoring*. Event time data can also be *interval censored*, meaning that individuals can come in and out of observation. Most event time data are right censored only, so the term "censoring" most commonly means "right censoring."

Additionally three types of censoring based on the nature of follow-up have been described. The usual type of censoring in clinical trials is random (or type III) because the staggered entry and losses to follow-up produce unequal times at risk and censoring. Some laboratory experiments place all animals on study at the same time and end after a fixed observation period. This produces type I censoring where all censored times are the same. Alternatively, if the investigator waits until a fixed proportion of the animals have had events, type II censoring is produced. Type I and II censoring are usually not seen in clinical trials.

We do not discard censored or incomplete observations but employ statistical methods to use all the available information about the failure rate contained in the follow-up time (time at risk). For example, if we followed a cohort of patients for 50 person-years of time and observed no deaths, we have learned something useful about the death *rate*, even though no deaths have been observed.

Using the information in censored observation times requires some special statistical procedures. Event time or "survival" distributions (e.g., life tables) might be used to summarize the data. Clinicians often use medians or proportions at a fixed time to summarize these outcomes. One of the most useful summaries is the hazard rate, which can be thought of as a proportion adjusted for follow-up time. The effect of prognostic factors or confounders on hazard rates can often be modeled using survival regression models.

Right censoring can occur administratively, as when the study observation period ends, or throughout a trial if study participants are lost to follow-up. The statistical methods for dealing with these are the same. The common methods that account for right censoring assume that the censoring mechanism is independent of the outcome. This means that there is no information about the outcome in the fact that the event time is censored. If this is not true, a situation called *informative censoring*, then the usual methods of summarizing the data may yield biased estimates. For example, suppose individuals are more likely to be censored just before they experience an event. Treating the censored observations as though they are independent of the event will underestimate the event rate.

8.3.11 Event Time Data Require Two Numerical Values

To capture the information in event times, whether censored or not, it is necessary to record two data elements for each individual. The first is the time at risk (follow-up or exposure time). It is a measure such as number of days, weeks, months, or years of time. The second item needed is an indicator variable that designates whether the event time represents the interval to an event or to a censoring point. The censoring indicator is usually given the value 1 for an event and 0 if the observation is censored. An example is shown in Table 8.3. If both left and right censoring are present, two indicator variables will be required to describe the risk interval fully.

The distribution of event times can be described in several ways. Often, it is summarized as the cumulative probability of remaining event free over time. Because this method is used so commonly in survival applications, the resulting curves are often called "survival curves" even when the outcome is not death. The most commonly employed method for estimating survival curves is the product-limit method (Kaplan and Meier, 1958). In many other situations, a more natural descriptive summary of the data is the overall hazard or event rate. Methods for describing such data are discussed in more detail in Chapter 16.

TABLE 8.3 Example of Censored Event Time Data

ID Number	Exposure Time	Censoring Indicator
1	141	0
2	153	1
3	124	0
4	251	0
5	160	1
6	419	0
7	257	1
8	355	0
⋮	⋮	⋮

8.3.12   Censoring and Lost to Follow-up Are Not the Same

It is easy to confuse the concepts of censoring and lost to follow-up, but important to understand the distinction. Not all censored observations are lost to follow-up. The difference is readily apparent for the commonest type of censoring in a survival study, administrative censoring at closeout. Subjects alive (event free) when the trial closes have censored event times even though they may be very actively followed until the end. They may be returning for regular follow-up, have all required exams, and be readily contacted. Thus they are in no sense lost to follow-up, but censored nonetheless. Censoring created by calendar termination of a study is not likely to be related to the event status of the subjects. Because subjects typically enter a trial over an extended accrual period, this type of censoring may occur for a sizable fraction of the study cohort.

Lost to follow-up means that the event status of a subject cannot be determined, even with an active follow-up effort. This usually occurs much less frequently than administrative censoring. Subjects lost to follow-up during a trial are frequently analyzed as being censored at the time of last contact. This makes the implicit assumption that being lost to follow-up does not carry information about the event; for example, subjects are not lost to follow-up as a consequence of impending death. If there is an association between the risk of being lost to follow-up and the risk of the event, then it is not appropriate to analyze the data under the usual assumption of independent censoring. More generally, events that compete with one another (e.g., different causes of mortality in the same cohort) cannot be analyzed as though they were independent.

An egregious confusion of administrative censoring with loss to follow-up happened in a meta-analysis of interferon treatment for multiple sclerosis (MS) (Filipini et al., 2003a). One study included in the analysis (Jacobs et al., 1996) had substantial administrative censoring because it was terminated before its planned conclusion when results of other similar trials became known. However, the meta-analysis investigators imputed worst-case values for the administratively censored subjects as part of a sensitivity analysis intended for lost to follow-up observations. Because the meta-analysis of interferon included relatively few studies, this method of handling censored data in a large trial may have clouded the results.

Although attention was drawn to this error (Rudick et al., 2003), the meta-analysts stated:

> With respect to lost to follow-up, an authoritative definition is that these are patients who "become unavailable for examinations at some stage during the study" for any reason, including "clinical decisions ... to stop the assigned interventions" (Jüni et al., 2001). Thus, the patients were lost to follow-up (Filipini et al., 2003b).

The exact quote is from the Jüni et al. (2001) discussion of attrition bias, which reads:

> Loss to follow-up refers to patients becoming unavailable for examinations at some stage during the study period because they refuse to participate further (also called drop outs), cannot be contacted, or clinical decisions are made to stop the assigned interventions.

This informal definition is ambiguous in light of my discussion here, and is probably irrelevant for the MS meta-analysis. A clinical decision to stop or change treatment during the trial is invariably based on how the subject is doing. Therefore it likely represents informative censoring and can contribute to the bias that Jüni et al. discuss. Such subjects might be censored when treatment changes but are not necessarily lost to follow-up. To the contrary, their treatment might change because of findings at a follow-up visit.

Administrative censoring is different because it occurs at the end of the study, affects all subjects equally, and is generally uninformative (i.e., will not produce a bias). Statistical methods for censored data were developed explicitly to avoid discarding incomplete observations or imputing arbitrary values. As indicated above, these observations are also not necessarily lost to follow-up. Thus the methods applied in the MS meta-analysis were inappropriate.

One can imagine how problematic it would be to confuse the definitions of censoring and lost to follow-up in a setting such as a prevention trial. In a low-risk population such as that, a large majority of subjects will be administratively censored (event free) at the close of the study. Analytic methods that treat the event-free subjects in a prevention trial as lost to follow-up (and impute worst-case values) rather than censored will produce erroneous results.

8.3.13 Survival and Disease Progression

Survival time is often taken to be a prototypical definitive outcome for clinical trials in serious chronic diseases. There is usually little ambiguity about a subject's vital status, and it is therefore very reliable. Prolonging survival is often a worthy goal. Furthermore treatments that affect survival are very likely to have a fundamental biological action. Survival is a pure outcome—there is no sense in which it predicts anything. As a point of methodology, it is important therefore to understand survival as a clinical trial outcome.

There are a few potential limitations to using survival as an outcome. It may not be the most clinically relevant parameter for some settings. Symptoms or other patient reported outcomes might be more appropriate measures of therapeutic efficacy, especially for diseases that are not life-threatening. Although reliable, survival may be somewhat inaccessible because of the long follow-up that is required. Because of censoring, survival is often estimated with lower precision than other outcomes, leading to larger trials. All these factors should be considered when selecting survival as the primary outcome for a trial.

On balance, survival is seen as a relevant and definitive outcome for studies in many chronic diseases. This is especially true in cancer, where it is something of a gold standard. The regulatory process surrounding anticancer drugs has reinforced the idea that survival outcomes are paramount. However, if one looks carefully at the basis of approval of new anticancer drugs in recent years, it is evident that other outcomes are equally important (Johnson, Williams, and Pazdur, 2003).

Disease progression is an outcome that has properties similar to survival. Progression will usually be a composite outcome and, if ascertained reliably, could be an early and useful signal of therapeutic activity. Problems with using this outcome are that it is often vaguely defined, inconsistently evaluated and interpreted, or based on infrequent or imprecise clinical evaluations. To be useful, disease progression ascertainment should be based on active follow-ups required in the protocol, have relatively short intervals so time bias does not occur, and use reliable diagnostic evaluations. Even when rigorously done, the resolution afforded by the schedule of follow-up evaluations may not allow us to detect small treatment effects on progression reliably.

8.3.14 Composite Outcomes Instead of Censoring

Investigators would prefer to count only events that relate specifically to the condition under study. There is often a firm clinical rationale for censoring events unrelated to the target disease. For example, in a large trial with a new treatment trying to prolong survival after cancer diagnosis, some patients will invariably die of noncancer causes such as cardiovascular disease. It seems appropriate to censor these noncancer events rather than count them as deaths. However, this creates some serious unintended consequences.

When two or more failure processes affect a population (competing risks), we cannot expect to obtain unbiased estimates of risk for a single cause by censoring all other events. If the failure processes are not independent of one another, as is usually the case, events attributable to one cause contain some information about events of other types. This concern is often valid biologically. The reporting mechanisms for events can obscure the connections. For example, a myocardial infarction listed as the primary cause of death may be secondary to an advanced state of some other underlying disease, such as cancer. Therefore the censored cardiovascular events are not independent of the cancer events, creating a potential for bias if independence is assumed. As a further example, death from injury, such as a motor vehicle accident, may be partly a consequence of the patient's psychological state due to the underlying disease.

Rather than assuming independence, the study could count deaths from any cause as a composite outcome. This all-cause or overall mortality is not subject to many biases and has a straightforward interpretation. Consequently it is the preferred mortality outcome for many clinical trials.

8.3.15 Waiting for Good Events Complicates Censoring

In most event time studies the interval of interest is measured from a hopeful clinical landmark, such as treatment, to a bad event, such as disease progression or death. Our perspective on censoring is that events such as disease progression or death could be seen with additional follow-up. In a few studies, however, investigators measure the waiting time to a good event. Short event times are better than long ones and censoring could be a more difficult problem than it is in survival studies.

For example, suppose that patients undergoing bone marrow transplantation for hematologic malignancies are observed to see how long it takes the new bone marrow to "recover" or become functional. The restoration of bone marrow function is a good outcome, but some patients may never recover fully or may die. Death from complications of not having a functioning bone marrow is more likely early rather than late in follow-up. In this circumstance death censors time to recovery, but the censoring paradigm is different than that discussed above. A short event time is good if it terminates with recovery but bad if it terminates with death (censoring). Long event times are not good but may be unlikely to end in death (censoring). This example illustrates the importance of understanding the relationship between the censoring process and the event process.

8.4   SURROGATE OUTCOMES

A surrogate outcome is one that is measured in place of the biologically definitive or clinically most meaningful outcome. Typically a definitive outcome measures clinical benefit, whereas a surrogate outcome tracks the progress or extent of the disease. A good surrogate outcome needs to be convincingly associated with a definitive clinical outcome so that it can be used as a reliable replacement. Investigators choose a surrogate when the definitive outcome is inaccessible due to cost, time, or difficulty of measurement (Herson, 1989). The difficulty in employing surrogate outcomes is more a question of their validity or strength of association with definitive outcomes than trouble designing, executing, or analyzing trials that use them.

Surrogate outcomes are sometimes called surrogate endpoints, surrogate markers, intermediate, or replacement endpoints. The term "surrogate" may be the best overall descriptor and I will use it here. Some authors have distinguished auxiliary from surrogate outcomes (Fleming et al., 1994). Auxiliary outcomes are those used to strengthen the analysis of definitive outcome data when the latter are weak because of a lack of events. Such outcomes can be used statistically to recover some of the information that is missing because of unobserved events. Auxiliary outcomes may be measurements such as biomarkers or other manifestations of the disease process. Also intermediate outcomes can be distinguished from surrogates, particularly in the context of cancer prevention (Freedman and Schatzkin, 1992).

Prentice (1994) offered a rigorous definition of a surrogate outcome as

> a response variable for which a test of the null hypothesis of no relationship to the treatment groups under comparison is also a valid test of the corresponding null hypothesis based on the true endpoint.

An outcome meeting this definition will be a good surrogate for the definitive outcome. A measurement that is merely correlated with outcome will not be a useful surrogate, unless it also reflects the effects of treatment on the definitive outcome. Surrogates may exist for efficacy, but there can be no convincing surrogates for safety.

It is sometimes useful to distinguish types or levels of surrogates. One type of surrogate is a predictor of disease severity, and it may be useful independently of treatment. In other contexts, these might be seen as prognostic factors. They may be partially outcomes and partially predictors. Examples are measures of extent or severity of disease, such as staging in cancer. All prognostic factors are not surrogates in this sense because many do not change with the disease (e.g., sex).

A second type of surrogate is a disease marker that responds to therapy with a frequency and magnitude that relates to the efficacy of the treatment. Blood pressure, cholesterol level, and PSA are examples of this. These surrogates are likely to be useful in middle developmental trials to demonstrate the activity of the treatment and motivate comparative trials with definitive outcomes.

The third type of surrogate is the ideal one—a marker that captures the clinical benefit of the treatment. Such a marker would be useful for establishing comparative efficacy. These kinds of surrogates are the ones we wish we had routinely. They are difficult or impossible to find and require strong validation. From the sketches of these categories, it seems intuitively clear that surrogate outcomes will be most useful in middle developmental trials. Their use in comparative trials creates important concerns and limitations, discussed below.

8.4.1 Surrogate Outcomes Are Disease-Specific

Surrogate outcomes are disease-specific because they depend on the mechanism of action of the treatment under investigation. A universally valid surrogate for a disease probably cannot be found. Some examples of surrogate-definitive outcome pairs are listed in Table 8.4. Trialists are interested in surrogate outcomes like these because of their potential to shorten, simplify, and economize clinical studies. The potential gain is greatest in chronic diseases, where both the opportunity to observe surrogates and the benefit of doing so are high. However, surrogate outcomes are nearly always accompanied by questions about their validity. Trials with surrogate outcomes usually require independent verification, but this does not establish the validity of the surrogate.

Some important characteristics of surrogate outcomes can be inferred from Table 8.4. First, a good surrogate can be measured relatively simply and without invasive procedures. Second, a surrogate that is strongly associated with a definitive outcome will likely be part of, or close to, the causal pathway for the true outcome. In other words, the surrogate should be justified on biologically mechanistic grounds. Cholesterol level is an example of this because it fits into the model of disease progression: high cholesterol $\Rightarrow$ atherosclerosis $\Rightarrow$ myocardial infarction $\Rightarrow$ death. This is in contrast to a surrogate like prostatic specific antigen (PSA), which is a reliable marker of tumor burden but is not in the chain of causation. We might say that cholesterol is a direct, and PSA is an indirect, surrogate. However because of temporal effects, PSA rather than cholesterol may be more strongly associated with a definitive disease state.

Third, we would expect a good surrogate outcome to yield the same inference as the definitive outcome. This implies a strong statistical association, even though the definitive outcome may occur less frequently than the surrogate. Several authors have pointed out that this statistical association is not a sufficient criterion for a surrogate to be useful (e.g., Boissel et al., 1992). Fourth, we would like the surrogate to have a short latency with respect to the natural history of the disease. Finally, a good surrogate should be responsive to the effects of treatment.

Cancer

In testing treatments for cancer prevention, surrogate outcomes are frequently proposed because of the long latency period for the disease. Also we are most interested in applying preventive agents or measures to a population of patients without disease. Even populations at high risk may have only a small fraction of people developing cancer each year. These factors inhibit our ability to observe definitive events such as

TABLE 8.4 Examples of Surrogate Endpoints Frequently Used in Clinical Trials

	Definitive Endpoint	Surrogate Endpoint
HIV infection	AIDS (or death)	Viral load
Cancer	Mortality	Tumor shrinkage
Colon cancer	Disease progression	CEA level
Prostate cancer	Disease progression	PSA level
Cardiovascular disease	hemorrhagic stroke, myocardial infarction	Blood pressure, cholesterol level
Glaucoma	Vision loss	Intraocular pressure

new cases of cancer or deaths attributed to cancer. This situation provides strong motivation for using surrogate outcomes like biomarkers, provided they are valid (Bogoch
and Bogoch, 1994; Kelloff et al., 1994).

In studies of cancer therapy, investigators also find many reasons to be interested in
surrogate outcomes (Ellenberg and Hamilton, 1989). When the interval between treatment and a definitive outcome is long, there is an opportunity for intercurrent events
to confuse the assessment of outcomes. An example of this type of event is when the
patient receives additional active treatment during the follow-up period. Deaths due to
causes other than the disease under investigation can also confuse outcome assessment.

Tumor size reduction (tumor response) is used as definitive outcome and proposed
as a surrogate outcome in cancer clinical trials. In SA trials, measurable tumor response
is often taken as evidence that the therapy is active against the disease. The degree of
such activity determines whether or not the treatment is recommended for continued
testing in comparative trials. In CTE trials, tumor response is sometimes proposed
as a surrogate for improved survival or longer disease free survival. However, the
association between response and definitive event times is weak for most cancers.
Consequently tumor shrinkage (response) should not generally be used as the primary
outcome variable in CTE trials. The occasional exception might be for conditions in
which reduction in tumor size provides a clinically important improvement in quality
of life or reduced risk of complications.

There are situations where evidence from studies employing surrogate outcomes is
considered strong enough to establish convincing safety and efficacy, and consequently
regulatory approval of some treatments for cancer. The clinical context in which this
occurs is at least as important as the nature of the study and outcome measure. Recent
examples include trastuzumab (Herceptin®) for treatment of metastatic breast cancer in
1998 using a surrogate marker in 222 patients, leuprolide (Eligard®) for palliation of
advanced prostate cancer in 2003 using a surrogate marker in 140 patients, and temo-
zolomide (Temodar®) for treatment of recurrent glioblastoma in 1999 using response
rates in 162 patients. There are many other examples, but these illustrate the balance of
circumstance, treatment options, safety, and evidence of benefit inherent in such decisions.

"Cure" or "remission" are stronger types of tumor response that have been used
as surrogate outcomes in some cancer trials. For example, permanent or long-term
tumor shrinkage below the level of clinical detectability might be labeled a remission
or cure as is commonly the case in studies of childhood hematologic malignancies.
There are ample data to support the strong association between this type of outcome
and survival in these diseases. The connection is further supported by evidence that
the failure rate diminishes to near zero or that the survival curve has a plateau in
patients achieving remission, providing long-term survival for them. In other cancers a
disease-free interval of, say, five years after disease disappearance is often labeled as
a "cure," but the failure (recurrence) rate may not be near zero. Thus tumor response
is not a uniformly good surrogate outcome for all types of cancer.

Some cancer biomarkers have been considered reliable enough to serve as surrogate
outcomes. Two well-known ones are prostatic specific antigen (PSA) and carcinoembryonic antigen (CEA), produced by some gastrointestinal malignancies. These markers
are particularly useful for following disease status after treatment, when an elevation
is strongly associated with recurrence. The usefulness of these and other biomarkers
as surrogate outcomes remains to be established.

Cardiovascular Diseases

Studies of cardiovascular diseases present opportunities to use potentially valid surrogate outcomes for definitive outcomes like mortality (Wittes, Lakatos, and Probstfield, 1989). This is possible because we understand the mechanisms leading to many cardiovascular events fairly well and can measure entities in the causal path. For example, elevated blood pressure, serum cholesterol, left ventricular ejection fraction, and coronary artery patency are in the chain of events contributing to myocardial infarction and can be measured quantitatively. Sometimes these surrogates, or risk factors, are used as the primary outcomes in a trial, while in other cases they are secondary.

Because of the strong mechanistic connection between some cardiovascular surrogates and definitive outcomes, they may be used more effectively in trials in this setting than surrogates for cancer outcomes. However, interventions that modulate risk factors do not necessarily change definitive outcomes. If treatment modifies the risk factor through mechanisms unrelated to action on the definitive outcome, we can be misled by a study using the factor as a surrogate outcome.

HIV Infection

In patients with HIV infection, CD4 positive lymphocyte (CD4) count is a widely discussed candidate for a surrogate outcome for clinical AIDS and death. Unfortunately, the available data suggest that CD4 count is not reliable enough to serve as a valid surrogate outcome. See Fleming (1994) for a review of this point. AIDS, by the Centers for Disease Control (CDC) clinical criteria, could be considered a surrogate for death because of the severely compromised immune system that it implies. Other clinically valid measures of immune function such as P-24 antigen levels and plasma HIV viral load have been suggested as possible surrogate outcomes but are unproven. Like SA cancer trials that use tumor response as an outcome despite its poor utility as a definitive outcome, developmental trials in AIDS may be able to use measures of immune function to evaluate the potential benefit of new treatments.

Eye Diseases

In trials studying diseases of the eye, Hillis and Seigel (1989) discuss some possible surrogate outcomes. On example is retinal vein occlusion, which can lead to loss of vision. Hypertensive vascular changes are a precursor to vein occlusion and can be observed noninvasively. However, in an eye affected by vein occlusion, the blood vessel changes may not be observable because of tissue damage. Observations in the opposite eye may be useful as a surrogate for the affected eye. Thus the opposite eye is a surrogate for observing the state of the retinal vessels, which is a possible surrogate for vein occlusion. In this situation the eye least affected by hypertensive vascular changes is likely to be used as a surrogate, leading to a biased underestimate of the relationship between the surrogate and the definitive outcome.

A second example in eye diseases discussed by Hillis and Seigel (1989) is the use of intraocular pressure as a surrogate for long-term visual function in patients with glaucoma. The validity of this surrogate depends on certainty that the elevated pressure is a cause of optic nerve damage in glaucoma patients, that the pressure can be determined reliably with the type of measurements commonly used, and that lower intraocular pressure will result in better vision in the long term. Many recent trials of glaucoma therapy have followed this reasoning, correct or not, and used intraocular pressure as a surrogate outcome.

8.4.2 Surrogate Outcomes Can Make Trials More Efficient

Most clinical trials require an extended period of accrual and observation for each patient after treatment. This is especially true of comparative studies with event time as a primary outcome. Disease prevention studies, where event rates are low and event times are long because the study population is relatively healthy, are even more lengthy than most treatment trials. It can be impractical or very expensive to conduct studies that take such a long time to complete. Good surrogate outcomes can shorten such clinical trials, which explains why they are of particular interest in prevention trials. However, to be useful, a surrogate outcome needs to become manifest relatively early in the course of follow-up.

A simple example will illustrate the potential gain in efficiency using surrogate outcomes. Suppose that we wish to test the benefit of a new antihypertensive agent against standard therapy in a randomized trial. Survival is a definitive outcome and blood pressure is a surrogate. If it were practical and ethical to follow patients long enough to observe overall mortality, such a trial would need to be large. For example, using calculations discussed in detail in Chapter 11, the difference between 95% and 90% overall mortality at five years requires 1162 subjects to detect as statistically significant with 90% power and a two-sided 0.05 α-level test. By contrast, if we use diastolic blood pressure as the outcome, we could detect a reduction of as little as 1/2 of a standard deviation using 170 patients with a trial duration of a few weeks or months. Larger reductions in blood pressure could be detected reliably using fewer patients. This hypothetical example is not meant to equate this benefit in mortality with this degree of blood pressure reduction. The benefit from such a small reduction is probably smaller. It does illustrate the potential difference in the scope of trials using surrogate outcomes.

In some instances, trials using surrogate outcomes may provide a clearer picture of the effects of treatment than those employing the "definitive" outcome. For example, suppose that we are studying the effects of a treatment on the prevention of coronary occlusion in patients with high risk for this event. Clinicians react to these significant and morbid events when they occur by attempting to restore blood flow to the heart muscle using drugs, surgery, or other interventions and often make modifications in other aspects of the patient's treatment. Thus coronary occlusion is an important clinical milestone and can be used as a basis for establishing the efficacy of treatments. A trial that used a "definitive" outcome such as death can present a somewhat confusing picture of treatment efficacy. Some patients will live a long time after the first coronary occlusion, allowing noncardiac complications to intervene. Also the patient may change lifestyle or therapy after a coronary occlusion and there may be comorbidities that influence the course of treatment. It may be difficult to describe or account for the effects of these changes on the natural history of the disease.

A similar situation is not uncommon in cancer, where effective second-line treatments (if they exist) will be chosen to intervene if patients start to do poorly after primary therapy. A trial with survival as a definitive outcome may not yield the correct treatment effect in such a case because the outcome will be confounded by the secondary therapy. A better outcome to evaluate the primary treatment might be time to treatment failure. Overall survival would essentially be an unsatisfactory surrogate for this.

In other cases the ethical acceptability of a trial can be enhanced by using surrogate outcomes. For example, suppose that the definitive outcome becomes apparent only after a long period of follow-up. A rigorous trial design comparing two therapies would likely require control of ancillary treatments administered during the follow-up period. However, restrictions on such treatments may be ethically problematic. By contrast, a valid surrogate outcome that can be measured early in the post-treatment period allows the comparison of interest to proceed and permits physicians to respond with fewer constraints to changes in the patients' clinical status.

8.4.3 Surrogate Outcomes Have Significant Limitations

Although surrogate outcomes are used frequently for developmental trials and are occasionally helpful for comparative studies, they often have serious limitations. Sources of difficulty include determining the validity of the surrogate, coping with missing data, having the eligibility criteria depend on the surrogate measurement, and the fact that trials using these outcomes may be too small to reliably inform us about uncommon but important events. Of these, the validity of the surrogate outcome is of most concern. For a review of surrogate outcomes and their limitations in various diseases, see Fleming and DeMets (1996).

Many surrogates are proposed because they appear to represent the biological state of disease, and observational data suggest that they are convincingly associated with the definitive outcome. It this setting it seems counterintuitive that a treatment effect on the surrogate could not yield clinical benefit. The problem is that treatment effects on the definitive outcomes may not be predicted accurately by treatment effects on the surrogate. This can occur for two reasons. First is the imperfect association between the surrogate and the true outcome, which may not reflect the effects of treatment. Second is the possibility that treatment affects the true outcome through a mechanism that does not involve the surrogate.

Some explanations are suggested by Figure 8.2. In phase Ia, the treatment affects the surrogate, but there is a second pathway by which the definitive outcome is produced by the disease. In phase IIb, the surrogate does indeed lie on a causal pathway for the outcome, but there is a second pathway that could invalidate the treatment effect at b. In phase IIc, the treatment will affect the true outcome, but the surrogate will

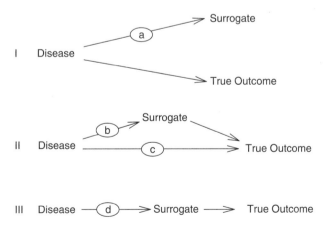

Figure 8.2 Possible relationships between surrogate outcomes, definitive outcomes, and treatment effects.

not capture the effect. Phase IIId appears to be an ideal situation, but difficulties can still arise. The action on the surrogate may produce unanticipated effects on the true outcome. For example, the direction of the effect could be reversed. The time interval between the effect on the surrogate and the final outcome may be so long as to dampen the utility of the surrogate measurement.

Some Particular Problems

A problem in the development of cancer treatments arises when trying to evaluate cytostatic (rather than cytotoxic) drugs. The traditional outcome for SA trials is response, which although a surrogate for benefit, is objective and widely understood. Some investigators have proposed new "clinical benefit" response criteria that could be useful for cytostatic drugs (Rothenberg et al., 1996). (See Verweij, 1996, and Gelber, 1996, for discussions.) However, new outcomes will require validation with respect to disease status, treatment effects, and existing outcomes before being used to evaluate, and possibly discard, new drugs.

In cardiovascular diseases there are sobering accounts of the use of surrogate outcomes in clinical trials. One example is the Cardiac Arrhythmia Suppression Trial (CAST), which was a randomized, placebo-controlled, double masked treatment trial with three drug arms (Cardiac Arrhythmia Suppression Trial Investigators, 1989). The drugs employed, encainide and flecainide, appeared to reduce arrhythmias, and therefore were promising treatments for reducing sudden death and total mortality. After a planned interim analysis, CAST was stopped early in two arms because of a convincing *increase* in sudden deaths on the treatments. Control of arrhythmia, although seemingly justified biologically, is not a valid surrogate outcome for mortality.

A randomized placebo-controlled trial of milrinone, a phosphodiesterase inhibitor used as a positive inotropic agent in chronic heart failure, in 1088 patients also showed benefit on surrogate outcomes (measures of hemodynamic action) with increased long-term morbidity and mortality (Packer et al., 1991). A similar failure of surrogate outcomes was seen in a randomized trial of the vasodilator flosequinan, compared with captopril in 209 patients with moderate to severe chronic heart failure (Cowley et al., 1994). Flosequinan had similar long-term efficacy and mortality compared with captopril, but a higher incidence of adverse events.

Another example that does not speak well for the use of surrogate outcomes is the randomized study of fluoride treatment on fractures in 202 women with osteoporosis (Riggs et al., 1990). Although bone mass was increased with fluoride therapy, the number of nonvertebral fractures was higher. Measures such as bone mass and bone mineral density are not valid surrogates for the definitive clinical outcomes.

A surrogate outcome in a specific disease may be useful for some purposes but not others. For example, surrogate outcomes in drug development may be appropriate for verifying the action of a drug under new manufacturing procedures or for a new formulation of an existing drug. Their use for a new drug in the same class as an existing drug may be questionable and their use for testing a new class of drugs may be inappropriate. For illnesses in which patients have a short life expectancy (e.g., advanced cancer and AIDS), it may be worthwhile to use treatments that improve surrogate outcomes, at least until efficacy can be verified definitively in earlier stages of disease.

8.5   SOME SPECIAL ENDPOINTS

8.5.1   Repeated Measurements Are not Common in Clinical Trials

It is occasionally necessary in clinical trials to summarize outcomes repeatedly over some interval of time. An example is when comparing control of blood pressure on two or more experimental antihypertensive therapies. A trial could be designed using measurements before and after treatment to assess treatment differences. Because of the variability in blood pressure recordings, a better strategy might be to monitor blood pressure at frequent intervals, and/or with the patient in different positions, during the treatment period.

A major difficulty after implementing such a scheme is using all of the information collected for analysis. It can be difficult to implement statistical techniques that simultaneously use all the longitudinal information collected, are robust to the inevitable missing data, and are flexible enough to permit valid inferences concerning a variety of questions. Repeated measures analyses of variance and other methods of analyzing longitudinal data can be used. However, their complexity and higher cost and administrative needs for trial designs that require repeated assessments make these outcomes relatively uncommon.

8.5.2   Patient Reported Outcomes

Patient reported outcomes (PRO) are obvious sources of information on which to base therapeutic evaluations. Practitioners are accustomed to this because symptoms can be reliable indicators of disease and its mitigation. PRO are widely used in clinical trials where they can yield varying impressions of their reliability. Pain is an example of an important PRO that has been extensively used in trials.

The essential issues surrounding the use of PRO are the same as for any outcome as listed in Table 8.2. Not all PRO perform well with regard to the needed characteristics, especially validity and objectivity, leading many investigators to disfavor them. But a given patient reported outcome can be studied and standardized the same as any laboratory measurement. The nature of communicating with sick and frightened patients may cause these outcomes to seem less rigorous than those reported by observers or machines, but PRO may often be more relevant.

Quality of life assessments are a special category of PRO that are broadly used in clinical trials as secondary outcomes, especially in chronic diseases like cancer. These outcomes attempt to capture psychosocial features of the patient's condition, symptoms of the disease that may be distressing, and/or functional (physical) status. See Testa and Simonson (1996) for a review, and Chow and Ki (1994, 1996) for a discussion of statistical issues. It is easy to imagine situations where the underlying disease is equally well controlled by either of two treatments, as measured by objective clinical criteria, but the quality of life might be superior with one treatment. This could be a consequence of the severity and/or nature of the side effects, duration of treatment, or long-term consequences of the therapy. Thus there is also a strong rationale for using quality of life assessments as primary outcomes in the appropriate setting.

Quality of life assessments are often made by summarizing items from a questionnaire (instrument) using a numerical score. Individual responses or assessments on the quality of life instrument might be summed, for example, for an overall score. In other circumstances quality of life assessments can be used as "utility coefficients" to weight or adjust other outcomes. This has been suggested in cancer investigations to create "quality-adjusted" survival times. Usually these quality-adjusted analyses require subjective judgments on the part of investigators in assigning utilities to various outcomes or conditions of the patient. For example, one could discount survival after chemotherapy for cancer by subtracting time with severe side effects from the overall survival time. This corresponds to a utility of zero for time spent in that condition. Because of the subjectivity required, quality-adjusted measurements often are not considered as reliable or rigorous as more objective measurements by many investigators.

A perpetual issue with quality of life is construct validity: Does the measurement instrument actually measure "quality of life," or can it even be defined? Overall quality of life is usually explicitly defined as a sum of subscale measurements, and there is no guarantee that a given treatment will have an impact on it. Furthermore, how to measure or define quality of life appears to depend on the disease, and to a certain extent, on the treatment. Finally longitudinally measured quality of life scores are subject to informative missingness because sick or dead subjects do not yield measurements. Although there are approaches to fixing these problems, it has proved challenging to use quality of life as a primary outcome in clinical trials.

There is an additional problem with quality of life and other PRO in longitudinal studies of progressive chronic diseases. As a disease advances, patients may become sicker and less willing to report subjective outcomes that require a task such as an extensive questionnaire. Thus the missing data in studies that use PRO may be informative because of progressive disease or survival. This creates an informative censoring bias that can make the average quality of life in a study cohort appear to improve as the subjects feel worse. This effect can limit the usefulness of such outcomes.

8.6   SUMMARY

The primary statistical objective of a clinical trial is usually to estimate some clinically important quantity, such as a treatment effect or difference. There may be numerous secondary objectives employing different outcomes, but the properties of the trial can usually only be controlled for one (primary) objective. Some trials are not designed to provide unbiased estimates of treatment effects, but instead select the "best" treatment from among several being tested.

Different scales of measurement may be required depending on the outcome being used. Scales of measurement include nominal or categorical, ordered, interval (for estimating differences), and ratio. Aside from scales of measurements, outcomes can be classified as measures, categorical, counts, and event times. Event times are widely used outcomes for clinical trials, especially in chronic diseases like cancer, cardiovascular disease, and AIDS. Special statistical methods are required to cope with censored event times, which are frequently present in clinical trial data.

Trial methodologists continue to examine and debate the merits of surrogate outcomes, essentially on a study-by-study basis. Plausible surrogate outcomes have been proposed or used in cancer (e.g., tumor size), AIDS (e.g., viral load), cardiovascular disease (e.g., blood pressure), and other disease trials. These types of outcomes can potentially shorten and increase the efficiency of trials. However, they may be imprecisely associated with definitive outcomes such as survival and can, therefore, yield misleading results.

8.7 QUESTIONS FOR DISCUSSION

1. Rank the different scales of measurement in order of their efficiency in using the available information. Discuss the pros and cons of each.

2. Repeated measurements on the same study subjects increase precision. Because repeated measurements are correlated with one another, adding new study subjects may increase precision more. Discuss the merits and weaknesses of each approach.

3. Surrogate or intermediate outcomes are frequently used in prevention trials. Discuss reasons why they may be more appropriate in studies of disease prevention than in studies of treatment.

4. Read and comment on the study by the Chronic Granulomatous Disease Cooperative Study Group (*N. Engl. J. Med.* 324: 509–516, 1991).

9

TRANSLATIONAL CLINICAL TRIALS

9.1 INTRODUCTION

A large number of therapeutic ideas originate in the laboratory, but only a small fraction of them survive and progress to developmental clinical trials. The transition from lab to clinic, and frequently back again, is guided by small targeted studies rather than large clinical trials. These small early human experiments, which I will call translational trials, form a bridge between therapeutic ideas developed in the laboratory and clinical development, and are among the most common types of clinical trials performed.

The methodology of translational clinical trials has not been fully formulated in the literature. Ideas presented in this chapter regarding methodology are not universal, but are based on my previous discussion of the topic (Piantadosi, 2005). Despite vagaries in definitions and methods, there is intense interest among researchers and sponsors regarding translational research because nearly every new therapy depends on a successful transition from laboratory to clinic. Such trials are especially relevant in fields where therapeutic development is highly active, like oncology. But translational research is equally important in settings where new therapeutic ideas occur less frequently and consequently may be even more valuable. Translational studies can sometimes be performed as part of later developmental trials, provided that the subjects and the questions are compatible with it.

The traditional dividing line between basic research and clinical development is often said to be the "phase I" study. Some investigators define phase I on this basis, as the earliest human trials of a new therapy. In fact the interface between laboratory research and clinical development is formed by translational clinical trials. Failure to recognize this has probably impeded both the formalization of translational trials as well as the rigor of dose-finding methodology. Dose-finding and related experimental designs are discussed in Chapter 10.

I have indicated elsewhere in this book that clinical trials can be viewed as empirical devices that test or select treatments, or as biological experiments that reveal truths of nature. Trials are both simultaneously. How best to view a trial depends on the nature and setting of the question and the study design. Many randomized trials or large-scale studies fit better in the empirical end of the spectrum, while translational trials are typical of the biological end.

Another major impediment to defining translational trials may be skepticism among methodologists that such small studies, as these typically are, can provide useful information. I believe that this issue is a critical one, and I will devote considerable discussion to it. Translational trials do form a coherent class, whose utility does not depend on size or disease focus. Consider, for example, the relatively large amount, and important nature, of information that dose-finding/ranging trials provide with minimal sample size. Similarity of purpose and design for dose-finding for various cytotoxic agents contribute to the specification of the phase I design. From that, the more general notion of dose optimization can be derived (with much methodologic development left to be done). It is similarity of purpose and structure that codifies trial designs into a class. Translational trials can be sufficiently stereotyped across disciplines to justify their being defined as a class.

A great deal of translational research is done by the pharmaceutical and biotechnology industry. Because much of their process is hidden from view, it is not clear how efficient they are at doing it. It is also not clear if academic models of translational research are the same as private industry's. It is evident that translational research is successful if revenue streams are large enough and properly directed.

9.1.1 Translational Setting and Outcomes

After developing a new therapeutic concept in the laboratory, there is a need for human studies that assess the effects of treatment on an in vivo biological target, to learn if the therapy is worthy of additional development. The treatment is changeable and is likely to be modified based on the results of these translational trials. In contrast, clinical trials from dose-finding/ranging forward use either definitive outcomes or surrogate outcomes that bear on the question of clinical benefit (or risk). Examples of these outcomes are toxicity, safety, and survival. During these developmental trials the treatment under investigation is not changeable as a consequence of results (except for dose or schedule).

The outcome used for a translational trial is a target or biological marker, which itself may require additional validation as part of the study. This outcome is not a surrogate outcome as defined in Chapter 8 because it is not used to make inferences about clinical benefit. The outcome used for a translational trial might anticipate later questions of clinical benefit, but it has more urgent purposes. The activity of the treatment on the target defines the next experimental steps to be taken. The outcome has to be measurable with reasonable precision following treatment, and the absence of a positive change can be taken as reliable evidence of inactivity of the treatment.

A carefully selected biological outcome is the linchpin of the experiment because it provides definitive mechanistic evidence of effect—an irrefutable signal—within the working paradigm of disease and treatment. The signal has to reveal promising changes in direction and magnitude for proof of principle, and to support further clinical development. Good biological signals might be mediated through, or the direct result of, a change in levels of a protein or gene expression, or the activity of some enzyme. Such a finding by itself would not prove clinical benefit but might lay a foundation for continued development.

As an example consider a new drug for the secondary prevention of cancer. The translational goal might require that our new drug reduce the presence of specific biomarkers in the tissue (biopsy specimens) or increase other biomarkers. The absence of these effects would necessitate discarding or modifying the drug, or reconsidering the appropriate target. Weakly positive biomarker changes might suggest additional preclinical improvements. Neither outcome could reliably establish clinical efficacy.

9.1.2   Character and Definition

The basic characteristics of a translational trial are the following:

- The trial is predicated on promising preclinical evidence, which creates a need to evaluate the new treatment in human subjects.
- The treatment and/or its algorithm can be changed.
- The treatment and method of target evaluation are fully specified in a written protocol.
- The evaluation relies on one or more biological targets that provide definitive evidence of mechanistic effect within the working paradigm of disease and treatment.
- The outcome is measurable with small uncertainty relative to the effect size. Target validation is sometimes also an objective, in which case imprecision in the outcome measurement may also represent a failure.
- Large effects on the target are sought.
- There is a clear definition of "lack of effect" or failure to demonstrate effects on the target.
- The protocol specifies the next experimental step to be taken for any possible outcome of the trial. Thus the trial will be informative with respect to a future experiment.
- The study is sized to provide reliable information to guide additional experiments but not necessarily to yield strong statistical evidence.
- The study is typically undertaken by a small group of investigators with limited resources.

These considerations lead to the following formal definition of a translational trial:

A clinical trial where the primary outcome: (1) is a biological measurement (target) derived from a well established paradigm of disease, and (2) represents an irrefutable signal regarding the intended therapeutic effect. The design and purposes of the trial are to guide further experiments in the laboratory or clinic, inform treatment modifications, and validate the target, but not necessarily to provide reliable evidence regarding clinical outcomes. (Piantadosi, 2005)

Translational trials imply a circularity between the clinic and the laboratory, with continued experimentation as the primary immediate objective. We are likely to use more than one biological outcome in such studies, and the trial may also provide evidence about the utility of each. Many therapeutic ideas will prove useless or not feasible during this cycle. For others, the lab–clinic iteration will beget the familiar linear development of a new therapy, perhaps after numerous false starts.

9.1.3   Small Does Not Mean Translational

My attempt to add rigor to translational trial design is not a general justification for small clinical trials. Although translational studies often use small sample sizes, small trials are not necessarily translational. The important distinctions are the setting, purpose, nature of the outcome, and how the studies are designed, conducted, and interpreted. For example, small comparative trials using clinical outcomes are seldom justified, almost always being the product of resource limitations. Translational clinical trials need no special justification. Investigators are convinced that they learn important information from such studies and the long run evidence supports the truth of this view.

Having said this, it is important to be aware of the deficiencies of these designs. The most obvious limitations are the lack of proven clinical validity for the outcome and the poor statistical properties of estimates. Not so obvious is the fact that the translational trial paradigm partially confounds three things: (1) the correctness of the disease paradigm, (2) the selection of a relevant biological outcome, and (3) the action of the therapy. Errors in any of these components can masquerade as either a positive or negative treatment effect. However, the correctness of the disease paradigm and the selection of a relevant outcome can be strongly supported using evidence from earlier studies.

9.2   INFORMATION FROM TRANSLATIONAL TRIALS

Introducing a new treatment into humans is usually characterized by high uncertainty about both beneficial effects and risks. In this circumstance even a fairly small experiment can substantially increase information for the purposes of guiding subsequent studies. The reduction of uncertainty (gain in information) can be quantified. The combination of biological knowledge with information from formal observation is a powerful tool for designing early experiments.

9.2.1   Parameter Uncertainty versus Outcome Uncertainty

The usual statistical perspective associates uncertainty with unknown parameters related to the problem at hand. Examples of parameters that are of interest in clinical trials include relative risks of failure or death, probability of benefit, and mean plasma level of a drug. The appropriate size of many experiments may be driven by the precision with which such parameters must be determined.

There is a second source or level of uncertainty that is a consequence of different possible outcomes. Information regarding the parameter(s) may not greatly reduce this overall outcome uncertainty. For example, we could be very sure that the probability, p, of one of two outcomes following treatment A is 0.5. Despite knowledge of the parameter there is still much uncertainty associated with the use of treatment A because either outcome is equally likely. In contrast, if we were sure that treatment *A* produces one outcome with probability 0.95 the overall uncertainty would be lower. Hence even if we know probability parameters with high precision, outcome (overall) uncertainty remains. Furthermore outcome uncertainty depends on the value of the parameter but not as much on the precision of the parameter.

Entropy (defined below) measures overall uncertainty. Specifically it formalizes and quantifies the idea of *information*. This is a useful concept for translational trials because they guide early development more from the perspective of overall uncertainty than from precision of estimation. There is no proof of this assertion—only the general behavior of translational investigators. Later developmental trials are appropriately focused on precision of estimation, as discussed elsewhere in this book.

9.2.2   Entropy

Suppose that a new treatment can yield one of five possible outcomes on a target biological marker: (1) strong benefit, (2) weak benefit, (3) no change, (4) weak harm, or (5) strong harm. The frequency of outcomes in each category is a response from a multinomial distribution. The information in a random variable can be measured by the entropy (Shannon, 1948). The entropy of a discrete random variable is

$$H = - \sum_{i=1}^{m} p_i \log p_i, \qquad (9.1)$$

where there are *m* possible states, each with probability p_i, and $\Sigma p_i = 1$. Gain in information is measured by change in entropy. If the probabilities change by virtue of acquiring information (q_i), the gain is

$$\Delta H = \sum_{i=1}^{m} q_i \log q_i - \sum_{i=1}^{m} p_i \log p_i. \qquad (9.2)$$

After our translational trial, suppose we take the observed frequencies of outcomes as the "true" q_i. We then have a basis for estimating the empirical entropy, or information apparently gained from the trial, which I will denote by ΔH^*.

Many results yield the same amount of information because the ordering of the q_i is immaterial. Large sample sizes will not produce proportionate gains in information because the proportions in the outcome categories will tend to stabilize after modest sample sizes. More data will increase the precision in the estimated frequencies but may not decrease entropy.

These properties seem to reflect investigator behavior with respect to translational trials. Many clinical outcomes are equally informative about the treatment effect, though not equally promising. Large studies are seen as unnecessary to accomplish the immediate goals. Definitive clinical inferences are neither made nor needed from such trials, but the information gathered can be used to guide subsequent experiments. I suspect that investigators are instinctively reacting to reduced uncertainty, or gain in information (as opposed to statistical precision), when they design and assess translational trials.

9.2.3   Empirical Entropy Is Biased

Equation (9.2) was applied as though the q_i from the trial were the true frequencies. However, small studies may tend to overrepresent some responses, so equation (9.2) overestimates the gain in information from the study. The expected value of H^*, using the observed data from n subjects, where $q_i = r_i/n$, is

$$E\{H^*\} = -E\left\{\sum_{i=1}^{m} \frac{r_i}{n} \log \frac{r_i}{n}\right\}$$

$$= -\sum_{i=1}^{m} \frac{1}{n}\left(E\{r_i \log r_i\} - E\{r_i\}\log n\right).$$

The responses are multinomial and the number in a given category is binomial. Using the definition of expectation for the binomial distribution, we have

$$E\{H^*\} = -\sum_{i=1}^{m}\left(\frac{1}{n}\sum_{k=0}^{n} k \log k \binom{n}{k} p_i^k (1-p_i)^{n-k} - p_i \log n\right), \qquad (9.3)$$

where the p_i are the true classification probabilities. In other words, $E\{H^*\} \neq H = -\Sigma p_i \log p_i$ from equation (9.3). $E\{H^*\}$ is a function of sample size, the number of categories, and the true p_i.

   The bias in H^* or ΔH^* is the difference between equations (9.3) and (9.1),

$$b\{H^*\} = \sum_{i=1}^{m} p_i \log_2 p_i - \sum_{i=1}^{m}\left(\frac{1}{n}\sum_{k=0}^{n} k \log k \binom{n}{k} p_i^k (1-p_i)^{n-k} - p_i \log n\right). \quad (9.4)$$

The bias calculated from equation (9.4) is illustrated in Figure 9.1. The bias is positive, indicating that for small sample sizes, the empirical entropy overestimates the information gained from a study. Some bias is present even for large sample sizes but is reduced substantially for sample sizes above 15.

9.2.4   Variance

The variance of H^* can be calculated from

$$\text{var}\{H^*\} = E\{H^{*2}\} - E\{H^*\}^2.$$

The first term yields

$$E\{H^{*2}\} = \sum_{i=1}^{m}\sum_{j=1}^{m} E\left\{\left(\frac{r_i}{n}\log\frac{r_i}{n}\right)\left(\frac{r_j}{n}\log\frac{r_j}{n}\right)\right\}.$$

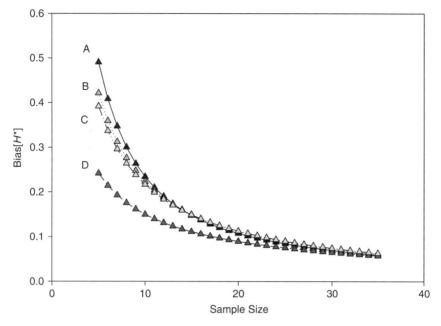

Figure 9.1 Bias versus sample size for empirically calculated entropy. The true probabilities used to generate each curve are $A = \{0.20, 0.20, 0.20, 0.20, 0.20\}$; $B = \{0.46, 0.26, 0.15, 0.08, 0.05\}$, $C = \{0.01, 0.15, 0.80, 0.03, 0.01\}$, and $D = \{0.50, 0.30, 0.10, 0.05, 0.05\}$.

The outcomes r_i and r_j are not independent. The expectation inside the sum can be calculated using $E\{XY\} = E_X\{XE_Y\{Y|X\}\}$. For $m \geq 3$,

$$E\left\{\left(\frac{r_i}{n}\log\frac{r_i}{n}\right)\left(\frac{r_j}{n}\log\frac{r_j}{n}\right)\right\} = \sum_{k=0}^{n}\frac{k}{n}\log\frac{k}{n}$$

$$\times\left(\sum_{r=0}^{n-k}\frac{r}{n}\log\frac{r}{n}\binom{n-k}{r}\left(\frac{p_j}{1-p_i}\right)^r\left(1-\frac{p_j}{1-p_i}\right)^{n-k-r}\right)\times\binom{n}{k}p_i^k(1-p_i)^{n-k},$$

$$\text{for } i \neq j,$$

$$\sum_{k=0}^{n}\left(\frac{k}{n}\log\frac{k}{n}\right)^2\binom{n}{k}p_i^k(1-p_i)^{n-k} \qquad \text{for } i = j.$$

Therefore

$$\text{var}\{H^*\} = \sum_{i=1}^{m}\sum_{\substack{j=1\\j\neq i}}^{m}\sum_{k=0}^{n}\frac{k}{n}\log\frac{k}{n} \tag{9.5}$$

$$\times\left(\sum_{r=0}^{n-k}\frac{r}{n}\log\frac{r}{n}\binom{n-k}{r}\left(\frac{p_j}{1-p_i}\right)^r\left(1-\frac{p_j}{1-p_i}\right)^{n-k-r}\right)$$

$$\times \binom{n}{k} p_i^k (1 - p_i)^{n-k} + \sum_{i=1}^{m} \left(\frac{k}{n} \log \frac{k}{n} \right)^2 \binom{n}{k} p_i^k (1 - p_i)^{n-k}$$

$$- \left[\sum_{i=1}^{m} \left(\frac{1}{n} \sum_{k=0}^{n} k \log k \binom{n}{k} p_i^k (1 - p_i)^{n-k} - p_i \log n \right) \right]^2 .$$

For $m = 2$ (binomial), the conditional expectation simplifies so that

$$E \left\{ \left(\frac{r_i}{n} \log \frac{r_i}{n} \right) \left(\frac{r_j}{n} \log \frac{r_j}{n} \right) \right\} =$$

$$\sum_{k=0}^{n} \left(\frac{k}{n} \log \frac{k}{n} \right) \left(\frac{n-k}{n} \log \frac{n-k}{n} \right) \binom{n}{k} p_i^k (1 - p_i)^{n-k} \text{ for } i \neq j,$$

$$\sum_{k=0}^{n} \left(\frac{k}{n} \log \frac{k}{n} \right)^2 \binom{n}{k} p_i^k (1 - p_i)^{n-k} \quad \text{for } i = j.$$

and equation 9.5 becomes

$$\text{var}_2 \{ H^* \} \tag{9.6}$$

$$= \sum_{i=1}^{m} \sum_{\substack{j=1 \\ j \neq i}}^{m} \sum_{k=0}^{n} \left(\frac{k}{n} \log \frac{k}{n} \right) \left(\frac{n-k}{n} \log \frac{n-k}{n} \right) \binom{n}{k} p_i^k (1 - p_i)^{n-k}$$

$$+ \sum_{i=1}^{m} \left(\frac{k}{n} \log \frac{k}{n} \right)^2 \binom{n}{k} p_i^k (1 - p_i)^{n-k}$$

$$- \left[\sum_{i=1}^{m} \left(\frac{1}{n} \sum_{k=0}^{n} k \log k \binom{n}{k} p_i^k (1 - p_i)^{n-k} - p_i \log n \right) \right]^2 .$$

9.2.5 Sample Size for Translational Trials

The relationship between var$\{H^*\}$, bias$\{H^*\}$, and sample size can be used to help select an appropriate sample size for a translational trial. The relationship is plotted in Figure 9.2 for several representative true multinomial response vectors and sample sizes ranging from 5 to 35. Figures 9.3 and 9.4 show the same for multinomial responses of lengths 3 and 2, respectively. Similar behavior results in most circumstances. Small sample sizes leave large bias and high variance in the empirical entropy. Modest increases in sample size reduce the bias and variance substantially. Large increases in sample size reduce the bias and variance to negligible levels, but most of the benefit can be achieved by sample sizes around 10 to 20.

It makes sense to balance bias and gain in information. We generally cannot afford to eliminate bias completely, and will be satisfied to get a useful reduction in uncertainty. A specific hypothetical response vector should be studied quantitatively, but in

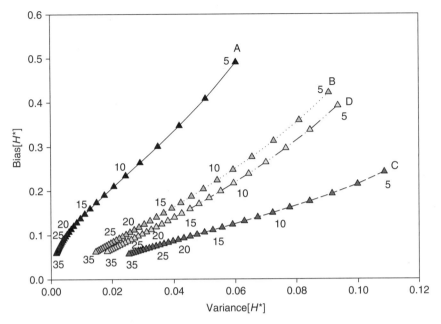

Figure 9.2 Relationship of bias, variance, and sample size for empirically calculated entropy. The true probabilities used to generate each curve are $A = \{0.2, 0.2, 0.2, 0.2, 0.2\}$; $B = \{0.46, 0.26, 0.15, 0.08, 0.05\}$, $C = \{0.01, 0.15, 0.80, 0.03, 0.01\}$, and $D = \{0.50, 0.30, 0.10, 0.05, 0.05\}$.

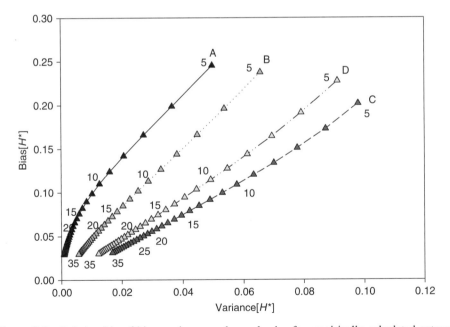

Figure 9.3 Relationship of bias, variance, and sample size for empirically calculated entropy. The true probabilities used to generate each curve are $A = \{0.33, 0.34, 0.33\}$, $B = \{0.15, 0.35, 0.50\}$, $C = \{0.20, 0.75, 0.05\}$, and $D = \{0.10, 0.25, 0.65\}$.

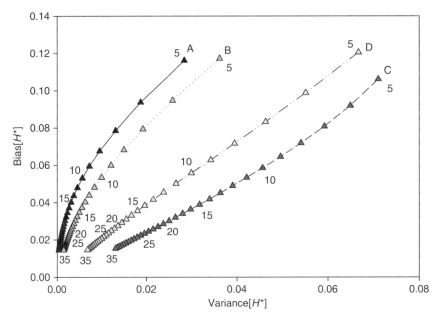

Figure 9.4   Relationship of bias, variance, and sample size for empirically calculated entropy. The true probabilities used to generate each curve are $A = \{0.5, 0.5\}$; $B = \{0.1, 0.9\}$, $C = \{0.25, 0.75\}$, and $D = \{0.6, 0.4\}$.

many circumstances relatively small sample sizes will provide adequate information for guiding the next experiments.

Example 5   Say we plan a translational trial with a dichotomous outcome (success or failure) and assume maximal uncertainty with regard to the outcome (50% chance of success). This circumstance corresponds to Figure 9.4, curve A. Substantial reductions in the variance and bias of the empirical entropy can be attained with a sample size of 10 to 15 subjects. Under these assumptions that sample size will provide a large fraction of the available information.

Example 6   Suppose that a translational trial is planned with an outcome classified into one of five ordinal categories, and the probabilities of the outcomes are expected to be $\{0.46, 0.26, 0.15, 0.08, 0.08\}$. Then Figure 9.2, curve B applies. A large fraction of the available information will be gained with 15 subjects.

9.3   SUMMARY

Translational studies represent the interface between therapeutic ideas emerging from the laboratory and clinical development. A translational clinical trial is a study where the primary outcome (1) is a biological measurement (target) derived from a well-established paradigm of disease and (2) represents an irrefutable signal regarding the intended therapeutic effect. The design and purposes of the trial are to guide further experiments in the laboratory or clinic, inform treatment modifications, and validate the target, but not necessarily to provide reliable evidence regarding clinical outcomes. Typically these studies are small compared to developmental trials and are often performed by resource limited investigators.

Translational trials are done in a setting where there is substantial uncertainty regarding the biological effects of the treatment. Hence a relatively small trial can provide a critical gain in information to guide subsequent studies. The gain in information can be formalized using the concept of entropy. Small studies yield a biased overestimate of information gained (entropy) that can be shown to be a function of the sample size and outcome probabilities. The sample size of a translational trial can therefore be determined based on relative reductions in the bias and variance of estimated entropy.

9.4   QUESTIONS FOR DISCUSSION

1. The mean squared error (MSE) is *variance* $+$ *bias*^{2}. Plot this function of the empirical entropy versus sample size for a few cases of your choosing. What does it suggest regarding sample size?

2. Suppose that reducing the variance of H^* is more important than reducing the bias (and vice versa). Suggest an appropriate modification of the sample size approach presented in this chapter.

10

DOSE-FINDING DESIGNS

10.1 INTRODUCTION

This chapter discusses a broad class of clinical trial designs in which the essential purpose is to determine an optimal biological dose (OBD) of drug. The optimum, described further in Section 10.2.3, is often derived from the relationship between dose and safety, and is established early in development. Dose-finding (DF) studies are a type of treatment mechanism (TM) trial, and focus on the OBD or schedule of drug, estimating pharmacokinetic parameters that relate to dose and safety, and the frequency of side effects. One might say that for the purposes of early development, drugs are "devices" and their mechanisms (e.g., absorption, biodistribution, elimination, and side effects) need to be explored. These studies are not necessarily the first introduction of an agent into humans, as is often stated, but are the earliest trials to employ clinical outcomes, and they form a basis for subsequent therapeutic development.

Our perspective on dose-finding has been heavily influenced, if not damaged, by phase I oncology studies, which are a subset of DF trials. I say this because the dose versus safety concerns for cytotoxic agents are rather narrow, artificial, and simpler than general dose optimization questions. But many clinical investigators have difficulty breaking the "phase I" paradigm and terminology despite its limitations for more general questions. Because early drug trial designs in oncology are stereotypical, relatively highly evolved from years of experience with cytotoxic agents, and reflect important design principles, they will be discussed here in detail. Many new treatments, even in oncology, do not fit the paradigm of cytotoxic drug development. Examples are many molecular targeted cancer drugs, vaccines and other preventives, analgesics, antibiotics, and antihypertensives. I will try to indicate how the ideas that derive from cytotoxic drug dose finding need to be modified or extended to be more broadly useful.

Dose-finding studies are conducted with sequentially rising doses in successive cohorts and terminate when certain predefined clinical outcomes are observed. Because of their sequential nature these designs tend to yield a biased underestimate of the targeted dose. One dose-finding design that is not as subject to this bias is the continual reassessment method (CRM). This design, with suitable modifications, is the method of choice for dose escalations of cytotoxic drugs in oncology, for reasons provided below. However, the CRM has not been as widely employed by clinical investigators as it should be.

DF designs raise important ethics issues. One difficulty is that the principal purpose of such trials is often to cause side effects. Patient benefit is a secondary goal, but it is often mistaken to be primary. Cancer patients who participate in these trials are often terminally ill, which has implications for obtaining valid informed consent.

These concerns are offset by the fact that patients participate in these trials with new agents only when there are no other therapeutic options. Informed consent documents are crafted attentively, and patients usually consider their decisions carefully with family members. Perhaps altruism is an important motivation, but there is also the occasional patient who benefits tangibly. It is often claimed that this occurs less than 5% of the time in oncology studies.

A recent publication from NIH indicates that, over the last decade, oncology dose-finding trials may have been associated with a higher degree of benefit than was the case historically (Horstmann et al., 2005). Higher response rates (up to 17%) appear to be the result of studies that also include an agent known to be effective. Classic single agent trials still yield response rates under 5%. The safety of these trials in the recent decade is similar to that seen historically with serious toxicities occurring in about 14% of subjects. Deaths due to toxic events occur with a frequency of 0.5%. Despite their focus on safety, these trials remain an appropriate option for patients with many types of cancer.

10.2 PRINCIPLES

This chapter is constructed mainly around a conceptual relationship between drug dose and a safety response. I will refer to the safety outcome either as response or toxicity, usually measured on a probability scale. This discussion could also pertain to responses other than probability (e.g., measured effects on a target) and efficacy outcomes. I will usually refer to dose as though it were literally a drug, say, on a milligram basis. But dose is also a more general concept that might refer to amount, number, intensity, or exposure, depending on the therapy. Further it is not essential to assume that dose is measured on a linear scale.

10.2.1 What Does "Phase I" Mean?

Investigating the relationship between dose and safety is variably named as "phase I," dose escalation, dose-ranging, or dose-finding. These terms are not equivalent, and I distinguish between them in Table 10.1. Phase I clinical trials, to use the term properly, are dose-finding or dose-ranging studies for cytotoxic agents. This concept is often overgeneralized to say that phase I trials are the first studies in which a new drug is administered to human subjects. A similar mistake labels as phase I any early developmental design, such as a translational trial. I will employ this term to refer narrowly to the dose–safety relationship in cytotoxic drug trials. In common discussions, the term "phase I" has become so diffuse as to be nearly useless.

TABLE 10.1   Characteristics of Early Developmental Trials

Term	Meaning
TM	Early developmental trial that investigates mechanism of treatment effect. Pharmacokinetics is an example of mechanism in dose–safety trials.
Phase I	Imprecise term for dose ranging designs in oncology, especially for cytotoxic drugs. Such studies are a subset of dose-ranging designs.
Dose escalation	Design or component of a design that specifies methods for increases in dose for subsequent subjects.
Dose-ranging	Design that tests some or all of a prespecified set of doses, meaning fixed design points.
Dose-finding	Design that titrates dose to a prespecified optimum based on biological or clinical considerations, such as frequency of side effects or degree of efficacy.

The primary purposes of classic phase I studies are to investigate toxicity and organ systems involved, establish an optimal biological dose, estimate pharmacokinetics, and assess tolerability and feasibility of the treatment. Secondary purposes are to assess evidence for efficacy, investigate the relation between pharmacokinetics and pharmacodynamics of the drug, and targeting (Ward, 1995). Not all these goals can be met completely in any phase I trial. In part this is because the number of patients treated is usually small. However, well-conducted phase I studies can achieve substantial progress toward each of these goals and provide reasonable therapeutic alternatives for some patients (ASCO, 1997). Phase I trials are not synonymous with dose–response studies, but they have many characteristics in common. For a discussion of dose–response designs, see Ruberg (1995a, b) or Wong and Lachenbruch (1996). Some statistical and ethical issues are discussed by Ratain et al. (1993).

I will use the term *dose-ranging* to refer to designs that specify in advance the design points, or doses, to be employed in the investigation. The investigator specifies the doses to try and some simple decision rules for moving from level to level based on the outcomes in small patient cohorts. Some criteria based on safety are also given for ending the dose jumps and selecting one of them for subsequent studies. Even when such a design uses quantitative decision rules for escalating, de-escalating, or terminating, it is not a true dose optimization because of its poor statistical characteristics discussed below. This is the most common type of design employed for addressing early developmental dose questions for both drugs and biologicals. There is no standard or universally employed dose-ranging design for clinical trials. A frequently used approach is discussed below (see also Storer, 1989).

I will use the term *dose-finding* for designs that attempt to locate the optimum biological dose based on prespecified criteria. Dose-finding designs iteratively employ doses that are themselves determined as intermediate outcomes of the experiment, and they are evaluated on a continuum rather than from a small set. By this definition, only a few dose-finding designs exist. The difference is that dose-ranging operationally tries a small set of doses that are specified in advance, whereas dose-finding employs an algorithm that queries a large or infinite number of doses. Thus only the latter can truly optimize or titrate dose to a specific outcome criterion.

10.2.2   Distinguish Dose–Safety from Dose–Efficacy

The relationship between dose and safety should be addressed early in drug development. It is important to understand pharmacokinetic and pharmacodynamic characteristics of a new agent early because safety will be largely a consequence of absorption, distribution, metabolism, and elimination of the drug. It may also be true that the dose–safety relationship is more important than the dose–efficacy one, meaning efficacy is irrelevant if the dose (or treatment) is unsafe. Both risk and benefit are functions of dose, but it is more efficient and logical to assess safety before investigating efficacy.

Sometimes it becomes clear later in development that higher doses can (and should) be safely employed. Revisiting dose–safety questions a second time in late development is inefficient and expensive. Studies with efficacy outcomes usually need to be large. Dose–efficacy trials should be performed only when the extra time and resources needed are likely to yield essential information. Such studies are logically the same as those that compare different treatments, and the discussion elsewhere in this book also applies to them.

Vaccines may be a potential exception to this rule. The relationship between dose (or schedule) and efficacy of vaccines may need to be routinely investigated early in development. Preventive agents are usually quite safe, at least compared to drugs. Because they are administered to individuals at risk, but free of disease, the dose versus efficacy question may be as imperative as dose versus safety. For vaccines and other preventives, safety questions may persist late in development because of an interest in uncommon events. Many other putative preventive agents, such as vitamins or trace elements, are likely to be safe even when given in slight excess. Hence efficacy trials for these agents may not need to address the question of dose, but they wind up being large and complex for other reasons.

In early development it occasionally happens that a very broad range of doses appears to be safe, but information regarding efficacy is not yet available. Efficacy information may be lacking because relatively few subjects have been treated. In this situation there may be little basis to select a dose for upcoming safety and activity trials. One strategy is to choose a dose based primarily on pharmacokinetic studies. Efficacy could be a consequence of factors such as peak concentration or time above a threshold, for example. Another possibility is to conduct SA trials at more than one dose, perhaps using a randomized design, and choose the dose that appears to have highest activity. This is somewhat inefficient, but developmentally it may be a good option.

10.2.3   Dose Optimality Is a Design Definition

Before a clinical investigator can design dose finding trials, he or she must have some notion of what characterizes an optimum dose. For example, the best dose might be the one with the highest therapeutic index (maximal separation between benefit and risk). In other circumstances the best dose might be one that maximizes benefit, provided that risk is below some prespecified threshold. I refer to the broad concept of an ideal dose as *optimality,* specifically the optimal biological dose (OBD) or just optimal dose. Apart from the exact milligrams or concentration to employ, the OBD needs to be characterized conceptually as part of the study design. The OBD also depends on the clinical circumstances and purposes of treatment.

OBD Examples

The optimal dose of a new analgesic might be the lowest dose that completely relieves mild to moderate pain in 90% of recipients. This is a type of "minimum effective dose" (MED). In contrast, the optimal dose of an new antibiotic to treat serious infections and discourage resistance might be the highest dose that causes major side effects in less than 5% of patients. This is a type of "maximum nontoxic dose" (MND). For cytotoxic drugs to shrink tumors, it has been thought historically that more is better, leading investigators to push doses as high as can be tolerated. We might choose the dose that yields serious (even life-threatening) but reversible toxicity in no more than 30% of patients. This is a "maximum tolerated dose" (MTD).

The optimal dose for development of a molecular targeted agent might be the dose that suppresses 99% of the target activity in at least 90% of patients. Until definitive clinical data are obtained, this might be viewed as a most likely to succeed dose (MLS). For drug combinations or adjuvants, the optimal joint dose might require a substantially more complex definition and correspondingly more intricate design, depending for example on interactions or synergy. For specific agents, there is no guarantee that any of these optima actually exist or have acceptable safety profiles. Whatever the clinical circumstances, the nature of the OBD must be defined explicitly or dose finding will be ambiguous at best.

10.2.4 The General Dose-Finding Problem Is Unsolved

Characterizing the OBD is a critical conceptual step that will suggest specific experimental design features for dose-finding. The narrow question of finding the MTD for cytotoxic drugs is facilitated by several lucky features. Many of the concepts in this chapter are reflective of that specific setting. First, the relationship between dose and toxicity is essentially biologically predetermined to be a nondecreasing function that saturates at high doses; that is, the chance of toxicity is 100% if the dose increases sufficiently. The curve labeled *A* in Figure 10.1 shows this qualitative behavior. Second, preclinical studies usually provide reliable guidance regarding the specific side effects and the quantitative dose–toxicity relationship, at least for a starting dose. Third, cytotoxic drugs are often members of a broad class about which much is known pharmacologically. Thus important properties of the dose–toxicity function are known in advance (i.e., its shape), and the problem can be reduced to determining the location and steepness of the shaped curve on the dose axis.

Experimental designs for the general dose optimization question (i.e., a drug with arbitrary properties) have not been developed. For general dose optimization the experiment must provide information about the shape, steepness, and location of the dose response function. Of these characteristics, shape is probably the most difficult to determine empirically. Curves *B*, *C*, and *D* in Figure 10.1 demonstrate general dose–response possibilities. Curve *B* shows a response that does not completely saturate, at least at low doses. In such a case the optimum might be on the shoulder of the curve.

Curve *C* shows a maximal response with declining response at higher doses. This behavior seems unlikely for pharmacologic agents, but it might be seen with vaccines or biologicals. Curve *D* shows minimal activity at low doses with essentially a threshold, after which response increases rapidly. These simple examples show how difficult dose-finding might be in the absence of knowledge about the qualitative behavior of the dose–response function. The practical implications of the general dose-finding problem

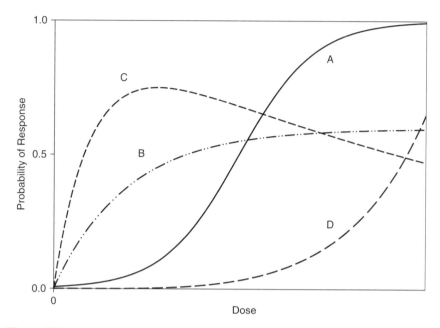

Figure 10.1   Some example dose–response curves. The dose axis is in arbitrary units.

are that a wider range of doses need to be tested and the optimal dose may depend on subtle features of the dose–response function. It is always possible that early trials will reveal a previously unknown property of the drug that strongly influences the definition of optimal dose. A flexible design for dose-finding would have to accommodate such changes.

Because trialists have not developed methods to approach the general dose optimization problem, the remainder of this chapter will emphasize methods used to study dose–safety derived from cytotoxic drug development. Many of these concepts are useful broadly. I will return to discuss some aspects of the general dose optimization question in Section 10.5.

10.2.5   Unavoidable Subjectivity

Any design used for early developmental questions requires a number of subjective judgments. In classic dose-ranging studies, subjectivity enters in (1) the design points chosen to study, including the starting dose, (2) the between patient or cohort escalation/de-escalation rules, (3) the assessment and attribution of side effects critical to connecting dose and safety, (4) decisions to employ within-patient dose increases (although the data from higher doses in the same subject is often disregarded in escalation algorithms), (5) reacting to unexpected side effects, and (6) the size of the cohort treated at each dose. Subjectivity is absolutely unavoidable early in development, but it is often denied or overlooked when investigators design trials based on tradition or dogma.

Some dose-finding designs, such as the CRM, encapsulate subjectivity explicitly. This means that some of the formal components of the design that require subjective assessment are made explicit and self-contained. A standard implementation of the

CRM is a Bayesian method and uses a prior probability distribution for unknown dose–toxicity model parameters. Because of the overt role of subjectivity in the CRM, the method is sometimes criticized, when in fact the same characteristics are present in other designs where the subjectivity is more diffuse and obscure. The CRM is discussed in detail below.

10.2.6   Sample Size Is an Outcome of Dose-Finding Studies

Usually it is not possible to specify the exact size of a dose-finding clinical trial in advance. Except in bioassay or other animal experiments, the size of the dose-finding trial depends on the outcomes observed. If the starting dose has been chosen conservatively, the trial will require more patients than if the early doses are near the MTD. Because the early doses are usually chosen conservatively, most dose-ranging trials are less efficient (larger) than they could be if more were known about the MTD.

Regardless of the specific design used for a dose-ranging trial, sample size, power, and the type I error rate are not the usual currency of concern regarding the properties of the study. Instead, such studies are dominated by clinical concerns relating to pharmacokinetic parameters, properties of the drug, and choosing a dose to use in later developmental trials. In the last few years investigators have been paying more attention to the quantitative properties of dose-ranging designs, particularly how well they inform us about the true dose–toxicity or dose–response function for the agent.

For DF trials sample size is not a major statistical design concern. Convincing evidence regarding the relationship between dose and safety can often be obtained after studying only a few patients. But the principal reason that sample size is not a major issue is that it is an outcome of a sequential study rather than a design characteristic. However, small sample size is not a rationale for conducting the entire DF study informally.

10.2.7   Idealized Dose-Finding Design

The ideal design for a DF study in humans might resemble a bioassay experiment. The investigator would employ a range of doses, $D_1 \ldots D_k$, bracketing the anticipated optimal one. Trial participants would be randomly assigned to one of the dose levels, with each level tested in n patients. At the completion of the experiment, the probabilities of response (or toxicity) would be calculated as $p_1 = r_1/n_1, \ldots, p_k = r_k/n_k$, where r_i is the number of responses at the ith dose. Then the dose response curve could be modeled with an appropriate mathematical form and fit to the observed probabilities. Model fitting facilitates interpolation because optima may not reside at one of the design points used. Estimating the MTD would not depend strongly on the specific dose–response model used.

To implement this design, one would have only to choose an appropriate sample size, mathematical form for the dose–response model, and range of doses. Designs such as this are used extensively in quantitative bioassay (Govindarajulu, 2001). Because experimental subjects would be randomly assigned to doses, there would be no confounding of dose with time, as would occur if doses were tested in increasing or decreasing order. One could even use permuted dose blocks of size k (or a multiple of k) containing one of each dose level.

One of the advantages of using such an ideal design is the ability to estimate reliably and without bias the dose of drug associated with a specific probability of response. For example, suppose that investigators are interested in using a dose of drug that yields a 50% chance of having serious but reversible side effects. One could estimate this from the dose–toxicity (dose–response) function after it is fitted to the data described above. Assume that the probability of toxicity as a function of dose, d, is $f(d; \boldsymbol{\theta})$, where $\boldsymbol{\theta}$ is a vector of parameters that characterize the dose–toxicity function, and the target dose is d_0, such that $0.5 = f(d_0; \widehat{\boldsymbol{\theta}})$, where $\widehat{\boldsymbol{\theta}}$ is an estimate of $\boldsymbol{\theta}$ obtained by curve fitting, maximum likelihood, Bayesian methods, or some other appropriate method. In other words,

$$d_0 = f^{-1}(0.5; \widehat{\boldsymbol{\theta}}),$$

where $f^{-1}(p; \boldsymbol{\theta})$ is the inverse function of $f(d; \boldsymbol{\theta})$ and p is the probability of response. For suitable choices of $f(d; \boldsymbol{\theta})$, the inverse will be simple, and an ideal design will assure a good estimate of $\boldsymbol{\theta}$.

For example, for a logistic dose–response function

$$f(d; \boldsymbol{\theta}) = \frac{1}{1 + e^{-\lambda(d-\mu)}},$$

where $\boldsymbol{\theta}' = \{\lambda, \mu\}$. In this parameterization, μ is the dose associated with a response probability of $\frac{1}{2}$. The inverse function is

$$f^{-1}(p; \lambda, \mu) = \mu + \frac{1}{\lambda} \log\left(\frac{p}{1-p}\right).$$

Another widely used mathematical model for the relationship between dose and response is the probit model

$$f(d; \boldsymbol{\theta}) = \int_{-\infty}^{d} \frac{1}{\sqrt{2\pi}} e^{-(d-\mu)/\sigma} \, dx,$$

where $\boldsymbol{\theta}' = \{\mu, \sigma\}$. This model is complicated by the integral form for f. Nevertheless, the inverse function can be found using approximations or numerical methods.

10.3   FIBONACCI AND RELATED DOSE-RANGING

The ideal bioassay type design is not feasible in human studies because ethics considerations require us to treat patients at lower doses before administering higher doses. Also we require designs that minimize the number of patients treated at both low, ineffective doses and at excessively high, toxic doses. Dose-ranging designs strike a compromise among the various competing needs of ethics, safety, simplicity, and dose optimization.

There are four components to the typical dose ranging design: (1) selection of a starting dose, (2) specification of the dose increments (design points) and cohort sizes, (3) definition of dose limiting toxicities, and (4) decision rules for escalation and de-escalation. It is worth noting, again, that the choices for these design features are substantially subjective. A starting dose is selected based on preclinical data and experience with similar drugs or the same drug in different patient populations. One method for choosing a starting dose is based on one-tenth of the dose that causes 10% mortality in rodents on a mg/kg basis—the so-called LD_{10}.

10.3.1 Some Historical Designs

A classic design based on this method was proposed by Dixon and Mood (1948). Their exact method is not used routinely, however. In most dose escalation designs, investigators do not fit the probit (cumulative normal) dose–response model as originally proposed, and patients are often treated in groups of three rather than individually before deciding to change the dose. This grouped design was suggested by Wetherill (1963) and a widely used modification by Storer (1989). Bayesian methods have been suggested (Gatsonis and Greenhouse, 1992), as have those that use graded responses (Gordon and Wilson, 1992). Dose increments of this type may not be suitable for studies of vaccines and biologicals, where logarithmic dose increments may be needed.

Another design potentially useful for phase I studies is the stochastic approximation method (Robbins and Monro, 1951). As with other approaches a set of fixed dose levels is chosen, and groups of patients can be used. If we denote the ith dose level by D_i and the ith response by Y_i, we then choose

$$D_{i+1} = D_i - a_i[Y_i - \theta],$$

where θ is the target probability of toxicity (or response) and a_i is a sequence of positive numbers converging to 0. For example, we might choose the a_i such that $a_i = c/i$, where c is the inverse of the slope of the dose–toxicity curve. It is clear that when the response is close to θ, the dose level does not change (this is why the a_i are chosen to be decreasing in absolute value). The recommended dose (MTD) is the last design point. Although it has some desirable properties, this design has not been widely used.

10.3.2 Typical Design

A common recent technique for specifying the doses to be tried is the "modified Fibonacci" scheme. Fibonacci (Leonardo of Pisa) was a thirteenth-century Italian number theorist, for whom the sequence of numbers, 1, 1, 2, 3, 5, 8, 13, 21, ... was named, where each term is the sum of the preceding two terms. The ratio of successive terms approaches $(\sqrt{5} - 1)/2 = 0.61803\ldots$ – the so-called golden ratio, which has some optimal properties with regard to mathematical searches (Wilde, 1964). An example of how this scheme might be used to construct dose levels in a dose-ranging trial is shown in Table 10.2. The ideal doses from the Fibonacci sequence are usually modified to increase more slowly.

In cytotoxic drug development, a dose-limiting toxicity (DLT) is a serious or life-threatening side effect. Death is also dose limiting. However, most DLTs are reversible side effects. The exact definition of a DLT depends on the clinical setting and the nature of the therapy. For non–life-threatening diseases and safe therapies, tolerance for side effects might be quite low. Headache, gastrointestinal side effects, and skin rash, for example, might be substantial dose limitations for many drugs to treat self-limited conditions.

Rather than treating a fixed number of patients at every dose, as in bioassay, dose-ranging studies are adaptive. This means that the decision to use a next higher or lower dose depends on the results observed at the current dose, according to the prespecified rules. If unacceptable side effects or toxicity are seen, higher doses will not be used. Conversely, if no DLTs or side effects are seen, the next higher dose will be tested.

**TABLE 10.2   Fibonacci Dose Escalation Scheme
(with Modification) for DF Trials**

Step	Ideal Dose	Actual Dose	Percent Increment
1	D	D	—
2	$2 \times D$	$2 \times D$	100
3	$3 \times D$	$3.3 \times D$	67
4	$5 \times D$	$5 \times D$	50
5	$8 \times D$	$7 \times D$	40
6	$13 \times D$	$9 \times D$	29
7	$21 \times D$	$12 \times D$	33
8	$34 \times D$	$16 \times D$	33

Note: D represents the starting dose employed and is selected
based on preclinical information.

**TABLE 10.3   Simulated Phase I Study Using
Modified Fibonacci Dose Escalation Scheme**

Step	Dose Used	Patients Treated	Number of Responses
1	100	3	0
2	200	3	0
3	330	3	0
4	500	3	0
5	700	3	0
6	900	3	1
7	700	3	1
⋮	⋮	⋮	⋮

Note: The MTD is operationally defined as the 700 unit dose.

Example 7   An example of a dose-ranging trial using Fibonacci escalations, cohorts
of size 3, and simple decision rules is shown in Table 10.3.

10.3.3   Operating Characteristics Can Be Calculated

The principal weakness of the Fibonacci dose escalation and related designs can be
seen in their operating characteristic (OC), which is the expected performance of the
design under specified conditions. A good design should have a high probability of
terminating at or near the optimal dose, such as the true MTD, regardless of the design
points chosen for study. The cumulative probability of stopping before the true MTD
should be low, and the probability of escalating beyond the MTD should be small. A
poor design will have a high chance of stopping at a dose other than the true MTD
(bias), at a highly variable stopping point, or both.

We can already anticipate that dose-ranging designs will have some difficulties with
bias. The first reason is because of their sequential nature, Second, and perhaps more

important, when investigators choose the doses to test, it is highly unlikely that they will include the unknown true MTD as a design point. The study is required to stop on one of the design points.

Aside from that heuristic argument, the OC can be explicitly calculated for designs and decision rules like those put forward in Section 10.3.2. The OC for more complex designs can always be determined by simulation. From the decision rules we can calculate the conditional probability of passing from the current dose to the next higher dose, and the unconditional probability of passing any given dose. Such calculations will yield the *chance of stopping at or before a given dose*, the operating characteristic.

Denote the binomial probability of observing k responses (toxicities) out of n subjects as $b(k; n, p)$, where the probability of response is p. Then define

$$B(a, c; n, p) = \sum_{i=a}^{c} b(i; n, p)$$

to be the cumulative binomial mass function between a and c inclusive and $0 \le a \le c \le n$. B represents the probability of having between a and c responses out of n subjects.

Suppose that the true probability of response at the ith dose is p_i. The ith cohort size is n_i, and the observed number of responses (toxicities) is r_i. The decision rule to escalate is if $r_i \le u_i$ and to de-escalate if $r_i \ge d_i$. If $u_i < r_i < d_i$, then m_i additional subjects will be placed on the same dose. After m_i subjects are added to the cohort, the decision rule becomes: escalate if $r'_i \le u'_i$ and de-escalate if $r'_i \ge d'_i$.

The conditional probability of escalating past the ith dose is

$$P_i = B(0, u_i; n_i, p_i) + \sum_{j=u_i+1}^{d_i-1} b(j; n_i, p_i)B(0, u'_i - j; m_i, p_i).$$

In other words, either the ith dose is passed when it is first visited (first term), or after the cohort is expanded (second term). The unconditional probability of passing the ith dose, Q_i, depends on what happens at lower doses,

$$Q_i = \prod_{k=1}^{i} \left[B(0, u_k; n_k, p_k) + \sum_{j=u_k+1}^{d_k-1} b(j; n_k, p_k)B(0, u'_k - j; m_k, p_k) \right].$$

It is worth noting that the OC for these designs does not depend on the dose(s) in natural units, only on the underlying probability of toxicity at the dose(s) selected (design points). A good OC would demonstrate a high chance of terminating at the optimal dose, independent of the underlying probabilities of the design points. As will be seen below, designs based on this method have a generally poor OC. A consequence of the dependency on the p_i is that different investigators using different design points for the same drug and question may conduct trials with very different properties. I have heard some CRM designs criticized from the perspective of "lack of repeatability" when, in fact, the traditional designs are deficient in this regard.

Written in generality, the OC is complicated, but usually all cohorts are the same size and employ exactly the same decision rules. For example, cohorts are often taken to be size 3. The dose is escalated if 0 toxicities are observed and de-escalated if 2 or more are seen. If 1 toxicity is seen, 3 additional patients are placed on the same dose. Then, if 2 out of 6 at that dose have toxicity, the trial is stopped. Fewer than 2 out of 6 increases the dose to the next level, and more than 2 out of 6 reduces the dose to a previous level. If 2 out of 3 toxicities are seen in the original cohort, the dose level is reduced.

The operating characteristics of this design can be calculated from the general formulas. Using the cohort sizes and numerical rules in the previous paragraph, the conditional probability of escalating past the ith dose, given that the ith dose has been reached, is

$$P_i = b(0; 3, p_i) + b(1; 3, p_i)b(0; 3, p_i),$$

where p_i is the true probability of toxicity at the ith dose. The ith dose is passed when there are no toxicities in the first 3 subjects, or when 1 toxicity occurs but there are no additional ones in the expanded cohort. The unconditional probability of passing the ith dose must account for the probability of passing all earlier doses, meaning it is the product

$$Q_i = \prod_{k=1}^{i} \left[b(0; 3, p_k) + b(1; 3, p_k)b(0; 3, p_k) \right].$$

The OC is $1 - Q_i$.

Numerical Example

Returning to the simple numerical case given above, the OC can be illustrated under several scenarios. To do so, it is necessary to hypothesize true values for $p_1, p_2, \ldots, p_k$. Again, the values of the doses are irrelevant—only the response probabilities at the selected doses matter. True probabilities of response for two cases are shown in Figure 10.2, in the order they will be employed in the escalation. For set A, the doses are conservative; that is, they yield a chance of toxicity that increases significantly for later doses. For set B, the probabilities increase steadily. Both of these are plausible based on the way doses are set out in advance of these experiments.

Assume that the optimal dose is defined as that which yields a 30% chance of toxicity. Then, for the conservative set of probabilities/doses (curve A), the trial should terminate between the 7th and 8th dose with a high probability. For the second set (curve B), the trial should terminate near the 5th dose with a high probability.

The calculated OCs for these cases are shown in Figure 10.3. For set A, the expected stopping point is the third dose, and the design has an 85% chance of stopping before the 5th dose (curve A, Figure 10.3). When such a set of toxicity probabilities is encountered, this design will most likely yield a biased estimate of the MTD. For set B, the expected stopping point is between the 6th and 7th dose, and the design has a nearly 90% chance of stopping before the 8th dose (curve B, Figure 10.3). Aside from the bias, in both cases the designs are not likely to produce repeatable values for the MTD. This performance is typical for dose-ranging designs when they are pressed into dose-finding service.

10.3.4 Modifications, Strengths, and Weaknesses

Over the years a large number of modifications have been proposed to the basic up-and-down Fibonacci type design sketched above. These modifications are a testament

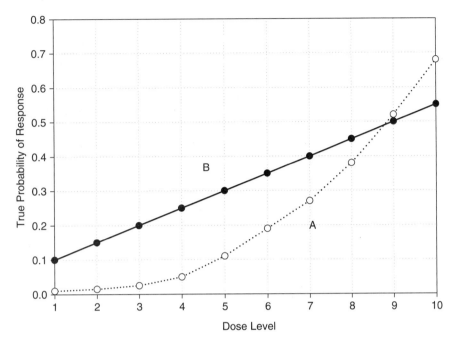

Figure 10.2 Hypothetical dose–response probabilities.

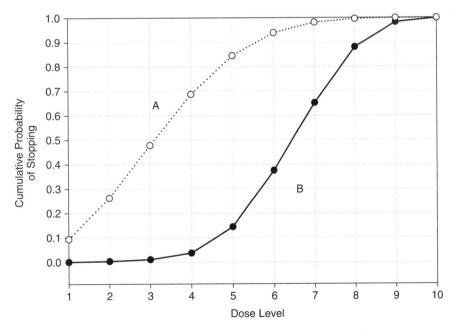

Figure 10.3 Operating characteristics for designs *A* and *B*.

to the flexibility and subjective nature of the design, and the fact that dose-ranging studies have been primarily the domain of clinical researchers rather than statistical methodologists.

Although certain limitations of these designs are evident, they offer some important advantages. The decision points and definitions are defined in advance and do not require elaborate statistical considerations. This is an important advantage in the clinic. The sample size, typically under 20 patients, is small so that the study can be completed quickly. Finally the design is familiar to many drug developers, allowing them to focus on clinical endpoints that are important when investigating new drugs, vaccines, or biologicals.

The principal advantage of dose-ranging designs is simplicity. They are easy to construct and carry out, and special methods of analysis are not needed. Furthermore, when modifications are necessary, either during a particular trial or when designing a new study, the changes can often be made by investigators based on their intuition. Statistical input is not required, and many studies are designed and conducted by example. When clinical protocols using such designs are peer reviewed, it is simple to see the doses to be employed and the plans of the investigators.

The principal weaknesses of these designs are the variability and bias in the estimate of the optimal dose. A design that converges quickly to the wrong answer is not useful. Because of their poor performance as discussed below, these designs should be used only for dose-ranging and not for dose-finding. An additional disadvantage is the illusion of objectivity. Investigators are always required to make subjective judgments about the trial, as discussed earlier. Simple designs used by rote may falsely appear objective.

In recent years various modifications and improvements on the basic Fibonacci dose-ranging design have been proposed. These are generally designed to increase efficiency, particularly use fewer subjects at low doses where the chance of clinical benefit is lower. A popular example is the accelerated titration method (Simon et al., 1997), which essentially escalates based on treatment of a single patient at low doses until some toxicity is seen. Such designs are appropriate in many circumstances, especially in oncology drug development, but also retain some of the fundamental shortcomings of the simple Fibonacci design.

10.4 DESIGNS USED FOR DOSE-FINDING

10.4.1 Mathematical Models Facilitate Inferences

The relationship between dose and response can be described by a mathematical function, model, or curve. This quantitative relationship is the subject of our investigation, and it may be helpful or necessary to make some reasonable biological assumptions about it. A flexible mathematical shape with qualitatively correct behavior (model) could be assumed, for example. A model that approximates the true state of nature is a powerful descriptive and analytic tool. Provided the model is accurate and flexible enough, our conclusions will depend minimally on its mathematical details.

Early DF trial designs in humans also employed a modeling step that improved dose titrations (e.g., probit and logit) by *interpolating* the data. Probably because of computational difficulties before the widespread availability of computers, model-based interpolation was dropped. The resulting simplicity facilitated the use of these designs by clinical researchers, but with the unfortunate consequence that the MTD (in the case of oncology studies) was usually taken to be the last design point (dose) visited according to the escalation/de-escalation rules adopted. This feature contributes to the poor operating characteristics of these designs.

As I indicated previously, the CRM is the best available method for dose-finding, at least in the classic cytotoxic drug domain. It also might be adaptable more broadly. The strengths and efficiencies of the CRM follow, in part, from its use of a mathematical model for dose *extrapolation* to guide the choice of a new dose. The method is inseparable from modeling the dose–toxicity function. The model also provides a formal way for outcomes at low doses to influence directly decisions made at higher doses. The Fibonacci method, for example, is purely local in the sense that once a step is taken, the outcomes below that step are no longer directly relevant.

In addition to the use of mathematical models for the design and execution of these trials, other models are essential for analyzing the pharmacokinetic data (Chapter 16) that routinely arises from phase I studies.

10.4.2   Continual Reassessment Method

The continual reassessment method (CRM) is a dose finding design that derives its advantages from an explicit mathematical model describing the relationship between dose and toxicity (O'Quigley, Pepe, and Fisher, 1990; O'Quigley and Chevret, 1991; O'Quigley, 1992). In the CRM algorithm the mathematical model is used to fit the available data and extrapolate to the optimal dose (in this case the MTD). The next cohort of subjects is administered the dose that appears to be optimal, based on the currently available data and the model extrapolation. As originally described, the CRM is a Bayesian method. In particular, Bayesian methods describing the probability distribution of model parameters are required. A broader view of dose escalation methods is discussed by Whitehead et al. (2001).

The CRM method begins with the quantitative specification of a dose–response relationship. The information that clinical investigators need to make this specification is essentially the same that they require to choose a starting dose and design points for a more traditional dose-ranging trial. An essential difference between the CRM and traditional dose-ranging methods is that the CRM does not necessarily specify the limited set of doses at the beginning of the experiment. The CRM is best viewed as an algorithm for updating the best guess regarding the optimal dose. It can be used to select (escalate or de-escalate) from among a set of the specified doses. However, this reduces its efficiency at dose-finding.

The CRM algorithm makes no assumptions about (1) the actual dose used, (2) the cohort sizes at the doses, (3) ordering of the doses, or (4) integer responses. Investigators are free to use the recommended dose or one thought to be more appropriate. They can vary the cohort size to reflect clinical circumstances, and the dose can and will go up or down depending on responses. Finally fractional responses are appropriate. The response from each subject is not restricted to be 0 or 1 (yes or no) but can be any numerical value in the interval [0,1].

By this method a starting dose is chosen, and after the response is assessed, a one-parameter dose–toxicity model (e.g., logit curve with one free parameter) is fit to the data. This fitting process, or parameter estimation, can be accomplished by Bayesian or other standard statistical methods. From the estimated dose–toxicity curve, we calculate the dose associated with the target probability of response. This estimated dose is the one used to treat the next patient. The process of treating, assessing response, model-fitting, and dose estimation is repeated until it converges (no additional dose changes) or until a preset number of patients have been treated. Simulations show that the CRM is somewhat more efficient (i.e., reaches the actual MTD sooner) than the other designs discussed above.

Example 8   An example of how the CRM works is shown in Figure 10.4 and Table 10.4. Small technical details are omitted here to focus on the example. Figure 10.4 shows the dose–toxicity models at each step of the CRM iterations with the data in Table 10.4. Assume that we intend to select the dose of a new drug that yields a target probability of 0.3 of toxicity (horizontal dotted lines in Figure 10.4). The postulated dose–response model at the initiation of the trial (which in this case has the same mathematical form as

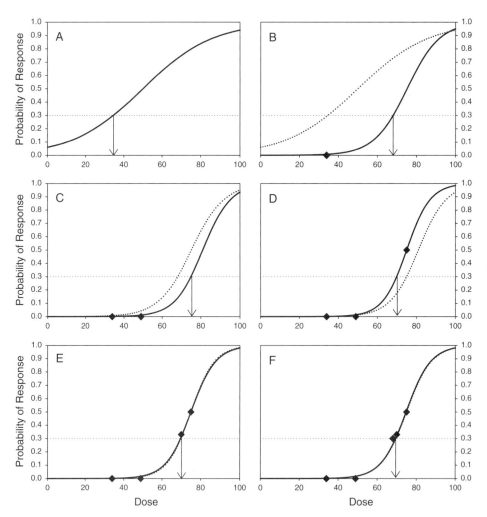

Figure 10.4   Simulated CRM dose-finding. Each panel shows the current data, updated model fit (*solid curve*), and the previous iteration's model (*dotted curve*).

TABLE 10.4   Simulated Dose-Finding Study Using the CRM

Step	CRM Dose	Dose Used	Patients Treated	Score of Responses
A	35	35	3	0
B	68	50	3	0
C	75	75	4	2
D	70	70	3	1
E	69	69	3	0.9
F	69	—	—	—

Note: See Figure 10.4 for the model fits.

curve *A* in Figure 10.1) is specified by the investigators, and explicitly provides a value of 35 for the starting guess for the optimum dose (Figure 10.4A). Investigators must be satisfied with this model and starting dose, as they are free to adjust the initial parameters of the model to locate it in accord with their intuition and knowledge. Suppose that 3 subjects are treated at the dose of 35 with no dose-limiting toxicity (DLT). This data point and model fit are shown in Figure 10.4B. The model now indicates that the optimal dose is 68. Investigators might be unwilling to administer a dose increment more than 1.5 times the previous dose. Suppose that they use a dose of 50 instead and treat 3 new subjects with no DLTs. This step is shown in Figure 10.4C, and yields a new recommended dose of 75. Investigators now feel that they are close to knowing the MTD, so 4 subjects are treated at 75, with 2 DLTs being observed (Figure 10.4D). The MTD appears to be overshot, and the CRM indicates 70 as the correct dose. Three subjects are treated at 70 with 1 DLT (Figure 10.4E). The iterative process has now nearly converged, with the next dose recommendation being 69. Three more subjects are treated at 69, with the total DLT score of, say, 0.9 (Figure 10.4F). The new recommended dose is also 69. Assuming this meets prespecified convergence criteria (e.g., two recommended doses within 10%), investigators could recommend 69 as the MTD based on all the available data.

This simple example illustrates some important features of the CRM method. It accommodates a mixture of clinical judgment and statistical rigor. All the available data are used at each step (i.e., not just the outcomes at the current dose). It is fairly efficient, the process having converged after 5 dose escalations. Also the CRM accommodates noninteger responses. Outcomes can be graded on a scale from 0 to 1, and different types of outcomes can be scored differently.

Additional advantages of the CRM are that it requires only a starting dose as opposed to an entire set of doses specified in advance, the estimation method is unbiased, and it does not depend strongly on the starting dose. It has been claimed that the CRM has a tendency to treat a larger fraction of patients at higher doses, increasing the chances for serious toxicity from new drugs (Korn et al., 1994). These deficiencies can be corrected by grouping patients (e.g., groups of three) and not allowing the escalation to skip to high dose levels too quickly (Goodman, Zahurak, and Piantadosi, 1995). Other modifications have been suggested by Faries (1994). More recently a likelihood, rather than a Bayesian, approach has been proposed for the CRM (O'Quigley and Shen, 1996). Despite the promise of this design, it has not yet been widely used because many investigators are unfamiliar with it.

There are a few disadvantages to the CRM. It is complex compared to dose-ranging designs, requiring special computer software and the aid of a statistical expert. In oncology clinical research these are not serious limitations, but they may be in some other disciplines. An additional potential disadvantage in the CRM can be seen from the perspective of a peer reviewer or regulator. The CRM requires that we review and approve an algorithm rather than a set of prespecified doses, leaving some uncertainty as to the design points that will be visited during the trial. This uncertainty, although relatively small, is sometimes consequential in the minds of reviewers. Much can be done to alleviate these concerns at the initiation of the study by enumerating all possible outcomes for the first one or two cohorts. By assuming integer responses a fairly small number of outcome possibilities can be obtained. Reviewers can then examine the doses that will be visited early in the escalation, which provides some reassurance that the escalations will be safe.

10.4.3 Pharmacokinetic Measurements Might Be Used to Improve CRM Dose Escalations

There is potential for further increasing the efficiency of the CRM using the information in pharmacokinetic (PK) measurements of the type that are routinely obtained during phase I investigations (Piantadosi and Liu, 1996). This method may be helpful when the PK measurements, usually derived by fitting compartmental models to time-concentration data, carry information beyond that conveyed by dose because of complexities in drug absorption, distribution, or elimination.

For example, suppose that toxicity is thought to be a consequence of both the dose of drug administered and the area under the time-concentration curve (AUC) for blood. For relatively simple physiological distributions of drug, such as one- or two-compartment models with first-order transfer rates, dose of drug and AUC are directly proportional and therefore carry redundant information. In such a situation we would not expect the addition of AUC to an escalation scheme based on dose of drug to improve the efficiency of the dose finding. However, if dose and AUC become "uncoupled" because of complex pharmacokinetics, random variability, or other reasons, and AUC is a strong determinant of toxicity, then information carried by AUC about toxicity may improve the efficiency of escalation schemes based only on dose.

The CRM facilitates using ancillary information in the AUC or other PK measurements. Specifically, the basic CRM employs a parametric model of the probability of toxicity as a function of dose. For example, if X_i is a random response variable for the ith patient that takes the value 1 for a toxicity and 0 otherwise, a logistic dose–toxicity function can be parameterized as

$$\Pr[X_i = 1] = \frac{1}{1 + e^{b_0 - b_1 D_i}},$$

where D_i is the dose of drug administered and b_0 and b_1 are population parameters that characterize the function. As dose increases, the probability of a toxic response also increases.

Usually b_0 is fixed at the start of the dose escalations and b_1 is estimated from patient data during the escalations. After sufficient information becomes available, both parameters are estimated from the data. A Bayesian procedure can be used, by which the parameter estimate is taken to be the posterior mean of a distribution based on the hypothesized prior distribution and the likelihood. For example,

$$\widehat{\beta}_1 = \frac{\int_0^\infty \beta_1 f(\beta_1)\mathcal{L}(\beta_1)\,d\beta_1}{\int_0^\infty f(\beta_1)\mathcal{L}(\beta_1)\,d\beta_1},$$

where $f(\beta_1)$ is the prior distribution for β_1 and $\mathcal{L}(\beta_1)$ is the likelihood function. These Bayesian estimates have desirable properties for the purposes at hand.

By this formalism, the effect of AUC on toxicity is incorporated as

$$\Pr[X_i = 1] = \frac{1}{1 + e^{b_0 - b_1 D_i - b_2 AUC_i}},$$

where b_2 is an additional population parameter. Thus large AUCs increase the chance of toxicity and low AUCs decrease it in this model. Bayesian estimates for the parameters can be obtained using a joint prior distribution. An approach similar to this shows the possibility of increasing the efficiency of dose escalation trials based on statistical simulations (Piantadosi and Liu, 1996). However, practical applications and experience with this method are currently lacking.

10.4.4 The CRM Is an Attractive Design to Criticize

The CRM focuses very sharply the nature of dose-finding trials. These studies by any method are characterized by subjective designs and assessments, high variability, use of ancillary data from outside the study design, and the need for efficiency. Because the CRM makes these characteristics so explicit, is fairly aggressive, and uses a visible dose response model, the design is often criticized. For example, some investigators complain that the CRM is not much more efficient than Fibonacci escalations. The dose–response model has been criticized as being falsely precise or unnecessary.

I am not aware of any substantive criticisms of the CRM that are not simply problems with dose-finding/ranging designs generally. It is demonstrably more efficient and less biased than classic designs when properly used. The fitted model is central to the method, allowing both interpolation and extrapolation of the available data. It is not clear how this can be falsely precise, or how the absence of a model is "better."

10.4.5 CRM Example

A prototypical example of the use of the CRM for dose-finding can be seen in the study of the antineoplastic agent irinotecan for the treatment of malignant gliomas (Gilbert et al., 2003). This trial accrued 40 subjects who received a total of 135 cycles of the study drug. Other than the starting dose, levels were not specified in advance. The CRM implementation selected target doses on a continuum, which were administered to the nearest milligram/meter^2. Dose cohorts were size 3, and parallel groups were studied for patients receiving enzyme-inducing anticonvulsants (EIA) or not.

The trial provided estimates of the MTD for both subsets of patients, which were found to differ by a factor of 3.5 (411 mg/m^2/week in the EIA$^+$ subset versus 117 mg/m^2/week in the EIA$^-$ subset). As is typical, the trial also provided extensive information regarding the pharmacokinetics of the study drug, and preliminary evidence regarding clinical efficacy. This trial illustrates a successful seamless incorporation of the CRM method into multicenter collaborative dose-finding trials (Piantadosi, Fisher, and Grossman, 1998).

10.4.6   Can Randomization Be Used in Phase I or TM Trials?

Early developmental trials are not comparative and selection bias is not much of a problem when using primarily treatment mechanism and pharmacological endpoints. Randomization seems to have no role in such studies. However, consider the development of new biological agents, which are expensive to manufacture and administer and may have large beneficial treatment effects. An example of such a treatment might be genetically engineered tumor vaccines, which stimulate immunologic reactions to tumor cells and can immunize laboratory animals against their own tumors. DF study objectives for these agents are to determine the largest number of cells (dose) that can be administered and to test for side effects of the immunization.

Because these treatments give live (but lethally irradiated) tumor cells with new genes to the patient, side effects can arise from two sources: the new genes in the transfected tumor cells or the cells themselves. One way to establish the safety of the transfected cells is to compare their toxicity with that from untransfected cells. This can be done during DF trials using a randomized design. Such considerations led one group of investigators to perform a randomized dose escalation trial in the gene therapy of renal cell carcinoma (Simons, Jaffee, Weber et al., 1997). An advantage of this design was that the therapy had the potential for large beneficial effects, which might have been evident (or convincing) from such a design.

10.4.7   Phase I Data Have Other Uses

For all phase I studies, learning about basic clinical pharmacology is important and includes measuring drug uptake, metabolism, distribution, and elimination. This information is vital to the future development and use of the drug, and is helpful in determining the relationship between blood levels and side effects, if any. These goals indicate that the major areas of concern in designing phase I trials will be selection of patients, choosing a starting dose, rules for escalating dose, and methods for determining the MTD or safe dose.

If basic pharmacology were the only goal of a phase I study, the patients might be selected from any underlying disease and without regard to functioning of specific organ systems. However, phase I studies are usually targeted for patients with the specific condition under investigation. For example, in phase I cancer trials, patients are selected from those with a disease type targeted by the new drug. Because the potential risks and benefits of the drug are unknown, patients often are those with relatively advanced disease. It is usually helpful to enroll patients with a normal cardiac, hepatic, and renal function. Because bone marrow suppression is a common side effect of cytotoxic drugs, it is usually helpful to have normal hematologic function as well when testing new drugs in cancer patients. In settings other than cancer, the first patients to receive a particular drug might have less extensive disease or even be healthy volunteers.

10.5   MORE GENERAL DOSE-FINDING ISSUES

The general dose-finding problem involves potential relationships between dose and response that are more complex than those typical of cytotoxic drugs. Some potential features are threshold, background, nonsaturating response, decreasing dose–response, schedule effects, and interactions. Threshold implies that dose must exceed a critical

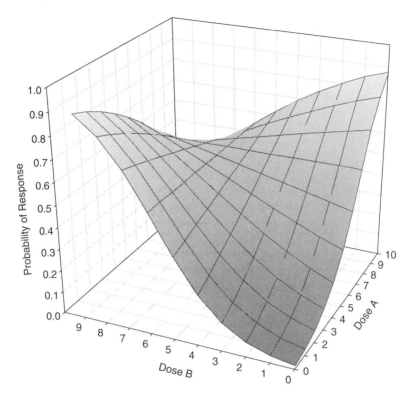

Figure 10.5 Joint dose–response surface for two drugs.

level before producing any response. This might be true of many drugs such as enzyme inhibitors. Background means that some level of response is produced at zero dose. Pain relief and cure of infections might show this phenomenon. A nonsaturating response means that high doses cannot produce the response in 100% of subjects.

Interaction means that a specific agent, or dose of an agent, has an effect that is modulated by other factors. Schedule effects are a type of interaction between time or sequence and dose. Interactions between different drugs or modalities of treatment are probably more familiar. In many clinical circumstances we attempt to take advantage of such interactions to enhance treatment effects. But interactions between different treatments can produce very complex relationships. For example, in Figure 10.5, the presence of Agent *A* alters the dose–response of Agent *B*. Even when beneficial interactions for efficacy are intended or desired, dose-finding trials often ignore the implications with regard to safety. This will be discussed more fully below.

10.5.1 Dose-Finding Is Not Always One Dimensional

Nearly always, the formulation and dose of a drug can be manipulated to yield a safe therapeutic effect when administered on a simple schedule. However, the general dose optimization problem can involve questions of scheduling. For many drugs, most vaccines and biologic agents, and radiotherapy, the question of treatment timing and schedule can be critical for both maximal safety and efficacy. The potential effects of scheduling on safety and efficacy must be treated as an additional dimension or factor, just as we would investigate jointly the effects of a second therapy.

A related area in which contemporary dose-finding designs remain primitive is when studying combinations, particularly those that intend to take advantage of a therapeutic interaction. Examples are radiosensitizers or two antitumor drugs in combination for treatment of cancers. Drugs that alter the metabolism, distribution, or excretion of an active agent are another example. A common approach in this circumstance is to keep the dose fixed for one drug, that is, for the agent about which most is known, and conduct a standard dose escalation for the other. This approach somewhat minimizes ethics concerns. While it is simple and understandable, this method is not guaranteed to yield a good estimate of the global optimum for the combination, particularly when interactions are expected. In fact, even independent one-dimensional optimizations are not guaranteed to yield the joint optimal dose. The solution to the problem requires a two-dimensional search; that is, more design points than are needed for a single factor. A more thorough exploration of the design space using multiple doses of each drug is necessary to identify optima reliably. Some approaches to this problem are suggested in the next section.

These types of questions increase design complexity very rapidly for multiple agents, modalities, and adjuvants. No general dose-finding solution is offered here, except to make the reader aware of the shortcomings of simplistic designs. However, similar problems historically in industrial production and agriculture have been approached efficiently, but requiring more sophisticated designs than many clinical investigators have used. Such principles can be adapted readily to dose-finding questions.

10.5.2 Dual Dose-Finding

This discussion will pertain specifically to two drug combinations, except by implication. The typical dose-finding approach for adding a new agent to an existing drug is to use a "one-dimensional" optimization after fixing the dose of the standard agent. In other words, dose-finding is performed only with respect to the new drug. Occasionally some ad hoc or informal changes in the dose level of the existing drug are added to the design. Such strategic decisions are usually handled entirely by clinical intuition.

This approach is unfavorable in several regards. Drug combinations are often studied primarily with the expectation that there will be synergistic therapeutic effect of the two drugs. If so, the potential for synergistic toxic effect must also be very likely to be enhanced. In the case of therapeutic (and potentially toxic) synergy, the best strategy is to explore the design space (combination doses) more fully than typical one-dimensional dose-finding allows. This will satisfy both a therapeutic and safety perspective. A principal reason why clinical investigators seem reluctant to explore a sizable number of dose combinations is concern that a large sample size required. This concern stems, in part, from thinking narrowly about such designs in the style of the Fibonacci dose escalation.

This problem has arisen several times recently in drug development settings in the Hopkins Oncology Center. When circumstances allow, we employ a safe and efficient strategy that explores the design space more thoroughly. The method is based on considerations depicted in Figure 10.6. Many design points could be tested, only to discover that the optimal combination exists at none of the joint doses actually utilized (gray region in Figure 10.6). It seems unlikely that investigators can lay out a simple

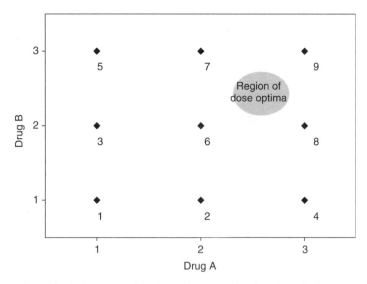

Figure 10.6 Possible design space for a two-drug combination dose-finding experiment. The gray area indicates that the optimal dose combination may not exist at a design point. The grid numbers indicate the sequence of dosing.

design prior to obtaining any data and choose a dose combination that is precisely optimal. It is more likely that some two-dimensional modeling procedure (analogous to the one-dimensional dose response modeling of bioassay) will accurately and reliably locate the optimal dose combination. For achieving a close optimum, the principal issues become (1) selecting the number of patients to test at each dose combination, (2) working through the dose combinations in a safe fashion, and (3) modeling the relationship between the joint doses and outcome.

Workable solutions to these problems have been described in the statistical literature. The problem appears amenable to response surface methodology, a well-studied statistical problem. The main idea in a response surface is to model response as a function of multiple experimental factors (linear in the parameters), and to optimize that response. Here we have only two experimental factors, so the problem is relatively simple. The two factors (drug doses) are laid out on the X–Y plane, with the response measured in the Z direction. The response surface is mathematically a two-dimensional "manifold" in three-dimensional space—it is similar to a plane but does not have a flat surface.

Plackett and Burman (1946) described the design of optimal multifactorial experiments. Subsequent investigators have refined this methodology. For example, see Box and Draper (1987). Short et al. (2002) discussed some of these design concerns in the setting of anesthetic drugs. The basic idea behind the Plackett and Burman designs is that the response surface can be described using relatively few parameters compared to the number of design points in the experiment. Thus the number of experimental subjects at each design point can be kept small. Classically, the number of design points and number of parameters were set equal, and a single experimental unit was treated at each design point. This stretches the statistical information to the limit.

In the present context a similar design could be employed, but with fewer parameters than design points. Figure 10.7 shows a possible design for a two-drug combination,

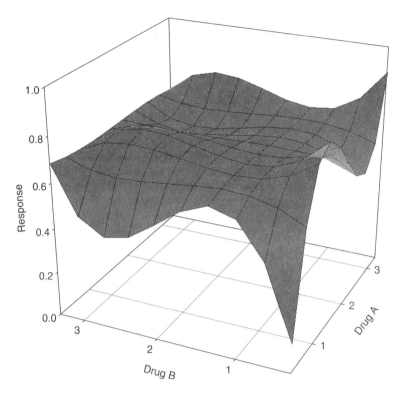

Figure 10.7 A hypothetical response surface for the two-drug combination dose-finding design from Figure 10.6. The interior minimum of the surface corresponds approximately to the region of dose optima from Figure 10.6.

where each drug is administered at three doses as discussed above and illustrated in Figure 10.6. A hypothetical response surface is shown in the Z direction. Although a response surface could be quite convoluted, it seems biologically reasonable that the surface would be relatively smooth. We might assume that the surface can be parameterized by a cubic polynomial, for example. There are nine design points, and the number of parameters describing a complete cubic response surface is 10, including an intercept and all first and second-order terms.

In this case a flexible model (but not saturated) theoretically could be fit using a single experimental unit (patient) at each design point. Nevertheless, larger cohorts at each design point are appropriate, particularly in the center of the design space where the optimal is likely to be (assuming he investigators have chosen the dose combinations insightfully). One could use cohorts of size 2 at the "corners" and cohorts of size 3 at all other design points, for example, allowing a fully saturated cubic response surface model to be fit. Such a plan would require only 23 subjects for nine dose combinations. The safest dose at which to begin an escalation is the one in the lower left corner. In special circumstances additional design points could be added. The dose combinations would be visited sequentially in a manner that escalates only one of the drugs relative to the previous design point. Thus the path through the dose space might be diagonal in the fashion indicated by the sequence numbers in Figure 10.6. The design points in the upper right corner may never be reached if dose-limiting toxicities are encountered.

After the data are acquired, a parsimonious but flexible response surface can be fit. The response surface should permit reasonable but smooth nonlinearities to help define the optimum. Investigators might treat additional subjects at the model-defined optimum to acquire more clinical experience. However, this is not required under the assumptions of the design. The first dose-finding study employing this design has yet to be completed. However, the ideas are well worked out statistically and the design seems compatible with the clinical setting and assumptions that typically enter studies of drug combinations. Variations of this basic design will likely follow from early experiences.

It is also possible to generalize the CRM to accommodate two drugs and a relatively smooth response function. However, to do so might require four parameters for the typically employed logistic model. In addition to the intercept, one parameter is required for the "slope" of each drug effect, and the last parameter would describe the interaction or synergy between the two drugs. This is the simplest generalization, and even more complex models could be necessary. Because of the difficulties in specifying prior values or information for model parameters (a relatively simple task in the one-dimensional case) when employing complex models, such generalizations of the CRM are not likely to be practical.

10.5.3 Optimizing Safety and Efficacy Jointly

Studies of cytotoxic agents are greatly aided by a priori knowledge regarding the form of the relationship between dose and toxicity. Investigators know that for such drugs there is a dose that saturates the response. For example, one can administer a high enough dose of a cytotoxic agent to create toxicity in 100% of recipients. This implies that there is a monotonically increasing dose–response function. This knowledge can be used implicitly in the dose escalation design. It can also be used explicitly (such as in a probit or logit CRM model), which yields even more efficiency in the design.

Suppose, however, that relatively little is known about the shape of the dose–response function. This might be the case with drugs where the effect saturates, or when the behavior of the response at low doses does not qualitatively mimic a cytotoxic drug. Such assumptions are completely reasonable if investigators are interested in the maximum nontoxic dose (antibiotics) or minimum effective dose (analgesics), as discussed above.

When designing dose-finding experiments in the absence of qualitative knowledge about the shape of the dose–response function, the investigator has little option except to study a range and number of doses that permits assessing shape, while simultaneously locating the position that is considered optimal. For example, if the dose–response function has a plateau, the OBD might be a dose just on the shoulder of the curve. Higher doses would not substantially increase the response but could increase the chance of undesirable side effects. In essence, there are two unknowns: the level of the plateau and the dose at which the plateau begins. Extra information is needed from the experimental design to optimize in such a circumstance.

A potentially more difficult problem arises when balancing the toxicity of a new drug against its biological marker response or intermediate outcome. We may need to select a dose on this basis before estimating definitive clinical activity. This situation is shown hypothetically in Figure 10.8, where both the probability of toxicity (right-hand axis) and biological response (left-hand axis) are plotted against dose. The two effect curves are not directly comparable because the respective vertical axes are different.

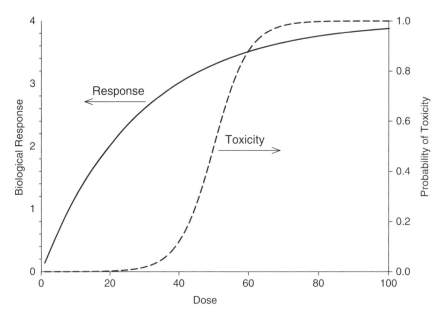

Figure 10.8   Relationship of toxicity and biological response to dose for a hypothetical drug.

An optimum dose would produce low toxicity while nearly maximizing the biological response. Such a dose is not guaranteed to exist. After the data are collected and modeled, the region of the optimum dose might be fairly obvious. However, it might be difficult to specify in advance either a definition of, or a strategy for finding, the optimum.

When the dose–response relationships are well defined for both effects, an approach to an optimum might be to consider the parametric space of the outcomes. A curve parameterized by dose exists in the plane whose *X* and *Y* axes are toxicity response and biological response. Suppose that the toxicity and biological responses are given by the curves drawn in Figure 10.8. The corresponding parametric curve is shown in Figure 10.9 along with a region of clinical utility defined by high biological response and low toxicity. The parametric curve defines the only biological possibilities that we can observe. If the OBD is an actual dose rather than a hypothetical, it would have to be a point on this curve. The response–toxicity curve might pass through the region of clinical utility, in which case it is straightforward to select the OBD. Alternatively, it might be necessary to select as the OBD a point in the parametric space closest to the region of clinical utility (e.g., Figure 10.9, dose = 50). If the parametric curve does not pass through or close to some clinically appropriate joint outcomes, the drug will not be useful. This might be the case in Figure 10.9 if higher doses are unusable for other reasons.

For smooth dose–toxicity and dose–response functions, the parametric space could be nearly linear over a wide range of effects. Figure 10.9 has this property. This could lead to an efficient design for such studies. The region of clinical utility would be prespecified on a toxicity–response parametric plot. As outcomes are gathered from small experimental cohorts at each dose, points in the parametric space can be plotted and extrapolated linearly (e.g., Figure 10.10). If the resulting projections convincingly

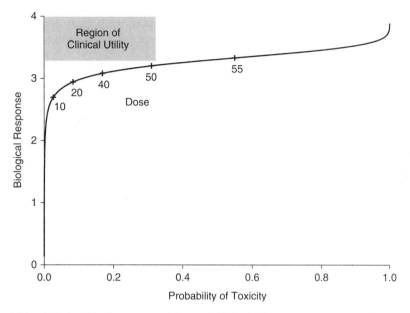

Figure 10.9 Relationship between toxicity and biological response (parameterized by dose) for the example functions from Figure 10.8.

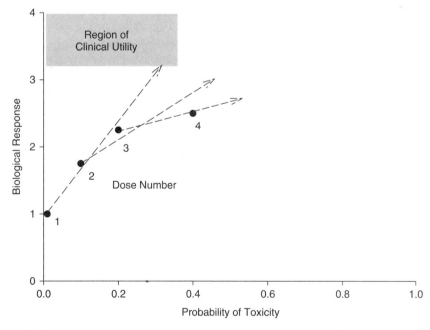

Figure 10.10 Possible experimental design to assess two outcomes that determine clinical utility. The black circles are design points, not necessarily taken in order, and the gray area is thought to be that of clinical utility. The dashed lines show extrapolations made by investigators based on a sequential experiment. At dose number 4, the lack of utility becomes convincing.

exclude the region of clinical utility, the drug can be abandoned. Such a situation is shown in Figure 10.10, where extrapolations from early data (black circles) suggest the chance of clinical utility (gray region), but data at higher doses demonstrate that the drug will not be useful. No particular ordering of the design points is required, although increasing doses will most likely be required as in other dose-finding trials. In practice, there would be some uncertainty associated with each outcome point in Figure 10.10 in both dimensions, which could be reduced by increasing the sample size at that dose.

10.6   SUMMARY

The purpose of a dose-finding or dose-ranging clinical trial is to locate an optimal biological dose, usually defined with respect to safety. The notion of a dose optimum depends on the purpose of the agent and the appropriate balance between safety and efficacy. It could be quite different for analgesics, antibiotics, and cytotoxic agents, for example. There are no clinical trial designs that can locate optimal doses in general. However, for specific purposes, such as toxicity titration for cytotoxic anticancer agents, broadly useful study designs have been developed. All such designs are strongly subjective but are necessary and useful to obtain reliable dosing parameters for subsequent developmental trials.

For cytotoxic drugs, a serviceable dose-ranging design is the (modified) Fibonacci dose escalation or one of its derivatives. Although simple to execute and analyze, such designs have poor operating characteristics with respect to true dose-finding. The continual reassessment method (CRM) is a more efficient and precise design for cytotoxic dose-finding. It comes at a slightly higher cost in terms of aggressiveness of dose escalation but almost universally employs fewer subjects.

Two more general dose-finding problems remain unsolved. One is balancing beneficial response against toxicity for drug doses and the other is optimizing joint doses of two or more agents. New experimental designs that target a region of clinical utility might be helpful in the first case. To investigate interactions, some type of modified response surface method might be helpful. Whatever complications arise in dose-finding, thoroughness is needed to resolve all or most of the dose versus safety questions before initiating safety and activity trials, where remaining dose questions will be inefficient to investigate.

10.7   QUESTIONS FOR DISCUSSION

1. Suppose that drug A is an established treatment administered at a fixed dose. Investigators wish to study the combination of drugs A and B but are unsure of the best dose of either one in the presence of the other. What type of design would permit studying this question? Compare it with a typical phase I trial. What if the correct doses are known but the order of A and B is in question?

11

SAMPLE SIZE AND POWER

11.1 INTRODUCTION

Questions regarding the quantitative properties of clinical trial designs, especially the correct precision or power, best sample size, and optimal study duration, are among those most frequently posed by clinicians to statisticians. The concern over sample size and power is appropriate because studying an adequate number of subjects is the only way to align clinical and statistical inferences. This concern is critical for CTE and SA studies and they will be emphasized in this chapter. For translational trials, these questions are not as important because findings from them are not used to support therapeutic decisions. For dose-finding and dose-ranging trials, sample size is often an outcome of the study.

The literature regarding sample size and power in clinical trials is large, although not especially well organized. Exceptions are the works by Desu and Raghavarao (1990), Kraemer and Thiemann (1987), and Machin and Campbell (1987). Shuster (1990) has also provided a compendium of methods and tables. Other useful references will be provided in context below. The widespread availability of flexible computer programs with good interfaces and graphics has made the use of tables nearly unnecessary. Even so, there are overall relationships that a well organized table can highlight, but that may be obscured by a one-at-a-time calculation from a computer program.

There are two widely used frequentist perspectives for determining the appropriate sample size for a clinical trial. The first is based on confidence intervals around the effect that we expect to observe. By this method the sample size is chosen to make the confidence interval suitably narrow. This approach is usually fairly simple. The second approach is based on the ability of the study to reject the null hypothesis when a specified treatment effect is hypothetically present (i.e., power). This perspective is very common, especially for comparative trials. I will also discuss a third perspective based on likelihood ratios, a very basic biostatistical idea that is very appealing theoretically

Clinical Trials: A Methodologic Perspective, 2E, by S. Piantadosi
Copyright © 2005 John Wiley & Sons, Inc.

TABLE 11.1   Brief Description of Quantitative Design Parameters Frequently Used in Clinical Trials

Power:	$1 - \beta$
β level:	Type II error probability
α level:	Type I error probability
Likelihood ratio:	Relative strength of evidence
Sample size:	Number of experimental subjects
Effect size:	Treatment difference expressed as number of standard deviations
Number of events:	Number of experimental subjects who have a specific outcome
Study duration:	Interval from beginning of trial to end of follow-up
Percent censoring:	Percentage of study participants left without an event by the end of follow-up
Allocation ratio:	Ratio of sample sizes in the treatment groups
Accrual rate:	New subjects entered per unit of time
Loss to follow-up rate:	Rate at which study participants are lost before outcomes can be observed
Follow-up period:	Interval from end of accrual to end of study
Δ :	Smallest treatment effect of interest based on clinical considerations

Note: For exact definitions of these terms, see Appendix B.

but has not been used widely. This approach is closest to formal methods regarding strength of evidence. The idea common to all approaches is precision of estimation, so it is not surprising that they are closely related or interchangeable.

Specification and study of the quantitative properties of a clinical trial requires both a conceptual and a computational effort. The conceptual framework involves the interplay of several factors: (1) the nature of the outcome variable, (2) the framework for specifying precision, (3) accrual dynamics, and (4) the design of the trial. A number of design parameters are, to varying degrees, under the control of the investigator (Table 11.1)

The required computations make the sample size problem seem even more complex than it really is. There are currently several good computer programs for alleviating the computational burden. I present a brief discussion of several programs later in the chapter. Many of the calculations can be performed using methods described in Appendix A.

11.2   PRINCIPLES

Because it is not possible to specify a universal approach to power and sample size, I will discuss some basic ideas and examples on a case-by-case basis, trying to synthesize whenever possible. Although some abbreviated sample size tables are presented later in the chapter, it will be necessary to use the formulas for direct calculation of specific cases or to consult the original references for more extensive tabulations. For a more statistically oriented review of fixed sample size calculations, see Donner (1984) or Lachin (1981). Tables and equations for specific purposes are presented in Machin and Campbell (1987), Shuster (1990), and Kraemer and Thiemann (1987). Quantitative aspects of group sequential designs for comparative trials are discussed in Chapter 14.

11.2.1 What Is Precision?

The underlying theme of sample size considerations in all clinical trials is *precision*. Precision is the reproducibility of multiple measurements. High precision implies little variation. Precision is therefore usefully described by the confidence interval around, or standard error of, the estimate. For this discussion, I will always assume that our estimates are unbiased, meaning accurate.

Precision of estimation is the characteristic of an experiment that is most directly a consequence of the size of the investigation. In contrast, other important features of the estimates, such as validity, unbiasedness, and reliability, do not necessarily relate to study size. Precision is a consequence of measurement error, person-to-person (and other) variability, number of replicates (e.g., sample size), experimental design, and methods of analysis. By specifying quantitatively the precision of measurement required in an experiment, the investigator is implicitly outlining the sample size, and possibly other features of the study.

There are several approaches to specifying the precision required for an experiment, including (1) direct specification on the scale of measurement, (2) indirect specification through confidence limits, (3) the power of a statistical hypothesis test, (4) relative evidence through a likelihood ratio, (5) a guess based on ordinal scales of measuring effect sizes, (6) ad hoc use of designs similar to those employed by other investigators, and (7) simulations of the design and analysis assuming different sample sizes. In this chapter, I will discuss primarily the second and third approaches.

Direct specification of precision might be possible for some studies. For example, we may need a sample size sufficient to estimate a mean diastolic blood pressure with a standard error of ± 2 mm Hg. Such an objective is unusual in clinical trials. More typically we specify precision through the absolute or relative width of a confidence interval. This is a useful approach in single cohort studies. For example, our sample size may need to be large enough for the 95% confidence interval around the mean diastolic blood pressure to be ± 4 mm Hg.

The size of comparative trials is often described by the power of a hypothesis test. We may need a sample size sufficient to yield 90% power to detect a 5 mm Hg difference as being statistically significant (using a specified type I error level). It may not be obvious how this captures the notion of precision, but it will be made more clear later. In all of these circumstances a convenient specification of precision combined with knowledge of variability can be used to determine the sample size.

11.2.2 What Is Power?

Many clinical investigators have difficulty with the concept of power and the way it should be used to design and interpret studies (or *if* it should be used to interpret studies). A principal reason for this is that power is a somewhat unnatural idea. However, it is an idea that is central to thinking about clinical trials and therefore critical to understand. The main utility of power is that it requires investigators to address precision of estimation while a study is being designed.

One obstacle is that power is defined in terms of a hypothetical—namely that a treatment effect of a certain size is actually present. (This contrasts with the more ubiquitous null hypothesis of no treatment difference.) For making clinical inferences, we would like our study to yield the probability that the true treatment effect is at least a given size. However, power *assumes* a treatment effect of a specified size and provides the probability that a result as large or larger would be statistically significant. Power is therefore neither the chance that the null hypothesis is true nor a probability statement about the treatment effect, both common errors of inference.

A third source of difficulty with the idea of power is that we tend to conceptualize variation at the level of the individual (the experimental unit), whereas power refers to variation at the level of the experiment. The two levels are connected fundamentally, but understanding variation from study to study does not seem to be intuitive. We take only a single sample from the space of all possible experiments, meaning we perform our study once. It is not obvious what probability statements can be made under such conditions. Precision also refers to variation at the level of the experiment, but may be easier to understand.

Finally we cannot separate power from either the size of the study or the magnitude of the hypothetical treatment effect. All three values must be discussed simultaneously. However, statements about power are often made in isolation. For example, someone might say "The trial has 90% power." By itself, such a statement is ambiguous or meaningless. The frequency of such errors is high and contributes to misunderstanding power.

The size of a comparative study is intertwined with the intended error rates, in the frequentist paradigm. By convention, most power equations are written in terms of the corresponding normal quantiles for the type I and type II error rates (Z_α and Z_β), rather than the error probabilities themselves. The definitions of Z_α and Z_β follow from that for the cumulative normal distribution, which gives "lower" or "left" tail areas as

$$\Phi(Z) = \int_{-\infty}^{Z} \frac{1}{\sqrt{2\pi}} e^{-x^2/2} \, dx. \tag{11.1}$$

Then $1 - \alpha/2 = \Phi(Z_{1-\alpha/2})$ and $1 - \beta = \Phi(Z_{1-\beta})$. To simplify the notation slightly in the remainder of the chapter, I will define $Z_\alpha = \Phi^{-1}(1 - \alpha/2)$ and $Z_\beta = \Phi^{-1}(1 - \beta)$. For $\alpha = 0.05$, $Z_\alpha = 1.96$, and for $\beta = 0.10$, $Z_\beta = 1.282$.

11.2.3 What Is Evidence?

In an informal sense we like our clinical trial to provide *evidence* regarding the treatment effect sufficient for the decisions at hand. Statistically it is difficult or impossible to define evidence in support of a single hypothesis. However, when assessing competing hypotheses, relative evidence can be defined in a useful way. Evidence in support of one hypothesis compared to another is captured in the statistical likelihood function—specifically the likelihood ratio (LR), which can be interpreted as relative strength of evidence (Royall, 1997).

Likelihoods form a basis for a wide class of both estimation methods and statistical tests, and it is perhaps not surprising that the LR can be interpreted as relative evidence. The likelihood is the probability of observing the data under an assumed probability model. The parameters of the probability model are the entities about which we would like to make inferences. Clinical trialists have not been accustomed to assessing and designing studies in terms of LRs, but they are extremely useful statistical tools. More detail is provided in Chapter 5.

For methods based on the normal distribution of some statistic, as many of the common power and sample size techniques are, a simple perspective on likelihood-based thinking can be obtained by considering the following: The LR can be calculated from a standard Z-score as $R = e^{\Lambda} = \exp(Z^2/2)$. This follows directly from the normal distribution and likelihood. Power formulas derived from hypothesis tests almost always contain a term of the form $(Z_{\alpha} + Z_{\beta})^2$ arising from quantification of the type I and II errors. Therefore this term can be replaced by $2\log(R)$, where R is an appropriately chosen standard of relative evidence for the comparison being planned. With a small basis of experience, it is no more difficult to choose an appropriate standard for relative evidence than to choose sensible values for type I and type II errors. For example, a two-sided 5% type I error and 90% power would result in $(Z_{\alpha} + Z_{\beta})^2 = (1.96 + 1.282)^2 = 10.5$, for which $R = \exp(10.5/2) = 190$. This LR corresponds to strong evidence, as discussed below.

Evidence can be weak or strong, misleading or not, as discussed extensively by Royall (1997, ch. 2). It is relatively easy to reduce the probability of misleading evidence by an appropriate choice for the likelihood ratio. For example, the chance of misleading evidence is 0.021, 0.009, and 0.004 for $R = 8$, 16, and 32, respectively. But producing strong evidence in favor of the correct hypothesis requires sample sizes somewhat larger than those yielded by the typical hypothesis testing framework. With sample sizes based on hypothesis tests in the usual paradigm, we will fail to produce likelihood ratios in excess of 8 about 25% to 30% of the time. The sample sizes required by an evidentiary approach can be 40% to 75% greater. This emphasizes that choosing between competing hypotheses in the typical framework is an easier task than generating strong evidence regarding the correct hypothesis. Some quantitative consequences of this can be seen later in this chapter (e.g., Section 11.5.9).

11.2.4 Sample Size and Power Calculations Are Approximations

There are several ways in which power and sample size calculations are approximations. First, the equations themselves are often based on approximations to the exact statistical distributions. For example, a useful and accurate equation for comparing means using the t-test can be based on a normal approximation with known variance (a so-called z-test). Second, the idea of predicting the number of patients or study subjects required in a study is itself an approximating process because it depends on guesses about some of the parameters. For example, the accrual rate, censoring rate, or variance may be known only approximately. The difference in samples sizes produced by hypothesis testing approaches and those based on relative evidence also indicate how solutions to this problem depend on our framework and assumptions.

One cannot take the estimated sample size or power that results from such approximations rigidly. We hope that our errors are small and the required sample size is accurate, generally within 5 or 10 percent of what is truly needed. However, we must be prepared to make the design more robust if some parameters are not known reliably. Usually this means increasing the sample size to compensate for optimistic assumptions. Furthermore, we should not allow small differences in sample size to dictate important scientific priorities, one way or the other. If it is essential clinically to achieve a certain degree of precision, then it is probably worthwhile to accrue 10% or 20% more participants to accomplish it. Finally, some subjects are usually lost from the experiment. We can inflate the sample size to compensate for this, but the exact amount by which to do so is often only an educated guess.

When solved for sample size, the equations that follow do not necessarily yield whole numbers. The answers must be rounded to a nearby integer. For comparative trial designs, it is probably best to solve for the sample size in one group, round to the next higher integer, and multiply by the number of treatment groups. The tables presented should be consistent with this policy. I have not rounded the answers for many of the calculations in the text because the important concepts do not rely on it.

11.2.5   The Relationship between Power/Precision and Sample Size Is Quadratic

Investigators are often surprised at how small changes in some design parameters yield large changes in the required sample size or study duration for experiments. At least for SA and CTE clinical trials, sample size tends to increase as the square, or inverse square of other design parameters. In general, the sample size increases as the square of the standard deviation of the treatment difference and the (sum of) normal quantiles for the type I and II error rates. Reducing the error rates increases Z_α and Z_β and increases the required sample size. Larger variability in the endpoint measurement increases the sample size. Also the sample size increases as the inverse square of the treatment difference; that is, detecting small treatment effects increases the required sample size greatly. These and other quantitative ideas will be made clear in the following sections.

11.3   EARLY DEVELOPMENTAL TRIALS

Early developmental trials do not present difficult issues with regard to sample size or precision, but they are somewhat unconventional. These studies are not designed to provide the strong statistical evidence or decision properties of later trials. I include a short discussion here for completeness, but the reader should refer to the appropriate chapters for more details regarding design considerations for these trials.

11.3.1   Translational Trials

Translational clinical trials are small, almost by definition, and nearly always exist below the sample size threshold for any reasonable degree of purely statistical certainty. However, as I indicate in Chapter 9, with a relatively small numbers of subjects and biological knowledge, these trials can reduce uncertainty enough to guide subsequent experiments. Their size can be motivated formally by considering the information gained, but practical constraints will usually make them smaller than 20 subjects.

A problem in the design of some translational studies, as well as other circumstances, is taking a large enough sample to estimate reliably the mean of some measurement. Let us assume that the measurements are samples from a normal distribution with mean μ and known variance σ^2. The sample mean $\overline{X}$ is an unbiased estimate of μ, and the absolute error is $|\overline{X} - \mu|$.

The sample size could be chosen to yield a high probability that the absolute error will be below some tolerance d. In other words, we intend

$$\Pr\left(|\overline{X} - \mu| \le d\right) \ge 1 - \alpha, \tag{11.2}$$

where α is the chance of exceeding the specified tolerance. The sample size required to satisfy this is

$$n \ge \left(Z_{1-\alpha/2}\frac{\sigma}{d}\right)^2. \tag{11.3}$$

In practice, we round n up to the next highest integer.

To test a hypothesis regarding the mean, this equation must be modified slightly to account for the type II error. Specifically, we obtain

$$n \geq \left(\frac{(Z_{1-\alpha/2} + Z_\beta)\sigma}{(\mu_1 - \mu_0)} \right)^2,$$

where Z_β is the normal quantile for the intended type II error (Section 11.2.2) and μ_1 and μ_0 represent the alternative and null means. This formula will arise later in the discussion of comparative trials (Section 11.5.2).

If σ^2 is unknown, we cannot use this formula directly. However, if d is specified in units of σ, the formula is again applicable. In that case $d = m\sigma$ and

$$n \geq \left(\frac{Z_{1-\alpha/2}}{m} \right)^2.$$

If d cannot be specified in terms of σ, a staged procedure is needed where the first part of the sample is used to estimate σ^2 and additional samples are taken to estimate the mean. See Desu and Raghavarao (1990) for a sketch of how to proceed in this circumstance.

Example 9   Suppose that the variance is unknown, but we require a sample size such that there is a 95% chance that the absolute error in our estimate of the mean is within $\frac{1}{2}$ a standard deviation (0.5σ). We then have

$$n = \left(\frac{1.96}{0.5} \right)^2 \approx 16,$$

so 16 observations are required. The sample size is very strongly affected by the precision. If we intend to be within 0.1σ, for example, almost 400 samples are required.

A similar circumstance arises in a sequence of Bernoulli trials when estimating the proportion of successes. If p is the probability of success with each trial, r, the number of successes in n trials, follows a binomial distribution with mean np and variance $np(1 - p)$. Equation (11.2) then becomes

$$\Pr \left(\left| \frac{r}{n} - p \right| \leq d \right) \geq 1 - \alpha,$$

and we can use the normal approximation to the binomial with $\sigma^2 = p(1 - p)$ so that equation (11.3) yields

$$n \geq \left(Z_{1-\alpha/2} \frac{\sqrt{p(1 - p)}}{d} \right)^2. \tag{11.4}$$

The variance is maximal when $p = \frac{1}{2}$, so conservatively

$$n \geq \left(\frac{Z_{1-\alpha/2}}{2d} \right)^2.$$

Further use of this equation will be illustrated below.

11.3.2 Dose-Finding Trials

Dose-finding/ranging trials are commonly not designed to meet criteria of precision in estimation. Instead, they are designed to select or identify a dose that meets certain operational criteria, and the resulting sample size is an outcome of the study. The precision with which important biological parameters are determined is usually not a formal part of the design of such trials. For example, in classic DF oncology trials we usually do not focus on the precision in the estimate of the MTD or of pharmacokinetic parameters, even though some such assessment might be available from the data at the end of the study. Perhaps this lack of attention to precision is an oversight.

Numerous methods might be applied to DF trials to assess their precision. For example, from the model-fitting that is integral to the CRM, the variance of the parameters can be estimated. It would be reduced by increasing cohort sizes or additional dose points. The precision in estimates of PK parameters might be assessed by simulation or bootstrapping and increased in similar ways. Study of such methods would have to provide strong evidence of inadequacy in current designs to change the habits of clinical investigators.

Sampling issues can arise in pharmacokinetic studies. For example, when estimating the area under the blood time–concentration curve (AUC) after administration of a drug, what is the best number and spacing of sample points. Under an assumed model for the time–concentration curve, the precision in the estimated AUC as a function of the number of samples and their spacing could be studied by simulation. This is not routinely done, perhaps for several reasons. First, great precision in the estimated AUC is not usually required. Second, the optimal design will necessarily depend on unknown factors such as the half-life of the drug and its peak concentration. Finally, the sample points are often largely dictated by clinical considerations, such as patient comfort and convenience.

11.4 SAFETY AND ACTIVITY STUDIES

Safety and activity studies have one nearly universal feature—they are single cohort studies in which the treatment effect is compared to a standard external to the experiment. The external standard may be derived from other trials or may be based on the clinical intuition and experience of the investigator, suitably quantified. A useful general way of conceptualizing such studies is that they estimate the unconditional probability of benefit (or lack of benefit), and thus form a basis for deciding whether or not to investigate the treatment in a larger, lengthier, and more expensive trial with an internal control (i.e., randomized study). These trials are almost always inadequate with respect to choice of outcome, size, precision, and control of bias to form a sole basis for changing therapeutic practice.

A basic approach to the precision of a simple SA trial is shown in Figure 11.1. The reference probability of benefit is depicted by p_0. A hypothetical improvement is given by p_1, that which we might expect on the trial. Also shown are two 95% confidence intervals—one for a small sample size (N_1) and one for a large study (N_2). In one case, the 95% confidence interval includes the reference standard, suggesting that the typical statistical test of the null hypothesis would not reject. In the other case, the 95% confidence interval convincingly excludes p_0. Thus the larger sample size might be preferred on that basis. To be technically precise, the formal test of the hypothesis

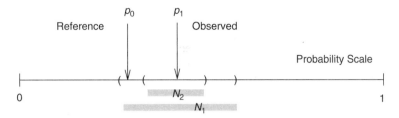

Figure 11.1 Conceptual framework for comparing the proportion of successes on a new therapy with a reference value external to the experiment.

that $p_0 = p_1$ would be performed under the null rather than the alternative hypothesis as implied by the figure. The following sections provide technical details.

11.4.1   Simple SA Designs Can Use Fixed Sample Size

The goal of SA studies is typically to estimate a clinical endpoint with a specified precision. Examples of endpoints with clinical interest are average blood or tissue levels of a drug, the proportion of patients responding (i.e., with side effects, or meeting other pre-defined criteria), and population failure rates. Here I consider a dichotomous outcome, success or failure, so that the trial results are summarized as a proportion. A useful measure of precision in this circumstance is the confidence interval around the estimated proportion. Narrow 95% confidence intervals indicate a higher degree of certainty about the location of the true effect than wide 95% confidence intervals do.

Because confidence intervals depend on the number of subjects studied, precision specified in this way can be translated into sample size. There are many methods for estimating confidence intervals for a proportion. See Vollset (1993) for a review. Here I first consider a simple method based on the normal approximation to the binomial. This is essentially the approached described above in equation (11.4).

An approximate $100(1 - \frac{\alpha}{2})\%$ confidence interval for an estimated proportion, $\widehat{p}$, is

$$\widehat{p} \pm Z_\alpha \times \sqrt{\frac{p(1-p)}{n}} \, , \tag{11.5}$$

where n is the number of patients tested and Z_α is the quantile from the normal distribution corresponding to a two-sided probability (equation 11.1). Because p is unknown, we usually substitute $\widehat{p}$ on the right-hand side of equation (11.5). This employs the normal distribution idea of mean $\pm$ standard deviation, and yields symmetric confidence intervals. It will not yield accurate results for extreme values of $\widehat{p}$ and small sample sizes.

Example 10   Consider a trial in which patients with esophageal cancer are treated with chemotherapy prior to surgical resection. A complete response is defined as the absence of macroscopic and microscopic tumor at the time of surgery. We suspect that this might occur 35% of the time and would like the 95% confidence interval of our estimate to be $\pm 15\%$. The formula yields $0.15 = 1.96 \times \sqrt{0.35(1 - 0.35)/n}$ or $n = 39$ patients required to meet the stated requirements for precision. Because 35% is just an estimate of the proportion responding and some patients may not complete the study, the actual sample size to use might be increased slightly. Expected accrual rates can be used to estimate the duration of this study in a straightforward fashion.

A rough but useful general guide for estimating sample sizes needed for proportions may be derived in the same way. Because $p(1 - p)$ is maximal for $p = 0.5$, and $Z_{0.975} \approx 2$, an approximate and conservative relationship between n, the sample size, and w, the width of the 95% confidence interval, is $n = 1/w^2$. To achieve a precision of $\pm 10\%$ (0.10) requires 100 patients, and a precision of $\pm 20\%$ (0.20) requires 25 patients. The quadratic nature of this relationship greatly increases the sample size needed to obtain high precision. This rule of thumb is not valid for proportions that deviate greatly from 0.5. For proportions less than about 0.2 or greater than about 0.8, exact binomial methods should be used.

11.4.2 Exact Binomial Confidence Limits Are Helpful

The method just sketched for SA studies uses 95% confidence intervals based on the normal approximation to the binomial distribution. This approximation may not be valid when p is extreme ($p < 0.2$ or $p > 0.8$), when the sample size is small, or when we require a high degree of confidence (e.g., 99% confidence intervals). Quantitative safety guidelines for SA trials are often concerned with probabilities that fall into these ranges. For example, we might want to stop a trial if the incidence of serious side effects is greater than 10% or the success rate is less than 20%. In these circumstances, exact confidence limits based on the tail area of the binomial distribution are more accurate than those based on a normal approximation.

Tail areas of a discrete distribution like the binomial can be obtained by summing probability terms. For example, the probability of obtaining exactly r successes out of n independent trials when the success probability for each trial is p is

$$\Pr[X = r] = \binom{n}{r} p^r (1 - p)^{n-r},$$

and the chance of r *or fewer* successes is the lower tail area

$$\Pr[X \leq r] = \sum_{k=0}^{r} \binom{n}{k} p^k (1 - p)^{n-k}.$$

Our confidence bounds exclude values of p that are not consistent with $\widehat{p} = r/n$. Therefore a $100\left(1 - \frac{\alpha}{2}\right)\%$ lower confidence bound for the true proportion is the value of p that satisfies

$$\frac{\alpha}{2} = \sum_{k=0}^{r} \binom{n}{k} p^k (1 - p)^{n-k}. \tag{11.6}$$

An upper $100\left(1 - \frac{\alpha}{2}\right)\%$ confidence bound is the value of p that satisfies

$$\frac{\alpha}{2} = \sum_{k=r}^{n} \binom{n}{k} p^k (1 - p)^{n-k}. \tag{11.7}$$

Note the resemblance of these formulaes to equation (11.1). When $r = 0$, the lower limit is defined to be 0, and when $r = n$, the upper limit is defined to be 1. The confidence limits from these equations have the properties that

$$p_L \text{ for } \frac{r}{n} = 1 - p_U \text{ for } \frac{n-r}{n}$$

and

$$p_U \text{ for } \frac{r}{n} = 1 - p_L \text{ for } \frac{n-r}{n}.$$

In general, these equations have to be solved numerically. Alternatively, some values can be found tabulated (Diem and Lentner, 1970). Because this calculation is correct and useful even when the normal approximation can be used, a flexible computer program for it is available, as explained in Appendix A. Some values are provided in Tables 11.2 and 11.3.

Note that the confidence interval, like the binomial distribution, need not be symmetric around p, even when both tails contain the same fraction of the distribution. It is customary to refer to confidence limits based on equations (11.6) and (11.7) as being "exact" because they use the correct probability distribution for $\hat{p}$, namely the binomial. In reality, they can still be approximate because of the discreteness of the distribution. These binomial confidence limits are known to be conservative, meaning wider than necessary, in general. A slightly improved confidence interval can be based on a Bayesian method (Walters, 1985). This is discussed below.

Example 11 Suppose that 3 of 19 patients respond to α-interferon treatment for multiple sclerosis. Exact 95% binomial confidence limits on the proportion responding

TABLE 11.2 Exact Binomial 95% Confidence Limits for r Responses out of n Subjects

				r		
n	0	1	2	3	4	5
5		0.005	0.053			
	0.451	0.716	0.853			
6		0.004	0.043	0.118		
	0.393	0.641	0.777	0.882		
7		0.004	0.037	0.099		
	0.348	0.579	0.710	0.816		
8		0.003	0.032	0.085	0.157	
	0.312	0.526	0.651	0.755	0.843	
9		0.003	0.028	0.075	0.137	
	0.283	0.482	0.600	0.701	0.788	
10		0.003	0.025	0.067	0.122	0.187
	0.259	0.445	0.556	0.652	0.738	0.813
11		0.002	0.023	0.060	0.109	0.167
	0.238	0.413	0.518	0.610	0.692	0.766
12		0.018	0.055	0.099	0.152	0.211
	0.221	0.385	0.484	0.572	0.651	0.723

Note: For each n, the first row is the lower bound; the second row is the upper bound. One-sided confidence interval for $r = 0$, two-sided for all other cases.

TABLE 11.3 Exact Binomial 95% Confidence Limits for r Responses out of n Subjects

n	0	1	2	3	4	5	6	7	8	9	10
13		0.002	0.019	0.050	0.091	0.139	0.192				
	0.206	0.360	0.454	0.538	0.614	0.684	0.749				
14		0.002	0.018	0.047	0.084	0.128	0.177	0.230			
	0.193	0.339	0.428	0.508	0.581	0.649	0.711	0.770			
15		0.002	0.017	0.043	0.078	0.118	0.163	0.213			
	0.181	0.320	0.405	0.481	0.551	0.616	0.677	0.734			
16		0.002	0.016	0.040	0.073	0.110	0.152	0.198	0.247		
	0.171	0.302	0.383	0.456	0.524	0.587	0.646	0.701	0.753		
17		0.001	0.015	0.038	0.068	0.103	0.142	0.184	0.230		
	0.162	0.287	0.364	0.434	0.499	0.560	0.617	0.671	0.722		
18		0.001	0.014	0.036	0.064	0.097	0.133	0.173	0.215	0.260	
	0.153	0.273	0.347	0.414	0.476	0.535	0.590	0.643	0.692	0.740	
19		0.001	0.013	0.034	0.060	0.091	0.126	0.163	0.203	0.244	
	0.146	0.260	0.331	0.396	0.456	0.512	0.566	0.616	0.665	0.711	
20		0.001	0.012	0.032	0.057	0.087	0.119	0.154	0.191	0.231	0.272
	0.139	0.249	0.317	0.379	0.437	0.491	0.543	0.592	0.639	0.685	0.728

Note: For each n, the first row is the lower bound; the second row is the upper bound. One-sided confidence interval for $r = 0$, two-sided for all other cases.

are [0.03–0.40] (Table 11.3). Approximate 95% confidence limits based on the normal approximation are [0.00–0.32]. In this case the normal approximation yielded a negative number for the lower 95% confidence bound.

An interesting special case arises when $r = 0$, that is, when no successes (or failures) out of n trials are seen. For $\alpha = 0.05$, the one-sided version of equation (11.7) is

$$0.95 = \sum_{k=1}^{n} \binom{n}{k} p^k (1 - p)^{n-k} = 1 - \binom{n}{0} p^0 (1 - p)^n$$

or

$$0.05 = (1 - p)^n.$$

Thus $\log(1 - p) = \log(0.05)/n \approx -3/n$. This yields

$$p \approx 1 - e^{-3/n} \approx \frac{3}{n} \qquad (11.8)$$

for $n \gg 3$. Actually, n does not need to be very large for the approximation to be useful. For example, for $r = 0$ and $n = 10$, the exact upper 95% confidence limit calculated from equation (11.7) is 0.26. The approximation based on equation (11.8) yields 0.30. When $n = 25$, the exact result is 0.11 and the approximation yields 0.12. So, as a general rule, to approximate the upper 95% confidence bound on an unknown proportion which yields 0 responses (or failures) out of n trials, use $3/n$. This estimate is sometimes helpful in drafting stopping guidelines for toxicity or impressing colleagues with your mental calculating ability. Some one-sided values are given in the first column of Tables 11.2 and 11.3.

Example 12   The space shuttle flew successfully 24 times between April 12, 1981, and the January 28, 1986, Challenger disaster. What is the highest probability of failure consistent with this series of 24 successes?

We could calculate a one-sided 95% confidence interval on 0 events out of 24 trials. The solution to the exact equation (11.7) yields $\hat{p} = 0.117$. Equation (11.8) gives $\hat{p} = 3/24 = 0.125$, close to the exact value. The shuttle data did not rule out a high failure rate before the Challenger accident.

Up to 2003 at the time of the Columbia accident, there were 106 successful space shuttle flights. If no improvements in launch procedures or equipment had been made (a highly doubtful assumption), an exact two-sided 95% binomial confidence interval on the probability of failure would have been (0.002–0.05). Thus we could not exclude a moderately high failure rate before Columbia (e.g., the data are not inconsistent with a 5% failure rate). Two events in 106 trials yield $\hat{p} = 0.02$ (0.006–0.066). (Some other interesting problems related to the Challenger can be found in Agresti, 1996, p. 135, and Dalal, Fowlkes, and Hoadley, 1989.)

11.4.3   Bayesian Binomial Confidence Intervals

Bayesian intervals have been much less frequently used than the classical formulation. However, they generally have good statistical properties, are flexible, and may be more appropriate for some purposes (Jaynes, 1975; Walters, 1985, 1997). A Bayesian formulation for a binomial confidence interval can be derived as discussed below. We should be a little careful regarding terminology, because the Bayesian typically refers to credible intervals rather than confidence intervals. This arises from the Bayesian view that the parameter of interest is a random quantity, rather than a fixed constant of nature as in the frequentist formulation. Thus, a Bayesian credible interval is a true probability statement about the parameter of interest.

Before any observations are taken, we might represent our knowledge of p as a uniform prior distribution on the interval $[0, 1]$. In other words, any value of p is equally likely. This assumption is subjective but leads to a workable solution. With a uniform prior, the posterior cumulative distribution for p is

$$F(p) = \frac{\int_0^p u^r (1-u)^{n-r} du}{\int_0^1 u^r (1-u)^{n-r} du} = \frac{(n+1)!}{r!(n-r)!} \int_0^p u^r (1-u)^{n-r} du.$$

Therefore, by the same reasoning as for the frequentist approach, the lower confidence bound, p_L, satisfies

$$\frac{\alpha}{2} = \frac{(n+1)!}{r!(n-r)!} \int_0^{p_L} u^r (1-u)^{n-r} du = \sum_{k=r+1}^{n+1} \binom{n+1}{k} p_L^k (1-p_L)^{n+1-k}. \quad (11.9)$$

Similarly the upper confidence bound satisfies

$$\frac{\alpha}{2} = \frac{(n+1)!}{r!(n-r)!} \int_{p_U}^1 u^r (1-u)^{n-r} du = \sum_{k=0}^{r} \binom{n+1}{k} p_U^k (1-p_U)^{n+1-k}. \quad (11.10)$$

**TABLE 11.4   Exact Bayesian Binomial 95%
Confidence Limits for *r* Responses out of *n* Subjects**

				r		
n	0	1	2	3	4	5
5		0.043	0.118			
	0.393	0.641	0.777			
6		0.037	0.099	0.184		
	0.348	0.579	0.710	0.816		
7		0.032	0.085	0.157		
	0.312	0.527	0.651	0.755		
8		0.028	0.075	0.137	0.212	
	0.283	0.482	0.600	0.701	0.788	
9		0.025	0.067	0.122	0.187	
	0.259	0.445	0.556	0.652	0.738	
10		0.023	0.060	0.109	0.167	0.234
	0.238	0.413	0.518	0.610	0.692	0.766
11		0.021	0.055	0.099	0.152	0.211
	0.221	0.385	0.484	0.572	0.651	0.723
12		0.019	0.050	0.091	0.139	0.192
	0.206	0.360	0.454	0.538	0.614	0.684

Note: For each *n*, the first row is the lower bound; the second
row is the upper bound. One-sided confidence interval for $r = 0$,
two-sided for all other cases.

Confidence intervals based on this approach tend to be less conservative (narrower)
than those based on the "classical" formulation. The difference is not large, and some
may consider the additional conservativeness of the usual method to be desirable for
designing trials. In any case it is important to see the different frameworks and to
recognize the technically superior performance of the Bayesian intervals. Tables 11.4
and 11.5 show Bayesian confidence limits for some cases that can be compared to
those from the classical approach. A useful tabulation of Bayesian confidence limits
for a binomial proportion is given by Lindley and Scott (1995).

11.4.4   A Bayesian Approach Can Use Prior Information

Consider the problem of planning a SA study when some information is already
available about efficacy, such as the success probability (response rate) for the new
treatment. In the case of cytotoxic drug development, objective sources for this evidence
may be animal tumor models, in vitro testing, DF studies, or trials of pharmacologically related drugs. The evidence could be used to plan a SA trial to yield more
precise information about the true response rate for the treatment. Thall and Simon
(1994) discuss an approach for the quantitative design of such trials. Here I consider
a rudimentary case using binomial responses to illustrate further a Bayesian approach.

At the start of the trial, the Bayesian paradigm summarizes prior information about
response probability in the form of a binomial probability distribution. For the confidence intervals presented above, the prior distribution on *p* was taken to be uninformative, meaning any value for *p* was considered equally likely. Suppose that the prior is

TABLE 11.5   Exact Bayesian Binomial 95% Confidence Limits for r Responses out of n Subjects

n	0	1	2	3	4	5	6	7	8	9	10
13		0.018	0.047	0.084	0.128	0.177	0.230				
	0.193	0.339	0.428	0.508	0.581	0.649	0.711				
14		0.017	0.043	0.078	0.118	0.163	0.213	0.266			
	0.181	0.319	0.405	0.481	0.551	0.616	0.677	0.734			
15		0.016	0.040	0.073	0.110	0.152	0.198	0.247			
	0.171	0.302	0.383	0.456	0.524	0.587	0.646	0.701			
16		0.015	0.038	0.068	0.103	0.142	0.184	0.230	0.278		
	0.162	0.287	0.364	0.434	0.499	0.560	0.617	0.671	0.722		
17		0.014	0.036	0.064	0.097	0.133	0.173	0.215	0.260		
	0.153	0.273	0.347	0.414	0.476	0.535	0.590	0.643	0.692		
18		0.013	0.034	0.060	0.091	0.126	0.163	0.203	0.244	0.289	
	0.146	0.260	0.331	0.396	0.456	0.512	0.566	0.616	0.665	0.711	
19		0.012	0.032	0.057	0.087	0.119	0.154	0.191	0.231	0.272	
	0.139	0.249	0.317	0.379	0.437	0.491	0.543	0.592	0.639	0.685	
20		0.012	0.030	0.055	0.082	0.113	0.146	0.181	0.218	0.257	0.298
	0.133	0.238	0.304	0.363	0.419	0.472	0.522	0.570	0.616	0.660	0.702

Note: For each n, the first row is the lower bound; the second row is the upper bound. One-sided confidence interval for $r = 0$, two-sided for all other cases.

taken to be equivalent to evidence that there have been r_1 responses out of n_1 patients on the treatment. This information about the number of responses, r, or equivalently the true response rate, p, could be summarized as a binomial distribution

$$f(r) = \binom{n_1}{r_1} p^{r_1} (1 - p)^{n_1 - r_1}.$$

The current best estimate of the response rate is $\widehat{p} = r_1/n_1$. The goal of the SA trial is to increase the precision of the estimate of $\widehat{p}$ using additional observations.

When the experiment is completed, the response rate can be estimated by $\widehat{p} = r/n$, where there are r responses out of n trials, and a confidence interval can be calculated (e.g., from equations 11.6 and 11.7). To assist with sample size determination, an alternate parameterization of equations (11.6) and (11.7) for binomial tail areas could be used,

$$\frac{\alpha}{2} = \sum_{k=0}^{r} \binom{n}{k} \left(\frac{[np]}{n} - w \right)^k \left(1 - \left(\frac{[np]}{n} - w \right) \right)^{n-k} \tag{11.11}$$

and

$$\frac{\alpha}{2} = \sum_{k=r}^{n} \binom{n}{k} \left(\frac{[np]}{n} + w \right)^k \left(1 - \left(\frac{[np]}{n} + w \right) \right)^{n-k}, \tag{11.12}$$

where w is the width of the confidence interval and $[\cdot]$ denotes the nearest integer function. Specifying a value for w at the completion of the study and assuming a value for $\hat{p}$, will allow calculation of a total sample size by solving equations (11.11) and (11.12) for n.

This problem is somewhat more tractable when using the incomplete beta function,

$$B_p(a, n - a + 1) = \sum_{k=a}^{n} \binom{n}{k} p^k (1 - p)^{n-k},$$

and its inverse, which I will denote by $B^{-1}(p, a, b)$. Equations (11.11) and (11.12) then become

$$p - w = B^{-1}\left(\frac{\alpha}{2}, r, n - r + 1\right) = B^{-1}\left(\frac{\alpha}{2}, np, n - np + 1\right) \tag{11.13}$$

and

$$p - w = B^{-1}\left(1 - \frac{\alpha}{2}, r + 1, n - r\right) = B^{-1}\left(1 - \frac{\alpha}{2}, np + 1, n - np\right), \tag{11.14}$$

which are more obscure but easier to solve numerically using standard software.

Example 13 Suppose that our prior information is $n_1 = 15$ observations with $r_1 = 5$ responses so that $\hat{p} = 1/3$. Assume that our final sample size should yield a 99% credible interval with width $w = 0.15$ and the true response rate is near $1/3$. Equations (11.13) and (11.14) will be satisfied by $n = 63$ and $n = 57$ (with $r \approx 20$). This solution will also be approximately valid for $\hat{p}$ near $\frac{1}{3}$. Therefore 42 to 48 additional observations in the SA study will satisfy the stated requirement for precision. An approximate solution can also be obtained from equation (11.5) with $Z_\alpha = 2.57$. Then

$$n \approx 0.33 \times 0.66 \times \left(\frac{2.57}{0.15}\right)^2 = 64,$$

which compares favorably with the calculation based on exact confidence limits.

11.4.5   Likelihood-Based Approach for Proportions

The binomial likelihood function is

$$e^{\mathcal{L}(p)} = p^k (1 - p)^{n-k},$$

where p is the probability of success, k is the number of successes, and n is the number of subjects. The likelihood ratio (relative evidence) for comparing the observed success rate, p_1, to a hypothetical value, p_0, is

$$e^\Lambda = \frac{p_1^k (1 - p_1)^{n-k}}{p_0^k (1 - p_0)^{n-k}}$$

so that

$$\Lambda = k \log \frac{p_1}{p_0} + (n - k) \log \frac{1 - p_1}{1 - p_0}.$$

TABLE 11.6 Sample Sizes for One-Sample Binomial Comparisons Using Relative Evidence and Equation (11.15)

p_1	Λ	p_0 0.05	0.1	0.15	0.2	0.25	0.3	0.35	0.4	0.45
0.5	8	3	4	6	9	14	24	44	102	414
	32	4	7	10	16	24	40	73	170	690
	64	5	8	12	19	29	48	88	204	828
	128	6	9	14	22	34	56	103	238	966
0.4	8	4	7	11	20	38	92	386		
	32	6	11	19	33	64	153	643		
	64	7	13	23	40	77	184	772		
	128	9	16	26	46	90	215	901		
0.3	8	6	14	29	74	325				
	32	11	23	48	123	541				
	64	13	27	58	148	650				
	128	15	32	67	172	758				
0.2	8	15	47	230						
	32	25	78	384						
	64	30	94	460						
	128	35	109	537						

As a likelihood ratio test, k and n represent data in this equation (see also Section 14.4.2), but a slight change in perspective allows us to imagine the data required to produce some specified relative evidence. Solving for n yields

$$n = \frac{\Lambda}{p_0 \log(\Theta) + \log \frac{1-p_1}{1-p_0}} = \frac{\Lambda}{\log \frac{p_1}{p_0} - (1 - p_1) \log(\Theta)}, \qquad (11.15)$$

where Θ is the odds ratio for p_1 versus p_0. Examples are shown in Table 11.6.

11.4.6 Confidence Intervals for a Mean Provide a Sample Size Approach

When the treatment effect of interest is the mean of a distribution, an approach similar to that for proportions can be used to estimate sample size. We assume that the sample mean has a normal distribution. This leads to two essential differences compared to the method sketched above for proportions. First, both the mean and standard deviation must be specified for the normal distribution, whereas the binomial mean and standard deviation could both be expressed in terms of p. Preliminary data may be needed to estimate the standard deviation. Here I assume that the standard deviation is known.

Second, there is no bounded range for the mean and standard deviation for the normal distribution, whereas the binomial p was bounded by the interval $[0, 1]$. Therefore there is no universal choice of scale for specifying the width of confidence intervals for the normal distribution. However, one could express precision either in absolute terms, or relative to the mean or variance.

If $\widehat{\mu}$ is the estimated mean from n observations, the $1 - \frac{\alpha}{2}$ confidence interval for the true mean is $\widehat{\mu} \pm Z_{1-\alpha/2} \times \sigma/\sqrt{n}$, where σ is the standard deviation. If our tolerance for the width of the confidence interval is $w = Z_{1-\alpha/2} \times \sigma/\sqrt{n}$, the sample size must satisfy

$$n = \left(\frac{Z_{1-\alpha/2}\sigma}{w} \right)^2.$$

See also Section 11.3.1 and equation (11.3).

Example 14   Suppose that we are measuring the serum level of a new drug such that we require a precision of ± 0.5 mg/dl in the 95% confidence interval, and the standard deviation of the distribution of drug levels is $\sigma = 5$ mg/dl. The required sample size is then $n = \left(\frac{1.96 \times 5}{0.5} \right)^2 = 384.2$ or 385 subjects.

The tolerance, w, specifies one-half the width of the confidence interval in absolute terms. It is not a percentage and relates to the scale of the distribution only in absolute terms. In some cases this may be appropriate. However, precision could be expressed relative to either μ or σ. If we specify the width of the confidence interval in terms of the standard deviation, $w = w'\sigma$,

$$n = \left(\frac{Z_\alpha \sigma}{w'\sigma} \right)^2 = \left(\frac{Z_\alpha}{w'} \right)^2.$$

Similarly in terms of the mean, $w = w''\mu$,

$$n = \left(\frac{Z_\alpha \sigma}{w''\mu} \right)^2 = \left(\frac{Z_\alpha}{w''} \right)^2 \left(\frac{\sigma}{\mu} \right)^2,$$

when the mean is not zero. The parameters w' and w'' have convenient interpretations: w' is the desired tolerance expressed as a fraction of a standard deviation and w'' has a similar interpretation with respect to the mean. The ratio of standard deviation to mean, σ/μ, is the "coefficient of variation."

Example 15   To narrow the 95% confidence interval to $\pm 5\%$ of the mean, set $w'' = 0.05$,

$$n = \left(\frac{1.96}{0.05} \right)^2 \left(\frac{\sigma}{\mu} \right)^2 \approx 1536 \times \left(\frac{\sigma}{\mu} \right)^2.$$

Unless the coefficient of variation is quite small, a large sample size will be required to attain this precision. From the other perspective, if the 95% confidence interval for the true mean is one-fourth of a standard deviation, $w' = 0.25$, then

$$n = \left(\frac{1.96}{0.25} \right)^2 \approx 62.$$

11.4.7   Confidence Intervals for Event Rates Can Determine Sample Size

Some safety and efficacy trials employ time-to-event measurements with censoring, such as the death, recurrence, or overall failure rate in the population under study. This type of study design is becoming more common in cancer trials as new agents are developed that do not work through a classical cytotoxic mechanism. For example, some new drugs target growth factor receptors or inhibit angiogenesis. These may not shrink tumors but might improve survival. The classical SA trial using response (tumor shrinkage) as the primary design variable will not work for such agents. However, many of the concepts that help determine sample size for proportions also carry over to event rates.

Absolute Precision

As in the discussion above for proportions, the confidence interval (precision) required for a failure rate on an absolute scale can be used to estimate sample size. This situation is shown in Figure 11.2—note the similarity to Figure 11.1 for proportions. The reference failure rate is λ_0, which might be obtained from earlier cohort studies or clinical knowledge. On treatment, we expect the failure rate to be λ_1. If the improvement is real, we would like for the trial to reliably resolve the difference between the two failure rates; that is, the confidence interval around λ_1 should exclude λ_0.

A study of this design will most likely include a fixed period of accrual and then a period of follow-up, during which the cohort can be observed for additional events. Sometimes accrual can continue to the end of the study, which shortens the length of the trial. However, patients accrued near the end of the study may not contribute much information to the estimate of the failure rate.

Assume that accrual is constant at rate a per unit time over some interval T, and that a period of continued follow-up is used to observe additional events. Also assume that the failure rate is constant over time (exponential event times) and that there are no losses to follow-up. At the end of the study, we summarize the data using methods presented in Chapter 16, so that the estimated failure rate is

$$\widehat{\lambda} = \frac{d}{\sum t_i} \, ,$$

where d is the number of events or failures and the sum is over all follow-up times t_i. An approximate confidence interval for the failure rate is

$$\widehat{\lambda} \pm Z_{1-\alpha/2} \frac{\lambda}{\sqrt{d}} \, .$$

Alternatively, an approximate confidence interval for $\log(\lambda)$ is

$$\log(\widehat{\lambda}) \pm \frac{Z_{1-\alpha/2}}{\sqrt{d}} \, .$$

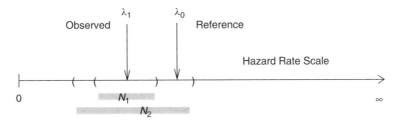

Figure 11.2 Study design for a SA trial with a failure time outcome.

Because of the design of this study, d is actually a function of time. Anticipating results from section 12.3.2, the number of events depends on time such that

$$
d(t) = \begin{cases} a_0 t - \dfrac{a_0}{\lambda}(1 - e^{-\lambda t}) & \text{if } t \le T, \\[2mm] a_0 T - \dfrac{a_0}{\lambda}(e^{-\lambda(t-T)} - e^{-\lambda t}) & \text{if } t > T, \end{cases}
\tag{11.16}
$$

where t is the total study duration. Thus the confidence interval for λ depends on time. Note that for fixed T, as $t \to \infty$, the number of failures approaches $a_0 T$, the total number of subjects accrued. We could express the approximate confidence interval as

$$
\widehat{\lambda}\left(1 \pm \frac{Z_{1-\alpha/2}}{\sqrt{d(t)}}\right) .
$$

If w is the desired width of the confidence interval expressed as a fraction of λ, the study design parameters must satisfy

$$
d(t) = \left(\frac{Z_{1-\alpha/2}}{w}\right)^2 ,
$$

which has a familiar form. A similar derivation could be used for $\log(\lambda)$.

Relative Precision

We can express the precision of an estimated event rate either in relative terms (e.g., $\pm 10\%$), or in absolute terms. Working with precision expressed on a percentage basis is relatively straightforward. Suppose that the hypothesized event rate is λ and we would like to provide an estimate that is $\pm P\%$. In other words, we would like to be fairly certain that the event rate lies between $\lambda_L = \lambda(1 - P)$ and $\lambda_U = \lambda(1 + P)$. On a logarithmic scale, an approximate confidence interval is $\log(\lambda) \pm Z_{1-\alpha/2}/\sqrt{d}$, where d is the number of events observed. This means that $\log(1 - P) = Z_{1-\alpha/2}/\sqrt{d}$ and $\log(1 + P) = Z_{1-\alpha/2}/\sqrt{d'}$, where the same d will not necessarily satisfy both equations. Thus, we are led to choosing a sample size based on the maximum of

$$
d = \left(\frac{Z_{1-\alpha/2}}{\log(1 \pm P)}\right)^2 .
$$

For example, if we intend for a two-sided 95% confidence interval on the failure rate to yield $\pm 20\%$, then $d = \left[1.96/\log(1.2)\right]^2 = 116$ events. The number of study subjects required to yield 116 events is likely to be substantially higher, and the following discussion regarding accrual dynamics is also relevant here.

Example 16   In patients with malignant gliomas, the historical median failure time is 8 months ($\lambda = \log(2)/8 = 0.087$ events per patient-month). A safety and efficacy trial using a new agent intends to estimate the failure rate with a precision of $\pm 25\%$ (95% confidence interval) so that is can be compared to the historical rate. A promising improvement in the failure rate will motivate a CTE trial. The required number of events for this degree of precision is $(1.96/0.25)^2 = 61$. The accrual rate for the SA trial is expected to be 9 patients per month. For the expected accrual and failure rates, one can calculate that the following combinations of accrual periods and total study durations yield 61 events (all times are in months): $(6.8, \infty)$; $(9, 21)$; $(12, 16)$; $(15.25, 15.25)$. Many other combinations would also satisfy the design equation, but the shortest total trial duration possible is 15.25 months.

11.4.8 Likelihood-Based Approach for Event Rates

Assuming a normal model for $\log(\lambda)$, the likelihood is

$$e^{\mathcal{L}(\lambda)} = \exp\left(-\frac{\left(\log(\widehat{\lambda}) - \log(\lambda)\right)^2}{2/d}\right),$$

where λ is the true hazard and $\widehat{\lambda}$ is the observed hazard. This represents the relative evidence for $\widehat{\lambda}$ versus λ so that

$$\Lambda = \frac{\left(\log(\widehat{\lambda}) - \log(\lambda)\right)^2}{2/d}$$

or

$$d = \frac{2\Lambda}{\left(\log(\widehat{\lambda}) - \log(\lambda)\right)^2} = \frac{2\Lambda}{\log(\Delta)^2},$$

where Δ is the hypothesized hazard ratio. The previous discussion regarding accrual dynamics pertains to this equation also. To connect this with hypothesis tests, the evidence produced by our trial can be partitioned into any appropriate allocation of type I and type II errors,

$$d = \frac{2\Lambda}{\log(\Delta)^2} = \frac{Z^2}{\log(\Delta)^2} = \frac{(Z_\alpha + Z_\beta)^2}{\log(\Delta)^2}.$$

However, the focus here is on controlling the total error, a distinction that has some practical consequences.

Example 17 Returning to the glioma example, if we expect a new treatment to reduce the hazard of failure by a factor of 1.75 (a large effect given the clinical circumstances), modest evidence ($\Lambda = 8$) regarding this hazard ratio can be produced by $d = 16/\log(1.75)^2 = 51$ events.

11.4.9 Ineffective or Unsafe Treatments Should Be Discarded Early

The purpose of middle development is to discard unsafe treatments or those that are inactive or less active than existing therapy (Section 6.5.2). SA studies in oncology assess the response (tumor shrinkage) rate of patients treated with new drugs or modalities. If the response rate is low or there is too high an incidence of side effects, investigators would like to terminate the trial as early as possible. This has led to the widespread use of "early stopping rules" for these studies that permit accrual to be terminated before the fixed sample size end. A detailed look at this issue is given in Chapter 14, but some designs are introduced here because they are often used to determine the size of SA trials.

An historic approach to this problem is the following: Suppose that 0.2 (approximately) is the lowest response rate that investigators consider acceptable for a new treatment. The exact binomial upper 95% confidence limit on 0 responses out of 12 tries is 0.22. Therefore, if none of the first 12 study participants respond, the treatment can be discarded because it most likely has a response rate less than 0.22. This rule was proposed by Gehan (1961) and is still used in some SA trials because it is simple to employ and understand. It is easily modified using exact binomial confidence limits for target proportions other than 0.2 (Tables 11.2 and 11.3). A more flexible approach to deal with this problem is outlined in the next section.

11.4.10 Two-Stage Designs Are Efficient

Optimal two-stage designs for SA trials are discussed by Simon (1989). The suggested design depends on two clinically important response rates p_0 and p_1 and type I and II error rates α and β. If the true probability of response is less than some clinically uninteresting level, p_0, the chance of accepting the treatment for further study is α. If the true response rate exceeds some interesting value, p_1, the chance of rejecting the treatment should be β. The study terminates at the end of the first stage only if the treatment appears ineffective. The design does not permit stopping early for efficacy.

The first stage (n_1 patients) is relatively small. If a small number ($\leq r_1$) of responses are seen at the end of the first stage, the treatment is abandoned. Otherwise, the trial proceeds to a second stage (n patients total). If the total number of responses after the second stage is large enough ($>r$), the treatment is accepted for further study. Otherwise, it is abandoned. The size of the stages and the decision rules can be chosen optimally to test differences between two response rates of clinical interest.

The designs can be specified quantitatively in the following way: Suppose that the true probability of response using the new treatment is p. The trial will stop after the first stage if r_1 or fewer responses are seen in n_1 patients. The chance of this happening is

$$B(r_1; p, n_1) = \sum_{k=0}^{r_1} \binom{n_1}{k} p^k (1-p)^{n_1-k}, \tag{11.17}$$

which is the cumulative binomial mass function given above in equation (11.6). At the end of the second stage the treatment will be rejected if r or fewer responses are seen. The chance of rejecting a treatment with true response probability p is the chance of stopping at the first stage plus the chance of rejecting at the second stage if the first stage is passed. We must be careful to count all of the ways of passing the first stage but not passing the second. The probability of rejecting the treatment is

$$Q = B(r_1; p, n_1) + \sum_{k=r_1+1}^{\min(n_1,r)} \binom{n_1}{k} p^k (1-p)^{n_1-k} B(r-k; p, n_2), \tag{11.18}$$

where $n_2 = n - n_1$.

When the design parameters, p_0, p_1, α, and β are specified, values of r_1, r_2, n_1, and n_2 can be chosen that satisfy the type I and type II error constraints. The design that yields the smallest *expected* sample size under the null hypothesis is defined as "optimal." Because a large number of designs could satisfy the error constraints, the search

TABLE 11.7   Optimal Two-Stage Designs for SA (Phase II) Trials for $p_1 - p_0 = 0.20$ and $\alpha = 0.05$

p_0	p_1	β	r_1	n_1	r	n	$E\{n \mid p_0\}$
0.05	0.25	0.2	0	9	2	17	12
		0.1	0	9	3	30	17
0.10	0.30	0.2	1	10	5	29	15
		0.1	2	18	6	35	23
0.20	0.40	0.2	3	13	12	43	21
		0.1	4	19	15	54	30
0.30	0.50	0.2	5	15	18	46	24
		0.1	8	24	24	63	35
0.40	0.60	0.2	7	16	23	46	25
		0.1	11	25	32	66	36
0.50	0.70	0.2	8	15	26	43	24
		0.1	13	24	36	61	34
0.60	0.80	0.2	7	11	30	43	21
		0.1	12	19	37	53	30
0.70	0.90	0.2	4	6	22	27	15
		0.1	11	15	29	36	21

Note: The last column gives the expected sample size when the true response rate is p_0.

for optimal ones requires an exhaustive computer algorithm. Although many useful designs are provided in the original paper and some are tabulated here, a computer program for finding optimal designs is described in Appendix A.

Trial designs derived in this way are shown in Tables 11.7 and 11.8. Multistage designs for SA studies can be constructed in a similar way. Three-stage designs have recently been proposed (Ensign et al., 1994). However, two-stage designs are simpler and nearly as efficient as designs with more stages, or those that assess treatment efficacy after each patient. These designs are discussed in Chapter 14.

It is important to recognize that many combinations of sample sizes and decision rules will satisfy the error criteria specified for a two-stage design. Sensible definitions of "optimal" (e.g., minimum expected sample size) help us select a single design. It is possible that the best definition of "optimal" would change under special circumstances.

Example 18   Suppose that we intend to detect a success rate of 40% compared to a background rate of 20% with $\alpha = 0.05$ and $\beta = 0.1$, Table 11.7 indicates that the decision rules and stage sizes should be 4 out of 19 and 15 out of 54. For total sample sizes less than or equal to 54, there are 610 designs that satisfy the error criteria, and for total sample sizes less than or equal to 75, there are 4330 designs that satisfy the error criteria. Designs with decision rules based on larger denominators will have smaller type I and type II error rates. The design that allows a decision with the smallest number of successes (not listed in Table 11.7) is 2 out of 15 and 13 out of 45.

11.4.11   Randomized SA Trials

As discussed in Chapter 6, the real purpose of randomization in middle development is to facilitate selection of the best-performing therapy rather than to estimate the

**TABLE 11.8   Optimal Two-Stage Designs for SA
(Phase II) Trials for $p_1 - p_0 = 0.15$ and $\alpha = 0.05$**

p_0	p_1	β	r_1	n_1	r	n	$E\{n \mid p_0\}$
0.05	0.20	.2	0	10	3	29	18
		.1	1	21	4	41	27
0.10	0.25	.2	2	18	7	43	25
		.1	2	21	10	66	37
0.20	0.35	.2	5	22	19	72	35
		.1	8	37	22	83	51
0.30	0.45	.2	9	27	30	81	42
		.1	13	40	40	110	61
0.40	0.55	.2	11	26	40	84	45
		.1	19	45	49	104	64
0.50	0.65	.2	15	28	48	83	44
		.1	22	42	60	105	62
0.60	0.75	.2	17	27	46	67	39
		.1	21	34	64	95	56
0.70	0.85	.2	14	19	46	59	30
		.1	18	25	61	79	43
0.80	0.95	.2	7	9	26	29	18
		.1	16	19	37	42	24

Note: The last column gives the expected sample size when the
true response rate is p_0.

magnitude of treatment differences. Selection is an easier task (statistically more efficient) than estimation because it determines only relative ranking and not actual effect size. The probability of selecting the superior treatment (correct selection) provides a quantitative approach to determining sample size for randomized SA trials.

Two Groups
Suppose that there are two treatment groups of size n and the treatment group mean is the basis of the selection. Assume that the true means are μ and $\mu + \delta$, the observed means are $\widehat{\mu}_1$ and $\widehat{\mu}_2$ respectively, and that the variance is known. For every possible value of the first mean, we must calculate the chance that the second mean is greater, and sum all the probabilities. Therefore the chance that $\widehat{\mu}_2 \geq \widehat{\mu}_1$ is

$$\Pr(\textit{correct ordering}) = \int_{-\infty}^{\infty} \phi(u) \int_{u}^{\infty} \phi(u + \delta) du \; du = \int_{-\infty}^{\infty} \phi(u) \Phi(u + \delta) \; du,$$

(11.19)

where $\phi(\cdot)$ is a normal density function with mean 0 and variance $1/n$ and $\Phi(\cdot)$ is the cumulative normal distribution function. The structure of this formula is quite general, but for two groups a shortcut can be used.

The probability of a correct ordering of two groups is also the chance that the ordered difference between the observed means is positive. The difference $\widehat{\mu}_2 - \widehat{\mu}_1$ has mean δ and variance $2/n$, so that

$$\Pr(\textit{correct ordering}) = \int_{0}^{\infty} \phi(u + \delta) du,$$

**TABLE 11.9   Sample Size Needed for Reliable
Correct Ordering of Two Means**

δ	P				
	0.8	0.85	0.9	0.95	0.99
0.10	142	215	329	542	1083
0.15	63	96	146	241	482
0.20	36	54	83	136	271
0.25	23	35	53	87	174
0.30	16	24	37	61	121
0.35	12	18	27	45	89
0.40	9	14	21	34	68
0.45	7	11	17	27	54
0.50	6	9	14	22	44
0.55	5	8	11	18	36
0.60	4	6	10	16	31
0.65	4	6	8	13	26
0.70	3	5	7	12	23
0.75	3	4	6	10	20
0.80	3	4	6	9	17
0.85	2	3	5	8	15
0.90	2	3	5	7	14
0.95	2	3	4	6	12
1.00	2	3	4	6	11

Note: P is the probability of correct ordering.

where $\phi(\cdot)$ has mean 0 and variance $2/n$. Thus we can choose n to make this probability as high as needed, or conversely, solve numerically for n, given a specified probability. For example, to have 90% chance of a correct ordering when $\delta = 0.50$, 14 subjects are needed in each group (Table 11.9).

For a binary outcome the same theory can be used after transforming the proportions p and $p + \delta$ using the arcsin–square root,

$$\theta = 2 \left(\arcsin \sqrt{p + \delta} - \arcsin \sqrt{p} \right),$$

which is approximately normal with mean θ and variance $2/n$. For example, when $\delta = 0.1$ and $p = 0.4$, $\theta = 0.201$. Note that in Table 11.9, 83 subjects are required in each group to yield 90% chance of correct ordering.

There is a small problem with sample sizes determined for binary outcomes using this method. The equations are derived for a continuous distribution whereas binomial outcomes are discrete. Suppose that r_1 and r_2 are the number of responses in the treatment groups with true success probabilities p and $p + \delta$ respectively. For $r_2 > r_1$, the binomial distribution yields

$$\Pr(\textit{correct ordering}) = \sum_{i=0}^{n-1} b(p) \sum_{j=i+1}^{n} b(p + \delta),$$

whereas for $r_2 \geq r_1$,

$$\Pr(\textit{correct ordering}) = \sum_{i=0}^{n} b(p) \sum_{j=i}^{n} b(p + \delta).$$

From these "exact" equations, to have 90% chance of correct ordering defined strictly as $r_2 > r_1$, 91 subjects are needed in each group. To have 90% chance of correct ordering defined as $r_2 \geq r_1$, 71 subjects are needed in each group. The normal approximation yielded 83, a compromise between these numbers.

For correctly ordering hazards of failure from time-to-event outcomes, the derivation for the normal case applies with the following modifications. The log hazard is approximately normal with variance $1/d$, where d is the number of events observed in the group. To simplify, assume that the number of events in each of two groups is d. An appropriate transformation for hazards is

$$\eta = \log(\lambda + \delta) - \log(\lambda) = \log\left(\frac{\lambda + \delta}{\lambda}\right) = \log(\Delta),$$

which is approximately normal with mean η and variance $2/d$. For example, when $\delta = 0.05$, $\lambda = 0.1$, and $\eta = 0.405$. To have a 90% chance correctly ordering the hazards, Table 11.9 shows that $d = 21$ *events* are required in each group. The same accrual dynamics discussed in Section 11.4.7 apply, so more than 21 subjects must be placed on study to yield the required number of events. A more elaborate approach to this type of problem is given by Liu et al., (1993).

More Than Two Groups
See Desu and Raghavarao (1990, ch. 6) for a general discussion of ranking and selection. The essential sample size problem is to make a reliable selection of the best out of $k \geq 2$ treatments. We assume that the sample size in each group is the same and that the variances are equal and known. The best treatment can be chosen by chance with probability $1/k$ so that $P > 1/k$, where P is the probability of correct selection.

For more than two groups, equation (11.19) generalizes to

$$\int_{-\infty}^{\infty} \Phi^{k-1}(u + \gamma)\phi(u)\,du \geq P. \tag{11.20}$$

Values of γ that satisfy this equation for some k and P have been tabulated by Bechhofer (1954) and Desu and Sobel (1968). They can be used to determine sample size according to

$$n \geq \left(\gamma\frac{\sigma}{\delta}\right)^2 = \left(\frac{\gamma}{\xi}\right)^2, \tag{11.21}$$

where σ is the standard deviation, δ is the minimum difference between the largest mean and all others, and ξ is the effect size of that difference. Equation (11.21) looks similar to other precision formulae given earlier in this chapter. It is relatively easy to calculate the necessary γ values that satisfy equation (11.20) using current software.

TABLE 11.10   Values of γ That Satisfy Equation (11.20)

			k	
P	2	3	4	5
0.80	1.190	1.652	1.893	2.053
0.85	1.466	1.908	2.140	2.294
0.90	1.812	2.230	2.452	2.600
0.95	2.326	2.710	2.916	3.055
0.99	3.290	3.617	3.797	3.920

Some values are given in Table 11.10 that can also be used to determine the sample sizes for the two group selection cases given above.

Example 19   Suppose there are four treatment groups, the effect size (δ/σ) is 0.5, and we intend to select the correct group with probability 99%. From Table 11.10, $\gamma = 3.797$. From equation (11.21), the required sample size is $n = (3.797/0.5)^2 = 58$ per group.

11.5   COMPARATIVE TRIALS

For CTE studies the discussion that follows emphasizes an approach based on a planned hypothesis test when the trial is completed. This is a convenient and frequently used perspective for determining the size of comparative trials, and motivates the use of the term "power." The null hypothesis usually represents equivalence between the treatments. The alternative value for the hypothesis test is chosen to be the smallest difference of clinical importance between the treatments. Following this, the size of the study is planned to yield a high probability of rejecting the null hypothesis if the alternative hypothesis is true. Therefore the test statistic planned for the analysis dictates the exact form of the power equation. Even though the power equation depends on the test being employed, there is a similar or generic form for many different statistics.

11.5.1   How to Choose Type I and II Error Rates

Convention holds that most clinical trials should be designed with a two-sided α-level set at 0.05 and 80% or 90% power ($\beta = 0.2$ or 0.1, respectively). In practice, the type I and II error rates should be chosen to reflect the consequences of making the particular type of error. For example, suppose that a standard therapy for a certain condition is effective and associated with few side effects. When testing a competing treatment, we would probably want the type I error rate to be small, especially if it is associated with serious side effects, to reduce the chance of a false positive. We might allow the type II error rate to be higher, indicating the lower seriousness of missing an effective therapy because a good treatment already exists. Circumstances like this are commonly encountered when developing cytotoxic drugs for cancer treatment.

   In contrast, suppose that we are studying prevention of a common disease using safe agents such as diet or dietary supplements. There would be little harm in the widespread application of such treatments, so the consequences of a type I error are not severe. In fact some benefits might occur, even if the treatment was not preventing the target condition. In contrast, a type II error would be more serious because a safe, inexpensive, and possibly effective treatment would be missed. In such cases there is a rationale for using a relaxed definition of "statistical significance," perhaps $\alpha = 0.10$, and a higher power, perhaps $\beta = 0.01$.

Special attention to the type I and II error rates may be needed when designing trials to demonstrate equivalence (noninferiority) of two treatments. This topic is discussed in Section 11.5.9.

11.5.2   Comparisons Using the t-Test Are a Good Learning Example

Suppose that the endpoint for a comparative clinical trial is a measurement, so the treatment comparison consists of testing the difference of the estimated means of the two groups. Assume that the true means in the treatment groups are μ_1 and μ_2 and the standard deviation of the measurement in each patient is σ. Let the treatment difference be $\Delta = \mu_1 - \mu_2$. The null hypothesis is $H_0 : \Delta = 0$. Investigators would reject the null hypothesis if $|\Delta|$ exceeds the critical value, c, where

$$c = Z_\alpha \times \sigma_\Delta$$

and σ_Δ is the standard deviation of Δ (Figure 11.3). In other words, if the estimated difference between the treatment means is too many standard deviations away from 0, we would disbelieve that the true difference is 0. Under the alternative hypothesis, the distribution of Δ is centered away from 0 (the right-hand curve in Figure 11.3).

The *power* of our statistical test is the area under the alternative distribution to the right of c, i.e., the probability of rejecting H_0 when the alternative hypothesis is

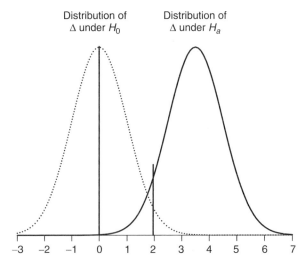

Figure 11.3   Sampling distribution of an estimate under the null and alternative hypotheses. Vertical lines are drawn at the null $\Delta = 0$, and the critical value $c = 1.96$ as explained in the text.

true. This area can be calculated by standardizing c with respect to the alternative distribution

$$-Z_\beta = \frac{c - \Delta}{\sigma_\Delta} = \frac{Z_\alpha \times \sigma_\Delta - \Delta}{\sigma_\Delta} \; ,$$

that is, by subtracting the mean of the alternative distribution and dividing by its standard deviation. The minus sign for Z_β comes from the fact that areas are tabulated from the left tail, whereas we are taking the area of the right tail of the distribution (equation 11.1). Thus

$$-Z_\beta = \frac{Z_\alpha \times \sigma_\Delta - \Delta}{\sigma_\Delta} = Z_\alpha - \frac{\Delta}{\sigma_\Delta}$$

or

$$Z_\alpha + Z_\beta = \frac{\Delta}{\sigma_\Delta} \; . \tag{11.22}$$

Now

$$\sigma_\Delta = \sqrt{\frac{\sigma^2}{n_1} + \frac{\sigma^2}{n_2}} = \sigma \sqrt{\frac{1}{n_1} + \frac{1}{n_2}} \; ,$$

assuming that the groups are independent and of sizes n_1 and n_2. Substituting into (11.22) and squaring both sides yields

$$\frac{1}{n_1} + \frac{1}{n_2} = \frac{\Delta^2}{(Z_\alpha + Z_\beta)^2 \sigma^2}. \tag{11.23}$$

Now suppose that $n_1 = r n_2$ (r is the *allocation ratio*) so that equation (11.23) becomes

$$\frac{1}{r n_2} + \frac{1}{n_2} = \frac{1}{n_2} \frac{r + 1}{r} = \frac{\Delta^2}{(Z_\alpha + Z_\beta)^2 \sigma^2}$$

or

$$n_2 = \frac{r + 1}{r} \frac{(Z_\alpha + Z_\beta)^2 \sigma^2}{\Delta^2}. \tag{11.24}$$

The denominator of the right-hand side of (11.24), expressed as $(\Delta/\sigma)^2$, is the square of the number of standard deviations between the null and alternative treatment means. All of the factors that affect the statistical power are evident from equation (11.24): the variance of an observation, the difference we are trying to detect, the allocation ratio, and the type I error level for the statistical comparison.

Some convenient values that solve equation (11.24) are shown in Table 11.11. A small amount of ambiguity remains, however. Equation (11.24) gives the sample size in one group, and its solution is not necessarily an integer. Table 11.11 rounds to the nearest integer and multiplies by 2 to obtain a total sample size. Other strategies are possible, such as rounding to the next higher integer. It is probably best to round first so that the total sample size is divisible by 2.

TABLE 11.11   Approximate Total Sample Sizes for Comparisons Using the t-Test and Equal Group Sizes

	$\beta = 0.1$		$\beta = 0.2$	
Δ/σ	$\alpha = 0.05$	$\alpha = 0.10$	$\alpha = 0.05$	$\alpha = 0.10$
0.25	672	548	502	396
0.50	168	138	126	98
0.75	75	62	56	44
1.00	42	34	32	24
1.25	28	22	20	16
1.50	18	16	14	12

Note: Δ is the difference in the treatment group means and σ is the standard deviation. See Equation (11.24).

Equation (11.24) can be solved for the type II error,

$$Z_\beta = \frac{\Delta\sqrt{N}}{2\sigma} - Z_\alpha,$$

where I have set $r = 1$ and $n_1 = n_2 = N/2$. Then the power is

$$1 - \beta = \Phi\left(\frac{\Delta\sqrt{N}}{2\sigma} - Z_\alpha\right),$$

where I have used the notational conventions indicated at the end of Section 11.2.2.

Although this power equation has been derived assuming normal distributions (so called z-test), it yields values nearly correct for the t-test. One could increase accuracy by using quantiles from the t-distribution in place of Z_α and Z_β. However, when using the t-test, the power calculations are made more difficult by the need to evaluate the noncentral t-distribution. Also we have assumed that the variance is known. In some situations the variance will be estimated from the observed data. This slightly increases the sample size required. However, the effect is small, amounting to an increase of only one or two subjects for sample sizes near 20 (Snedecor and Cochran, 1980). Consequently the correction can be ignored for most clinical trials.

Although many important test statistics for clinical trials fit assumptions of normality, at least approximately, some important cases do not. For example, power and sample size equations for analyses of variance involve F-distributions that are computationally more cumbersome than Gaussian distributions. Because these are uncommon designs for clinical trials, details are not given here. Approaches to this problem can be found in Winer (1971). Some important nonnormal cases more commonly encountered in clinical trials are discussed below.

11.5.3   Likelihood-Based Approach

Suppose that we observe n independent values that arise from a normal distribution with unknown mean μ. For simplicity, the variance, σ^2, will be assumed known. The likelihood, $e^{\mathcal{L}(\mathbf{X}|\mu)}$ (where $\mathcal{L}(\mathbf{X}|\mu)$ is the log likelihood) is the probability of observing the data under the normal model,

$$e^{\mathcal{L}(\mathbf{X}|\mu)} = \frac{1}{\sqrt{2\pi}\sigma} \prod_{i=1}^{n} \exp\left(-\frac{(x_i - \mu)^2}{2\sigma^2}\right).$$

The relative evidence comparing two hypothetical values for μ is the likelihood ratio

$$e^{\Lambda} = \frac{\prod_{i=1}^{n} \exp(-(x_i - \mu_a)^2/2\sigma^2)}{\prod_{i=1}^{n} \exp(-(x_i - \mu_b)^2/2\sigma^2)} = \frac{\exp(-\sum_{1}^{n}(x_i - \mu_a)^2/2\sigma^2)}{\exp(-\sum_{1}^{n}(x_i - \mu_b)^2/2\sigma^2)}$$

$$= \exp\left(\frac{1}{2\sigma^2} \sum_{i=1}^{n}(x_i - \mu_b)^2 - (x_i - \mu_a)^2\right).$$

After some rearranging of the right-hand side, it can be shown that

$$\Lambda = \frac{n(\mu_a - \mu_b)}{\sigma^2}\left(\bar{x} - \bar{\mu}_{ab}\right),$$

where Λ is the $\log(LR)$, $\bar{x}$ is the observed mean, and $\bar{\mu}_{ab}$ is the midpoint of the hypothetical means. Therefore

$$n = \frac{\Lambda\sigma^2}{(\mu_a - \mu_b)(\bar{x} - \bar{\mu}_{ab})}.$$

The preceding equation has all of the intuitive properties of a sample size relationship for testing a mean. Sample size is increased by a higher standard of evidence (larger Λ), hypothetical values that are close to one another, a larger person-to-person variance, and an observed mean that lies close to the middle of the range in question. If we compare the observed mean to some value as in a typical hypothesis test, we can set $\mu_a = \bar{x}$ and obtain

$$n = \frac{2\Lambda\sigma^2}{(\mu_a - \mu_b)^2}. \tag{11.25}$$

From the fact derived above that $2\Lambda = Z^2$, it can be seen that

$$n = \frac{(Z_\alpha + Z_\beta)^2\sigma^2}{(\mu_a - \mu_b)^2}, \tag{11.26}$$

which is the common sample size formula for testing a mean when the variance is known.

For a two sample problem as in a comparative trial, the same basic approach applies. Then x_i represents the difference of two means, the variance of which is $2\sigma^2$, so equations (11.25) and (11.26) become

$$n = \frac{4\Lambda\sigma^2}{(\mu_a - \mu_b)^2} \tag{11.27}$$

and

$$n = \frac{2(Z_\alpha + Z_\beta)^2 \sigma^2}{(\mu_a - \mu_b)^2}, \tag{11.28}$$

which is identical to equation (11.24).

11.5.4 Dichotomous Responses Are More Complex

When the outcome is a dichotomous response, the results of a comparative trial can be summarized in a 2×2 table:

	Treatment	
Success	A	B
Yes	a	b
No	c	d

The analysis essentially consists of comparing the proportion of successes or failures in the groups, for example, $a/(a+c)$ versus $b/(b+d)$. The full scope of methods for determining power and sample size in this situation is large. A review of various approaches is given by Sahai and Khurshid (1996). Here I discuss only the basics.

The usual analysis of such data would employ Fisher's exact test or the χ^2 test, with or without continuity correction. The exact test assumes that $a+b$, $c+d$, $a+c$, and $b+d$ are fixed by the design of the trial. However, in a trial, a and b are random variables, indicating that the χ^2 test without continuity correction is appropriate. However, a fixed sample size with random treatment assignment leads to the exact test or χ^2 test with continuity correction. The sample size required for a particular trial can be different, depending on which perspective is taken (Pocock, 1982; Yates, 1984).

A derivation similar to the t-test above for comparing two proportions, π_1 and π_2, without continuity correction yields

$$n_2 = \frac{\left(Z_\alpha \sqrt{(r+1)\bar{\pi}(1-\bar{\pi})} + Z_\beta \sqrt{r\pi_1(1-\pi_1) + \pi_2(1-\pi_2)} \right)^2}{r\Delta^2}, \tag{11.29}$$

where $\bar{\pi} = (\pi_1 + r\pi_2)/(r+1)$ is the (weighted) average proportion and $\Delta = \pi_1 - \pi_2$. Convenient values that solve equation (11.29) for $r = 1$ are given in Tables 11.12 and 11.13. When $r = 1$, equation (11.29) can be approximated by

$$n_2 = \frac{(Z_\alpha + Z_\beta)^2 (\pi_1(1-\pi_1) + \pi_2(1-\pi_2))}{\Delta^2}. \tag{11.30}$$

The calculated sample size must be modified when planning to use the χ^2 test with continuity correction. The new sample size must satisfy

$$n_2^* = \frac{n_2}{4} \left(1 + \sqrt{1 + \frac{2(r+1)}{rn_2\Delta}} \right)^2,$$

where n_2 is given by equation (11.29).

It is noteworthy that equation (11.24) could be solved algebraically for any single parameter in terms of the others. However, equation (11.29) cannot be solved simply

TABLE 11.12   Sample Sizes per Group for Comparisons Using the χ^2 Test without Continuity Correction with Equal Group Sizes Determined from Equation (11.29)

π_1	$\Delta = \pi_2 - \pi_1$					
	0.05	0.10	0.15	0.20	0.25	0.30
0.05	435	141	76	49	36	27
	582	188	101	65	47	36
0.10	686	199	100	62	43	32
	918	266	133	82	57	42
0.15	905	250	121	73	49	36
	1212	335	161	97	65	47
0.20	1094	294	138	82	54	39
	1465	393	185	109	72	52
0.25	1251	329	152	89	58	41
	1675	440	203	118	77	54
0.30	1376	356	163	93	61	42
	1843	477	217	125	81	56
0.35	1470	376	170	96	62	43
	1969	503	227	128	82	57
0.40	1533	388	173	97	62	42
	2053	519	231	130	82	56
0.45	1565	392	173	96	61	41
	2095	524	231	128	81	54
0.50	1565	388	170	93	58	39
	2095	519	227	125	77	52

Note: In all cases, $\alpha = 0.05$. For each pair of π_1 and π_2, the upper number corresponds to $\beta = 0.20$ and the lower number corresponds to $\beta = 0.10$.

for some parameters. For example, if we wish to determine the treatment difference that can be detected with 90% power, a sample size of 100 per group, equal treatment allocation, and a response proportion of 0.5 in the control group, equation (11.29) must be solved using iterative calculations—starting with an initial guess for π_2 and using a standard method such as Newton's iterations to improve the estimate. The need for iterative solutions is is a general feature of sample size equations. Consequently good computer software is essential for performing such calculations that may have to be repeated many times before settling on the final design of a trial.

11.5.5   Hazard Comparisons Yield Similar Equations

Comparative clinical trials with event time endpoints require similar methods to estimate sample size and power. To test equality between treatment groups, it is common to compare the ratio of hazards (defined below) versus the null hypothesis value of 1.0. In trials with recurrence or survival time as the primary endpoint, the power of such a study depends on the number of events (e.g., recurrences or deaths). Usually there is a difference between the number of *patients* placed on study and the number of *events* required for the trial to have the intended statistical properties.

In the following sections, I give several similar appearing sample size equations for studies with event time outcomes. All will yield similar results. The sample size

TABLE 11.13   Sample Sizes per Group for Comparisons Using the χ^2-Test without Continuity Correction with Equal Group Sizes Determined from Equation (11.29)

π_1	\multicolumn{8}{c}{$\Delta = \pi_2 - \pi_1$}
	0.35	0.40	0.45	0.50	0.55	0.60	0.65	0.70
0.05	22	18	15	12	11	9	8	7
	28	23	19	16	14	12	10	8
0.10	25	20	16	14	11	10	8	7
	33	26	21	17	15	12	10	9
0.15	27	22	17	14	12	10	8	7
	36	28	23	19	15	13	11	9
0.20	29	23	18	15	12	10	8	7
	39	30	24	19	16	13	11	9
0.25	31	24	19	15	12	10	8	7
	40	31	24	19	16	13	10	8
0.30	31	24	19	15	12	10	8	6
	41	31	24	19	16	12	10	8
0.35	31	24	18	14	12	9	7	6
	41	31	24	19	15	12	9	7
0.40	31	23	17	14	11	8	7	5
	40	30	23	27	14	11	8	6
0.45	29	22	16	12	10	7	6	4
	39	28	21	16	12	9	7	5
0.50	27	20	15	11	8	6	4	3
	36	26	19	14	10	8	5	3

Note: In all cases, $\alpha = 0.05$. For each pair of π_1 and π_2, the upper number corresponds to $\beta = 0.20$ and the lower number corresponds to $\beta = 0.10$.

equations can be classified into those that use parametric forms for the event time distribution and those that do not (nonparametric). The equations use the ratio of hazards in the treatment groups, $\Delta = \lambda_1/\lambda_2$, where λ_1 and λ_2 are the individual hazards.

Exponential

If event times are exponentially distributed, some exact distributional results can be used. Suppose that d observations are uncensored and $\widehat{\lambda}$ is the maximum likelihood estimate of the exponential parameter (see Chapter 16),

$$\widehat{\lambda} = \frac{d}{\sum t_i} \, ,$$

where the denominator is the sum of all follow-up times. Then $2d\lambda/\widehat{\lambda}$ has a chi-square distribution with $2d$ degrees of freedom (Epstein and Sobol, 1953; Halperin, 1952). A ratio of chi-square random variables has an F-distribution with $2d_1$ and $2d_2$ degrees of freedom (Cox, 1953). This fact can be used to construct tests and confidence intervals for the hazard ratio. For example, a $100(1 - \alpha)\%$ confidence interval for $\Delta = \lambda_1/\lambda_2$ is

$$\widehat{\Delta} F_{2d_1,2d_2,1-\alpha/2} < \Delta < \widehat{\Delta} F_{2d_1,2d_2,\alpha/2} \, .$$

See Lee (1992) for some examples. Power calculations can be simplified somewhat using a log transformation, as discussed in the next section.

Other Parametric Approaches

Under the null hypothesis, $\log(\Delta)$ is approximately normally distributed with mean 0 and variance 1.0 (George and Desu, 1974). This leads to a power/sample size relationship similar to equation (11.24):

$$D = 4\frac{(Z_\alpha + Z_\beta)^2}{[\log(\Delta)]^2}, \tag{11.31}$$

where D is the total number of events required and Z_α and Z_β are the normal quantiles for the type I and II error rates. It is easy to verify that to have 90% power to detect a hazard ratio of 2.0 with a two-sided 0.05 α-level test, approximately 90 total events are required. It is useful to remember one or two such special cases because the formula may be difficult to recall.

A more general form for equation (11.31) is

$$\frac{1}{d_1} + \frac{1}{d_2} = \frac{[\log(\Delta)]^2}{(Z_\alpha + Z_\beta)^2}, \tag{11.32}$$

which is useful because it shows the number of events required in each group. Note the similarity to equation (11.23). Ideally the patients should be allocated to yield equal numbers of events in the two groups. Usually this is impractical and not much different from allocating equal numbers of patients to the two groups. Equations (11.31) and (11.32) are approximately valid for nonexponential distributions as well, especially those with proportional hazards such as the Weibull.

Nonparametric Approaches

To avoid parametric assumptions about the distribution of event times, a formula given by Freedman (1982) can be used. This approach is helpful when it is unreasonable to make assumptions about the form of the event time distribution. Under fairly flexible assumptions, the size of a study should satisfy

$$D = \frac{(Z_\alpha + Z_\beta)^2(\Delta + 1)^2}{(\Delta - 1)^2}, \tag{11.33}$$

where D is the total number of events needed on the study.

Example 20 Based on this formula, to detect a hazard rate of 1.75 as being statistically significantly different from 1.0 using a two-sided 0.05 α-level test with 90% power requires

$$\frac{(1.96 + 1.282)^2(1.75 + 1)^2}{(1.75 - 1)^2} = 141 \; events.$$

A sufficient number of patients must be placed on study to yield 141 events in an interval of time appropriate for the trial. Suppose that previous studies suggest that approximately 30% of subjects will remain event free at the end of the trial. The total number of study subjects would have to be

$$n = \frac{141}{1 - 0.3} = 202.$$

This number might be further inflated to account for study dropouts.

11.5.6 Parametric and Nonparametric Equations Are Connected

Interestingly equations (11.32) and (11.33) can be connected directly in the following way: For $\Delta > 0$, a convergent power series for the logarithmic function is

$$\log(\Delta) = \sum_{i=1}^{\infty} \frac{2}{2i-1} \psi^{2i-1}, \tag{11.34}$$

where $\psi = (\Delta - 1)/(\Delta + 1)$. Using only the first term of equation (11.32), we have

$$\log(\Delta) \approx 2\psi.$$

Substituting this into equation (11.34), we get equation (11.33). The quantity $2(\Delta - 1) /(\Delta + 1)$ gives values closer to 1.0 than $\log(\Delta)$ does, which causes equation (11.33) to yield higher sample sizes than equation (11.32).

The ratio of sample sizes given by the two equations, R, is

$$\sqrt{R} = \frac{\frac{1}{2}\log(\Delta)}{\psi} = \sum_{i=1}^{\infty} \frac{1}{2i-1} \psi^{2i-2}.$$

Taking the first two terms of the sum,

$$\sqrt{R} \approx 1 + \tfrac{1}{3}\psi^2.$$

For $\Delta = 2$, $\psi = \frac{1}{3}$, and

$$\sqrt{R} = 1 + \frac{1}{27}$$

or $R \approx 1.08$. Thus equation (11.33) should yield sample sizes roughly 8% larger than equation (11.32) for hazard ratios near 2.0. This is in accord with Table 11.14.

11.5.7 Accommodating Unbalanced Treatment Assignments

In some circumstances it is useful to allocate unequally sized treatment groups. This might be necessary if one treatment is very expensive, in which case the overall cost of the study could be reduced by unequal allocation. In other cases we might be interested in a subset of patients on one treatment group. If d_1 and d_2 are the the required numbers of events in the treatment groups, define $r = d_2/d_1$ to be the allocation ratio. Then equation (11.32) becomes

$$d_1 = \frac{r+1}{r} \frac{(Z_\alpha + Z_\beta)^2}{\left[\log(\Delta)\right]^2}. \tag{11.35}$$

From equation (11.33) the corresponding generalization is

$$d_1 = \frac{r+1}{r} \frac{(Z_\alpha + Z_\beta)^2 (\Delta + 1)^2}{4(\Delta - 1)^2}$$

TABLE 11.14 Number of Total Events for Hazard Rate Comparisons Using the Log Rank Test

Δ	$\beta = 0.1$		$\beta = 0.2$	
	$\alpha = 0.05$	$\alpha = 0.10$	$\alpha = 0.05$	$\alpha = 0.10$
1.25	844	688	630	496
	852	694	636	500
1.50	256	208	192	150
	262	214	196	154
1.75	134	110	100	80
	142	116	106	84
2.00	88	72	66	52
	94	78	70	56
2.25	64	52	48	38
	72	58	54	42
2.50	50	42	38	30
	58	48	42	34

Note: Δ is the hazard ratio. The upper row is for the exponential parametric assumption (equation 11.31) and the lower row is a nonparametric assumption (equation 11.33). All numbers are rounded so as to be divisible evenly by 2, although this is not strictly necessary.

and

$$d_1 + d_2 = \frac{(r+1)^2}{r} \frac{(Z_\alpha + Z_\beta)^2 (\Delta + 1)^2}{4(\Delta - 1)^2}. \tag{11.36}$$

When $r = 1$, we recover equation (11.33).

The effect of unequal allocations can be studied from equation (11.35), for example. Suppose that the total sample size is held constant:

$$D = d_1 + d_2 = d_1 + rd_1 = d_1(r + 1)$$

or $d_1 = D/(r + 1)$. Then equation (11.35) becomes

$$D = \frac{(r+1)^2}{r} \frac{(Z_\alpha + Z_\beta)^2}{\left[\log(\Delta)\right]^2}. \tag{11.37}$$

From this we have

$$\text{Power} = \Phi\left(\frac{\sqrt{Dr}}{r+1} \log(\Delta) - Z_\alpha\right).$$

A plot of power versus r using this equation is shown in Figure 11.4.

As the allocation ratio deviates from 1.0, the power declines. However, this effect is not very pronounced for $0.5 \leq r \leq 2$. Thus moderate imbalances in the treatment

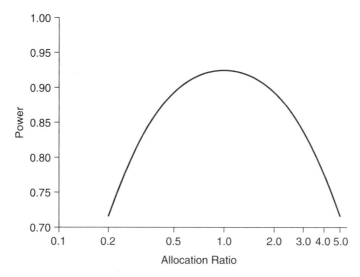

Figure 11.4   Power versus allocation ratio for event time comparisons.

group sizes can be used without great concern about loss of power or the need to increase total sample size.

11.5.8   A Simple Accrual Model Can Also Be Incorporated

Often it is important to estimate the length of time required to complete a clinical trial with event time endpoints. This is true in studies of many chronic diseases like cancer and AIDS. Consider a trial that accrues patients over an interval of time from 0 to T. After time T the study does not terminate but continues without new accruals while those patients already on study are followed for events (Figure 6.1). This scheme is advantageous because individuals accrued close to the end of the study ordinarily would not be followed long enough to observe many events. Using a period of additional follow-up permits subjects accrued near T to contribute information to the study.

Under the assumptions of Poisson accrual rates and exponentially distributed failure times, the study parameters in this case should satisfy

$$\sum_{i=1}^{2} \frac{2\left(\lambda_i^*\right)^2}{n\lambda_i} \frac{1}{\lambda_i^* T - e^{-\lambda_i^* \tau}(1 - e^{-\lambda_i^* T})} = \frac{\left[\log(\Delta)\right]^2}{(Z_\alpha + Z_\beta)^2} \tag{11.38}$$

where the subscript indicates group 1 or 2, λ is the event rate, $\lambda_i^* = \lambda_i + \mu$, which is the event rate plus μ, a common loss to follow-up rate, T is the accrual period, τ is the period of additional follow-up, and the other parameters are as above (Rubinstein, Gail, and Santner 1981). This equation must be used thoughtfully to be certain that the time scales for the event rates and accrual periods are the same (years, months, etc.). Numerical methods are required to solve it for the event rates, accrual time, and follow-up time parameters. The other parameters can be found algebraically. Values of accrual time that satisfy equation (11.38) for various other parameter values are shown in Table 11.15.

TABLE 11.15   Accrual Times That Satisfy Equation (11.38) with Equal Group Sizes

Event Rates		Accrual Rate						
λ_1	λ_2	20	30	40	50	75	100	200
0.1	0.15	20.4	15.6	13.0	11.3	8.9	7.5	5.1
	0.20	9.9	7.8	6.6	5.8	4.6	3.9	2.7
	0.25	6.9	5.5	4.7	4.1	3.3	2.8	2.0
0.15	0.20	31.2	22.7	18.3	15.6	11.8	9.8	6.4
	0.25	12.9	9.9	8.2	7.2	5.6	4.8	3.2
	0.30	8.5	6.6	5.5	4.9	3.9	3.3	2.2
0.20	0.30	16.9	12.5	10.2	8.8	6.8	5.7	3.8
	0.35	10.4	7.9	6.6	5.7	4.5	3.8	2.5
	0.40	7.7	5.9	4.9	4.3	3.4	2.9	2.0

Note: $\alpha = 0.05$ (two-sided), $\beta = 0.10$, and $\tau = 0$.

Example 21   A randomized clinical trial is planned to detect a twofold reduction in the risk of death following surgery for non–small cell lung cancer using adjuvant chemotherapy versus placebo. In this population of patients, the failure rate is approximately 0.1 per person-year of follow-up on surgery alone. Investigators expect to randomize 60 eligible patients per year. Assuming no losses to follow-up and type I and II error rates of 5% and 10%, respectively, equation (11.38) is satisfied with $T = 2.75$ and $\tau = 0$. This is the shortest possible study because accrual continues to the end of the trial, meaning it is maximized. The total accrual required is $60 \times 2.75 = 165$ subjects. However, those individuals accrued near the end of the study may contribute little information to the treatment comparison because they have relatively little time at risk and are less likely to have events than those accrued earlier. As an alternative design, investigators consider adding a two-year period of follow-up on the end of the accrual period, so that individuals accrued late can be followed long enough to observe events. This type of design can be satisfied by $T = 1.66$ and $\tau = 2$, meaning these parameters will yield the same number of events as those above. Here the total accrual is $60 \times 1.66 = 100$ patients and the total trial duration is 3.66 years. Thus this design option allows trading follow-up time for up-front accruals. This strategy is often possible in event time studies. Follow-up time and accruals can be exchanged to yield a favorable balance while still producing the necessary number of events in the study cohort.

In its original form equation (11.38) is somewhat difficult to use and the sample size tables often do not have the needed entries. If the exponential terms in equation (11.38) are replaced by first-order Taylor series approximations, $e^{\lambda x} \approx 1 - \lambda x$, we obtain

$$n \approx \tilde{\lambda} \frac{(Z_\alpha + Z_\beta)^2}{T \tau [\log(\Delta)]^2} ,$$

where $\tilde{\lambda}$ is the harmonic mean of λ_1 and λ_2. This approximation yields accrual rates that are slightly conservative. Nevertheless, it is useful in situations where the first-order approximation is valid, for example, when the event rates are low. This might be the case in disease prevention trials where the population under study is high risk but the event rate is expected to be relatively low.

11.5.9   Noninferiority

Noninferiority or equivalence questions and some of their design issues were introduced in Section 6.6.2. Because there is a tendency to accept the null whenever we fail to reject it statistically, noninferiority decisions made under the typical hypothesis testing paradigm can be anticonservative. A trial with low power might fail to reject the null, leading to a declaration of equivalence and adoption of a useless therapy.

Based on this, there is some sense in designing such trials with the null hypothesis being "the treatments are different" and the alternative hypothesis being "the treatments are the same," which is a reversal of the usual situation where the null is the hypothesis of no difference (Blackwelder, 1982). Reversing the null and alternative hypotheses reduces the chance of adopting a therapy simply because it was inadequately studied. Such tests are naturally one-sided because we don't wish to prove that the new treatment is significantly better or worse, only that it is equivalent. Thus the roles of α and β are reversed, and they must be selected with extra care.

The impact on sample size of choosing α and β in this way does not have to be great. It was shown above that the quantiles for type I and II error probabilities add together directly in the numerator of many power formulas. However, the operational definition of "significant" may change depending on α and β, which can be consequential. A careful consideration of the consequences of type I and II errors is useful before planning any noninferiority trial. Approaches to power and sample size for these trials are discussed by Roebruck and Kühn (1995) and Farrington and Manning (1990). Some of the conventional regulatory perspectives on the statistical aspects of these studies are discussed and challenged by Garrett (2003).

The sample size for noninferiority trials depends strongly on the quantitative definition of equivalence, just as that for superiority trials depends on the effect size. Equivalence defined with high precision requires an extraordinarily large sample size. In equation (11.30), for example, the sample size increases without being bound as $\pi_1 \rightarrow \pi_2$. If we are willing to declare two proportions equivalent when $|\pi_1 - \pi_2| \leq \delta$, then equation (11.30) can be modified as

$$n_2 = \frac{(Z_\alpha + Z_\beta)^2 [\pi_1(1 - \pi_1) + \pi_2(1 - \pi_2)]}{[\delta - (\pi_1 - \pi_2)]^2} . \tag{11.39}$$

Most frequently in this context, we also assume that $\pi_1 = \pi_2$.

Example 22   Suppose that the success rate for standard induction chemotherapy in adults with leukemia is 50% and that we would consider a new treatment equivalent to it if the success rate was between 40% and 60% ($\delta = 0.1$). The null hypothesis assumes that the treatments are unequal. Then, using an $\alpha = 0.10$ level test, we would have 90% power to reject non-equivalence with 428 patients assigned to each treatment group. In contrast, if equivalence is defined as $\delta = 0.15$, the required sample size decreases to 190 per group.

Confidence Limits

From the above discussion it is evident that the usual hypothesis-testing framework is somewhat awkward for noninferiority questions. Either confidence intervals or likelihood methods may be a more natural way to deal with such issues. See Fleming (2000) for a nice summary of this approach. To simplify the following discussion, assume that safety is not at stake, and that the benefits of standard treatment have already been established. Noninferiority will be assessed in terms of a survival hazard ratio. Other outcomes can be handled in an analogous way.

Suppose that a trial of A versus placebo shows a convincing reduction in risk of some magnitude, and a later comparison of A versus B yields a hazard ratio $\widehat{\Delta}_{AB} = 1.0$. Although the point estimate from the second study suggests that B is not inferior to A, we need to be certain that the data are not also consistent with substantial loss of benefit, for example, if the study were underpowered. One assurance is to check that the confidence interval around $\widehat{\Delta}_{AB}$ does not strongly support low values less than 1, perhaps below some tolerance, b. A small study with a wide confidence interval around the estimated hazard ratio might not convincingly rule out values below b. As an example, b might be taken to be 0.75, that is, requiring confidence that three-quarters of the treatment benefit is preserved to declare noninferiority.

An approximate confidence interval for $\widehat{\Delta}_{AB}$ can be obtained from

$$\log \widehat{\Delta}_{AB} \pm \frac{Z_\alpha}{\sqrt{D/2}},$$

where D is the total number of events observed on study (assumed to be evenly divided between the two treatment groups) and Z_α is the two-sided normal distribution quantile. If we require that the entire confidence interval exceed the tolerance, b,

$$\log b = \log \widehat{\Delta}_{AB} - \frac{Z_\alpha}{\sqrt{D/2}}.$$

From this we obtain

$$D = \left(\frac{2Z_\alpha}{\log(\widehat{\Delta}_{AB}/b)} \right)^2 \tag{11.40}$$

as the required number of events in our study, assuming that it yields a hazard ratio $\widehat{\Delta}_{AB}$. Studies that satisfy such a design criterion could end up being quite large for several reasons. First, we would require relatively high precision (narrow confidence limits). Second, a high fraction of observations could be censored. Third, we cannot always expect $\widehat{\Delta}_{AB} \geq 1$. Figure 11.5 shows the relationship between total events required, the hazard ratio observed in the study cohort, and the width of the confidence interval. Table 11.16 provides the number of events calculated from this equation for a few cases. The sample size does not have to be large (relative to a superiority trial), as Table 11.16 indicates. The bottom two panels show more modest sample sizes when $\widehat{\Delta}_{AB} \geq 1$. As for most event time outcomes, the number of patients needed for the study will be substantially larger than the number of events required.

This method has some deficiencies. If the trial yields a confidence interval for the hazard ratio that extends slightly below b, the decision rule indicates that we should

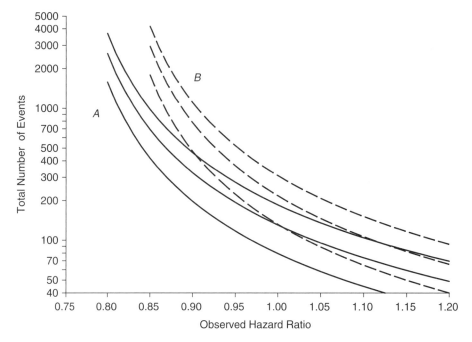

Figure 11.5   Relationship between the observed hazard ratio and the total number of events required for a two-sided lower confidence bound to exceed 0.75 (set A, solid) or 0.80 (set B, dashed) in a noninferiority trial. Note the logarithmic vertical axis. In each set the upper curve is for 95%, the middle curve for 90%, and the lower curve for 80% confidence bounds.

reject noninferiority. The principal flaw in such an approach is that not all points in the confidence interval are equally supported by the data. The data support the center of the interval most strongly, but the decision rule is sensitive to the behavior of points at the end that are not well supported by the data. Furthermore there is as much support for the upper end of the confidence interval as for the lower, but only the lower end strongly affects the decision.

This situation is logically consistent with hypothesis tests and the common interpretation of p-values, as a cutoff of 0.05. Rigidity is inappropriate, as stated, because it presupposes the equivalence of all points in the confidence (or rejection) interval. In reality the choice between nearly equivalent treatments will depend on additional outcomes such as toxicity, quality of life, or convenience. The discussion here is intended to be illustrative, particularly of the large sample sizes required, and not necessarily an endorsement of this as being the optimum approach to noninferiority.

Likelihood Based

An approach to sample size for noninferiority trials based on likelihood ratios may be slightly more natural than confidence intervals. Simply stated, we would require the same relative evidence (measured by the likelihood ratio) that $\widehat{\Delta} \geq b$ in a noninferiority trial as for $\widehat{\Delta} \geq 1$ in a superiority trial. Using the fact that $\log(\widehat{\Delta})$ is approximately distributed as $N\left(\log(\Delta), \frac{1}{d_1} + \frac{1}{d_2}\right)$, where d_1 and d_2 are the numbers of events in the

TABLE 11.16   Approximate Total Number of Events Required to Satisfy Equivalence as Defined by Confidence Intervals and Equation (11.40)

Δ	b			
	0.75	0.8	0.85	0.9
0.85	(981)	(4181)		
	(691)	(2946)		
	(420)	(1789)		
0.90	(463)	(1108)	(4704)	
	(326)	(781)	(3314)	
	(198)	(474)	(2013)	
0.95	275	521	1243	(5257)
	194	367	875	(3703)
	118	223	532	(2249)
1.00	186	309	582	1385
	131	218	410	976
	80	133	249	593
1.05	136	208	345	647
	96	147	243	456
	59	89	148	277
1.10	105	152	232	382
	74	107	163	269
	45	65	99	164

Note: Δ is the observed hazard ratio and b is the value of the lower confidence bound that must be exceeded to satisfy equivalence. Within each Δ grouping, the first row is for $Z = 1.96$ (e.g., a two-sided 95% confidence interval), the second row for $Z = 1.645$, and the third row for $Z = 1.282$. Numbers in parentheses indicate that the *upper* confidence bound is less than 1.0.

two treatment groups, the normal likelihood yields

$$\frac{L(\Delta_1)}{L(\Delta_2)} = \exp\left(\frac{\left(\log(\widehat{\Delta}) - \log(\Delta_2)\right)^2}{8/D} - \frac{\left(\log(\widehat{\Delta}) - \log(\Delta_1)\right)^2}{8/D}\right),$$

where we have assumed that $d_1 \approx d_2$ and $d_1 + d_2 = D$. In the usual superiority situation we would reject the null value ($\Delta = 1$) if we observed sufficient evidence that $\widehat{\Delta}$ is different. Thus

$$R = \frac{L(\widehat{\Delta})}{L(1)} = \exp\left(\frac{\log(\widehat{\Delta})^2}{8/D}\right),$$

where R is a specified likelihood ratio. The number of events required is

$$D = \frac{8\log(R)}{\log(\widehat{\Delta})^2}.$$

For example, suppose that we need moderate to strong evidence against the null if $\widehat{\Delta} = 2$. When $R = 128$ (strong evidence), $D = 82$, and when $R = 32$ (moderate evidence), $D = 58$, in accord with sample sizes produced by conventional hypothesis test based designs.

For noninferiority, the relative evidence for $\widehat{\Delta}$ versus the tolerance b defined above yields

$$D = \frac{8\log(R)}{\left(\log(\widehat{\Delta}) - \log(b)\right)^2} = \frac{8\log(R)}{\left(\log(\widehat{\Delta}/b)\right)^2}, \qquad (11.41)$$

which could have been more directly derived from equation (11.40) because $R = \exp(Z^2/2)$. Again, the principal difficulty for trial size will arise when $b \le \widehat{\Delta} \le 1$. The relative evidence implied by the 95% confidence interval is moderate to weak: $\exp(1.96^2/2) = 6.8$. Hence the sample sizes implied by Table 11.16 are actually relatively small. Table 11.17 is based on likelihood ratios of 8, 32, and 128 (up to strong evidence), resulting in numbers of events to demonstrate noninferiority that are very large.

TABLE 11.17   Approximate Total Number of Events Required to Satisfy Equivalence as Defined by Likelihood Ratios and Equation (11.41)

Δ	b 0.75	0.8	0.85	0.9
0.85	2478	10561		
	1770	7544		
	1062	4526		
0.90	1168	2798	11881	
	834	1999	8486	
	500	1199	5092	
0.95	695	1314	3138	13278
	496	939	2241	9485
	298	563	1345	5691
1.00	469	780	1470	3497
	335	557	1050	2498
	201	334	630	1499
1.05	343	525	869	1634
	245	375	621	1167
	147	225	373	700
1.10	265	382	584	964
	189	273	417	689
	113	164	250	413

Note: Δ is the observed hazard ratio and b is the value of the relative hazard that must be exceeded to satisfy equivalence. Within each Δ grouping, the first row is for $R = 128$, the second row for $R = 32$, and the third row for $R = 8$.

11.6   ES TRIALS

ES (phase IV) studies are large safety trials, typified by postmarketing surveillance, and are intended to uncover and accurately estimate the frequency of uncommon side effects that may have been undetected in earlier studies. The size of such studies depends on how low the event rate of interest is in the cohort under study and how powerful the study needs to be.

11.6.1   Model Rare Events with the Poisson Distribution

Assume that the study population is large and that the probability of an event is small. Also assume that all subjects are followed for approximately the same length of time. The number of events, D, in such a cohort would follow the Poisson distribution. The probability of observing exactly r events is

$$\Pr[D = r] = \frac{(\lambda m)^r e^{-\lambda m}}{r!},$$

where λ is the event rate, and m is the size of the cohort. The chance of observing r or fewer events is the tail area of this distribution or

$$\Pr[D \le r] = \sum_{k=0}^{r} \frac{(\lambda m)^k e^{-\lambda m}}{k!}.$$

The study should be large enough to have a high chance of seeing at least one event when the event rate is λ.

Example 23   The chance of seeing at least one event in the Poisson distribution is

$$\beta = 1 - \Pr[D = 0] = 1 - e^{-\lambda m},$$

or $m = -\log(1 - \beta)/\lambda$. If $\lambda = 0.001$ and $\beta = 0.99$, then $m = 4605$. In other words, if we studied 4605 patients and observed no events, we would have a high degree of certainty that the event rate was lower than 0.1%.

Confidence intervals for λ can be calculated from tails areas of the distribution in a fashion similar to those for the binomial above. Some results are given in Table 11.18, where d represents the number of events observed. The numerical values for the upper and lower limits provided in Table 11.18 must be divided by the cohort size to obtain the actual confidence bounds.

Example 24   If 1 event is observed in a cohort of size 10, from Table 11.18 the 95% confidence bounds on λ are $(0.024, 0.557)$. If 2 events are observed in a cohort of 100, the 95% confidence bounds on λ are $(0.006, 0.072)$. Finally, if 11 events are observed in a cohort of 250, the bounds are $(6.201/250, 19.68/250) = (0.025, 0.079)$.

In some cases comparative trials are designed with discrete outcomes having a Poisson distribution. Power computations in these cases are discussed by Gail (1974).

TABLE 11.18   Exact Two-Sided 95% Poisson Confidence Limits

d	Lower	Upper	d	Lower	Upper	d	Lower	Upper
1	0.242	5.572	26	17.79	38.10	51	38.84	67.06
2	0.619	7.225	27	18.61	39.28	52	39.70	68.19
3	1.090	8.767	28	19.42	40.47	53	40.57	69.33
4	1.623	10.24	29	20.24	41.65	54	41.43	70.46
5	2.202	11.67	30	21.06	42.83	55	42.30	71.59
6	2.814	13.06	31	21.89	44.00	56	43.17	72.72
7	3.454	14.42	32	22.72	45.17	57	44.04	73.85
8	4.115	15.76	33	23.55	46.34	58	44.91	74.98
9	4.795	17.08	34	24.38	47.51	59	45.79	76.11
10	5.491	18.39	35	25.21	48.68	60	46.66	77.23
11	6.201	19.68	36	26.05	49.84	61	47.54	78.36
12	6.922	20.96	37	26.89	51.00	62	48.41	79.48
13	7.654	22.23	38	27.73	52.16	63	49.29	80.60
14	8.395	23.49	39	28.58	53.31	64	50.17	81.73
15	9.145	24.74	40	29.42	54.47	65	51.04	82.85
16	9.903	25.98	41	30.27	55.62	66	51.92	83.97
17	10.67	27.22	42	31.12	56.77	67	52.80	85.09
18	11.44	28.45	43	31.97	57.92	68	53.69	86.21
19	12.22	29.67	44	32.82	59.07	69	54.57	87.32
20	13.00	30.89	45	33.68	60.21	70	55.45	88.44
21	13.79	32.10	46	34.53	61.36	71	56.34	89.56
22	14.58	33.31	47	35.39	62.50	72	57.22	90.67
23	15.38	34.51	48	36.25	63.64	73	58.11	91.79
24	16.18	35.71	49	37.11	64.78	74	58.99	92.90
25	16.98	36.90	50	37.97	65.92	75	59.88	94.01

Note: When $r = 0$, the table entries for one-sided upper bounds are 3.689 (95%) and 4.382 (97.5%).

11.6.2   Likelihood Approach for Poisson Rates

Some useful insights can be obtained by considering the relative evidence for an observed Poisson event rate, λ. The likelihood ratio for λ versus μ is

$$e^{\Lambda} = \left(\frac{\lambda}{\mu}\right)^{r} e^{-m(\lambda-\mu)},$$

where r events are observed with cohort size (exposure) m. Then

$$\Lambda = r \log\left(\frac{\lambda}{\mu}\right) - m(\lambda - \mu) \tag{11.42}$$

or

$$m = \frac{r \log\left(\frac{\lambda}{\mu}\right) - \Lambda}{\lambda - \mu}. \tag{11.43}$$

Equation (11.43) can be viewed as an event-driven sample size relationship and might have some utility if our design called for observing a fixed number of events.

TABLE 11.19   Sample Sizes for Comparison of λ versus 1.0 from Equation (11.44)

λ	\Lambda			
	8	32	64	128
0.1	6	23	46	92
0.2	10	40	80	159
0.3	16	64	127	254
0.4	26	102	203	405
0.5	42	166	332	663
0.6	73	289	578	1155
0.7	142	565	1130	2259
0.8	346	1383	2766	5531
0.9	1493	5970	11940	23879

For example, if we anticipate observing 0 events, we can only consider hypothetical rates $\mu > \lambda$. If we need to generate relative evidence against μ close to λ, more events are needed (and consequently a larger cohort).

Equation (11.42) also yields

$$m = \frac{\Lambda}{\lambda \log(\lambda/\mu) + (\mu - \lambda)}, \tag{11.44}$$

which has a more familiar structure and is rate-driven in the sense that $\lambda = r/m$. Some example sample sizes from equation (11.44) rounded to the next highest integer are shown in Table 11.19. In all cases the reference event rate is $\mu = 1$.

11.7   OTHER CONSIDERATIONS

11.7.1   Cluster Randomization Requires Increased Sample Size

The effect of clustered experimental units is to reduce efficiency (increase variance) in measured summaries. Intuitively this seems justified because observations within a cluster are more similar than those between clusters. This correlation acts to reduce the effective sample size relative to independent units. In the case of comparing means, as in Section 11.5.2, for k clusters of m individuals in each treatment group, the mean of either treatment group has variance

$$\text{var}\{\overline{X}\} = \frac{\sigma^2}{km}\left(1 + (m-1)\rho\right),$$

where ρ is the intracluster correlation coefficient.

The inflation factor, $1 + (m-1)\rho$, can be applied in a straightforward way to obtain sample sizes for clustered designs. Also σ^2 can be replaced by $p(1-p)$ for dichotomous outcomes to yield an appropriate inflation factor for the usual sample size formulas for those outcomes. Many other considerations must also be applied for cluster randomized studies besides sample size. For example, the analysis of observations that are not independent of one another requires special statistical procedures.

See Donner and Klar (2000) and Murray (1998) for thorough discussions of cluster randomized trials.

Example 25   If we randomize in clusters of size 5, and the intracluster correlation coefficient is 0.25, the inflation factor is $1 + 4 \times 0.25 = 2$. Thus, for an effect size of 1.0 when comparing means (90% power, two-sided $\alpha = 0.05$), the sample size must be increased from 21 per group to 42 per group to compensate for the clustering.

11.7.2   Simple Cost Optimization

It is possible to perform a simple cost optimization using unbalanced allocation and information about the relative cost of two treatment groups. For the t-test case as an example, equation (11.24) indicates that the total sample size must satisfy

$$n_2 + n_1 = \left(\frac{r+1}{r} + r + 1\right)\frac{(Z_\alpha + Z_\beta)^2 \sigma^2}{\Delta^2},$$

where r is the allocation ratio and $n_1 = rn_2$. Suppose that the relative cost of subjects on group 2 (treatment) compared to group 1 (control) is C. Then the total cost of the study, M, is proportional to

$$M \propto C\frac{r+1}{r} + \frac{r^2+r}{r}.$$

The minimum cost with respect to the allocation ratio can be found from

$$\frac{\partial M}{\partial r} = 1 - \frac{C}{r^2},$$

which, after setting the derivative equal to zero, yields

$$r_{\min} = \sqrt{C}.$$

Thus the allocation ratio that minimizes total cost should be the square root of the relative costs of the treatment groups. This result is fairly general and does not depend explicitly on a particular statistical test. Rather than allocation ratio, r, we could consider the fractional allocation, $f = r/(r+1)$, for which the minimum is

$$f_{\min} = \frac{1}{1 + \frac{1}{\sqrt{C}}}.$$

Example 26   Suppose that a treatment under study costs 2.5 times as much as control therapy. The allocation ratio that minimizes costs is $r = \sqrt{2.5} = 1.58$. Therefore 1.6 times as many subjects should be enrolled on the control therapy compared to the experimental treatment. The fractional allocation is $f = 1/(1 + 1/\sqrt{2.5}) = 0.61$ or approximately 61:39 control to experimental. Note that the power loss for this unbalanced allocation would be quite modest.

11.7.3   Increase the Sample Size for Nonadherence

Frequently patients on randomized clinical trials do not comply with the treatment to which they were assigned. Here we consider a particular type of nonadherence, called drop-in, where patients intentionally take the other (or both) treatment(s). This might happen in an AIDS trial, for example, where study participants want to receive all treatments on the chance that one of them would have a large benefit. A more general discussion of the consequences of nonadherence is given by Schechtman and Gordon (1994). Nonadherence, especially treatment crossover, always raises questions about how best to analyze the data. Such questions are discussed in Chapter 15.

Drop-ins diminish the difference between the treatment groups, requiring a larger sample size than one would need to detect the same therapeutic effect in perfectly complying subjects. Investigators frequently inflate the sample size planned for a trial on an ad hoc basis to correct for this problem. For example, if 15% of subjects are expected to fail to comply, one could increase the sample size by 15% as though the analysis would be based on compliant patients only. This strategy is helpful because it increases the precision of the trial, but gives credence to the incorrect idea that noncompliers can be removed from the analysis.

It is possible to approximate quantitatively the consequence of nonadherence on the power of a comparative clinical trial. Under somewhat idealized conditions, suppose that the trial endpoint is survival and that a fixed proportion of study participants in each group cross over to the opposite treatment. For simplicity, we assume that the compliant proportion is p, and is the same in both groups, and that patients only drop-in to the other treatment. Denote the hazard ratio in compliant subjects by Δ and in partially compliant subjects by Δ'. Then, for the same type I and II error rates, the required number of events are D and D', respectively.

Hazards

By equation (11.33), the sample size inflation factor, R, that corrects for different hazard ratios satisfies

$$ R = \frac{D'}{D} = \frac{(\Delta - 1)^2}{(\Delta + 1)^2} \frac{(\Delta' + 1)^2}{(\Delta' - 1)^2} . $$

Expressing Δ' in terms of Δ will put this equation in a usable form.

To accomplish this, assume that $\Delta' = \lambda'_1 / \lambda'_2$, where λ'_1 and λ'_2 are the composite hazards in the two treatment groups after noncompliant patients are taken into account. In the first treatment group, p subjects have hazard λ_1 and $1 - p$ have hazard λ_2. The "average" hazard in this group is the weighted harmonic mean,

$$ \lambda'_1 = \frac{2}{(p/\lambda_1) + (1 - p)/\lambda_2}, $$

and in the second treatment group,

$$ \lambda'_2 = \frac{2}{(1 - p)/\lambda_1 + (p/\lambda_2)}. $$

Therefore

$$\Delta' = \frac{1 + p(\Delta - 1)}{\Delta - p(\Delta - 1)}.$$

From this, it is straightforward to show that

$$\frac{\Delta' + 1}{\Delta' - 1} = \frac{\Delta + 1}{\Delta - 1} \frac{1}{2p - 1}.$$

Substituting into the equation defining R yields

$$R = \frac{D'}{D} = \frac{(\Delta - 1)^2 (\Delta' + 1)^2}{(\Delta + 1)^2 (\Delta' - 1)^2} = \frac{1}{(2p - 1)^2},$$

which gives the inflation factor as a function of the adherence rate.

It is not obvious why the harmonic mean rather than an ordinary mean should be used for aggregate failure rates. A brief justification for this is as follows: Suppose that d failure times are ranked from smallest to largest. The total exposure time is

$$\sum_{i=1}^{d} t_i = \sum_{i=1}^{d} (t_i - t_{i-1})(d - i + 1) = \sum_{i=1}^{d} \delta_i (d - i + 1),$$

where $t_0 = 0$ and $\delta_i = t_i - t_{i-1}$. There is one event in each interval so that the interval hazard rates are

$$\lambda_i = \frac{1}{\delta_i (d - i + 1)}.$$

The overall hazard is

$$\lambda = \frac{d}{\sum_{i=1}^{d} t_i} = \frac{d}{\sum_{i=1}^{d} \delta_i (d - i + 1)} = \frac{d}{\sum_{i=1}^{d} (1/\lambda_i)}.$$

Thus the overall hazard is the harmonic mean of the interval hazards.

Alternatively, the problem can be viewed from the perspective of the mean exposure time (exposure time per event). Exposure times can be averaged in the usual manner. However, the mean exposure time is the reciprocal hazard, again indicating the need for the harmonic mean of the hazards.

Means

Returning to the underlying problem of nonadherence, the same inflation factor also applies to power and sample size considerations for differences in means. Suppose that when all participants adhere to their assigned treatments, the treatment difference is $\delta = \mu_1 - \mu_2$. Same as above, assume that a fraction, p, of patients in each group crossover to the other treatment. The observed treatment difference would then be a consequence of the weighted average means in the treatment groups, which would yield

$$\delta' = \mu_1' - \mu_2'$$
$$= \mu_1(1 - p) + \mu_2 p - \mu_1 p - \mu_2(1 - p)$$
$$= (\mu_1 - \mu_2)(1 - 2p).$$

The inflation factor, R, is the ratio of sample sizes required by δ versus δ' (e.g., from equation 11.28), or

$$R = \frac{1}{(2p - 1)^2}.$$

Proportions

For dichotomous outcomes or proportions, the same inflation factor applies, at least approximately. To see this, note that the weighted proportions in the treatment groups, accounting for noncompliance, are

$$\pi_1' = \pi_1 p + \pi_2(1 - p)$$

and

$$\pi_2' = \pi_1(1 - p) + \pi_2 p,$$

so $\Delta' = \pi_1' - \pi_2' = (\pi_1 - \pi_2)(2p - 1)$. As can be seen from equation (11.30), if differences in the numerator as a consequence of π_1' and π_2' are ignored, the inflation factor will again be approximately $R = (2p - 1)^{-2}$.

When all patients comply, $p = 1$ and $R = 1$. If 95% of patients in each group comply (5% "drop-in" to the opposite group), $R = 1/(1.9 - 1)^2 = 1.23$. In other words, to preserve the nominal power when there is 5% nonadherence (crossover) of this type in each treatment group, the number of events has to be inflated by 23%. If the nonadherence is 10% in each group, the sample size inflation factor is 56%, illustrating the strong attenuation of the treatment difference that crossover can cause. Similar equations can be derived for drop-outs or when the drop-in rates are different in the two groups.

11.7.4 Simulated Lifetables Can Be a Simple Design Tool

Sometimes one does not have a computer program available but still wishes to have a rough idea of how long a trial might take to complete. Assuming that we can calculate the number of events needed from an equation like (11.31) or (11.33), a basic lifetable can be constructed to show how long such a study might take. The only additional piece of information required is the accrual rate.

The procedure is fairly simple. For each interval of time, calculate (1) the number of patients accrued, (2) the total number at risk, (3) the number of events in the interval produced by the overall event rate, and (4) the cumulative number of events. When the cumulative number of events reaches its target, the trial will stop. As an example, suppose that we wish to know how long a trial will take that requires 90 events and

TABLE 11.20   Simulated Lifetable to Estimate Study Size and Duration Assuming Exponential Event Times

Time Interval	Number Accrued	Number on Study	Events in Interval	Number Event Free	Cumulative Events
1	30	30	4	26	4
2	30	56	8	48	12
3	30	78	11	67	23
4	30	97	14	83	37
5	30	113	16	97	53
6	30	127	18	109	71
7	30	139	19	120	90
⋮	⋮	⋮	⋮	⋮	⋮

has event rates of 0.1 and 0.2 per person-year in the two treatment groups. Accrual is estimated to be 30 patients per year. Table 11.20 shows the calculations. Here I have assumed the overall event rate is

$$\bar{\lambda} = \frac{2}{(1/\lambda_1) + (1/\lambda_2)} = \frac{2}{10 + 5} = 0.14,$$

which is the harmonic mean of the two event rates. In the seventh year of the lifetable, the cumulative number of events reaches 90. Thus this study would require approximately seven years to complete. In the event that accrual is stopped after five years, a similar calculation shows that the trial would take nine years to yield 90 events.

This method is approximate because it assumes that all patients accrued are at risk for the entire time interval, there are no losses from the study, and both the event rate and accrual rate are constant. Also not all accruals are always accounted for at the end of the table because of round-off error. These problems could be corrected with a more complex calculation, but this one is useful because it is so simple.

11.7.5   Sample Size for Prognostic Factor Studies

The generalization of sample size to cases where the treatment groups are unbalanced suggests a simple method to estimate sample sizes for prognostic factor studies (PFS). Prognostic factor studies are discussed in Chapter 17. Somewhat simplistically, consider designing a prognostic factor study to detect reliably a hazard ratio, Δ, attributable to a single dichotomous variable. If there is more than one factor, we might consider Δ to be the hazard ratio adjusted for all the other covariate effects.

If somehow we could make our PFS like a comparative trial, we would "assign" half of the subjects at random to "receive" the prognostic factor. The power of such a study would then be assessed using the methods just outlined. The distribution of the prognostic factor of interest will not be 1:1 in the study cohort in general. Instead, it will be unbalanced, perhaps severely. However, equation (11.36) might be used, for example, to estimate the size of the cohort required. When the imbalance is severe, we know to expect sample sizes in excess of those required to detect the same size effect in a designed comparative trial.

Example 27   Consider the surgical staging of lung cancer, which is based on both the size of the tumor and lymph node involvement. Suppose that a molecular marker can detect tumor DNA in lymph nodes that appear negative using conventional pathology evaluation. Up-staging such tumors would place the subject in a higher risk category for recurrence or death following surgical resection. Assume that the frequency of a positive molecular marker in a subject thought to be negative by conventional pathology is 15%. What size cohort would be needed to reliably detect a twofold increase in risk attributable to the up-staging? The imbalance in the risk factor is 6:1. Assuming that we want 90% power to detect a hazard ratio of 2 using a one-side 0.025 α level test, then equation (11.36) indicates that approximately 193 events are required. Here the relative hazard is assumed to be constant across all levels of other risk factors. Such a study could be done retrospectively on banked specimens, assuming that clinical follow-up is available. The number of specimens needed might exceed this, if some follow-ups are censored. The study could also be done prospectively, where considerably more subjects might need to be entered in the study cohort to yield the required number of events in a reasonable time. For a prospective study the production of the required events depends on the absolute event rate in the study cohort, the accrual rate, and the length of follow-up. In either case the sample size exceeds that required for a balanced trial to detect the same magnitude effect, where approximately 90 events are needed.

11.7.6 Computer Programs Simplify Calculations

The discussion in this chapter demonstrates the need for flexible and fast, user-friendly computer software to alleviate hand calculations. Most statisticians who routinely perform such calculations have their own programs, commercial software, tables, or other labor-saving methods. Good software packages can greatly simplify these calculations, saving time, increasing flexibility, and allowing the trialist to focus on the conceptual aspects of design. Most commercial programs are periodically updated. It is important to weigh ease of use and cost. Often the easier a program is to use, the more costly it is. A computer program to perform some of the calculations described in this chapter is available as described in Appendix A. Some other programs are described below.

First is *Power and Sample Size* (PASS), sold by NCSS Software (Hintze, 2002). It also performs most of the calculations outlined here and permits simple power curves to be graphed. The price of this program is under US$500. It is a useful learning device and the price may be discounted for students.

Another program for power and sample size is *nQuery Advisor* by Statistical Solutions (Elashoff, 1995). It is *Windows* based and easy to use. This program performs many sample size calculations, but permits solving only for power, sample size, or effect size. Other parameters are assumed to be fixed. This program can also graph power versus sample size.

The Statistical Analysis System (SAS; SAS Institute, 2004) has a power procedure in Version 9.1 of its software. The program is quite flexible and fairly complete. The integration of such software with data and graphics opens the door to diverse sample size approaches, including trial simulations.

A review of sample size programs for survival applications is given by Iwane, Palensky, and Plante (1997). Other computer programs are available for (1) general power calculations (see Goldstein, 1989, for a review) and (2) specialized purposes (e.g., group sequential designs). Some complexities such as time varying drop-out and event rates can be handled (Wu, Fisher, and DeMets, 1980). Other special purpose programs are sketched in Appendix A.

11.7.7   Simulation Is a Powerful and Flexible Design Alternative

It is not uncommon to encounter circumstances where one of the numerous sample size equations is inadequate or based on assumptions known to be incorrect. In these situations it is often possible to study the behavior of a particular trial design by simulating the data that might result and then analyzing it. If this process is repeated many times with data generated from the appropriate model with a random component, the distribution of various trial outcomes can be seen. The process of simulating data, sometimes called the "Monte Carlo method," is widely used in statistics to study analytically intractable problems.

An example of a situation in which the usual assumptions might be inaccurate occurs in prevention clinical trials studying the effects of treatments such as diet or lifestyle changes on the time to development of cancer. In this circumstance it is likely that the effect of the treatment gradually phases in over time. Thus the risk (hazard) ratio between the two treatment groups is not constant over time. However, a constant hazard ratio is an assumption of all of the power and sample size methods presented above for event time endpoints. It is also difficult to deal analytically with time-varying hazard ratios. Additionally the loss rates or drop-in rates in the treatment groups are not likely to be constant over time. An appropriate way to study the power of such a trial might be to use simulation.

Simulations can facilitate studies in other situations. Examples include studying dose escalation algorithms in DF trials, the behavior of flexible grouped designs, and making decisions about early stopping. Reasonable quantitative estimates of trial behavior can be obtained with 1000 to 10,000 replications. Provided the sample size is not too great, these can usually be performed in a reasonable length of time on a microcomputer.

11.7.8   Power Curves are Sigmoid Shaped

I emphasized earlier the need to include the alternative hypothesis when discussing the power of a comparative trial. It is often helpful to examine the power of a trial for a variety of alternative hypotheses (Δ) when the sample size is fixed. Alternatively, the power for a variety of sample sizes for a specific Δ is also informative. These are "power curves" and generally have a sigmoidal shape with increasing power as sample size or Δ increases. It is important to know if a particular trial design is a point on the plateau of such a curve, or lies more near the shoulder. In the former case, changes in sample size, or in Δ, will have little effect on the power of the trial. In the latter case, small changes can seriously affect the power.

As an example, consider the family of power curves shown in Figure 11.6. These represent the power of a clinical trial using the log rank statistic to detect differences in hazard ratios as a function of Δ. Each curve is for a different number of total events and all were calculated from equation (11.33) as

$$1 - \beta = \Phi(Z_\beta) = \Phi\left(\sqrt{d}\,\frac{\Delta - 1}{\Delta + 1} - Z_\alpha\right).$$

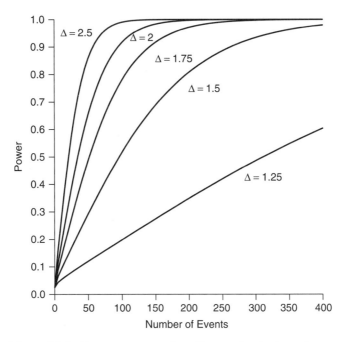

Figure 11.6   Power versus number of events for the logrank test.

For any sample size, the power increases as Δ increases. In other words, even a small study has a high power against a sufficiently large alternative hypothesis. However, only large trials have a high power to detect smaller (and clinically realistic) alternative hypotheses, and retain power, even if some sample size is lost.

Suppose that we find ourselves having designed a study with a power of 0.8 on the $\Delta = 2$ curve (Figure 11.6). If difficulties with the trial reduce the number of events observed (losses to follow-up, for example), the power loss will be disproportionately severe compared to if we had designed the trial initially with a power of over 0.9. Thus reduction of the type II error is not the only benefit from high power. It also makes a trial less sensitive to inevitable inaccuracies in other design assumptions.

11.8   SUMMARY

The size of a clinical trial must be motivated by precision, power, or relative evidence, all of which can be formally related. Size quantifications, although approximate, are useful and necessary for designing trials and planning for resource utilization. A hypothesis-testing framework is often adopted for sample size considerations. Important design parameters besides sample size and power include the intended type I error rate, the number of events in an event–time study, accrual rate and duration, losses to follow-up, allocation ratio, total study duration, and the smallest treatment difference of clinical importance. The practical use of each of these depends on the type of trial and its specific design.

The sample size needed for a phase I trial is usually an outcome of the study, although it is frequently less than 20 to 25 patients. The exact sample size cannot usually be specified in advance because the study stops only after prespecified clinical outcome criteria are met. For dose-finding, fairly simple and familiar study designs typically suffice, although newer methods such as the continual reassessment method have better performance characteristics. Designs that explicitly incorporate pharmacokinetic information into the dose escalation may make these studies more efficient.

Developmental studies, like SA trials, which look for evidence of treatment efficacy, can employ a fixed sample size. The sample size can be determined as a consequence of the precision required to estimate the response, success rate, or failure rate of the treatment. These trials usually require 25 to 50 subjects. When faced with evidence of low efficacy, investigators wish to stop a SA trial as early as possible. This motivates the use of quantitative early stopping rules, sequential, or staged designs that minimize the number of patients given an unpromising treatment. When evidence about efficacy is available from previous studies, a Bayesian SA design may be useful.

Sample size and power relationships for CTE trials depend on the particular test statistic used to compare the treatment groups. Sample size increases as the type I and type II error rates decrease. Sample size decreases as the treatment difference (alternative hypothesis) increases or as the variance of the treatment difference decreases. For event–time studies, the currency of design is the number of events required to detect a particular hazard ratio. The actual sample size needed to yield the required number of events depends on the degree of censoring and the duration of the trial and its follow-up. Nonadherence with assigned treatment may increase the required number of study participants dramatically.

Noninferiority designs often require very large sample sizes because evidence is needed to demonstrate that at least a high fraction of the treatment benefit from standard therapy is preserved. Operationally this requires relatively narrow confidence intervals (high precision), increasing the sample size compared to superiority designs.

Besides analytic power and sample size equations, statistical simulation may be a useful way to study the quantitative properties of some trial designs. These can be as simple as a single hypothetical lifetable constructed under the assumptions of the trial, accrual, and treatment alternatives. More complex simulations could help quantify the effects of nonadherence or other analytically intractable problems in trials. Specialized, flexible computer programs are necessary for performing the required calculations efficiently. Depending on the shape of the power curve, small changes in design parameters can have large or small effects on the power of the study against a fixed alternative.

11.9 QUESTIONS FOR DISCUSSION

1. How large a study is required to estimate a proportion with a precision of ± 0.1 (95% confidence limits)? How large a study is required to estimate a mean value with precision ± 0.1 standard deviation (95% confidence limits)?

2. How many space shuttle flights, in the presence of two failures, would it take to provide a high degree of confidence that the true failure rate is $\leq 0.1\%$? Assume that you intend for the upper two-sided 95% confidence bound to have the value 0.001.

3. What is the largest proportion consistent with 0 successes out of 5 binomial trials? Define what you mean by "consistent with." What about 0 out of 10, 12, 15, and 20 trials?

4. Consider equation (11.8) and the upper confidence bound when the point estimate for a proportion is 0. Derive an analogous rule for hazard rates, assuming that the rate is the intensity of a Poisson distribution.

5. With the normal approximation, confidence limits for a proportion are symmetric around the point estimate. Exact binomial confidence limits can also be constructed to be symmetric around the point estimate. Give some numerical examples for $p \neq 0.05$, and discuss the advantages and disadvantages of this approach.

6. Suppose that a series of binomial trials is stopped when the first "success" occurs. What distribution should be used for confidence intervals on the probability of success? Give some numerical examples.

7. A SA trial is planned with a fixed sample size of 25 patients, expecting 5 (20%) of them to "respond." Investigators will stop the study if 0 of the first 15 patients respond. What is the chance that a treatment with a true response rate of 20% will be discarded by such an approach? What about a treatment with a true response rate of 5%? Discuss the pros and cons of this simple rule for early stopping.

8. Standard treatment produces successes in 30% of patients. A new treatment will be pursued only if it seems to increase the success rate by 15%. Write a formal statistical methods section for a protocol using an appropriate design for this situation.

9. A staged SA trial design uses the following stopping rules: 0/5, 1/10, 2/15 and 5/5, 9/10, 13/15. These correspond approximately to lower and upper 95% confidence bounds on $p = 0.5$. If the study stops early, the estimated proportion of successes will be biased. For example, when the true $p = 0.5$ and the 5/5 upper boundary is hit, the apparent p is 0.75. When the lower 2/15 boundary is hit, the apparent p is 0.13. Discuss ways to systematically investigate this bias.

10. Suppose that preliminary data are available regarding the mean value of some clinically important measurement from patients in a SA trial. If a and b are the minimum and maximum data values, assume a uniform distribution for the preliminary data and derive a sample size equation based on requiring a precision of $w\%$ of the mean. If $a = 5$, $b = 35$, and $w = 10\%$, what sample size is required?

11. A CTE randomized trial is investigating the reduction in mortality attributable to reduced dietary fat. Investigators choose $\alpha = 0.05$ (two-sided) and $\beta = 0.10$ for the trial design. Discuss the appropriateness of these choices.

12. Suppose the trial in question 8 is studying the use of a synthetic fat substitute instead of dietary modification. Does your opinion of $\alpha = 0.05$ and $\beta = 0.10$ change? Why or why not?

13. A CTE randomized trial will compare the means of the two equally sized treatment groups using a t-test. How many patients are required to detect a difference of 0.5 standard deviations using $\alpha = 0.05$ (two-sided) and $\beta = 0.20$? How does the

sample size change if twice as many patients are assigned to one treatment as the other?

14. Suppose that investigators wish to detect the difference between the proportion of successes of $p_1 = 0.4$ and $p_2 = 0.55$, using equal treatment groups with $\alpha = 0.05$ (two-sided) and $\beta = 0.10$. How large must the trial be? Suppose that the allocation ratio is 1.5:1 instead of 1:1?

15. Investigators will compare survival on two treatments using the logrank statistic. What sample size is required to detect a hazard ratio of 1.5, assuming no censoring?

16. Five-year survival on standard treatment is 40%. If a new treatment can improve this figure to 60%, what size trial would be required to demonstrate it using $\alpha = 0.05$ (two-sided) and $\beta = 0.10$? How does this estimated sample size compare with one obtained using equation (11.29) for the difference of proportions? Discuss.

17. Compare the sample sizes obtained from equation (11.33) with those obtained by recasting the problem as a difference in proportions and using equation (11.29). Discuss.

18. Considering equation (11.32), what is the "optimal" allocation of subjects in a trial with a survival endpoint?

19. One could employ a Bayesian or frequentist approach to specifying the size of a SA clinical trial. Discuss the strengths and weaknesses of each and when they might be used.

12

THE STUDY COHORT

12.1 INTRODUCTION

There is a well-known report by Moertel et al. (1974) describing response rates to
5-fluorouracil treatment for colon cancer in phase II (SA) trials. The authors show that
essentially the same treatment for the same extent of disease as tested by 21 different
investigators varied between 8% and 85%. Many of the response rates were estimated
in studies with patient populations over 40 subjects, suggesting that random variation
in the response rate is not the only reason for the observed differences. (See also
Moertel and Reitemeier 1969; Moertel and Thynne 1982; and Leventhal and Wittes
1988.) How can rigorously performed studies of the same treatment yield such large
differences in outcome?

Differences in outcome like these can be a consequence of many factors. However,
given the same eligibility criteria for the various studies, only a few factors are likely
to yield effects on outcome as strong as those seen: patient selection (or, equivalently,
uncontrolled prognostic factors), different definitions of response, and differences in
analysis such as post-treatment exclusion of some patients. Of these, patient selection
factors may be the most difficult for investigators to control. I will use the term "eligibility criteria" to denote formal written specifications of inclusion/exclusion factors
for a study, and "selection factors" to describe uncontrolled selection effects.

Even when eligibility criteria are drafted carefully and followed closely by different
investigators, they usually allow enough heterogeneity in the treatment groups so that
the results can vary considerably at different institutions (e.g., Toronto Leukemia Study
Group, 1986). Although each patient individually meets the eligibility criteria, other
prognostic or selection factors may be influential and systematically different at institutions, causing variability in outcomes. As in the studies cited above, the differences
in outcomes, which are really differences in selection, can be large enough to affect
clinical inferences.

Apart from defining prognosis and encouraging homogeneity, another practical effect of eligibility criteria is to define the accrual base (i.e., accrual rate) for the trial. Every criterion tends to restrict the composition of the study cohort and prolong accrual. Because of this the eligibility criteria can have important consequences for all the resources that a trial will require. Also the utility of a homogeneous study cohort produced by restrictive eligibility criteria must be balanced against greater empirical external validity resulting from looser criteria and a more heterogeneous population.

12.2 DEFINING THE STUDY COHORT

The composition of the study cohort has a dominant effect on results from case series, uncontrolled cohort studies, and database analyses. Interpretation of such studies depends substantially on the difficult task of comparing selection effects and anticipating selection biases that may not be so obvious. An interesting recent example of this was the results of the WHI hormone replacement trial discussed in Section 6.4.2. There may be reasons why well-designed and well-executed developmental trials may not be as sensitive to selection effects as nonexperimental studies, but this seems to be more of a theoretical than practical point.

12.2.1 Active Sampling or Enrichment

An ideal experiment would use an active sample, meaning that the characteristics of the study cohort would be controlled and representative of the target population. Familiar examples of active sampling are surveys used to predict the outcome of elections. Such surveys are useless if the sample is not drawn appropriately. Active sampling with respect to biological characteristics has been termed "enrichment." If we are testing a targeted agent, we might need to be sure that subjects in the cohort actually have the target or that their disease is of an appropriate subtype. Without this control, a treatment can appear to be ineffective only because it was applied in the incorrect biological setting.

There are several reasons why clinical trials do not routinely employ active samples. One reason is that an active sample presupposes knowledge of the major determinants of outcome to control by the sampling design. Many therapeutic and prevention questions either are predicated on universal efficacy or lack sufficient background knowledge to expect differently. Examples of this might be treatments such as analgesic drugs and antihypertensives. Testing of an antibiotic or antiviral would require active sampling, but with regard to the causative agent rather than patient characteristics. Some newer drugs based on genomic features of either the disease or the recipient would also have to be tested in active samples.

A second reason why active samples are infrequently used in clinical trials is expense. In the absence of knowledge or reasonable expectation of subject differences and the interaction of those differences with treatment effects, it is not worth the expense to plan for them. Trials that do so would have to be several times larger than those that do not. Lack of knowledge about interactions is not the same as knowledge that interactions are absent. But experience and expense suggest strongly what the default approach should be.

A third reason why active samples might be unnecessary is that many developmental trials use outcomes that can be ascertained in essentially any subject. Examples of this might be studies of drug distribution, metabolism, excretion, and safety. A fourth reason is that relative treatment effects often generalize outside the study sample on the basis of biological knowledge. For example, a drug may target a universal receptor or pathway.

A final reason why active samples may not be needed routinely in trials may be the most important. It is that trials universally employ a sensible approximation to active sampling—eligibility (inclusion) and exclusion criteria. Exclusions are set primarily for safety. Eligibility criteria are set as delimiters to define the study cohort approximately. I say approximately because the inclusion criteria are usually permissions rather than requirements, and the actual subjects entered do not necessarily cover the multidimensional space implied by the inclusion limits. If we require study subjects to represent that space, we create an active sample. If we allow but do not guarantee them to represent that space, we essentially have an approximate active sample.

12.2.2 Participation May Select Subjects with Better Prognosis

Subjects who participate in a clinical trial tend to have more favorable outcomes than those who refuse, even if the treatment is not effective (Tygstrup, Lachin, and Juhl, 1982). This effect of selection can be quite strong, and could be called the "trial participant effect," much like the "healthy worker effect," a name for the observation that the general health of individuals who are employed is better than average. It is likely to occur in situations where one treatment requires better baseline health and organ system function than alternative therapies. Studies that require subject to travel to a referral center or have more intensive follow-up visits to clinic can also create the effect.

Such an effect is common for surgical treatments where, on average, patients must be healthier to undergo an operative procedure than to receive medical therapy. However, even within surgical treatment, selection effects can be pronounced (Davis et al., 1985). This circumstance arose in the trial of lung volume reduction surgery discussed in Section 4.6.6. Similar issues surround "ventricular remodeling," a proposed surgical treatment for end-stage congestive heart failure in which a portion of heart muscle is removed to improve cardiac function.

The trial participant effect is largely unavoidable in noncomparative designs, although as mentioned above, eligibility criteria can partially control it. In randomized comparative trials, it is less of a problem, provided that the selection effects do not affect the treatment groups differentially. Treatment *differences* are unlikely to be affected by selection effects that operate prior to randomization. However, if selection effects operate differentially in the treatment groups because of either poor design or patient exclusions, the estimated treatment effect can be biased.

Selection could in principle yield a study cohort with either a more favorable or less favorable prognosis, although the former seems to be more common. It is usually not possible to find a single cause (risk factor) for what appears to be a strong effect. It is likely that several factors, each with modest effects, combine to produce an overall bias. Consider factors that produce relative risks in the range of 1.2 (20% increase or decrease). This risk ratio is relatively small and would probably be below the resolution of all but the largest prognostic factor or epidemiologic studies. However, only four such factors together could produce a net relative risk over 2.0 (1.2^4), or under 0.5, if operating in the opposite direction. This is quite substantial relative to the size of most treatment effects, and it is perhaps not surprising that we see selection biases frequently.

Idealized Analytic Examples

The quantitative impact of selection effects can be seen in a theoretical example. Suppose that we are interested in the average chance that individuals will benefit from a particular treatment (response rate) and not everyone has the same probability of benefit. A simple case is when the population consists of two types of individuals—those with response rate p_1 and those with p_2. Assume arbitrarily $p_1 < p_2$. The overall response rate from a sample with n_1 and n_2 such subjects respectively will be the weighted average of the two rates,

$$\widehat{p} = \frac{n_1 p_1 + n_2 p_2}{n_1 + n_2}.$$

The expected success rate depends on the composition of the study cohort with

$$p_1 \leq \widehat{p} \leq p_2.$$

If the study cohort is enriched with one subtype or the other, the observed success rate will be influenced accordingly. The study results will be representative of the overall population only if the sample is unbiased.

A more complex mixture arises if there is a distribution of response probabilities in the population. We denote it by $f(p)$, and assume that it follows a standard beta distribution (Johnson and Kotz, 1970),

$$f(p) = \frac{p^{r-1}(1-p)^{s-1}}{B(r,s)},$$

for $0 \leq p \leq 1$, where $B(r,s) = \Gamma(r)\Gamma(s)/\Gamma(r+s)$ is the beta function. The reason for choosing this functional form for the distribution of response probabilities is purely convenience. The function ranges between 0 and 1 (as p must), is flexible, and is mathematically tractable. It was used in a similar way in Sections 11.4.3 and 5.4.2. When $r = s$, the beta distribution is symmetric and the mean equals 0.5. High values of r and s make the distribution more peaked. For this discussion we assume $r, s \geq 1$. The expected value of p in the population is

$$\begin{aligned}
E(p) &= \int_0^1 p \frac{p^{r-1}(1-p)^{s-1}}{B(r,s)} \, dp \\
&= \frac{B(1+r,s)}{B(r,s)} \\
&= \frac{r}{r+s}.
\end{aligned}$$

The composition of the study cohort is determined by a weighting function $w(p)$ analogous to $n_1/(n_1 + n_2)$ and $n_2/(n_1 + n_2)$ above. For convenience, assume that the weighting of the study sample can also be summarized as a beta distribution,

$$w(p) = \frac{p^{a-1}(1-p)^{b-1}}{B(a,b)},$$

with $a, b \geq 1$. This does not mean that the average response probability in individuals selected for the study cohort is $a/(a+b)$. That quantity will be calculated below. It means that individuals are selected such that those with response probability $a/(a+b)$ have the highest chance of being included. However, individuals with any response probability could be included because the selection distribution is continuous and covers the entire interval $(0,1)$.

In a sample with this type of selection, the expected response rate will be an average *weighted* by the selection distribution. Thus the probability distribution for p in the study sample is also beta,

$$g(p) = \frac{p^{a+r-2}(1-p)^{b+s-2}}{B(a+r-1, b+s-1)}.$$

The expected value of p is

$$
\begin{aligned}
E_w(p) &= \frac{\int_0^1 p p^{a+r-2}(1-p)^{b+s-2}\, dp}{B(a+r-1, b+s-1)} \\
&= \frac{a+r-1}{a+r+b+s-2}.
\end{aligned}
\tag{12.1}
$$

This expected response rate in the study cohort is not necessarily equal to $r/(r+s)$, the mean value in the population, illustrating the potential for bias due to the selection effects. Such a selection effect could arise from nonrandom sampling of the population or could be constructed intentionally to choose those with a higher (or lower) response probability.

A special case of the weighting function is the uniform distribution or uniform selection, $a = b = 1$. If individuals have a uniform chance of being included in the study cohort without regard to p, the expected value of the observed response rate will be $r/(r+s)$. This is what we would expect intuitively: uniform selection provides an unbiased view of the population. There are many other cases for which $a \neq b \neq 1$ but the study cohort yields the "correct" population success rate. For an arbitrary chosen value of b,

$$\frac{a+r-1}{a+r+b+s-2} = \frac{r}{r+s}$$

whenever

$$a = \frac{br+s-r}{s}.$$

Aside from the fact that this example is highly idealized, investigators would never have the control over selection effects that would guarantee an unbiased estimate from a single cohort study.

Example 28   Suppose $r = s = 5$; that is, the average probability of response in the population is 0.5, but there is substantial heterogeneity. Suppose also that the sample is biased, e.g., $a = 4$, and $b = 2$ (Figure 12.1). In the study cohort equation (12.1) yields

$$\hat{p} = \frac{4+5-1}{4+5+2+5-2} = \frac{8}{14} = 0.57,$$

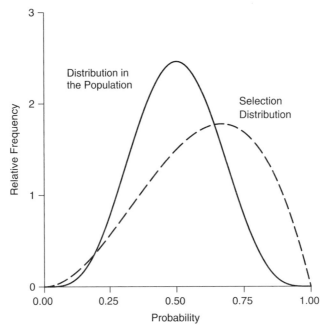

Figure 12.1   Distribution of true response probabilities in a population and selection probabilities for a study cohort.

demonstrating the bias that is possible from nonuniform selection. Similarly, if the selection effect were characterized by $a = 2$, $b = 4$, then $\hat{p} = 0.43$. Selection criteria that define a cohort narrowly in terms of response probability can produce a larger bias than broad selection. It is also possible for nonuniform selection to yield no bias. For example, consider the previous case with $a = b = 5$.

A randomized comparative design will yield an unbiased estimate of the treatment *difference* in this situation unless the treatment effect depends on the true response probability (e.g., treatment–covariate interaction). This illustrates again the benefit of a randomized concurrent control group.

12.2.3   Define the Study Population Using Eligibility and Exclusion Criteria

Because of the possible effects of prognostic and selection factors on differences in outcome, the eligibility criteria for the study cohort need to be defined thoughtfully. In practice, there are two general philosophies that seem to be followed (Yusuf et al., 1990). One approach is to define eligibility criteria with considerable restricting detail, increasing the homogeneity of the study cohort, reducing the variability in the estimated treatment effect, and possibly minimizing the size of the study. Studies designed in this way can yield clear and convincing evidence of biological effects but may seem to have less external validity than other strategies. An alternative approach is to simplify and expand the eligibility criteria, permitting heterogeneity in the study cohort, and control the extra variability by increasing the sample size. This approach is suggested by advocates of large-scale trials (LST), CTE studies that appear to increase external validity because of their large and diverse study cohort (Chapter 6).

Each approach to defining the study cohort is useful in some situations but may have serious shortcomings in others. Broadly defined cohorts of the type used in LS trials may be inappropriate for developmental studies, where it is important to demonstrate a biological effect efficiently and convincingly. A heterogeneous cohort in a large trial can provide additional empirical reassurance regarding general validity. Thus both internal and external validity can depend substantially on the composition of the study cohort. In turn the cohort is defined, at least operationally, by the eligibility criteria *and their interpretation* by the investigators. In a real trial, eligibility criteria may be interpreted by less experienced investigators or study administrators. This somewhat subjective nature of eligibility criteria needs to be considered when planning the study.

Although a perspective on external validity is an interesting one for eligibility criteria, there is a more fundamental view. Eligibility criteria constitute a safety screen for trial participants. Stressful treatments or those with significant risks will be tested first in cohorts restricted by eligibility criteria to minimize risk, or more specifically to have an appropriate risk–benefit ratio. Treatments found or known to be safer might be studied in less restricted populations. Thus we can expect that developmental trials will generally use study cohorts that must be free of comorbidities or other factors that elevate risk, that is, will use more restrictive eligibility criteria.

An example of the relationship between the required sample size of a comparative study, the standard deviation of the outcome measure, and the treatment difference of clinical importance is shown in Figure 12.2. This figure was plotted using points obtained from equation (11.24). Small changes in the standard deviation (actually the standard deviation divided by the treatment difference) do not strongly affect the sample size. However, the relationship between these two factors is an inverse square, so that larger differences substantially affect the sample size required. This motivates using a homogeneous cohort to minimize the size of the trial.

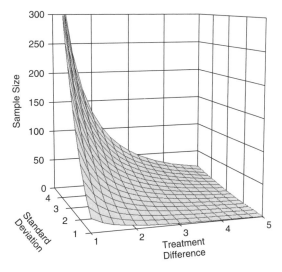

Figure 12.2 Sample size as a function of standard deviation and treatment difference.

In many instances endpoints may be evaluated more easily if certain complicating factors are prevented by patient exclusions. For example, if patients with recent non-pulmonary malignancies are excluded from a lung cancer trial, evaluation of tumor recurrences and second primaries are made simpler. If and when disease progresses or recurs, investigators are more certain that it was a consequence of the index cancer, rather than a preexisting condition.

Ethical considerations also suggest that patients who are unlikely to benefit from the treatment (e.g., because of organ system dysfunction) not be allowed to participate in the trial. Presumably the risk–benefit ratio will not be favorable for these individuals if the treatment carries much risk, as it often does in cancer trials. Whenever possible, quantitative parameters such as laboratory values should be used to make inclusionary and exclusionary definitions rather than qualitative clinical assessments. For example, some studies in patients with advanced cancer might call for a "life expectancy of at least six months." A more useful and reproducible criteria would be Karnofsky performance status greater than, say, 70.

12.2.4   Quantitative Selection Criteria versus False Precision

Potential vagaries of interpretation are reduced by stating inclusion and exclusion criteria quantitatively. In many clinical trials, especially those using drugs or cytotoxic therapy, the trial eligibility criteria require demonstrating good major organ system reserve, such as "normal" renal and hepatic function. Normal lab values often differ from hospital to hospital, and some sense of institutional norms is needed when specifying the criteria. For surgical treatment trials the same principle of using quantitative selection criteria may extend to other organ systems.

One quantitative measure that should not routinely be used as an exclusionary criterion is age. Although advanced age is a good surrogate for declining functional reserve in many organ systems, one should not automatically exclude older persons from the study unless specific physiologic, anatomic, or other criteria require it. In studies of advanced illnesses such as AIDS, cancer, and some heart diseases, it may be sensible to exclude patients who have a very poor prognosis. Given the difficulties of prognosticating, criteria should be based on objective functional scores of the type that could be derived from quantitative measures.

A mistake that investigators frequently make is to specify inclusion criteria to an unnecessary degree of precision. This happens in two circumstances. The first is when an excessive number of criteria are specified, each with its own upper and lower limits. This is typical for laboratory criteria in an attempt to validate organ reserve and enhance safety. Invariably clinic personnel will screen an individual who meets all but one criterion and just misses the cutoff. Both investigator and patient desire inclusion because the spirit of the study makes it appropriate and routine clinical practice suggests no incremental risk. When the call comes into the coordinating center to generate an exception to the protocol specifications, the answer will likely be to exclude the patient.

The difference between what is written in the protocol and what is intended clinically can create considerable friction in the study. The problem can be minimized by not asking nonmedical personnel to override medical specifications, and by writing the specifications in a way that is consistent with clinical practice. Therapeutic decisions are made by a mixture of judgment and rule-based reasoning, so eligibility criteria should reflect this. In a couple of trials I have suggested eliminating rule-based criteria entirely. While investigators were not comfortable going that far, it did permit them to see the problem and increase the clinical judgment employed.

A second way in which false precision can enter the eligibility criteria is in the specification of diagnostic classifications, such as pathology reviews. Routine clinical practice most often accommodates a single pathology reading by the institutional pathologist. Difficult, ambiguous, or complex cases may have a second independent pathology opinion. In a clinical trial investigators sometimes insist on validating all pathology readings, which leads to a couple of strategies. The simplest is to use a single highly regarded pathology opinion centrally, and delay eligibility determination until such is obtained. This removes a possible institutional effect.

However, the institutional reading may be the basis for routine therapeutic decisions, and might be differ from the central reading. To help resolve such discrepancies, a third opinion could be used, provided the logistical difficulties can be tolerated. Such a process might be unassailable when the manuscript is reviewed but may also be far removed from clinical practice. In a randomized trial, particularly a sizable one with experienced participating centers, I believe that a single pathology reading for eligibility is sufficient. If a serious methodologic problem exists with this approach, it is mostly in the minds of the consumers of such a study. The homogeneity afforded by other mechanisms is not essential, and is rarely worth the cost.

12.2.5 Comparative Trials Are Not Sensitive to Selection

Regardless of how carefully eligibility and exclusion criteria are crafted, a clinical trial cohort is self-selected, and therefore unlikely to be representative of the population with the disease. Some critics make much of this fact, refusing to generalize the results because the sample is atypical in some ways. As I have indicated elsewhere, this ignores the principal justification for external validity, that being biological knowledge regarding mechanism.

To the extent that this concern is valid, it is less so for comparative (randomized) trials than for single cohort studies. The robustness of inferences from randomized trials with regard to selection effects is one of their strengths. Although selection factors typically render the trial cohort somewhat different from the general disease population, the relative treatment effect is more likely to be externally valid than in developmental trials. Furthermore the treatment effect from a RCT is likely, in absolute terms, to be externally valid independently of the composition of the study cohort.

A properly designed and executed RCT will yield an incorrect estimate of the (relative) treatment effect only if two criteria are met. First, the treatment effect has to depend on certain characteristics of the subject. In other words, there must be a treatment–covariate interaction such that individuals with certain characteristics benefit, whereas others do not. Second, the study cohort must disproportionately contain one type of such individuals that will produce a nongeneralizable result. Because treatment–covariate interactions are relatively uncommon, and must be further enhanced by unlucky sampling to be damaging, RCTs are quite reliable externally. The circumstance is even more favorable toward RCTs than it may seem at first because only large interactions would create clinically troublesome results.

This reasoning at first may seem counterintuitive. However, most of the selection parameters that tend to frustrate us, such as sex, age, and race, are biologically irrelevant with respect to many therapies. It is true that some of these factors are surrogates for biological differences, such as sex and age, but those differences often do not interact with treatment effects. Other factors are not biological constructs at all but are political/sociodemographic, such as race, and we would therefore not expect them to modulate treatment effects. More details regarding these points is given in Section 12.4.4.

Thus our expectations are that RCTs produce externally valid estimates of treatment effect. It is probably not a wise use of resources to behave otherwise or to expect that study designs should accommodate interactions without strong a priori evidence. In fact, if we believed that interactions were commonplace, then the frequent null results that we obtain would have to be investigated more fully, to exclude that possibility, or that selection effects and a qualitative interaction conspired to obscure treatment benefits.

12.3   ANTICIPATING ACCRUAL

Often in clinical trials we think of measuring or anticipating accrual and determining other trial parameters, such as study duration, from it. In some circumstances accrual may be under the control of the investigators. This can happen, for example, in multicenter trials, where additional institutions can be added to increase accrual.

One unfortunate and preventable mishap in conducting clinical trials is to plan and initiate a study, only to have it terminate early because of low accrual. This is a waste of resources and time for all those involved in the study. This situation can be avoided with some advanced planning and assumptions about the trial. First, investigators should know the accrual rate that is required to complete the study in a certain fixed period of time. Most researchers would like to see comparative treatment trials completed within five years and pilot or feasibility studies finished within a year or two. Disease prevention trials may take longer. In any case, the accrual rate required to complete a study within the time allowed can usually be estimated easily from the total sample size required.

Second, investigators need to obtain *realistic* estimates of the accrual rate. The number of patients with a specific diagnosis can often be determined from hospital or clinic records, but it is a large overestimate of potential study accrual. It must be reduced by the proportion of subjects likely to meet the eligibility criteria, and again by the proportion of those willing to participate in the trial (e.g., consenting to randomization). This latter proportion is usually less than half. Study duration can then be projected based on this potential accrual rate, which might be one-quarter to one-half of the patient population.

Third, investigators can project trial duration based on a worst-case accrual. The study may still be feasible under such contingencies. If not, plans for terminating the trial because of low accrual may be needed so as not to waste resources. Accrual estimates from participating institutions other than the investigator's own are often overly optimistic. How long will the study take as a single-institution trial?

12.3.1   Using a Run-in Period

Occasionally accrual of eligible patients is not a concern, but the trial and its interpretation are likely to be complicated by the failure of many patients to adhere to the treatment. When there is a potential for this to occur, a run-in period could be used to increase the fraction of individuals on the study who comply with the treatment.

During the run-in period eligible patients are monitored for treatment compliance, using a placebo, for example. This monitoring can be complicated to implement. Methods suggested for compliance monitoring include simple assessments like self-reports, pill or medication counts, or more complex assessments such as those based on blood or urine samples. In any case, patients who are found to have poor compliance with the treatment can then be dropped from the study before the treatment period begins.

It is possible that such a run-in period could decrease the external validity of the trial. In the real world patients regularly fail to comply with recommended treatments. A trial done under different circumstances with highly compliant patients may overestimate the effectiveness of the treatment. This criticism is potentially valid. However, it may be more important to establish the efficacy of the treatment before deciding if it can be applied broadly.

In some SA studies, a run-in period is used to assess the safety of a new agent, particularly if it is combined with standard therapy. The purpose is not to select adherent subjects but to be sure that there is not a strong unexpected safety signal from the combination. This might be a concern, for example, if the dose of the new agent was established in the absence of standard therapy. Run-in period is a misnomer in this circumstance because the design does not actually use a calendar interval. Instead a small number of subjects receive the combination. If the frequency and intensity of side effects are low, the full trial proceeds.

12.3.2 Estimate Accrual Quantitatively

To estimate potential accrual more accurately, a formal survey of potential participants can be done before the trial begins. As patients are seen in clinic over a period of time, investigators can keep a record to see if potential participants match the eligibility criteria. To estimate the proportion of eligible patients willing to give consent, one could briefly explain the proposed study and ask if they would (hypothetically) be willing to participate. These accrual surveys are most useful in multicenter collaborative groups that have a track record for performing studies. Often databases exist in these groups that can be used effectively to help estimate accrual.

The Expected Number of Events on the Study Can Be Modeled
Modeling the cumulative number of events in the study cohort is helpful for predicting the required duration of a trial with time to event endpoints. It can also be used to predict cost, paper flow, or other resource requirements (Piantadosi and Patterson, 1987). Suppose that subjects are accrued over an interval of time $[0, T]$, at the end of which accrual is terminated. Denote the accrual rate, survival function, and hazard function by $a(t)$, $S(t)$, and $\lambda(t)$ respectively.

The number of events observed at an arbitrary time t depends on individuals accrued at all earlier times. To be counted as an event at t, an individual must have been accrued at $u < t$, survived $t - u$ units of time, and had an event at t. If $n(t)$ represents the number of events at t,

$$n(t) = \int_0^\tau a(u)\, S(t - u)\, \lambda(t - u)\, du = \int_0^\tau a(u) f(t - u)\, du, \qquad (12.2)$$

where $f(t)$ is the density function,

$$f(t) = \lambda(t)S(t) = -\frac{dS(t)}{dt},$$

and $\tau = \min(t, T)$, meaning the upper limit of integration extends only to the minimum of t and T. The cumulative number of events observed at or before t is

$$D(t) = \int_0^t n(u)\, du = \int_0^t \int_0^\tau a(v) f(u - v)\, dv\, du.$$

Although numerical integration renders these formulae usable as given, it will often be the case that

$$a(t) = \begin{cases} a_0 & \text{if } t \le T, \\ 0 & \text{if } t > T. \end{cases}$$

Then

$$n(t) = \begin{cases} a_0 (1 - S(t)) & \text{if } t \le T, \\ a_0 (S(t - T) - S(t)) & \text{if } t > T, \end{cases}$$

and

$$D(t) = \begin{cases} a_0 t - a_0 \int_0^t S(u)\, du & \text{if } t \le T, \\[2mm] D(T) + \int_T^t S(u - T)\, du - \int_T^t S(u)\, du & \text{if } t > T. \end{cases}$$

Note that

$$D(0) = 0$$

and

$$\lim_{t \to \infty} D(t) = a_0 T,$$

where $a_0 T$ is simply the total accrual. An example of $n(t)$ and $D(t)$ is shown in Figure 12.3, where $T = 10$, $a_0 = 25$ with a Weibull survival function $S(t) = e^{-\mu t^k}$, with $\mu = 0.1$ and $k = 1.5$.

If we make the additional simplifying assumption of exponential survival, equation (12.2) would yield

$$n(t) = a_0 \lambda \int_0^\tau e^{-\lambda \cdot (t - u)}\, du.$$

Then

$$n(t) = \begin{cases} a_0 (1 - e^{-\lambda t}) & \text{if } t \le T, \\ a_0 (e^{-\lambda(t - T)} - e^{-\lambda t}) & \text{if } t > T. \end{cases} \tag{12.3}$$

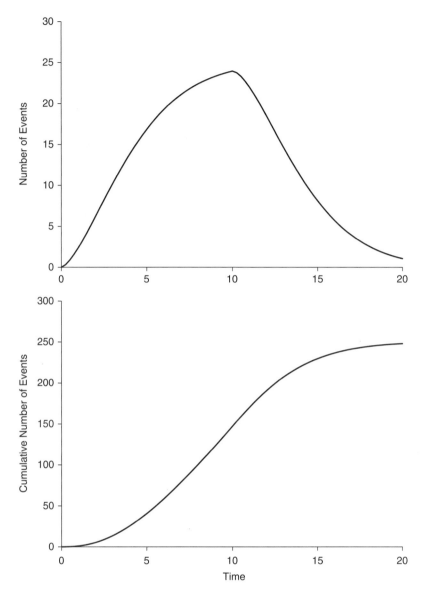

Figure 12.3 Accumulating events in a cohort with constant accrual and Weibull survival: $T = 10$, $a_0 = 25$, $\mu = 0.1$, $k = 1.5$.

From equation (12.3), the cumulative number of events is

$$D(t) = \begin{cases} \frac{a_0}{\lambda}(\lambda t + e^{-\lambda \cdot t} - 1) & \text{if } t \leq T, \\ \frac{a_0}{\lambda}(\lambda T + e^{-\lambda \cdot t} - e^{-\lambda \cdot (t-T)}) & \text{if } t > T. \end{cases} \tag{12.4}$$

Example 29   Suppose that a clinical trial requires 180 events to achieve its planned power. If accrual proceeds at a constant rate of 80 subjects per year for 4 years and the event rate is 0.13 per person-year of follow-up, how many events will have taken place after 4 years? We assume exponential survival. Substituting into equation (12.4) yields

$$D(4) = \frac{80}{0.13} \times (0.13 \times 4 + e^{-0.13 \cdot 4} - 1) = 70 \; events.$$

The study will be only 40% finished, with respect to events, after 4 years. However, additional events will accumulate more quickly after 4 years due to the size of the study cohort. The number of subjects remaining on-study at 4 years is $320 - 70 = 250$ and the length of time required to observe 180 events is just under 7 years.

12.4   INCLUSIVENESS, REPRESENTATION, AND INTERACTIONS

12.4.1   Inclusiveness Is a Worthy Goal

Having a broadly inclusive study cohort is a good idea from at least two perspectives. It allows an opportunity to investigate biological heterogeneity if it truly is present. This is a more valid point with respect to genotypic rather than phenotypic differences in study subjects. Nevertheless, it is applied mostly in the latter case.

Second, a broadly inclusive study cohort helps justly distribute the risks and burdens of research in society. The removal of barriers to participation in research is an important sociopolitical goal that scientists must assist. Even so, participation in research is voluntary and reflects the personal and cultural attitudes and beliefs of subjects. Medical investigators have only limited opportunity to influence those factors.

I did not say that an inclusive study cohort distributes the benefits of research justly. As indicated elsewhere in this chapter, knowledge from clinical trials generalizes on the basis of biology rather than on the superficial characteristics of the cohort (empiricism). We are all more similar than different with regard to biology. Clinical trials clearly have the potential to provide benefit to individual participants, but only indirectly to others. The societal benefits of this research depend almost entirely on the structure and delivery of health care, which is determined by political, social, and economic factors.

12.4.2   Barriers Can Hinder Trial Participation

Among adult cancer patients in the United States, less than 3% participate in clinical trials (Gotay, 1991). A recent study of disparities in cooperative oncology groups shows the same low participation rate and poor acceptance by minorities, women, and the elderly (Murthy, Krumholz, and Gross, 2004). Similar levels of participation in research studies would probably be found for many diseases. Exceptions might include AIDS and some uncommon pediatric conditions. Although the reasons for lack of participation are complex, they fall into three general categories: physician, patient, and administrative.

The reasons that physicians give for failing to enter eligible patients onto clinical trials include the perception that trials interfere with the physician–patient relationship and difficulties with informed consent (Taylor, Margolese, and Soskolne, 1984). In some cases participation in a trial may threaten the supposed expertise and authority of the physician. This is likely to be less of a problem in the future as patients become more active in choosing from among treatment options. The increasing use of second opinions instituted by patients and insurers and patients' assertive participation in multiple AIDS treatments are examples.

Informed consent procedures can be cumbersome, intimidating, and time-consuming. Often patients who give consent do not retain the detailed knowledge about the treatments or the study that investigators would like them to. This suggests difficulties with the consent process that may discourage patients and their families from completing it. Many consent forms are not written at an appropriate reading level. The process tends to overstate risks and understate benefits from participating in research studies as a way of minimizing the liability of investigators and their institutions. An accurate portrayal of these details would require excessively long and technical consent documents.

Some investigators are implicitly or explicitly formulating the idea that patients have a *right* to participate in research studies (Elks, 1993). This new right is seen to arise as an extension of the principle of autonomy. I believe this notion is incorrect because it presupposes benefits from participation that may not exist. This is true on the individual participant level and perhaps more so on a group or societal level. This point will be expanded below. Furthermore this right cannot be exercised universally, even if it exists, because the number of clinical trials is too small.

Many patients are mistrustful of the medical establishment even if they trust their individual physicians. Recent public reactions to proposed changes in health care, such as managed care, indicate this. Stronger reactions are often evident among racial and ethnic minority groups regarding participation in clinical trials. Of the small amount of research that has been done in this area, results related to cancer suggest three points: (1) patients are not well informed about trials, (2) they believe that trials will be of personal benefit, and (3) patients are mistrustful (Roberson, 1994). Such factors partially explain the low minority participation on many research studies.

12.4.3 Efficacy versus Effectiveness Trials

There are two views of the role of clinical trials in evaluating medical therapies today. The views overlap considerably with regard to the methodology that is required but can generate differences of opinion about the appropriate study sample to enroll, coping with data imperfections (Chapter 15), and the best settings for applying clinical trials. The first is that trials are primarily developmental tools used to make inferences about biological questions. These types of studies tend to be smaller than their counterparts (discussed below) and employ study cohorts that are relatively homogeneous. Studies with these characteristics are sometimes called "efficacy" trials, a term and distinction incorrectly attributed to Cochrane (1972). With this perspective, investigators tend to emphasize the internal validity of the study and generalize to other settings based primarily on biological knowledge.

A second perspective on trials is that they are evaluation tools used to test the worth of interventions that should be applied on a large scale. They are motivated by the fact that a biologically sound therapy may not be effective when applied outside the controlled setting of a developmental trial. These trials are large, applied in heterogeneous populations, and use simple methods of assessment and data capture. In Chapter 6, I referred to these studies as large-scale (LS) trials. They have also been called "effectiveness trials," "large simple," and "public health" trials. Investigators conducting

TABLE 12.1   General Characteristics and Orientation of Efficacy and Effectiveness Trials

Characteristic	Efficacy Trial	Effectiveness Trial
Purpose	Test a biological question	Assess effectiveness
Number of participants	Less than 1000	Tens of thousands
Cost	Moderate	Large
Orientation	Treatment	Prevention
Cohort	Homogeneous	Heterogeneous
Data	Complex and detailed	Simple
Focus of inference	Internal validity	External validity
Eligibility	Strict	Relaxed

these trials tend to emphasize the external validity of the studies and expect that their findings will have immediate impact on medical practice.

A concise summary of the difference between these types of studies is shown in Table 12.1. These characterizations are approximations and simplifications. It is easy to find studies that fit neither category well or have elements of both. As we have already seen, some treatments become widely accepted and validated without using any structured methods of evaluation. Other treatments are developed using both types of studies at different times. Viewed in this way, we can see that the proper emphasis of clinical trials is not an "either-or" question about effectiveness or efficacy. Both pathways and types of studies have useful roles in developing and disseminating medical therapies.

The importance of these distinctions for the study cohort primarily relates to the breadth of the eligibility criteria. For any particular trial these criteria should be broad enough to permit inclusion of enough patients to answer the research question quickly. However, the criteria should not permit more heterogeneity than is clinically useful. For example, the bioavailability of a compound may be in question and could be established using a small number of study subjects who meet fairly narrow eligibility criteria. Questions about the bioavailability in different individuals are important; they might be answered by combining the study result with other biological knowledge. It may not be efficient to try to determine differences in bioavailability empirically.

12.4.4   Representation: Politics Blunders into Science

In the recent decade the clinical trials community has focused a great deal of attention on the composition of study cohorts, particularly with regard to their gender and minority makeup. A perception developed early in the discussion that women and minorities were "under represented" in clinical studies of all types and that this was a problem for scientific inference regarding treatment effects (e.g., Roberson, 1994; Schmucker and Vesell, 1993; Sechzer et al., 1994). One of the strongest contributing factors to this perception might have been that women "of childbearing potential" were routinely considered ineligible for many drug trials.

Ignoring Benjamin Franklin's admonition that "the greatest folly is wisdom spun too fine," many politicians thought it necessary to have scientists correct this perceived injustice. The legislative solution was contained in the NIH Revitalization Act of 1993 (Public Law 103-43, 1993):

The Director of NIH shall ensure that the trial is designed and carried out in a manner sufficient to provide for a valid analysis of whether the variables being studied in the trial affect women or members of minority groups, as the case may be, differently than other subjects in the trial.

Some exceptions to this requirement are permitted, although the law explicitly excludes cost as a reason for noncompliance. There are scientific reasons why one would occasionally need to know if sex or ethnicity modulate the efficacy of treatment. More specifically, do the biological components of sex or ethnicity interact with a given treatment in a way that affects therapeutic decisions? However, to require *always* testing for them while ignoring the consequences of known effect modifiers such as major organ function, and to increase the cost of performing trials to do so, requires something other than a scientific perspective.

As for the underrepresentation belief, in the case of women, reliable and convincing data to support or refute such a claim are simply not available (Mastroianni, Faden, and Federman, 1994). Data from published studies suggest that female:male participation of individuals in clinical trials was about 2:3. Published female-only trials tend to be larger than male-only studies, and there were probably as many female-only trials as male-only trials before the law was passed (Gilpin and Meinert, 1994). Recent data from NIH sponsored trials in cancer show a firm female preponderance (Sateren et al. 2002). It would not be a surprise to see the same in other diseases. Possibly there never was a women's participation problem (as I suspect), and according to the political calculus of 1993, we might now be neglecting men's health.

The sex question now has a life of its own, there being a growing body of scientific evidence that males and females are different (e.g., Gesensway, 2001). However, it remains difficult to tell politically correct differences from politically incorrect ones. Aside from reproductive biology and its immediate hormonal consequences, there are other male–female differences in metabolism as well as risk and consequences of chronic disease. What is not clear is how many such differences are important therapeutically (we know that many are not) and if we should try to uncover them all.

For ethnic subsets relatively few data are available, but it is a universal experience that minorities join clinical trials less often than, say, whites. This is true even in diseases like cancer where some minorities have higher incidence rates and higher death rates. Many factors contribute to this, including history and cultural perceptions, underserved primary medical needs, and comorbidities that render potential participants ineligible for some research studies. Comorbidities were an important factor in low participation in a recent survey even at a minority institution (Adams-Campbell et al., 2004). Other barriers to participation in cancer studies have been described as more attributable to religion, education, and income than race (Advani et al., 2002). This last paper explicitly makes the mistake of stating generally that there is questionable applicability of research findings to ethnic groups when they are underrepresented. Aside from the biological weakness of such a claim, it is inconsistent with the conclusion of the investigators who find that race is largely a surrogate for religion, education, and income.

Convenience versus Active Samples

In issues of efficacy and efficiency it is important to understand the difference between convenience samples and cohorts that actively control one or more aggregate characteristics, such as sex or ethnic composition. Convenience samples accept the composition of the cohort passively. The nature of the cohort is dictated by local factors such as referral patterns, chance, selection effects, and the like. Only infrequently would such a sample resemble the population with the disease under investigation, and even then it most likely would not be a true random sample. However, even a convenience sample uses important filters such as eligibility and ineligibility criteria. Convenience samples have the advantage of simplicity and presumably the lowest cost.

A sample in which subject characteristics are controlled (e.g., demographics or risk factors) has potential advantages over a sample based largely on convenience. Active sampling offers the possibility of a truly random study cohort representative of the population with the disease. This could be a critical design feature if the study must provide an accurate estimate of an absolute rate—not a very frequent need in clinical trials. A common circumstance where an active sample is required is when predicting political elections. Active samples require a great deal of planning and expense compared to convenience samples.

A simple example will illustrate the potential inefficiency of an active sample. Suppose that a disease under study has an asymmetric sex distribution, 25% female versus 75% male, and we need to control the study cohort so it has the same sex composition. Suppose further that the convenience sample readily available to investigators has the inverse composition, 75% female versus 25% male. For simplicity, assume that all patients are actually eligible for the study. To obtain the target of three times as many males as females, we have to discard almost 90% of the female potential participants—well over half of the convenience sample. Thus the extra calendar time and resources required to actively control the sex composition would have to be well justified in terms of scientific gain.

This dichotomy is not perfect, but as a rule, clinical trials do well with samples that are closer to the passive end of the spectrum than the active end. Reasons why include the following: many biologically important characteristics are already controlled by routine eligibility criteria; accurate inferences about the disease population may not be needed; relative differences rather than absolute treatment effects might be the primary focus and these can be estimated from simpler designs; other factors typically controlled in active samples do not often have biological importance. It is this last reason that is most relevant in the present context.

NIH and FDA Guidelines

The response of scientists at NIH to the requirements of the law is evident in the guidelines published to implement them (DHHS, 1994). The guidelines essentially restrict the applicability of the law to phase III trials and require an assessment of existing evidence from animal studies, clinical observations, natural history, epidemiology, and other sources to determine the likelihood of "significant differences of clinical or public health importance in intervention effect." When such differences are expected, the design of the trial must permit answering the primary question in each subset. When differences in treatment effects are not expected, are known not to exist, or are unimportant for making therapeutic decisions, the composition of the study group is unimportant. In cases where the evidence neither supports nor refutes the possibility of different treatment effects, representation is required, although "the trial will not be required to provide high statistical power for each subgroup."

In view of the wording of the law, the NIH guidelines are helpful because they intend to restore a scientific perspective on the design of phase III trials. When no conclusive information about treatment–covariate interactions is available, a "valid analysis" is implicitly defined as being unbiased because trials with low statistical power are permitted. Presumably very large interactions would be evident from such designs. However, if "valid analysis" means "unbiased" as opposed to "high statistical power," then some clinically important interactions could remain undetected. This seems to be contrary to the intent of the law. Thus there does not appear to be a reasonable scientific perspective that can be superimposed on the law and the "fix-up" is inescapably empirical.

The FDA guidelines concerning women of childbearing potential (DHHS, 1993a) also attempt to impose scientific thinking on an empirical law. They deal with the issues in a more mechanistic fashion, emphasizing the role of pharmacokinetic analyses in detecting possible gender differences. Also the exclusion of women of childbearing potential from early trials has been eliminated. Additionally the FDA states

> ... representatives of both genders should be included in clinical trials in numbers adequate to allow detection of clinically significant gender-related differences in drug response. (DHHS, 1993a)

Although questions of "validity" and "power" are not addressed, this statement seems to be at odds with the NIH guidelines. Nowhere in the guidelines does it say explicitly that FDA will refuse drug approval on the basis of failure to study patients adequately representative of those with the disease. One can easily imagine some circumstances in which this would be reasonable and other circumstances in which it would not be.

An interesting example arose when the FDA approved the use of Tamoxifen for treatment of male breast cancer, virtually exclusively on the basis of biological similarity of the breast cancer disease, rather than on the results of well-performed clinical trials in men. Furthermore, during advisory committee review of the issue, there was little, if any, discussion of male–female differences in response to Tamoxifen. Outcomes such as this are consistent with a biology-based approach to the issues but are at odds with the current law and written policy.

Treatment–Covariate Interactions

A primary reason for requiring a cohort with a particular composition (e.g., age, sex, or ethnic background) on a clinical trial is to be able to study interactions between the treatment and the actively controlled factor (covariate). For example, if males and females are likely to have clinically significantly different treatment effects, the trial should probably be designed to permit studying the difference. To accomplish this efficiently, equal numbers of males and females should be studied (rather than proportions representative of the disease population).

Another important point regarding treatment–covariate interactions is the distinction between modifying the *magnitude* of the treatment effect as compared to changing its *direction* (Figure 12.4). The former is termed a quantitative interaction whereas the latter is termed qualitative. In principle, we would be interested in sizable interactions of either type. However, qualitative interactions are the ones of real therapeutic concern because they mean that one subset should receive treatment *A*, whereas the other should receive treatment *B*. Such circumstances are exceptionally rare. Quantitative interactions are less consequential because they inform us that both subsets should receive the same treatment, although one subset will benefit more than the other.

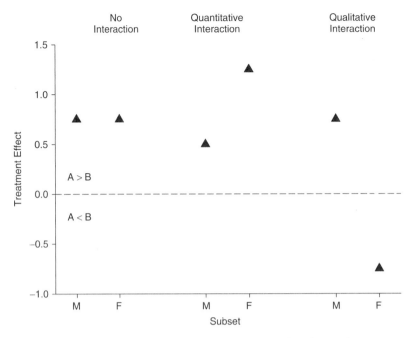

Figure 12.4 Different types of treatment–covariate interaction.

The cost of testing treatment–covariate interactions can be high, even for balanced covariates. As will be seen in Chapter 19 in some detail, the test of interaction in a balanced 2×2 factorial trial has variance 4σ, whereas the main effects have variance σ. This indicates the inefficiency of studying interactions. The variance of the difference of two means, analogous to a main effect, is $2\sigma^2/n$. The variance of the difference of differences, analogous to a treatment–covariate interaction, is $4\sigma^2/n$. This also indicates the inefficiency of testing treatment–covariate interaction. Covariate imbalances will decrease the precision. Ordinarily we would not design trials to detect such effects with high precision, unless it was very important to learn about them.

Final Comments

The U.S. government's fetish with racial and sex composition of study cohorts has important practical consequences and must be examined from a scientific perspective, despite the fact that it originated from political concerns. The only way to make sense out of the inconsistent thinking that plagues this issue both within and outside the government is to separate political considerations from scientific ones. I cannot say generally if the law and its consequent NIH policy make good politics or good science policy. My guess would be not. But it is reasonably certain that they do not generally make good science. Even so, there are many apologists who view the requirements as justified on scientific or ethical grounds. The rationale is that such differences are common and important. I believe that neither is the case—people are demonstrably more similar than they are different. In fact the legislation and the guidelines support my claim. If differences are common and important, then both the funding levels and valid analysis rules are terrible injustices. Gender differences in treatment effects are very uncommon. If real, they are mostly inconsequential clinically.

In 2004, plans to complete development of a new drug for heart failure targeted to African-Americans were announced with great fanfare. The drug is a combination of hydralizine and isosorbide dinitrite, neither of which is new. This combination was said to be the first "ethnic" drug and has been rationalized on the basis of epidemiologic and other differences in heart disease and its treatment in blacks and whites. Superficially, this circumstance appears to support the idea of consequential biological differences attributable to race. However, the driving forces behind it are sociopolitical, not scientific. The relevant history behind this drug combination and its proposed indication is deconstructed by Kahn (2004).

Race is not a biological construct but a sociodemographic one. The study of human genetic heterogeneity does not support the general attribution of biological differences to race. Genetic differences are generally larger, more varied, and more consequential within racial categories than between them (Bamshad and Olson, 2003). It would be a surprise if factors derived on this superficial basis turned out to have important biological consequences. Schwartz (2001) stated the problem especially well:

> ... Such research mistakenly assumes an inherent biological difference between black-skinned and white-skinned people. It falls into error by attributing a complex physiological or clinical phenomenon to arbitrary aspects of external appearance. It is implausible that the few genes that account for such outward characteristics could be meaningfully linked to multigenic diseases such as diabetes mellitus or to the intricacies of the therapeutic effect of a drug.

Thus therapeutic differences attributable to race rather than to a more fundamental biological construct are more likely to be errors. We can often extrapolate inferences from animal species (or even in vitro experiments) to humans (e.g., carcinogenicity and dose-testing). Why would we not think we can learn about all humans from studies in other humans?

On the ethics side, there is an interesting paradox that surrounds such concerns. The notion is that the principle of justice requires us to address the concerns of sex and minority subsets, the premise being that minorities need to be studied to derive the benefits of biomedical research. When health concerns have been identified, usually by means other than clinical trials, this is an important concern. For clinical trials broadly and abstractly, I believe it is exactly backward. Because we have learned much about the treatment of human disease, we have learned about the treatment of disease in particular subsets. If individuals from one of those subsets have not participated in clinical trials, they have derived benefit without incurring risk, very likely disproportionately.

Consider the diagnosis and treatment of prostate cancer as an example. Black men have been underrepresented in most trials, but we have learned a great deal about screening and treatment that applies directly to them. Disparities in outcome may be a consequence of failure to apply knowledge or health care access, but within the domain of clinical trials the justice argument does not apply. Of course, additional studies could target specific questions.

12.5 SUMMARY

A clinical trial cohort is defined by the eligibility and exclusion criteria. The criteria are chosen with a target population in mind, but often the study cohort will not closely resemble the intended population. This selection effect is a consequence of interpretations by the investigators and chance and limits the comparability of different single-arm studies, even though they have the same selection criteria. In randomized comparative trials the selection effects are the same in all treatment groups, validating the comparisons.

Realistic quantitative assessments of accrual are necessary when planning a clinical trial. Simple mathematical models can help predict the expected duration of a study and the number of events that will be observed over time.

In the last few years there has been a great deal of interest in the gender and ethnic composition of trial participants. This interest arises, in part, from concerns about the generalizability of trial results. Large heterogeneous study populations offer some advantages for generalizing results empirically, but may not be optimal for proving biological principles. Removing barriers to trial participation and employing representative cohorts is essential for the health of medical studies. However, required representation in clinical trials could be a hindrance to acquiring new knowledge if it consumes too many resources.

12.6   QUESTIONS FOR DISCUSSION

1. Suppose that the population consists of only two types of individuals. Half of all patients have a response probability of 0.25 and the other half have a response probability of 0.40. If both types of patients are equally likely to be accrued on a trial, what response probability can we expect to observe? How will this change if the eligibility criteria select 2:1 in favor of those with a higher response probability?

2. Suppose that the accrual rate onto a trial is constant and the event times (survival) are exponentially distributed with hazard λ. After the accrual period there is a fixed-length follow-up period. Derive expressions for the number of events and cumulative number of events as functions of time and these parameters. Illustrate how follow-up time (or total study duration) can be substituted for new accruals.

3. Response data from a clinical trial will be analyzed using a linear model. For the ith patient,

$$Y_i = \beta_0 + \beta_1 T_i + \beta_2 X_i + \gamma T_i X_i,$$

where T and X are binary indicator variables for treatment group and covariate, respectively, Y is the response, and the other parameters are to be estimated from the data. γ models the interaction between treatment and covariate. Using ordinary least squares (or maximum likelihood), what is the relative magnitude of $\mathrm{var}\{\beta_1\}$, $\mathrm{var}\{\beta_2\}$, and $\mathrm{var}\{\gamma\}$? Discuss.

4. One way to generalize the results from clinical trials is empirical. This perspective suggests that, if females are not in the study population, the results do not pertain to them. Another way to generalize is biological. This perspective suggests that learning about males informs us about females insofar as biological similarities will allow. Discuss the merits and deficiencies of each point of view.

5. Read the study report by Exner et al. (2001). Comment on the research design, methods, and conclusions.

13

TREATMENT ALLOCATION

13.1 INTRODUCTION

Practitioners can recommend treatments for patients on the basis of one of several perspectives. The most common way is when a physician uses the diagnosis, characteristics of the individual patient, medical knowledge, opinion, patient preference, and circumstantial constraints to recommend a treatment that he or she believes to be optimal. Treatment choices made using this type of judgment do not directly facilitate investigating questions of efficacy. They do not preclude it if differences in outcomes are very large or incremental improvements of existing therapies are made. This method of treatment selection is not a consideration in clinical trials, despite it being the way most practitioners are trained.

A second method for treatment assignment occurs when physicians use only a single therapy for a given condition. Often this happens because practitioners are explicitly taught to behave in this way. But exclusive use of one treatment can result from expert opinion, ignorance, tradition, dogma, economic incentives, or individual or institutional preferences. There are superficial similarities to treatment allocation in an SA trial, but the trial is performed with a high degree of standardization, explicit eligibility criteria, a defined outcome, and may explicitly invalidate the therapy. Practice-based use of a single therapy often does not control selection bias or confounders and almost never has the goals of a clinical trial. These shortcomings are evident when comparing case series from different times, institutions, or clinics. Treatment assignments made in this way are haphazard with respect to assessing treatment differences, and are also not a consideration in comparative clinical trials.

A third way of making treatment choices is based on active control of the alternatives and allocation process, with the purposes of evaluating treatment effects while avoiding biases and confounders. Treatment assignments made in this way are directly relevant to true experimental designs. Although nonrandom but unbiased treatment

Clinical Trials: A Methodologic Perspective, 2E, by S. Piantadosi
Copyright © 2005 John Wiley & Sons, Inc.

schemes (e.g., alternating assignments) can facilitate comparisons, they allow investigators to discover and possibly use information about future assignments. This opens the possibility for selection bias. In contrast, properly constructed and implemented random or dynamic schemes are not subject to discovery and are more reliable and convincing tools to eliminate bias. This chapter is only about the third method of treatment allocation.

When choosing an allocation scheme for a clinical trial, there are three technical considerations: reducing bias, producing a balanced comparison, and quantifying errors attributable to chance. The allocation scheme should satisfy all these objectives or the outcome of the trial may be questioned. Selection bias is reduced or eliminated when patient characteristics (prognostic factors) do not differentially affect the composition of comparison groups. Operationally this means that control over individual subject assignment is removed from the investigator. Credibility requires the assignment process to be nondiscoverable.

Balanced comparison groups are seen as a hallmark of lack of bias. This is not literally true because, as will be discussed below, we often *use* bias to induce balance on key factors. Moreover groups can appear to be balanced when compared marginally, but the joint or multivariate distribution of factors can remain biased. From a practical perspective, investigators tend to look almost exclusively at marginal measures and prefer balance more than the literal use of randomization.

When randomization is used in a comparative trial, it is always possible to construct a valid test of the null hypothesis. The validity of such a procedure does not depend on any outcome, such as prognostic factor balance. Despite this fact, balance is again seen as paramount and investigators would have a difficult time selling a valid comparison that appeared to ignore it.

Implementation of randomized treatment allocation appears as though it should be straightforward. Indeed, there are few circumstances where simple methods of treatment assignment in clinical trials are inappropriate. Problems with simple allocation schemes tend to be small and infrequent. However, there are several different methods of treatment allocation commonly labeled as "randomized" (Kalish and Begg, 1985). We refer to them collectively as randomization, although each may have different properties and advantages. Also, we tend to equate *impartial* assignment with *balanced* assignment and are often surprised at just how unbalanced impartial (simply randomized) assignments can be.

13.2 RANDOMIZATION

There are three potential problems that motivate the use of randomized treatment assignments in comparative clinical trials. The first is the wide variation in clinical outcomes relative to the magnitude of effects or differences that treatments are likely to produce. Good study design, including randomization, helps control and quantify this variability. The second motive arises because of a type of selection bias, or as Miettinen (1983) described it, "confounding by indication." This is a sort of reverse causality, wherein the disease determines the treatment, and means that individuals with certain characteristics are more likely to receive particular treatments. This confounds the effects of the therapy with prognostic factors. Without randomization, selection of patients for particular treatments could yield differences of a clinically important size but due only to bias.

Third, researchers are usually unable to characterize why individuals receive a particular treatment or define homogeneous risk strata for nonrandomized comparisons. If we could characterize quantitatively the relationship between risk factors and treatment choice outside a trial, the effect of selection could be undone, yielding an unbiased treatment comparison. Alternatively, if we knew all prognostic factors, we could define homogeneous risk strata and make treatment comparisons within each. The treatment differences could then be pooled across strata to yield a valid overall estimate. The most practical and convincing way to avoid these problems is to use random treatment assignment.

In some cultures, chance phenomena were a means of communicating with the gods or were associated with chaos and darkness. Unpredictability was associated with danger. In other cultures, randomness has been a time-honored method to apportion blame and remove human bias. As Barrow (1991) says in his brief but interesting discussion of this topic in a cosmological context,

> ... dabbling with random devices was serious theological business, not something to be trifled with or merely studied for the fun of it.

In contemporary Western society we use randomization more for demonstrating a convincing lack of bias than for superstitious reasons, such as coin flips before sporting contests and choosing straws for unpleasant tasks. In clinical trials researchers use randomization for its objectivity in removing bias.

In Western science randomization was suggested formally by R.A. Fisher in the 1920s (Box, 1980) and used in medical studies by Bradford Hill and Richard Doll in Great Britain in the 1940s. In the United States randomization was advocated by early trialists such as Tom Chalmers and Paul Meier. It continues to be widely used for preventing bias in allocating treatments in comparative clinical trials and is popular because of its simplicity and reliability.

In an early single-masked tuberculosis treatment clinical trial in the United States, randomization was used slightly differently than it is today. The investigators stated:

> The 24 patients were then divided into two approximately comparable groups of 12 each. The cases were individually matched, one with another, in making this division. ... Then, by a flip of the coin, one group became identified as group I (sanocrysin-treated) and the other as group II (control). (Amberson et al., 1931)

Today investigators seldom have treatment groups already composed and ready for randomization as a whole. Instead, each individual patient is usually randomized independently, although both procedures are logically equivalent. Even when the experimental unit is a group rather than an individual, we may not be able to randomize all clusters at once. Current practice and theory associated with randomization is discussed by Lachin, Matts, and Wei (1988) and Lachin (1988a, b). For a discussion of the timing of randomization, see Durrleman and Simon (1991).

13.2.1 Randomization Controls the Influence of Unknown Factors

Randomization is an effective means for reducing bias in treatment selection because it guarantees that treatment assignment will not be based on patients' prognostic factors.

This prevents the investigators from consciously or unconsciously assigning better prognosis patients to a treatment that they hope will be superior. Because selection bias can influence outcomes as strongly as many treatment effects, preventing it is an important benefit of randomization.

Selection bias is not the only one that plagues clinical studies. Treatment administration, outcomes assessment, and counting endpoints can all be biased, despite random treatment assignment.

Example 30   Consider an unblinded trial with randomization between treatment and no treatment. The placebo effect will contribute to the apparent efficacy of the treatment. In other words, it may bias the estimated treatment effect. In contrast, if the same trial employs a placebo control, the estimated treatment difference should not be biased.

Thus randomization does not guarantee complete objectivity in a trial and must be combined with other design strategies to reduce bias.

A second, and more far-reaching, benefit of randomization is that it prevents confounding, even if the investigator is unaware that the effects exist and/or has not measured them. One argument against randomization is that it is unnecessary because confounders can be controlled in the analysis by using statistical adjustment procedures. The extent to which confounding can be controlled reliably depends on two additional assumptions: (1) the investigators are aware of, and have measured, all the important confounders in the experimental subjects, and (2) the assumptions underlying the statistical models or other adjustment procedures are known to be correct. Randomization obviates these problems. Critics sometimes overlook this last point, which provides randomized studies with their high degree of inferential directness and credibility.

Thus randomization corrects many of the important limitations of studies with nonrandomized controls. It prevents effects from unknown prognostic factors and eliminates differential treatment selection effects, whether due to patient choice or physician judgment. Randomization facilitates defining start of treatment and controls time trends in the disease, patient population, diagnostic methods, and ancillary care. Interesting examples where investigators did not make the mistake of claiming definitive treatment differences in a nonrandomized comparison for reasons similar to those just outlined are given by Green and Byar (1984). One example is based on the report by Byar et al. (1979).

In Chapter 2, I noted the objections to randomization by Abel and Koch (1997) and Urbach (1993), and indicated the worth of studying their concerns and likely errors. They reject randomization as a (1) means to validate certain statistical tests, (2) basis for causal inference, (3) facilitation of masking, and (4) method to balance comparison groups. They omit the principle benefit of randomization often discussed in the methodologic literature and emphasized here, namely the control over unknown prognostic factors. In Section 13.5.3, I demonstrate how randomization alone does indeed validate an important class of statistical tests. In Section 13.3.2, I show how randomization combined with stratification and blocking induces balanced treatment groups. Facilitation of masking is obvious and needs no further clarification. Randomization as a basis for causal inference is an important and complex epistemological question discussed by many authors (Fisher, 1935; Holland, 1986; Rubin, 1974, 1990, 1991).

Although the benefits of randomization are considerable, there are methodologic problems and biases that it cannot eliminate. One example is limitation of the external validity of a trial. Eligibility restrictions that reduce variability in trials limit the spectrum of patients studied and consequently can seem to reduce the external validity of findings. A second problem is the limited treatment algorithm that most clinical trials study. The experimental structure simplifies and restricts interventions to control them. This may not reflect how the treatments are used in actual practice. A third problem that randomization alone cannot eliminate is bias in the ascertainment of outcomes. Treatment masking and other design features are needed to eliminate this bias.

13.2.2 Haphazard Assignments Are Not Random

In many nonexperimental study designs, the methods by which patients came to receive the treatments they actually got are unknown. Presumably the treating physicians used information at hand in the usual way to select treatments which they felt were best. At the worst the physicians were ineffective at this and the treatments were assigned in a haphazard fashion. Even in such a case treatment comparisons based on these data lack credibility and reliability compared with those from a randomized trial. Haphazard assignments are not random and cannot be relied upon to reduce bias. When reviewing the results of a comparative trial, we want to be convinced that accidents or selection and other biases are not responsible for differences observed.

If one could be certain that patients presented themselves for study entry in a purely random fashion and that investigators would use no prognostic information whatsoever to assign treatments, then virtually any allocation pattern would be effectively random. For example, a trial might employ alternating treatment assignments under the assumption that the patients available are a temporally random sample of those with the disease under study. While this might be credible, it relies on the additional assumption of random arrival in the clinic. Investigators would prefer to control randomization convincingly instead of leaving it to chance!

13.2.3 Simple Randomization Can Yield Imbalances

Simple randomization makes each new treatment assignment without regard to those already made. In other words, a simply randomized sequence has no memory of its past history. While this has important advantages, it can also produce certain unwanted effects in a clinical trial. The most common adverse consequence is an imbalance in the number of subjects assigned to one of the treatments. For example, suppose we are studying two treatments, A and B, and we randomize a total of N subjects with probability $p = 0.5$ to each treatment group. When the trial is over, the size of the two treatment groups will not, in general, be equal. In fact the chance that $N_A = N_B = N/2$ is only

$$\Pr\left[X = \frac{N}{2}\right] = \binom{N}{N/2} \frac{1}{2^N},$$

which can be quite small. For example, when $N = 100$, the chance of the randomization yielding exactly 50 subjects per group is only about 8%.

It is informative to consider the probability of imbalances greater than or equal to some specified size. If properties of the binomial distribution are used, the expected number of assignments to each of two groups is

$$E\{N_A\} = E\{N_B\} = Np,$$

and the variance of the number of assignments is

$$\text{var}\{N_A\} = \text{var}\{N_B\} = Np(1 - p).$$

When $N = 100$ and $p = 1/2$, the variance of N_A is 25. An approximate 95% confidence bound on the number of assignments to treatment A is $N_A \pm 1.96 \times \sqrt{25} \approx N_A \pm 10$. Thus we can expect more than a 60/40 imbalance in favor of A (or B) 5% of the time using simple randomization.

This problem of imbalance becomes more noticeable when we must account for influential prognostic factors. Even if the number of treatment assignments is balanced, the distribution of prognostic factors may not be. For example, suppose that there are k independent dichotomous prognostic factors, each with probability 0.5 of being "positive." We compare two treatment groups with respect to the proportion of subjects with a positive prognostic factor. If a type I error of 5% is used for the comparison, the chance of balance on any one variable is 0.95. The chance of balance on all k factors is 0.95^k (assuming independence). Thus the chance of finding at least one factor unbalanced is $1 - 0.95^k$. When $k = 5$, the chance of at least one statistically significant imbalance is 0.23, which explains the frequency with which this problem is noticed.

From a practical point of view, large imbalances in the number of treatment assignments or the distribution of covariates can lessen the credibility of trial results. Even though statistical methods can account for the effects, if any, of imbalances that occur by chance, we are usually left feeling uneasy by large or significant differences in baseline characteristics between the treatment groups. For large, expensive, and unique trials to be as credible as possible, the best strategy is to prevent the imbalances from occurring in the first place. One way to control the magnitude of imbalances in these studies is to use constrained, rather than simple, randomization. There are two general methods of constrained randomization—blocked randomization and minimization. These will be discussed in the next sections.

13.3 CONSTRAINED RANDOMIZATION

13.3.1 Blocking Improves Balance

Randomization in blocks is a simple constraint that improves balance in the number of treatment assignments in each group (Matts and Lachin, 1988). A "block" contains a prespecified number and proportion of treatment assignments. The size of each block must be an exact integer multiple of the number of treatment groups. A sequence of blocks makes up the randomization list. Within each block the order of the treatments is randomly permuted, but they are exactly balanced at the end of each block. After all the assignments are made in each block, the treatment groups are exactly balanced as intended.

To see that this scheme is constrained randomization, consider two treatment groups, A and B, and blocks of size $N_A + N_B$. During the randomization, the current number of assignments in a block made to treatment A is n_A and likewise for B. At the start of each new block, we reset $n_A = n_B = 0$. Then, the blocking constraint can be produced by setting the probability of assignment to treatment A to

$$\Pr[A] = \frac{N_A - n_A}{N_A + N_B - n_A - n_B} \tag{13.1}$$

TABLE 13.1   All Possible Permutations of Two Treatments in Blocks of Size 4

Within-block Assignment Number	Permutation Number					
	1	2	3	4	5	6
1	A	A	A	B	B	B
2	A	B	B	B	A	A
3	B	A	B	A	B	A
4	B	B	A	A	A	B

for each assignment. When N_A assignments have been made in a block, $n_A = N_A$ and the probability of getting A is 0. When N_B assignments have been made, the probability of getting A is 1. This is a useful way to generate blocked assignments by computer. To make each new assignment, we compare a random number, u, uniformly distributed on the interval (0, 1) to $\Pr[A]$. If $u \leq \Pr[A]$, the assignment is made to A. Otherwise, it is made to B.

Suppose that there are two treatments, A and B, and blocks of size 4 are used. There are six possible permutations of two treatments in blocks of size 4 (Table 13.1). The assignment scheme consists of a string of blocks chosen randomly (with replacement) from the set of possible permutations. If we are unlucky enough to stop the trial halfway through a block in which the first two assignments are A's, the imbalance will be only 2 extra in favor of treatment A. This illustrates that the maximum imbalance is one-half of the block size.

The number of orderings of the treatments within small blocks is not very large. Usually it makes sense to use relatively small block sizes to balance the assignments at frequent intervals. Suppose there are 2 treatments and the block size is $2b$. Then, there are $\binom{2b}{b}$ different permutations of the assignments within blocks. If the trial ends exactly halfway through a block, the excess number of assignments in one group will be no greater than b and is likely to be less. Blocking in a similar fashion can be applied to unbalanced allocations or for more than two treatments. For more than two treatments, the number of permutations can be calculated from multinomial coefficients and the assignments can be generated from formulae similar to equation (13.1).

In an actual randomization, all blocks do not have to be the same size. Varying the length of each block (perhaps randomly) makes the sequence of assignments appear more random and can help prevent discovery. These sequences are not much more difficult to generate or implement than fixed block sizes. In practice, one could choose a small set of block sizes from which to select. For example, for two treatments, we could randomly choose from among blocks of length 2, 4, 6, or any multiple of 2. For three treatments, the block sizes would be multiples of 3.

13.3.2   Blocking and Stratifying Balances Prognostic Factors

To balance several clinically important prognostic factors, the randomization can be blocked and stratified. Every relevant prognostic factor combination can define an individual stratum, with blocking in each. Then the treatment assignments will be balanced in each stratum, yielding balanced prognostic factors in each treatment group. To be most advantageous, stratification should be used only for prognostic factors with relatively strong effects, such as risk ratios over about 2.0.

The balance resulting from blocked stratification can improve the power of trials by reducing unwanted variation (Palta, 1985). Other advantages to the balance induced by stratification and blocking include reduced type I and II error rates, higher efficiency, and improved subgroup analyses. These have been emphasized by Kernan et al. (1999) who also present useful information on the current practice of stratification and suggestions for reporting its use.

An example using blocks of size 4 should help make the procedure clear (Table 13.2). Suppose that there are two prognostic factors, age (young versus old) and sex. To balance both, we would form four strata: young-female, young-male, old-female, and old-male. Blocked assignments are generated for each stratum. Each patient receives the next treatment assignment from the stratum into which he or she fits and the strata do not necessarily all fill at the same rate.

In a blocked stratified trial such as this, we could stop the trial when all strata are half filled, all with an excess of assignments on A (or B). For example, the sequence of assignments depicted in Table 13.2 could end halfway through all current blocks and with the last two assignments being AA (or BB). Block 2 in the old-female stratum is of this type. This situation creates the largest imbalance that could occur.

For a trial with b assignments of each treatment in every block (block size $= 2b$) and k strata, the size of the maximum imbalance using blocked strata is $b \times k$. In other words, each stratum could yield an excess of b assignments for the same treatment. The same is true if b is the *largest* number of assignments possible to one treatment in a block in each of k strata when a variable block size is used. If b_j represents the largest excess number of assignments possible in a block in the jth stratum (not necessarily the same for all strata), the largest imbalance (total excess number of assignments on one group) possible is $b_1 + b_2 + \cdots + b_k$.

Admittedly we would have to be very unlucky for the actual imbalance to be at the maximum when the study ends. For example, suppose that there are b assignments of each type in every block and the trial ends randomly and uniformly within a block. In other words, there is no increased likelihood of stopping on either a particular treatment

TABLE 13.2   Example of Blocked Stratified Treatment Assignments with Equal Allocation in Two Groups

Block Number	Strata			
	Males		Females	
	Young	Old	Young	Old
1	A	B	B	A
1	B	B	A	B
1	B	A	B	A
1	A	A	A	B
2	A	B	A	B
2	A	A	B	B
2	B	B	A	A
2	B	A	B	A
⋮	⋮	⋮	⋮	⋮

assignment or position within a block. Assume that all blocks are the same size. The chance of stopping in every stratum exactly halfway through a block is

$$u = \left(\frac{1}{2b}\right)^k,$$

and the chance that the first half of all k blocks has the same treatment assignment is

$$v = \left(\frac{2b}{b}\right)^{-k}.$$

Then uv is the chance of ending the trial on the worst imbalance. For 2 treatments and 4 strata with all blocks of size 4 ($b = 2$), as in the example above, the maximum imbalance is $2 \times 4 = 8$. The chance of stopping with an imbalance of 8 in favor of one treatment or the other is only

$$uv = 2 \times \left(\frac{1}{4}\right)^4 \times \left(\frac{4}{2}\right)^{-4} = 6.03 \times 10^{-6}.$$

The chance of a lesser imbalance is greater.

Stratified randomization *without* blocking will not improve balance beyond that attainable by simple randomization. In other words, simple randomization in each of several strata is equivalent to simple randomization overall. Because a simple random process has no memory of previous assignments, stratification serves no purpose unless the assignments are constrained within strata. Another type of constraint that produces balance using adaptive randomization is discussed below.

13.3.3 Other Considerations Regarding Blocking

One can inadvertently counteract the balancing effects of blocking by having too many strata. For example, if each patient winds up in his or her own stratum uniquely specified by a set of prognostic factors, the result is equivalent to having used simple randomization. Plans for randomization must take this possible problem into account and use only a few strata, relative to the size of the trial. In other words, most blocks should be filled because unfilled blocks permit imbalances.

If block sizes are too large, it can also counteract the balancing effects of blocking. Treatment assignments from a very large block are essentially equivalent to simple randomization, except toward the end of the block. For example, consider equation (13.1) for Pr[A] when N_A and N_B are very large. Then Pr[A] $\approx \frac{1}{2}$ for all the early assignments in a block.

Because at least the last assignment in each block (and possibly up to half of some blocks) is not random (i.e., determined from the previous assignments), the block size should not be revealed to the clinical investigators. Otherwise, some assignments are potentially discoverable. Random block sizes can be used, provided all possible sizes are small, to reduce or eliminate the chance that an investigator will acquire knowledge about future treatment assignments. For example, in a two-group trial, choosing randomly from among block sizes of 2, 4, and 6 yields assignments that would look almost purely random if revealed, except that they are fairly closely balanced.

Blocking can also be used with unequal treatment allocation. For example, suppose that we wish to have a 2:1 allocation in favor of treatment A. Block sizes that are multiples of three could be used with two-thirds of the assignments allocated to treatment A. At the end of each block the treatment allocation will be exactly 2:1 in favor of treatment A.

One advantage of blocked assignments, with or without strata, is that they can be generated on paper in advance of the trial and placed in notebooks accessible to the staff responsible for administering the treatment allocation process. An excess number of assignments can be generated and used sequentially as accrual proceeds. The resulting stream of assignments will have the intended properties and will be easy to prepare and use. Excess assignments not used at the end of the study can be ignored.

The use of stratified blocked randomization is an explicit acknowledgment that the variability due to treatment center and/or the prognostic factor is sizable but extraneous to the treatment effect. When analyzing a study that has used this method of treatment allocation, it is important to account for the stratification. This point has been discussed by several authors (Simon, 1982; Fleiss, 1986b). If the strata are ignored in the analysis, it is equivalent to lumping that source of variability into the error or variance term of the test statistic. As a result the variance will be larger than necessary and the efficiency of the test will be lower. Blocking and stratifying the randomization has virtually no drawbacks and properly accounting for the use of stratification in the analysis increases precision. Consequently these are very useful devices for controlling extraneous variation and should be considered whenever feasible.

13.4   ADAPTIVE ALLOCATION

Adaptive allocation is a process in which the probability of assignment to the treatments in a clinical trial does not remain constant but is determined by the current balance, composition, or outcomes of the groups. In a limited sense this same characteristic is true for blocked randomization. However, adaptive allocation is a more general idea and is often motivated by a desire to minimize the number of subjects entered on what will be shown to be the inferior treatment. General discussions of this method can be found in Hoel, Sobel, and Weiss (1975), Efron (1971), and Pocock and Simon (1975). Urn designs (biased coin randomization) and minimization are two types of adaptive treatment allocation.

13.4.1   Urn Designs Also Improve Balance

An alternative to blocking to prevent large imbalances in the numbers of assignments made to treatment groups is a technique based on urn models (Wei and Lachin, 1988). Urn models are ubiquitous devices in probability and statistics, which are helpful in illustrating important concepts. Imagine an urn containing one red and one white ball. To make a treatment assignment, a ball is selected at random. If the ball is red, the assignment is to treatment A. A white ball yields treatment B. After use, the ball is replaced in the urn. This procedure is equivalent to simple randomization with probability of $\frac{1}{2}$ of receiving either treatment.

To discourage imbalances, however, a slightly different procedure might be used. Suppose that we begin with one ball of each color in the urn and the first ball picked is red. Then the original ball plus an additional *white* ball might be placed in the urn. With the next choice there is one red ball and two white balls in the urn, yielding a higher probability of balancing the first treatment assignment. With each draw the chosen ball and one of the opposite color are placed in the urn. Suppose that n_A and

TABLE 13.3   Example of Stratified Treatment Assignments Using an Urn Model

	Stratum 1				Stratum 2		
n_a	n_b	Pr[A]	Treatment	n_a	n_b	Pr[A]	Treatment
1	0	0.500	A	1	0	0.500	A
1	1	0.333	B	1	1	0.333	B
2	1	0.500	A	1	2	0.500	B
3	1	0.400	A	1	3	0.600	B
4	1	0.333	A	1	4	0.667	B
4	2	0.286	B	2	4	0.714	A
5	2	0.375	A	3	4	0.625	A
5	3	0.333	B	4	4	0.556	A
5	4	0.400	B	4	5	0.500	B
6	4	0.455	A	5	5	0.545	A
7	4	0.417	A	5	6	0.500	B
8	4	0.385	A	6	6	0.538	A
8	5	0.357	B	6	7	0.500	B
8	6	0.400	B	7	7	0.533	A
8	7	0.438	B	7	8	0.500	B
9	7	0.471	A	7	9	0.529	B
10	7	0.444	A	7	10	0.556	B
10	8	0.421	B	8	10	0.579	A
11	8	0.450	A	8	11	0.550	B
11	9	0.429	B	9	11	0.571	A
⋮	⋮	⋮	⋮	⋮	⋮	⋮	⋮

Note: Pr[A] is the probability of assignment to treatment A calculated prior to the assignment.

n_B represent the current number of red and white balls, respectively, in the urn. Then, for each assignment,

$$\Pr[A] = \frac{n_B}{n_A + n_B} .$$

When the number of draws is small, there tends to be tighter balance. As the number of draws increases, the addition of a single ball is less important and the process begins to behave like 1:1 simple randomization. At any point, when equal numbers of assignments have been made for each group, the probability of getting a red or white ball on the next draw is $\frac{1}{2}$. This process could also be used for stratified assignments.

This urn procedure might be useful in trials where the final sample size is going to be small and tight balance is needed. These assignments can also be generated in advance and stratified, if necessary, to balance a prognostic factor. For example, Table 13.3 shows the first few treatment assignments from a two-group, two-stratum sequence. Any sequence is possible, but balanced ones are much more likely.

13.4.2   Minimization Yields Tight Balance

Other schemes to restrict randomization to yield tighter balance in treatment groups can be based on minimization, as suggested by Pocock and Simon (1975) and Begg and Iglewicz (1980). To implement minimization, a measure of imbalance based on the current number of treatment assignments (or the prognostic factor composition of the treatment groups) is used. Before the next treatment assignment is made, the imbalance that will result is calculated in two ways: (1) assuming the assignment is made to A, and (2) assuming it is made to B. The treatment assignment that yields the smallest imbalance is chosen. If the imbalance will be the same either way, the choice is made randomly.

The strengths of this scheme are that it can produce a tighter balance than blocked strata. However, it has the drawback that one must keep track of the current measure of imbalance. This is not a problem for computer-based randomization schemes, which are frequently used today. Also, when minimization is employed, *none* of the assignments (except perhaps the first) is necessarily made purely randomly. This would seem to prevent the use of randomization theory in analysis. However, treating the data as though they arose from a completely randomized design is commonly done and is probably inferentially correct. At least in oncology, many multi-institutional "randomized" studies actually employ treatment allocation based on minimization.

13.4.3 Play the Winner

Other methods for making treatment allocations in clinical trials have been suggested to meet certain important constraints. For example, Zelen (1969) and Wei and Durham (1978) suggested making treatment assignments according to a "play the winner" rule, to minimize the number of patients who receive an inferior treatment in a comparative trial. One way of implementing this scheme is as follows: Suppose that randomization to one of two groups is made by the logical equivalent of drawing balls labeled A or B from an urn, with replacement. For the first patient, the urn contains one ball of each type (or an equal number of each). The first patient is assigned on the basis of a random draw. If the assigned treatment "fails," a ball of the other type is added to the urn. The next patient therefore has a higher probability of receiving the other treatment. If the first treatment "succeeds," a ball of the same type is added to the urn. Thus the next patient has a higher probability of receiving the same therapy.

This method has the advantage of preferentially using what appears to be the best treatment for each assignment after the first. However, implementing it in the manner described requires being able to assess the final outcome in each study subject before placing the next patient on the trial. This may not always be feasible. A more flexible approach is to use a multistage design with adjustment of the allocation ratio, to favor the treatment arm that appears to be doing the best at each stage. This approach can also be used with outcomes that are not dichotomous. Some investigators believe that trials designed in this way may be viewed more favorably than conventional designs by patients and Institutional Review Boards.

A design of this type was used in a clinical trial of extracorporeal membrane oxygenation (ECMO) versus standard therapy for newborn infants with persistent pulmonary hypertension (Bartlett et al., 1985). Based on developmental studies over 10 years, ECMO appeared to be a potentially promising treatment for this fatal condition when it was tested in a clinical trial in 1985. With the Wei and Durham biased coin randomization, the trial was to be stopped when the urn contained 10 balls favoring one treatment. This stopping rule was chosen to yield a 95% chance of selecting the best treatment when $p_1 \geq 0.8$ and $p_1 - p_2 > 0.4$, where p_1 and p_2 are the survival probabilities on the best and worst treatments, respectively.

TABLE 13.4   Results from an ECMO Clinical Trial Using Play the Winner Randomization

Patient Number	Treatment Assignment	Outcome
1	ECMO	Success
2	Control	Failure
3	ECMO	Success
⋮	⋮	⋮
10	ECMO	Success

Some of the data arising from the trial (in the stratum of infants weighing more than 2 kilograms) are shown in Table 13.4. The first patient was assigned to ECMO and survived. The second patient received conventional treatment and died. All subsequent patients were assigned to ECMO. The trial was terminated after 10 patients received ECMO, but 2 additional patients were non-randomly assigned to ECMO. Because of the small sample size, one might consider testing the data for statistical significance using Fisher's exact test, which conditions on the row and column totals of the 2×2 table. However, the analysis should not condition on the row and column totals because, in an adaptive design such as this, they contain information about the outcomes. Accounting for the biased coin allocation, the data demonstrate a marginally significant result in favor of ECMO (Wei, 1988).

This trial was unusual from a number of perspectives: the investigators felt fairly strongly that ECMO was more effective than conventional therapy before the study was started, an adaptive design was used, consent was sought only for patients assigned to experimental therapy (Section 13.7), and the stopping criterion was based on selection or ranking of results. Ware and Epstein (1985) discussed this trial, stating that from the familiar hypothesis testing perspective, the type I error rate for a trial with this design is 50%. The methodology has also been criticized by Begg (1990).

The ECMO trial is also unusual because of the subsequent controversies that it sparked. Not all of the discussions stem from the design of the trial; much is a consequence of the clinical uncertainties. However, alternative designs may have helped resolve the clinical questions more clearly. For example, a design with an unequal treatment allocation (Section 13.6) might have reduced some of the uncertainty. For a clinical review of the ECMO story, see the paper by O'Rourke (1991). Ethical issues were discussed by Royall (1991). A confirmatory ECMO trial was performed (Section 14.1.4), with resulting additional controversy (Ware, 1989).

13.5   OTHER ISSUES REGARDING RANDOMIZATION

13.5.1   Administration of the Randomization

The beneficial effects of randomization for reducing bias can be undone if future treatment assignments are discoverable by investigators. Potentially this information could be available if all treatment assignments are listed in advance of the study in a notebook or folder and dispensed as each patient goes on study, as is commonly done. This problem could arise if clinical investigators responsible for placing patients on study and selecting treatment are involved with, or have access to, the randomization process. To convincingly demonstrate that this has not occurred, the administration of the randomization process should be physically separated from the clinical investigators.

In multicenter trials, this is commonly accomplished by designating one of the institutions, or an off-site location, to serve as a trial coordinating center. Treatment assignments are then made by telephone or by secure computer access. The concern about access to treatment assignments may be more relevant in single-institution studies, where the clinical investigators often administratively manage their own trials. It is sometimes necessary for the investigators to manage treatment assignment for reasons of accessibility or convenience. When this is required, extra care should be taken to keep the assignments secure.

Example 31 Investigators conducted a multicenter randomized trial of low tidal volume versus standard ventilation in patients with adult respiratory distress syndrome. Because the patients were all ventilator-dependent and hospitalized in intensive care units (ICU), it was necessary to have the capability to randomize treatment assignments 24 hours each day. This could not be accomplished easily using an independent study coordinator. Instead, it was more efficient to designate one of the participating ICUs as the randomization center and have the nursing staff always available there perform an occasional randomization. This ICU environment is potentially "hostile" to performing such tasks rigorously. Nevertheless, a suitably written computer program helped the randomization proceed routinely. The program contained password protection for each randomization and safeguards to prevent tampering with either the program itself or the sequence of treatment assignments. Initially stratified randomized assignments from the computer program were made with a block size of 8. Interestingly, after 3 of the first 4 assignments in one stratum were made to the same treatment group, the ICU nurses became convinced that the program was not functioning "randomly." This illustrates the scrutiny that the sequence of assignments is likely to get. Reassuring investigators that the program was operating correctly was a strong suggestion about the (fixed) block size. Consequently the randomization program was modified to use a variable block size and the staff was reassured that all was well.

This example illustrates that under special circumstances, treatment assignments can be protected from tampering, while still permitting flexibility and convenience. Of course, no safeguards are totally effective against a determined unscrupulous investigator. For example, one might be able to defeat seemingly secure randomization programs by running two copies. The second copy would permit experimentation that could help to determine the next treatment assignment. The most secure randomization for 24-hour-per-day availability is provided by centralized, automated, dial-in computer applications kept at a location remote from the clinic.

The use of sealed envelopes managed by clinical investigators is sometimes proposed as a method of randomization. Although expedient, it should be discouraged as a method of treatment allocation because it is potentially discoverable, error prone, and will not provide as convincing evidence of lack of bias as treatment allocation methods that are independent of the clinicians. Even so, sometimes sealed envelopes are the only practical allocation method. This may be the case when trials are conducted in settings with limited resources or research infrastructure, such as some developing countries or in some clinical settings where 24-hour-per-day treatment assignment is needed. Assignments using sealed envelopes should be periodically audited against a master list maintained centrally.

13.5.2 Computers Generate Pseudorandom Numbers

I am always fascinated that some investigators seem reluctant to employ truly random methods of generating numbers in preference to computer-based (pseudorandom) methods. Consider the familiar Plexiglas apparatus filled with numbered ping-pong balls, which is often used to select winning numbers for lottery drawings. This is a reliable and convincingly random method for selecting numbers and could well be used for treatment assignments in clinical trials, except that it is slow and inconvenient.

Computers, in contrast, generate pseudorandom numbers quickly. The numbers are generated by a mathematical formula that yields a completely predictable sequence if one knows the constants in the generating equation and the starting seed. The resulting stream of numbers will pass most important tests for randomness, despite its being deterministic, provided that the seed and constants are chosen wisely. Excellent discussions of this topic are given by Knuth (1981) and Press et al. (1992).

The most commonly used algorithm for generating a sequence of pseudorandom numbers, $\{N_i\}$, is the linear congruential method. This method uses a recurrence formula to generate N_{i+1} from N_i, which is

$$N_{i+1} = aN_i + c \ (\text{mod} \ k),$$

where a is the *multiplier*, c is the *increment*, and k is the *modulus*. Modular arithmetic takes the remainder when N is divided by k. This method can produce numbers that appear random and distributed uniformly over the interval $(0,1)$.

There are some important limitations to this method of generating pseudorandom numbers. The stream of numbers can be serially correlated, and it will eventually repeat after no more than k numbers. The sequence will have a short period if a, c, and k are not carefully chosen, so one must be cautious about using untested random number generators built-in to computer systems. Also the formula and number stream are sensitive to the hardware of the particular computer system. These are important points for statistical simulation, but, perhaps, not so critical for clinical trial treatment assignments because the operative benefit of randomization is that the assignments are made impartially.

In any case, the reader should realize that (1) computers generate pseudorandom numbers, (2) the method of generation is subject to mistakes that could lead to pseudo-random number streams with very undesirable properties, and (3) extra care is needed to have good pseudorandom number generators or ones that yield identical results on all computer systems. It seems we have come full circle now in terms of treatment allocation. The "randomization" of computer-based methods is actually a nondiscoverable, seemingly random, deterministic allocation scheme. Used with care, it is perfectly appropriate for "randomized" assignments in clinical trials.

13.5.3 Randomization Justifies Type I Errors

In addition to eliminating bias in treatment assignments, a second benefit of randomization is that it guarantees the correctness of certain statistical procedures and control of type I errors in hypothesis tests. For example, the important class of statistical tests known as permutation (or randomization) tests of the null hypothesis are validated by the random assignment of patients to treatments. To understand this, we must discuss some aspects of analysis now, even though it is jumping ahead in the scheme of our study of clinical trials. An in-depth discussion of permutation tests is given by Good (1994), and randomization tests by Edgington (1987).

The validity of randomization tests is motivated by the exchangeability of treatment assignments in a clinical trial. Under a population model, two samples from the same population are exchangeable. This might be the view of a randomized clinical trial where, under the null hypothesis, the sample observations from the two treatments are exchangeable. Alternatively, we could motivate such tests by a randomization model as suggested by R.A. Fisher. Under this model, the outcomes from the experiment are viewed as fixed numbers and the null hypothesis states that treatment has no effect on these numbers. Then the group assignments are exchangeable under this null hypothesis.

Suppose that there are two treatments, A and B, N patient assignments in total, $N/2$ patients in each treatment group, and treatment assignments have been made by simple randomization. Let Y_i be the response for the ith patient, $\overline{Y}_A$ and $\overline{Y}_B$ be the average responses in the two treatment groups, and $\Delta = \overline{Y}_A - \overline{Y}_B$. The null hypothesis states that there is no difference between the treatment groups, $H_0 : \Delta = 0$. Stated another way, the treatment assignments and the responses are independent of one another and exchangeable under the null hypothesis. If so, any ordering of the treatment assignments is equally likely under randomization and all should yield a "similar" Δ.

Following this idea, the null hypothesis can be tested by forming all permutations of treatment assignments independently of the Y_i's, calculating the distribution of the Δ's that results, and comparing the observed Δ to the permutation distribution. If the observed Δ is atypical (i.e., if it lies far enough in the tail of the permutation distribution), we would conclude that the distribution arising from the null hypothesis is an unlikely origin for it. In other words, we would reject the null hypothesis. This procedure is "exact," in the sense that the distribution generated is the only one that we need to consider and can be completely enumerated by the permutation method.

Number of Permutations

The number of permutations for even modest sized studies is very large. For example, the number of permutations for two treatment groups, A and B, using simple random allocation with N_A and N_B assignments in each group, is

$$M = \binom{N_A + N_B}{N_A} = \binom{N_A + N_B}{N_B} = \frac{(N_A + N_B)!}{N_A! N_B!} ,$$

which is the binomial coefficient for N objects (treatment assignments) taken N_A (or N_B) at a time. The formula is valid when $N_A \neq N_B$. This number grows very rapidly with $N_A + N_B$. For more than two treatments, the number of permutations is given by the multinomial coefficient

$$M = \binom{N}{N_A, N_B, \ldots} = \frac{N!}{N_A! N_B! \ldots},$$

where $N = N_A + N_B + \cdots$ is the total number of assignments. This number also grows very rapidly with N.

TABLE 13.5   Number of Possible Permutations of Two Treatment Assignments for Different Block and Sample Sizes

Total Sample Size	Block Size			
	2	4	8	∞
4	4	6	6	6
8	16	36	70	70
16	256	1296	4900	12870
24	4096	46656	343000	2.7×10^6
56	2.68×10^8	7.84×10^{10}	8.24×10^{12}	7.65×10^{15}
104	4.50×10^{15}	1.71×10^{20}	9.69×10^{23}	1.58×10^{30}

Note: The calculation assumes all blocks are filled. An infinite block size corresponds to simple randomization.

When blocking is used, the number of permutations is smaller, but still becomes quite large with increasing sample size. For two treatments with b assignments in each fixed block and all blocks filled, the number is

$$M = \binom{2b}{b}^k,$$

where the number of blocks is k. A more general formula for blocks of variable size is

$$M = \prod_{i=1}^{r} \binom{N_{A_i} + N_{B_i}}{N_{A_i}}^{k_i},$$

where there are r different sized blocks and k_i of each one. Here also the formula is valid when $N_{A_i} \neq N_{B_i}$. This last formula can also be adapted to cases where the last block is not filled. Examples of the number of permutations arising from two-treatment group studies of various sizes are shown in Table 13.5.

Implementation

Because the number of permutations from even modest size studies is so large, it is impractical to use exact permutation tests routinely in the analysis of clinical trials, despite the desirable characteristics of the procedure. A more economical approach, which relies on the same theory, is to sample randomly (with replacement) from the set of possible permutations. If the sample is large enough, although considerably smaller than the total number of possible permutations, the resulting distribution will accurately represent the exact permutation distribution and will be suitable for making inferences. For example, one might take 1000 or 10,000 samples from the permutation distribution. Such a procedure is called a "randomization test," although this term could be applied to permutation tests as well. The terminology is not important, as long as one realizes that the exact permutation distribution can be approximated accurately by a sufficiently large sample.

An example of a permutation test is shown in Table 13.6. Here there were 5 assignments in each of two groups, yielding a total of 252 possible permutations, conditional on the number in each group being equal. For each possible permutation,

TABLE 13.6 Lower Half of the Permutation Distribution for 10 Balanced Treatment Assignments in Two Groups

Permutation	Δ	Permutation	Δ	Permutation	Δ
AAABABBABB	2.4	AAABBBBABA	1.0	ABBAAABBBA	0.4
ABAAABBABB	2.2	BAAAABBBAB	1.0	BAABAABBBA	0.4
AAAAABBBBB	2.2	AABBAABBAB	1.0	BBAAABAABB	0.4
ABABAABABB	2.0	AAABBABBAB	1.0	BBBAAABAAB	0.4
AAAABBBABB	2.0	BBABAABAAB	0.8	ABABBAAABB	0.4
AABAABBABB	2.0	ABABAABBBA	0.8	ABBBAAAABB	0.4
AAABAABBBB	2.0	AAAABBBBBA	0.8	BAAAABABBB	0.4
ABAAAABBBB	1.8	AABAABBBBA	0.8	BAAABBBABA	0.4
BAAAABBABB	1.8	ABAAABABBB	0.8	BABAABBABA	0.4
AAABBABABB	1.8	ABAABABBAB	0.8	BBAABABAAB	0.4
AABBAABABB	1.8	ABAABBBABA	0.8	AABABBAABB	0.4
ABAABABABB	1.6	ABBAABBABA	0.8	AABBAAABBB	0.4
ABABABBAAB	1.6	BAABABBABA	0.8	AAABBAABBB	0.4
ABBAAABABB	1.6	AABBABAABB	0.8	BAAABABBAB	0.4
BAABAABABB	1.6	BAAABBBAAB	0.8	BABAAABBAB	0.4
AAAABABBBB	1.6	BABAABBAAB	0.8	ABABBBBAAA	0.2
AABAAABBBB	1.6	AAABBBAABB	0.8	ABBBABBAAA	0.2
AAABABBBAB	1.6	AABBBABAAB	0.8	ABAABAABBB	0.2
BAAAAABBBB	1.4	ABBAAABBAB	0.8	ABBABABABA	0.2
ABAAABBBAB	1.4	BAABAABBAB	0.8	BAABBABABA	0.2
BBAAAABABB	1.4	AABABBBABA	0.6	BBAAAABBBA	0.2
AAABBBBAAB	1.4	BBAAABBABA	0.6	AABBBAAABB	0.2
AABABABABB	1.4	ABAABBAABB	0.6	ABABABABAB	0.2
AABBABBAAB	1.4	ABABBABABA	0.6	BABABABAAB	0.2
ABABABBABA	1.2	ABBBAABABA	0.6	BBABAAAABB	0.2
AAAABBBBAB	1.2	AAAABBABBB	0.6	ABBAAAABBB	0.2
AABAABBBAB	1.2	AAABBABBBA	0.6	BAAABBAABB	0.2
ABAABBBAAB	1.2	AABAABABBB	0.6	BAABAAABBB	0.2
ABBAABBAAB	1.2	ABBABABAAB	0.6	BABAABAABB	0.2
BAAABABABB	1.2	BAAAABBBBA	0.6	BABBAABABA	0.2
BAABABBAAB	1.2	BAABBABAAB	0.6	AABABABBBA	0.2
BABAAABABB	1.2	BABBAABAAB	0.6	AAABBBBBAA	0.2
ABABAABBAB	1.2	BBAAAABBAB	0.6	AABBABBBAA	0.2
AAABABBBBA	1.2	AABBAABBBA	0.6	AAABBBABAB	0.0
ABAAABBBBA	1.0	ABABAAABBB	0.6	AABABAABBB	0.0
ABABBABAAB	1.0	ABBAABAABB	0.6	ABAABBBBAA	0.0
ABBBAABAAB	1.0	BAABABAABB	0.6	ABBAABBBAA	0.0
ABABABAABB	1.0	AABABABBAB	0.6	BAAABABBBA	0.0
BBAAABBAAB	1.0	BBABAABABA	0.4	BAABABBBAA	0.0
AAABABABBB	1.0	AABBBABABA	0.4	BABAAABBBA	0.0
AABABBBAAB	1.0	ABAABABBBA	0.4	BBAAAAABBB	0.0
AABBABBABA	1.0	ABABABBBAA	0.4	BBAABABABA	0.0

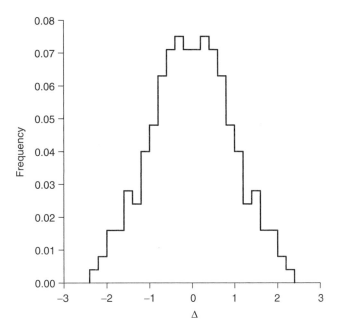

Figure 13.1 Example permutation distribution from 10 treatment assignments.

let $\Delta = \overline{X}_B - \overline{X}_A$. The 5 responses on A and B were $\mathbf{X} = \{1.1, 1.3, 1.2, 1.4, 1.2, 1.5, 2.0, 1.3, 1.7, 1.9\}$. From these the Δ's that would be obtained for each permutation of treatment assignments are shown in Table 13.6, ordered by the size of Δ. Only the upper half of the permutations are listed—the mirror-image half (i.e., A and B switching places) has Δ's that are the negative of those in the table. Figure 13.1 shows the permutation distribution. Suppose the observed Δ was 2.2, the sequence of assignments actually being $BABBBAABAA$, (from permutation number 2), which lies above the 0.012 percentile of the distribution. Thus we would reject the null hypothesis of no treatment difference. The validity of this procedure relies on nothing except the use of randomization in allocating treatment assignments: no assumptions about statistical distributions or models have been made. Similar procedures can be applied to event data from clinical trials (Freedman, Sylvester, and Byar, 1989; Ohashi, 1990).

Permutation and randomization tests are appealing because they can be applied to virtually any test, they do not depend on the sample size, and they require few assumptions beyond the use of randomization. They have not been as widely applied as many other tests, probably because of the computing resources required. However, recent improvements in computing performance on all platforms make randomization tests very attractive.

13.6 UNEQUAL TREATMENT ALLOCATION

Usually comparative trials employ an equal allocation of subjects to each of the treatments under study. To maximize the efficiency (power) of the primary comparison, this is often the best approach. However, the trial may be designed intentionally to employ unequal group sizes to meet important secondary objectives. Then unbalanced designs can be more efficient than equal allocation for meeting all study objectives. Examples where this can be useful include important subset analyses, large differences in the cost of treatments, and when the responses have unequal variances. For a review of unequal allocation, see Sposto and Krailo (1987).

Some investigators feel that unbalanced random assignment reflects a preference for the more frequently employed treatment. This view is not correct (at least with any sensible allocation ratio). Randomization is employed in the presence of equipoise to eliminate selection bias. This same rationale applies to unequal allocation. Furthermore one cannot correct a lack of equipoise by appealing to a collective ethic that does less harm on average. It is either appropriate to randomize or not. The need to use an unbalanced allocation should be separated from the decision to randomize, and is more properly based on other study objectives as outlined below.

13.6.1 Subsets May Be of Interest

It may be important to acquire as much experience with a new treatment as possible, while also comparing it with standard therapy. In this case an unequal allocation of subjects, such as 2:1 in favor of the new treatment, may allow important additional experience with the new treatment without substantially decreasing the efficiency (power) of the comparison. Similarly subsets of patients who receive the new treatment (e.g., older patients) may be of interest and unequal allocation may increase their number. The sensitivity of power to the allocation ratio was discussed in Chapter 11.

13.6.2 Treatments May Differ Greatly in Cost

When one treatment is much more expensive than the other in a two-group clinical trial, the total cost of the study will be minimized by an unequal allocation of subjects. Fewer patients would be allocated to the more expensive treatment. This circumstance was discussed in Section 11.7.2. There it was shown that under fairly general circumstances the cost is minimized by an allocation ratio of

$$r = \sqrt{C},$$

where C is the relative cost of the more expensive therapy. If $C = 10$, for example, the cost-minimizing allocation ratio will be approximately 3:1 in favor of the less expensive treatment. When cost differences are very large, we must be attentive to the potential power loss that this unbalanced allocation implies.

13.6.3 Variances May Be Different

Suppose the means, μ_1 and μ_2, of two treatment groups are being compared using a test, for which the statistic is

$$t = \frac{\mu_1 - \mu_2}{\sqrt{\frac{\sigma_1^2}{n_1} + \frac{\sigma_2^2}{n_2}}}$$

where the variances, σ_1^2 and σ_2^2, are not necessarily equal. Because of this, one may wish to change the allocation ratio (i.e., n_1 and n_2) to make the test as efficient as possible. For fixed μ_1 and μ_2, the test statistic is maximized when the denominator is minimized, or when $\sigma_1^2/n_1 + \sigma_2^2/n_2$ is minimized. For $N = n_1 + n_2$, the minimum satisfies

$$0 = \frac{\partial}{\partial n_1}\left(\frac{\sigma_1^2}{n_1} + \frac{\sigma_2^2}{N - n_1}\right)$$

or

$$\frac{\sigma_2^2}{\sigma_1^2} = \left(\frac{N - n_1}{n_1}\right)^2,$$

from which the optimal fraction in group 1 is found to be

$$\frac{n_1}{N} = \frac{\sigma_1}{\sigma_1 + \sigma_2}.$$

Therefore the optimal allocation ratio is

$$r = \frac{n_1}{n_2} = \frac{\sigma_1}{\sigma_2}.$$

In other words, more subjects should be placed in the group where the measured response is less precise.

13.7 RANDOMIZATION BEFORE CONSENT

Partly in response to the difficulties of getting patients to accept randomization, Zelen (1979, 1990) suggested a "pre-randomized" scheme that works in the following way: Suppose that a new therapy is being compared with standard treatment. Eligible patients are randomized before being approached to participate in the trial. If the treatment assignment is standard therapy, the patient is offered the standard as a matter of routine and randomization and consent need not be discussed. If the randomized assignment is the new therapy, the patient is approached for consent. If the patient refuses, he or she is offered standard therapy. The final comparison is between the groups based on randomized assignment. A double randomized consent has also been suggested, in which those initially assigned to standard therapy are also asked if they will accept the new treatment. A review of these designs is given by Parmar (1992).

The randomization before consent strategy can increase the number of trial participants. However, it creates some difficulties that limit its use. For example, in the analysis by treatment assigned, the patients who refuse their assignment dilute the treatment difference. If the trial is analyzed by treatment received, there is potential for bias. Most important, the design has difficulty passing ethical review because patients assigned to standard therapy are part of a research study without being properly informed (highlighting a double standard for consent on research studies). For example, the patients on standard therapy are not given the chance to obtain the new treatment. Some may not wish to participate in the study for reasons not directly related to treatment.

This design was used in the ECMO trial discussed above, although only a single patient was assigned to conventional therapy. A second trial of ECMO was conducted using this consent procedure (O'Rourke et al., 1989). This pre-randomization was thought to be necessary because of the lack of equipoise on the part of the investigators. This trial, like the one before it and nonrandomized studies, supported the efficacy of ECMO compared with conventional therapy. This second ECMO study also raises a number of important issues about trial conduct as discussed by Meinert (1989) and Chalmers (1989). A few other trials have used randomization before consent, including some collaborative studies in the United States and Europe (Parmar, 1992).

Because of its practical and ethical difficulties, pre-randomization should probably not be routinely used for treatment assignments in clinical trials. Although there may be special circumstances that warrant its use and it can increase accrual, traditional randomized designs seem to be better suited to the objectives of comparative trials and the limits of ethical standards.

13.8   SUMMARY

Treatment allocation in clinical trials and other true experimental designs is characterized by active control of the treatments and the process used to make assignments. The practical concerns in choosing an allocation scheme for comparative studies are reducing bias, quantifying random errors, and increasing the credibility of results. Simple or constrained randomization satisfies the practical concerns and offers important theoretical advantages over other methods of allocation. Randomization reduces or eliminates biases in treatment assignment, guarantees the expectation that unobserved confounders will be controlled, validates control over type I error, and motivates an important class of analyses.

Despite its appeal the actual realization of a simple randomization scheme can result in chance imbalances in influential prognostic factors or the size of treatment groups. This can be prevented using constrained randomization. Constraints frequently used include blocking and stratifying, adaptive randomization, and minimization. These methods, while theoretically unnecessary, encourage covariate balance in the treatment groups, which tends to enhance the credibility of trial results. Unequal group sizes might be used when it minimizes the cost of a trial or facilitates secondary objectives.

Administering treatment assignments may be as important as using an unbiased allocation method. The assignments should be supervised and conducted by individuals who have no vested interest in the results of the trial. The method employed should be convincingly tamper-proof and distinct from the clinic site. The beneficial effects of randomization in reducing bias can be undone if investigators can discover future treatment assignments.

13.9   QUESTIONS FOR DISCUSSION

1. Simple randomization is used in a two-group CTE trial with 100 participants. The probability of being assigned to treatment A is $p = 0.5$. What is the chance that treatment A will end up having between 48 and 52 assignments? Using a 0.05 α-level test, what is the chance that the groups will be so unbalanced that investigators would conclude that $p \neq 0.5$?

2. Simple randomization is used in a two-group CTE trial. The probability of being assigned to either treatment is 0.5. How many independent binary covariates would one need to test to be 50% certain of finding at least one "significantly" unbalanced in the treatment groups? How about being 95% certain?

3. A clinical trial with four groups uses treatment assignments in blocks of size 4 in each of 8 strata. What is the maximum imbalance in the number of treatment assignments that can occur? What is the chance of this happening? What is the chance that the trial will end exactly balanced?

4. Investigators have written a computer program to generate random integers between 0 and 32767. The output is tested repeatedly and found to satisfy a uniform distribution across the interval. Even integers result in assignment to group A and odd ones to group B. When the actual assignments are made, group B appears to have twice as many assignments as group A. How can this happen? Can this generator be used in any way? Why or why not?

5. A new random number generator is tested and found to satisfy a uniform distribution. Values below the midpoint are assigned to treatment A. When the actual assignments are made, there are very few "runs" of either treatment (i.e., AA, AAA, ..., BB, BBB, ...). How would you describe and quantify this problem? Can this generator be used after any modifications? If so, how?

6. A trial with 72 subjects in each of two groups uses randomized blocks of size 8. The data will be analyzed using a permutation test. You find a fast computer that can calculate 10 permutation results each second and start the program. How long before you have the results? Can you suggest an alternative?

14

TREATMENT EFFECTS MONITORING

14.1 INTRODUCTION

Continuing a clinical trial should entail an active affirmation by the investigators that the scientific and ethical milieu requires and permits it, rather than a passive activity based only on the fact that the study was begun. As the trial proceeds, investigators must consider ongoing aspects of ethics (risks and benefits), data quality, the precision of results, the qualitative nature of treatment effects and side effects, resource availability, and information from outside the trial in deciding whether or not to continue the study. Circumstances may require ending a treatment arm or discontinuing accrual of a subset of patients rather than stopping the entire trial.

The appropriate source of information on which to base most aspects of these decisions is the accumulating data, although sometimes information from other studies may be relevant. This explains the idea of "data-dependent stopping," also an appropriate terminology for this activity in which decisions about continuing the trial are based on the evidence currently available. However, monitoring a clinical trial touches more issues than only the decision to stop.

Correctly gathering and interpreting information accumulating during a clinical trial has been described using a variety of other terms: interim analysis, data monitoring, data and safety monitoring, treatment effects monitoring, or early stopping. It is possible to construct nuances of meaning among all these terms, but is not necessary for the purposes here, where I will use them interchangeably. Strictly speaking, an interim analysis need not address the question of study termination for reasons of efficacy. It might be done for administrative or quality control reasons. However, the most interesting and difficult questions are those surrounding early termination because of treatment effects or their lack, so that will be the focus of my discussion. Statistical reasoning plays an essential, but not exclusive, role in such deliberations. In any case, preplanned and structured information gathering during a trial is the most reliable way to proceed for all purposes.

In past years most clinical trials were monitored principally for data quality and safety rather than treatment effects. The need to minimize exposure of patients to inferior treatments in comparative studies, safety, and the development of statistical methods for interim inferences have facilitated the wide application of monitoring for treatment effects. Now there is the expectation that all comparative clinical trials and many other designs be monitored formally. Sponsors such as the NIH have formal written policies for data and safety monitoring (Appendixes F and G). The FDA also has a draft guidance on the subject (www.fda.gov/cber/gdlns/clindatmon.htm).

In this chapter, I discuss ways of minimizing errors that can result from interim looks at accumulating trial data and ways of improving decision making. Both a statistical perspective and organizational and procedural considerations are presented. Methods for data-dependent stopping fall into four general categories: likelihood, Bayesian, frequentist, and others. There is a large literature base related to this topic. The best place to start an in-depth study is the recent book by Ellenberg, Fleming, and DeMets (2003). An issue of *Statistics in Medicine* contains papers from a workshop on early stopping rules in cancer clinical trials (Souhami and Whitehead, 1994). For older but useful reviews, see Gail (1982, 1984), Rubinstein and Gail (1982), Freedman and Spiegelhalter (1989), Freedman, Spiegelhalter, and Parmar (1994), Fleming and DeMets (1993), Berry (1985), Pocock (1993), or Fleming, Green, and Harrington (1984).

The focus of most methodologic work on monitoring is termination of the entire trial. However, practical issues sometimes force us to consider terminating a treatment arm or a subset of subjects who may be entering both treatments. This latter circumstance arose in a recent trial of lung volume reduction surgery (e.g., Section 4.6.6) and is discussed by Lee et al. (2004).

The discussion that follows pertains only to two-group randomized trials and a single major endpoint. There is much less in the clinical trials literature on the monitoring and stopping of trials with more than two treatment groups. Exceptions are the papers by Hughes (1993) and Proschan, Follmann, and Geller (1994). In addition to statistical guidelines for interim analyses, I discuss some computer programs that can assist with the necessary computations.

14.1.1 Motives for Monitoring

The primary purpose for monitoring a clinical trial is to provide an ongoing evaluation of risk–benefit that addresses the uncertainty necessary to continue. This requires explicit views of both safety data (risk) and efficacy data (benefit). This might seem obvious, but there are occasional requests from study sponsors to monitor trials solely for safety. This is operationally possible, but it is seldom appropriate to isolate safety concerns from efficacy ones. Rarely it might happen that accrual and treatment proceed very quickly relative to the accumulation of definitive efficacy outcomes, meaning that predominantly safety issues are relevant during the trial.

Aside from terminating a study early because of treatment effects, there are other reasons why investigators may want a view of trial data as they accumulate. Problems with patient accrual or recruitment may become evident and require changes in the study protocol or investigator interactions with prospective participants. In some cases incentives or recruitment programs may need to be initiated. Additionally monitoring and judicious interim reporting can help maintain investigator interest in, and enthusiasm for, the study. It is natural for highly motivated investigators to wonder about study progress and be curious about results. Not all such curiosity can or should be satisfied by interim reports, but they can serve to maintain active interest. Finally it may be necessary to make other adjustments in the trial design, such as sample size, on the basis of information from the early part of the study (Herson and Wittes, 1993).

14.1.2 Components of Responsible Monitoring

Making good decisions on an interim basis is not primarily a technical problem. It requires a mixture of advanced planning, skill, and experience that must consider the context of the trial, ethics, data quality, statistical reasoning, committee process and psychology, and scientific objectives. Deficiencies in any of these components can lead to unserviceable decisions.

In all that follows, I assume that timely and accurate data are available on which to base decisions. The production, management, and quality control of complex dynamic data systems is an important subject in itself, and is made even more difficult in multicenter collaborations typical of clinical trials. Good data result from a smooth integration of planning, trained personnel, system design, experience, quality control, and diligence. Interim deliberations often require unexpected views of the data and unanticipated analyses that put additional stresses on the data system and staff. Doubts about data quality will undermine critical interim decisions. Planning, guidelines, and formal decision process cannot substitute for confidence in the data sources. Thus reliable and timely data are a foundation for all of the decision aids discussed here.

The components of monitoring are brought together in the form of a monitoring committee, often called a data and safety monitoring board (DSMB) committee (DSMC or DMC), or a treatment effects monitoring committee (TEMC). The last term is probably the most descriptive. The name used is inconsequential, but the structure, composition, competence, objectivity, and role assigned to the committee are vital. These aspects of data-dependent stopping are discussed below.

14.1.3 Trials Can Be Stopped for a Variety of Reasons

Some practical reasons for terminating a trial early are shown in Table 14.1. My use of the term "data-dependent stopping" to describe *all* of these aspects of trial monitoring is broader than usual because the term is applied often only to the statistical aspects of planning for study termination. Because one cannot easily separate the statistical aspects of trial monitoring from the administrative ones, I will use the term to refer to both.

The simplest approach to formalizing the decision to stop a clinical trial is to use a fixed sample size. Fixed sample size designs are easy to plan, carry out, stop, and analyze, and the estimated treatment effects will be unbiased if the trial is performed and analyzed properly. Often, however, investigators have important reasons to examine and review the accumulating data while a study is ongoing. If convincing evidence becomes available during the trial about either treatment differences, safety, or the quality of the data, there may be an imperative to terminate accrual and/or stop the study before the final fixed sample size has been reached. Assessment of study progress from an administrative perspective, updating collaborators, and monitoring treatment compliance also require that the data be reviewed during the study.

TABLE 14.1   Some Reasons Why a Clinical Trial Might Be Stopped Early

- Treatments are found to be convincingly different by impartial knowledgeable experts.
- Treatments are found to be convincingly not different by impartial knowledgeable experts.
- Side effects or toxicity are too severe to continue treatment in light of potential benefits.
- The data are of poor quality.
- Accrual is too slow to complete the study in a timely fashion.
- Definitive information about the treatment becomes available (e.g., from a similar trial), making the study unnecessary or unethical.
- The scientific questions are no longer important because of other medical advances.
- Adherence to the treatment is unacceptably poor, preventing an answer to the basic question.
- Resources to perform the study are lost or no longer available.
- The study integrity has been undermined by fraud or misconduct.

There are two decision landmarks when monitoring a study: when to terminate accrual and when to disseminate results. Some investigators view these as one and the same, but it is often useful to separate them in studies with prolonged follow-up (e.g., survival studies). As sketched in Chapter 11, one can advantageously trade accrual rate, accrual time, or follow-up time for events in these types of trials.

It is routine to view the stopping decision as symmetric with respect to the outcome of the trial. For example, whether A is superior to B, and vice versa, the quantitative aspects of the stopping guidelines are often taken to be the same or similar. Clinical circumstances are usually not symmetric. If A is standard and B is a new treatment, we would likely wish to stop a CTE trial (1) when B cannot be superior to A rather than when it is convincingly worse, or (2) when B is convincingly better than A. Thus stopping guidelines for efficacy should be drafted considering the asymmetric consequences of the differences implied, and supplemented with expert opinion. The presence of additional important outcomes will further complicate the asymmetry.

14.1.4   There Is Tension in the Decision to Stop

Reviewing, interpreting, and making decisions about clinical trials based on interim data is necessary, but error prone. Before the data are finalized, monitors will get an imprecise view of most aspects of the study. If information is disseminated prematurely to study investigators, it may affect their objectivity about the treatments, leading to unsupported conclusions or inappropriate changes in behavior. If statistical hypothesis tests are performed repeatedly on the accumulating data, the chance of making a type I error is increased. Despite these potential problems, investigators are vitally interested in the ongoing status of a trial and they or their agents need summaries of the data in a useful format and with an appropriate level of detail to stay informed without corrupting the study.

There are pressures to terminate a trial at the earliest possible moment. These include the need to minimize the size and duration of the trial and the number of patients receiving an inferior treatment. The sponsor and investigators are concerned about the economy of the study, and timeliness in disseminating results. These pressures to shorten a trial are opposed by reasons to continue the study as long as possible. Benefits of longer, larger trials include increasing precision and reducing errors of inference, obtaining sufficient power to account for the effects of prognostic factors, the ability to examine clinically important subgroups, observing temporal trends, and gathering information on secondary endpoints, which may be obtained only in the context of an ongoing trial. Thus, there is a natural tension between the needs to stop a study and those tending to continue it.

Those responsible for monitoring must be cognizant of the sponsor's (and investigators') view of the trial and the anticipated use of the evidence generated. Will the study have the needed impact on clinical practice if it is shortened? For example, a trial used to support a regulatory decision could require a somewhat higher standard of evidence than one conducted for less contentious reasons. In any case, there is an ethical mandate to maintain the integrity of the trial until it provides a standard of evidence appropriate for the setting. In this sense those who watch the data and evidence evolve throughout the trial must be somewhat concerned with the collective good (what will we learn from the trial) as well as the individual good (is the risk–benefit appropriate for participants). The tension or balance between these perspectives is different early in the trial than later, and the trade-off between them cannot be avoided.

Example 32   Tension can be illustrated by the second clinical trial of extracorporeal membrane oxygenation (ECMO) versus standard treatment for newborn infants with persistent pulmonary hypertension (O'Rourke et al., 1989). (This study, and the first trial of ECMO, were briefly discussed in Chapter 13.) Thirty-nine newborn infants were enrolled in a trial comparing ECMO with conventional medical therapy. In this study randomization was terminated on the basis of 4 deaths in 10 infants treated with standard therapy compared with 0 of 9 on ECMO ($p = 0.054$). Regardless of how one evaluates the data, the evidence from such a small trial is not an ideal basis on which to make changes in clinical practice. If ECMO is a truly effective treatment, the failure to adopt it broadly because of weak evidence favoring it is also an ethical problem.

Example 33   The Second International Study of Infarct Survival (ISIS-2) is an example of a trial that continued past the point at which many practitioners would have been convinced by the data. In this study the streptokinase versus placebo comparison was based on 17,187 patients with myocardial infarction (ISIS-2, 1988). After five weeks, the death rate in the placebo group was 12.0% and in the streptokinase group was 9.2% ($p = 10^{-8}$). With regard to this outcome, convincing evidence in favor of streptokinase was available earlier in the trial. However, the impact of this study on practice is largely related to its precision, which would have been diminished by early stopping. An earlier overview of 22 smaller trials had little effect on practice, even though the estimate it produced in favor of streptokinase was similar. See also Section 16.7.4.

14.2   ADMINISTRATIVE ISSUES IN TRIAL MONITORING

Investigators use interim analyses to assess several important aspects of an ongoing trial. These include accrual, data quality, safety, and other parameters listed in Table 14.1. Deficiencies in any area could be a reason to stop the trial. Thus, the statistical guidelines for efficacy discussed below are only part of a broader trial monitoring activity. A number of authors have discussed the broad issues and offer useful perspectives for the clinician and trialist alike (Geller, 1987; Green, Fleming, and O'Fallon,

1987; O'Fallon, 1985). An interesting discussion in the context of cardiovascular disease is given by the Task Force of the Working Group on Arrhythmias of the European Society of Cardiology (1994).

14.2.1 Monitoring of Single-Center Studies Relies on Periodic Investigator Reporting

The practical issues in trial monitoring may be somewhat different for multicenter clinical trials than for single-institution studies. The funding and infrastructure for multicenter trials usually permits a separate monitoring mechanism for each study or group of related studies. For single academic institutions conducting a variety of trials, such as cancer centers, the same oversight mechanism may be required to deal with many varied clinical studies. In such settings the responsibility for study conduct and management may reside more with the individual investigator than it does in collaborative clinical trial groups.

A typical mechanism to assist single institutions in monitoring clinical trials sponsored by their faculty is an annual report. It is the responsibility of the study principal investigator to prepare the report and submit it to the oversight committee, which may be the IRB or a similar body. This report should contain objective and convincing evidence that the study is safe and appropriate to continue. At a minimum the renewal report should address the following concerns:

- *Compliance with governmental and institutional oversight.* The investigator should document that all of the necessary regulations have been satisfied.
- *Review of eligibility.* Document that there is a low frequency of ineligible patients being placed on study.
- *Treatment review.* Most or all patients should have adhered to the intended treatment. Frequent nonadherence may be a sign of problems with the study design, eligibility, or conduct.
- *Summary of response.* For diseases like cancer where response to treatment is a well defined and widely used measure of efficacy, investigators should summarize responses and the criteria used to judge them.
- *Summary of survival.* For most serious chronic diseases, the survival (or disease-free survival) experience of the study cohort is important. In any case, events possibly related to death on treatment need to be summarized.
- *Adverse events.* Convincing evidence of safety depends on accurate, complete, and timely reporting of all adverse events related to the treatment. Investigators sometimes attribute adverse events to causes other than the investigational treatment, such as to the underlying disease. This aspect of the annual report merits careful review.
- *Safety monitoring rules.* Many study protocols have formal guidelines or statistical criteria for early termination. The data must be summarized appropriately and compared with these guides.
- *Audit and other quality assurance reviews.* Data audits examine some of the issues outlined above and other aspects of study performance. The results of such audits and other reviews of process should be available for consideration by the oversight committee.

14.2.2   Composition and Organization of the TEMC

There are several reasons why TEMCs have become widely used. First, they frequently provide a workable mechanism for coping with the problem of protecting the interests and safety of study participants, while preserving the scientific integrity of the trial. Second, the TEMC is intellectually and financially independent of the study investigators, thus providing objective assessments. Therefore investigators and observers can have confidence that decisions will be made largely independent of academic and economic pressures. Finally, many sponsors of trials (e.g., NIH) support or require such a mechanism. For example, the National Heart Lung and Blood Institute (NHLBI) has explicit guidelines for data quality assurance that emphasize the TEMC mechanism (NHLBI, 1994a, b). Similar policies have recently been adopted by the National Cancer Institute (2001). Advice regarding the TEMC charter is given by the DAMOCLES Study Group (2005).

Relationship to the Investigators
One of the most important aspects of the TEMC is its structural relationship to the investigators. Two models are commonly used (Figure 14.1). The first is having the TEMC be advisory to the trial sponsor (Figure 14.1*A*); the second is making the TEMC advisory to the study investigators, perhaps through a Steering Committee (Figure 14.1*B*). This difference is not a nuance and can have consequences for the quality of the recommendation and ethics responsibilities. Neither model is more advantageous with respect to optimal selection of committee members, internal functioning, or the requirements for accurate and timely data.

When the TEMC is advisory to the study sponsor, the opinions and judgments rendered are likely to be as independent as possible from investigator opinion and bias; that is, there is potentially a premium on objectivity. This model is typified by studies performed under contract, particularly those sponsored by the NHLBI and some pharmaceutical companies. Recommendations might be more efficiently implemented, although this is not always the case. The principal disadvantage of this model is that the study sponsor may be an imperfect filter for recommended actions and transmission of information that is important to the investigators. In other words, sponsors have conflicts of interest of their own. Thus, to be optimal, this design must be supplemented by a written policy requiring investigators to be notified of relevant TEMC recommendations.

This issue of reporting relationships is particularly important for ethics. Consider what might occur if the TEMC makes an ethically significant recommendation to

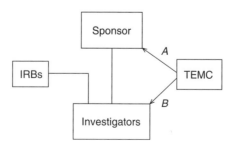

Figure 14.1   Important relationships in treatment effects monitoring.

a sponsor, who in turn disagrees and does not deem it necessary to transmit the recommendation to the investigators. The investigators are the ones who incur the ethical responsibility for representing risks and benefits to the study participants, but they can be denied relevant information from their own data. But the investigators' obligations to the patients cannot be alleviated or abdicated to a TEMC or sponsor. Thus a structure that does not require TEMC reporting to investigators is potentially unethical.

A better model is to make the TEMC advisory to the investigators, specifically through a steering or executive committee. This model is equally simple, objective, and effective as the alternative, and is employed by most cancer cooperative clinical trials groups and many academic institutions. It assures that recommendations that bear on the ethics of the study will find their way reliably and efficiently to the individuals responsible for patient safety. After due process, investigators may take a course of action different than the TEMC recommendation. This is fine if appropriately justified and documented. But it is not appropriate for the sponsor to follow a similar course without the concurrence of the investigators.

My discussion has articulated a principle of obligation for the TEMC to the investigators/patients, which I will now make explicit. The TEMC has an obligation to inform the investigators of their opinions and recommendations about actions that carry ethics implications. In the first model discussed above, the TEMC obligations do not end with the sponsor. The investigators have their own widely recognized and discussed obligations to the patients. The sponsor has its own obligations but is irrelevant with respect to the connection between the TEMC and the investigators. Suppose, for example, that the TEMC produces a non-unanimous vote to continue a clinical trial, meaning some members vote to terminate the trial. (The reason is irrelevant for this discussion.) We can see the potential for a double standard about TEMC function in this situation if this information is withheld from the investigators. On the one hand, we are interested in what objective knowledgeable experts have to say about the trial. On the other hand, we might refuse to represent a difference of opinion to the investigators from such experts because it might alter their thinking.

Relationship to IRBs

Another important but often overlooked aspect of TEMC structure is its relationship to the IRBs at participating institutions. The roles of these two entities in protecting patient safety are somewhat overlapping. However, their roles are often seen as separate and independent, and there is usually a reluctance of the TEMC to provide information to the IRB because of the emphasis on confidentiality. This can be more of a problem if the TEMC is advisory to the sponsor. I have seen this be an acute problem in a large NHLBI sponsored trial when the IRB of the institution supporting the coordinating center requested interim data on treatment risks. The sponsor resisted furnishing the data, and the IRB request was honored only when it signaled its intent to refuse renewal of the study.

There is little doubt that the IRB has the right and responsibility to examine interim data relating to risks and benefits as presented to the patients in the consent documents. IRBs can easily review such data if they are properly summarized, while preserving confidentiality. Furthermore IRBs are not intended or equipped for, nor do they desire, a more active role in the details of study monitoring.

TEMC Membership

TEMCs are usually composed of a combination of clinical, statistical, epidemiological, laboratory, data management, and ethical experts. All members of the TEMC should be knowledgeable about the circumstances surrounding the trial and should not have any vested interest in the outcome. Of course, the committee members will have a scientific interest in the outcome, but only to the extent that they would like to see a valid and well-performed trial. Individuals with any financial or other conflicts of interest should not be members of the TEMC.

Monitoring committees often consist of 3 to 10 members, depending on the size of the trial and complexity of the issues. It is good practice to include some experienced members on every TEMC. A new committee should not be constructed entirely from inexperienced members, even if they are all experts. Multidisciplinary or multinational membership may be helpful because some issues may require these varying perspectives. Individuals who would make good TEMC members are usually known to the trial investigators. There is no formal source for such expertise except for personal contacts. Experienced trial methodologists, clinical experts, ethicists, and investigators who have conducted similar studies are good sources for TEMC members.

Investigators and trialists do not agree universally on whether or not the TEMC should include an investigator from the trial. To the extent that the role of the TEMC is to protect the interests of the patients, a clinical investigator can be a member. To the extent that the TEMC should preserve the scientific integrity of the trial, a clinical investigator may be unnecessary or inappropriate. This question need not be very difficult to resolve. For example, a clinical investigator could participate in safety and data quality related aspects of the TEMC deliberations but be excluded from discussions regarding efficacy and possible early termination of the trial because of convincing treatment differences. An investigator or other individual with particular expertise can be a nonvoting member of the committee. There are many circumstances where detailed knowledge of study conduct, protocol, and participants is essential to TEMC deliberations.

In institutions that conduct many clinical trials simultaneously, such as large comprehensive cancer centers, a single TEMC may review multiple studies or may even monitor all ongoing trials. The role or emphasis of this committee may be broader than simply safety and efficacy monitoring. For example, such a committee may help investigators with protocol development and review, setting institutional priorities for potentially competing studies, monitoring adverse events, and reviewing protocols and other administrative matters. These functions are distinct from those usually performed by the Institutional Review Board.

There are other questions that an impartial view of the data can help answer besides evaluating treatment or side effects. For example, the need for ancillary studies, secondary questions, or requests for the data might be evaluated by the TEMC. The committee will meet once or twice each year (more often if the study progress or issues require it) and review formal reports of study outcomes. Because of the potential for this interim information to affect the clinicians' interactions with patients, the TEMC's meetings and the data presented to it are kept confidential. Reports typically address a number of issues vital to decision making about the trial and are not disseminated.

Masking

The TEMC should not be masked as to treatment assignment. Doing so can inhibit interpretation of important details. For example, if clinically significant differences in safety or efficacy between treatment groups are evident, the TEMC could wrongly attribute them to one treatment or another. Thus unexpected results can be missed until the end of the trial. Masking of the TEMC can facilitate some aspects of the scientific integrity of the trial. However, masking probably cannot improve patient safety and should be discouraged. Because the TEMC will be aware of treatment assignment and must make important decisions based on limited data, the members should be selected thoughtfully.

Meeting Format and Reporting Responsibilities

The TEMC will meet usually at the same time that trial investigators review the progress of their studies. This often occurs semiannually, for example. The frequency of meetings depends on how quickly information in the trial accumulates. The deliberations of the TEMC are confidential and there should be no discussion or dissemination of any results outside the committee meetings. A fairly standard but flexible format for meetings has evolved in many trial settings. It includes a graded degree of confidentiality, depending on the aspect of the study being discussed. Topics from nonconfidential to most confidential are (1) relevant information from outside the trial, (2) accrual and study progress, (3) safety data, (4) efficacy differences, and (5) TEMC speculations.

For example, the general progress of the study (e.g., accrual) can be discussed in an open forum. Additionally scientific information from outside the study can be brought to the attention of the committee and investigators. This might include results or progress on related trials. Attendance at this portion of the meeting can include the sponsor, study chair (PI) or other investigators, and statistical center personnel in addition to the TEMC members.

A more restricted portion of the meeting follows this. Discussions and summary data concerning safety should probably be restricted to the sponsor, study chair, study statistician, and the TEMC. Discussions and data presented by treatment group should be closed to investigators, except for the trial statistician or designee who presents quantitative findings to the TEMC. During this portion of the meeting, outcome data on safety and efficacy by treatment group are presented and discussed. No formal decisions are made by the TEMC during this part of the meeting.

Following the presentation of safety and efficacy data by treatment group, the TEMC meets alone in executive session. During these deliberations, decisions or recommendations regarding changes in the study are made. These decisions can then be transmitted to the appropriate group representative. In some clinical trials, the TEMC reports to the sponsor. This is the case, for example, in many NIH-sponsored trials. Other times, the committee reports to the principal investigator or study chair. In the AIDS Clinical Trial Group, the TEMC reports to the executive committee of the group. There are many workable models, provided that the recommendations of the TEMC are considered seriously by those ultimately responsible for the conduct of the study. In rare cases the investigators may choose not to follow the course of action suggested by the TEMC. This can lead to a great deal of friction between the investigators and the TEMC, sometimes resulting in resignations from the monitoring committee.

14.2.3 Complete Objectivity Is Not Ethical

A basic dilemma when obtaining advice of any type is achieving a balance between objectivity and expertise. This problem arises in several places when monitoring clinical trials—the relationship between the monitoring committee and the investigators, the composition of the monitoring committee itself, the use of masking, and the crafting and implementation of statistical guidelines. To avoid certain criticisms, most clinical trial investigators tend to err on the side of more objectivity and less expertise, tipping the balance in the wrong direction. When this balance is lost, ethics obligations to the patients cannot be adequately met.

As a simple example, consider having a masked TEMC in a study comparing standard treatment to a new therapy. From a purely theoretical view, it may seem appropriate to consider evaluating evidence regarding treatment differences as a symmetric, objective decision. This would endorse the complete masking of the study—patients, physicians, and TEMC. From a patient safety perspective, however, the decision is not symmetric. We would not want to prove that a new treatment is significantly *worse* than standard therapy. Being convinced that it is not better than standard therapy is the appropriate stopping point. Thus the TEMC needs to know the treatment group assignments to interpret the data appropriately and make the most expert risk–benefit assessment. A recent illustration of the importance of this was the trial of hormone replacement therapy discussed in Section 6.4.2. Most of the decisions faced by a TEMC are not symmetric, and it is my opinion that masking such committees is unethical, despite it being more objective.

Hypertrophied objectivity can enter the monitoring process in other structural and procedural ways. Consider the optimal composition of the TEMC. Complete objectivity can only be provided by members who have little or no knowledge of the investigators, the study, the competition, the sponsor, market factors, and the regulatory issues. Such persons could not offer expert advice. This indicates that we do not, in reality, seek complete objectivity. Similarly it is unrealistic to expect an ideal balance of objectivity and expertise in every member of the TEMC.

A more appropriate goal is that the TEMC should comprise experts who are reasonably objective in a collective sense and employ objective methods. Prior opinions can be balanced in the same way that expertise is. Also bear in mind that expertise is a characteristic of the *membership* of the TEMC, whereas objectivity is partially a property of the membership and substantially a characteristic of the *process* of review. Investigator expertise is a fundamental prerequisite for human experimentation, so we should preferentially admit expertise over objectivity at every point in monitoring. I am not arguing against objectivity but against the imbalance between expertise and objectivity that is prevalent in this field today. We permit attitudinal biases in TEMC members and the sponsor but are reluctant to cope with the biases of study investigators. Such thinking is inconsistent, and we should be prepared to utilize the expertise of investigators while counterbalancing any lack of objectivity. At the least this means making the TEMC advisory to the investigators, and perhaps making a membership position on the TEMC open to a study investigator.

The most serious deficiency of expertise on TEMCs as typically constituted today is lack of knowledge of the study protocol, trial conduct, and patient interaction. The best source for the missing expertise is one of the principal investigators. Current dogma indicates that it is inappropriate for the TEMC to include an investigator. For example the FDA Draft Guidance states flatly:

> Knowledge of unblinded interim comparisons from a clinical trial is not necessary for those conducting or those sponsoring the trial; further, such knowledge can bias the outcome of the study by inappropriately influencing its continuing conduct or the plan of analyses. Therefore, interim data and the results of interim analyses should generally not be accessible by anyone other than DMC members.

Some have gone further to suggest that firewalls be constructed so that even those who interact with the investigators (e.g., coordinating center staff) not be privy to interim data. I have heard of a circumstance where investigators were discouraged by NIH sponsors from looking at data from their RCT for grant-writing purposes even though recruitment was over, the treatment period was finished for all patients, and those looking were not treating or evaluating patients on the study. A philosophy that forces this degree of "objectivity" is nonsense.

Clinical trials seem to be the only science where the prevalent view is that investigators should not see their own data. Some sponsors have put forth guidelines for trial monitoring that objectify the process by excluding investigators. There are two problems with such guidelines. First, they tend to sacrifice expertise as I have discussed. Second, they are not necessarily produced by a reasoned process of debate and review. Guidelines put forward by NIH (Appendix F and G) are often taken by inexperienced investigators without criticism. The FDA draft regarding monitoring is given for public comment, but the debate is not extensive and all such regulations have a strong ossifying effect because pharmaceutical companies have so much at stake.

A landmark to help navigate these difficulties is the Institutional Review Board (IRB). The conduct of research is based on the relationships between the investigator, patient, and IRB. If these relationships are disrupted, the study may lose its ethical foundation. Although assuring an appropriate balance of risk and benefit is the responsibility of the investigator as overseen by the IRB, the practicalities are now usually abdicated to a TEMC. But the sponsor and the TEMC typically do not answer to any IRB. Thus it is not appropriate for the TEMC/sponsor to withhold knowledge from the investigator (and from the IRB) but require that investigator to assure the IRB that the study remains ethically appropriate. Furthermore, the IRB cannot be completely comfortable with the continuation of a study with only passive confirmation by the TEMC. In many cases, IRBs have requested and obtained active confirmation based on data provided by the TEMC. This is a somewhat paradoxical situation in which the investigator role is usurped by the TEMC.

14.3   ORGANIZATIONAL ISSUES RELATED TO DATA

14.3.1   The TEMC Assesses Baseline Comparability

Before recommending termination of a clinical trial because of treatment or side effects, the TEMC will establish that any observed differences between the treatment groups are not due to imbalances in patient characteristics at baseline. When such differences are present and thought to be of consequence for the outcome, the treatment effect typically would be "adjusted" for the imbalance, perhaps using methods discussed in Chapter 17. Without the exploratory adjusted analyses, the trial might continue under the assumption that the differences are, in fact, a consequence only of baseline imbalances. This is likely to be a risky assumption. Familiar statistical methods will usually suffice to assess baseline balance and determine its impact on the observed outcome.

Baseline imbalances by themselves are not likely to be a cause for much concern. For example, in a well-conducted randomized trial, they must be due to chance. However, because imbalances can undermine the appearance or credibility of a trial, some interventions might be proposed to correct them. These could include modifications of blocking or other methods employed to create balance.

14.3.2 The TEMC Reviews Accrual and Expected Time to Study Completion

Design Assumptions
Another consideration for the TEMC that reflects on the decision to continue is the rate of accrual onto the trial and the projected duration of the study based on the current trends. Often the accrual rate at the beginning of a trial is slower than that projected when the study was designed or slower than that required to finish the trial in a timely fashion. In many studies the accrual rate later increases and stabilizes. In any case, after accrual has been open at all centers for several months, investigators can get a reliable estimate of how long it will take to complete the study under the original design assumptions and the observed accrual and event rates.

At this early point in the trial it may be evident that design assumptions are inaccurate. For example, the dropout rate may be higher than expected, the event rate may be lower than planned (because trial participants are often at lower risk than the general population with the same disease), or the intervention may not be applied in a manner sufficient to produce its full effect (e.g., some dietary interventions). Each of these circumstances could significantly prolong the length of time necessary to complete the study. Unless remedial actions are taken, such as increasing accrual by adding study centers, the trial may be impossible to complete. The TEMC may recommend that these studies terminate prematurely.

Resource Availability
It may also happen that some of the limited resources available to conduct the clinical trial are being consumed more rapidly than planned. Money is likely to be at the top of this list, but one would have to consider shrinking human resources as well. Difficulties obtaining rare drugs are another example. Occasionally irreplaceable expertise is lost, for example, because of the death of an investigator. Any of these factors may substantially impede the timely completion of a study.

14.3.3 Timeliness of Data and Reporting Lags

The TEMC cannot perform its duties without up-to-date data and the database cannot be updated without a timely and complete submission of data forms. Monitoring this activity is relatively routine for the coordinating center and is a reliable way of spotting errors or sloppy work by clinics or investigators. While there are numerous reasons why forms might be submitted behind schedule, there are not so many why they would be chronically delayed or extremely late. The percentage of forms submitted to the coordinating center within a few weeks or months of their due date should be quite high, such as over 90%.

There can be ambiguities in exactly how timely the data are at the time of a TEMC review. Many interim analyses are done on a database with a given cutoff date, perhaps a month or so prior to the formal data review. This gives the impression that the data are up to date as of the cutoff date. However, because of the natural lag between the occurrence of an event and the time it is captured in the database, analyses on data as of a given cutoff date may not include events immediately prior to the cutoff date. For long-term ongoing reviews, this may not be much of an issue.

In other cases it may be essential to know that all events up to the specified cutoff date are included in the interim report. Extra data sweeps after the cutoff date are required to assure this. The TEMC should understand which circumstance pertains, especially for reviews early in a trial. Although it makes extra demands on the trial infrastructure to know that all events are captured as of the cutoff date, the TEMC may be reassured to operate under that model.

14.3.4 Data Quality Is a Major Focus of the TEMC

Even when the design assumptions and observed accrual rate of an ongoing clinical trial suggest that the study can be completed within the planned time, problems with data quality may cause the TEMC to recommend stopping. Evidence of problems with the data may become available after audits or as a result of more passive observations. An audit can be a useful device for the TEMC to assure them that data errors do not contribute to any treatment differences that are observed.

The TEMC will routinely review patient eligibility. Minor deviations from the protocol eligibility criteria are common and are not likely to have any consequence for the internal or external validity of the trial. These minor errors are study dependent, but they might include things such as errors of small magnitude in the timing of registration or randomization or baseline laboratory values that are out of bounds. More serious eligibility violations undermine the internal validity of the trial and might include things such as randomizing patients who have been misdiagnosed or those with disease status that prevents them from benefiting from the treatment. Usually in multicenter trials, eligibility violations occur in less than 10% of those accrued. If the rate is much higher than this, it may be a sign of internal quality control problems.

In nearly all trials there is a set of laboratory or other tests that needs to be completed either as a part of eligibility determination or to assess the baseline condition of the patient. Investigators should expect that the number of such tests would be kept to the minimum required to address the study question and that 100% of them are performed on time. One must not assume that missing data are "normal." Failure of study centers or investigators to carry out and properly record the results of these tests may be a sign of serious shortcomings that demand an audit or closing accrual at the clinic. The TEMC will be interested in the completion rate of these required tests.

Treatment compliance or adherence is another measure of considerable importance in the evaluation of data quality. Viewing trials from the intention-to-treat (ITT) perspective (see Chapter 15), the only way to ensure that the ITT analysis yields an estimate of treatment effect that is also the effect of actually receiving the treatment is to maintain a high rate of treatment adherence. When adherence breaks down it might be a sign of one or more of the following: (1) serious side effects, (2) patients too sick to tolerate the therapy or therapies, (3) poor quality control by the investigators, or (4) pressures from outside the study (e.g., other treatments). In any case, poor adherence threatens the validity of the study findings and may be cause for the TEMC to stop the trial.

14.3.5 The TEMC Reviews Safety and Toxicity Data

Safety and toxicity concerns are among those most carefully considered by the TEMC. There are three characteristics of side effects that are relevant: the frequency of side effects, their intensity or seriousness, and whether or not they are reversible. Frequent side effects of low intensity that are reversible by dose reduction or other treatment modifications are not likely to be of much consequence to the patients or concern to the investigators. In contrast, even a rarely occurring toxicity of irreversible or fatal degree could be intolerable in studies where patients are basically healthy or have a long life expectancy.

In diseases like cancer or AIDS, serious but reversible toxicities are likely to be common side effects of treatment and are frequently accepted by patients because of the life-threatening nature of their illness. In fact, in the cytotoxic drug treatment of many cancers, the therapeutic benefit of the treatment may depend on, or be coincident with, serious but reversible toxicity. This idea is especially important when considering the design of phase I trials in oncology. Here experienced investigators may realize that the implications of serious toxicity are not as far-reaching as they would be in many other diseases.

14.3.6   Efficacy Differences Are Assessed by the TEMC

Efficacy, or the lack of it, is the most familiar question to evaluate when considering the progress of a trial. The importance of learning about efficacy as early as possible has ethical and resource utilization motivations. Before attributing any observed differences in the groups to the effects of treatment, the TEMC will be certain that the data being reviewed are of high quality and that baseline differences in the treatment groups cannot explain the findings. Statistical stopping rules like the ones outlined later in this chapter may be of considerable help in structuring and interpreting efficacy comparisons.

It is possible that efficacy and toxicity findings during a trial will be discordant, making the question of whether or not to stop the study more difficult. For example, suppose that a treatment is found to offer a small but convincing benefit in addition to a small but clinically important increase in the frequency and severity of side effects. On balance, these findings may not be persuasive for terminating the trial. In this situation it may be important to continue the study to gather more precise information or data about secondary endpoints.

14.3.7   The TEMC Should Address a Few Practical Questions Specifically

From a practical perspective the TEMC needs to answer only a few questions to provide investigators with the information required to complete the experiment.

Should the Trial Continue?

The most fundamental question, and the one most statistical monitoring guidelines are designed to help answer, is "should the study be stopped?" When the efficacy and safety information is convincing, the TEMC will likely recommend stopping. However, there is room for differences of opinion about what is, to a large degree, a subjective assessment.

Clinical trials yield much more information than just a straightforward assessment of treatment and side effect differences. There are many secondary questions that are often important motivations for conducting the trial. Also the database from the study is a valuable resource that can be studied later to address ancillary questions and generate new hypotheses. Much of this secondary gain from a completed clinical trial can be lost when the study is stopped early. If the window of opportunity for performing comparative studies closes, as it might if the TEMC terminates the trial, some important questions will remain unanswered. Therefore the TEMC should weigh the decision to stop carefully and in the light of the consequences of losing ancillary information.

Should the Study Protocol Be Modified?

If the trial is to continue, the TEMC may ask if the study protocol should be modified on the basis of the interim findings. A variety of modifications may be prudent, depending on the clinical circumstances. For example, side effects may be worrisome enough to adjust the frequency or timing of diagnostic tests, but not serious (or different) enough to stop the trial. If more than one treatment comparison is permitted by the study design (e.g., in a factorial trial), convincing stopping points may be reached for some treatment differences but not for others. Thus the structure of part of the trial could be changed, while still allowing the remaining treatment comparisons to be made.

Numerous other types of modifications may be needed after an interim look at the data. These include changes in the consent documents or process, improvements in quality control of data collection, enhancing accrual resources, changes in treatment to reduce dropouts or nonadherence, or changes in the eligibility criteria or their interpretation. In some cases one or more treatments or treatment schedules will have to be modified, hopefully in ways that preserve the integrity of the biological question being asked.

Does the TEMC Require Other Views of the Data?

Another practical question for the TEMC is whether or not the data have been presented in sufficient detail and proper format to examine the monitoring questions. Remaining questions could be answered by additional views of the data, analysis of certain subsets, or presentation of supporting details. These could be important considerations for two reasons. First, preparation of additional analyses and supporting documentation may not be a trivial task for the statistical office. It could require extra quality control effort and verification of interim data. Second, continuation of the trial may strongly depend on subtle findings or those in a small percentage of the study participants. In such cases it is likely that some ways of viewing the results will not display the influential findings. An unfortunate encounter with fraudulent or fabricated data would illustrate this.

Should the TEMC Meet More/Less Frequently?

Interesting trends in the data might prompt the TEMC to meet more often than originally planned. Such a decision could create problems for some statistical monitoring plans, which is a reason to use as flexible an approach as possible. In the same way slower accrual or a lower event rate than originally projected could prolong the interval between TEMC meetings. In terms of "information time" for the trial, the meeting may occur at the recommended intervals. However, in terms of calendar time, the frequency could change.

Are There Other Recommendations by the TEMC?

Finally, there are other recommendations that may result from the TEMC review. An example would be changes in the committee itself, perhaps adding expertise in specific areas. If recruitment goals are not being met, the TEMC might recommend initiating efforts to improve it, such as interactions with target populations through community leaders or organizations. Sometimes for multicenter trials a meeting of the investigators may increase enthusiasm and clear up minor problems with accrual. In any case, the TEMC should be viewed and should act as objective experts who have in mind the best interests of all parties associated with the study. The committee should thus feel free to recommend any course of action that enhances the safety and quality of the clinical trial.

14.3.8 The TEMC Mechanism Has Potential Weaknesses

Although TEMCs can satisfy the need for informed impartial decision making in clinical trials, the mechanism has potential shortcomings, some of which were discussed earlier in this chapter. Trial investigators are often not permitted to be part of the TEMC so that this vital perspective may be underrepresented during deliberations. Although clinical investigators are in a state of equipoise when a trial begins, experience, knowledge, and opinion gained during the study can eliminate it. The TEMC may preserve the state of equipoise, either because members do not have the clinical experiences of working with the treatments and patients, or because they employ somewhat abstract, statistically based termination criteria (discussed below). The stopping guidelines typically consider only a single outcome, whereas the decision to stop a trial may depend on several factors. The partially isolated character of the TEMC can sometimes be advantageous, but it is generally undesirable.

The relationship between the TEMC and trial sponsor can sometimes be an issue as discussed earlier. In some trials sponsored by NIH, for example, the TEMC reports formally to the sponsoring institute only, and not to the study investigators or local IRBs. The sponsor has no written obligation to inform the investigators about TEMC recommendations. It is possible (though unlikely) that the sponsor would not agree with TEMC concerns and not convey recommendations to the study investigators. However, the investigators are the ones who carry ethical obligations to the patients. Thus a policy that does not require TEMC recommendations to be transmitted explicitly and formally to study investigators does not honor all ethical obligations.

The relationship between TEMCs and IRBs or Ethics Boards is sometimes strained. IRBs may operationally abdicate to a TEMC some of their responsibilities to assure patient safety, creating a potential for dissonance. The multidisciplinary expertise needed to monitor a large trial may not be present on IRBs, requiring them to rely on a TEMC. However, TEMC confidentiality often prevents details of interim safety analyses from reaching the IRB. This circumstance can be quite contentious but can usually be managed by transmitting adequate information from the TEMC to the IRB to satisfy the latter's obligations.

Because of their composition and function, TEMCs can emphasize impartiality over expertise. Overzealous adherence to statistical stopping criteria could be a manifestation of this. Some TEMCs remain masked to treatment assignments to increase impartial assessments of differences that may become evident. As mentioned previously, it is hard to see how a TEMC can adequately assure patient safety while masked. Finally there is some concern that studies employing TEMCs tend to stop too soon, leaving important secondary questions unanswered.

14.4   STATISTICAL METHODS FOR MONITORING

14.4.1   There Are Several Approaches to Evaluating Incomplete Evidence

To facilitate decision making about statistical evidence while a trial is still ongoing, several quantitative methods have been developed. These include likelihood based methods, Bayesian methods, decision-theoretic approaches, and frequentist approaches (e.g., fixed sample size, fully sequential, and group sequential methods). Evaluating incomplete evidence from a comparative trial is one area of statistical theory and practice that tends to highlight differences between all of these, but particularly Bayesian and frequentist approaches. No method of study monitoring is completely satisfactory in all circumstances, although certain aspects of the problem are better managed by one method or another. Frequentist methods appear to be the most widely applied in practical situations. For a general discussion of monitoring alternatives, see Gail (1984). Some practical issues are discussed in DeMets (1990).

Statistical guidelines for early stopping can be classified further in two ways: those based only on the current evidence at the time of an interim analysis and those that attempt to predict the outcome after additional observations have been taken (e.g., if the trial continued to its fixed sample size conclusion). Frequentist procedures that use only the current evidence include sequential and group sequential methods, alpha spending functions, and repeated confidence intervals. Stochastic curtailment is a frequentist method that uses predictions based on continuation of the trial. Corresponding Bayesian methods are based on the posterior distribution or a predictive distribution.

Regardless of the method used to help quantify evidence, investigators must separate the statistical guidelines and criteria used for stopping boundaries from the qualitative and equally important aspects of monitoring a clinical trial. Although the statistical criteria are useful tools for addressing important aspects of the problem, they suffer from several shortcomings. They tend to oversimplify the information relevant to the decision and the process by which it is made. The philosophy behind, and the results of, the various statistical approaches are not universally agreed upon, and they are sometimes inconsistent with one another. These problems set the stage for formal ways of introducing expert opinion into the monitoring process, discussed below.

Similarities

These different statistical methods to facilitate early termination of trials do have several things in common. First, all methods require the investigators to state the questions clearly and in advance of the study. This demands a priority for the various questions that a trial will address. Second, all methods for early stopping presuppose that the basic study is properly designed to answer the research questions.

Third, all approaches to early stopping require the investigator to provide some structure to the problem beyond the data being observed. For example, fixed sample size methods essentially assume that the interim data will be inconclusive. The frequentist perspective appeals to the hypothetical repetition of a series of identical experiments and the importance of controlling type I errors. Bayesian approaches require specifying prior beliefs about treatment effects, and decision methods require constructing quantitative loss functions. Any one of these approaches may be useful in specific circumstances or for coping with specific problems.

Fourth, except for using a fixed sample size, the various approaches to early stopping tend to yield quantitative guidelines with similar practical performance on the same data. Perhaps this is not surprising because they are all ways of reformulating evidence from the same data. In any specific study, investigators tend to disagree more about the non-statistical aspects of study termination, making differences between quantitative stopping guidelines relatively unimportant.

Fifth, investigators pay a price for terminating trials early. This is true whether explicit designs are used to formalize decision making or ad hoc procedures are used.

Unless investigators use designs that formalize the decision to stop, it is likely that the chance for error will be increased or the credibility of the study will suffer. Studies that permit or require formal data-dependent stopping methods as part of their design are more complicated to plan and carry out, and when such trials are stopped early, the estimates of treatment differences can be biased. Also accurate data may not be available as quickly as needed. Good formal methods are not available to deal with multiple endpoints such as event times, safety, and quality of life or to deal effectively with new information from outside the trial itself. Finally there is a difficult to quantify but very important question of how convincing the evidence will be if the study is terminated early.

No method is a substitute for judgment. Investigators should keep in mind that the sequential boundaries are *guidelines* for stopping. The decision to stop is considerably more complex than capturing information in a single number for a single endpoint would suggest. Although objective review and consensus regarding the available evidence is needed when a stopping boundary is crossed, the decision cannot be made on the basis of the statistic alone. Mechanisms for making such decisions are discussed later in this chapter.

Differences and Criticisms

There are substantial differences between the various approaches to study monitoring and stopping, at least from a philosophical perspective. Decision-theoretic approaches require specifying a loss function that attempts to capture the consequences of incorrect conclusions. This usually requires an oversimplification of losses and the consequences of incorrect conclusions. These approaches use the idea of a "patient horizon," which represents those who are likely to be affected by the trial. The magnitude of the patient horizon has a quantitative effect on the stopping boundary.

Bayesian approaches rely on choosing a prior distribution for the treatment difference, which summarizes prior evidence and/or belief. This can be a weakness of the method because a prior probability distribution may be difficult to specify, especially for clinicians. Apart from this, a statistical distribution may be a poor representation of one's knowledge (or ignorance). Even if we ignore these problems, different investigators are likely to specify a different prior distribution. Its exact form is somewhat arbitrary.

The frequentist approach measures the evidence against the null hypothesis using the α-level of the hypothesis test. However, hypotheses with the same α-level may have different amounts of evidence against them and Bayesian methods reject this approach (Anscombe, 1963; Cornfield, 1966a, b). For monitoring, frequentist methods have three additional problems. First, they lack the philosophical strength and consistency of Bayesian methods. Second, they are difficult to use if the monitoring plan is not followed. Finally, the estimated treatment effects at the end of the trial are biased if the study is terminated early according to one of these plans. This is discussed below.

14.4.2 Likelihood Methods

Likelihood methods for monitoring trials are based principally or entirely on the likelihood function constructed from an assumed probability model. The likelihood function is frequently used for testing hypotheses, but it also provides a means for summarizing evidence from the data without a formal hypothesis test. Here we consider a likelihood method originally developed for testing, but use it as a pure likelihood-based assessment of the data. Likelihood methods represent nice learning examples, although they are not used in practice as often as frequentist methods discussed below.

Fully Sequential Designs Test after Each Observation

One could assess the treatment difference after each experimental subject is accrued, treated, and evaluated. Although this approach would allow investigators to learn about treatment differences as early as possible, it can be impractical for studies that require a long period of observation after treatment before evaluating the outcome. (An alternative perspective on this view is offered below.) This approach is definitely useful in situations where responses are evident soon after the beginning of treatment. It is broadly useful to see how quantitative guidelines for this approach can be constructed.

The first designs that used this approach were developed by Wald (1947) to improve reliability testing and are called sequential probability (or likelihood) ratio tests (SPRT). As such, they were developed for testing specific hypotheses using the likelihood ratio test, and could be classified as frequentist procedures. More generally as suggested above, the likelihood function is a summary of all the evidence in the data and the likelihood ratio does not have to be given a hypothesis testing interpretation. Therefore I have classified the SPRT as a likelihood method, despite it being a "test." Frequentist methods are discussed below.

SPRT Boundaries for Binomial Outcomes   Consider a binary response from each study subject that becomes known soon after receiving treatment. We are interested in knowing if the estimated probability of response favors p_1 or p_2, values of clinical importance chosen to facilitate decisions about early stopping. We will test the relative evidence in favor of p_1 or p_2, using the likelihood ratio test (Stuart and Ord, 1991), a commonly used statistical procedure.

Because the response is a Bernoulli random variable for each subject, the likelihood function is binomial,

$$e^{\mathcal{L}(p,\mathbf{X})} = \prod_{i=1}^{m} p^{r_i}(1-p)^{1-r_i} = p^s(1-p)^{m-s},$$

where $\mathbf{X} = \{r_1, r_2, r_3, \ldots r_m\}$ is the data vector, r_i is 1 or 0 depending on whether the ith subject responded or not, s is the total number of responses, and m is the number of subjects tested. The likelihood is a function of both the data, $\mathbf{X}$, and the unknown probability of success, p. The relative evidence in favor of p_1 or p_2 is the likelihood ratio, which was also given in Section 11.4.5 as

$$\Lambda = \frac{\mathcal{L}(p_1, \mathbf{X})}{\mathcal{L}(p_2, \mathbf{X})} = \left(\frac{p_1}{p_2}\right)^s \left(\frac{1-p_1}{1-p_2}\right)^{m-s}.$$

If $\Lambda \gg 1$, the evidence favors p_1 and if $\Lambda \ll 1$, it favors p_2. When $\Lambda = 1$, the evidence is balanced. Thus we can define two critical values for Λ: Λ_1 causes us to favor p_1 and Λ_2 favors p_2. The likelihood ratio summarizes the relative evidence in the data and could be used at any point in the trial.

When analyzing interim data, we can calculate the likelihood ratio and stop the study only if the evidence in favor of p_1 or p_2 is the same as it would have been for the final (fixed) sample size. Thus the stopping criteria, expressed as the likelihood ratio, are constant throughout the study. This means that to stop early, we must satisfy either

$$\left(\frac{p_1}{p_2}\right)^k \left(\frac{1 - p_1}{1 - p_2}\right)^{n-k} \geq \Lambda_1 \qquad (14.1)$$

or

$$\left(\frac{p_1}{p_2}\right)^k \left(\frac{1 - p_1}{1 - p_2}\right)^{n-k} \leq \Lambda_2, \qquad (14.2)$$

where we *currently* have k responses out of n subjects tested. It is more convenient to express the stopping criteria either in terms of k versus n or $\pi = k/n$ versus n, where π is the observed proportion of responders. Values of π (and therefore k and n) that satisfy equality in equations (14.1) and (14.2) are "boundary values."

Equation (14.1) or (14.2) yields

$$\log(\Lambda) = k \log\left(\frac{p_1}{p_2}\right) + (n - k) \log\left(\frac{1 - p_1}{1 - p_2}\right),$$

where $\Lambda = \Lambda_1$ or $\Lambda = \Lambda_2$. Thus

$$k = \frac{\log(\Lambda) - n \log\left(\frac{1-p_1}{1-p_2}\right)}{\log(\Theta)}, \qquad (14.3)$$

where Θ is the odds ratio. Also

$$\pi = \frac{k}{n} = \frac{\log(\Lambda) - n \log\left(\frac{1-p_1}{1-p_2}\right)}{n \log(\Theta)}. \qquad (14.4)$$

A similar derivation has been given by Wald (1947) and Duan-Zheng (1990).

Equation (14.3) is helpful because it shows that the values of k and n for the boundary are linearly related. We will obtain an upper or lower boundary, depending on whether we substitute Λ_1 or Λ_2 into equations (14.3) or (14.4). Choosing Λ_U and Λ_L is somewhat arbitrary, but for "strong" evidence, we might take $\Lambda = \pm 32$. A weaker standard of evidence might use $\Lambda = \pm 8$.

To make a tighter analogy with frequentist testing procedures, choices which attempt to satisfy controlling the type I and type II error level of the comparison (Wald, 1947) might be

$$\Lambda_1 = \frac{1 - \alpha}{\beta}$$

and

$$\Lambda_2 = \frac{\alpha}{1 - \beta},$$

where $\alpha = 0.05$ and $\beta = 0.20$, for example. Plots of π versus n for several values of Λ_1 and Λ_2 are shown in Figure 14.2. Early in the trial only extreme values of π (or k) will provide convincing evidence to stop the study.

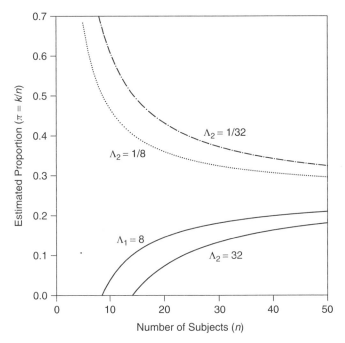

Figure 14.2 SPRT boundaries for a binomial outcome calculated from equation 14.4: $p_1 = 0.17$, $p_2 = 0.35$.

SPRT Boundaries for Event Rates   Similar equations can be used to derive the boundaries for a sequential likelihood ratio test for deciding if evidence favors either of two event rates, λ_1 and λ_2. Again, the hypothesis testing perspective is convenient but unnecessary. For event times with a Weibull distribution with shape parameter v and possible right censoring, the likelihood ratio test for $H_0: \lambda_1 = \lambda_2$ is

$$\Lambda = \frac{\prod_{i=1}^{N} e^{-\lambda_1 t_i^v} \lambda_1^{z_i} (v t_i^{v-1})^{z_i}}{\prod_{i=1}^{N} e^{-\lambda_2 t_i^v} \lambda_2^{z_i} (v t_i^{v-1})^{z_i}} = \frac{e^{-\lambda_1 \sum_{i=1}^{N} t_i^v} \lambda_1^d}{e^{-\lambda_2 \sum_{i=1}^{N} t_i^v} \lambda_2^d}, \tag{14.5}$$

where N is the total number of subjects under study, and z_i is a binary indicator variable for each subject such that $z_i = 1$ if the ith subject has failed and $z_i = 0$ if the ith subject is censored. The exponential distribution is a special case of the Weibull with $v = 1$. Note that $d = \sum z_i$ is the total number of failures. Equation (14.5) can be written

$$\Lambda = e^{-(\lambda_1 - \lambda_2)T} \Delta^d, \tag{14.6}$$

where $\Delta = \lambda_1/\lambda_2$ and $T = \sum_{i=1}^{N} t_i^v$ is the total "effective" exposure time for all study subjects. Taking logarithms and solving equation (14.6) for d yields

$$d = \frac{\log(\Lambda) + (\lambda_1 - \lambda_2)\,T}{\log(\Delta)}, \tag{14.7}$$

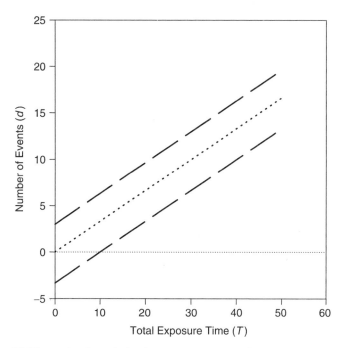

Figure 14.3   SPRT stopping boundaries for event rate comparisons. Note that the vertical axis is number of events and not the failure rate.

which shows the linear relationship between d and T. A plot of number of events versus total exposure time using this equation with $v = 1$ is shown in Figure 14.3. If λ had been plotted instead of d, the graph shape would resemble Figure 14.2.

Alternatively, solving for the boundary value $d/T = \lambda$ yields

$$\lambda = \frac{d}{T} = \frac{(\Lambda/T) + \lambda_1 - \lambda_2}{\log(\Delta)}. \tag{14.8}$$

These equations express the stopping boundary value, d or λ, in terms of T, the current "total" exposure time in the trial cohort. T is an ordinary total exposure time when $v = 1$. For $v \neq 1$, it is similar to a "weighted" total. For the upper boundary, one would choose $R = R_U$ and λ_1 and λ_2 would be given values that reflect clinically relevant high and low failure rates. For the lower boundary, $R = R_L$. After observing T units of total follow-up time in the cohort, the observed failure rate can be compared with the corresponding boundary point to help decide whether or not to stop the trial.

Using SPRT Boundaries   The SPRT is an appealing method for constructing stopping guidelines for a number of reasons. First, it is based on the likelihood function, an efficient summary of the data and one of the few points of agreement for all of the data-dependent stopping methods commonly used. Second, it is conceptually and mathematically relatively simple to employ. Third, it permits continuous monitoring of study outcomes, which is a natural and flexible use for most clinicians. Finally, it is simple to interpret, both when designing the guidelines and in the event that a boundary is reached. These points assume that the statistical model is correct.

These types of designs, although optimal from certain perspectives, have been criticized as impractical for clinical trials because the outcome or response must be known quickly and because the boundary is "open," meaning there is a chance that the interim estimate of the event rate will never touch a boundary and the trial will not terminate. The first criticism is not as problematic as it might seem at first because, if the outcome is not evident immediately for each subject in the trial, one can use instead the subjects who have been followed long enough to be assessed.

For binary outcomes that require a prolonged period of observation, it is as though study subjects are in a "pipeline" with a delay, during which a response cannot occur. Only when they emerge from the other end of the pipeline can they be evaluated for response. The longer the delay, the less ability there is to stop the trial immediately because patients remain in the pipeline. Even so, the basic methodology of the fully sequential design remains applicable. For event rates every increment in follow-up time contributes to the estimate of the failure rate. Therefore all patients accrued contribute continuously to the information needed to monitor the study. In fact, even after accrual stops, the follow-up time is increasing and can lead to having the ratio reach a boundary.

The problem of never encountering a stopping boundary in an SPRT trial is harder to deal with. The accumulating evidence from a trial may never favor either parameter within a reasonable period of time. In this case the study must be interpreted as yielding evidence, which does not distinguish between p_1 and p_2. This may be a perfectly appropriate assessment of the evidence, so this characteristic is not necessarily a drawback. Also we may be interested in more than p_1 and p_2.

At least three possible solutions to the "problem" exist. After a fixed number of patients have been accrued (or a fixed total follow-up time), the trial could be declared finished with the evidence favoring neither alternative. This is probably the most correct representation of the evidence. Alternatively, the inconclusive circumstance could be taken conservatively as evidence in favor of the null hypothesis. This would increase the type II error from the frequentist point of view. Finally the design could be replaced by one that is guaranteed to terminate at a stopping boundary at some point. Such designs are termed "closed" and are discussed in the section on frequentist methods (below).

14.4.3   Bayesian Methods

Bayesian methods for statistical monitoring of trials seem to have much appeal to clinicians and have a strong philosophical foundation, but have not been widely used. Bayesian methods for trials, in general, are discussed by Berry (1985) and Freedman and Spiegelhalter (1989). A concise review of Bayesian methods for monitoring clinical trials is given by Freedman, Spiegelhalter, and Parmar (1994). Robust Bayesian methods are discussed by Carlin and Sargent (1996). Additional Bayesian methods for single-arm trials are discussed by Thall, Simon, and Estey (1995, 1996).

Problems of inference can arise when the assessment of evidence about the treatments depends on the plans for analyses in the future, as it does in frequentist methods. This problem is acute when considering, for example, a trial "result" with a statistical significance level of 0.03. If this is the final analysis of a fixed sample size study, we might conclude that the findings are significant and be willing to stand by them. However, if we learn that the result is only the first of several planned interim analyses (especially one using a very conservative early stopping boundary), the frequentist will be unwilling to embrace the same evidence as being convincing. A similar situation might arise if different investigators are monitoring the same study using different frequentist stopping boundaries. Problems increase if a trial is extended beyond its final sample size. The strict frequentist, having "spent" all of the type I error during the planned part of the trial, cannot find any result significant if the study is extended indefinitely.

Bayesian monitoring methods avoid these problems because the assessment of evidence does not depend on the plan for future analyses. Instead, the interim data from the trial are used to estimate the treatment effect or difference and Bayes rule is used to combine the prior distribution and the experimental evidence, in the form of a likelihood, into a posterior distribution. The posterior distribution is the basis for recommendations regarding the status of the trial. If the posterior distribution is shifted far in one direction, the probability that the treatment effect lies in one region will be high and the inference will favor that particular treatment. If the posterior distribution is more centrally located, neither treatment will be convincingly superior. Thus these posterior probabilities can be used to make decisions about stopping the trial or gathering additional evidence.

Bayesian Example

To illustrate the Bayesian monitoring method, consider a randomized controlled trial with survival as the main endpoint. We suppose that investigators planned to detect a hazard ratio of 2.0 with 90% power using a fixed sample size of 90 events. Suppose also that investigators plan one interim analysis approximately halfway through the trial and a final analysis. The logarithm of the hazard ratio is approximately normally distributed, making it simpler to work on that scale. The variance of the log-hazard distribution is approximately $1/d_1 + 1/d_2$, where d_1 and d_2 are the numbers of events in the treatment groups. Under the null hypothesis of no-treatment difference, the log-hazard ratio is 0 and for a hazard ratio of 2.0, the logarithm is 0.693.

Skeptical Prior   For a prior distribution, investigators summarize the evidence as being equivalent to a distribution with zero mean and a standard deviation of 0.35. If correct, random samples from this distribution would exceed 0.693 only 2.5% of the time. This corresponds to having observed approximately 32 events on the treatments with an estimated risk ratio of 1.0. This particular prior distribution, if based only on strength of belief, expresses some skepticism about the treatment benefit.

At the time of the interim analysis, there are 31 events in one group and 14 in the other, with the hazard ratio estimated to be 2.25. We can summarize this evidence as a normal distribution centered at $\log(2.25) = 0.811$ with a standard deviation of $\sqrt{1/31 + 1/14} = 0.32$. Then the mean of the posterior distribution (Figure 14.4) is

$$\mu_p = \frac{(0 \times 32 + 0.811 \times 45)}{(32 + 45)} = 0.474$$

and the standard deviation of the posterior distribution is

$$\sigma_p \approx \sqrt{\frac{4}{32 + 45}} = 0.228.$$

These quantities can be used to calculate probabilities about the current estimate of the treatment effect (hazard ratio). For example, the probability that the true hazard ratio

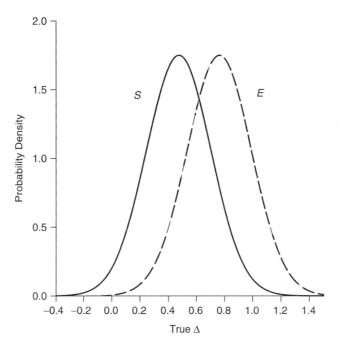

Figure 14.4 Bayesian posterior distributions for monitoring a trial with event rate outcomes based on a skeptical (*S*) or enthusiastic prior (*E*).

exceeds 2.0 is

$$\Pr\{\Delta > 2.0\} = 1 - \Phi\left(\frac{\log(2.0) - 0.474}{0.228}\right) = 1 - \Phi(0.961) = 0.168,$$

where $\Phi(\cdot)$ is the cumulative normal distribution function and Δ is the true hazard ratio. In other words, there is not convincing evidence to terminate the trial on the basis of this skeptical prior distribution.

Enthusiastic Prior For comparison, consider what happens when investigators are subjectively enthusiastic about the new treatment. They quantify their opinion in the form of a different prior distribution, centered at the alternative hypothesis, but with the same variance as the skeptical prior. If true, random samples from this distribution would yield hazard ratios in excess of 2.0, 50% of the time. Following the calculations now, the posterior mean (Figure 14.4) is

$$\mu'_p = \frac{(0.693 \times 32 + 0.811 \times 45)}{(32 + 45)} = 0.762,$$

with the same standard deviation as that obtained above. Investigators would then estimate the chance that the true hazard ratio exceeds 2.0 as

$$\Pr\{\Delta > 2.0\} = 1 - \Phi\left(\frac{\log(2.0) - 0.762}{0.228}\right) = 1 - \Phi(-0.302) = 0.682.$$

Furthermore the chance that the true hazard ratio exceeds 1.75 is 81%, which might lead some investigators to want the trial terminated because of efficacy.

This example illustrates both the Bayesian method and its dependence on the choice of a prior distribution for the treatment effect. In many actual cases one would not choose either of these distributions as a prior for monitoring purposes, and instead let the evidence from the trial stand more on its own by employing a "less informative" density function. An example of this might be a normal distribution that is made to be very flat by virtue of having a large variance. This would correspond to having few prior observations. Such a distribution might be chosen by investigators who prefer not to influence monitoring decisions greatly with information from outside the trial.

14.4.4 Decision-Theoretic Methods

One can approach the problem of stopping or continuing a clinical trial at the time of an interim analysis as a decision-theory question. Given a prior distribution for the treatment difference and a utility function, it is possible to construct optimal group sequential tests for terminating the trial. Similarly any particular group sequential test is optimal under some assumed prior and utility, which can be determined using this approach. Like the likelihood and Bayesian methods, decision-theoretic approaches do not control the type I error properties of the stopping procedure. Fixing the probability of a type I error in advance does not lead to stopping guidelines with optimal decision-theoretic properties.

Many of the designs that have been developed using this approach have been impractical and have not been applied to actual clinical trials. A principal difficulty is that the exact designs are sensitive to the "patient horizon," which is the number of patients who stand to benefit from the selected therapy. It is usually difficult or impossible to specify this number. Despite this shortcoming, it is useful to examine the general properties of this approach. With it, subjective judgments are isolated and formalized as utility functions and prior distributions. The utility function quantifies the benefit from various outcomes and the prior distribution is a convenient way to summarize knowledge and uncertainty about the treatment difference before the experiment is conducted. Using these tools, the behavior of decision rules can be quantified and ranked according to their performance.

Standard group sequential stopping rules (discussed below) have been studied from a decision-theoretic view and found to be lacking (Heitjan, Houts, and Harvey, 1992). When the utility functions and prior distributions are symmetric, the group sequential methods commonly employed are optimal. However, when a trial is judged by how well it improves the treatment of future patients, symmetric utilities may not be applicable, and standard group sequential stopping rules perform poorly.

Lewis and Berry (1994) discussed trial designs based on Bayesian decision theory but evaluated them as classical group sequential procedures. They show that the clinical trial designs based on decision theory have smaller average costs than the classical designs. Moreover, under reasonable conditions, the mean sample sizes of these designs are smaller than those expected from the classical designs.

14.4.5 Frequentist Methods

Sequential and group sequential clinical trial designs have been developed from the perspective of hypothesis testing, which permits trial termination when a test of the null hypothesis of no difference between the treatments rejects. As interim assessments of the evidence (e.g., tests of the null hypothesis) are carried out, the test statistic is compared with a pre-specified set of values called the "stopping boundary." If the test statistic exceeds or crosses the boundary, the statistical evidence favors stopping the trial.

By repeatedly testing accumulating data in this fashion, the type I error level can be increased. Constructing boundaries that preserve the type I error level of the trial, but still permit early termination after multiple tests, is part of the mechanics of frequentist sequential trial design. A simplified example will illustrate the effect.

Example 34   Suppose that we take the data in n nonoverlapping batches (independent of one another) and test treatment differences in each one using $\alpha = 0.05$. If the null hypothesis is true, the chance of not rejecting each batch is $1 - 0.05 = 0.95$. Because of independence the chance of not rejecting *all* batches is 0.95^n. Thus the chance of rejecting one or more batches (i.e., the overall type I error) is $1 - 0.95^n$. The overall type I error, α^*, would be

$$\alpha^* = 1 - (1 - \alpha)^n,$$

where α is the level of each test.

This example is only slightly deficient. In an actual trial the test is performed on overlapping groups as the data accumulate so that the tests are not independent of one another. Therefore the type I error may not increase in the same way as it would for independent tests, but it does increase nonetheless. One way to correct this problem is for each test to be performed using a smaller α to keep α^* at the desired level.

Triangular Designs Are "Closed"

A complete discussion of the theory of sequential trial designs is beyond the scope of this book. This theory is extensively discussed by Whitehead (1992) and encompasses fixed sample size designs, SPRT designs, and triangular designs. A more concise summary is given by Whitehead (1994). Computer methods for analysis are discussed by Duan-Zheng (1990). Here I sketch sequential designs for censored survival data while trying to avoid some of the complexities of the theory. For a recent clinical application of the triangular design, see Moss, Hall, Cannom, et al. (1996).

As in the case of fixed sample size equations, assume that the measure of difference between treatment groups is $\Delta = \log(\lambda_1/\lambda_2)$, the logarithm of the hazard ratio. At the time of the ith failure, we have observed d_{i1} and d_{i2} failures in the treatment groups from n_{i1} and n_{i2} subjects who remain at risk. Consider the logrank statistic, which measures the excess number of events on the control treatment over that expected:

$$Z = \sum_{i=1}^{N} \frac{n_{i1}d_{i2} - n_{i2}d_{i1}}{n_i},$$

where $n_i = n_{i1} + n_{i2}$. The variance of this quantity is

$$V = \sum_{i=1}^{N} \frac{d_i(n_i - d_i)n_{i1}n_{i2}}{(n_i - 1)n_i^2},$$

where $d_i = d_{i1} + d_{i2}$. V is a measure of the amount of information in the data. For equal sized treatment groups and a small proportion of events, $V \approx D/4$ where D is the total number of events.

To monitor the trial, Z_j and V_j will be calculated at the jth monitoring point. It is necessary to assume that V_j does not depend on $Z_1, \ldots Z_{j-1}$ so that the stopping rules cannot be manipulated. This can be accomplished by using a fixed schedule of interim analysis times. At the jth interim analysis, if Z_j is more extreme than a specified boundary point, the trial will be stopped. The boundaries depend on $V_1, \ldots V_j$. It is customary to draw them on a plot of Z versus V. Figure 14.5 shows an example for the triangular test. For discrete rather than continuous monitoring, the boundaries must be modified slightly, yielding a "Christmas tree" shape.

The mathematical form of the boundaries can be derived from two considerations. The first is the usual power requirement which states that the probability of rejecting the null hypothesis when the hazard ratio is Δ_a should be $1 - \beta$. The second consideration relates the upper boundary, Z_u and the lower boundary Z_ℓ, to V as

$$Z_u = a + cV,$$

$$Z_\ell = -a + 3cV.$$

In the very special case when $\beta = \alpha/2$, we have

$$a = \frac{-2\log(\alpha)}{\Delta_a}$$

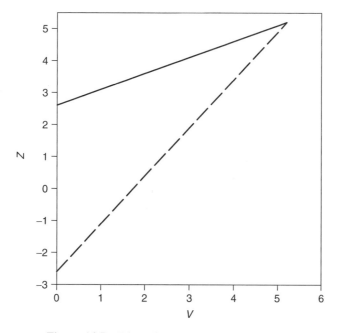

Figure 14.5 Triangular test stopping boundaries.

and

$$c = \frac{\Delta_a}{4}.$$

However, the restriction on α and β that produces this simplification is not generally applicable, and one must resort to numerical solutions for other cases.

Group Sequential Designs May Be Easier to Construct and Apply

In many circumstances investigators do not need to assess treatment differences after every patient is accrued. When monitoring large (multicenter) trials, it is more common for data about efficacy to be available only at discrete times, perhaps once or twice each year. When assessments are made at discrete intervals but frequently enough, the trial can terminate nearly as early as if it were monitored continually. This is the idea of group sequential methods. Typically only a handful of interim looks need to be performed.

The statistical approach to group sequential boundaries is to define a critical value for significance at each interim analysis so that the overall type I error criterion will be satisfied. Suppose that there are a maximum of R interim analyses planned. If the values of the test statistics from the interim analyses are denoted by $Z_1, Z_2, \ldots, Z_R$, the boundary values for early stopping can be denoted by the points $B_1, B_2, \ldots, B_R$. At the jth analysis, the trial stops with rejection of the null hypothesis if

$$Z_j \geq B_j \qquad \text{for } 1 \leq j \leq R.$$

Many commonly used test statistics have "independent increments" so that the interim test statistic can be written

$$Z_j = \frac{\sum_{i=1}^{j} Z_i^*}{\sqrt{j}},$$

where Z_i^* is the test statistic based on the ith group's data. This greatly simplifies the calculation of the necessary interim significance levels. Although the boundaries are constructed in a way that preserves the overall type I error of the trial, one may construct boundaries of many different shapes, which satisfy the overall error criterion. However, there are a few boundary shapes that are commonly used. These are shown in Figure 14.6 and the numerical values are given in Table 14.2.

Specific Group Sequential Boundaries   The Pocock boundary uses the same test criterion for all tests performed, $Z_i = Z_c$ for $1 \leq i \leq R$, where Z_c is calculated to yield the desired overall type I error rate. With this boundary it is relatively easy to terminate a trial early. However, the procedure suffers from the undesirable feature that the final test of significance is made with a larger critical value (smaller p-value) than that conventionally used for a fixed sample size trial. In theory, the final significance level after three analyses could be between, say, 0.05 and 0.0221 (Table 14.2). If one had not used a group sequential procedure, the trial would show a significant difference. However, because the sequential procedure was used, the result is not "significant." This is an uncomfortable position for investigators.

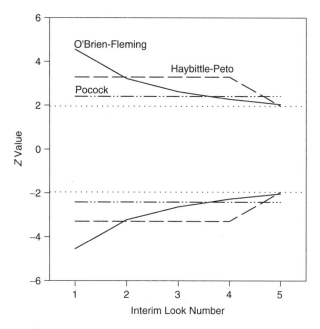

Figure 14.6 Group sequential stopping boundaries from Table 14.2.

TABLE 14.2 Some Frequently Used Group Sequential Stopping Boundaries with Z Scores and Significance Levels for Different Numbers of Interim Analyses

Interim Analysis Number	O'Brien-Fleming		Haybittle-Peto		Pocock	
	Z	p	Z	p	Z	p
			$R = 2$			
1	2.782	0.0054	2.576	0.0100	2.178	0.0294
2	1.967	0.0492	1.960	0.0500	2.178	0.0294
			$R = 3$			
1	3.438	0.0006	2.576	0.0100	2.289	0.0221
2	2.431	0.0151	2.576	0.0100	2.289	0.0221
3	1.985	0.0471	1.960	0.0500	2.289	0.0221
			$R = 4$			
1	4.084	5×10^{-5}	3.291	0.0010	2.361	0.0158
2	2.888	0.0039	3.291	0.0010	2.361	0.0158
3	2.358	0.0184	3.291	0.0010	2.361	0.0158
4	2.042	0.0412	1.960	0.0500	2.361	0.0158
			$R = 5$			
1	4.555	5×10^{-6}	3.291	0.0010	2.413	0.0158
2	3.221	0.0013	3.291	0.0010	2.413	0.0158
3	2.630	0.0085	3.291	0.0010	2.413	0.0158
4	2.277	0.0228	3.291	0.0010	2.413	0.0158
5	2.037	0.0417	1.960	0.0500	2.413	0.0158

The Haybittle-Peto boundary corrects this problem because the final significance level is very close to that conventionally employed, such as it is near 0.05. It is harder to terminate the trial early, but the final analysis resembles the hypothesis test that would have been used in a fixed sample size design. It is a very convenient and workable design. Similarly the O'Brien-Fleming design uses boundaries that yield nearly conventional levels of significance for the final analysis but makes it hard to terminate the trial early. For these designs, $Z_j = Z_c / \sqrt{R/j}$, where again, Z_c is calculated to control the overall type I error rate (O'Brien and Fleming, 1979). One needs very strong evidence to stop early when using this boundary.

Interim analyses of accumulating data are usually spaced evenly in calendar time or information time. This means that some analyses are planned when the trial data are relatively immature. It is quite difficult to meet stopping guidelines with less than about 50% of the information available. It may not be worthwhile to look prior to that unless we firmly believe that the truth lies outside the range of the null and alternative hypotheses. For example, suppose that survival is the primary outcome and the trial is powered for an alternative hazard ratio of 1.75. Very early analyses are not likely to be useful (result in stopping) unless the true hazard ratio is less than 1.0 or greater than 1.75.

Alpha Spending Designs Permit More Flexibility

One of the principal difficulties with the strict application of group sequential stopping boundaries is the need to specify the number and location of points for interim analysis in advance of the study. As the evidence accumulates during a trial, this lack of flexibility may become a problem. It may be more useful to be able to adjust the timing of interim analyses and, for the frequentist, to use varying fractions of the overall type I error for the trial. Such an approach was developed and called the "alpha spending function" (DeMets and Lan, 1994; Lan and DeMets, 1983). See also Hwang, Shih, and DeCani (1990).

Suppose that there are R interim analyses planned for a clinical trial. The amount of information available during the trial is called the information fraction and will be denoted by τ. For grouped interim analyses, let i_j be the information available at the jth analysis, $j = 1, 2, \ldots, R$, and let $\tau_j = i_j / I$ be the information fraction, where I is the total information. For studies comparing mean values based on a final sample size of N, $\tau = n/N$, where n is the interim accrual. For studies with time-to-event endpoints, $\tau \approx d/D$, where d is the interim number of events and D is the planned final number. The alpha spending function, $\alpha(t)$, is a smooth function on the information fraction such that $\alpha(0) = 0$ and $\alpha(1) = \alpha$, the final type I error rate desired. The alpha spending function must be monotonically increasing.

If the values of the test statistics from the interim analyses are denoted by $Z_1, Z_2, \ldots, Z_j$, the boundary values for early stopping are the points $B_1, B_2, \ldots, B_j$, such that

$$\Pr\{|Z_1| \geq B_1, \ or \ |Z_2| \geq B_2, \ldots, \ or \ |Z_j| \geq B_j\} = \alpha(\tau_j).$$

By this definition, the O'Brien-Fleming boundary discussed above is

$$\alpha_{OF}(\tau) = 2\left[1 - \Phi\left(\frac{Z_\alpha}{\sqrt{\tau}}\right)\right],$$

where $\Phi(x)$ is the cumulative standard normal distribution function. The Pocock boundary discussed above is

$$\alpha_P(\tau) = \alpha \log(1 + \tau(e - 1)).$$

Many other spending functions have been constructed (Hwang, Shih, and DeCani, 1990).

Early Stopping May Yield Biased Estimates of Treatment Effect

One drawback of sequential and group sequential methods is that, when a trial terminates early, the estimates of treatment effect will be biased. This can be understood intuitively. If the same study could be repeated a large number of times, chance fluctuations in the estimated treatment effect in the direction of the boundary would be more likely to result in early termination than fluctuations away from the boundary. Therefore, when the boundary is touched, these variations would not average out equally, biasing the estimated treatment effect. The sooner a boundary is hit, the larger is the bias in the estimated treatment effect. This effect can create problems with the analysis and interpretation of a trial that is terminated early (e.g., Emerson and Banks, 1994).

For example, consider a randomized comparative group sequential trial with 4 analyses using an O'Brien-Fleming stopping boundary. If the true treatment effect is 0.0, there is a small chance that the stopping boundary will be crossed at the third analysis. This would be a type I error. This probability is 0.008 for the third analysis, upper boundary. Among trials crossing the upper boundary there, the average Z score for the treatment effect is approximately 2.68. When the true treatment effect is 1.0, 50% of the trials will cross the upper boundary. Among those crossing at the third interim analysis, the average Z score is 2.82. When the true treatment effect is 2.0, 98% of trials will terminate on the upper boundary. Among those crossing at the third interim analysis, the average Z score is 3.19. These results, each determined by statistical simulation of 10,000 trials, illustrate the bias in estimated treatment effects when a boundary is crossed.

One could consider correcting for this bias using a Bayesian procedure that modifies the observed treatment effect by averaging over all possible "prior" values of the true treatment effect. However, such a procedure will be strongly affected by the properties of the assumed prior distribution and therefore may not be a good solution to the problem of bias. In other words, the true prior distribution of treatment effects is unknown, making it impossible to reliably correct for the bias.

Because of this bias the trialist must view these sequential designs as "selection designs" that terminate when a treatment satisfies the stopping criteria. In contrast, fixed sample size trials might be termed "estimation designs" because they will generally provide unbiased estimates of the treatment effect or difference. One should probably not employ a selection design when it is critical to obtain unbiased estimates of the treatment effect. A fixed sample size design would be more appropriate in such a situation. Similarly, when early stopping is required for efficient use of study resources or ethical concerns, a selection design should be used in place of the fixed sample size alternative.

14.4.6   Other Monitoring Tools

Futility Assessment with Conditional Power

The monitoring methods discussed above assume that the trial should be terminated only when the treatments are significantly different. However, we should also consider stopping a clinical trial when the interim result is unlikely to change after accruing more subjects (futility). The accumulated evidence could convincingly demonstrate that the treatments are equivalent, meaning it is virtually certain to end without a rejection of the null hypothesis, even if carried to the planned fixed sample size. It is possible to calculate the power of the study to reject the null hypothesis given the current results. This approach is called *conditional power* (Lan, DeMets, and Halperin, 1984; Lan, Simon, and Halperin, 1982). Some additional practical aspects are discussed by Andersen (1987).

Example 35   Suppose that we are testing a questionable coin to see if it yields heads 50% of the time when flipped. If π is the true probability of heads, we have H_0: $\pi =$ Pr[heads] $= 0.5$ or H_a: $\pi =$ Pr[heads] > 0.5. Our fixed sample size plan might be to flip the coin 500 times and reject H_0 if

$$Z = \frac{n_h - 250}{\sqrt{500 \times 0.5 \times 0.5}} \geq 1.96,$$

where n_h is the number of heads obtained and $\alpha = 0.025$. Stated another way, we reject H_0 if $n_h \geq 272$. Suppose that we flip the coin 400 times and observe 272 heads. It would be futile to conduct the remaining 100 trials because we are already certain to reject the null hypothesis.

Alternatively, suppose that we observe 200 heads after 400 trials. We can reject the null only if 72 or more of the remaining 100 trials yield heads. From a fair coin, which this appears to be, this event is unlikely, being more than 2 standard deviations away from the null. Thus, with a fairly high certainty, the overall null will not be rejected and continuing the experiment is futile.

Partial Sums   To illustrate how conditional power works in a one-sample case, suppose that we have a sample of size N from a $N(\mu, 1)$ distribution and we wish to test H_0: $\mu = 0$ vs. H_a: $\mu > 0$. Furthermore assume that we have a test statistic, $Z_{(N)}$, which is based on the mean of a sample from a Gaussian distribution,

$$Z_{(N)} = \frac{\sum_{i=1}^{N} x_i}{\sqrt{N}},$$

where we reject H_0 if $Z_{(N)} > 1.96$. Using a standard fixed sample size approach (equation 11.24), we solve $\theta = \sqrt{N}\mu = (Z_\alpha + Z_\beta)$ for N to obtain the sample size. Part way through the trial when n subjects have been accrued, the test statistic is

$$Z_{(n)} = \frac{\sum_{i=1}^{n} x_i}{\sqrt{n}} = \frac{S_n}{\sqrt{n}},$$

where $1 \leq n \leq N$, and S_n is the partial sum. It is easy to show that $E\{S_n\} = n\mu$ and var$\{S_n\} = n$. More important, using S_n, the increments are independent, meaning $E\{S_N \mid S_n\} = S_n + E\{S_{N-n}\}$ and var$\{S_N \mid S_n\} = N - n$. The expected value of S_n increases linearly throughout the trial. Therefore we can easily calculate the probability

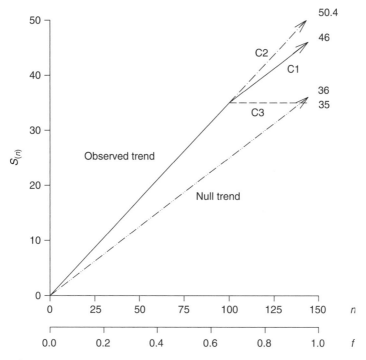

Figure 14.7 Projected partial sums during a trial.

of various outcomes conditional on observing certain interim results. This is represented in Figure 14.7 by the plot of S_n versus n (and the information fraction, f) and the following example:

Example 36 Suppose $\mu = 0.25$ and $\sqrt{N}\mu = (1.96 + 1.04) = 3$ so that $N = 144$. After $n = 100$ patients, suppose $S_{100} = 35$ ($Z_{100} = 35/\sqrt{100} = 3.5$). There are at least three scenarios under which we would like to project the end of the trial given the interim results: (C1) the original alternative hypothesis continues to the end of the trial, (C2) the current trend continues to the end, and (C3) the original null hypothesis continues to the end of the trial. These are shown in Figure 14.7. Other scenarios could also be important to consider. The method of calculation is shown in Table 14.3. For C1,

$$\Pr\{Z_{(144)} \geq 1.96 \mid Z_{(100)} = 3.5, \ \mu = 0.25\}$$
$$= \Pr\{S_{144} \geq 23.52 \mid S_{100} = 35, \ \mu = 0.25\}$$
$$= \Pr\left\{\frac{S_{144} - 46.0}{\sqrt{44}} \geq \frac{23.52 - 46.0}{\sqrt{44}} \mid S_{100} = 35, \ \mu = 0.25\right\}$$
$$= \Pr\{\Phi(x) \geq -3.39\}$$
$$= 0.99965.$$

TABLE 14.3   Projected Partial Sums during a Trial under Three Assumptions about the Trend from Example 36

Case	μ	$E\{S_{144} \mid S_{100}\}$	$\text{var}\{S_{144} \mid S_{100}\}$
C1	0.25	$35 + (144 - 100) \times 0.25 = 46.0$	$144 - 100 = 44$
C2	0.35	$35 + (144 - 100) \times 0.35 = 50.4$	44
C3	0.00	$35 + (144 - 100) \times 0.00 = 35.0$	44

Similarly, for C2,

$$\Pr\{Z_{(144)} \geq 1.96 \mid Z_{(100)} = 3.5, \ \mu = 0.35\}$$

$$= \Pr\left\{\frac{S_{144} - 50.4}{\sqrt{44}} \geq \frac{23.52 - 50.4}{\sqrt{44}} \ \middle| \ S_{100} = 35, \ \mu = 0.35\right\}$$

$$= \Pr\{\Phi(x) \geq -4.05\}$$

$$= 0.99997,$$

and for C3,

$$\Pr\{Z_{(144)} \geq 1.96 \mid Z_{(100)} = 3.5, \ \mu = 0.0\}$$

$$= \Pr\left\{\frac{S_{144} - 35.0}{\sqrt{44}} \geq \frac{23.52 - 35.0}{\sqrt{44}} \ \middle| \ S_{100} = 35, \ \mu = 0.0\right\}$$

$$= \Pr\{\Phi(x) \geq -1.73\}$$

$$= 0.95818.$$

Because of the high chance of rejecting the null hypothesis regardless of assumptions about the trend, investigators would probably terminate this study early.

B-Values   Conditional power can be made independent of the sample size (Lan and Wittes, 1988). To do so, we define the *information fraction* of the trial as $f = n/N$. Then for a comparative study we define

$$Z_n = \frac{\overline{X}_1 - \overline{X}_2}{s_n\sqrt{1/n_1 + 1/n_2}},$$

where $n_1 \approx n_2$, and

$$B(f) = Z_n\sqrt{f}.$$

The expected value of $B(f)$ is $E[B(f)] = E(Z_N)f = \theta f$, a linear function of f that is useful for extrapolations. Also $B(1) = B(f) + [B(1) - B(f)]$, from which we can see: $B(f)$ and $(B(1) - B(f))$ are independent and normal; $E[B(1) - B(f)] = \theta(1 - f)$; $\text{var}[B(f)] = f$, and $\text{var}[B(1) - B(f)] = 1 - f$. From these facts the conditional power given θ can be expressed as

$$Z_{CP} = \frac{Z_\alpha - B(f) - \theta(1 - f)}{\sqrt{1 - f}} \tag{14.9}$$

and

$$P(\theta) = 1 - \Phi(Z_{CP}).$$

If the null trend pertains to the remainder of the trial,

$$P(0) = 1 - \Phi\left(\frac{Z_\alpha - B(f)}{\sqrt{1-f}}\right),$$

whereas if the observed trend continues,

$$P(\theta) = 1 - \Phi\left(\frac{Z_\alpha - B(f)/f}{\sqrt{1-f}}\right).$$

Example 36 above can be re-worked in terms of B-values as follows.

Example 37   After $n = 100$ patients, $f = 100/144 = 0.694$, $Z_{100} = 3.5$, and $B(0.694) = 3.5\sqrt{0.694} = 2.92$. The three scenarios under which we would like to project the end of the trial are (C1) $\theta = 3$, (C2) $\theta = 2.92/0.694 = 4.20$, and (C3) $\theta = 0$. These can be calculated from equation (14.9). For C1,

$$Z_{CP} = \frac{Z_\alpha - B(f) - \theta(1-f)}{\sqrt{1-f}} = \frac{1.96 - B(0.694) - 3 \times 0.306}{\sqrt{0.306}}$$
$$= \frac{1.96 - 2.92 - 0.918}{0.553} = -3.396.$$
$$P(3) = 1 - \Phi(-3.396)$$
$$= 0.99965.$$

Similarly, for C2,

$$Z_{CP} = \frac{1.96 - 2.92 - 4.2 \times 0.306}{0.553}$$
$$= -4.05.$$
$$P(4.2) = 1 - \Phi(-4.05)$$
$$= 0.99997,$$

and for C3,

$$Z_{CP} = \frac{1.96 - 2.92 - 0 \times 0.306}{0.553}$$
$$= -1.73.$$
$$P(0) = 1 - \Phi(-1.73)$$
$$= 0.95818.$$

The conclusions are as above.

14.4.7   Some Software

A recurring problem in implementing new or difficult methods of design and analysis is the availability of reliable, flexible, and reasonably priced computer software to facilitate the computations. For the Bayesian and likelihood-based methods discussed above, the calculations are simple enough to be performed by hand. To my knowledge, there is no commercially available software to accomplish those calculations.

For calculating group sequential boundaries, the program *EaSt* (Cytel Software, 1992) is available for IBM-compatible microcomputers. Although the user interface is somewhat awkward, the program is accurate and reasonably priced. It also permits including stopping boundaries for both the null hypothesis and the alternative hypothesis. For fully sequential frequentist designs, illustrated above by the triangular design, there is *PEST* (MPS Research Unit, 2000), a commercially available program also for IBM PCs. It requires a fairly high level of expertise from the user and is very expensive. Because of these features it is a workable solution only for the investigator who needs to perform such calculations frequently.

Conditional power calculations using B-values can be performed using the software that is described in Appendix A.

14.5   SUMMARY

It is appropriate, if not ethically imperative, for investigators to examine the accumulating results from a clinical trial in progress. Information from the trial may convince investigators to close the study early for reasons of patient safety. Reasons why a trial might be stopped earlier than initially planned include: the treatments are convincingly different (or equivalent); side effects are too severe or adherence is too low; the data are of poor quality; or needed resources are lost or are insufficient. The tendency to stop a study for these or other reasons is balanced by an imperative to gather complete and convincing evidence about all the objectives.

The process used to make these decisions must balance expertise and objectivity. It is relatively easy to compose an advisory committee with the required expertise, and somewhat harder to assure objectivity of the monitoring process. By their nature, experts tend to be less than completely objective, but monitoring policies and procedures can compensate for it. Strong separation of the monitoring process from the expertise of the investigators (e.g., firewalling) is seldom truly necessary or wise.

Making inferences about treatments based on incomplete data can be error prone. Investigators must place additional structure on the interim analyses beyond the usual end-of-study plans to minimize errors. There are several approaches to accomplishing this, including frequentist methods that control the type I error, likelihood-based methods, Bayesian approaches, and decision theory. The product of each of these approaches to early stopping is a set of quantitative guidelines that help investigators evaluate the strength of the available evidence and decide if the trial should be terminated. All methods require timely and accurate interim reporting of data.

Frequentist methods for constructing early stopping guidelines (boundaries) have gained widespread acceptance. They control the overall type I error of the trial and are relatively simple and flexible to implement and interpret. However, they have the drawback that the same data can yield different inferences depending on the monitoring plans of the investigators. Likelihood methods base the decision to stop a trial early on achieving the same strength of evidence (measured by likelihood ratios) as one would obtain for the final analysis. Bayesian methods have similar characteristics but allow subjective notions of strength of evidence to play a part in the decision to stop.

In the commonly used group sequential boundaries, the decision to stop or not is taken at a small number of discrete points during the trial, typically once or twice each year. Additional flexibility in the timing and type I error control can be gained by using an alpha spending function approach. Fully sequential frequentist methods, likelihood methods such as the SPRT, and most Bayesian monitoring methods permit assessment of interim data continuously. Besides stopping when the "null hypothesis" appears to be false, trials may be terminated early if, after some point, there is little chance of finding a difference. Conditional power calculations are one means to facilitate this.

Statistical guidelines for the primary endpoint are not the only consideration when assessing the interim evidence from a clinical trial. Side effects, unanticipated events, data quality, secondary objectives, and evidence from outside the trial may all reflect on the decision to continue. Also, the sponsor or investigators may have other obligations or interests that can conflict with objective evaluation of trial data. Because monitoring can be complicated, many clinical trials use a formal Treatment Effects Monitoring Committee (TEMC) to assist with the task and assure patient safety. After reviewing interim data, the TEMC could recommend a variety of actions including stopping the trial, modifying the study protocol, examining additional data, or making a new assessment of the evidence sooner than originally planned.

14.6   QUESTIONS FOR DISCUSSION

1. Investigators plan a SA clinical trial to study the efficacy and side effects of a new genetically engineered treatment vaccine against prostate cancer. The toxicity is likely to be low and the potential treatment effect, based on animal studies, could be high. The clinicians are anxious to finish the trial as quickly as possible because of cost and scientific priorities. Sketch and defend a trial design that you believe would be appropriate under these circumstances.

2. The probability of success on standard therapy for a disease is 50%. A new treatment is being tested and investigators would like to end the trial if evidence supports a success rate of 75%. Construct, graph, and explain what you believe to be appropriate statistical monitoring guidelines for this trial.

3. In a clinical trial comparing treatments to improve survival following AIDS, investigators plan to adopt a new treatment if the evidence suggests a hazard ratio of 1.75 (new treatment is superior). Investigators prefer a Bayesian approach to monitoring and analysis. Prior to the trial, some data are available indicating a hazard ratio of 1.6 based on 50 events. During the trial, a hazard ratio of 2.0 is observed with 60 and 30 events in the treatment groups. What is your quantitative assessment of the evidence and would you favor continuing the trial?

4. Two statisticians are helping to conduct, monitor, and analyze a trial with a single planned interim analysis and a final analysis. At the interim, the statisticians perform some calculations independently and agree that the study should be continued. (Because they agree to recommend continuing the trial, the statisticians do not discuss details of their interim analyses.) At the final analysis, the p-value turns

out to be 0.035. To their consternation the statisticians discover that they have used different boundaries for the analyses and one can declare the final result "significant" while the other cannot. Discuss how you can help them out of this problem.

5. A trial is planned to accrue a maximum sample size of 80 patients. After 60 have been accrued, $Z = 25/\sqrt{60} = 3.23$. Should the trial be continued? Justify your recommendation.

6. For the previous problem, calculate the information fraction, f, and the B-value, B_f.

15

COUNTING SUBJECTS AND EVENTS

15.1 INTRODUCTION

The findings from a clinical trial can sometimes be sensitive to how investigators resolve imperfections in the data. Imperfections arise mainly when experimental subjects do not adhere precisely with some aspect of the protocol. Nonadherence is usually not an extensive problem in experiments other than clinical trials. For example, when it occurs in laboratory and animal studies, lack of adherence to the study plan tends to be minimal so that it has relatively little impact on the analysis and interpretation of the experiment.

In contrast, clinical trials are characterized, even dominated on occasion, by imperfect data arising from nonadherence with the protocol. Patients fail to follow the experimental protocol precisely because of their independence and autonomy and because of unforeseen clinical circumstances. The consequences of imperfect data depend on where the problems occur in the experimental design and how the investigators resolve them. It is a general principle that the designs and outcomes employed in clinical trials should be chosen to minimize the impact of protocol nonadherence.

Problems with data can be classified as *protocol nonadherence*, *missing or incomplete observations*, and *methodologic errors*. This categorization is similar to commonly used terms such as missing and incomplete observations, treatment dropouts, treatment crossovers, eligibility errors, uncounted events, and lack of adherence to planned schedules (Gail, 1985). Data imperfections are either correctable or permanent. Correctable mistakes are those derived through faulty definitions from source data that are free of error. Permanent, or uncorrectable, imperfections occur when the source data are lost or are in error. Both correctable and permanent errors can occur at any point in the experimental paradigm.

The difficulty with imperfections is how best to reconcile them with the experimental structure upon which reliable inference depends. This problem has been approached from two perspectives. One view has been called "explanatory" and the other "pragmatic," terms coined 35 years ago (Schwartz and Lellouch, 1967; Schwartz, Flamant, and Lellouch, 1980). Other terms used to describe the same distinction are "efficacy" versus "effectiveness" (Chapter 12) and "explanatory" versus "management" (Sackett, 1983). The explanatory perspective emphasizes acquiring information, while the pragmatic perspective focuses on making decisions. These perspectives have practical implications for how the analyst deals with trial design issues, as well as data imperfections. This chapter emphasizes the pragmatic perspective.

Suppose that investigators are studying the effect of pre-operative chemotherapy on disease recurrence and survival in patients with early stage non–small cell lung cancer. They plan a randomized trial with the treatment groups consisting of surgery (S) versus chemotherapy (C) plus surgery. A biologist might view the question like a laboratory experiment, and could isolate the effect of chemotherapy by scheduling surgery at the same time following randomization in both treatment groups. Thus patients in the S group would wait a period of time before having surgery. The waiting time in the S group corresponds to the chemotherapy period in the $C + S$ group. This design attempts to estimate the effect of chemotherapy.

In contrast, a clinician might view this design as unrealistic, noting that physicians and patients will be unwilling to wait for surgery. In practice, patients receiving surgery alone would have their operations immediately after diagnosis. This leads to a (different) practical design, where surgery is scheduled at once in the S group but after chemotherapy in the $C + S$ group. Although this design cannot isolate the effect of chemotherapy strictly, it is pragmatic and attempts to select the superior treatment as used in actual practice.

The two trial designs are superficially the same, but demonstrate an important conceptual difference. The explanatory trial attempts to estimate what has been called "method-effectiveness" (Meier, 1991; Sheiner and Rubin, 1995), while the pragmatic one addresses "use-effectiveness." One cannot say which is the superior or correct approach, only that both questions are relevant and important and that both types of queries cannot always be answered by the same trial. Similarly, when resolving data imperfections arising from nonadherence with the study protocol, explanatory and pragmatic views may each suggest methods based on relevant and important questions. However, like the design problem just outlined suggests, there is no guarantee that the questions can be answered from the existing data. The final resolution depends on the specific circumstances.

Distinguishing between explanatory and pragmatic approaches is a useful device to investigate different philosophies about coping with protocol nonadherence and other data imperfections, as well as some design issues. However, it is not possible to label all approaches to such problems in this way. Moreover it is not possible to say exactly what methods the explanatory or pragmatic views will emphasize in every circumstance. In any case, I will continue to use the labels in this chapter as a descriptive device.

15.2   NATURE OF SOME SPECIFIC DATA IMPERFECTIONS

Not all imperfections produce missing data. Imperfections can result from inappropriately "correcting" data that are properly missing. These and other types of data imperfections are avoidable. For example, patients who are too ill to comply with a demanding course of treatment should not be included in the study population because they are not likely to adhere to the treatment. In contrast, eligible subjects entered into a trial who then become too ill to complete either their treatment or evaluations cannot be discarded as though they were never present. Unavoidable imperfections are those due to human error, patients lost to follow-up, and some types of missing observations such as when patients refuse to undergo tests or clinic visits because they feel ill. Data imperfections can also be a consequence of poor study methodology, chance, or lack of protocol adherence. These topics are discussed below.

15.2.1   Evaluability Criteria Are a Methodologic Error

Protocols sometimes contain improper plans that can create or exacerbate imperfections in the data. A common example is *evaluability criteria*, by which some investigators attempt to define what it means to receive treatment. (Already there is a potential problem because the definition of "receiving treatment" would not be important unless exclusions based on it were planned.) Investigators might define "evaluable" patients as those who receive most or all planned courses of therapy. A biological rationale permits removing inevaluable patients from the analysis because the treatment did not have an opportunity to work. This circumstance may be typical in SA trials.

For example, suppose that among patients accrued to a SA trial, N_E are evaluable and R_E of those have a favorable outcome. N_I are inevaluable, of whom R_I have a favorable outcome (usually $R_I = 0$). The estimate of benefit among all patients is $P = (R_E + R_I)/(N_E + N_I)$, and that for evaluable patients is $P_E = R_E/N_E$. It appears that P_E has a firm biological basis because the exclusions are described in advance and are predicated on the fact that treatment cannot work unless a patient receives it. Attempts to isolate and estimate biological effects of treatment are explanatory, as compared with pragmatic analyses for which evaluability is immaterial. Evaluability criteria can create missing data, although the explanatory perspective sees them as irrelevant.

There are fundamental problems with this approach. Evaluability criteria define inclusion retroactively, that is, on the basis of treatment adherence, which is an outcome. Although treatment adherence is also a predictor of subsequent events, exclusions based on it incorrectly assume that it is a baseline factor. These exclusions create bias, as do other *retroactive definitions*. For example, suppose that comparison groups are defined using outcome or future events such as tumor response in cancer studies. Patients who respond must have lived long enough to do so, whereas patients who did not respond may have survived a shorter time. Therefore survival or other event time comparisons based on such categorizations can be biased. This problem is discussed by Anderson et al. (1985).

The pragmatic perspective is a better one for coping with questions of evaluability. The pragmatist recognizes that the trial does not guarantee an assessment of biological effect. Even if it did, P_E does not necessarily estimate anything because it is confounded with adherence. Furthermore the concept of biological effect degenerates if patients cannot adhere to the treatment. The trial does assure an unbiased estimate of treatment benefit if all eligible patients are included in the analysis. The way to assure that the pragmatic estimate of benefit is close to the explanatory estimate of biological effect is to select treatments and design and conduct the trial so that there is a high adherence.

Another perspective on this issue can be gained by considering the difference between conditional and unconditional estimates of treatment effect. Unconditional estimates of effect do not rely on adherence (nor any other factor) and are always applicable to a new patient at baseline. Effect estimates that condition on other factors, even those measurable at baseline, create potential problems. If the estimate conditions on events from the patient's future (adherence), it will not represent the true biological effect and is likely to be uninterpretable.

Evaluability criteria may be defined only implicitly by the analytic plan. This mistake can be subtle and hard to correct. Exactly this type of error plagues many uncontrolled comparisons of a stressful therapy compared with a nonstressful one (e.g., surgery versus medical management), which can create a survivors' bias. However, the same bias can occur in a randomized comparison unless an outcome is *defined for every patient* regardless of vital status. A good example of the problem and its fix can be seen in the discussion of lung volume reduction surgery in Section 4.6.6. In that randomized trial, comparison of mean functional measures (FEV_1, exercise capacity, quality of life) by treatment group were potentially biased because the surgical therapy produced a higher short-term mortality capable of spuriously raising the group average. To correct this effect, an outcome was defined based on achieving a prespecified degree of improvement. Missing or deceased subjects were classified as unimproved, removing the survivors' bias. It is surprising how resistant some clinicians were to this remedy.

15.2.2   Statistical Methods Can Cope with Some Types of Missing Data

In most circumstances, *unrecorded data* imply that a methodologic error has occurred. This may not be the case if measurements were omitted during follow-up because investigators did not know that the information was important. When this occurs frequently, especially in baseline or other clinically important assessments, it may be evidence of a fundamental problem with study design or conduct. Sometimes data may be missing for reasons associated with other patient characteristics or outcomes. Other times, it is essentially a random occurrence, for example, because of human error during data entry.

The explanatory way of coping with uncounted events because they are seemingly unrelated to the outcome of interest (e.g., noncardiac deaths in a cohort with time to myocardial infarction as the primary study outcome) might be to censor such observations. If events seemingly unrelated to the outcome of interest are in fact independent of the outcome, the explanatory approach might be appropriate. In other words, not counting or censoring the seemingly unrelated events would address a useful question about the *cause-specific* event rate and would not be subject to bias.

In clinical trials with longitudinal components such as time-to-event studies, some patients are likely to be *lost to follow-up*. In other words, the follow-up period may not be long enough for investigators to observe events in all patients. This can happen when a study participant is no longer accessible to the investigators because the patient does not return for clinic visits or has moved away. It can also happen when the follow-up period is shortened due to limited resources or error. Follow-up information is also lost when the earliest event time is the patient's death, which prevents observing all later event times such as disease progression or recurrence. In this case the competing risks of death and disease progression result in lost data. This can be a problem in studies of chronic disease, especially in older populations.

If losses to follow-up occur for reasons not associated with outcome, they have little consequence for affecting the study result, except to reduce precision. If investigators know that losses to follow-up are independent of outcome, the explanatory and pragmatic views of the problem are equivalent. In a trial with survival time as the main outcome, patients might stop coming to clinic for scheduled follow-up visits because they are too ill or have, in fact, died. In this case, being lost to follow-up is not a random event but carries information about the outcome. Such problems could affect the treatment groups of a randomized trial differently, producing a bias. In any case, studies designed with active follow-up and active ascertainment of endpoints in the participants will be less subject to this problem than those that rely on passive methods of assessing individual outcomes.

When losses occur frequently, even if they are not associated with the outcome, the external validity of the trial might be open to question. In studies using survival or disease recurrence or progression time as the major outcome, these losses to follow-up should occur in less than 5% of the trial participants when the study is conducted by experienced researchers. Often these types of missing data are not correctable but are preventable by designs and infrastructure that use active follow-up and ascertainment of events.

Investigators cannot reliably assume that losses are random events and conduct analyses that ignore them, particularly if they occur often. Every effort should be made to recover lost information rather than assuming that the inferences will be correct "as though all patients were followed completely." Survival status can sometimes be updated through the *National Death Index* or local sources. Active efforts to obtain missing information from friends or family members are also frequently successful.

There are many clinical rationalizations why investigators permit *uncounted events* in the study cohort. Suppose that a drug trial is being conducted in patients at high risk of death due to myocardial infarction. Because these patients are likely to be older than low-risk individuals, some study participants might die from other diseases before the end of the trial. Some investigators might prefer not to count these events because they do not seem to carry information about the effect of treatment on the target cause of death. Failure to count all such events can bias estimates of treatment effect. Sackett and Gent (1979) have some sensible advice on this point. It is also worth reading the discussion of the National Emphysema Treatment Trial in Section 4.6.6 for an illustration of the potential consequences of ignoring missing data that arise from differential stresses on the treatment groups in a randomized trial.

Different causes of failure compete with one another and failure of one kind usually carries some information about other types of failure. Being lost to follow-up may be associated with a higher chance of disease progression, recurrence, or death, for example. Death from suicide, infection, or cardiovascular causes may be associated with recurrence or progression of chronic disease such as cancer or AIDS. In cases such as these one cannot expect to estimate a cause-specific event rate without bias by censoring the seemingly unrelated events. Instead, the pragmatist would rely on a composite and well-defined endpoint such as time to disease recurrence or death from any cause (disease-free survival). Alternatively, time to death from any cause (overall survival) might be used. Censoring seemingly unrelated event times is usually a correctable error.

Imputation of Missing Values

There are only three generic analytic approaches to coping with missing values: (1) disregard the observations that contain a missing value, (2) disregard the variable if it has a high frequency of missing values, or (3) replace the missing values by some appropriate value. This last approach, called imputation of missing values, may be necessary or helpful in some circumstances. It sounds a lot like "making up data," but may be the most sensible strategy when done properly. Censored survival data is a type of missingness for which very appropriate and reliable strategies exist. The literature on missing data is large and technically difficult. I can only introduce the topic here, and the reader should consult a reference such as Rubin and Schenker (1991) for more depth.

The following general circumstances might support imputation as a worthwhile analytic strategy: (1) the frequency of missingness is relatively small (e.g., 10–15% of the data), (2) the variable containing missing values is especially important clinically or biologically to the research question, (3) reasonable (conservative) base assumptions and technical strategy for imputation exist, and (4) sensitivity of the conclusions to different imputation strategies can be determined. Then the effect of imputation of missing values will be to shift the debate away from concern about the possible influence of missing observations to concern about the effect of assumptions and methods. Because the latter can be studied, this shift of focus can be extremely helpful.

A familiar circumstance amenable to imputation of missing values is a longitudinal study with repeated quality of life assessments of subjects. Outcome measures are more likely to be missing for those with longer follow-up (attrition). A simple imputation method often used in this setting is *last observation carried forward* (LOCF) where the most recent observation replaces any subsequent missing ones. There are obvious limitations to such an approach, but it does permit certain analytic methods to be mechanically applied.

For missing values more generally, many other strategies come to mind. For example, we could replace the missing values by the mean value of the variable. Alternatively, we could replace missing values by *predicted* values derived from some reasonable data model. The data model itself might depend on imputed values, leading to an iterative procedure. Alternatively, we might randomly resample from similar or matched observations to obtain a replacement value. Obviously none of these approaches can correct a biased relationship between the fact of missingness and the outcome of interest.

In practice more than one approach would be tried, and the sensitivity of conclusions to each can be assessed. If all approaches yield similar answers, we would have more confidence in the conclusions. Otherwise we might not consider any approach to be reliable. Imputation is not a good solution for data with a high frequency of missing values, and is certainly not a substitute for thoughtful design, active ascertainment of outcomes, and an adequate infrastructure to support the study.

15.2.3   Protocol Nonadherence Is Common

Ineligible patients are a form of missing data with respect to the external validity of the trial. We do not usually think about eligibility in these terms but it is precisely the concern raised regarding inclusiveness or representation in the study cohort. Potentially this problem can be ameliorated by using a large heterogeneous study cohort. However, this may not always be possible because of resource limitations. This problem is interesting but peripheral to the focus of this chapter.

In most clinical trials it is common to find errors that result in ineligible patients being placed on study. These *eligibility errors* are a relatively common type of protocol nonadherence. If the eligibility criteria are defined objectively and are based only on information available before study entry or randomization, the consequences of such errors will be minimal. In contrast, if the criteria are partly subjective, or are defined improperly using outcome information, errors and the methods of resolving them carry greater consequences.

Objective eligibility criteria are less prone to error than subjective ones. Objective criteria include quantitative measurements such as age, the results of laboratory tests, and some categorical factors such as sex. Partly subjective criteria include some measures like extent of disease, histologic type, and the patient's functional capacity. These criteria are clinically well defined but require some element of expert opinion to specify them. This subjectivity can sometimes produce an error. Subjective criteria are those based on patient self-report or solely on physician judgment. These may not be appropriate eligibility criteria, although they may be the gold standard for some assessments such as quality of life and pain.

Patients can fail to comply with nearly any aspect of treatment specification. Common treatment nonadherence problems are reduced or missed doses and improper scheduling. All treatment modalities are subject to such errors including drugs, biologicals, radiotherapy, surgery, lifestyle changes, diet, and psychosocial interventions. Failure to comply with, or complete, the intended treatment can create a strong, but incorrect, rationale to remove patients from the final analysis. Evaluability criteria, mentioned above, are a type of non-adherence. Removing eligible but noncompliant patients from the analysis can create serious problems. In general, criteria that permit such removals are a methodologic error. Approaches to the analysis in the presence of treatment nonadherence are discussed in the section regarding intention to treat (below).

Ineligible patients who were mistakenly placed on study can be analyzed in two ways: included with the eligible patients from the trial cohort (pragmatic approach), or removed from the analysis (explanatory approach). In a randomized comparative trial, if the eligibility criteria are objective and determined from pre-study criteria, neither of these approaches will create a bias. However, excluding patients from any study can diminish the external validity of the results.

If there is any potential for the eligibility determination to be applied retroactively because of subjective interpretation of the criteria or methodologic error in the protocol design, the pragmatic and explanatory approaches have quite different properties. In particular, excluding patients in this situation can affect the treatment groups in a randomized comparison differently and produce a bias. The principal difficulty is that patient exclusions produce the possibility that eligibility will be confounded with outcome.

For example, suppose that a randomized trial compares medical versus surgical therapy for the same condition. Patients randomized to surgery may need evaluations and criteria beyond those required for patients to receive medical therapy. Simple examples are normal EKGs and good pulmonary function. If retroactive exclusions are based on such tests, they can differentially affect the treatment groups, yielding patients with a better prognosis in the surgical arm. This problem could be avoided by requiring all trial participants to pass the same (pre-surgical) criteria. However, such a policy would exclude many patients eligible for medical therapy, reducing the generalizations that can be made from the trial. Thus it is not necessarily the best course of action.

Treatment nonadherence is a special type of protocol nonadherence that has received a great deal of attention in the clinical trials literature because it is a common problem and there are different ways to cope with it when analyzing most trials. These are discussed in detail in the next section.

15.3   TREATMENT NONADHERENCE

There has been considerable investigation into the problems surrounding treatment non-adherence. Most work has focused on ways of measuring or preventing nonadherence and proposed ways of improving statistical estimates of treatment effects when adherence is a problem. There is a chronic debate about the advantages and problems of analyses based on treatment assigned compared with those based on treatment received in comparative trials. See Newell (1992), Lewis and Machin (1993), or Gillings and Koch (1991) for reviews. This section will highlight some of the issues surrounding this "intention-to-treat" debate.

15.3.1   Intention to Treat Is a Policy of Inclusion

Intention to treat (ITT) is the idea, often stated as a principle, that patients on a randomized clinical trial should be analyzed as part of the treatment group to which they were assigned, even if they did not actually receive the intended treatment. The term "intention to treat" appears to have been originated by Hill (1961). It can be defined generally as the analysis that

> includes all randomized patients in the groups to which they were randomly assigned, regardless of their adherence with the entry criteria, regardless of the treatment they actually received, and regardless of subsequent withdrawal from treatment or deviation from the protocol. (Fisher et al., 1990)

Thus ITT is an approach to several types of protocol nonadherence. "Treatment received" (TR) is the idea that patients should be analyzed according to the treatment actually given, even if the randomization called for something else. I can think of no other issue which perpetually generates so much disagreement between clinicians, some of whom prefer treatment received analyses, and statisticians, who usually prefer intention-to-treat approaches.

Actually, there is a third type of analysis that often competes in this situation, termed "adherers only." This approach discards all those patients who did not comply with their treatment assignment. Most investigators would instinctively avoid analyses that discard information. I will not discuss "adherers only" analyses for this reason and also because doing so will not help illuminate the basic issues.

Suppose that a trial calls for randomization between two treatments, A and B. During the study some patients randomized to A actually receive B. I will denote this group of patients by B_A. Similarly patients randomized to B who actually receive A will be designated A_B. ITT calls for the analysis groups to be $A + B_A$ compared with $B + A_B$, consistent with the initial randomization. TR calls for the analysis groups to be $A + A_B$ compared with $B + B_A$, consistent with the treatment actually received. Differences in philosophy underlying the approaches and the results of actually performing both analyses in real clinical trials fuel debate as to the correct approach in general.

With some careful thought and review of empirical evidence, this debate can be decided mostly in favor of the statisticians and ITT, at least as an initial approach to the analysis. In part, this is because one of the most useful perspectives of an RCT is as a test of the null hypothesis of no-treatment difference, a circumstance in which the ITT analysis yields the best properties. However, it is a mistake to adhere rigidly to either perspective. It is easy to imagine circumstances where following one principle or the other too strictly is unhelpful or inappropriate. The potential utility of TR analyses should probably be established on a study-by-study basis as part of the trial design (analysis plan). Then differences between TR and ITT results will not influence one's choice.

15.3.2 Coronary Drug Project Results Illustrate the Pitfalls of Exclusions based on Nonadherence

The potential differences between ITT and TR analyses have received attention in the clinical trials literature since the early work by the Coronary Drug Project Research Group (CDP) (Coronary Drug Project Research Group, 1980). The CDP trial was a large randomized, double-blind, placebo-controlled, multicenter clinical trial, testing the efficacy of the cholesterol lowering drug clofibrate on mortality. The overall results of the CDP showed no convincing benefit of clofibrate (effect size $= 0.60$, $p = 0.55$) (Table 15.1, bottom row). There was speculation that patients who adhered to the clofibrate regimen received a demonstrable benefit, while those who did not comply would have a death rate similar to the placebo group. The relevant analysis is also shown in Table 15.1, where adherence is defined as those patients taking 80% or more of their medication during the first 5 years or until death. Additional mortality rates in Table 15.1 have been adjusted for 40 baseline prognostic factors. Within the clofibrate group, the results appear to support the biological hypothesis that treatment adherence yields a benefit from the drug (effect size $= 3.86$, $p = 0.0001$).

However, an identical analysis of patients receiving placebo demonstrates why results based on adherence can be misleading. Patients who adhered to placebo show an even larger benefit than those adhering to clofibrate (effect size $= 8.12$, $p \ll 0.0001$). Comparison of mortality in the placebo group adjusted for the baseline factors does not explain the difference. Thus we cannot attribute the reduced mortality in the adhering patients to the drug. Instead, it must be a consequence of factors associated with patients adhering to medication but not captured by the baseline predictors. These could be other treatments, disease status, lifestyle, or unknown factors. In any case, comparison of groups based on adherence is potentially very misleading.

15.3.3 Statistical Studies Support the ITT Approach

Several authors have studied ITT and alternative approaches using analyses of existing data or simulation (Lee et al., 1991; Peduzzi et al., 1991; Peduzzi et al., 1993). There

TABLE 15.1 Five-Year Mortality in the Coronary Drug Project by Treatment Group and Adherence

Treatment Adherence	Clofibrate			Placebo		
	N	% Mortality	Adjusted	N	% Mortality	Adjusted
<80%	357	24.6 ± 2.3	22.5	882	28.2 ± 1.5	25.8
$\geq 80\%$	708	15.0 ± 1.3	15.7	1813	15.1 ± 0.8	16.4
Total	1065	18.2 ± 1.2	18.0	2695	19.4 ± 0.8	19.5

Source: Coronary Drug Project Research Group (1980).

is a consensus supporting ITT analyses. Lee et al. (1991) studied several types of analyses in a trial of phenobarbital in children with febrile seizures. They concluded that the ITT analysis was preferred and that analyses based on treatment received may confuse rather than help the interpretation of results.

Peduzzi et al. (1991) analyzed data from the Veterans Administration coronary bypass surgery study using TR and four other approaches. The ITT analysis yielded a nonsignificant treatment difference ($p > 0.99$), whereas the TR approach showed a survival advantage for the surgical therapy group ($p < 0.001$). Based on these findings and simulations, the authors conclude that ITT analyses should be the standard for randomized clinical trials. Similar conclusions were obtained in a second study (Peduzzi et al., 1993).

Lagakos, Lim, and Robins (1990) discuss the similar problem of early treatment termination in clinical trials. They conclude that ITT analyses are best for making inferences about the unconditional distribution of time to failure. The size of ITT analysis tests is not distorted by early treatment termination. However, a loss of power can occur. They propose modifications to ordinary logrank tests that would restore some of the lost power without affecting the size of the test.

15.3.4   Trials Can Be Viewed as Tests of Treatment Policy

It is unfortunate that investigators conducting clinical trials cannot guarantee that the patients who participate will definitely complete (or even receive) the treatment assigned. This is, in part, a consequence of the ethical principle of respect for individual autonomy. Many factors contribute to patients' failure to complete the intended therapy, including severe side effects, disease progression, patient or physician strong preference for a different treatment, and a change of mind. In nearly all circumstances failure to complete the assigned therapy is partially an outcome of the study and therefore can produce a bias if used to subset patients for analysis. From this perspective a clinical trial is a test of treatment policy, not a test of treatment received. ITT analyses avoid bias by testing policy or programmatic effectiveness.

From a clinical perspective, postentry exclusion of eligible patients is analogous to using information from the future. For example, when selecting a therapy for a new patient, the clinician is primarily interested in the probability that the treatment will benefit the new patient. Because the clinician has no knowledge of whether or not the patient will complete the treatment intended, he or she has little use for inferences that depend on events in the patient's future. By the time treatment adherence is known, the clinical outcomes may also be known. Consequently at the outset the investigator will be most interested in clinical trial results that do not depend on adherence or other events in the patient's future. If the physician wishes to revise the prognosis when new information becomes available, then an analysis that depends on some intermediate success might be relevant.

15.3.5   ITT Analyses Cannot Always Be Applied

The limitations of the ITT approach have been discussed by Feinstein (1991) and Sheiner and Rubin (1995). A breakdown in the experimental paradigm can render an analysis plan based on ITT irrelevant for answering the biological question. This does not mean that TR analyses are the best solution because they are subject to errors of their own. There may be no entirely satisfactory analysis when the usual or ideal procedures are inapplicable because of unanticipated complications in study design or conduct. On the other hand, when the conduct of the trial follows the experimental paradigm fairly closely, as is often the case, analyses such as those based on ITT are frequently the most appropriate.

Example 38 Busulfan is a preparative regimen for the treatment of hematologic malignancies with bone marrow transplantation. In a small proportion of patients, the drug is associated with veno-occlusive disease (VOD) of the liver, a fatal complication. Clinical observations suggested that the incidence of VOD might be eliminated with appropriate dose reduction to decrease the area under the time–concentration curve (AUC) of the drug, observable after patients are given a test dose. Furthermore the efficacy of the drug might be improved by increasing the dose in patients whose AUC is too low. These considerations led to a randomized trial design in which patients were assigned to two treatments consisting of the same drug but used in different ways. Group A received a standard fixed dose, while group B received a dose adjustment, up or down, to achieve a target AUC, based on the findings of a test dose. Partway through the trial the data were examined to see if dose adjustment was reducing the incidence of VOD. Interestingly none of the patients assigned to B actually required a dose adjustment because, by chance, their AUCs were all within the targeted range. On treatment A, some patients had high AUCs and a few experienced VOD. The intention-to-treat analysis would compare all those randomized to B, none of whom were dose adjusted, with those on A. Thus the ITT analysis could not carry much information about the efficacy of dose adjustment. On the other hand, when the trial data were examined in conjunction with preexisting data, the clinical investigators felt ethically compelled to use dose adjustment, and the trial was stopped.

In special circumstances TR analyses can yield estimated treatment effects that are closer to the true value than those obtained from the ITT analyses. However, this improved performance of the TR approach depends upon also adjusting for the covariates responsible for crossover. Investigators would have to know factors responsible for patients failing to get their assigned treatment, and incorporate those factors in correct statistical models describing the treatment effect. This is usually not feasible because investigators do not know the reasons why, or the covariates associated with, patients failing to complete their assigned treatment. Even if the factors that influence nonadherence are known, their effect is likely to be more complex than simple statistical models can capture. Thus the improved performance of TR methods in this circumstance is largely illusory.

Sommer and Zeger (1991) present an alternative to TR analyses that permits estimating efficacy in the presence of nonadherence. This method employs an estimator of biological efficacy that avoids the selection bias that confounds the comparison of compliant subgroups. This method can be applied to randomized trials with a dichotomous outcome measure, regardless of whether a placebo is given to the control group. The method compares the compliers in the treatment group to an inferred subgroup of controls, chosen to avoid selection bias.

Efron and Feldman (1991) discuss a statistical model that uses adherence as an explanatory factor and apply their method to data from a randomized placebo-controlled trial of cholestyramine for cholesterol reduction. Their method provides a way to reconstruct dose–response curves from adherence data in the trial. This and similar approaches based on models are likely to be useful supplements to the usual conservative ITT analyses and can recover valid estimates of method effectiveness when nonadherence is present (Sheiner and Rubin, 1995).

Treatment effects in the presence of imperfect adherence can be bounded using models of causal inference that do not rely on parametric assumptions. Examples of this is the work by Robins (1989), Manski (1990), and Balke and Pearl (1994). These methods permit estimating the extent to which treatment effect estimates based on ITT can differ from the true treatment effect. These methods show promise but have not been widely applied in clinical trials.

Another promising method for coping with treatment nonadherence is based on the idea of principal stratification (Frangakis and Rubin, 1999, 2002). This method permits a recovery of the effect of treatment, as opposed to the effect of randomization, under fairly general assumptions. For an example of its application, see the study of exercise in cancer patients by Mock et al. (2004).

15.3.6 Trial Inferences Depend on the Experimental Design

The best way to reconcile the legitimate clinical need for a good biological estimate of treatment efficacy and the statistical need for unbiased estimation and correct error levels is to be certain that patients entered on the trial are very likely to complete the assigned therapy. In other words, the eligibility criteria should exclude patients with characteristics that might prevent them from completing the therapy. This is different from excluding patients solely to improve homogeneity. For example, if the therapy is lengthy, perhaps only patients with good performance status should be eligible. If the treatment is toxic or associated with potentially intolerable side effects, only patients with normal function in major organ systems would be likely to complete the therapy.

It is a fact that the potential inferences from a clinical trial and the potential correctness of those inferences are a consequence of both the experimental design and the methods of analysis. One should not employ a design with particular strengths and then undo that design during the data analysis. For example, if we are certain that factors associated with treatment decisions are known, there might be very little reason to randomize. One could potentially obtain correct inferences from a simple database. However, if there are influential prognostic factors (known), then a randomized comparative trial offers considerable advantages. These advantages should not be weakened by ignoring the randomization in the analysis. Finally there may be legitimate biological questions that one cannot answer effectively using rigorous designs. In these circumstances it is not wise to insist on an ITT analysis. Instead an approximate answer to a well-posed biological question will be more useful than the exact answer to the wrong question.

15.4 SUMMARY

Clinical trials are characterized by imperfect data as a consequence of protocol nonadherence, methodologic error, and incomplete observations. Inferences from a trial can depend on how the investigators resolve data imperfections. Two approaches to such questions have been called "explanatory" and "pragmatic." The pragmatic approach tends to follow the statistical design of an experiment more closely. The explanatory approach may make some assumptions to try to answer biological questions.

Data imperfections that have the most potential for influencing the results of a trial arise from patients who do not adhere with assigned treatment or other criteria. Nonadherence encourages some investigators to remove patients from the analysis. If the reasons for exclusion are associated with prognosis or can affect the treatment groups differently, as is often the case, the trial results based on exclusions may not be valid.

Intention to treat is the principle that includes all patients for analysis in the groups to which they were assigned, regardless of protocol adherence. This approach is sometimes at odds with explanatory views of the trial, but usually provides a valid test of the null hypothesis. Approaches based on analyzing patients according to the treatment they actually received may be useful for exploring some clinical questions but should not be the primary analysis of a randomized clinical trial. Investigators should avoid any method that removes eligible patients from the analysis.

15.5   QUESTIONS FOR DISCUSSION

1. In some retrospective cohort studies patients who have not been followed for a while in clinic can be assumed to be alive and well. Discuss how such an assumption can create bias.

2. Generalizing the idea of intention to treat to developmental trials, one might require that all patients who meet the eligibility criteria should be analyzed as part of the treatment group. Discuss the pros and cons of this approach.

3. Discuss specific circumstances in a CTE trial that might make it difficult to apply the intention-to-treat approach.

4. Read the paper by Stansfield et al. (1993) and discuss the inclusion/exclusion properties of the analysis.

5. Apply the method of Sommer and Zeger (1991) to the data from the Coronary Drug Project, assuming that the placebo compliance data were not available. How does the method compare to the actual estimated relative risk? Discuss.

16

ESTIMATING CLINICAL EFFECTS

16.1 INTRODUCTION

A perspective on summarizing data from trials is necessary to understand some important experimental design issues. Because this book is not primarily about analyzing trials, the discussion here will be brief and will emphasize clinically relevant summaries of the data. Thorough accounting of the data from a clinical trial usually requires technical background and skills beyond the scope of this book. The references cited can provide some help, but there is little substitute for experience, good design, and expert statistical collaborators. For the reader with sufficient background and ambition, additional technical details can be found in Armitage and Berry (1994), Everitt (1989), and Campbell and Machin (1990). Reporting guidelines are discussed in Chapter 18. Readers needing a more complete introduction to the analysis of survival data can consult Kleinbaum (1996). The difficult issue of simultaneous analysis of treatment effects and covariates (adjusted analyses) is discussed in Section 17.3.

Most clinical trials use more than one of the types of outcomes discussed in Chapter 8 and usually require several qualitatively different views of the data to address the basic research question. Informative analyses of data are based on many factors, including the state of the clinical context in which questions are embedded, specific biological knowledge, the design of the study, proper counting of subjects and events, numeracy (understanding of numbers), familiarity with probability and statistical models, and methods of inference. Applying these various concepts to clinical trials is usually not difficult, but it does require a broad range of knowledge.

Exactly the same methods of data analysis are often used for designed experiments as for nonexperimental studies. But the greater validity of inference from experimental studies relies more on design than analysis, as emphasized elsewhere in this book. Data analysis has a certain fascination because it is more proximate to the final results than study design, and sometimes appears more important. However, the opposite is true. Experimental design and execution are not supplemental to analysis—a better description is that analysis is subordinate to design.

We must also understand the dichotomy between descriptive summaries of data and methods intended to be analytical. In this context "analytical" refers not to "data analysis," but to quantifying the relationships of observables in the data that have direct biological meaning. Statistics embodies a spectrum of methods from the descriptive to the analytical. But seemingly descriptive methods can be analytical when superimposed on good design. Methods commonly employed to be analytical may be only descriptive when applied to observations lacking a foundation of design.

16.1.1 Structure Aids Data Interpretation

The design of a clinical trial imposes structure on the data that facilitates analysis and interpretation. Other components of structure (e.g., a biological or population model) may also contribute to understanding. Provided that the structure supporting the data is reasonable and flexible, the results of the analysis are likely to be accurate, clinically useful, and externally valid. The appropriate structure supports interpretation *of* the data as well as interpretation *beyond* the data, meaning interpolation and extrapolation.

For example, in pharmacologic studies blood samples are often used to construct time concentration curves, which describe drug distribution and/or metabolism. When the data are supplemented by additional structure in the form of a pharmacokinetic model, statistical methods allow estimating the values of the parameters (e.g., half-life, elimination rate) that best fit the data. The resulting model fits and parameters allow estimation of physiologic effects, smoothing and interpolation of data, quantitative associations with side effects, and inferences about subjects yet to be treated. The extrapolation of knowledge to individuals not actually studied is based on biological knowledge embodied the model, which is external to the actual experiment.

In SA trials of cytotoxic drugs, investigators are often interested in tumor response and toxicity of the drug or regimen. The usual study design and structure permits estimating the unconditional probability of response or toxicity in patients who met the eligibility criteria. These probabilities, or measures of risk, summarize the observed effects of treatment and also generalize outside the study cohort to the extent that biology permits. With the additional structure of models, patient characteristics can be connected quantitatively to the observed outcomes. SA trials could not have a central role developmentally if their structure did not permit at least a moderate degree of generalizability.

Comparative trials can yield formal tests of statistical hypotheses but also provide more descriptive information about both absolute and relative probabilities of events and risks. We readily recognize design components that increase the internal validity of these trials. Such studies are also intended to generalize externally, validated by the structured internal comparison and control of extraneous factors. Sometimes it is helpful or necessary to impose an additional framework on analyses of comparative trials to address new questions or generate hypotheses. Such a framework could come in the form of risk models, subset definitions, or aggregated data from several studies. All of these, when appropriately done, can enhance the interpretation of the data from a trial.

For all types of studies, investigators must distinguish among those analyses, tests, or other summaries of the data that are specified a priori and justified by the design of the trial and those that are exploratory or post hoc. In many cases the data suggest certain comparisons that, although biologically justifiable, may not be statistically reliable because they were not anticipated in the study design. We would not want to focus on findings from such analyses as the principal result of a trial.

Example 39   Suppose that a series of patients meeting pre-defined eligibility criteria are given a six-week fixed-dose course of a new antihypertensive drug. The drug produces side effects in some individuals, which require dose reduction or discontinuation of the treatment. The design of this study suggests that the investigators are primarily interested in estimating the unconditional probability of benefit from the new drug and the proportion of patients who cannot tolerate the therapy. Suppose that investigators also observe that patients who remain on the new drug have "significantly" lower blood pressure during the six-week study than those who are forced to resume their previous treatment. Although there may be biological reasons why the new treatment is superior, the study design and events do not permit reliable conclusions by comparing it with standard therapy. The patients who experience side effects on the new treatment may be different in clinically important ways from those who tolerate the drug.

16.1.2   Estimates of Risk Are Natural and Useful

Risk is always important to the patient. Safety and efficacy can both usually be summarized in terms of the risk of clinically relevant events. Often this is the best or only way to quantify safety and efficacy. The clinician scientist is often interested in direct comparisons of risk—estimating the magnitude of differences, or relative differences. Depending on the setting, risk differences can be expressed in absolute magnitude or on a relative scale. These are discussed further below.

The populist view of risk is usually deficient. A major flaw is the confounding of quantification with risk perception. This confounding applies both to understanding a specific therapy as well as risks inherent in clinical trials. A second populist flaw is the distorted metric applied to assess risk. In horse racing, for example, the ordinary betting odds is a commonly understood measure of risk and is used as a wagering summary or guide. Unfortunately, it does not formally relate to the chances of a particular horse winning the race but merely reflects the current allocation of money. In football and some other sports contests, team differences are often quantified as a "point spread." It is generally understood that the "point spread" equalizes the probability of winning a bet on either side. It indirectly quantifies risk and can be converted into an odds of winning an even bet, though this is seldom done. Although widely understood, such measures of risk are unhelpful in medical studies.

Often in the clinical setting, we are interested in knowing the probability of a successful outcome, or the risk of failure, perhaps measured over time. When comparing treatment effects, changes or relative changes in risk are of primary interest. When comparing event times such as survival or disease recurrence/progression, group differences could be summarized as the difference in the absolute probabilities of failure. Such an effect can also be expressed as differences in median event times. When the hazard of failure is constant across time (or nearly so), a more efficient method is to summarize group differences as a ratio of hazards. These risk summaries may seem unfamiliar at first, but it is essential to understand them to be conversant with clinical trials. Additional details are provided later in this chapter.

A little discipline is required to generate the most useful summaries of risk from clinical trials. When absolute risks are of interest, as in safety assessments or the chance of benefit from the treatment in a single cohort study, unconditional estimates are needed. As a general principle, eligible study subjects should not be removed from the denominator of such calculations, even if doing so appears to address some useful post hoc clinical question. Risk estimates that condition on other outcomes, such as treatment adherence, also violate this principle and are often clinically meaningless for that and other reasons.

16.2   DOSE-FINDING AND PK TRIALS

Obtaining clinically useful information from DF trials such as phase I studies depends on (1) evaluation of preclinical data, (2) knowledge of the physical and chemical properties of the drug and related compounds, (3) modeling drug absorption, distribution, metabolism, and elimination, and (4) judgment based on experience. Instead of a comprehensive mathematical presentation, I will discuss a few interesting points related to modeling and obtaining estimates of important pharmacokinetic parameters. Complete discussions of modeling and inferences can be found in Carson, Cobelli, and Finkelstein (1983) or Rubinow (1975).

16.2.1   Pharmacokinetic Models Are Essential for Analyzing DF Trials

One of the principal objectives of phase I studies is to assess the distribution and elimination of drug in the body. Some specific parameters of interest are listed in Table 16.1. Pharmacokinetic (PK) models (or compartmental models) are a useful tool for summarizing and interpreting data from DF trials such as phase I studies. These models are helpful in study design, for example, to suggest the best times for blood or other samples to be taken. During analysis the model is essential as an underlying structure for the data and permitting quantitative estimates of drug elimination. PK models have limitations because they are idealizations. No model is better than the data on which it is based. However when properly used, the model may yield insights that are difficult to gain from raw data. Some other objectives of phase I studies may not require modeling assumptions. Examples of these include secondary objectives, such as exploring the association between plasma drug levels and severity of side effects, or looking for evidence of drug efficacy. In any case, PK models are at the heart of using quantitative information from DF studies.

TABLE 16.1   Outcomes of Clinical Interest in Dose-Finding Studies

- Maximal tolerated dose
- Absorption rate
- Elimination rate
- Area under the curve
- Peak concentration
- Half life
- Correlation between plasma levels and side effects
- Proportion of patients who demonstrate evidence of efficacy

Drug absorption, distribution, metabolism, and excretion is not generally a simple process. However, relatively simple PK models can yield extremely useful information, even though they do not capture all of the complexities of the biology. In the next section, I consider a simple but realistic PK model that facilitates quantitative inferences about drug distribution and elimination. An important consideration in assessing toxicity, especially in oncologic studies, is the area under the time–concentration curve in the blood (AUC). The model discussed below illustrates one method of studying the AUC.

16.2.2   A Two-Compartment Model Is Simple but Realistic

Consider a drug administered at a constant rate by continuous intravenous infusion. We assume that the drug is transferred from blood to a tissue compartment (and vice-versa) with first order kinetics, and eliminated directly from the blood, also by first order kinetics (Figure 16.1). This situation can be described by a two-compartment linear system in the following way. Suppose that $X(t)$ and $Y(t)$ are the drug concentrations within blood and tissue, respectively, at time t and the drug is infused into the blood compartment with a rate of $g(t)$. The drug is transported from compartment X to Y at a rate λ, from Y back to X at a rate μ, and eliminated from X at a rate γ. All coefficients are positive. The rate equations for the system are

$$\frac{dX(t)}{dt} = -(\lambda + \gamma)X(t) + \mu Y(t) + g(t),$$

$$\frac{dY(t)}{dt} = \lambda X(t) - \mu Y(t), \tag{16.1}$$

where d/dt denotes the derivative with respect to time. This is a nonhomogeneous linear system of differential equations and can be solved using standard methods (Simmons, 1972). The general solution of this system is

$$X(t) = c_1(t)e^{\xi_1 t} + c_2(t)e^{\xi_2 t},$$

$$Y(t) = c_1(t)\frac{\xi_1 + \lambda + \gamma}{\mu}e^{\xi_1 t} + c_2(t)\frac{\xi_2 + \lambda + \gamma}{\mu}e^{\xi_2 t}, \tag{16.2}$$

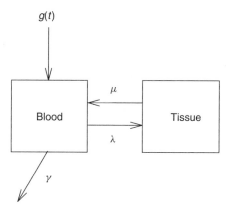

Figure 16.1   A simple two-compartment model for drug distribution.

where

$$\xi_{1,2} = \frac{-(\lambda + \mu + \gamma) \pm \sqrt{(\lambda + \mu + \gamma)^2 - 4\mu\gamma}}{2} \tag{16.3}$$

and $c_1(t)$ and $c_2(t)$ satisfy

$$\frac{dc_1(t)}{dt} = \frac{\xi_2 + \lambda + \gamma}{\xi_2 - \xi_1} g(t) e^{-\xi_1 t},$$

$$\frac{dc_2(t)}{dt} = -\frac{\xi_1 + \lambda + \gamma}{\xi_2 - \xi_1} g(t) e^{-\xi_2 t}. \tag{16.4}$$

To get an explicit solution for the system, we must specify $g(t)$, the infusion rate as a function of time, and initial conditions $X(0)$ and $Y(0)$. An interesting special case is constant infusion for a fixed time period,

$$g(t) = \begin{cases} g_0, & t \le t_0, \\ 0, & t > t_0. \end{cases} \tag{16.5}$$

Then, substituting equations (16.3) through (16.5) into equation (16.2),

$$X(t) = \begin{cases} \dfrac{g_0}{r} + \dfrac{\xi_2 + \lambda + \gamma}{\xi_2 - \xi_1}\left(\dfrac{g_0}{\xi_1} + X(0)\right)e^{\xi_1 t} \\ \quad - \dfrac{\xi_1 + \lambda + \gamma}{\xi_2 - \xi_1}\left(\dfrac{g_0}{\xi_2} + X(0)\right)e^{\xi_2 t}, & t \le t_0, \\[2ex] \dfrac{\xi_2 + \lambda + \gamma}{\xi_2 - \xi_1}\left(\dfrac{g_0}{\xi_1} + X(0) - \dfrac{g_0}{\xi_1}e^{-\xi_1 t_0}\right)e^{\xi_1 t} \\ \quad - \dfrac{\xi_1 + \lambda + \gamma}{\xi_2 - \xi_1}\left(\dfrac{g_0}{\xi_2} + X(0) - \dfrac{g_0}{\xi_2}e^{-\xi_2 t_0}\right)e^{\xi_2 t}, & t > t_0, \end{cases} \tag{16.6}$$

where the initial condition $Y(0) = 0$ has been incorporated. Here we have used the facts that $\xi_1 + \xi_2 = -(\lambda + \mu + \gamma)$ and $\xi_1\xi_2 = \mu\gamma$. Hence the area under the curve (AUC) for the first compartment is

$$AUC_x = \int_0^{t_0} X(t)dt + \int_{t_0}^{\infty} X(t)dt = \frac{g_0 t_0 + X(0)}{\gamma}. \tag{16.7}$$

Sometimes it is helpful to express these models in terms of amount of drug and volume of distribution. Equation (16.7) can be rewritten

$$AUC_x = \frac{Dt_0 + W(0)}{\gamma V},$$

where D is the dose of drug, V is the volume of distribution (assumed to be constant), $W(t) = V \times X(t)$ is the amount of drug, and γV is the "clearance." However, expressing AUC_x this way represents only a change of scale, which is not necessary for this discussion. Because drug "dose" is commonly expressed as weight of drug, weight of drug per kilogram of body weight, or weight of drug per square meter of body surface area, we can refer to $g_0 t_0 + X(0)$ as a dose, even though formally it is a concentration.

When the drug is infused constantly from time 0 to t_0 and $X(0) = 0$, $AUC_x = g_0 t_0 / \gamma$. This is the ratio of total dose to the excretion rate. Another interesting case is when the drug is given as a single bolus, meaning $t_0 = 0$, in which case $AUC_x = X(0)/\gamma$, which is also a ratio of total dose over excretion rate. The transport parameters λ and μ do not affect the AUC in the first compartment.

With similar calculations we can find the solution for the second compartment,

$$Y(t) = \begin{cases} \dfrac{g_0 \lambda}{\mu \gamma} - \dfrac{\lambda}{\xi_2 - \xi_1} \left(\dfrac{g_0}{\xi_1} + X(0) \right) e^{\xi_1 t} \\ \quad + \dfrac{\lambda}{\xi_2 - \xi_1} \left(\dfrac{g_0}{\xi_2} + X(0) \right) e^{\xi_2 t}, \qquad t \leq t_0, \\ \\ - \dfrac{\lambda}{\xi_2 - \xi_1} \left(\dfrac{g_0}{\xi_1} + X(0) - \dfrac{g_0}{\xi_1} e^{-\xi_1 t_0} \right) e^{\xi_1 t} \\ \quad + \dfrac{\lambda}{\xi_2 - \xi_1} \left(\dfrac{g_0}{\xi_2} + X(0) - \dfrac{g_0}{\xi_2} e^{-\xi_2 t_0} \right) e^{\xi_2 t}, \quad t > t_0, \end{cases} \tag{16.8}$$

and

$$AUC_Y = \frac{\lambda g_0 t_0 + \lambda X(0)}{\mu \gamma} = \frac{\lambda}{\mu} AUC_X. \tag{16.9}$$

It is directly related to AUC in the first compartment and the transport rates λ and μ.

The behavior of this model is shown in Figure 16.2. Values for both compartments determined by numerical integration are shown with the tissue compartment peak lagging behind the vascular compartment peak as one would expect. When the infusion

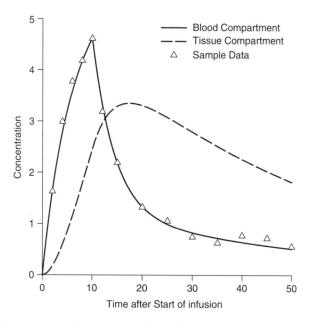

Figure 16.2 Time concentration curves and data from a two-compartment model. $T_0 = 10$, $\mu = 0.1$, $\gamma = 0.1$, $\lambda = 0.1$, and $g_0 = 1.0$.

stops, the concentration in the blood begins to decline abruptly, whereas the tissue curve shows an inflection point. Data points (with simulated error) that might be obtained from such a system are also shown in Figure 16.2.

16.2.3 PK Models Are Used by "Model Fitting"

It is common in actual DF studies to have data that consist of samples from only the vascular compartment taken at various time points during the uptake or infusion of drug and during its elimination. The model in the form of equations (16.6) and (16.8) can be fitted to such data to obtain estimates of the rate constants which are the clinical effects of interest. An example of this for a continuous intravenous infusion of the antitumor drug cyclophosphamide in four patients is shown in Figure 16.3. For each study subject, the infusion was to last 60 minutes although the exact time varied. The value of g_0 was fixed at 10.0. Because of measurement error in the infusion time and dosage, it is sometimes helpful to estimate both t_0 and g_0 (see Appendix A). The fitted curves show a good agreement with the measured values. The estimated parameter values are shown in Table 16.2, along with the estimated AUCs. Of course, the quality of inference about such a drug would be improved by examining the results from a larger number of study subjects.

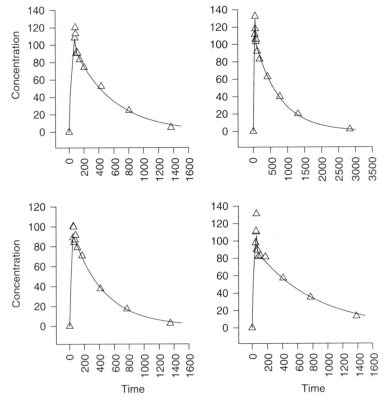

Figure 16.3 Sample data and model fits to time–concentration values from four patients in a DF trial. The triangles are observed serum levels and the solid lines are model fits.

TABLE 16.2   Estimated Rate Constants from Phase I Clinical Trial Data

#	t_0	$\widehat{\lambda}$	$\widehat{\gamma}$	$\widehat{\mu}$	$\widehat{\text{AUC}}$
1	85	0.284	0.015	0.042	53,475
		(0.0436)	(0.0014)	(0.0040)	
2	60	0.250	0.007	0.057	78,876
		(0.0190)	(0.0004)	(0.0065)	
3	50	0.782	0.012	0.196	40,251
		(0.9200)	(0.0068)	(0.1027)	
4	50	0.239	0.007	0.061	63,509
		(0.1684)	(0.0044)	(0.0275)	

Note: Estimated standard errors are shown in parentheses.

16.3   SA STUDIES

A common primary objective of SA studies is to estimate the frequency of side effects and the probability of success in treating patients with a new drug or combination. In oncology the investigator is also usually interested in grading the toxicity seen and estimating overall length of survival. Often the outcome assessments for response and toxicity are dichotomous, involving yes–no variables. Meeting these types of objectives requires estimating absolute probabilities.

16.3.1   Mesothelioma Clinical Trial Example

To illustrate some clinically useful summaries of information from SA trials, consider data of the kind shown in Table 16.3. The patients on this (phase II) trial were all

TABLE 16.3   Data from Mesothelioma SA (Phase II) Clinical Trial

Age	Sex	PS	Hist	Wtchg	Surg	PFS	Prog	Surv	Event
60	1	1	136	1	3	394	1	823	1
59	1	0	136	2	3	1338	0	1338	0
51	0	0	130	1	1	184	1	270	1
73	1	1	136	1	3	320	0	320	1
74	1	0	136	2	1	168	0	168	1
39	0	0	136	1	1	36	1	247	1
46	1	1	131	1	3	552	1	694	0
71	1	0	136	1	1	133	1	316	1
69	1	0	136	1	1	175	1	725	0
49	1	0	131	1	1	327	0	327	1
69	1	0	131	1	2	0	0	0	1
72	1	0	131	1	1	676	1	963	0
⋮	⋮	⋮	⋮	⋮	⋮	⋮	⋮	⋮	⋮

Note: PFS is progression free time; Surv is survival time. Prog and Event are censoring indicator variables for progression and death, respectively. Performance status (PS) is dichotomized as high versus low.

diagnosed with malignant mesothelioma, an uncommon lung tumor strongly related to asbestos exposure. Depending on the extent of disease at diagnosis, patients underwent one of three types of surgery: biopsy, limited resection, or extrapleural pneumonectomy (EPP), a more extensive operative procedure (Rusch, Piantadosi, and Holmes, 1991). The goals of the SA trial were to determine the feasibility of performing EPP and to document the natural history of the disease. The complete data on 83 subjects are presented in Appendix A.

Possibly important prognostic factors in patients with mesothelioma include sex, histologic subtype (hist), weight change at diagnosis (wtchg), performance status (ps), age, and type of surgery (surg). Disease progression is both an outcome of treatment and a potential predictor of survival. The progression and survival times are censored, meaning some patients remained progression free or alive at the end of the study or cutoff date for analysis.

16.3.2 Summarize Risk for Dichotomous Factors

Estimating the overall proportion of patients who progress is straightforward and will not be detailed here. Instead, we consider the relationships between disease progression as an intermediate outcome and the prognostic factors sex and performance status. In reality, progression is an event time, but will be treated as a dichotomous factor, temporarily, for simplicity. Both factors can be summarized in 2×2 tables (Table 16.4).

The probabilities and odds of progression are shown in Tables 16.5 and 16.6. Here the 95% confidence limits on the proportions are based on the binomial distribution. For dichotomous factors like those in Table 16.4, we are sometimes interested in absolute probabilities (or proportions). However, we are often interested in estimates of risk ratios such as the relative risk or odds ratio. If p_1 and p_2 are the probabilities of events in two groups, the odds ratio, θ, is estimated by

$$\widehat{\theta} = \frac{p_1}{1 - p_1} \div \frac{p_2}{1 - p_2} = \frac{ad}{bc},$$

where $p_1 = a/(a + c)$, $p_2 = b/(b + d)$, and so on from the entries in a 2×2 table. For example, the odds ratio for progression in males versus females (Table 16.4) is $15 \times 14/(49 \times 5) = 0.857$. This can also be seen in Table 16.5. For odds ratios, calculating confidence intervals on a log scale is relatively simple. An approximate confidence interval for the log odds ratio is

$$\log\{\widehat{\theta}\} \pm Z_\alpha \times \sqrt{\frac{1}{a} + \frac{1}{b} + \frac{1}{c} + \frac{1}{d}},$$

TABLE 16.4   Progression by Sex and Performance Status (PS) for an SA Mesothelioma Trial

Progression	Overall	Sex		PS	
		Male	Female	0	1
No	20	15	5	10	10
Yes	63	49	14	42	21

TABLE 16.5 Probabilities and Odds of Progression by Sex for an SA Mesothelioma Trial

	Overall	Sex	
		Male	Female
Pr[progression] 95% CL	0.24 (0.154–0.347)	0.23 (0.138–0.357)	0.26 (0.092–0.512)
Odds of progression	0.317	0.306	0.357

TABLE 16.6 Probabilities and Odds of Progression by Performance Status for an SA Mesothelioma Trial

	PS	
	0	1
Pr[progression] 95% CL	0.19 (0.096–0.325)	0.32 (0.167–0.514)
Odds of progression	0.238	0.476

where Z_α is the point on normal distribution exceeded with probability $\alpha/2$ (e.g., for $\alpha = 0.05$, $Z_\alpha = 1.96$). For the example above, this yields a confidence interval of $[-1.33, 1.02]$ for the log odds ratio or $[0.26, 2.77]$ for the odds ratio. Because the 95% confidence interval for the odds ratio includes 1.0, the male–female difference in risk of progression is not "statistically significant" using the conventional criterion.

The odds ratio is a convenient and concise summary of risk data for dichotomous outcomes and arises naturally in relative risk regressions such as the logistic model. However, the odds ratio has limitations and should be applied and interpreted with thought, especially in circumstances where it is important to know the absolute risk. For example, if two risks are related by $p' = p + \delta$, where δ is the difference in absolute risks, the odds ratio satisfies

$$\frac{p}{1-p} = \theta \frac{(p+\delta)}{1-(p+\delta)}$$

or

$$\delta = \frac{p(\theta-1)(1-p)}{1+p(\theta-1)}.$$

This means that many values of p and θ are consistent with the same difference in risk. For example, all of the following (p, θ) pairs are consistent with an absolute difference in risk of $\delta = -0.2$: (0.30, 3.86), (0.45, 2.45), (0.70, 2.33), and (0.25, 6.33). As useful as the odds ratio is, it obscures differences in absolute risks that may be biologically important for the question at hand. Similar comments relate to hazard ratios (discussed below).

16.3.3   Nonparametric Estimates of Survival Are Robust

Survival data are unique because of censoring. Censoring means that some individuals have not had the event of interest at the time the data are analyzed or the study is over. For these people we know only that they were at risk for a measured period of time and that the event of interest has yet to happen. Thus their event time is censored. Using the incomplete information from censored observations requires some special methods that give "survival analysis" its statistical niche. The basic consequence of censoring is that we summarize the data using cumulative probability distributions rather than probability density functions, which are so frequently used in other circumstances.

In many clinical trials involving patients with serious illnesses like cancer, a primary clinical focus is the overall survival experience of the cohort. There are several ways to summarize the survival experience of a cohort. However, one or two frequently used methods make few, if any, assumptions about the probability distribution of the failure times. These "nonparametric" methods are widely used because they are robust and simple to employ. See Peto (1984) for a helpful review. For censored failure time data, the commonest analytic technique is the lifetable. Events (failures or deaths) and times at which they occur are grouped into convenient intervals (e.g., months or years) and the probability of failure during each interval is calculated.

Here I review the product limit method (Kaplan and Meier, 1951) for estimating survival probabilities with individual failure times. It is essentially the same as lifetable methods for grouped data. To avoid grouping of events into (arbitrary) intervals or to handle small cohorts, individual failure times are used. The method is illustrated in Table 16.7 using the SA mesothelioma trial data introduced above. First, the observed event or failure times are ranked from shortest to longest. At each event time, indexed by i, the number of subjects failing is denoted by d_i and the number of subjects at risk just prior to the event is denoted by n_i. By convention, censoring times that are tied with failure times are assumed to rank lower in the list. Unless there are tied failure times, the number of events represented by each time will be 0 (for censorings) or 1 (for failures).

For an arbitrary event time, the probability of failure in the interval from the last failure time is

$$p_i = \frac{d_i}{n_i}.$$

The probability of surviving the interval is $1 - p_i = 1 - (d_i/n_i)$. Therefore the cumulative probability of surviving all earlier intervals up to the kth failure time is

$$\widehat{S(t_k)} = \prod_{i=0}^{k-1} \left(1 - \frac{d_i}{n_i}\right) = \prod_{t_i} \left(\frac{n_i - d_i}{n_i}\right), \tag{16.10}$$

where the product is taken over all distinct event times. This calculation is carried through a few observations in Table 16.7.

For censored events, the previous product will be multiplied by 1 and, in the absence of an event, the survival estimate remains constant, giving the curve its characteristic step-function appearance. Although more complicated to derive, the variance of the product limit estimator can be shown to be (Greenwood, 1926)

$$\text{var}\{\widehat{S(t_k)}\} = \widehat{S(t_k)}^2 \sum_{i=0}^{k-1} \frac{d_i}{n_i(n_i - d_i)}. \tag{16.11}$$

TABLE 16.7 Product Limit Estimates of Survival for Data from an SA Mesothelioma Clinical Trial (All Patients)

Event Time t_i	Number of Events n_i	Number Alive d_i	Survival Probability $\widehat{S(t_i)}$	Failure Probability p_i	Survival Std. Err.
0.0	1	83	1.0000	0	0
0.0	1	82	0.9880	0.0120	0.0120
4.0	1	81	0.9759	0.0241	0.0168
6.0	1	80	0.9639	0.0361	0.0205
17.0	1	79	0.9518	0.0482	0.0235
20.0	1	78	0.9398	0.0602	0.0261
22.0	1	77	0.9277	0.0723	0.0284
28.0	1	76	0.9157	0.0843	0.0305
⋮	⋮	⋮	⋮	⋮	⋮
764.0	1	12	0.2081	0.7919	0.0473
823.0	1	11	0.1908	0.8092	0.0464
948.0	1	10	0.1734	0.8266	0.0453
963.0	0	9	.	.	.
1029.0	0	8	.	.	.
1074.0	0	7	.	.	.
1093.0	0	6	.	.	.
1102.0	0	5	.	.	.
1123.0	0	4	.	.	.
1170.0	0	3	.	.	.
1229.0	1	2	0.1156	0.8844	0.0560
1265.0	1	1	0.0578	0.9422	0.0496
1338.0	0	0	.	.	.

The square roots of these numbers are shown in the last column of Table 16.7. Usually one plots $\widehat{S(t)}$ versus time to obtain familiar "survival curves," which can be done separately for two or more groups to facilitate comparisons. For the mesothelioma data, such curves are shown in Figure 16.4.

Investigators are frequently interested in the estimated probability of survival at a fixed time, which can be determined from the calculations sketched above. For example, the probability of surviving (or remaining event free) at one year is approximately 0.50 (Figure 16.4). When this estimate is based on a lifetable or product limit calculation, it is often called "actuarial survival." Sometimes clinicians discuss "actual survival," a vague and inconsistently used term. It usually means the raw proportion surviving, for example $25/50 = 0.5$ or 50% at one year in the data above.

16.3.4 Parametric (Exponential) Summaries of Survival Are Efficient

To discuss nonparametric estimates of survival quantitatively, it is necessary to agree on a reference point in time or have the entire survival curve available. This inconvenience can often be avoided by using a parametric summary of the data, for example, by calculating the overall failure rate or hazard. If we assume the failure rate is constant over time (i.e., the failure times arise from an exponential distribution), the hazard can

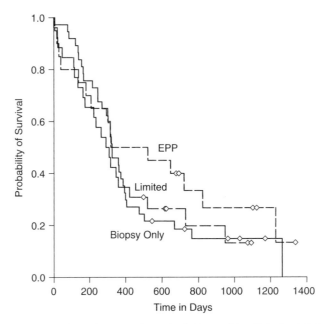

Figure 16.4 Nonparametric estimates of survival for a SA clinical trial in patients with mesothelioma.

be estimated by

$$\widehat{\lambda} = \frac{d}{\sum_{i=1}^{N}} t_i, \tag{16.12}$$

where d is the total number of failures and the denominator is the total follow-up or exposure time in the cohort. This estimate of the hazard was introduced in Chapter 11. It is the event rate per person-time (e.g., person-year or person-month) of exposure and summarizes the entire survival experience. More complicated procedures are necessary if the hazard is not constant over time or such an assumption is not helpful.

Because $2d\lambda/\widehat{\lambda}$ has a chi-square distribution with $2d$ degrees of freedom (Halperin, 1952), a $100(1 - \alpha)\%$ confidence interval for λ is

$$\frac{\widehat{\lambda}\chi^2_{2d,1-\alpha/2}}{2d} < \lambda < \frac{\widehat{\lambda}\chi^2_{2d,\alpha/2}}{2d}.$$

When the sample size is large, $\widehat{\lambda}$ has an approximate normal distribution with mean λ and variance $\lambda^2/(d-1)$. Then an approximate confidence interval is

$$\widehat{\lambda} - \frac{\widehat{\lambda}Z_{\alpha/2}}{\sqrt{d-1}} < \lambda < \widehat{\lambda} + \frac{\widehat{\lambda}Z_{\alpha/2}}{\sqrt{d-1}}.$$

Approximate confidence limits for the failure rate can also be calculated in the following way. It is easier to put confidence limits on the logarithm of the failure rate. In particular,

$$\mathrm{var}\{\log(\widehat{\lambda})\} = \frac{1}{d}$$

TABLE 16.8   Hazard Summary Data from a SA Mesothelioma Trial

Variable	Group	Exposure Time[a]	Number of Events	Hazard Rate	Approximate 95% CL
Overall	All	96.3	68	0.706	0.557–0.895
Sex	Males	74.8	52	0.695	0.530–0.912
	Females	21.5	16	0.743	0.455–1.213
PS	0	55.7	46	0.826	0.617–1.103
	1	40.6	22	0.541	0.356–0.822
Surgery	Biopsy	41.2	32	0.776	0.549–1.098
	Limited	26.5	21	0.792	0.516–1.215
	EPP	28.6	15	0.524	0.316–0.870

[a] Exposure time is measured in years.

so that an approximate $(1 - \alpha)\%$ confidence interval is

$$\log(\widehat{\lambda}) \pm \frac{Z_{1-\alpha}}{\sqrt{d}}, \tag{16.13}$$

where $Z_{1-\alpha}$ is the normal probability quantile for the width of the confidence interval (e.g., $Z_{0.95} = 1.96$). The ends of the confidence interval on $\log(\widehat{\lambda})$ can be transformed back to the natural scale to produce the desired confidence limits. These simple parametric methods have been used on the data from Table 16.3 and the results are given in Table 16.8.

In many circumstances one cannot reasonably assume that the hazard is constant over time. Certain parametric methods can still be employed if a good choice for an overall survival model is made or if there are intervals or epochs in which the hazard is approximately constant. Parametric methods can be very useful when studying subtle features of the data, for example, the tail of a survival distribution. In many cases specialized computer software or methods of estimating model parameters are needed. A detailed discussion of this topic is given by Gross and Clark (1980). Large or complex SA trials will likely employ descriptive or analytic methods similar to those given in the next section for comparative studies or in Chapter 17 for prognostic factor studies.

16.4   COMPARATIVE EFFICACY TRIALS (PHASE III)

Developmental studies like DF and SA trials primarily require descriptive analysis. Comparative trials usually require analyses that describe the data, quantify the possible effects of chance on the observed treatment difference, and assess extraneous influences. A frequentist approach to the question of the effects of chance might be to test for "statistical significance," meaning formally determine to what extent the observed differences between the treatment groups could be attributable to random variation. A Bayesian approach would be to derive a probability distribution for the treatment difference based on both prior information and the results of the trial. A likelihood approach would quantify the relative evidence for the observed treatment effect.

Questions of *clinical* significance depend partly on the outcomes being used and many considerations from outside the trial. There is not necessarily a correspondence between the objectivity of an outcome and the strength of its clinical significance. For example, most trialists and practitioners would agree that an improvement in length of survival is clinically significant in almost all circumstances. Survival is perhaps the most objective outcome. However, many patient reported outcomes (e.g., quality of life, pain relief, symptom control) are also clinically significant, but rank among the most subjective outcomes that are used. It is a major challenge in comparative trials to design and analyze studies in a way that uses clinically important outcomes without damaging their utility because of statistical or other methodologic weaknesses.

Outcomes employed most frequently in comparative trials can be classified as continuous measures, binary outcomes, or event times. Each of these can be used to estimate absolute or relative effects and assess those effects for the play of chance. Most often continuous measures are used to estimate treatment differences as opposed to risk ratios. Binary outcomes and event rates are measures of risk and are most frequently used in comparative studies to estimate risk ratios.

16.4.1   Examples of CTE Trials Used in This Section

To discuss points of analysis about CTE trials, it is useful to have examples. I have selected two small but interesting trials that illustrate relevant concepts. One is a randomized trial testing whether or not a nonsteroidal antiinflammatory agent (sulindac) can reduce the number and size of colonic polyps in patients with Familial Adenomatous Polyposis (FAP) (Giardiello et al., 1993). A second useful example, in which survival and disease recurrence are the primary statistical outcomes, is a randomized clinical trial comparing the benefit of cytoxan, doxorubicin, and platinum (CAP) as an adjuvant to radiotherapy for treatment of locally advanced non–small cell lung cancer following surgical resection (Lad, Rubinstein, Sadeghi, et al., 1988).

FAP Prevention Trial
FAP is an autosomal dominant genetic defect that predisposes those affected to develop large numbers of polyps in the colon, frequently at a young age. These polyps are prone to malignant changes and, if untreated, affected individuals may develop colon cancer. In addition to frequent screening and biopsy of polyps that form, patients may require surgical procedures such as colectomy to prevent the formation of cancers. Serendipitous observations suggested that the drug employed in this clinical trial might benefit FAP patients. (This illustrates the role that observational studies can have in the early development of disease prevention strategies.) Sulindac and related drugs inhibit the enzyme cyclo-oxygenase and may act beneficially to reduce polyp formation by enhancing pre-programmed cell death (apoptosis) as opposed to reducing proliferative rates.

In this trial patients with FAP were randomly assigned to receive the drug as a preventive measure or a placebo. The primary outcome was to examine the number and size of colonic polyps, as determined by colonoscopy, at 3, 6, 9, and 12 months after starting treatment. The analysis originally performed used relative changes over baseline in polyp number and size as the clinically relevant outcomes. This study was stopped at a planned interim analysis after approximately one-half of the fixed sample

TABLE 16.9 Data from a CTE Clinical Trial in Familial Adenomatous Polyposis

ID	Sex	Polyp Number at Month		Polyp Size at Month		Age	Treatment
		0	12	0	12		
1	0	7	—	3.6	—	17	1
2	0	77	—	3.8	—	20	0
3	1	7	4	5.0	1.0	16	1
4	0	5	26	3.4	2.1	18	0
5	1	23	16	3.0	1.2	22	1
6	0	35	40	4.2	4.1	13	0
7	0	11	14	2.2	3.3	23	1
8	1	12	16	2.0	3.0	34	0
9	1	7	11	4.2	2.5	50	0
10	1	318	434	4.8	4.4	19	0
11	1	160	26	5.5	3.5	17	1
12	0	8	7	1.7	0.8	23	1
13	1	20	45	2.5	3.0	22	0
14	1	11	32	2.3	2.7	30	0
15	1	24	80	2.4	2.7	27	0
16	1	34	34	3.0	4.2	23	1
17	0	54	38	4.0	2.9	22	0
18	1	16	—	1.8	—	13	1
21	1	30	57	3.2	3.7	34	0
22	0	10	7	3.0	1.1	23	1
23	0	20	1	4.0	0.4	22	1
24	1	12	8	2.8	1.0	42	1

size was accrued because the clinicians and trial methodologists felt that the evidence was ethically compelling in favor of treatment. Here we consider only the one-year data (Table 16.9). A complete listing of data is available as described in Appendix A.

Lung Cancer Chemotherapy Trial
This clinical trial tested the benefit of adding adjuvant chemotherapy to patients with surgically resected lung cancer. A total of 172 patients were randomized between 1979 and 1985. The study illustrates an interesting collaboration between investigators applying the major modes of cancer treatment: radiotherapy, surgery, and chemotherapy. It was among the first to show the benefit of platinum based chemotherapy for non–small cell lung cancer.

In this trial patients were followed for several years and substantial amounts of data were collected. Here we can only consider a portion of it, related to the primary outcomes and major prognostic factors. The data and analyses here differ slightly from the primary publication cited above because of additional follow-up information collected subsequently. However, the basic conclusions remain as previously reported. A data listing is given in Appendix A. The methodology of this trial is noteworthy because of the careful intraoperative anatomic staging of the patients, a process that improves knowledge of the extent of disease and permits more accurate prognostication.

16.4.2   Continuous Measures Estimate Treatment Differences

Continuous measures on each study subject are most often summarized as group averages or medians. A trial of blood pressure reducing drugs might compare the average diastolic blood pressure on each treatment. The actual level of the diastolic blood pressure might be less interesting than the differences between the groups. For example, a trial of human growth hormone in children with a deficiency as manifested by extremely short stature might compare the average height of the treatment groups. Here one would likely be interested in the group differences as well as the actual average heights in the treatment groups.

For these types of continuously distributed response measurements, an estimate of the effect of clinical interest might be $\widehat{\Delta}_{AB} = \overline{Y}_A - \overline{Y}_B$, where $\overline{Y}_A$ and $\overline{Y}_B$ are the average outcomes in the treatment groups. Not only are we interested in Δ_{AB}, the magnitude of the treatment difference, but we would also like to estimate its variability through a standard deviation or confidence interval. Although a p-value or other quantification of the play of chance will be interesting, it is likely that we would continue to study the treatment if $\widehat{\Delta}_{AB}$ is large enough, even if it is not "significantly different from zero."

In the FAP clinical trial the average numbers of polyps after 12 months of treatment were 78 and 13 in the placebo and treatment groups, respectively. The average polyp sizes were 3.1 and 1.8 in the placebo and treatment groups, respectively. The difference in number of polyps between the treatment groups is not statistically significant using the t-test ($p = 0.15$) but the polyp size is significantly decreased ($p = 0.02$). Although the t-test applied to 12-month measurements is a valid procedure for testing the treatment effect based on the design of the trial, it does not use all of the available information from the study. Both baseline measurements and longitudinal assessments of polyp number and size are available and could contribute to information about the treatment effect. These are discussed below.

16.4.3   Baseline Measurements Can Increase Precision

There are many instances where we measure outcomes both at the beginning (baseline) and end of treatment on each patient. This occurs often in behavioral research, for example. In these circumstances we could still estimate $\widehat{\Delta}_{AB}$ using only the values at end of treatment, as in the FAP analysis above. However, it might be better to calculate each patient's *increment* over baseline after the treatment period, $\delta_i = Y_{i2} - Y_{i1}$, where the subscript i denotes patient and 1 and 2 denote baseline and end of treatment, respectively. Using the δ_i is sometimes called a gain-score analysis. The average increments in the treatment groups are $\overline{\delta}_A$ and $\overline{\delta}_B$, so the effect of most clinical interest might be $\widehat{\Delta}_{AB} = \overline{\delta}_A - \overline{\delta}_B$, which is the difference of increments.

This approach, or one similar to it, is often more efficient than using only a single post-treatment measurement because the standard deviation of $\widehat{\Delta}_{AB}$ is probably reduced as the result of using two measurements from each study subject. Suppose that the variance of each of the measurements, Y_{ij}, is σ^2. Then

$$\text{var}(\widehat{\Delta}_{AB}) = \text{var}(\overline{\delta}_A) + \text{var}(\overline{\delta}_B)$$

and

$$\text{var}(\overline{\delta}_A) = \frac{\text{var}(\delta_i)}{n} = \frac{\sigma^2 + \sigma^2 - 2 \times \text{cov}(Y_{i1}, Y_{i2})}{n} = 2\frac{\sigma^2}{n}(1 - \rho), \qquad (16.14)$$

where ρ is the within-subject correlation between baseline and post-treatment measurements. The covariance (correlation) of Y_{i2} and Y_{i1} is most likely positive (i.e., $\rho > 0$), reducing the variance of $\bar{\delta}_A$. The same applies to $\bar{\delta}_B$.

Example

Using the gain-score approach on the data from the FAP clinical trial, the average difference in the number of polyps between baseline and 12 months were 26.3 and -18.7 in the placebo and treatment groups, respectively. That is, on placebo, the number of polyps tended to increase, whereas treatment with sulindac decreased the number of polyps. This difference was statistically significant ($p = 0.03$). For polyp size, the average change on placebo was -0.19 and on treatment was -1.52. This difference was also statistically significant ($p = 0.05$).

Analysis of Covariance

An alternative approach using baseline information, which is probably better than a gain-score analysis (Laird, 1983), is an analysis of covariance (ANCOVA). This method employs the baseline measurements as covariates, which are used to adjust the end of treatment values using a linear regression model. In a simple case the model takes the form

$$Y_{i2} = \beta_0 + \beta_1 Y_{i1} + \beta_2 T_i + \epsilon_i, \tag{16.15}$$

where T_i is the treatment indicator variable for the ith person. β_2 is the treatment effect and the focus of interest for the analysis. One can show that the relative efficiency of the gain-score versus the linear model, as measured by the ratio of variances, is $2/(1 + \rho)$ in favor of the ANCOVA (Bock, 1975). In any case, proper use of additional measurements on the study subjects can improve the precision or efficiency of the trial.

Example

An analysis of covariance for the FAP clinical trial is shown in Table 16.10. The linear model used is as given above with the variables coded, as in Table 16.9. The interpretation of the parameter estimates depends on the coding and on which terms are included in the model. Here β_0 is the sulindac effect, β_1 is the effect of the baseline measurement, and β_2 is the difference between the treatments. For polyp number the first model is equivalent to the t-test above: the average number of polyps at 12 months is 13.0 on sulindac and the average *difference* between the treatments is 64.9 polyps ($p = 0.15$).

The second model shows that the number of polyps at 12 months depends strongly on the number at baseline ($p = 0.0001$) and that the difference between the treatments after accounting for the baseline number of polyps is 43.2, which is significant ($p = 0.04$). The results for polyp size are different, showing that the average polyp size at 12 months in the sulindac group is 1.83 and the *difference* in size of 1.28 attributable to the treatment is significant ($p = 0.02$). Also the effect of treatment on polyp size does not depend on the baseline size.

16.4.4   Nonparametric Survival Comparisons

Above, I discussed the utility of nonparametric estimates of event time distributions. These are useful summaries because they are relatively simple to calculate, can be generalized to more than one group, can be applied to interval grouped data, and require no assumptions about the distribution giving rise to the data. Methods that

TABLE 16.10   Analyses of Covariance for the Familial Adenomatous Polyposis Clinical Trial

Dependent Variable	Model Terms	Parameter Estimate	Standard Error	P-Value
Polyp number	β_0	13.0	30.8	—
	β_2	64.9	42.5	0.15
Polyp number	β_0	−21.5	14.2	—
	β_1	1.1	0.1	0.0001
	β_2	43.2	18.9	0.04
Polyp size	β_0	1.83	0.37	—
	β_2	1.28	0.51	0.02
Polyp size	β_0	1.13	0.90	—
	β_1	0.21	0.24	0.40
	β_2	1.29	0.51	0.02

Note: All dependent variables are measured at 12 months.

share some of these properties are widely used to *compare* event time distributions in the presence of censored data. One of the most common methods is the logrank statistic (Mantel and Haenszel, 1959). To understand the workings of this statistic, consider a simple two-group (A versus B) comparison of the type that might arise in a randomized clinical trial.

As in the product limit method discussed above, the data are sorted by the event time from smallest to largest. At each failure time, a 2×2 table can be formed:

		Status:	
		Event	No Event
Group:	A	d_{iA}	$n_{iA} - d_{iA}$
	B	d_{iB}	$n_{iB} - d_{iB}$

where i indexes the failure times, the d's represent the numbers of events in the groups, the n's represent the number of subjects at risk, and either d_{iA} or d_{iB} is 1 and the other is 0 if all of the event times are unique. These tables can be combined over all failure times in the same way that 2×2 tables are combined across strata in some epidemiologic studies. In particular, we can calculate an overall "observed minus expected" statistic for group A (or B) as a test of the null hypothesis of equal event rates in the groups. This yields

$$O_A - E_A = \sum_{i=1}^{N} \frac{n_{iA} d_{iB} - n_{iB} d_{iA}}{n_i}, \tag{16.16}$$

where $n_i = n_{iA} + n_{iB}$ and $d_i = d_{iA} + d_{iB}$. The variance can be shown to be

$$V_A = \sum_{i=1}^{N} \frac{d_i (n_i - d_i) n_{iA} n_{iB}}{(n_i - 1) n_i^2}. \tag{16.17}$$

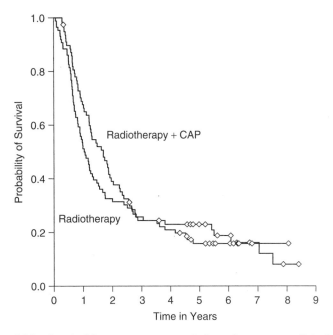

Figure 16.5 Survival by treatment group during a lung cancer clinical trial.

Then the test statistic can be calculated as

$$Z = \frac{O_A - E_A}{\sqrt{V_A}}.$$

Z will have a standard normal distribution under the null hypothesis.

Example

For the lung cancer clinical trial introduced above, investigators were interested in testing the difference in disease recurrence rates and survival on the two treatment arms (Figures 16.5 and 16.6). When eligible patients are analyzed, the logrank statistic for survival calculated from equations (16.16) and (16.17) is 1.29 ($p = 0.26$) and for recurrence is 9.18 ($p = 0.002$).

16.4.5 Risk (Hazard) Ratios and Confidence Intervals Are Clinically Useful Data Summaries

Although nonparametric methods of describing and comparing event time distributions yield robust ways of assessing statistical significance, they may not provide a concise clinically interpretable summary of treatment effects. For example, product-limit estimates of event times require us to view the entire recurrence curves in Figure 16.6 to have a sense of the magnitude of benefit from CAP. The problem of how to express clinical differences in event times concisely can be lessened by using hazard ratios (and confidence intervals) as summaries. These were introduced in the power and sample size equations in Chapter 11. The ratio of hazards between the treatment groups can be considered a partially parametric summary because it shares characteristics of both

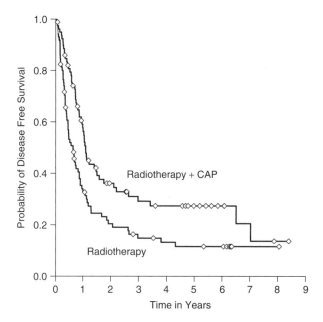

Figure 16.6 Disease-free survival by treatment group during a lung cancer clinical trial.

parametric and nonparametric statistics. The hazard ratio is usually assumed to be constant over the course of follow-up, which is a parametric assumption. However, the ratio does not depend on the actual magnitude of the event times, only on their ranking. This is typical of nonparametric methods.

The hazard ratio is a useful descriptor because it summarizes the magnitude of the treatment difference in a single number. Hazard ratios that deviate from 1.0 indicate increasing or decreasing risk depending on the numerical coding of the variable. It is also relatively easy to specify the precision of the hazard ratio using confidence intervals. Assuming a constant hazard ratio is likely to be at least approximately correct for the period of observation, even in many situations where the hazards are changing with time. In other words, the ratio may remain constant even though the baseline risk fluctuates. A fixed ratio is often useful, even when it is not constant over time. Furthermore the ratio has an interpretation in terms of relative risk and is connected to the odds ratio in fundamental ways. Finally the effects of both categorical and continuous prognostic factors can usually be expressed in the form of a hazard ratio, making it widely applicable.

Confidence intervals are probability statements about an estimate and not about the true treatment effect. For example, suppose that the true hazard ratio has exactly the value we have estimated in our study. Then a 95% confidence interval indicates the region in which 95% of hazard ratio estimates would fall if we repeated the experiment. Informally, a confidence interval is a region in which we are confident that a true treatment effect or difference lies. Although incorrect, this notion is not too misleading. The value of confidence intervals is that they convey a sense of the precision with which an effect is estimated.

As indicated in Chapter 11, the estimated hazard from exponentially distributed event times has a chi-square distribution with $2d$ degrees of freedom. A ratio of chi-square random variables the hazard ratio—has an F distribution with $2d_1$ and $2d_2$ degrees of freedom (Cox, 1953). Therefore a $100(1 - \alpha)\%$ confidence interval for $\Delta = \lambda_1/\lambda_2$ is

$$\widehat{\Delta} F_{2d_1, 2d_2, 1-\alpha/2} < \Delta < \widehat{\Delta} F_{2d_1, 2d_2, \alpha/2}.$$

Here again, the calculations can be made more simple by using the approximate normality of $\log(\Delta)$.

Example

In the CAP lung cancer clinical trial we can summarize the survival difference by saying that the hazard ratio is 1.22 with a 95% confidence interval of 0.87 to 1.70. The fact that the hazard ratio is near 1.0 tells us that the treatment offers little overall benefit on survival. The fact that this confidence interval includes 1.0 tells us that, even accounting for the minimal improvement in survival, the difference is not statistically significant at conventional levels. The hazard ratio for disease free survival is 1.72 with 95% confidence interval 1.21 to 2.47. Thus the benefit of CAP for recurrence is clinically sizeable and statistically significant at conventional levels.

16.4.6 Statistical Models Are Helpful Tools

Statistical models are extremely helpful devices for making estimates of treatment effects, testing hypotheses about those effects, and studying the simultaneous influence of covariates on outcome (Section 17.3). All models make assumptions about the data. Commonly used survival or relative risk models can be parametric, in which case they assume that some specific distribution is the source of the event times, or partially parametric, in which case some assumptions are relaxed. Models cannot be totally nonparametric because this would be something of an oxymoron.

Here we view a small set of models principally as devices to facilitate estimating and comparing hazard ratios. The proportional hazards (PH) model (Cox, 1972) is probably the best-known and most useful device. The essential feature of the model is that time, t, and the covariate vector (predictor variables), $\mathbf{X}$, enter the hazard function, $\lambda(t; \mathbf{X})$, in a way that conveniently factors,

$$\lambda(t; \mathbf{X}) = \lambda_0(t) e^{\beta \mathbf{X}},$$

where $\lambda_0(t)$ is the baseline hazard and β is a vector of regression coefficients to be estimated from the data. In other words, for an individual characterized by $\mathbf{X}$, the ratio of their hazard to the baseline hazard is $e^{\beta \mathbf{X}}$. We could write this relationship as

$$\log \left\{ \frac{\lambda(t)}{\lambda_0(t)} \right\} = \beta \mathbf{X} = \beta_1 X_1 + \beta_2 X_2 + \cdots$$

to reveal its similarity to other covariate models. Thus β is a vector of log hazard ratios. The model assumes that the hazard ratio is constant over time, and the covariates are also assumed to be constant over time. The effect of covariates is to multiply the baseline hazard, even for $t = 0$.

Estimating β is technically complex and not the subject of primary interest here. Computer programs for parameter estimation are widely available, enabling us to focus directly on the results. One advantage of the PH model is that the estimation can be stratified on factors that we have no need to model. For example, we could account for the effects of a risk factor by defining strata based on its levels, estimating the treatment effect separately within each level of the factor, and pooling the estimated hazard ratios over all strata. In this way we can account for the effects of the factor without having to assume that its hazard is proportional because it is not entered into the model as a covariate.

When interpreting β, we must remember that it represents the hazard ratio per unit change in the predictor variable. If the predictor is a dichotomous factor such as an indicator variable for treatment group, then a unit change in the variable simply compares the groups. However, if the variable is measured on a continuous scale such as age, then the estimated hazard ratio is per year of age (or other unit of measurement). For variables that are measured on a continuous scale, the hazard ratio associated with an n-unit change is Δ^n, where Δ is the hazard ratio. For example, if age yields a hazard ratio of 1.02 per year increase, then a 10-year increase will have a hazard ratio of $1.02^{10} = 1.22$.

For the CAP lung cancer trial, estimated hazard ratios and 95% confidence limits for predictor variables calculated from the PH model are shown in Table 16.11. Because of differences in the exact method of calculation, these might be slightly different from estimates that could be obtained by the methods outlined earlier in this chapter.

16.4.7   *P*-Values Do Not Measure Evidence

Not even a brief discussion of estimation and analysis methods for clinical trials would be complete without an appropriate de-emphasis of p-values as the proper currency for conveying treatment effects. There are many circumstances in which p-values are useful, particularly for hypothesis tests specified a priori. However, p-values have properties which make them poor summaries of clinical effects (Royall, 1986). In particular, p-values do not convey the magnitude of a clinical effect. The size of the p-value is a consequence of two things: the magnitude of the estimated treatment difference and its estimated variability (which is itself a consequence of sample size). Thus the p-value partially reflects the size of the experiment, which has no biological importance. The p-value also hides the size of the treatment difference, which does have major biological importance.

When faced with an estimated effect that is not statistically significant at customary levels, some investigators speculate "the effect might be statistically significant in a larger sample." This, of course, misses a fundamental point because any effect other than zero will be statistically significant in a large enough sample. What the investigators should really focus on is the size and clinical significance of an estimated treatment effect rather than its p-value. In summary, p-values only quantify the type I error and incompletely characterize the biologically important effects in the data.

One of the weaknesses of p-values as summaries of strength of evidence can be illustrated in a simple way using an example given by Walter (1995). Consider the following 2×2 tables summarizing binomial proportions:

$$
\begin{array}{c c c}
 & A & \overline{A} \\
B & 1 & 7 \\
\overline{B} & 13 & 7
\end{array}
$$

TABLE 16.11 Estimated Hazard Ratios from the Proportional Hazards Model for the CAP Lung Cancer Trial

Variable	Hazard Ratio	95% Confidence Bounds	P-Value
		Survival Results	
treat='2'	1.22	0.867–1.70	0.26
cell type='2'	1.28	0.907–1.81	0.16
karn='2'	0.84	0.505–1.40	0.51
t='2'	0.94	0.558–1.57	0.80
t='3'	0.94	0.542–1.63	0.82
n='1'	1.09	0.542–2.20	0.81
n='2'	1.26	0.691–2.30	0.45
age	1.00	0.984–1.02	0.73
sex='1'	1.09	0.745–1.58	0.67
wtloss='1'	1.09	0.602–1.98	0.78
race='1'	1.21	0.734–1.98	0.46
		Recurrence Results	
treat='2'	1.73	1.21–2.47	0.003
cell type='2'	1.68	1.16–2.43	0.006
karn='2'	0.72	0.433–1.21	0.22
t='2'	0.99	0.576–1.71	0.98
t='3'	0.89	0.495–1.60	0.70
n='1'	1.06	0.499–2.26	0.88
n='2'	1.30	0.674–2.49	0.44
age	1.00	0.980–1.02	0.96
sex='1'	0.93	0.629–1.37	0.71
wtloss='1'	1.48	0.814–2.67	0.20
race='1'	0.89	0.542–1.48	0.67

and

$$
\begin{array}{c|cc}
 & A & \overline{A} \\
\hline
B & 1 & 6 \\
\overline{B} & 13 & 6 \\
\end{array}
$$

The column proportions are the same in both tables, although the first has more data and should provide stronger evidence than the second. For both tables we can compare the proportions $\frac{1}{14}$ versus $\frac{1}{2}$ using Fisher's exact test (Agresti, 1996). Doing so yields two-sided p-values of 0.33 and 0.26, respectively. In other words, the second table is more "statistically significant" than the first, even though it provides less evidence. This outcome is a result of discreteness and asymmetry. One should probably not make too much of it except to recognize that p-values do not measure strength of evidence.

To illustrate the advantage of estimation and confidence intervals over p-values, consider the recent discussion over the prognostic effect of peri-operative blood transfusion in lung cancer (Piantadosi, 1992). Several studies (not clinical trials) of this phenomenon have been performed because of firm evidence in other malignancies and diseases that blood transfusion has a clinically important immunosuppressive effect.

TABLE 16.12   Summary of Studies Examining the Peri-operative Effect of Blood Transfusion in Lung Cancer

Study	Endpoint	Hazard Ratio	95% Confidence Limits
Tartter et al., 1984	Survival	1.99	1.09–3.64
Hyman et al., 1985	Survival	1.25	1.04–1.49
Pena et al., 1992	Survival	1.30	0.80–2.20
Keller et al., 1988	Recurrence		
	Stage I	1.24	0.67–1.81
	Stage II	1.92	0.28–3.57
Moores et al., 1989	Survival	1.57	1.14–2.16
	Recurrence	1.40	1.01–1.94

Note: All hazard ratios are transfused versus untransfused patients and are adjusted for extent of disease.

Disagreement about results of various studies has stemmed, in part, from too strong an emphasis on hypothesis tests instead of focusing on the estimated risk ratios and confidence limits. Some study results are shown in Table 16.12. Although the authors of the various reports came to different qualitative conclusions about the risk of blood transfusion because of differing p-values, the estimated risk ratios, adjusted for extent of disease, appear to be consistent across studies. Based on these results, one might be justified in concluding that peri-operative blood transfusion has a modest adverse effect on lung cancer patients. Interestingly a randomized trial of autologous versus allogeneic blood transfusion in colorectal cancer has been reported recently (Heiss et al., 1994). It showed a 3.5-fold increased risk attributable to the use of allogeneic blood transfusion with a p-value of 0.10.

16.5   STRENGTH OF EVIDENCE THROUGH SUPPORT INTERVALS

16.5.1   Support Intervals Are Based on the Likelihood Function

Strength of evidence can be measured using the likelihood function (Edwards, 1972; Royall, 1997). We can obtain a measure of the relative strength of evidence in favor of our best estimate versus a hypothetical value using the ratio of the likelihoods evaluated with each parameter value. Support intervals are quantitative regions based on likelihood ratios that quantify the strength of evidence in favor of particular values of a parameter of interest. They summarize parameter values consistent with the evidence without using hypothesis tests and confidence intervals. Because they are based on a likelihood function, support intervals are conditional on, or assume, a particular model of the data. Like confidence intervals, they characterize a range of values that are consistent with the observed data. However, support intervals are based on values of the likelihood ratio rather than on control of the type I error.

The likelihood function, $L(\theta|\mathbf{x})$, depends on the observed data, $\mathbf{x}$, and the parameter of the model, θ (which could be a vector of parameters). We can view the likelihood as a function of the unknown parameter, conditional on the observed data. Suppose that our best estimate of the unknown parameter is $\widehat{\theta}$. Usually this will be the "maximum likelihood estimate." We are interested in the set of all values of θ that are consistent with $\widehat{\theta}$, according to criteria based on the likelihood ratio. We could, for example, say that the data (evidence) support any θ for which the likelihood ratio relative to $\widehat{\theta}$ is less than R. When the likelihood ratio for some θ exceeds R, it is not supported by the data. This defines a support interval or a range of values for θ which are consistent with $\widehat{\theta}$. An example is given below.

16.5.2 Support Intervals Can Be Used with Any Outcome

Support intervals can be constructed from any likelihood using any outcome. If the purpose of our clinical trial is to estimate the hazard ratio on two treatments, we could employ a simple exponential failure time model. Let i denote an arbitrary study subject. We define a binary covariate, X_i, which equals 1 for treatment group A and 0 for treatment group B. The survival function is $S(t_i) = e^{-\lambda_i t_i}$ and the hazard function, λ_i, is assumed to be constant. Usually we model the hazard as a multiplicative function of covariates, $\lambda_i = e^{\beta_0 + \beta_1 X_i}$ (or $\lambda_i = e^{-\beta_0 - \beta_1 X_i}$), where β_0 and β_1 are parameters to be estimated from the data, i.e., $\theta' = \{\beta_0, \beta_1\}$. The hazard for a person on treatment A is $\lambda_A = e^{\beta_0 + \beta_1}$ and for a person on treatment B is $\lambda_B = e^{\beta_0}$. The hazard ratio is

$$\Delta_{AB} = e^{\beta_0 + \beta_1} / e^{\beta_0} = e^{\beta_1}.$$

Thus β_1 is the log hazard ratio for the treatment effect of interest. (β_0 is a baseline log hazard, which is unimportant for the present purposes.)

To account for censoring, we define an indicator variable Z_i that equals 1 if the ith person is observed to have an event and 0 if the ith person is censored. The exponential likelihood is

$$L(\beta_0, \beta_1 \mid \mathbf{x}) = \prod_{i=1}^{N} e^{-\lambda_i t_i} \lambda_i^{Z_i} = \prod_{i=1}^{N} e^{-t_i e^{\beta_0 + \beta_1 X_i}} (e^{\beta_0 + \beta_1 X_i})^{Z_i}.$$

If $\widehat{\beta}_0$ and $\widehat{\beta}_1$ are the MLEs for β_0 and β_1, a support interval for $\widehat{\beta}_1$ is defined by the values of ζ that satisfy

$$
\begin{aligned}
R &> \frac{\prod_{i=1}^{N} e^{-t(_i e^{\widehat{\beta}_0 + \widehat{\beta}_1 X_i})} (e^{\widehat{\beta}_0 + \widehat{\beta}_1 X_i})^{Z_i}}{\prod_{i=1}^{N} e^{-t_i (e^{\widehat{\beta}_0 + \zeta X_i})} (e^{\widehat{\beta}_0 + \zeta X_i})^{Z_i}} \qquad (16.18) \\[2mm]
&= \frac{e^{-\sum_{i=1}^{N} t(_i e^{\widehat{\beta}_0 + \widehat{\beta}_1 X_i})}}{e^{-\sum_{i=1}^{N} t(_i e^{\widehat{\beta}_0 + \zeta X_i})}} \prod_{i=1}^{N} \left(\frac{e^{\widehat{\beta}_1 X_i}}{e^{\zeta X_i}} \right)^{Z_i} \\[2mm]
&= e^{-\sum_{i=1}^{N} t_i (e^{\widehat{\beta}_0 + \widehat{\beta}_1 X_i} - e^{\widehat{\beta}_0 + \zeta X_i})} \prod_{i=1}^{N} (e^{(\widehat{\beta}_1 - \zeta) X_i})^{Z_i}.
\end{aligned}
$$

Example

Reconsider the CTE lung cancer trial for which the estimated hazard ratio for disease-free survival favored treatment with CAP chemotherapy. The log hazard ratio estimated

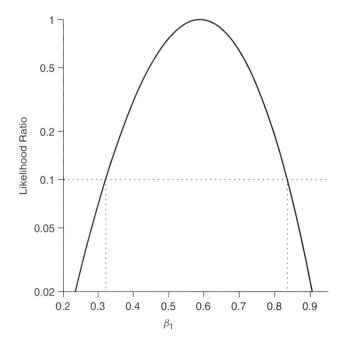

Figure 16.7   Support interval for the estimated hazard ratio for disease free survival in a lung cancer trial.

using the exponential model is $(\widehat{\beta}_1 =)$ 0.589 in favor of the CAP group. With some software for fitting this model, it may be necessary to re-code the treatment group indicator variable as 0 versus 1, rather than 1 versus 2. Note that $e^{0.589} = 1.80$, which is similar to the hazard ratio estimated above from the PH model. The estimated baseline hazard depends on the time scale and is unimportant for the present purposes. However, from the exponential model, $\widehat{\beta}_0 = -6.92$.

Using these parameter estimates and applying equation (16.18) to the data, we can calculate support for different values of β_1 (Figure 16.7). The vertical axis is the likelihood ratio relative to the MLE and the horizontal axis is the value of β_1. For these data, values of β_1 between 0.322 and 0.835 fall within an interval defined by $R = 10$ and are thus strongly supported by the data. This interval corresponds to hazard ratios of 1.38 to 2.30. The 95% confidence interval for β_1 is 0.234 –0.944, which corresponds approximately to a support interval with $R = 12$.

16.6   SPECIAL METHODS OF ANALYSIS

Some biological questions, whether on the underlying disease, method of treatment, or the structure of the data, require an analytic plan that is more complicated than the examples given above. This tends to happen more in nonexperimental studies because investigators do not always have control over how the data are collected. In true experiments it is often possible to measure outcomes in such a way that simple analyses are sufficient. Even so, special methods of analysis may be needed to address specific clinical questions or goals of the trial. Some examples of situations that may require special or more sophisticated analytic methods are correlated responses, such as those from repeated measurements on the same individual over time or clustering of study subjects, pairing of responses such as event times or binary responses, covariates that change their values over time, measurement errors in independent variables, and accounting for restricted randomization schemes. Each of these can require generalizations of commonly used analytic methods to fully utilize the data.

Sometimes the hardest part of being in these situations is recognizing that ordinary or simple approaches to an analysis are deficient in one important way or another. Even after recognizing the problem and a solution, computer software to carry out the analyses may not be readily available. All of these issues indicate the usefulness of consulting a statistical methodologist during study design and again early in the analysis.

16.6.1 The Bootstrap Is Based on Resampling

Often one of the most difficult tasks for the biostatistician is to determine how precise an estimate is. Stated more formally, determining the variance is usually more difficult than determining the estimate itself. For statistical models with estimation methods like maximum likelihood, there are fairly simple and reliable ways of calculating the approximate variance of an estimate. Sometimes, however, either the assumptions underlying the approximation are invalid or standard methods are not available. One example is placing confidence limits on the estimate of a median from a distribution.

In situations like this, one simple and reliable way to approximate the variance is to use a resampling method called the *bootstrap* (Efron and Tibshirani, 1986, 1993). In the bootstrap the observed data are resampled and the point estimate or other statistic of interest is calculated from each sample. The process is repeated a large number of times so that a distribution of possible point estimates is built up. This distribution, characterized by ordinary means, serves as a measure of the precision of the estimate. The sample at each step is taken "with replacement," meaning any particular datum can be chosen more than once in the bootstrap sample.

Like randomization distributions there are a large number of possible bootstrap samples. For example, suppose that there are N observed values in the data and each bootstrap sample consists of M values. Then there are N^M possible bootstrap samples. Some of these may not be distinguishable because of duplicate observations or sample points. Nevertheless, there are a large number of samples, in general, so complete enumeration of the sample space is difficult. For this reason the bootstrap distribution is usually approximated from a random sample of the N^M possibilities. A simple example should illustrate the procedure.

Suppose that we have a sample of 50 event times and we are interested in placing confidence intervals on our estimate of the median failure time. The observed data plotted as a survival curve are shown in Figure 16.8. There are no censored observations in this example. Standard lifetable methods yield a 95% confidence interval for the

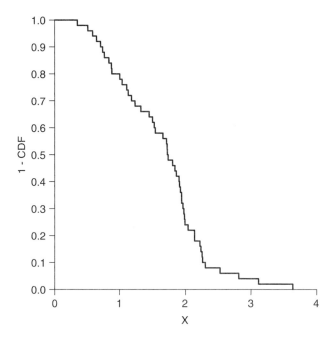

Figure 16.8   Survival distribution for bootstrap example.

median event time of (1.50–1.92). Samples of size 50 were taken with replacement and the bootstrap procedure was repeated 500 times. The medians and cumulative distribution from the bootstrap resamples actually obtained are shown in Table 16.13. Ninety-five percent of the bootstrap medians fall between 1.49 and 1.92, in excellent agreement with standard methods. Calculations such as these are greatly facilitated by computer programs dedicated to the purpose. One example of a general resampling program is *Resampling Stats* (Simon, 2004).

Bootstrap methods can be used to validate modeling procedures. An example of its use in pharmacodynamic modeling is given by Mick and Ratain (1994).

16.6.2   Some Clinical Questions Require Other Special Methods of Analysis

A book such as this can do little more than mention a few of the important statistical methods that are occasionally required to help analyze data from clinical trials. Here I briefly discuss some special methods that are likely to be important to the student of clinical trials. Most of these represent areas of active biostatistical research. For many of the situations discussed below, there are no widely applicable guidelines about the best statistical summary of the data to answer clinical questions. This results from the lower frequency with which some of these situations arise, and their greater complexity.

Longitudinal Measurements
Because most clinical trials involve or require observation of patients over a period of time following treatment, investigators often record longitudinal assessments of

TABLE 16.13   Cumulative Distribution of Medians Obtained from Bootstrap Simulation

Median	Number	Cumulative %
1.254	1	0.2
1.280	1	0.4
1.342	1	0.6
1.385	1	0.8
1.411	1	1.0
1.448	3	1.6
1.473	1	1.8
1.486	2	2.2
1.494	2	2.6
1.498	9	4.4
1.511	2	4.8
⋮	⋮	⋮
1.912	3	95.2
1.915	11	97.4
1.920	3	98.0
1.921	1	98.2
1.922	1	98.4
1.924	1	98.6
1.931	2	99.0
1.931	1	99.2
1.939	1	99.4
1.940	1	99.6
1.941	1	99.8
1.974	1	100.0

outcomes, endpoints, and predictor variables. Using the additional information contained in longitudinal measurements can be difficult. For example, survival and disease progression are familiar longitudinal outcomes that require special methods of analysis. To the data analyst, the relationship between measurements taken in a longitudinal study are not all equivalent. For example, there is a difference between measurements taken within the same individual, which are correlated, and those taken from different individuals, which are usually uncorrelated or independent. Because most simple statistical approaches to analysis rely on the assumption of independent observations, coping with correlated measurements is an important methodologic issue.

When the primary data analysis tool is the linear model, such as for analyses of variance or linear regressions, correlated measurements are analyzed according to "repeated measures" or "longitudinal data" models. These types of data and models used to analyze them are increasingly used in studies of HIV and other diseases with repeated outcomes such as migraine, asthma, and seizure disorders. An in-depth discussion of statistical methods is given by Diggle, Liang, and Zeger (1994). See also Zeger and Liang (1986), Zeger, Liang, and Albert (1988), and Liang and Zeger (1986). These references also pertain to the analysis of other correlated outcomes.

Correlated Outcomes

Longitudinal measures is not the only circumstance that gives rise to correlated observations. When the experimental unit is a cluster or group, outcomes may be correlated from individual to individual. This situation can arise if families are the experimental unit, as in some disease prevention trials. For example, consider a clinical trial to assess the efficacy of treatments to eliminate *helicobacter pylori* in patients living in endemic areas. *Helicobacter pylori* has a causal association with peptic ulcer disease. After treatment with antibiotics and/or bismuth, patients may become re-infected because of environmental exposure or family contact. Therefore it may be necessary to treat families as the experimental unit, in which case the outcomes between individuals in the same family will be correlated. In situations such as this, the analysis may need to employ methods to account for this dependency.

Other circumstances can lead to correlations, such as when individuals have more than one outcome. For example, suppose that recurrent pre-malignant or malignant lesions (e.g., skin cancers, bladder polyps, and colonic polyps) are the clinical outcome of interest. We might be interested in the interval between such events, recognizing that prolongation of the between-lesion time could be the sign of an effective secondary preventive agent. Thus each individual on study could give rise to two or more event times correlated with each other because they arise from the same person. Correlated dichotomous outcomes can arise in a similar way.

Individual study subjects can yield more than one outcome as a consequence of the experimental design. This is the case for crossover designs (discussed in Chapter 20) where each study participant is intentionally given more than one treatment. Estimating the difference in treatment effects within individuals while accounting for dependency is the major issue in the use of crossover designs.

Time-Dependent Covariates

Most of the prognostic factors (covariates) measured in clinical trials are fixed at baseline (start of treatment) and do not change during follow-up. Examples are severity of disease at diagnosis, age at diagnosis, sex, race, treatment group assignment, and pathologic type or class of disease. The methods of accounting for the influence of such prognostic factors in statistical regression models typically assume that their effects are constant over time. Statistical models usually assume that the effect of treatment as a prognostic factor is immediate following administration and constant over the follow-up period.

These assumptions are inadequate to describe the effects of all prognostic factors. First, disease intensity or factors associated with it can fluctuate following any clinical landmark such as beginning of treatment. Second, long-term prognosis may be a direct consequence of disease intensity or other time-varying prognostic factors measured at an earlier time. Third, some time-varying prognostic factors may be associated with an outcome without being causal. In any case, we require statistical models that are flexible enough to account for the effects of predictor variables whose value changes over time. These type of variables are called "time-dependent covariates" (TDC). TDCs are discussed in depth by Kalbfleisch and Prentice (2002) and Marubini and Valsecchi (1995).

It is helpful to distinguish between two basic types of TDCs. The first is *external*, implying that the factor can affect the individual's prognosis but does not carry information about the event time. External TDCs can be *defined* by a particular mechanism. An example is age, which is usually regarded as a fixed covariate (age at time 0) but may be considered as defined and time varying when prognoses change with the age of the study subject. TDCs can be *ancillary*, which indicates that they arise from a process not related to the individual. A second type of TDC is *internal*, a term that implies that a process within the individual, perhaps even the disease itself, gives rise to the prognostic factor. Extent of disease and response to treatment are examples of this. Cumulative dose of drug may also be an internal TDC if it increases because the patient survives longer.

Investigators should interpret the results of analyses employing TDCs with care. For internal TDCs in a treatment trial, the therapy may determine the value of the covariate. Adjusting on the TDC in such a case can "adjust away" the effect of treatment, making the prognostic factor appear significant and the treatment appear ineffective. This is almost never the correct description of the trial outcome.

Measurement Error

Another usual assumption of most statistical modeling methods is that the predictor variables are measured without error. Although such models explicitly incorporate the effects of random error, it is usually assumed to be associated with the response measurement rather than with the predictor variables. This paradigm is clearly not applicable in all circumstances. For example, suppose that we attempt to predict the occurrence of cancer from dietary factors, such as fat, calorie, and mineral content. It is likely that the predictor variables will be subject to measurement error as a consequence of recall error and inaccuracies converting from food substances to the components of interest.

Accounting for errors in predictor variables complicates statistical models and is not usually needed when analyzing clinical trials. Nonlinear models with predictor variables subject to measurement error are discussed by Carroll, Ruppert, and Stefanski (1995).

Random versus Fixed Effects

Most widely used statistical models assume that the effects of predictor variables are fixed (nonrandom). Sometimes it makes more sense to model the influence of a predictor variable as a random effect. For example, in a multicenter study, the effect of interest may vary from institution to institution. Comparing two institutions may not be of interest. Testing the average treatment effect in different institutions also may not be of interest because of systematic differences. In such a situation investigators may be more interested in the relative size of the within and between institution variability.

A random effects model regards the study centers as a random sample of all possible institutions and accounts for variation both within and among centers. When using linear models or analyses of variance, such models are also called *variance components*. It will almost never be the case that centers participating in a trial can realistically be regarded as randomly chosen from a population of centers. However, this perspective may be useful for assessing the relative sizes of the sources of variation. An overview of random and mixed effects linear models is given by McLean, Sanders, and Stroup (1991). Such models have been applied to longitudinal data (Laird and Ware, 1982). See Taylor, Cumberland, and Sy (1994) for an interesting application to AIDS.

16.7 EXPLORATORY OR HYPOTHESIS-GENERATING ANALYSES

16.7.1 Clinical Trial Data Lend Themselves to Exploratory Analyses

The data from clinical trials can and should be used to address questions in addition to those directly related to the primary objectives of the study. Such questions may be specified in advance of the trial, but some are often suggested by investigators after seeing the data. The structure of the experiment is usually not perfectly suited to answering these questions, so the quality of evidence that results is almost always inferior to that for the primary objectives. Even so, it is scientifically important for investigators to use clinical trial data to explore new questions.

Usually investigators engage in this activity only after the primary questions have been answered, often in secondary publications or reports. Mistakes arise not because of exploratory analyses, but when the findings are represented as the primary results of the trial. It is essential to acknowledge the hypothetical nature of exploratory findings and recognize that the usual calculation of type I errors may be incorrect, especially when the data themselves suggest the hypothesis test. As a general rule, the same data should not be used both to generate a new hypothesis and to test it. Apart from the statistical pitfalls, investigators must guard against forcing the data to fit the hypotheses (data torturing) (Mills, 1993).

16.7.2 Multiple Tests Multiply Type I Errors

Data, in sufficient quantity and detail, can be made to yield nearly any effect desired by the adventuresome analyst performing hypothesis tests. They will almost certainly yield *some* effect if studied diligently.

> *There once was a biased clinician,*
> *Who rejected the wise statistician.*
> *By flogging his data*
> *With α and β,*
> *He satisfied all his ambition.*

A small thought experiment will illustrate effects of concern. Suppose that we generate N observations sampled from a Gaussian distribution with mean 0 and variance 1. With each observation we also randomly generate 100 binary indicator variables, $x_1, x_2, \ldots, x_{100}$ that can be used to assign the observation to either of two groups. We then perform 100 "group" comparisons defined by the x's, each at a specified α-level (the nominal level), such as $\alpha = 0.05$. Thus the null hypothesis is true for all comparisons. Using this procedure, we expect 5% of the tests to reject simply by chance. Of course, the type I error rate for the entire testing procedure greatly exceeds 5%. It equals $\alpha^* = 1 - (1 - \alpha)^{100} \approx 0.99$. In other words, we are virtually certain to find at least one "significant" difference based on partitioning by the x's.

What if we put aside the problem of multiplicity of tests and restrict our attention to only those differences that are large in magnitude? This corresponds to performing significance tests suggested by findings in the data. If we test only the 10 largest group differences, presumably all 5 of the expected "significant" differences will be in this group. Thus the expected type I error rate for each test will increase from the nominal 5% to $5/10 = 50\%$.

Investigators must be aware of these types of problems when performing exploratory analyses. Findings observed in this setting, even if supported by biological rationale, have a high chance of being incorrect. Independent verification is essential, perhaps through another clinical trial.

16.7.3 Kinds of Multiplicity

There are four well-described sources of multiple tests that can contribute to an increased frequency of type I errors. They are multiple endpoints in a single trial, multiple interim analyses of accumulating data, tests performed on a large number of subsets, and analyses of prognostic factors (covariates) and their interactions. We can refer to these circumstances collectively as multiplicity. An extreme of any can increase the type I error arbitrarily high, as indicated in the numerical example above. It is sometimes surprising how rapidly the error rate can increase above the nominal level with seemingly few such tests.

Reducing errors due to a multiplicity of tests requires mainly discipline, which is to say that it can be quite difficult. The best conceptual approach to control such errors is illustrated in Chapter 14, where the frequentist framework for interim analyses is discussed. This framework is designed to control type I errors at the nominal level in interim analyses by (1) planning for multiplicity in advance, (2) restricting the number of tests performed, (3) performing each test in a structured fashion at a more stringent critical value, and (4) tempering the technical recommendations with good biological judgment. These same points can be used to reduce errors when examining multiple endpoints or subset analyses. Some sensible advice on this subject in the context of cardiovascular trials is given by Parker and Naylor (2000).

Applying these ideas to subset analyses implies that we should specify such tests in advance of obtaining the data and tightly constrain the number of comparisons performed. The nominal type I error level for such tests should not be interpreted uncritically or represented as a primary finding of the trial. If it is important to control the overall error rate, each test can be restricted using some appropriate correction factor. This point deserves careful consideration because there are disparate technical opinions about it. Last, seemingly significant findings from subset analyses must be interpreted cautiously or dismissed entirely if they are the product of abusive analyses. It is too much to ask the same data to generate a new hypothesis and simultaneously provide convincing evidence supporting it.

16.7.4 Subgroup Analyses Are Error Prone

One of the easiest ways for the analysis of a clinical trial to follow an inappropriate direction occurs when investigators emphasize the findings from a particular subset of patients, especially when the results are different from the overall findings (i.e., an analysis of all randomized patients). These interesting results may be found in a particular subset of patients after an extensive search that is not based on any a priori biological hypothesis. Other times an accidental observation in a subset may suggest a difference, which is then tested and found to be "statistically significant." If the investigators have prejudices or reasons from outside the trial to believe the findings, these circumstances could lead to a fairly firmly held belief in the validity of the results. Besides type I errors, subset analyses are not protected by the randomization. As a result a small bias can become amplified.

A classic cautionary example regarding subset analyses was derived from the second International Study of Infarct Survival (ISIS-2) (ISIS-2 Collaborative Group, 1988). In that trial (see also Section 14.1.4), 12 subset analyses we performed by zodiacal sign. Although, overall, aspirin was strongly superior to placebo ($p < 0.00001$), subjects born under the Gemini and Libra signs appeared to do better with placebo ($9\% \pm 13\%$). This example has become so familiar to trialists that it has lost some of its ability to surprise and amuse. In any case, the lesson cannot be forgotten.

Another example of how subgroup analyses help create and perpetuate errors is illustrated by the enthusiasm some clinicians had in the 1980s and 1990s (and beyond) for the treatment of cancer patients with hydrazine sulfate. Hydrazine, H_2N-NH_2 (diamide or diamine), was discovered in 1887 and synthesized in 1907. However, it was not until World War II that interest developed in the compound as a rocket fuel. It is a powerful reducing agent and readily reacts with acids to form salts. Hydrazine is a known carcinogen in rodents and probably in humans (IARC, 1974). It is metabolized by N-acetylation (Colvin, 1969) with varying rapidity according to acetylator phenotypes (Weber, 1987). A condensed but well-referenced review of hydrazine sulfate in cancer treatment can be found in the National Cancer Institute's final IND report to the FDA (NCI, 1993).

In the 1970s there were suggestions that hydrazine sulfate could improve the survival of patients with advanced cancer (Gold, 1975). Investigators suggested that the mechanism of action was by normalizing glucose metabolism in patients with cachexia. Cachexia is common in cancer patients and is a sign of poor prognosis. It may be due, in part, to tumor glycolysis. Although blocking glycolysis systemically is undesirable, inhibiting gluconeogenesis might be beneficial. Hydrazine sulfate is a noncompetitive inhibitor of phosphoethanol pyruvate carboxykinase, the enzyme that catalyzes the conversion of oxaloacetate to phosphoenolpyruvate. This mechanism may explain observations that hydrazine sulfate inhibits tumor growth in animals (Gold, 1975).

Based on the possible biological actions suggested above and anecdotal case reports of human benefit, early uncontrolled trials of hydrazine sulfate were undertaken in the Soviet Union. The results were mixed. In 1990 a group of investigators reported results from a randomized clinical trial testing the effects of hydrazine sulfate in patients with advanced lung cancer (Chlebowski et al., 1990). Although no statistically significant overall effect was found, a subset analysis revealed a group of patients that appeared to benefit from the drug. The subset of patients who seemed to have improved survival after treatment with hydrazine included those with the most advanced disease. It was not made clear how many subset analyses had to be conducted to discover the truth.

The analysis and report of this study emphasized the subset findings as primary results of the trial. Criticism of the trial report, suggesting that it might represent a type I error (Piantadosi, 1990), annoyed proponents of hydrazine sulfate. Commentary on the question in the well-known scientific forum afforded by *Penthouse* magazine suggested that patients were being denied a virtual cure for cancer and that unscrupulous researchers were profiting from continued denial of hydrazine's salutary effects:

> If you ... come down with cancer or AIDS ... you will probably be denied the one drug that may offer the best possibility of an effective treatment with the least side effects. It works for roughly half of all the patients who have received it and it's being deliberately suppressed ... [A] million Americans alone are being denied lifesaving benefits each

year .... [T]he apparent sabotaging of federally funded clinical trials of the drug deny the public access to it. (Kamen, 1993)

The author of this article was said to be working on a book and a documentary film regarding hydrazine.

Scientists with no financial interests in the outcome were performing large randomized clinical trials testing the effects of hydrazine in patients with lung and colon cancer. One clinical trial was conducted by the Cancer and Leukemia Group B (CALGB) and two others were done by the North Central Cancer Treatment Group (NCCTG) based at the Mayo Clinic. These studies were all published in the same issue of the *Journal of Clinical Oncology* in 1994 (Kosty et al., 1994; Loprinzi et al., 1994a, b). All of these studies showed hydrazine sulfate to be no better than placebo. Measurements of quality of life using standard methods of assessment also showed trends favoring placebo.

The findings at a planned interim analysis from one randomized trial suggested that hydrazine sulfate was nearly significantly *worse* than placebo (Loprinzi et al., 1994b). This finding could have been due to chance imbalances in the treatment groups. In any case, the trial was terminated before its planned accrual target of 300 patients because there was little chance that given the interim results, hydrazine sulfate would be found to be beneficial. In addition to the rigorous design and analyses of these trials, the final reports emphasized only findings that were protected by the randomization.

Even with these methodologically rigorous findings in evidence, the proponents of hydrazine would not be silenced. Returning to *Penthouse*, they said:

> The contempt of the Romanovs for their people has almost been rivaled in the United States by key figures in the American cancer establishment. These senior physicians have demonstrated their own brand of imperviousness to the suffering of millions by systematically destroying the reputation of an anticancer drug that has already benefited thousands and made life longer and easier for many more .... (Kamen, 1994)

In 1993, because of the failure of hydrazine sulfate to demonstrate any benefit in large, randomized, double-masked, placebo-controlled clinical trials, the National Cancer Institute inactivated the IND for the drug (NCI, 1993). However, an editorial accompanying the publication of the results from the three NIH trials prophesied that the treatment might rise again (Herbert, 1994). This was true, for in 1994, the General Accounting Office (GAO) of the U.S. government began an investigation into the manner in which the three trials were conducted. The investigation was fueled by documents from the developer of hydrazine alleging that NCI had compromised the studies. In particular, questions were raised about the number of patients who might have received barbiturates or related drugs, said to interfere with the beneficial effects of hydrazine. A collection of new data (retrospectively) and re-analysis of the lung cancer trial supported the original conclusions (Kosty et al., 1995).

The final GAO report on the NCI hydrazine clinical trials appeared in September 1995 (GAO, 1995). After a detailed investigation of the allegations and research methods, the GAO supported conclusions by the study investigators and sponsors that hydrazine sulfate is ineffective. The GAO did criticize the failure to keep records of tranquilizer, barbiturate, and alcohol use, and the late analyses of such questions. These criticisms were dismissed by the Public Health Service in Appendix II of the GAO report. In any case, mainstream scientific and governmental opinion has converged on the view that hydrazine sulfate in not effective therapy for patients with cancer.

The findings of lack of efficacy continue to be handily explained away by proponents of hydrazine. The most recent statement (2004) by a leading advocate regarding the compound (www.hydrazinesulfate.org) has extensive rationalizations for all the study findings, positive and negative. However, there seems to be little or no enthusiasm among oncology clinical researchers for any additional testing of this compound. Aside from the greater weight of evidence assigned by most scientists to the negative studies, the tone of conspiracy and paranoia by its advocates will likely continue to discourage any quality investigations of hydrazine in the near future.

Lessons to Be Learned from Hydrazine Sulfate

The student of clinical trials can learn some very important general lessons by reading the scientific papers and commentaries cited above regarding hydrazine sulfate. With regard to subset analyses, power is reduced because of the smaller sample sizes on which comparisons are based. This tends to produce a high false negative rate (see the discussion in Section 7.2.7). If investigators perform many such comparisons, the overall false positive rate will be increased. Thus almost all findings are incorrect. Explorations of the data are worthwhile and important for generating new questions. However, one cannot test a hypothesis reliably using the same data that generated the question.

The hydrazine example also sketches implicitly some of the circumstances in which we need to maintain a cautious interpretation of clinical trial results. Even when there is *a priori biological justification for the hypothesis from animal studies*, interpretations and extrapolations of methodologically flawed human trials should be cautious before confirmatory studies are done. *Small trials* or those with *poor designs or analyses* should be treated skeptically. The *results of subset analyses should not be emphasized* as the primary findings of a trial nor should *hypothesis tests suggested by the data*. When exploratory analyses are conducted, *statistical adjustment* procedures can help interpret the validity of findings.

The reader of trial reports and other medical studies should be aware of the ways in which *observer bias* can enter trial designs or analyses. Aside from inappropriate emphasis, those analyzing or reporting a study should not have any *financial interests* in the outcome. Finally, apparent *treatment–covariate interactions* should be explored rigorously and fully.

16.8   SUMMARY

The clinical effects of interest in phase I trials are usually drug distribution and elimination parameters and the association, if any, between dose and side effects. Pharmacokinetic models are essential for quantifying kinetic parameters such as elimination rate and half-life. These models are often based on simple mass action distribution of drug in a few idealized compartments. The compartments can often be identified physiologically as intravascular, extravascular, body fat, or similar tissues. Phase I trials usually provide useful dosage recommendations and usually do not provide evidence of efficacy.

SA clinical trials often focus on treatment feasibility and simple estimates of clinical efficacy or toxicity such as success or failure rates. More formally, these studies estimate the probability of success or failure (according to pre-defined criteria) when new patients are treated. Phase II studies in patients with life-threatening diseases can also provide estimates of survival and other event rates. Many other clinical or laboratory effects of treatment can be estimated from SA trials.

Comparative trials estimate relative treatment effects. Depending on the outcome being used, the relative effect of two treatments might be a difference of means, a ratio, or a qualitative difference. Estimated risk ratios are important and commonly used relative treatment effects in phase III trials. Although not always required, statistical models can be useful when analyzing comparative trials to help estimate relative treatment effects and account for the possible influence of prognostic variables.

Besides statistical models, clinical trials often employ other special methods of analysis. These may be necessary for using the information in repeated measurements, when outcomes are correlated with one another, or when predictor variables change over time. The statistical methods for dealing with such situations are complex and require an experienced methodologist.

It is often important to conduct exploratory analyses of clinical trials, that is, analyses that do not adhere strictly to the experimental design. In comparative trials these explorations of the data may not be protected by the design of the study (e.g., randomization). Consequently they should be performed, interpreted, and reported conservatively. This is true of results based on subsets of the data, especially when many such analyses are performed.

16.9 QUESTIONS FOR DISCUSSION

1. A new drug is thought to be distributed in a single, mostly vascular, compartment and excreted by the kidneys. Drug administration will be by a single IV bolus injection. What rate (differential) equation describes this situation? What should the time–concentration curve look like? Four hours after injection, 65% of the drug has been eliminated. What is the half-life? If the drug is actually distributed in two or more compartments, what will be the shape of the time–concentration curve?

2. For the solution of the two-compartment pharmacokinetic model in Section 16.2.2, it was found that

$$\xi_{1,2} = \frac{-(\lambda + \mu + \gamma) \pm \sqrt{(\lambda + \mu + \gamma)^2 - 4\mu\gamma}}{2}.$$

Prove that the expression under the radical is positive.

3. The following event times are observed on a SA trial in patients with cardiomyopathy (censored times are indicated by a +): 55, 122, 135, 141+,144, 150, 153+, 154, 159, 162, 170, 171, 171+, 174, 178, 180, 180+, 200+, 200+, 200+, 200+, 200+. Construct a lifetable and estimate the survival at 180 days. Estimate the overall failure rate and an approximate 95% confidence interval.

4. In the mesothelioma SA trial, estimate the overall rate of progression and death rate. Are these rates associated with age, sex, or performance status? Why or why not?

5. In the CAP lung cancer trial, there are as many as $2 \times 2 \times 3 \times 3 \times 2 \times 2 \times 2 = 288$ subsets based on categorical factors (cell type, performance status, T, N, sex, weight loss, and race). The number of possible ways of forming subsets is even larger. Can you suggest a simple algorithm for forming subsets based on arbitrary classifications? Use your method to explore treatment differences in subsets in the CAP trial. Draw conclusions based on your analyses.

6. In the FAP clinical trial, test the treatment effect by performing gain score analyses on polyp number and polyp size. How do the results compare with analyses of covariance. Discuss. Can these analyses test or account for the effects of sex and age?

7. At one medical center, a new cancer treatment is classified as a success in 11 out of 55 patients. At another center using the same treatment and criteria, the success rate is 14 out of 40 patients. Using resampling methods, place a confidence interval on the *difference* in the success rates. What are your conclusions?

8. The data in Table 16.14 show the survival times in years measured from inauguration, election, or coronation for U.S. presidents, Roman Catholic popes, and

TABLE 16.14   Survival of Presidents, Popes, and Monarchs from 1690

Washington	10	Harding	2	Pius IX	32
J. Adams	29	Coolidge	9	Leo XIII	25
Jefferson	26	Hoover	36	Pius X	11
Madison	28	F. Roosevelt	12	Ben XV	8
Monroe	15	Truman	28	Pius XI	17
J.Q. Adams	23	Kennedy	3	Pius XII	19
Jackson	17	Eisenhower	16	John XXIII	5
Van Buren	0	L. Johnson	9	Paul VI	15
Harrison	20	Nixon	27	John Paul	0
Polk	4	Ford	29+	John Paul II	27
Taylor	1	Carter	26+		
Fillmore	24	Reagan	24	James II	17
Pierce	16	Bush 1	16+	Mary II	6
Buchanan	12	Clinton	12+	William III	13
Lincoln	4	Bush 2	4+	Anne	12
A. Johnson	10			George I	13
Grant	17	Alex VIII	2	George II	33
Hayes	16	Innoc XII	9	George III	59
Garfield	0	Clem XI	21	George IV	10
Arthur	7	Innoc XIII	3	William IV	7
Cleveland	24	Ben XIII	6	Victoria	63
Harrison	12	Clem XIII	11	Edward VII	9
McKinley	4	Clem XIV	6	George V	25
T. Roosevelt	18	Pius VI	25	Edward VIII	36
Taft	21	Pius VII	23	George VI	15
Wilson	11	Greg XVI	15	Elizabeth II	53+

Note: Survival in years from first taking office (see Lunn and McNeil, 1990). Updated by the author.

TABLE 16.15   Results of a Double-Masked Randomized Trial of Mannitol versus Placebo in Patients with Ciguatera Poisoning

	Signs and Symptoms		Treatment Group	Baseline Severity	Sex	Age	Time
ID	Baseline	2.5 Hours					
1	10	3	M	18	F	24	330
2	10	7	M	12	M	12	270
3	12	8	M	21	F	18	820
4	12	6	M	24	M	11	825
26	5	0	P	9	M	37	910
28	4	4	P	9	M	41	2935
33	7	7	P	12	M	44	1055
37	10	10	P	23	M	46	1455
42	5	5	P	6	M	54	780
43	6	6	M	6	M	44	9430
50	11	11	M	16	F	46	390
51	7	7	M	12	F	27	900
59	7	7	M	12	M	46	870
94	12	6	M	19	M	44	80
95	14	14	M	27	M	45	490
98	14	7	M	34	F	34	415
116	8	7	M	12	F	25	605
117	8	5	M	20	F	18	650
118	16	11	M	27	F	16	645
501	3	0	P	3	F	27	2670
503	15	14	M	23	M	57	5850

Source: Data from Palafox, Schatz, Lange et al. (1997).

British monarchs from 1690 to the present (Lunn and McNeil, 1990). (I have added censored observations to the list for individuals remaining alive since 1996.) Are there differences among the groups? Discuss your methods and conclusions.

9. Do this exercise without consulting the reference. The data in Tables 16.15 and 16.16 show the results of a double-masked randomized clinical trial of mannitol infusion versus placebo for the treatment of ciguatera poisoning (Palafox, Schatz, Lange, et al., 1997). The number of neurological signs and symptoms of poisoning at the start of treatment and 2.5 hours after treatment are given. Estimate the clinical effect and statistical significance of mannitol treatment. Discuss your estimate in view of the small size of the trial.

10. Do this exercise without consulting the reference. The data in Table 16.17 shows the results of a double-masked randomized clinical trial of implantable biodegradable polymer wafers impregnated with (BCNU) versus placebo for the treatment of patients with newly diagnosed malignant gliomas (Valtonen, Timonen, Toivanen, et al., 1997). Survival time (in weeks) following surgical resection is given for each patient. Estimate the clinical effect and statistical significance of BCNU wafer treatment. Discuss your estimate in view of the small size of the trial. You may wish to compare your findings with those of Westphal et al. (2003).

TABLE 16.16   Results of a Double-Masked Randomized Trial of Mannitol versus Placebo in Patients with Ciguatera Poisoning

ID	Signs and Symptoms		Treatment Group	Baseline Severity	Sex	Age	Time
	Baseline	2.5 Hours					
504	13	13	P	22	M	42	440
505	13	13	P	21	F	16	1460
512	13	1	P	31	F	14	545
513	5	0	M	29	F	30	670
514	17	8	P	37	M	40	615
515	25	24	M	54	F	19	645
516	21	20	M	27	F	29	1280
517	8	8	M	11	M	48	1500
613	12	1	P	34	M	9	740
615	11	8	M	45	M	9	415
701	11	0	P	11	M	40	540
705	12	12	M	14	F	38	1005
706	3	3	M	4	M	39	370
707	8	7	P	8	M	54	250
708	11	3	P	10	F	27	795
709	9	9	M	9	M	34	—
737	4	3	M	4	M	31	865
738	11	8	P	22	M	48	630
766	11	0	P	29	M	22	630
768	11	7	M	27	M	18	690
1000	6	6	P	11	M	50	795

Source: Data from Palafox, Schatz, Lange, et al. (1997).

TABLE 16.17   Results of a Randomized Double-Masked Trial of BCNU Polymer versus Placebo for Patients with Newly Diagnosed Malignant Gliomas

ID	Treat	Score	Sex	Age	Karn	GBM	Time	Status
103	1	25	M	65	1	0	9.6	1
104	1	25	M	43	1	0	24.0	0
105	1	19	F	49	0	1	24.0	0
107	1	17	M	60	0	1	9.2	1
203	1	21	F	58	0	0	21.0	1
204	1	26	F	45	1	1	24.0	0
205	1	27	F	37	1	0	24.0	0
301	1	23	M	60	1	1	13.3	1
302	1	26	F	65	0	1	17.9	1
306	1	24	F	44	0	1	9.9	1
307	1	24	F	57	0	1	13.4	1
402	1	27	M	68	1	1	9.7	1
404	1	27	M	55	0	1	12.3	1
407	1	10	M	60	0	1	1.1	1
408	1	23	M	42	0	1	9.2	1
412	1	27	F	48	1	0	24.0	0
101	2	26	F	59	1	1	9.9	1

TABLE 16.17   (*continued*)

ID	Treat	Score	Sex	Age	Karn	GBM	Time	Status
102	2	14	F	65	0	1	1.9	1
106	2	21	M	58	1	1	5.4	1
108	2	21	F	51	0	1	9.2	1
109	2	22	F	63	1	1	8.6	1
201	2	17	F	45	0	1	9.1	1
202	2	20	F	52	0	1	6.9	1
303	2	26	M	62	1	1	11.5	1
304	2	27	M	53	1	1	8.7	1
305	2	25	F	44	1	1	17.1	1
308	2	24	M	57	0	1	9.1	1
309	2	25	M	36	1	1	15.1	1
401	2	27	F	47	1	1	10.3	1
403	2	27	M	52	1	1	24.0	0
405	2	18	F	55	0	1	9.3	1
406	2	26	F	63	0	1	4.8	1

Source: Data from Valtonen et al. (1997).

17

PROGNOSTIC FACTOR ANALYSES

17.1 INTRODUCTION

Prognostic factor analyses (PFAs) are studies, often based on data that exist for other reasons, that attempt to assess the relative importance of several predictor variables simultaneously. The need to prognosticate is basic to clinical reasoning, but most of us are unable to account quantitatively for the effects of more than one or two variables at a time. Using the formality of a PFA, the additional structure provided by a statistical model, and thoughtful displays of data and effect estimates, one can extend quantitative accounting to many predictor variables. This is often useful when analyzing the data from clinical trials (Armitage, 1981).

The terms "predictor variables," "independent variables," "prognostic factors," and "covariates" are usually used interchangeably to describe the predictors. Prognostic factor analyses are closely related to methods that one might employ when analyzing a clinical trial, especially when adjusting for the effects of covariates on risk. In fact the treatment effect in a designed experiment is a prognostic factor in every sense of the word. The statistical models employed for PFAs are the same as those used in clinical trials and the interpretation of results is similar. Furthermore, advanced planning is needed to conduct PFAs without making errors. Useful references discussing this subject include Byar (1982, 1984), George (1988), Harris and Albert (1991), Collett (1994), Marubini and Valsecchi (1995), and Parmar and Machin (1995).

PFAs differ from clinical trial analyses is some significant ways. Most important, PFAs are usually based on data in which investigators did not control the predictor variables or confounders, as in experimental designs. Therefore the validity of prognostic factor analyses depends on the absence of strong selection bias, on the correctness of the statistical models employed, and on having observed, recorded, and analyzed appropriately the important predictor variables. Accomplishing this requires much stronger assumptions than those needed to estimate valid effects in true experimental designs.

Clinical Trials: A Methodologic Perspective, 2E, by S. Piantadosi
Copyright © 2005 John Wiley & Sons, Inc.

Not only are the patients whose data contribute to PFAs often subject to important selection effects, but the reasons why treatments were chosen for them may also relate strongly to their prognosis. In fact we expect that this will be the case if the physicians are effective in their work. A prognostic factor analysis does not permit certain control over unobserved confounders that can bias estimates of treatment effects. Similar concerns pertain to these types of analyses when used in epidemiologic studies, where they are also common.

After some preliminaries, I discuss methods for PFAs based on familiar statistical models. The discussion pertains to linear, logistic, parametric survival, and proportional hazards regression models. Finally, I discuss assessing prognostic factors using methods that are not based directly on statistical models.

17.1.1 Studying Prognostic Factors Is Broadly Useful

A well-conceived study of prognostic factors can yield useful information about the past, present, or future. PFAs can describe, explain, analyze, or summarize preexisting data, such as from a database or a comparative clinical trial. When results of PFAs are applied to new patients, they suggest information about the future of the individuals. All these contexts are clinically useful.

One reason for studying prognostic factors is to learn the relative importance of several variables that affect, or are associated with, disease outcome. This is especially important for diseases that are treated imperfectly such as AIDS, cardiovascular disease, and cancer. A large fraction of patients with these diseases will continue to have problems and even die from them. Therefore prognostication is important for both the patient and the physician. Examples of clinically useful prognostic factors in these settings are stage (or other extent of disease measures) in cancer and left ventricular function in ischemic heart disease. Predictions based on these factors can be individualized further, using other characteristics of the patient or the disease.

A second reason for studying prognostic factors is to improve the design of clinical trials. For example, suppose that we learn that a composite score describing the severity of recurrent respiratory papillomatosis is strongly prognostic. We might stratify on disease severity as a way of improving the balance and comparability of treatment groups in a randomized study comparing treatments in this condition. Other studies may select only high- or low-risk subjects as determined by prognostic factors.

Knowledge of prognostic factors can improve our ability to analyze randomized clinical trials. For example, imbalances in strong prognostic factors in the treatment groups can invalidate simple comparisons. However, analytic methods that adjust for such imbalances can correct this problem. This topic is discussed later in this chapter. The same methods are essential when analyzing nonrandomized comparisons.

Interactions between treatment and covariates or between prognostic factors themselves can be detected using the methods of PFAs (Byar, 1985). Large treatment–covariate interactions are likely to be important, as are those that indicate that treatment is helpful in one subset of patients but harmful in another, so-called qualitative interactions. Detecting interactions depends not only on the nature of the covariate effects on outcome but also on the scale of analysis. For example, if covariates truly affect the outcome in a multiplicative fashion with no interactions, as in most nonlinear models for survival and other endpoints, interactions are likely to be found if an additive scale of analysis is used. Conversely, if the effect of covariates is truly

**TABLE 17.1   Hypothetical
Effect Estimates Illustrating
Interaction between Two Factors
on Different Scales of
Measurement**

Factor B Present	Factor A Present	
	No	Yes
No	10	20
Yes	30	60(40)

additive, interactions are likely to be observed if a multiplicative scale of analysis is used. Thus interactions depend on the model being employed for analysis.

These effects are illustrated in Table 17.1, which shows hypothetical outcomes in 4 groups defined by 2 dichotomous prognostic factors. In the absence of both factors, the baseline response is 10. If the effect of factor A is to multiply the rate by 2 and the effect of B is to multiply the rate by 3, then a response of 60 in the yes–yes cell demonstrates no interaction. In contrast, if the effect of factor A is to add 10 to the baseline response and the effect of B is to add 20, then a response of 40 in the yes–yes cell demonstrates no interaction. In either case there will be interaction on *some* scale of measurement, illustrating that interaction is model dependent.

Prognostic factors are also useful in assessing clinical landmarks during the course of an illness and deciding if changes in treatment strategy are warranted. This could be useful, for example, when monitoring the time course of CD4+ lymphocyte count in HIV positive patients. When the count drops below some threshold, a change in treatment may be indicated. A threshold such as this could be determined by an appropriate PFA.

17.1.2   Prognostic Factors Can Be Constant or Time-Varying

The numerical properties of prognostic factor measurements are the same as those for the study endpoints discussed in Chapter 8. That is, prognostic factors can be continuous measures, ordinal, binary, or categorical. Most often prognostic factors are recorded at study entry or "time 0" with respect to follow-up time and remain constant. These are termed "baseline" factors. The measured value of a baseline factor, such as sex or treatment assigned, does not change over time. Usually, when studying the effects of baseline factors, we assume that differences attributable to them happen instantaneously and remain constant over time.

Some prognostic factors may change their value over time, as discussed in Chapter 16. They would be assessed both at baseline and at several points following treatment as longitudinal measurements. For example, prostatic specific antigen (PSA) is a reliable indicator of disease recurrence in patients with prostate cancer. The PSA level may decrease to near zero in patients with completely treated prostate cancer, and the chance of detecting future cancer recurrence or risk of death may depend on recent PSA levels. Prognostic factors that change value over time are termed "time-dependent covariates" (TDC) and special methods are needed to account for their effects on most outcomes.

There are two types of TDCs: intrinsic or internal, and extrinsic or external. Intrinsic covariates are those measured in the study subject, such as the PSA example above. Extrinsic TDCs are those that exist independently of the study subject. For example, we might be interested in the risk of developing cancer as a function of environmental levels of toxins. These levels may affect the patient's risk but exist independently. Both types of TDCs can be incorporated in prognostic factor models with appropriate modifications in the model equations and estimation procedures.

17.2   MODEL-BASED METHODS

One of the most powerful and flexible tools for assessing the effects of more than one prognostic factor simultaneously is a statistical model. These models describe a plausible mathematical relationship between the predictors and the observed endpoint in terms of one or more model parameters which have handy clinical interpretations. To proceed with this approach, the investigator must be knowledgeable about the subject matter and interpretation and (1) collect and verify complete data, (2) consult an experienced statistical expert to guide the analysis, and (3) make a plan for dealing with decision points during the analysis.

Survival or time-to-event data constitute an important subset of prognostic factor information associated with clinical trials. Because of the frequency with which such data are encountered, statistical methods for analyzing them have been highly developed. Sources for the statistical theory dealing with these types of data are Lee (1992), Cox and Oakes (1984), and Kalbfleisch and Prentice (2002). Recently a new journal, *Lifetime Data Analysis*, has appeared as an international journal devoted to statistical methods for time-to-event data.

17.2.1   Models Combine Theory and Data

A model is any construct which combines theoretical knowledge (hypothesis) with empirical knowledge (observation). In mathematical models the theoretical component is represented by one or more *equations* that relate the measured or observed quantities. Empirical knowledge is represented by *data*, that is, the measured quantities from a sample of individuals or experimental units. The behavior of a model is governed by its structure or functional form and unknown quantities or constants of nature called "parameters." One goal of the modeling process might be to estimate, or gain quantitative insight, into the parameters. Another goal might be to see if the data are consistent with the theoretical form of the model. Yet another goal might be to summarize large amounts of data efficiently, in which case the model need not be precise.

Statistical models generally have additional characteristics. First, the equations represent convenient biological constructs but are usually not fashioned to be literally true as in some biological or physical models. Second, statistical models often explicitly incorporate an error structure or a method to cope with the random variability that is always present in the data. Third, the primary purpose of statistical models is often to facilitate estimating the parameters so that relatively simple mathematical forms are most appropriate. Even so, the models employed do more than simply smooth data (Appleton, 1995). A broad survey of statistical models is given by Dobson (1983).

If it has been constructed so that the parameters correspond to clinically interpretable effects, the model can be a way of estimating the influence of several factors simultaneously on the outcome. In practice, statistical models usually provide a concise way for estimating parameters, obtaining confidence intervals on parameter estimates, testing hypotheses, choosing from among competing models, or revising the model itself. Models with these characteristics include linear, generalized linear, logistic, and proportional hazards models. These are widely used in clinical trials and prognostic factor analyses and are discussed in more detail below.

17.2.2   The Scale of Measurements (Coding) May Be Important

The first step in a PFA is to code the measurements (variable values) in a numerically appropriate way. All statistical models can be formulated by the numerical coding of variables. Even qualitative variables can be represented by the proper choice of numerical coding. There are no truly qualitative statistical models or qualitative variable effects. For this reason the coding and scale of measurement that seems most natural clinically may not be the best one to use in a model. If an ordinal variable with three levels is coded 1, 2, 3 as compared with 10, 11, 12, the results of model fitting might be different. If a coding of 1, 10, 100 is used, almost certainly the results will be different. This effect can often be used purposefully to transform variables in ways that are clinically sensible and statistically advantageous.

Simple summary statistics should always be inspected for prognostic factor variables. These may reveal that some factors have highly skewed or irregular distributions, for which a transformation could be useful. Predictor variables do not have to have "normal," or even symmetric, distributions. Categorical factors may have some levels with very few observations. Whenever possible, these should be combined so that subsets are not too small.

Ordinal and qualitative variables often need to be re-coded as binary "indicator" or dummy variables that facilitate group comparisons. A variable with N levels requires $N - 1$ binary dummy variables to compare levels in a regression model. Each dummy variable implies a comparison between a specific level and the reference level (which is omitted from the model). For example, a factor with three levels A, B, and C, would require two dummy variables. One possible coding is for the first variable to have the value 1 for level A and 0 otherwise. The second could have 1 for level B and 0 otherwise. Including both factors in a model compares A and B versus the reference group C. In contrast, if a single variable is coded 1 for level A, 2 for level B, and 3 for level C, for example, then including it in a regression will model a linear trend across the levels. This would imply that the $A - B$ difference was the same as the $B - C$ difference, which might or might not be appropriate. A set of three dummy variables for a factor with four levels is shown in Table 17.2.

17.2.3   Use Flexible Covariate Models

The appropriate statistical models to employ in PFAs are dictated by the specific type of data and biological questions. Some models widely used in oncology (and elsewhere) are discussed by Simon (1984). Regardless of the endpoint and structure of the model, the mathematical form will always contain one or more submodels that describe the effects of multiple covariates on either the outcome or a simple function of it. For example, most relative risk regression models used in failure time data and

**TABLE 17.2   Dummy
Variables for a Four Level
Factor with the Last Level
Taken as a Reference**

Factor Level	Dummy Variables		
	x_1	x_2	x_3
A	1	0	0
B	0	1	0
C	0	0	1
D	0	0	0

epidemiologic applications use a multiplicative form for the effects of covariates on a measure of relative risk. In the case of the proportional hazards regression model, the logarithm of the hazard ratio is assumed to be a constant related to a linear combination of the predictor variables. The hazard rate, $\lambda_i(t)$, in those with covariate vector $\mathbf{x}_i$ satisfies

$$\log\left\{\frac{\lambda_i(t)}{\lambda_0(t)}\right\} = \sum_{j=1}^{k} \beta_j \mathbf{x}_{ij} \, .$$

Because the covariates are additive on a logarithmic scale, the effects multiply the baseline hazard.

Similarly the widely used logistic regression model is

$$\log\left\{\frac{p_i}{1-p_i}\right\} = \beta_0 + \sum_{j=1}^{k} \beta_j \mathbf{x}_{ij},$$

where p_i is the probability of "success," $\mathbf{x}_{ij}$ is the value of the jth covariate in the ith patient or group, and β_j is the log-odds ratio for the jth covariate. This model contains an intercept term, β_0, unlike the proportional hazards model where the baseline hazard function, $\lambda_0(t)$, is arbitrary. In both cases a linear combination of covariates multiplies the baseline risk.

Generalized linear models (GLMs) (Nelder and Wedderburn, 1972; McCullagh and Nelder, 1989) are powerful tools that can be used to describe the effects of covariates on a variety of outcome data from the general exponential family of distributions. GLMs have the form

$$\eta_i = \beta_0 + \sum_{j=1}^{k} \beta_j \mathbf{x}_{ij},$$

$$E\{y_i\} = g(\eta_i),$$

where $g(\cdot)$ is a simple "link function" relating the outcome, y, to the linear combination of predictors. Powerful and general statistical theory facilitates parameter estimation in this class of models, which includes one-way analyses of variance, multiple linear regression models, log-linear models, logit and probit models, and others. For a comprehensive review, see McCullagh and Nelder (1989).

For both common relative risk models and GLMs, the covariate submodel is a linear one. This form is primarily a mathematical convenience and yields models whose parameters are simple to interpret. However, more complex covariate models can be constructed in special cases such as additive models and those with random effects.

Models with Random Effects

All statistical models contain an explicit error term that represents a random quantity, perhaps due to the effects of measurement error. Almost always, the random error is assumed to have mean zero and a variability that can be estimated from the data. Most other effects (parameters) in statistical models are assumed to be "fixed," which means that they estimate quantities that are fixed or constant for all subjects. In some circumstances we need to model effects that, like the error term, are random but are attributable to sources other than random error. Random effects terms are a way to accomplish this and are used frequently in linear models. For a recent discussion of this topic in the context of longitudinal data, see Rutter and Elashoff (1994).

An example of a circumstance where a random effect might be necessary is a clinical trial with a large number of treatment centers and a patient population at each center that is somewhat different from one another. In some sense the study centers can be thought of as a random sample of all possible centers. The treatment effect may be partly a function of study center, meaning there may be clinically important treatment by center interactions. The effect of study center on the trial outcome variable might best be modeled as a random effect. A fixed-effect model would require a separate parameter to describe each study center.

17.2.4   Building Parsimonious Models Is the Next Step

Quantitative prognostic factor assessment can be thought of as the process of constructing parsimonious statistical models. These models are most useful when (1) they contain a few clinically relevant and interpretable predictors, (2) the parameters or coefficients are estimated with a reasonably high degree of precision, (3) the predictive factors each carry independent information about prognosis, and (4) the model is consistent with other clinical and biological data. Constructing models that meet these criteria is usually not a simple or automatic process. We can use information in the data themselves (data-based variable selection), clinical knowledge (clinically based variable selection), or both. A useful tutorial on this subject is given by Harrell, Lee, and Mark (1996).

Don't Rely on Automated Procedures

With modern computing technology and algorithms, it is possible to automate portions of the model-building process. This practice can be dangerous for several reasons. First, the criteria on which most automated algorithms select or eliminate variables for inclusion in the model may not be appropriate. For example, this is often done on the basis of significance levels (p-values) only. Second, the statistical properties of performing a large number of such tests and re-fitting models are poor when the objective is to arrive at a valid set of predictors. The process will be sensitive to noise or chance associations in the data. Third, an automated procedure does not provide a way to incorporate information from outside the model building mechanics. This information may be absolutely critical for finding a statistically correct and clinically interpretable final model. Fourth, we usually want more sophisticated control over the consequences of missing data values (an inevitable complication) than that afforded by automated procedures.

There are ways of correcting the deficiencies just noted. All of them require thoughtful input into the model-building process from clinical investigators and interaction with a biostatistician who is familiar with the strengths and pitfalls. One statistical approach to prevent being misled by random associations in the data is to pre-screen prognostic factors based on clinical criteria. Many times prognostic factors are simply convenient to use rather than having been collected because of a priori interest or biological plausibility. Such factors should probably be discarded from further consideration.

An example of the potential difficulty with automated variable selection procedures is provided by the data in Table 17.3. The complete data and computer program to analyze them are available as described in Appendix A. The data consist of 80 observed event times and censoring indicators. For each study subject, 25 dichotomous predictor variables have been measured. These data are similar to those available from many exploratory studies.

The first approach to building a multiple regression model employed a step-up selection procedure. Variables were entered into the regression model if the significance level for association with the failure time was less than 0.05. This process found X_1 and X_{12} to be significantly associated with the failure time.

A second model-building approach used step-down selection, retaining variables in the regression only if the p-value was less than 0.05. At its conclusion this method found X_3, X_{10}, X_{11}, X_{14}, X_{15}, X_{16}, X_{18}, X_{20}, X_{21}, and X_{23} to be significantly associated with the failure time (Table 17.4). It may be somewhat alarming to those unfamiliar with these procedures that the results of the two model building techniques do not overlap. The results of the step-down procedure suggest that the retention criterion could be strengthened. Using step-down variable selection and a p-value of 0.025

TABLE 17.3   Simulated Outcome and Predictor Variable Data from a Prognostic Factor Analysis

#	Dead	Time	$X_1 - X_{25}$
1	1	7.38	1111111001110100010100101
2	1	27.56	1001100111001011001001110
3	1	1.67	1000110000111000011010010
4	1	1.82	1101001111011101100101101
5	0	10.49	0001110011110011111100101
6	1	14.96	1011111011000100101001011
7	1	0.63	0111001011101000010001111
8	1	10.84	1001010111111011110100101
9	1	15.65	1110010000001001110001100
10	1	4.73	0100100000011100101100000
11	1	14.97	0010111110010111110000010
12	1	3.47	0010111111001100000000000
13	1	4.29	1100110000101100010000001
14	1	0.11	1111110110111101000011000
15	1	13.35	0010101010110011010010011
⋮	⋮	⋮	⋮

**TABLE 17.4   Multiple Regressions Using Automated Variable Selection Methods
Applied to the Data from Table 17.3**

Model	Variables	Parameter Estimate	Standard Error	Wald χ^2	Pr > χ^2	Risk Ratio
1	X_1	0.479	0.238	4.07	0.04	1.615
	X_{12}	−0.551	0.238	5.35	0.02	0.576
2	X_3	0.573	0.257	4.96	0.03	1.773
	X_{10}	−0.743	0.271	7.53	0.006	0.476
	X_{11}	0.777	0.274	8.06	0.005	2.175
	X_{14}	−0.697	0.276	6.38	0.01	0.498
	X_{15}	−0.611	0.272	5.06	0.02	0.543
	X_{16}	−0.670	0.261	6.57	0.01	0.512
	X_{18}	−0.767	0.290	7.01	0.008	0.465
	X_{20}	−0.610	0.262	5.42	0.02	0.544
	X_{21}	0.699	0.278	6.30	0.01	2.012
	X_{23}	0.650	0.261	6.20	0.01	1.916

to stay in the model, *no* variables were found to be significant. This seems even more unlikely than the fact that nine variables appeared important after the step-down procedure.

These results are typical and should serve as a warning as to the volatility of results when automated procedures, particularly those based on significance levels, are used to build models without the benefit of biological knowledge. A second lesson from this example extends from the fact that the data were simulated such that all predictor variables were both independent from one another and independent from the failure time. Thus all "statistically significant" associations in Table 17.4 are due purely to chance. Furthermore the apparent joint or multivariable association, especially the second regression, disappears when a single variable is removed from the regression.

Resolve Missing Data

There are three alternatives for coping with missing covariate values. One could disregard prognostic factors with missing values. This may be the best strategy if the proportion of unknowns is very high. Alternatively, one can remove the individuals with missing covariates from the analysis. This may be appropriate for a few individuals who have missing values for a large fraction of the important covariates. A third strategy is to replace missing observations with some appropriate value. Usually statistical packages have no way to cope with records that contain missing values for the variable being analyzed, other than to discard them.

We rarely have the luxury of analyzing perfectly complete data. Nearly all patients may have some missing measurements so that the number of records with complete observations is a fraction of the total. If data are missing at random, that is, the loss of information is not associated with outcomes or other covariates, it is appropriate to disregard individuals with the missing variable. Loss of precision is the only consequence of this approach, and we can still investigate the covariate of interest.

If missing data are more likely in individuals with certain outcomes (or certain values of other covariates), there may be no approach for coping with the loss of information that avoids bias. Removing individuals with missing data from the analysis in such a circumstance systematically discards influences on the outcome. For example, if individuals with the most severe disease are more likely to be missing outcome data, the aggregate effect of severity on outcome will be lessened. When we have no choice but to lose some information, decisions about how to cope with missing data need to be guided by clinical considerations rather than automated statistical procedures alone.

There is a third method of coping with missing values that can be useful. It is called *imputation*, or replacing missing data with values calculated in a way that allows other analyses to proceed essentially unaffected. This can be a reasonable alternative when relatively few data are missing, precisely the circumstance in which other alternatives are also workable. For example, missing values could be "predicted" from all the remaining data and replaced by estimates that preserve the overall covariance structure of the data. The effect of such procedures on the inferences that result must be studied on a case-by-case basis.

The way that missing data are resolved should not be driven by the effect it produces on the outcome. For example, discarding certain incomplete observations or variables may influence the outcome differently than imputation. The best strategy to use should be decided on principle rather than results. I have seen circumstances where investigators (or regulatory officials) create missing data by disregarding certain observations, regardless of completeness. While there are occasional circumstances in which this might be the correct thing to do, it can produce very strong effects on the outcome and should be discussed carefully ahead of time.

Screen Factors for Importance in Univariable Regressions

The next practical step in assessing prognostic factors is to screen all the retained variables in univariable regressions (or other analyses). It is important to examine the estimates of clinical effect (e.g., relative hazards), confidence intervals, and significance levels. Together with biological knowledge, these can be used to select a subset of factors to study in multivariable models. The most difficult conceptual part of this process is deciding which factors to discard as unimportant. The basis of this decision is threefold: prior clinical information (usually qualitative), the size of the estimated effect, and the statistical significance level. Strong factors, or those known to be important biologically, should be retained for further consideration regardless of significance levels.

The findings at this stage can help investigators check the overall validity of the data. For example, certain factors are known to be strongly prognostic in similar groups of patients. Investigators should examine the direction and magnitude of prognostic factor effects. If the estimates deviate from those previously observed or known, the data may contain important errors.

It is appropriate to use a relaxed definition of "significance" at this screening stage. For example, one could use $\alpha = 0.15$ to minimize the chance of discarding an important prognostic factor. This screening step usually reduces the number of factors to about $\frac{1}{4}$ of those started. The potential error in this process is the tendency to keep variables that are associated with the outcome purely by chance and/or that the modeling procedure may overestimate the effect of some factors on the outcome. The only way to prevent or minimize these mistakes is to use clinical knowledge to augment the screening process.

Build Multiple Regressions

The next practical step in a PFA is to build a series of multivariable regression models and study their relative performance. To decrease the chance of missing important associations in the data, one should try more models rather than fewer. Consequently step-down model-building procedures may be better than step-up methods. In step-down, all prognostic factors that pass the screening are included in a single multiple regression. The model is likely to be overparameterized at the outset, in which case one or more factors will have to be removed to allow the estimation process to converge.

Following model-fitting, we use the same evaluation methods outlined above to include or discard prognostic factors in the regression. After each change or removal of variables, the parameters and significance levels are re-estimated. The process stops when the model makes the most biological sense. It is satisfying when this point also corresponds to a local or global statistical optimum, such as when the parameter estimates are fairly precise and the significance levels are high.

Usually step-down variable selection methods will cause us to try more models than other approaches. In fact there may be partially (or non-) overlapping sets of predictors that perform nearly as well as each other using model-fitting evaluations. This is most likely a consequence of correlations between the predictor variables. Coping with severe manifestations of this problem is discussed below.

Some automated approaches can fit all possible subset regressions. For example, they produce all one-variable regressions or all two-variable ones. The number of such regressions is large. For r predictor variables the total number of models possible without interaction terms is

$$N = \sum_{k=1}^{r} \binom{r}{k}.$$

For 10 predictor variables, $N = 1023$. Even if all of these models can be fitted and summarized, the details may be important for distinguishing between them. Thus there is usually no substitute for guidance by clinical knowledge.

Correlated Predictors May Be a Problem

Correlations among predictor variables can present difficulties during the model-building and interpretation process. Sometimes the correlations among predictor variables are strong enough to interfere with the statistical estimation process, in the same way that occurs with familiar linear regression models. Although models such as logistic and proportional hazards regressions are nonlinear, the parameter estimates are most often obtained by a process of iterated solution through linearization of estimating equations. Furthermore covariates often form a linear combination in statistical models, as discussed above. Therefore collinearity of predictors can be a problem in nonlinear models also.

Even when the estimation process goes well, correlated variables can create difficulties in model building and interpretation. Among a set of correlated predictors, any one will appear to improve the model prediction, but if more than one is included, all of them may appear unimportant. To diagnose and correct this situation, a number of models will have to be fitted and one of several seemingly "good" models selected as "best."

Clinicians are sometimes disturbed by the fact that statistical procedures are not guaranteed to produce one regression model that is clearly superior to all others. In fact, even defining this model on statistical grounds can be difficult because of the large number of regressions that one has to examine in typical circumstances. In any case, we should not be surprised that several models fit the data and explain it well, especially when different variables carry partially redundant information. Models cannot always be distinguished on the basis of statistical evidence because of our inability to compare nonnested models formally and the inadequacies of summary statistics like significance levels for variable selection. Nested models are those that are special cases of more complex ones. Nested models can often be compared with parent models using statistical tests such as the likelihood ratio (see Chapter 14).

The only reasonable way for the methodologist to solve these difficulties is to work with clinicians who have expert knowledge of the predictor variables, based on either other studies or preclinical data. Their opinion, when guided by statistical evidence, is necessary for building a good model. Even when a particular set of predictors appears to offer slight statistical improvement in fit over another, one should generally prefer the set with the most clear clinical interpretation. Of course, if the statistical evidence concerning a particular predictor or set of predictors is very strong, then these models should be preferred or studied very carefully to understand the mechanisms, which may lead to new biological findings.

17.2.5 Incompletely Specified Models May Yield Biased Estimates

When working with linear models, it is well known that omission of important predictor variables will not bias the estimated coefficients of the variables included in the model. For example, if Y is a response linearly related to a set of predictors, $\mathbf{X}$, so that

$$E\{Y\} = \beta_0 + \beta_1 X_1 + \beta_2 X_2 + \cdots + \beta_n X_n$$

is the true model and we fit an incompletely specified model, for example,

$$Y = \beta_0 + \beta_1 X_1 + \beta_2 X_2 + \beta_3 X_3 + \epsilon,$$

the estimates of β_1, β_2, and β_3 will be unbiased. The effect of incomplete specification is to increase the variances but not to bias the estimates (Draper and Smith, 1981).

In contrast, when using certain nonlinear models, the proportional hazards model among them, omission of an important covariate will bias the estimated coefficients, even if the omitted covariate is perfectly balanced across levels of those remaining in the model (Gail et al., 1984). The same can be said for model-derived estimates of the variances of the coefficients (Gail et al., 1988). The magnitude of the bias that results from these incompletely specified nonlinear models is proportional to the strength of the omitted covariate. For example, suppose that we conduct a clinical trial to estimate the hazard ratio for survival between two treatments for coronary artery disease and that age is an influential predictor of the risk of death. Even if young and old patients are *perfectly* balanced in the treatment groups, failure to include age in a proportional hazards regression when using it to estimate the hazard ratio will yield a biased estimate of the treatment effect.

Although the existence of this bias is important for theoretical reasons, for situations commonly encountered in analyzing RCTs, there is not a serious consequence when important covariates are omitted, as they invariably are (Chastang et al., 1988). The lesson to learn from this situation is that models are important and powerful conveniences for summarizing data, but they are subject to assumptions and limitations that prevent us from blindly accepting the parameters or significance tests they yield. Even so, they offer many advantages that usually outweigh their limitations.

17.2.6 Study Second-Order Effects (Interactions)

One advantage to using models to perform PFAs is the ability to assess interactions. For example, if response depends on both age and sex, it is possible that the sex effect is different in young, compared with old, individuals. In a linear model an interaction can be described by the model

$$Y = \beta_0 + \beta_1 X_1 + \beta_2 X_2 + \gamma X_1 X_2 + \epsilon,$$

where γ represents the strength of the interaction between X_1 and X_2. Interactions of biological importance are relatively uncommon. Some interactions that seem to be important can be eliminated by variable transformations or different scales of analysis. However, large interactions may be important and regression models provide a convenient way to assess them.

One difficulty with using models to survey interactions is the large number of comparisons that have to be performed to evaluate all possible effects. If the model supports 6 prognostic factors, there are $6 + \binom{6}{2} = 21$ pairwise interactions possible (each variable can interact with itself). Even if all estimated interaction effects are due only to chance, we might expect one of these to be "significant" using p-values as test criteria. The number of higher order interactions possible is much larger. Usually we would not screen for interactions unless there is an a priori reason to do so or if the model does not fit the data well. Even then we should employ more strict criteria for declaring an interaction statistically significant than for main effects to reduce type I errors.

It is important to include the main effects, or low-order terms, in the model when estimating interaction effects. Otherwise, we may misestimate the coefficients and wrongly conclude that the high-order effect is significant when it is not. This can be illustrated by a very simple example. Suppose that we have the data shown in Figure 17.1 and the two models

$$E\{Y\} = \beta_0 + \beta_1 X_1$$

and

$$E\{Y\} = \beta_1^* X_1.$$

The first model has both a low-order effect (intercept) and a high-order effect (β_1). The second model has only the high-order effect β_1^*, meaning the intercept is assumed to be zero. When fit to the data, the models yield $\beta_1 \approx 0$ and $\beta_1^* \neq 0$. In the first case, we obtain a correct estimate of the slope (Figure 17.1, dotted line). In the second case, we obtain an incorrect estimate of the slope because we have wrongly assumed that the intercept is zero (Figure 17.1, solid line). An analogous problem can occur when interaction effects are estimated assuming the main effects are zero (i.e., omitting them from the model). A clinical application of interaction analyses can be seen in Section 4.6.6.

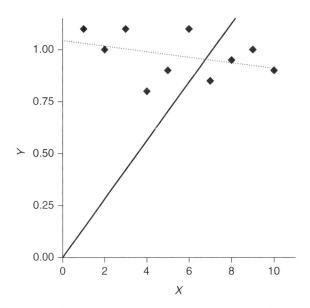

Figure 17.1 Hypothetical data for regression analyses with and without low-order effects.

17.2.7 PFAs Can Help Describe Risk Groups

Information from a PFA can help clinicians anticipate the future course of a patient's illness as a function of several (possibly correlated) predictors. This type of prognostication is often done informally on purely clinical grounds, most often using a few categorical factors such as functional classifications or anatomical extent of disease. Stage of disease is a good example of a measure of extent from the field of cancer. Prognostication based only on simple clinical parameters can be sharpened considerably by using quantitative model-based methods.

Suppose that we have measured two factors, X and Y that are either present or absent in each patient and relate strongly to outcome. Patients could be classified into one of the following four cells: X alone, Y alone, both, or neither. It is possible that the prognoses of individuals in each of the four cells are quite different and that knowing into which "risk group" a patient is categorized would convey useful information about his or her future. On the other hand, it might be that cells X and Y are similar, such that each represents the presence of a single risk factor and there are effectively only three risk levels. The three levels of risk are characterized by one risk factor, two risk factors, or none.

When there are several prognostic factors, some of them possibly measured on a continuous scale rather than being dichotomous or categorical, these ideas can be extended using the types of models discussed above. Rather than simply *counting* the number of factors, the actual variable *values* can be combined in a weighted sum to calculate risk. The resulting values can then be ranked and categorized. The best way to illustrate the general procedure is with an actual example, discussed in detail in the next section.

Example

In patients with HIV infection it is useful to have simple methods by which individual risk of clinical AIDS or death can be inferred. Many clinical parameters are known to carry prognostic information about time to AIDS and survival, including platelet count, hemoglobin, and symptoms (Graham et al., 1994). Here I illustrate a risk set classification that is a combination of clinical and laboratory values for HIV positive patients from the Multi-Center AIDS Cohort Study (MACS) (Piantadosi, 1994). The MACS is a prospective study of the natural history of HIV infection among homosexual and bisexual men in the United States. Details of the MACS study design and methods are described elsewhere (Kaslow et al., 1987).

From April 1984 through March 1985, 4954 men were enrolled in four metropolitan areas: Baltimore/Washington, DC, Chicago, Pittsburgh, and Los Angeles. There were 1809 HIV seropositive and 418 seroconverters in the MACS cohort. This risk set analysis included all HIV-1 seroprevalent men and seroconverters who used zidovudine prior to developing AIDS, and who had CD4+ lymphocyte data at study visits immediately before and after the first reported use of zidovudine. The total number of individuals meeting these criteria was 747.

Prognostic factors measured at baseline included age, CD4+ lymphocyte count, CD8+ lymphocyte count, hemoglobin, platelets, clinical symptoms and signs, and white blood cell count. The clinical symptoms and signs of interests were fever (greater than 37.9 degrees C) for more than two weeks, oral candidiasis, diarrhea for more than four weeks, weight loss of 4.5 kg, oral hairy leukoplakia, and herpes zoster. There were 216 AIDS cases among individuals treated with zidovudine prior to the cutoff date for this analysis and 165 deaths.

The initial step was to build to a multiple regression model for prediction of time to AIDS or time to death using the proportional hazards regression model. After model building, the log relative risk for each individual patient was calculated according to the equation

$$r_i = \sum_{j=1}^{p} \widehat{\beta}_j X_{ij}, \tag{17.1}$$

where r_i is the aggregate relative risk for the ith patient, $\widehat{\beta}_j$ is the estimated coefficient for the jth covariate from the regression model, and X_{ij} is the value of the jth covariate in the ith patient. Data values were then ordered from smallest to largest, based on the value of r_i. Following this ordering based on aggregate estimated relative risk, individuals were grouped or categorized into discrete risk sets. The cut points used for categorization and the number of groups were chosen empirically. That is, they were chosen so that the groups formed would display the full range of prognoses in the cohort. Usually this can be accomplished with three to five risk groups.

Following the formation of risk groups, survival curves were drawn for individuals in each group. Finally predicted survival curves were generated from the proportional hazards model and drawn superimposed on the observed curves as an informal test of goodness of fit. The predicted survival curves in the risk groups cannot be obtained directly from the original covariate regression. Instead, a second regression must be fit using a single covariate representing risk group. Because the risk group classification is based only on covariates from the first regression, the second regression a valid illustration of model performance based on the original predictors.

The units for expressing covariate values are shown in Table 17.5. The scale of measurement is important because it directly affects the interpretation of risk ratio estimates. For example, symptoms were coded as present or absent. Consequently

TABLE 17.5   Coding of Covariate Values for AIDS Prognostic Factor Analysis

Variable Name	Measurement Units
CD4 number	100 cells
CD8 number	100 cells
Neopterin	mg/dl
Microglobulin	mg/dl
Platelets	25,000 cells
Hemoglobin	gm/dl
Symptoms	1 = yes, 0 = no

TABLE 17.6   Proportional Hazards Multiple Regressions for Time to AIDS

Variable	Relative Risk	95% CI	P-Value
CD4	0.69	0.63–0.75	0.0001
CD8	1.02	1.00–1.04	0.04
Platelets	0.93	0.89–0.97	0.0009
Hemoglobin	0.85	0.78–0.92	0.0001
Symptoms	1.34	1.08–1.67	0.008

the risk ratio for symptoms is interpreted as the increase in risk associated with any symptoms. In contrast, change in CD4+ lymphocyte count was measured in units of 100 cells, so that its relative risk is per 100 cell increase.

For the time to AIDS outcome, the final regression model shows significant effects for CD4 count, platelets, hemoglobin and symptoms (Table 17.6). When risk set assignments are based on this model, the resulting time-to-event curves are shown in Figure 17.2. For time to AIDS, the first risk set was formed from individuals with the lowest risk of AIDS. This group constitutes 35% of the population. For the last risk set, the 15% with the highest risk was chosen. These individuals developed AIDS very rapidly, virtually all within the first 1.5 years after beginning zidovudine. The remaining risk sets were each formed from 25% fractions of the population. They indicate levels of risk intermediate between the most favorable and least favorable subsets. For the baseline variables model, equation (17.1) becomes

$$r_i = -0.373 \times CD4 + 0.020 \times CD8 - 0.071 \times Platelets \qquad (17.2)$$
$$- 0.168 \times Hemoglobin + 0.295 \times Symptoms.$$

The cut points used for these risk sets were: low risk, $r_i < -4.25$; low–intermediate risk, $-4.25 \le r_i < -3.70$; high–intermediate risk, $-3.70 \le r_i < -3.13$; and high risk, $r_i \ge -3.13$.

Regression models for survival time are shown in Table 17.7. For this outcome, the proportions of individuals assigned to each risk set are slightly different than for time

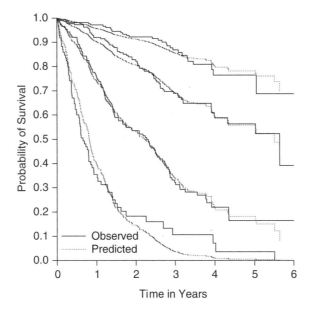

Figure 17.2 Observed and predicted survival of risk sets in the MACS cohort.

TABLE 17.7 Proportional Hazards Multiple Regressions for Time to Death

Variable	Relative Risk	95% CI	P-Value
CD4	0.64	0.59–0.71	0.0001
CD8	1.03	1.01–1.06	0.0009
Platelets	0.93	0.89–0.98	0.003
Hemoglobin	0.83	0.76–0.91	0.0001
Symptoms	1.35	1.08–1.69	0.01

to AIDS (Figure 17.2). This was done to separate the highest and lowest risk subsets as much as possible. The first risk set for the baseline variables model was formed from the 30% of individuals with the lowest risk of death. For the last risk set, the 15% with the highest risk was chosen. These individuals had the poorest survival, living generally less than 1.5 years after beginning AZT. The remaining risk sets were each formed from 30% and 25% fractions of the population. They indicate levels of risk intermediate between the most favorable and least favorable subsets. For the baseline variables model, equation (17.1) becomes

$$r_i = -0.440 \times CD4 + 0.034 \times CD8 - 0.070 \times Platelets \qquad (17.3)$$
$$- 0.182 \times Hemoglobin + 0.030 \times Symptoms.$$

The cut points used for the risk sets were: low risk, $r_i < -4.73$; low–intermediate risk, $-4.73 \leq r_i \leq -3.95$; high–intermediate risk, $-3.95 < r_i \leq -3.29$; and high risk, $r_i > -3.29$.

The utility of these results lies in our ability to classify individuals into one of the risk groups on the basis of covariate values. This is fairly easily accomplished using the calculation defined by the linear combination in equation (17.2) or (17.3). A particular risk set is *heterogeneous* with respect to individual covariate values, but *homogeneous* with respect to risk. Therefore the character of a risk set does not have a simple interpretation in terms of covariates, but it does have a simple clinical interpretation, that is, increased or decreased risk.

17.2.8   Power and Sample Size for PFAs

Sometimes we can conduct PFAs on a surplus of data and would like to know how few observations can be used to meet our objectives. Other times we wish to know in advance how large a covariate effect must be to yield statistical significance. Both of these questions can be addressed by methods for determining power and sample size in PFAs. Calculating power for PFAs is difficult in general, but a few guidelines can be given.

Suppose that we are studying a single binary predictor, denoted by X, in a time-to-event analysis analogous to the effect of treatment in a planned trial. The distribution of $X = 0$ and $X = 1$ in the study population will not be equal as it might have been if it actually represented an assignment to a randomized treatment group. Instead, X is probably going to be unbalanced. Furthermore we are likely to be interested in the effect of X adjusted for other variables in the regression.

Under these assumptions a sample size formula such as equation (7.21) might be used to determine the approximate number of observations needed in a proportional hazards regression model. To detect an adjusted hazard ratio of $\Delta = 1.5$ attributable to a binary covariate with 20% of the population having $X = 1$, with 90% power and using a type I error of 5%, we calculate

$$D = \frac{(r + 1)^2}{r} \frac{(Z_\alpha + Z_\beta)^2}{[\log(\Delta)]^2} = \frac{(4 + 1)^2}{4} \frac{(1.96 + 1.282)^2}{[\log(1.5)]^2} = 240.$$

Therefore we have a reasonable chance of detecting an adjusted risk ratio of 1.5 in a PFA using 240 subjects where 20% of them have $X = 1$ and 80% have $X = 0$. More precise methods for performing these types of calculations are available (Statistics and Epidemiology Research Co., 1993). However, the general method is complex and not likely to be used frequently by clinical trial investigators. Consequently it is not be discussed in detail here.

In any case, prognostic factor analyses most often utilize all of the data that bear on a particular question or analysis. Missing variables (incomplete records), interest in subsets, and highly asymmetric distributions of variable values tend to limit precision even when there appears to be a large database on which to perform analyses. For survival and time-to-event outcomes, the number of events is frequently the most restrictive characteristic of the data.

17.3   ADJUSTED ANALYSES OF COMPARATIVE TRIALS

Covariates are also important purveyors of ancillary information in designed clinical trials. They facilitate validating the randomization, allow improved prognostication, can generate or test new hypotheses because of their associations with each other and with outcome, and may be used to improve (reduce the variance of) estimates of treatment effect. The possibility that estimates of treatment effect can be influenced or invalidated by covariate imbalance is one of the main reasons for studying the results of adjusted analyses. However, analysis of covariates and treatment effects simultaneously in a clinical trial is typically not a well-conditioned exercise practically or inferentially and should be considered with care (Ford and Norrie, 2002; Grouin, Day, and Lewis, 2004).

In essence, covariate adjustment is a type of subset analysis, and the tests generated contribute to problems of multiplicity (Section 16.7). This problem can be made even worse when it becomes necessary to explore interactions. The number of possible combinations of factors, and therefore the potential number of significance tests, increases dramatically with the number of covariates.

Not all clinical trial statisticians agree on the need for adjusted analyses in comparative clinical trials. From a theoretical perspective, randomization and proper counting and analysis guarantee unbiasedness and the correctness of type I error levels, even in the presence of chance imbalances in prognostic factors. In a randomized experiment, variability may still influence the results. The distinction is between the *expectation* of a random process and its *realization*. Adjustment can increase the precision of estimated treatment effects or control for the effects of unbalanced prognostic factors. Also the difference in estimated treatment effects before and after covariate adjustment often conveys useful biological information. Thus, although not necessary for valid tests of the treatment effect, adjusted analyses may facilitate other goals of randomized clinical trials.

Suppose that the analyst thought that the data from an RCT arose from an observational study rather than from an experimental design. The analysis would likely proceed much as it was sketched above for PFAs. Other investigators, knowing the true origins of the data, might not perform covariate adjustment. The conclusions of the two analyses could be different. Although we would prefer the analysis that most closely follows the paradigm of the study, it is not guaranteed to yield the most efficient or informative estimate of the treatment effect. It seems we have little choice but to explore the consequences of covariate adjustment and emphasize the results that are most consistent with other knowledge.

17.3.1 What Should We Adjust For?

There are two sources of information to help in deciding which covariates should be used for adjusted analyses: the data and biological knowledge from outside the trial. An excellent discussion of using the observed data to decide which covariates should be studied in adjusted analyses is given by Beach and Meier (1989). They conclude on the basis of some real-world examples and statistical simulations that only covariates associated with disparity (i.e., distributed differently in the treatment groups) and influence (i.e., distribution of outcomes across the covariate levels) are likely candidates for adjusted analyses. Specifically, the product of Z statistics for influence and disparity appears to govern the need for covariate adjustment, at least in simple cases.

Investigators might use these ideas in adjusting estimated treatment effects for prognostic factors that meet one of the following criteria:

1. Factors that (by chance) are unbalanced between the treatment groups.
2. Factors that are strongly associated with the outcome, whether unbalanced or not.

3. To demonstrate that a prognostic factor does not artificially create the treatment effect.

4. To illustrate, quantify, or discount the modifying effects of factors known to be clinically important.

The philosophy underlying adjusting in these circumstances is to be certain that the observed treatment effect is "independent" of the factors. The quantitative measures of clinical interest after adjustment are changes in relative risk parameters rather than changes in p-values. Therefore some adjusted analyses will include statistically non-significant variables but will be informative in a broad context.

One should not adjust estimated treatment effects for all of the covariates that are typically measured in comparative trials. Not only are large numbers of uninteresting covariates often recorded, but model building in that circumstance can produce spurious results due to multiplicity and collinearities. However, clinical knowledge, preclinical data, and findings in the trial data, can contribute to covariate adjustment that can improve inferences from comparative trials and generate new hypotheses.

17.3.2 What Can Happen?

There are many possible qualitative results when treatment is adjusted for covariates. Some hypothetical examples are shown in Table 17.8, where I have assumed that balance of the prognostic factors in the treatment groups is immaterial. Table 17.8 is constructed for proportional hazards models but the behavior discussed here could be seen with other analytic models. All prognostic factors in Table 17.8 are dichotomous. Models 1 through 3 are univariable analyses of treatment or covariates. All effect estimates are statistically significant at conventional levels.

Model 4 gives the treatment hazard ratio (HR) adjusted for sex. There is an increase in the size and significance of the treatment HR, typical of circumstances where the covariate effect is important on its own and also to the treatment effect. Adjustment usually pushes the treatment HR toward the null, opposite of what model 4 indicates. If the reduction in the HR were severe and it was rendered nonsignificant, it might be taken as evidence that the treatment effect was spurious, namely due to an imbalance in a strong covariate. A shift in the adjusted HR away from the null seems to occur infrequently in practice. Model 5 shows a different sort of result. When adjusted for risk level, the treatment HR is virtually unaffected. The effects of treatment and that covariate are nearly independent of one another.

Model 6 demonstrates some typical behavior that occurs with multiple covariates. The presence of both covariates alters the treatment HR, in this case away from the null. There is evidence of some collinearity between the covariates because risk group is reduced in significance, although the effect estimates are all similar to univariable results. The univariable analyses already sketched are essential for distinguishing unimportant covariates from those that are collinear.

It is worth mentioning that the circumstances of Table 17.8 were all controlled by simulation of the underlying data, which does not make the observed behavior any less typical. The sample size was 500 subjects in each treatment group with 20% censoring, and the true multivariable hazard ratios were treatment, 2.0; sex, 1.25; and high risk, 0.75. The true effect values are most clearly seen in model 6, which resembles the way the data were generated. Although less important for this discussion and not evident

TABLE 17.8 Hypothetical Proportional Hazards Regression Models Illustrating Adjusted Treatment Effects

Model	Variable	Hazard Ratio	95% Confidence Limits	P-Value
1	Treatment group	1.74	1.51–2.01	<0.0001
2	Sex	1.46	1.27–1.69	<0.0001
3	High risk	0.84	0.73–0.97	0.01
4	Treatment group	1.93	1.66–2.24	<0.0001
	Sex	1.66	1.43–1.92	<0.0001
5	Treatment group	1.76	1.52–2.04	<0.0001
	High risk	0.82	0.71–0.94	0.004
6	Treatment group	1.94	1.67–2.25	<0.0001
	Sex	1.62	1.40–1.88	<0.0001
	High risk	0.88	0.76–1.01	0.07

form the results, the interaction effect of sex and risk group on treatment was a hazard ratio of 1.75, and the odds ratio of association between sex and risk group was 1.6.

The presence of additional covariates could complicate such findings greatly. Three problems are made much more difficult by a sizable number of covariates: multivariable collinearities, interactions among predictors, and selecting a parsimonious model. The principles already discussed are not different in such cases, but the scale of the problem requires more planning, disciplined analyses, and cautious interpretation.

Brain Tumor Case Study

The utility of conducting adjusted analyses of treatment effects in randomized clinical trials can be seen in a trial of BCNU impregnated implantable polymers for the treatment of recurrent malignant gliomas (Brem et al., 1995). Two-hundred and twenty-two patients were randomized with equal probability to receive either BCNU polymer or placebo polymer, implanted in the cavity remaining after surgical resection of recurrent tumors. The polymer released BCNU slowly over a three-week period to achieve higher local concentrations of drug than one could obtain with systemic administration. In this study the randomization was not stratified to balance strong prognostic factors.

When the trial was analyzed, 215 patients had died of recurrent disease. The overall effect of BCNU polymer was to reduce the risk of death by approximately 18% (Figure 17.3). The estimated hazard ratio was 0.82 and the 95% percent confidence interval on the hazard ration was 0.45 to 1.03. Thus an analysis adhering strictly to the design of the trial and using the conventional interpretation of statistical significance would conclude that BCNU polymer was not of convincing benefit.

Interestingly the pathological type of the tumor was a strong determinant of survival time, as were the previous use of nitrosureas, pathologically active versus quiescent cells, and other factors listed below. When the treatment effect was estimated stratified by pathological type and adjusted for other factors, the precision of the estimate was

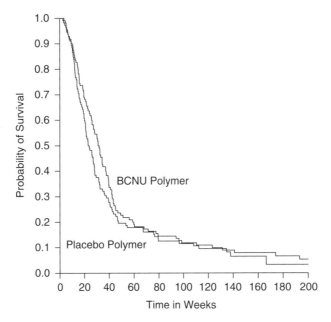

Figure 17.3   Observed survival following reoperation for malignant glioma.

improved and more convincing evidence of a beneficial effect of BCNU emerged. The adjusted regression analysis is shown in Table 17.9.

The primary purpose of the adjusted analysis was not to create statistical significance where none existed before. In fact considerable evidence regarding the likely beneficial effect of BCNU already existed. The agent has shown activity in many studies when administered systemically. Furthermore preclinical data for the polymer delivery system strongly supported its biological efficacy. Finally the estimated hazard ratio for BCNU polymer did not change substantially after adjustment for the prognostic factors listed. Based on the findings of this trial and supporting data, biodegradable polymers impregnated with BCNU were approved by the Food and Drug Administration in 1996 for treatment of recurrent brain tumors.

TABLE 17.9   Proportional Hazards Multiple Regression Stratified by Pathologic Type of Tumor

Variable	Hazard Ratio	95% Confidence Limits	P-Value
BCNU polymer vs. placebo	0.69	0.52–0.91	0.01
Performance status $\geq$70 vs.<70	0.66	0.49–0.91	0.01
Local vs. whole brain irradiation	0.59	0.42–0.83	0.003
Active vs. quiescent	1.93	1.26–3.78	0.02
Previous nitrosurea vs. none	1.53	1.13–2.08	0.006
White vs. other races	1.75	1.03–2.99	0.04
>75% resection vs. <75%	0.67	0.49–0.93	0.02
Age (per decade)	1.25	1.11–1.40	<0.001

This example illustrates the utility of adjusted analyses of comparative trials for reducing the variability of estimated treatment effects. Adjustment would also be useful for accounting for the effects of chance imbalances. Although there is a potential to use adjustment for data dredging and inappropriately enhancing statistical significance, it is a valuable tool that can assist with the data analysis of randomized studies.

17.4   NON–MODEL-BASED METHODS FOR PFAS

Statistical models are flexible and informative tools that help estimate useful biological effects from experimental (and nonexperimental) study designs. Models derive their strengths from knowledge or assumptions about the error structure, covariate effects, and other mathematical relationships in the data. When these assumptions are incorrect, models may yield the wrong answers or may not utilize the information in the data efficiently. There are other approaches useful for examining structure and associations in data that do not require all of the assumptions of statistical models. These approaches are more general and as simple to apply as statistical models.

No statistical procedures are free of assumptions or pitfalls, including those that do not use structural models. Although the methods discussed here are relatively assumption-free, they do not necessarily yield inferences that are more clinically useful, relevant, accurate, or simple than statistical models. In some cases they yield less information. These limitations will be discussed below. Interesting reviews and comparisons of model and non–model-based methods are given by Hand (1992) and Hadorn et al. (1992).

17.4.1   Recursive Partitioning Uses Dichotomies

Recursive partitioning is a method for developing clinical prediction rules that uses dichotomies in prognostic variables to divide and subset the cohort. Its purpose is to form a small set of classes or categories, defined on several simple criteria, that have clinically significantly different outcomes. A new patient could be classified using the resulting scheme into a single category, and this classification would carry useful prognostic information.

Process
Recursive partitioning uses a stepwise approach to developing prediction rules. At each step a bivariate analysis is performed to determine which predictive factor from the ones under consideration yields a partition of the data with the lowest false-positive and false-negative rates. In some algorithms the different types of errors can be assigned different weights. Then each step partitions the data according to the lowest weighted probability of error. For each of the partitions that results, the same process is repeated using other prognostic variables. The result is a type of "tree," where each branch splits into two daughter branches based on a dichotomization of one predictor variable.

If this process is repeated enough times, the outcomes in one of the groups will all be the same, and no further partitioning of that group will be possible. Eventually this will be true for all the remaining branches and the partitioning process stops. At this point the classification tree can be "pruned" by recombining branches that do not increase the classification error. This process is called *amalgamation*, giving the procedure its common name, "recursive partitioning and amalgamation" (RPA).

Practical Use

The probabilities of correct classification from the partitioning tree can be used to guide therapeutic (or diagnostic) decisions. The trees that result from this process are clinically appealing because they are defined directly in terms of simple dichotomies of predictor variables. They require no computations to categorize patients and can yield reliable classifications.

Examples

Recursive partitioning has been used in a variety of clinical applications and is comparable to similar predictive methods based on discriminant or other models. RPA has been used to develop improved diagnostic capability for upper extremity fractures in children, where it appears to perform as well as a logistic discriminant function (McConnochie, Roghmann, and Pasternack, 1993). Seleznick and Fries (1991) used RPA to detect prognostic factors for systemic lupus erythematosus. The technique has also been applied to prognosticating hospitalization for asthma (Li et al., 1995). The classification methods have been extended to censored survival data (Clark et al., 1994; LeBlanc and Crowley, 1992; Segal, 1988) and have been used to study prognostic factors in breast cancer (Albain et al., 1992), ovarian cancer (Ansell, 1993), malignant gliomas (Curran et al., 1993), and childhood neuroblastoma (Shuster et al., 1992), to name a few examples.

17.4.2   Neural Networks Are Used for Pattern Recognition

Neural networks are logically defined structures and associated mathematical models that are modeled after simple neuronal interconnections and were developed to assist with complex classification processes such as pattern recognition. Using data with known associations between the input variables and the outcomes (or output or classification variables), the network can be "trained" to yield a high probability of correct classification. This process involves modifying the strengths of internal associations (weights) to reinforce correct classifications and de-emphasize incorrect ones. A variety of algorithms can be used to optimize and error function. After training, the network is applied to new data, where it uses its internal rules to classify or predict outcomes from the input variables. In this way it can be used to make prognostic classifications similar to the model-based methods and recursive partitioning described above. See Warner and Misra (1996) for a recent review.

Neural networks are well suited to large complex problems such as pattern recognition. They have been used for speech synthesis and identifying military targets. However, they have also been successfully applied to diverse areas in the biological sciences such as predicting protein structure, diagnosis, and prognostication for patients with breast cancer. They can even be used to model censored survival data (Faraggi and Simon, 1995). For a recent review of neural network applications to clinical data, see Minor and Namini (1996).

Structure

A simple neural network has three layers: input, hidden, and output (Figure 17.4). The input layer has one node for each input factor (variable, in the case of prognostication). Each of these nodes is connected to each node in the hidden layer, which typically contains fewer nodes than the input layer. Finally each node in the hidden layer is

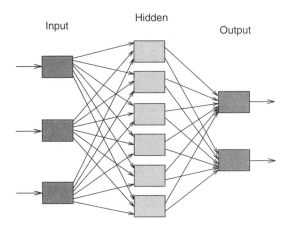

Figure 17.4   Connections in a basic neural network.

connected to each output node. For simple prognostication (e.g., a binary response) the output layer can contain a single node.

A consequence of this high degree of connectivity is that the network is complex and a large number of constants or parameters are required to specify its state. For example, a network with N inputs and N nodes in the hidden layer would have N^2 connections or weights that must be adjusted during training. This is considerably more parameters than a typical statistical model would require.

Practical Use

Some practical aspects of using neural networks for prognostication are discussed by Clark et al. (1994) and Ravdin and Clark (1992). Networks have been used for prognostication in cancer patients and appear to perform nearly as well as discrimination procedures such as logistic regression for predicting binary outcomes. However, neural networks do not seem to have been compared systematically and quantitatively to statistical models with a comparable number of parameters. For example, the interconnections in a neural network can only be described by using a relatively large number of parameters. This number may approximate n^2, where n is the number of nodes. In contrast, the typical statistical regression model usually employs much fewer than k parameters, where k is the number of variables.

Although developing prediction algorithms with neural networks is fairly simple, investigators may sacrifice some useful information when they give up more traditional model-based approaches. For example, it is difficult to quantify and formalize the degree of improvement that results from adding additional predictors to network models. There are no simple ways to quantify variability or precision in neural networks and no way to estimate risks or other clinical parameters attributable to individual prognostic factors. Thus there is a trade-off between lack of assumptions and utility of the information that is produced by such methods.

17.5   SUMMARY

An organized, detailed, quantitative study of prognostic factors usually provides valuable clinical information. These analyses help clinicians understand disease and select the best treatment for patients. They also provide a structure that can help to analyze clinical trials and design new studies. In some circumstances there are biologically important interactions between risk factors or between treatment and predictors that can be illuminated by PFAs.

Statistical models, especially multiple regressions, provide a convenient and powerful way to conduct prognostic factor analyses. These models perform several useful services. They combine theory and data, provide structure to the data, allow estimation of clinically important quantities, and permit testing and revising of the model itself. To be useful, statistical models must be built on a foundation of good data, statistical expertise, and clinical judgment. Investigators should take particular care with numerical coding of predictors, selecting variables for multiple regressions, and assessing interactions.

The same methods used for PFAs can often help analyze and interpret the results of randomized clinical trials. In particular, these so-called adjusted analyses of treatment effect may be helpful when strong prognostic factors are not controlled in the execution of the study, to demonstrate that known predictors have not confounded the outcome, and to explore the data. It is not necessary to analyze comparative trials in this way routinely, but it can often help explain and illuminate the findings.

Besides statistical models there are other useful ways for assessing the influence of prognostic factors on outcomes. These include recursive partitioning and neural network modeling. These methods tend to be operationally simple and yield clinically interpretable information about the effects of prognostic variables. The drawbacks to these techniques are that they do not quantify the effects of predictors or yield formal methods for revising the selection of variables.

17.6 QUESTIONS FOR DISCUSSION

1. When two continuous covariates are studied on their natural scales, they appear to interact significantly. However, when they are both "transformed" using 0, 1, or 2 for levels <1, $1-10$, and $10-100$, respectively, no interaction is observed. How do you explain this?

2. Two prognostic factors have been previously observed to be strong predictors of outcome. In your analysis, however, they are only individually almost statistically significant. When multiple regressions are constructed, the two predictors enter the model together (with opposite algebraic signs) and are highly significant. What might be going on, and what should you do next?

3. Investigators using neural networks for assessing prognostic factors suggest that the method is superior than logistic regression for predicting outcome. Design and discuss a convincing method to compare the two methodologies in the same data. Assume that the data are plentiful and of high quality.

18

REPORTING AND AUTHORSHIP

18.1 INTRODUCTION

Reporting the results of a trial is one of the most important aspects of clinical research. Investigators have an obligation to each other, the study participants, and the scientific community to disseminate results in a competent and timely manner. Chalmers (1990) suggested that failure to publish a well-designed trial is a type of scientific misconduct. The goal of a research report is to provide useful clinical information while allowing the reader to evaluate the investigators' conclusions. This chapter will cover some basics for accomplishing this task.

Authorship is an issue that superficially appears to take care of itself, particularly from the perspective of the junior clinical trialist. However, it is essential that authorship issues be addressed during the initiation of a collaboration, especially the kind that typically produce clinical trials. Failure to do so before study results are generated or draft manuscripts are prepared will invite disputes, even among well-intentioned and otherwise congenial colleagues. Authorship disputes can consume unimaginable amounts of energy and have lasting effects on one's career. Controversy can be minimized or prevented only by a deliberate plan early in the study. Small collaborations (e.g., early developmental trials) are no less prone to authorship disputes than multicenter studies.

With regard to reporting clinical trials, there are relatively few formal landmarks. One is the CONSORT (Consolidated Standards of Reporting Trials) statement that targets randomized trials (Moher, Schulz, and Altman, 2001; also www.consort-statement.org). Broader reporting issues have been addressed by the International Committee of Medical Journal Editors (ICMJE) evolving from the so-called Vancouver Group beginning in 1978. Their statement regarding biomedical journal manuscripts is broad (www.icmje.org) and also covers ethics principles related to reporting (ICMJE, 1997). Although not aimed at journal publication, ICH (International Conference on Harmonisation) Efficacy Guidelines (www.mcclurenet.com/ICHefficacy.html) provide a comprehensive framework for thinking about the content of clinical trial reports. The Guideline on Structure and Content of Clinical Study Reports (Federal Register, 1996; also www.mcclurenet.com/FedRegisterPDFs/E3.pdf) is especially useful, and is essential in the regulatory context. With regard to statistical content of biomedical reports, advice provided in Lang and Secic (1997) is excellent. The Council of Science Editors has addressed fundamental concerns regarding traditional authorship as opposed to a contributorship model (www.councilscienceeditors.org/ services/authorship.cfm).

The guidelines given in this chapter are typical of ones found in the literature for clinical trials and are flexible enough to apply to various trial types (e.g., Bailar and Mosteller, 1988). Additional details have been given by the Standards of Reporting Trials Group (1994) and the Working Group on Recommendations for Reporting Clinical Trials in the Biomedical Literature (1994). Also knowledge of good reporting practices can help with critical reading and reviewing. For an in-depth discussion of evaluating the medical literature (with an epidemiologic orientation), see Gehlbach (1988).

The guidance available for randomized trials and biomedical publications in general has probably not had a full effect on the reporting of developmental trials. The quality of reports for dose-finding studies and safety and activity (SA) trials is highly variable. For translational trials the structure and content of reports is so variable that it may be difficult to determine that they are translational trials at all. Some of this lack of coherence in reports of early developmental studies is due to the heterogeneity of the investigations. Much of it is a consequence of investigator laxity.

18.2 GENERAL ISSUES IN REPORTING

The overall utility of a trial report is a consequence of several components including (1) the importance of the scientific question, (2) the rigor of the study design and execution, (3) the quality of the written report, and (4) the external validity of the research design. It may not be possible to judge the quality of the trial and assess the importance of the results if the report is poorly written. We have all had the experience in our various journal clubs of reviewing a paper that might be very significant, but feeling uneasy about the results because of the poor quality of the report. Reporting methods and techniques can affect the reader's comprehension of results (e.g., Elting and Bodey 1991), aside from questions about validity. The way in which findings are described can have an impact on practitioner behavior (Bobbio, Demichelis, and Giustetto, 1994; Forrow, Taylor, and Arnold, 1992; Naylor, Chen, and Strauss, 1992).

Authoring a clinical trial report involves coordinating a number of interdependent tasks. These include resolving all questions about the data, finalizing the statistical analyses, interpreting the findings, building a consensus of investigator opinion, drafting the report, and responding to peer review and editorial feedback. These tasks can be extended in time as the data mature or additional questions arise. Each of these areas offers an opportunity for diverse opinions to be heard and possibly delay the reporting process. However, they also represent an important opportunity to improve the quality of the final product by accommodating different views and criticisms. A structured reporting plan can reduce error during this last stage of research.

Frequently results must be presented in more than a single forum such as an investigator's meeting, national society meeting, and journal. Reports must be tailored to the nature of the scientific question, the disease, and the audience. The important issues will be different for disease prevention studies than for therapeutic trials. Similarly papers written for a specialty journal may need to have a somewhat different emphasis from those for a more general audience.

The International Committee of Medical Journal Editors (ICMJE) announced a new requirement, effective July 2005, to register clinical trials in a public registry as a prerequisite for publishing results in their 11 member journals (De Angelis et al., 2004). The full impact of this may take some time to be evident, and it may reduce bias from underreporting or lack of awareness of negative results. It will also be an important factor for authors and investigators to consider, especially if other medical journals adopt the same or similar policies. See Steinbrook (2004b) for a perspective on this issue.

18.2.1 Uniformity of Reporting Can Improve Comprehension

The most informative summaries and the amount of detail to report from a clinical trial depend on the nature of the clinical hypotheses being studied. We would not expect reports of single-institution pharmacologic trials to be similar to those from multicenter randomized trials. Furthermore not all journal editors and reviewers will agree on the ideal content of reports. However, many features of good reports are similar for all types of trials and find widespread acceptance in journals. Some care must be taken so that uniform standards do not obscure important facts. See Walter (1995) for a discussion of this point in the context of confidence intervals.

One simple example of a uniform reporting criterion is a requirement to have all papers and presentations reviewed by the study principal investigator and the trial methodologist. This requirement is not likely to generate much controversy. A second simple example of uniformity relates to the protection of research subjects. Journals require affirmation that reports of research studies have been reviewed and approved by the investigators' Institutional Review Boards. Some other attempts at uniformity focus on the content of the report title and the content and structure of the abstract. Structured abstracts have been advocated by the Ad hoc Working Group for Critical Appraisal of the Medical Literature (1987; Haynes et al., 1990) but are not universally employed.

Uniformity is helpful to readers, particularly to those who are the least familiar with the details of the disease or intervention under investigation. Uniformity of reporting is likely to become more of a priority in situations where the historical record, corporate memory, and detailed findings of particular studies become more distant in time but remain vital to new research. The benefits of uniformity are evident in some chronic diseases like cancer, where standardized staging has improved trial design, reporting, and interpretation. Some cancer journals have called for a uniform reporting format (Weymuller and Goepfert, 1991) as have other specialty journals (e.g., Waldhausen and Localio, 1996).

Many journals have limitations on the length of research reports or at least discourage those that are long. Editors prefer short assertive papers without much speculation. Often the methods section of papers are trimmed to meet such restrictions, causing important details to be omitted. One possible solution to this problem is to publish a separate methods paper for large or important clinical trials. Then the results paper can contain only a summary of the key methodologic points. Because publishing a separate methods paper is usually not possible, investigators should provide complete and well-organized information in a single report.

18.2.2   Quality of the Literature

Despite many years of published randomized comparative trials, the content and quality of trial reports in the literature remains remarkably inconsistent on these points. A few systematic studies of published papers highlight the variable quality of trial reports. Mahon and Daniel (1964) examined 203 reports of drug trials published in the *Canadian Medical Association Journal* between 1956 and 1960. Only 11 of these reports were judged by the examiners to meet the criteria for a valid report. The most consistent deficiencies they noted were the absence of any type of control group (133 studies) and nonrandom allocation of controls (49 studies).

Pocock, Hughes, and Lee (1987) examined 45 trial reports published during 1985 in three leading medical journals and found that the large majority did not provide sample size, misused confidence intervals, or tended to overestimate the efficacy of the treatment under investigation. Similar examinations of the published literature have shown deficiencies in descriptions of randomization (Altman and Doré, 1990; Gøtzsche, 1989; Schulz et al., 1995). Although most attention regarding reporting has been given to randomized trials, phase I trials in oncology have also been found to be poorly described (Winget, 1996).

Traditionally the information contained in a published report of a clinical trial has been determined entirely by the investigators and the editorial process of the journal. Some authors have given guides for reporting trial results (Grant 1989; Meinert, 1986; Mosteller, Gilbert, and McPeek, 1980; Simon and Wittes, 1985; Zelen, 1983). Despite these suggestions trial reporting remains essentially unstructured. A few journals have also attempted to improve this area by publishing checklists and guidelines for reporting (Gardner, Machin, and Campbell, 1986; Gore, Jones, and Thompson, 1992; Murray, 1991b), but most have not.

18.2.3   Peer Review Is the Only Game in Town

The content of the medical literature reflects an imperfect editorial and peer review process. Limitations of peer review include unqualified, biased, tired, or inattentive referees and deficiencies in the management of submitted manuscripts. Some strengths of the process are discussed in Abby et al. (1994). Review of findings by one's peers is a fundamental value of the scientific process. Discussions of these and other aspects of peer review were undertaken at the Second International Congress on Peer Review in Biomedical Publication (Chicago, September 9–11, 1993). Many of the papers presented at the congress were published in Volume 272 of *JAMA* (No. 2). This journal issue provides a broad view of the peer review process. Several authors suggested quality rating instruments as a way to correct deficiencies in the peer review process (Cho and Bero, 1994; Feurer et al., 1994; Rochon et al., 1994).

Peer review has limitations and may be modified in the future because of its corruption, the growth of science, and electronic dissemination of information (Judson, 1994). However, despite the current and future limitations of peer review, currently there are no good alternatives for judging the relative merits of scientific papers. Modifications to the process, such as Internet publishing, may improve timeliness, decrease bias, and correct other shortcomings, but peer review will not soon disappear. Electronic publishing may provide a quick and more flexible alternative in the future, and can potentially correct some of the deficiencies in a paper-based system. It can improve retrieval and linkages to related information, both for the reader and the author. As a consequence papers can be more inclusive and detailed. Finally electronic media can shorten the editorial and review process, perhaps even eliminate it, as the practice of posting preprints on electronic bulletin boards has done in some physics and biology groups.

Until peer review is replaced, investigators have only the option of preparing reports in the traditional manner. A trial report may fail the initial test of peer review for a number of reasons. These include (see Kassirer and Campion, 1994): design flaws or consequential biases that prevent the question from being answered; deficiencies in presentation such as poor rationale or background, inappropriate omission or analysis of data, lack of objectivity, and poor writing; deficiencies in presentation of the results such as overemphasis on weak data, preliminary findings, or irrelevant data; and concerns about the importance or relevance of the work. Avoiding such errors usually permits the investigators to have an audience. What they do with the opportunity to have others read and review their work depends on some of the finer points of reporting.

18.2.4   Publication Bias Can Distort Impressions Based on the Literature

Publication bias is the name given to the tendency for studies with "positive" (i.e., non-null) findings to be preferentially selected for publication over those with "negative" findings. It is a *selection bias* for studies that yield a particular type of result. This phenomenon has been convincingly demonstrated by several authors (e.g., Begg and Berlin, 1988; Dickersin, 1990). An interesting study of publication bias has been discussed by Kemper (1991). Less rigorous research designs are not the only ones subject to publication bias—large randomized studies are subject to it as well (Krzyzanowska, Pintilie, and Tannock, 2003).

It is common to have trials criticized when they are "not strongly positive" and to struggle to have them published. At the time of this writing, I have on my desk the review from a collaborative multicenter SA trial. The associate editor of the *Journal of Clinical Oncology*, a major journal in the field, rejected the manuscript by stating, in part,

> The strengths of this study are that the premise . . . is reasonable, and it was well executed. However, this manuscript has several weaknesses. Most importantly, this was a negative study, and no data are presented that suggest that this approach or a subsequent iteration might be positive for a subset of patients.

This is a nearly perfect illustration of a source of publication bias. More formal research also supports the notion that nonsignificant results are less likely to be published (Dickersin et al., 1992). If the published literature is used as a basis for drawing conclusions about a treatment or group of related treatments from independent studies (as it should be), a bias in favor of the treatment could result. This effect is possible for both informal reviews of the literature and more quantitative ones such as meta-analyses or overviews (discussed in Chapter 21).

In principle, publication bias should not be a problem. If an important therapeutic question is addressed by a trial, and the study has been conducted rigorously, and the report is properly written, the information will be useful to other investigators and practitioners. Therefore it deserves to be published as much as studies that appear to show a therapeutic advance. However, investigators often lose enthusiasm for negative results because they seem less glamorous than positive ones and may be viewed as failures by the research team and sponsors. This can lead to weaker reports. Some journal editors prefer to publish positive studies because they seem to improve the stature of the journal. This is paradoxical because, if we assume that true treatment advances are uncommon, then positive reports are more likely to be in error than negative ones.

There is no completely satisfactory way for the reader to correct for publication bias when viewing a particular paper, although some analytic approaches to the problem have been suggested (Iyengar and Greenhouse, 1988). One must recognize that selection bias exists and retain a little skepticism about positive results. In a few cases readers may be aware of unpublished studies that contradict or fail to support a particular finding and these can be used to soften enthusiasm. It may be useful to ask if the paper in question would have been published if it was a negative trial. In other words, is the quality of the study good enough to convey externally valid information consistent with no effect? Even so, there is no way to separate true-positive results from false-positive ones without replicating the study (several times). When conducting exhaustive quantitative overviews (Chapter 21), it is important to include data from both published and unpublished trials to counteract this bias.

Journal editors and reviewers can reduce publication bias in two ways. First, they must weight methodologic rigor and thorough reporting ahead of statistical significance when judging the merits of an article. Second, they must be willing to report negative findings from sound studies with as much enthusiasm as positive reports. In the first edition of this work, I suggested that it would be helpful to have a peer-reviewed *Journal of Negative Trials* devoted to such results, an idea that may have originated among European colleagues. There is now an on-line *Journal of Negative Results in Biomedicine* (http://www.jnrbm.com/home/), although not devoted to clinical trials. Such journals should encourage authors to complete and publish negative studies because they would obtain the same academic credit from them as from positive trials.

Negative results are vitally important to clinical practice and research broadly. They especially have the potential to alter clinical practice immediately. Consider the estrogen replacement trial example from Section 6.4.2. Similar remarks pertain to negative prevention or vaccine trials. Many of the studies that have had a major impact on their respective disciplines have had results consistent with the null.

Journals should also consider publishing shortened trial protocols for impending, well-designed studies. This practice is already done for a few large, selected, expensive, or otherwise high-profile trials. Publication of protocols would (1) alert the community to the conduct of a study, (2) make that fact available to literature searches and later research, (3) encourage and shorten the results paper, and (4) provide educational experience for trialists.

18.3   CLINICAL TRIAL REPORTS

There is considerable overlap in good reporting form and content for translational and developmental studies. One comprehensive source for structure and content of trial reports is the International Conference on Harmonisation (ICH) consensus guideline (ICH, 1996). This guideline requires several dozen pages to describe and contains a level of detail in excess of what can be put in the peer-reviewed literature. Nevertheless,

TABLE 18.1   CONSORT Guidelines for Randomized Clinical Trials (first half)

Topic	Description
Title & abstract	How participants were allocated to interventions (e.g., "random allocation," "randomized," or "randomly assigned").
Introduction	
Background	Scientific background and explanation of rationale.
Methods	
Participants	Eligibility criteria for participants and the settings and locations where the data were collected.
Interventions	Precise details of the interventions intended for each group and how and when they were actually administered.
Objectives	Specific objectives and hypotheses.
Outcomes	Clearly defined primary and secondary outcome measures, and when applicable, any methods used to enhance the quality of measurements (e.g., multiple observations, training of assessors).
Sample size	How sample size was determined, and when applicable, explanation of any interim analyses and stopping rules.
Randomization	
Sequence generation	Method used to generate the random allocation sequence, including details of any restriction (e.g., blocking, stratification).
Allocation concealment	Method used to implement the random allocation sequence (e.g., numbered containers or central telephone), clarifying whether the sequence was concealed until interventions were assigned.
Implementation	Who generated the allocation sequence, who enrolled participants, and who assigned participants to their groups.
Blinding (masking)	Whether or not participants, those administering the interventions, and those assessing the outcomes were blinded to group assignment. When relevant, how the success of blinding was evaluated.
Statistical methods	Statistical methods used to compare groups for primary outcome(s); Methods for additional analyses, such as subgroup analyses and adjusted analyses.

it illustrates aspects of study reporting that should be familiar to all investigators, even though a published report will be less extensive.

The CONSORT guidelines for RCTs are shown in Tables 18.1 and 18.2. The principles underlying these guidelines serve as a sensible basis for reports of all types of trials. In the remainder of this section, I will emphasize some points related to the guidelines without reiterating them.

18.3.1   General Considerations

The basic issues that will concern readers of a clinical trial result are the importance of the scientific question addressed by the study, the strength of the research design, the appropriateness of the analytic methods, and the quality of the report. There are

TABLE 18.2   CONSORT Guidelines for Randomized Clinical Trials (second half)

Topic	Description
Results	
Participant flow	Flow of participants through each stage (a diagram is strongly recommended). Specifically, for each group report the numbers of participants randomly assigned, receiving intended treatment, completing the study protocol, and analyzed for the primary outcome. Describe protocol deviations from study as planned, together with reasons.
Recruitment	Dates defining the periods of recruitment and follow-up.
Baseline data	Baseline demographic and clinical characteristics of each group.
Numbers analyzed	Number of participants (denominator) in each group included in each analysis and whether the analysis was by "intention to treat." State the results in absolute numbers when feasible (e.g., 10/20, not 50%).
Outcomes and estimation	For each primary and secondary outcome, a summary of results for each group, and the estimated effect size and its precision (e.g., 95% confidence interval).
Ancillary analyses	Address multiplicity by reporting any other analyses performed, including subgroup analyses and adjusted analyses, indicating those prespecified and those exploratory.
Adverse events	All important adverse events or side effects in each intervention group.
Discussion	
Interpretation	Interpretation of the results, taking into account study hypotheses, sources of potential bias or imprecision and the dangers associated with multiplicity of analyses and outcomes.
Generalizability	Generalizability (external validity) of the trial findings.
Overall evidence	General interpretation of the results in the context of current evidence.

important nuances in each of these, but there is essentially only one opportunity in the primary publication to present each effectively.

Avoid Statistical Pitfalls
Standard notation should be used throughout the report. Although even good statistical notation can appear obscure to those unfamiliar with it, it usually helps clarify results. Using the proper summary or descriptive statistic is essential. For example, investigators sometimes use standard deviations and standard errors interchangeably. The standard deviation (or other measures of distributional spread such as interquartile range) should be used as a descriptive summary, whereas the standard error is intended to convey the uncertainty of an estimate, such as a mean. Confidence intervals, discussed in Chapter 16, also convey the precision of an estimate and are more informative than significance levels.

Reports of clinical trials often do not distinguish between clinical significance and statistical significance, or actively confuse the two. Clinical significance is the stronger notion and is independent of statistical significance (p-values). It can be expressed in terms of the magnitude and direction of treatment effects or differences. Therefore the currency of a trial report should be the size of treatment effects and the precision with which they are estimated. This is particularly important when assessing equivalence. Significance levels do not capture evidence of equivalence, whereas effect sizes do. Evidence of no difference is not the same as no evidence of difference (Altman and Bland, 1995).

To avoid overemphasis on p-values, it is instructive to prepare the study report without them, describing all findings in terms of effect sizes, confidence intervals, and clinical significance. Then p-values can be added in appropriate places, of which there might be surprisingly few.

Titles and Abstracts Should Inform

Titles and abstracts need special attention because they are the only part of many reports that some readers will bother reading. The title should name both the drug, agent, regimen, test, device, or therapy and the target disease or condition. Generic names for drugs should be given, along with trade names if a particular product has been used. The exact nature of the trial design (e.g., dose escalation, safety and efficacy, comparative efficacy) can often be included in a subtitle.

Example 40   To illustrate how important nuances in a title can be, consider the National Emphysema Treatment Trial Research Group paper describing a high-risk subset of patients, titled Patients at High Risk of Death after Lung-Volume-Reduction Surgery (NETT, 2001). This trial was high profile in the pulmonary, rehabilitation, and thoracic surgery communities, and the paper appeared in a high-impact journal. This resulted in some attention in the lay press and by national societies. Many reporters, members of the public, as well as health professionals read this title as a newspaper headline instead of as the title of a research article. The meaning is changed, if not nearly reversed, if we impute part of the verb "to be" in the title rather than leave it in its intended (and appropriate) adjectival form. The authors did not anticipate this misreading, which caused considerable consternation in the NETT Research Group.

Some journals require structured abstracts. Even when they do not, the abstract can usually be improved by paying attention to the following points. A good outline for an abstract recapitulates the text and includes the following points:

- Objectives of the study
- Design of the trial
- Setting or type of practices
- Characteristics of the study population
- Interventions used
- Primary outcome measures
- Principal result
- Conclusions

Abstracts generally should be shorter than 250 words. Statistical methods are not usually included in an abstract.

Dose-Finding Trial Reports Need Improvement

Although there is a large statistical literature on the merits of various designs for dose-finding trials, there has been little written about the methodological adequacy of published reports. One exception is the work by Winget (1996), who reviewed phase I studies of cytotoxic drugs for cancer therapy. Even in this highly specific and structured area, deficiencies are evident in the way that results are reported and drug doses selected for later clinical trials. From these observations in oncology, one would expect that early developmental trials in all disease areas are not documented optimally in the medical literature.

Keeping in mind that the basic objectives of dose-finding trials are to observe pharmacokinetics, find a good therapeutic dose, assess toxicity, and look for evidence of efficacy, it is not difficult to outline a good phase I study report. The categories to cover are (1) study design features, (2) characteristics of the study population, (3) estimates of clinically important pharmacokinetic parameters, (4) recommendations for the proper dose of drug to use in SA trials, (5) nature, severity, and reversibility of toxicity or side effects, and (6) evidence of treatment efficacy.

Design   Methods for dose-finding designs are still actively evolving. In recent years new designs have been suggested that appear to provide information efficiently and accurately about the dose–response or dose–toxicity function that underlies the observed outcomes. In reporting, it is important to describe the exact nature of the study design used because the quality of inference about the dose–response relationship can depend on it. The design description should also address selection criteria, the clinical and biological basis for the starting dose, reasons for stopping the trial, definition of the optimal biological dose, and the route and schedule of administration of the study drug and ancillary therapies.

Patient Characteristics   The outcome of developmental trials depends substantially on the characteristics of the patients being treated. In cancer trials patients who have been previously treated with cytotoxic therapy may have lower functional reserve in major organ systems due to toxicity. These patients may not experience side effects and responses similar to those in untreated patients. When major organ system toxicity is less of a concern, it may be satisfactory or advantageous to use a heterogeneous study population. In either case it is essential to provide a sufficient level of detail regarding the patients' characteristics so that the reader can assess the external validity of the trial.

Some specific points to report regarding the study population are the primary disease sites and extents of disease, extent of previous treatment, demographic characteristics, and the proportion of patients with poor prognosis based on objective criteria (e.g., performance status or functional index). It is standard practice in oncology to use toxicity and side effect criteria that are widely agreed upon and well understood. Although more difficult in other disease settings, there are many circumstances where quantitative criteria can be used to describe side effects. Total number of deaths in the study population and any attributable to the treatment under study should be reported.

Pharmacokinetics   Pharmacokinetic models are useful devices for describing drug distribution and excretion. Investigators are usually interested in certain parameters of clinical importance, such as drug excretion rate, half-life, peak concentration, or area under the time–concentration curve (AUC). For many agents, cytotoxic drugs in particular, the frequency and severity of toxicity may relate strongly to parameters such as the AUC or the peak blood concentration. Estimates of these important pharmacokinetic parameters should be reported in the publication, along with any empirical connections between them and actual toxicities observed. Pharmacokinetic estimates can depend on the model employed to analyze the data. Any such dependencies should also be reported.

Efficacy   It is relatively uncommon to have convincing objective evidence of efficacy during a dose-finding trial. However, it can occur and is an important finding when it does. For any sign of efficacy that might be attributable to the new therapy, investigators should report the type of (standard) response criteria used, the number of study subjects benefiting, and the duration of benefit. Many readers will also be interested in the cumulative dose (or dose intensity) used prior to observing responses, especially for cancer drugs. An important result from phase I studies is the recommended dose of drug to use in subsequent investigations. The study report should explicitly provide this recommendation, as well as the basis for it. In some cases side effects may be more likely in older, younger, or more frail patients, requiring the dose to vary on the basis of baseline characteristics. Recommendations related to dose differences should also be explicitly stated in the report.

SA Reports Should Minimize Biases and Overinterpretation

The clinical trials literature contains very little about reporting deficiencies, standards, or guidelines for SA clinical trials. However, investigators can anticipate these based on the purposes and potential weaknesses of these single cohort studies. The purposes of SA studies are to demonstrate treatment feasibility, to estimate success (efficacy), complication, and other event rates, and to facilitate a decision about proceeding with later development. SA trials also provide information about appropriate doses and scheduling of therapy for comparative trials.

Feasibility   With regard to feasibility goals, the SA report should present and quantify the frequency of problems that prevent the therapy from being carried out or require it to be modified (e.g., dosage reduced) because of side effects. To meet this goal effectively, investigators cannot permit "evaluability" or other clinical criteria to remove patients from the analysis who may represent evidence of nonfeasibility. The report must have a complete accounting of the treatment histories for all patients who met the eligibility criteria. It is sometimes helpful to track and report patients who met the eligibility criteria but did not enter the trial. If this happens frequently, it may be a sign that the therapy is unacceptable to patients. It may also be necessary to distinguish between short- and long-term feasibility using appropriate criteria.

Efficacy and Toxicity   Estimating success and complication rates requires a similar thorough accounting of patients. The criteria used to define a success should be reported. These criteria should use only information available at, or prior to, the specific clinical landmark. Successes defined in retrospect are suspect. The criteria used to establish a complication, toxicity, or side effect should also be contained in the report. Investigators should distinguish between major and minor side effects and define these terms in the report. In many chronic diseases, SA trials are designed to yield information about rates of disease progression, relapse, or death. Particularly for progression or relapse, the report should explain how they were defined. For relapse, presumably patients are found to be disease free by some criteria and then later found to have a relapse or recurrence. It is also helpful to describe the observed influence of prognostic factors on outcome because this can provide evidence that the study population is similar to others that have been reported.

Treatment Comparisons   SA trials permit, but usually do not encourage, comparisons with other treatments for the same population of patients. Some studies are carried out with the same or similar therapy at different institutions, again inviting comparisons. While such comparisons are error prone because of selection bias, they are natural and necessary. The SA trial report should recognize the potential for strong selection bias, qualify these types of comparisons (if they are made), and avoid definitive or overly enthusiastic statements about relative efficacy.

18.3.2   Employ a Complete Outline for CTE Reporting

Structured reports for randomized controlled trials have been discussed by the Standards of Reporting Trials Group (1994), who recommend a checklist to assist in preparing or evaluating a trial report. Although heavily weighted toward statistical issues, these guidelines are also helpful for assessing the clinical merits of trial reports. The Standards of Reporting Trials Group divides trial reports into four content areas: (1) treatment assignment, (2) treatment masking, (3) subject follow-up, and (4) statistical analysis (methods and results). Considering the necessary content of the entire report, we should extend these topics to include other areas such as introduction, design, discussion, and tables.

Considerations for writing titles and abstracts for CTE trials are the same as those given above for developmental trials. Additional points for CTE trial reports are given below. Many of the topics listed here may also pertain to some TM, DF, and SA trials. The format below roughly follows the outline of an article but is intended to be flexible.

Introduction and Background

The introduction serves two purposes. First, it informs readers who are less knowledgeable than the authors about the specific subject matter of the disease and the trial. Second, it justifies the setting and the scientific hypothesis tested. Both objectives require concise summaries and current references. The introduction, like that for a protocol, should anticipate the specific objectives of the trial. For investigators focused on more subtle aspects of the investigation, these points will seem uninteresting. However, for readers who are new to the study, this introduction is vital and may dictate whether or not they continue reading.

Study Sample and Sites   Clinically relevant descriptions of both the study and target population should be reported. This includes stating the experimental unit unambiguously. For example, is the experimental unit the individual, a cluster or family, or a larger group? In some phase I trials the unit of analysis may be the *treatment course* rather than the individual patient (appropriately or not). It may also be important to describe patients who met the eligibility criteria but chose not to participate in the trial, when this information is available. This may be essential when patients from a large group are asked to participate, but many refuse. It may be difficult for readers to generalize from these situations to clinic populations that are available at their institutions. For nonrandomized trials, referral and selection bias may make even detailed descriptions of the study group unsuitable or unconvincing to make comparisons with other studies. In any case, complete and accurate descriptions of the study population are essential so that the readers can assess the external validity of the results.

Study Design   There are many aspects of study design discussed throughout this book that are relevant to reporting the results of a CTE trial. Publications from long-standing collaborative groups may not require as much detail as those from single-institution or one-of-a-kind collaborations. However, when the trial is complex, controversial, high-profile, or otherwise of special interest, a separate design publication may be warranted and is likely to save space and increase clarity in the final report. A source of error is the monitoring plan, which must be very carefully described if the trial was terminated before the fixed sample size design indicated.

Methods to control bias such as type of internal control and treatment masking should be described. In randomized trials the method and administration of the randomization should be described to convince readers that it has been properly done.

Methods
Treatment and Eligibility Failures   Eligibility for a trial is determined in advance of study entry from pre-entry criteria specified in the protocol. When ineligible patients are mistakenly placed on study, investigators often analyze only the subset of eligible patients (perhaps in addition to other analyses). The estimate of treatment effect obtained in this way is not biased by patient exclusions because the selection factor (eligibility) affects both treatment groups equally. The trial report should describe those patients who were retrospectively found to have failed the eligibility criteria. Usually eligible patient analyses will be reported along with all randomized patient analyses. Similarly the report should describe patients who did not complete the assigned treatment.

Sometimes a large fraction of patients complete the assigned therapy but receive additional therapy not specified by the protocol or design of the trial. For example, patients with esophageal cancer may undergo resection and chemotherapy, and have a variety of second line treatments if signs of disease progression or recurrence are observed. If some of these latter treatments are effective, the results of an initial treatment comparison based on recurrence or survival may be skewed. In general, it is difficult or impossible to use the statistical information in studies that permit "crossovers" either to new treatments or to the other treatment arm, unless it is part of the study design. One might be able to use the information up to the time of stopping the assigned treatment using conventional methods of analysis.

Methods of Treatment Assignment and Masking   Although treatment assignment methods in clinical trials are generally straightforward, studies of trial reports indicate that assignment methods for many trials are inadequately reported (Williams and Davis, 1994). This seems to be true for both single-center and multicenter trials. To assure readers about the internal validity of the trial, investigators should report the methods by which treatment assignments were kept confidential.

Statistical Methods and Assumptions   Readers should be made aware of assumptions underlying the design and analysis of the trial. For a discussion of some practical issues, see DerSimonian et al. (1982). Many clinicians understand the assumptions and limitations of common statistical procedures. However, readers should be convinced that the data analyst has verified all important assumptions and reported methods in detail for less well-known statistical procedures. Examples of assumptions that are often made in analysis, sometimes violated by the data, and also likely to be consequential are distributional assumptions underlying parametric hypothesis tests, error distributions in regression analyses, and proportionality of hazards in lifetable regressions.

Results

Univariable Analyses   It is likely that the data analyst will test the effect of all potentially important prognostic variables on the major outcomes. For these univariable analyses, investigators should report estimated treatment effects (odds ratios or hazard ratios), confidence intervals, and significance levels of tests of no treatment effect (*p*-values). This does not preclude presenting other displays of univariable analyses (e.g., survival curves or 2 × 2 tables) if these analyses are especially relevant to the goals of the study. However, the investigators should keep in mind that univariable analyses, particularly in uncontrolled studies, are subject to confounding. Consequently these analyses need not be emphasized or presented in excessive detail.

Adjusted and Multiple Variable Analyses   In most randomized trials the univariable (unadjusted) comparison of treatment groups is a simple and valid summary of the treatment difference. However, many times it is useful to demonstrate that the treatment effect is not due to confounders. Evidence against confounding (by observed prognostic factors) can be obtained by using adjusted (multivariable) analyses. The best style of reporting multivariable analyses is the same or similar to that for univariable effects. However, the adjusted analyses reported are usually selected from a larger set of less informative or preliminary results. As an example, consider a lifetable regression model attempting to predict time to cancer recurrence. The "best" (most predictive but parsimonious) model might be built using a step-down procedure from a large set of potential prognostic factors. Each step in the analysis need not be reported, but the final model should be because it is a major objective of the analysis.

For multiple regression analyses, investigators usually report adjusted estimates of treatment effects, confidence intervals, and *p*-values. Not all prognostic factors retained in multiple regression models must be "statistically significant." It is often useful to keep nonsignificant effects in a multiple regression model to demonstrate convincingly that the treatment effect persists in their presence. See Chapter 17 for details regarding adjustment.

Discussion

Negative (Null) Findings   In noncomparative studies such as SA trials, investigators focus appropriately on the magnitude of clinical effects. Large or small effects can be judged in the context of the benefits of existing treatments, tempered by concerns about patient selection and confounding. One should not abandon this perspective in randomized comparisons. In comparative trials, the absence of a statistically significant difference is not the same as convincing evidence of no difference. In these studies it is also important to focus on estimated treatment differences rather than hypothesis tests.

Between 1960 and 1977 Freiman et al. (1978) reviewed 300 studies reported in 20 journals. Of these, 71 randomized trials found no difference between the treatment groups. The authors found that most negative trials had low statistical power to detect a 25% to 50% relative difference in the treatment groups. A study of 102 negative trials from a set of 383 studies in 1994 had similar findings (Moher, Dulberg, and Wells, 1994). A similar review between 1988 and 1998 of the surgical literature showed that null findings had low power and deficiencies in reporting (Dimick et al., 2001).

When no statistically significant treatment effect or difference is found in a comparative trial, readers sometimes ask about the power of the study (1) against the original alternative hypothesis and (2) to detect the observed difference. However, such power calculations are not usually helpful. This is true because, when the result is not significant, the treatment difference will usually be smaller than that on which the original power calculation was based. However, the study was not designed to detect this difference, making the power to detect the observed difference lower than the original power. Also the original alternative hypothesis is no longer supported by the data. Therefore the power against it, as well as other unsupported alternatives, is not very interesting.

Even so, it is sometimes helpful to report the motivations and assumptions behind the sample size employed in a trial. While knowing these may not affect interpretation of treatment differences when the study is complete, they may help in planning new or confirmatory trials. Helpful advice regarding these and other aspects of negative clinical trials is provided by Detsky and Sackett (1985).

Patient Exclusions   Although the intention-to-treat principle should be followed for the principal analyses from designed experiments (Chapter 15), it is often helpful to conduct many other exploratory analyses. Exploratory analyses that violate the intention-to-treat principle may be particularly informative in large comparative studies where the data are themselves a valuable resource. Examples of other types of analyses include eligible patients only, treatment received, and compliers only.

Most trial methodologists do not object to exploring the data but take a fairly conservative view of what should be represented as the final primary result of the study. This is particularly true of analyses that exclude patients from consideration. If the intention-to-treat approach is not followed, for example, one should report differences between it and the results that investigators believe are more relevant. Then readers of the literature can decide for themselves how consequential the different approaches are.

Figures and Tables

Statistical Significance   P-values do not determine clinical or biological significance and, as discussed elsewhere in this book, are poor measures of strength of evidence. When basing inferences predominantly on p-values, an arbitrary level should not be the only criterion for a declaration of "statistical significance." When biological or clinical support is strong, effect estimates are large, and confidence intervals or p-values indicate significance near conventional levels, it seems appropriate to label the result as "statistically significant." Conversely, results with no biological or clinical support, or those that seem paradoxical, should be reported and interpreted with caution. Even when p-values are smaller than 0.05, these results can be type I errors.

Exploratory Analyses   Exploratory or hypothesis-generating analyses should not be reported as the major findings of the trial. If subset or other exploratory analyses are performed, discrepancies between these and the major analyses of the clinical trial should be reported.

Biological Consistency   The most reliable guides to the external validity of both primary study results and exploratory analyses are (1) confirmation of findings by similar studies and (2) consistency with established biological theory or findings. For clinical trials exact replication is often not done, and even when supporting trials are undertaken, the results may take years to come out. Thus it is common to use preclinical data to support or refute the findings of trials. One should exercise caution in emphasizing findings that are unverified, seem counter to intuition, or contradict other seemingly well-established biological evidence. This is not to say that such findings should not be reported. In fact they may be vitally important. However, they should not generally be reported as the major findings of a study that was designed with other objectives in mind.

P-Values   In Chapter 16, I discussed the inadequacy of p-values as summaries of evidence from data. Their well-known deficiencies should lead investigators to de-emphasize p-values as the most relevant summaries for published reports. In particular, the fact that repetitions of comparative clinical trials are unlikely to yield significance levels that resemble the original reports and the undesirable combining of effect size and precision implied by p-values should encourage investigators to de-emphasize them.

18.4 AUTHORSHIP

Principles upon which authorship should be based have changed over time, and themselves may be the subject of occasional disagreement. The dominant model for authorship has been the laboratory model, which I describe below as "conventional." In the past, authorship on clinical trials has been seen as a gratuity for significant contribution to the study through sponsorship, accepting intellectual responsibility for the conduct of the study, clinical care of study subjects, analysis of data, or drafting and reviewing the study report. Because of issues at and beyond the point of publication, journal editors and others have steadily refocused the principle of authorship to mean contributing to the intellectual content of the study report and accepting responsibility for it. See Croll (1984) and Rennie et al. (1997) for some perspective on this issue.

I believe this is an appropriate separation of the act of conducting a trial from the reporting of the study design and results. It does mean, however, that clinicians who treat patients per study protocol, but do not participate in the writing of the report, may not be "granted" authorship. The justice of this is imperfect, but it is wholly consistent with viewing the study report as a stand-alone intellectual product. Furthermore the feasibility of large numbers of investigators to contribute to a relatively brief journal article is low, meaning that most papers from extensive collaborations would conventionally have only a handful of authors.

The tension surrounding authorship arises from the fact that there are relatively few sources of academic credit for clinical researchers. Furthermore the number of publications possible relative to the duration and intensity of the effort required to produce a clinical trial is probably lower than for many other fields (e.g., bench science). Possible sources of tangible credit include acknowledgments, citations of earlier work, and authorship. For young investigators, authorship is the most valuable, and even for seasoned investigators, other sources of credit depend on it. It essentially forms the foundation for appointments, promotion, and grant support and therefore is vital to an academic career.

There are various authorship models (discussed below). Some of these do not allow a conventional by-line. These should be used only by prior agreement, and may be suitable to resolve very contentious issues of the type that might occur in large collaborations with relatively few chances for academic recognition. Assuming a by-line is used, its content and ordering must be worked out before a manuscript materializes.

18.4.1 Inclusion and Ordering

The main basis for including authors should be contribution to the report and the study, in that order. This follows the editorial perspective emphasizing who has taken responsibility for the content of the article. The historical argument that "investigators who create the original idea and execute the trial should be authors" can be satisfied by also giving them the opportunity to create the research report as well. Those named in the by-line should be the writers and reviewers of the paper and also have the most knowledge of the study.

This opens the possibility that an individual would come relatively late to a collaboration and receive disproportionate credit by helping to prepare a manuscript without having done much of the "clinical" work. This is no more or less justified than giving authorship credit to someone who performed relatively routine patient care duties in the context of a study but did not contribute to the final stage of the project. In short, the valuable components of the research include the originating concept, experimental design, study execution, analysis, interpretation, and writing.

Ultimately assessing the worth of individuals' contributions for authorship falls to the lead author or writing committee. Those who make a significant contribution can usually be easily separated from those who do not. Many hard working individuals also do have the need or expectation of academic credit and can be appropriately treated by acknowledgment. This often includes support staff such as secretaries, programmers, technicians, sponsors, and administrative leaders of laboratories. The manner in which individuals are employed is not relevant for assessing contributions. For example, just because an individual is paid to perform support services, does not mean that they should not receive academic recognition.

Ideally every author should be able to take responsibility for content of the research report with regard to the design and conduct of the study. In practice, collaborations often involve multidisciplinary teams, members of which have to rely partially (on faith or experience) on the expertise of others. The ICMJE guidelines acknowledge this and suggest basing authorship on "contributions to (a) conception and design, or analysis and interpretation of data; and to (b) drafting the article or revising it critically for important intellectual content; and on (c) final approval of the version to be published" (ICMJE, 1997).

Aside from inclusion, the ordering of names in the by-line is an important issue. The reason why stems from various conventions including how citations are printed and abbreviated, the historical (and perhaps anomalous position of the lab, division, or section head), and assumptions made by readers and reviewers. It is not unusual for the "senior" author's name to come last, whereas the lead author is listed first. In some disciplines the names are arranged alphabetically.

Whatever mechanism is invoked to decide who the authors are must also establish principles for ordering. One simple method is to designate the lead author or chair of the writing committee, who then has responsibility for ordering all others according to their contributions to the study report. This has the advantage of simplicity. Provided there are a sufficient number of academic products coming out of a collaboration, it may be the most viable approach.

18.4.2   Responsibility of Authorship

The basic responsibilities of authorship are to draft and review the manuscript, to correct and refine it (re-draft), and to take public responsibility for the study. Authors must also be able to assure that (1) standards of ethical conduct of research have been followed, (2) the paper appropriately links to previous work regardless of whether it substantiates or contradicts, and (3) the report includes only and all observations actually made. Accepting these responsibilities allows the author the right to assure and maintain his or her professional reputation.

Members of the writing committee for a particular publication have more specific responsibilities. These include the following.

1. The writing committee chair is responsible for:
    (a) Recommending members of the writing committee
    (b) Assignment of tasks to members
    (c) Internal submission of draft manuscript
    (d) Distribution of drafts and managing the internal review process
    (e) Maintenance of a local archive of drafts
    (f) Determination of order of authors (footnote or masthead)
    (g) Choice of journal for submission
    (h) Correspondence with coauthors, group researchers, and journal editors
    (i) Exact copy of the paper at the time of submission for the study governance
    (j) Distribution of galleys for a final check of the paper
2. Writing committee members are responsible for tasks as assigned by the chair
3. Each writing committee member is expected to contribute substantively to the manuscript
4. Providing status reports to colleagues in the collaboration
5. Maintaining data confidentiality

In addition to these, senior members of the collaboration incur additional responsibilities with regard to investigators that they mentor. There can be a disadvantageous work-recognition balance for young investigators in larger clinical trial collaborations. It is an obligation of the senior investigators to see that young researchers have the opportunity to contribute to the final stages of the research in a way that gives them the credit they deserve. Many junior faculty members are disadvantaged with regard to this because the more established investigators get the grants and often have a fair amount of ego tied up in associating certain results with their own names. Senior investigators often do not need highly visible credit for career advancement and must be very active in advocating for their junior colleagues.

18.4.3   Authorship Models

Appropriate academic rewards for participating in multicenter clinical trials is difficult to distribute fairly. The conventional (laboratory) authorship model is interpreted by readers to give more credit to the first (and sometimes last) author. This perception coupled with limitations on space has prompted several authorship styles for these collaborations. They are (1) corporate, (2) modified corporate, (3) modified conventional, and (4) conventional. Specifications of each are as follows:

- Corporate: masthead author is listed as "_____ Research Group"; membership of writing committee is nowhere specified in the published article.
- Modified corporate: masthead author is listed as "_____ Research Group"; title page footnotes include a listing of the writing committee for the paper; the writing committee for the paper determines the order of the named authors in the footnote.
- Modified conventional: masthead authors are listed as "Name A, Name B, Name C, etc., for the _____ Research Group"; the writing committee for the paper determines the order of the named authors.
- Conventional: masthead authors are listed as "Name A, Name B, Name C,"etc.; the writing committee for the paper determines the order of the named authors.

The advantages of the first three authorship models is that credit is distributed appropriately to the many individuals who contributed to the study. Typically they are all listed in an appendix to the paper. The issue is which model should be applied to the various publications emanating from a study, particularly the primary outcome paper. Many research groups have adopted the corporate or modified corporate models for design and primary outcomes papers. Less important papers, such as ancillary studies, employ the modified conventional style. This avoids disputes and disproportionate academic credit.

My personal opinion is that corporate authorship models are almost always a poor compromise. For very high profile publications where it is impossible to distribute credit equitably, they solve a contentious problem. Otherwise, modified conventional should be considered first, for the following reasons: First, the need of journals to point to responsibility for the study report per se, rather than the study, is vital. Second, modified conventional allows some individuals to get direct academic credit. Third, it is impossible for any members of the writing team or study group to get appropriate credit for the work in today's high-pressure academics. Promotions committees usually cannot recognize the level or quality of work contributed by an individual from the appendix. Peer reviewers of credentials for promotions committees are likely to have a similarly difficult time. Fourth, and most important, the corporate style is contrary to the mentoring relationship that senior investigators should have with their junior colleagues.

It is unwise for investigators beginning their academic careers to spend large efforts on projects that do not show a clear path for academic credit. When such projects promise essentially no credit or ambiguous credit, as in the corporate model, the opportunity costs for junior faculty participation are much too high. Furthermore it is wrong for senior colleagues to take advantage of their mentoring relationships in this way. Senior colleagues always have ways to get academic credit for the work they do. They get research grants, give talks, teach, and advise in contexts that point to their role. Junior investigators have many fewer opportunities. Thus the most appropriate authorship policy is to employ modified conventional style, and insist that junior investigators do the reporting.

In any case, the need for a written policy early in the collaboration is clear. Additionally the investigators or publications committee should sketch the major papers that will arise from a collaboration early in the process, along with intended author lists. In this way disputes are minimized, enthusiasm can be maintained by junior investigators, mentoring is enhanced, and the academic credit can be distributed fairly.

In recent years, demands on authors have increased to the point where prior planning is necessary to decide on the details of the author list. In the past, it was common for authorship to be a "reward" for contributions to collaborative clinical trials. Physicians placing the most patients on study were often listed as authors of the report, even if they had little or nothing to do with the preparation or editing of the manuscript. This practice is no longer satisfactory. Most journal editors require all authors to have contributed substantially to the *report* and to agree with the material that it contains. Authors who do not help with the actual writing of the paper cannot give such assurances.

18.4.4 Some Other Practicalities

Policies and governance should be established for most of the issues surrounding publication and presentation of study results. The reason why such issues are more of a concern in clinical trials compared to conventional bench science collaborations is because the intellectual products of the typical clinical trial collaboration have a more diffuse shared ownership than those from laboratory research. Written policies should therefore cover the following points vital to the health of any collaboration:

- Ensure that abstracts, presentations, and publications are accurate, objective, and do not compromise the scientific integrity of the study.
- Prioritize presentations and publications.
- Expedite orderly and timely reports to the scientific community.
- Ensure that investigators have the opportunity to participate in analyses and the preparation of papers.
- Encourage and enhance interdisciplinary collaboration.
- Establish procedures for writing committees.
- Ensure that publications and presentations will be reviewed and approved in a timely fashion.
- Maintain and distribute the current list of publications.

The appropriate authorship model depends partly on the type of paper emerging from a collaboration. Following are some generic types:

- *Primary*. Dealing with the protocol-stated primary outcome, major hypotheses, or overall aspects of the trial (e.g., overall recruitment effort, population description). These often use some sort of corporate model.
- *Secondary*. Dealing with secondary outcomes of the trial, substudy, or a question that is selective. Modified conventional is often a good authorship model for such papers.
- *Tertiary*. Dealing with tertiary outcomes of the trial; ancillary study. Conventional authorship often suffices for these papers.

18.5   ALTERNATIVE WAYS TO DISSEMINATE RESULTS

Peer-reviewed publication has its limitations and alternatives have been tried, some on a small but visible scale. Presentation of results in preliminary form at national or international society meetings is used concurrently with manuscript submission by many authors. This practice has evolved from the research setting, where such meetings can be used effectively to gain needed input from colleagues during the project, to clinical settings where results are disseminated more rapidly and incompletely than by printed media. Sometimes investigators deliberately use the popular press either to augment presentation at a meeting or entirely independently. Because the popular media only disseminate sketchy clinical research results, they are not truly an alternative to peer-reviewed publication.

Another mechanism besides peer-reviewed publication for disseminating results is the so-called clinical alert. This mechanism has been used by the National Institutes of Health since 1988 to make practitioners and researchers aware of trial results that appear to have great public health importance. The clinical alert is an abbreviated non–peer-reviewed report distributed to health professionals through direct mail or other means. The communication summarizes the trial design and results (there may be more than one study covered) and carries an implication that the conclusions are convincing enough to warrant a change in practice. This mechanism was first used in May 1988 to describe the results of three large trials in women with node-negative breast cancer that showed improvement in relapse-free survival (DeVita, 1991) and several times since then. This way of disseminating results has not been uniformly endorsed by the research community (Borgen, 1991). See Hellman (1991), Macdonald (1990), Henderson (1990), Friedman (1990), and Wittes (1988) for different points of view.

18.6   SUMMARY

There are many reasons for clinical investigators to focus considerable attention on the written reports of their trials. Because of the importance of the report and the complexity of the scientific issues, it is helpful to adopt standards for content and structure of the paper. The main theme for doing so is to use a logical structure that facilitates a complete and comprehensible paper. The report should allow the quality of the investigation to be evident. Otherwise, trials may not carry the weight of evidence that they should. Investigators cannot rely only on our imperfect peer review mechanism to produce reports of high quality.

The quality of phase I reports has been nearly neglected. They should document a basis for all of the important inferences that need to be made. This includes judging if the recommended dose is appropriate, learning about the pharmacologic or other behaviors of the treatment, and assessing side effects and efficacy. Because it is natural for readers to try to generalize and compare knowledge represented in SA trials, it is important for investigators to minimize the potential for bias and misinterpretation in these reports. This goal is facilitated by a complete accounting of eligible patients and acknowledgment of the limitations of the study design.

Readers of CTE trials will be most interested in the internal and external validity of the findings. Reports that do not facilitate both inferences may leave readers confused or ambivalent about the trial. If so, the findings may have little impact. Because of the strengths of design for most randomized trials, well-written reports of "negative" (null) results frequently carry as much or more useful information as "positive" results.

Authorship issues must be addressed early in any collaboration. For large groups it is critical to have written policies to cover types of publications, processing manuscripts through the group, and authorship models. The statistical contribution to these complex trials is typically very substantial, and both the content and authorship should reflect it. In recent years the responsibilities and granting of authorship has become more formalized. The traditional style of small laboratory collaborations is not always adequate. The authorship model used for clinical trial collaborations should respect the efforts of researchers who contribute to the infrastructure of the study, balance it with the need for academic credit for young investigators who do much of the actual work, and assure that named authors will take responsibility for the report as written.

18.7   QUESTIONS FOR DISCUSSION

1. Read or scan the following papers: Prostate Cancer Trialists' Collaborative Group, 1995; Greenberg et al., 1994; Fisher et al., 1989; Non–small Cell Lung Cancer Collaborative Group, 1995; Mountain and Gail, 1981; Sadeghi, Lad, Payne and Rubinstein, 1988; ISIS-4 Collaborative Group, 1995. Comment on the strengths and weaknesses of the authorship policy implied by each of them for primary outcome papers and how they are referenced in MedLine.

2. A small consortium of investigators is discussing their authorship policies as part of their formal organization meeting. An opinion is voiced that a biostatistician should be a major contributor to, and author of, all SA and phase III studies from the group and that this policy should be stated in the group's Constitution. Some clinicians disagree. Give your own opinion and help the investigators resolve this question. Would your answer change depending on the experience and academic rank of the biostatistician?

3. Design a one-page checklist using the ideas in this chapter for assessing the quality of a published report from a CTE randomized trial. Try your checklist out on at least the following papers and discuss your findings: Rosell et al., 1994; Fisher et al., 1996; Villanueva et al., 1996; Spector et al., 1996; Bhasin et al., 1996.

4. Read Sokal (1996a, b) (in that order) and comment on the papers from a peer review perspective.

5. Examine one of the following publications and comment on the authorship policy: Aubert et al., 2001; Drutskoy et al., 2004; Oliver et al., 1992. There are numerous similar publications.

19

FACTORIAL DESIGNS

19.1 INTRODUCTION

Factorial clinical trials are experiments that test the effect of more than one treatment using a design that permits an assessment of interactions among the treatments. The name arises because historically the control variables have been called *factors* rather than *treatments* as in most medical applications. A factor can have more than one *level*. For example, a *factor* can be defined by the presence or absence of a single drug. Different doses of the same drug constitute *levels* of a factor.

A factor cannot be purely qualitative. For example, the choice between treatments *A* and *B* is not a factor (assuming that one is not a placebo). Many factors have only 2 levels (present or absent) and are therefore both ordinal and qualitative. The essential feature of factorial designs is that all factors are varied systematically (i.e., some groups receive more than one treatment) and the experimental groups are arranged in a way that permits testing if the combination of treatments is better (or worse) than individual treatments.

The technique of varying more than one factor or treatment in a single study was used in agricultural experiments in England before 1900. The method did not become popular until it was developed further by R.A. Fisher (1935, 1960) and Yates (1935), but since then it has been used to great advantage in both agricultural and industrial experiments. Influential and more recent discussions of factorial experiments are given by Cox (1958) and Snedecor and Cochran (1980). Factorial designs have been used relatively infrequently in medical trials, except recently in disease prevention studies.

Factorial designs offer certain advantages over conventional comparative designs, even those employing more than two treatment arms. The factorial structure permits certain comparisons to be made that cannot be achieved by any other design. In some circumstances two treatments can be tested using the same number of subjects ordinarily used to test a single treatment. Despite their potential advantages factorial designs have important limitations. These must be understood before deciding if a factorial experiment is the best design to employ for a particular therapeutic question. More complete discussions of factorial designs in clinical trials can be found in Byar and Piantadosi (1985) and Byar, Herzberg, and Tan (1993). For a discussion of such designs related to cardiology trials, particularly in the context of the ISIS-4 trial (Flather et al., 1994), see Lubsen and Pocock (1994) and McAlister et al. (2003).

19.2   CHARACTERISTICS OF FACTORIAL DESIGNS

19.2.1   Interactions or Efficiency, but Not Both Simultaneously

Factorial designs embody an essential dichotomy that causes them to be widely misunderstood. The dichotomy is that the same structural design can be used either to gain substantial efficiency in questions about individual treatments or to study treatment interactions. However, both objectives can never be met at the same time. In fact these two purposes of a factorial trial and the features required by each are in competition. From outside the design of a factorial trial can therefore appear conflicting or confusing.

A factorial design is capable of answering questions for two treatments with one sample size, but only if the treatments are known not to interact with one another (Section 19.2.3). This is a two-for-one efficiency. Conversely, the same structural design can estimate the interaction between two treatments, but the sample size required would be roughly four times as large (Section 19.3.1). Thus both objectives can not be met simultaneously. Firm biological knowledge is required to support the purpose for a factorial trial because the consequences for sample size are so great.

19.2.2   Factorial Designs Are Defined by Their Structure

The quickest way to learn the basic features of a factorial design is to study an example. The least complex factorial design has 2 treatments (A and B) and 4 treatment groups (Table 19.1). There might be n patients entered into each of the 4 treatment groups for a total sample size of $4n$ and a balanced design. One group receives neither A nor B, a second receives both A and B, and the other two groups receive one of A or B. This is called a 2×2 (two by two) factorial design. Although basic, this design illustrates many of the general features of factorial experiments. The design generates enough information to test the effects of A alone, B alone, and A plus B. The efficiencies in doing so will be presented below.

The 2×2 design generalizes to "higher order" designs in a straightforward manner. For example, a factorial design studying 3 treatments, A, B, and C is the $2 \times 2 \times 2$.

TABLE 19.1   Four Treatment Groups and Sample Sizes in a 2 × 2 Balanced Factorial Design

| Treatment | Treatment B | | |
A	No	Yes	Total
No	n	n	$2n$
Yes	n	n	$2n$
Total	$2n$	$2n$	$4n$

TABLE 19.2   Eight Treatment Groups in a Balanced 2 × 2 × 2 Factorial Design

Group	Treatments			Sample Size
	A	*B*	*C*	
1	No	No	No	*n*
2	Yes	No	No	*n*
3	No	Yes	No	*n*
4	No	No	Yes	*n*
5	Yes	Yes	No	*n*
6	No	Yes	Yes	*n*
7	Yes	No	Yes	*n*
8	Yes	Yes	Yes	*n*

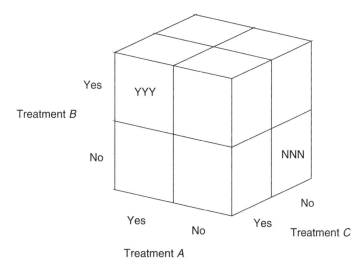

Figure 19.1   Structure of a 2 × 2 × 2 factorial design. The cell with no treatments and the cell with all three treatments are labeled.

Possible treatment groups for this design are shown in Table 19.2. The design can also be depicted as a cubic array of treatment cells (Figure 19.1). The total sample size is 8*n* if all treatment groups have *n* subjects.

Aside from illustrating the factorial structure, these examples highlight some of the prerequisites necessary for, and restrictions on, using a factorial trial. First, the treatments must be amenable to being administered in combination without changing dosage in the presence of each other. For example, in Table 19.1 we would not want to reduce the dose of *A* in the lower right cell where *B* is present. The reasons for this will become clearer below. This requirement implies that the side effects of the treatments cannot be cumulative to the point where the combination would be impossible to administer.

Second, it must be ethically acceptable not to administer the individual treatments, or administer them at lower doses as the case may be. In some situations, this means having a no-treatment or placebo group in the trial. In other cases, *A* and *B* may be

TABLE 19.3 Treatment Effects in a 2 × 2 Factorial Design

Treatment	Treatment B	
A	No	Yes
No	$\overline{Y}_0$	$\overline{Y}_B$
Yes	$\overline{Y}_A$	$\overline{Y}_{AB}$

administered in addition to a "standard," so all groups receive some treatment. An example of this circumstance was a factorial trial of chemotherapy and prophylactic brain radiotherapy in patients with non-small cell lung cancer, all of whom received chest radiotherapy (Miller et al., 1998). Third, we must be genuinely interested in learning about treatment combinations or else some of the treatment groups might be unnecessary. Alternatively, to use the design to achieve greater efficiency in studying two or more treatments, we must know that some interactions do not exist.

Fourth, the therapeutic questions must be chosen appropriately. We would not use a factorial design to test treatments that have exactly the same mechanisms of action (e.g., two ACE inhibitors for high blood pressure) because either would answer the question. Treatments acting through different mechanisms would be more appropriate for a factorial design (e.g., radiotherapy and chemotherapy for tumors). In some prevention factorial trials the treatments tested also target different diseases.

19.2.3 Factorial Designs Can Be Made Efficient

Although their scope is limited, factorial designs offer certain very important efficiencies or advantages when they are applicable. To illustrate this, consider the 2 × 2 design and the estimates of treatment effects that would result using an additive model for analysis (Table 19.3). Assume that the responses are group averages of some normally distributed response denoted by $\overline{Y}$. The subscripts on $\overline{Y}$ indicate which treatment group it represents. Note that half the patients receive one of the treatments (this is also true in higher order designs). For the moment further assume that the effect of A is not influenced by the presence of B.

There are two estimates of the effect of treatment A compared with placebo in the design, $\overline{Y}_A - \overline{Y}_0$ and $Y_{AB} - \overline{Y}_B$. If B does not modify the effect of A, it is sensible to combine (average) them to estimate the overall, or main, effect of A (denoted here by β_A),

$$\beta_A = \frac{(\overline{Y}_A - \overline{Y}_0) + (\overline{Y}_{AB} - \overline{Y}_B)}{2}. \tag{19.1}$$

Similarly

$$\beta_B = \frac{(\overline{Y}_B - \overline{Y}_0) + (\overline{Y}_{AB} - \overline{Y}_A)}{2}. \tag{19.2}$$

Thus in the absence of interactions (i.e., the effect of A is the same with or without B, and vice versa), the design permits the full sample size to be used to estimate two treatment effects.

Now suppose that each patient's response has a variance σ^2 and that it is the same in all treatment groups. We can calculate the variance of β_A to be

$$\text{var}(\beta_A) = \frac{1}{4} \times \frac{4\sigma^2}{n} = \frac{\sigma^2}{n}.$$

This is exactly the same variance that would result if A were tested against placebo in a single two-armed comparative trial with $2n$ patients in each treatment group. Similarly

$$\text{var}(\beta_B) = \frac{\sigma^2}{n}.$$

However, if we tested A and B separately, we would require $4n$ subjects in each trial or a total of $8n$ patients to have the same precision obtained from half as many patients in the factorial design. Thus, in the absence of interactions, these designs allow great efficiency in estimating main effects. In fact, in the absence of interaction, we get two trials for the price of one. Tests of both A and B can be conducted in a single factorial trial with the same precision as two single-factor trials using twice the sample size.

19.3   TREATMENT INTERACTIONS

The idea of interactions between treatments and discrete covariates was discussed in Section 12.4.4. We now consider more general circumstances where the effect of treatment A is influenced by the presence of treatment B (and vice versa). In such cases there is said to be a *treatment interaction*. Although the sample size efficiencies just discussed will be lost when this occurs, factorial designs become even more relevant.

19.3.1   Factorial Designs Are the Only Way to Study Interactions

One of the most consequential features of factorial designs is that they are the only type of trial design that permits study of treatment interactions. This is because the factorial structure has groups with all possible combinations of treatments, allowing the responses to be compared directly. Consider, again, the two estimates of the effect of A in the 2×2 design, one in the presence of B and the other in the absence of B. The definition of an interaction is that the effect of A in the absence of B is different from the effect of A in the presence of B. This difference can be estimated by comparing

$$\beta_{AB} = (\overline{Y}_A - \overline{Y}_0) - (\overline{Y}_{AB} - \overline{Y}_B) \tag{19.3}$$

with zero. If β_{AB} is near zero, we would conclude that no interaction is present. It is straightforward to verify that $\beta_{AB} = \beta_{BA}$.

An important principle of factorial trials is evident by examining the variance of β_{AB}. Under the same assumptions as in Section 19.2.3,

$$\text{var}(\beta_{AB}) = 4\frac{\sigma^2}{n},$$

which is four times larger than the variance for either main effect when an interaction is known to be absent. Therefore, to have the same precision for an estimate of an interaction effect as for a main effect, the sample size has to be four times larger. This illustrates again why both the efficiency and interaction objectives cannot be simultaneously met in the same factorial study.

When there is an AB interaction, we cannot use the estimators given above for the main effects of A and B (equations 19.1 and 19.2) because they assume that no interaction is present. In fact it is not sensible to talk about an overall main effect in the presence of an interaction because equation (19.1) or (19.2) would have us average over two quantities that are not expected to be equal. Instead, we could talk about the effect of A in the absence of B,

$$\beta'_A = (\overline{Y}_A - \overline{Y}_0),\qquad(19.4)$$

or the effect of B in the absence of A,

$$\beta'_B = (\overline{Y}_B - \overline{Y}_0).\qquad(19.5)$$

These are logically and statistically equivalent to what would be obtained from stand-alone trials.

In the $2 \times 2 \times 2$ design there are three main effects and four interactions possible, all of which can be estimated by the design. Following the notation above, the effects are

$$\beta_A = \tfrac{1}{4}\left[(\overline{Y}_A - \overline{Y}_0) + (\overline{Y}_{AB} - \overline{Y}_B) + (\overline{Y}_{AC} - \overline{Y}_C) + (\overline{Y}_{ABC} - \overline{Y}_{BC})\right]\qquad(19.6)$$

for treatment A,

$$\beta_{AB} = \tfrac{1}{2}\left[\left((\overline{Y}_A - \overline{Y}_0) - (\overline{Y}_{AB} - \overline{Y}_B)\right) + \left((\overline{Y}_{AC} - \overline{Y}_C) - (\overline{Y}_{ABC} - \overline{Y}_{BC})\right)\right]\qquad(19.7)$$

for the AB interaction, and

$$\beta_{ABC} = \left[(\overline{Y}_A - \overline{Y}_0) - (\overline{Y}_{AB} - \overline{Y}_B) - (\overline{Y}_{AC} - \overline{Y}_C) - (\overline{Y}_{ABC} - \overline{Y}_{BC})\right]\qquad(19.8)$$

for the ABC interaction. The respective variances are $\sigma^2/2n$, $2\sigma^2/n$, and $8\sigma^2/n$. Thus the precision of the two-way interactions relative to the main effect is $1/4$, and for the three-way interaction is $1/16$.

When certain interactions are present, here again it will not be sensible to think of the straightforward main effects. But the design can yield an alternative estimator for β_A or β_{BA} (or for other effects). Suppose that there is an ABC interaction. Then instead of β_A, an estimator of the effect of A in the absence of C would be

$$\beta'_A = \tfrac{1}{2}\left[(\overline{Y}_A - \overline{Y}_0) + (\overline{Y}_{AB} - \overline{Y}_B)\right],$$

which does not use β_{ABC} and implicitly assumes that there is no AB interaction. Similarly the AB interaction would be

$$\beta'_{AB} = (\overline{Y}_A - \overline{Y}_0) - (\overline{Y}_{AB} - \overline{Y}_B)$$

for the same reason. Thus, when high-order interactions are present, we must modify our estimates of lower order effects, losing some efficiency. However, factorial designs are the only ones that permit treatment interactions to be studied.

19.3.2 Interactions Depend on the Scale of Measurement

In the examples just given, the treatment effects and interactions have been assumed to exist on an additive scale. This is reflected in the use of sums and differences in the formulas for estimation. In practice, other scales of measurement, particularly a multiplicative one, may be useful. As an example, consider the response data in Table 19.4 where the effect of treatment A is to increase the baseline response by 5 units. The same is true of B, and there is no interaction between the treatments on this scale because the joint effect of A and B is to increase the response by $5 + 5 = 10$ units.

In contrast, Table 19.5 shows data in which the effects of both treatments are to multiply the baseline response by 2.0. Hence the combined effect of A and B is a fourfold increase, which is greater than the joint treatment effect for the additive case. If the analysis model were multiplicative, Table 19.4 would show an interaction, whereas if the analysis model were additive, Table 19.5 would show an interaction. Thus to discuss interactions, we must establish the scale of measurement.

19.3.3 The Interpretation of Main Effects Depends on Interactions

In the presence of an interaction in the 2×2 design, there is not an overall, or main, effect of either treatment. This is because the effect of A is different depending on the presence or absence of B. In the presence of a small interaction, where all patients benefit regardless of the use of B, we might observe that the magnitude of the "overall" effect of A is of some size and that therapeutic decisions are unaffected by the presence of an interaction. This is called a "quantitative" interaction because it does not affect the direction of the treatment effect. For large quantitative interactions it may not be sensible to talk about overall effects.

TABLE 19.4 Response Data from a Hypothetical Factorial Trial Showing No Interaction on an Additive Scale of Measurement

Treatment	Treatment B	
A	No	Yes
No	5	10
Yes	10	15

TABLE 19.5 Response Data from a Hypothetical Factorial Trial Showing No Interaction on a Multiplicative Scale of Measurement

Treatment	Treatment B	
A	No	Yes
No	5	10
Yes	10	20

In contrast, if the presence of B reverses the effects of A, then the interaction is "qualitative," and treatment decisions may need to be modified. Here we would not talk about an overall effect of A because it could be positive in the presence of B and negative in the absence of B, and could yield an average effect near zero.

19.3.4 Analyses Can Employ Linear Models

Motivation for the estimators given above can be obtained using linear models. There has been little theoretical work on analyses using other models. One exception is the work by Slud (1994) describing approaches to factorial trials with survival outcomes. Suppose that we have conducted a 2×2 factorial experiment with group sizes given by Table 19.1. We can estimate the AB interaction effect using a linear model of the form

$$E\{Y\} = \beta_0 + \beta_A X_A + \beta_B X_B + \beta_{AB} X_A X_B, \tag{19.9}$$

where the X's are indicator variables for the treatment groups and β_{AB} is the interaction effect. For example,

$$X_A = \begin{cases} 1 \ \textit{for treatment group } A, \\ 0 \ \textit{otherwise}. \end{cases}$$

The design matrix has dimension $4n \times 4$ and is

$$\mathbf{X}' = \begin{bmatrix} 1 & \ldots & 1 & \ldots & 1 & \ldots & 1 & \ldots \\ 0 & \ldots & 1 & \ldots & 0 & \ldots & 1 & \ldots \\ 0 & \ldots & 0 & \ldots & 1 & \ldots & 1 & \ldots \\ 0 & \ldots & 0 & \ldots & 0 & \ldots & 1 & \ldots \end{bmatrix},$$

where there are 4 blocks of n identical rows representing each treatment group and the columns represent effects for the intercept, treatment A, treatment B, and both treatments, respectively. The vector of responses has dimension $4n \times 1$ and is

$$\mathbf{Y}' = \{Y_{01}, \ldots, Y_{AI}, \ldots, Y_{BI}, \ldots, Y_{ABI}, \ldots\}.$$

By ordinary least squares estimation, the solution to equation (19.9) is

$$\widehat{\beta} = (\mathbf{X}'\mathbf{X})^{-1}\mathbf{X}'\mathbf{Y}.$$

When the interaction effect is omitted, the estimates will be denoted by $\widehat{\beta}^*$. The covariance matrix of estimates is $(\mathbf{X}'\mathbf{X})^{-1}\sigma^2$, where the variance of each observation is σ^2.
We have

$$\mathbf{X}'\mathbf{X} = n \times \begin{bmatrix} 4 & 2 & 2 & 1 \\ 2 & 2 & 1 & 1 \\ 2 & 1 & 2 & 1 \\ 1 & 1 & 1 & 1 \end{bmatrix}, \quad (\mathbf{X}'\mathbf{X})^{-1} = \frac{1}{n} \times \begin{bmatrix} 1 & -1 & -1 & 1 \\ -1 & 2 & 1 & -2 \\ -1 & 1 & 2 & -2 \\ 1 & -2 & -2 & 4 \end{bmatrix},$$

and

$$\mathbf{X}'\mathbf{Y} = n \times \begin{bmatrix} \overline{Y}_0 + \overline{Y}_A + \overline{Y}_B + \overline{Y}_{AB} \\ \overline{Y}_A + \overline{Y}_{AB} \\ \overline{Y}_B + \overline{Y}_{AB} \\ \overline{Y}_{AB} \end{bmatrix},$$

where $\overline{Y}_i$ denotes the average response in the ith group. Then

$$\widehat{\beta} = \begin{bmatrix} \overline{Y}_0 \\ -\overline{Y}_0 + \overline{Y}_A \\ -\overline{Y}_0 + \overline{Y}_B \\ \overline{Y}_0 - \overline{Y}_A - \overline{Y}_B + \overline{Y}_{AB} \end{bmatrix}, \tag{19.10}$$

which corresponds to the estimators given above in equations (19.3) through (19.5). However, if the test for interaction fails to reject and the β_{AB} effect is removed from the model,

$$\widehat{\beta^*} = \begin{bmatrix} \frac{3}{4}\overline{Y}_0 + \frac{1}{4}\overline{Y}_A + \frac{1}{4}\overline{Y}_B - \frac{1}{4}\overline{Y}_{AB} \\ -\frac{1}{2}\overline{Y}_0 + \frac{1}{2}\overline{Y}_A - \frac{1}{2}\overline{Y}_B + \frac{1}{2}\overline{Y}_{AB} \\ -\frac{1}{2}\overline{Y}_0 - \frac{1}{2}\overline{Y}_A + \frac{1}{2}\overline{Y}_B + \frac{1}{2}\overline{Y}_{AB} \end{bmatrix}.$$

The main effects for A and B are as given above in equations (19.1) and (19.2).

The covariance matrices for these estimators are

$$\widehat{\text{cov}\{\beta\}} = \frac{\sigma^2}{n} \times \begin{bmatrix} 1 & -1 & -1 & 1 \\ -1 & 2 & 1 & -2 \\ -1 & 1 & 2 & -2 \\ 1 & -2 & -2 & 4 \end{bmatrix} \tag{19.11}$$

and

$$\widehat{\text{cov}\{\beta^*\}} = \frac{\sigma^2}{n} \times \begin{bmatrix} \frac{3}{4} & -\frac{1}{2} & -\frac{1}{2} \\ -\frac{1}{2} & 1 & 0 \\ -\frac{1}{2} & 0 & 1 \end{bmatrix}. \tag{19.12}$$

In the absence of an interaction, the main effects of A and B are estimated independently and with higher precision than when an interaction is present. The interaction effect is relatively imprecisely estimated, indicating the larger sample sizes required to have a high power to detect such effects.

19.4   EXAMPLES OF FACTORIAL DESIGNS

Many trials conducted in recent years have used factorial designs. A few interesting examples are listed in Table 19.6. Factorial designs seem well suited to prevention trials for reasons outlined above, but many therapeutic trials have also needed factorial designs because of the questions being addressed. One important study using a 2×2 factorial design is the Physicians' Health Study (Hennekens and Eberlein, 1985; Stampfer et al., 1985). This trial has been conducted in 22,000 physicians in the United States and was designed to test the effects of (1) aspirin on reducing cardiovascular mortality, and (2) β-carotene on reducing cancer incidence. The trial is noteworthy in several ways, including its test of two interventions in unrelated diseases, use of physicians as subjects to report outcomes reliably, relatively low cost, and an all-male (high-risk) study population. This last characteristic has led to some criticism, which is probably unwarranted.

TABLE 19.6   Some Recent Randomized Clinical Trials Using Factorial Designs

Trial	Design	Cohort	Treatments	Outcomes	Reference
Physicians' Health Study	2×2	Healthy male physicians $n = 22,071$	Aspirin β-carotene	CHD; cancer	Hennekens and Eberlein, 1985
Linxian Nutrition Trial	2^4 *	4 Linxian communes $n = 29,584$	Retinol + zinc Riboflavin + niacin Ascorbic acid + molybdenum Selenium + β-carotene + α-tocopherol	Esophageal cancer; all cause mortality	Li et al., 1993
ISIS-4	2^3	Acute MI patients $n = 58,050$	Oral captopril Oral mononitrate IV magnesium sulphate	5 week mortality; 12 month mortality	ISIS-4 Collaborative Group, 1995
Prevention of Postoperative Nausea and Vomiting	2^6	Patients at high risk for nausea and vomiting $n = 5,199$	Ondansetron or no ondansetron Dexamethasone or no dexamethasone Droperidol or no droperidol Propofol or volatile anesthetic Nitrogen or nitrous oxide remifentanil or fentanyl	Nausea and vomiting within 24 hours postop	Apfel et al., 2004
Ipswich Childbirth Study	2×2	Women needing episiotomy repair $n = 793$	Repair: 2 stage or 3 stage suture: polyglactin or chromic	Pain or re-suturing	Grant et al., 2001
Thrombosis Prevention Trial	2×2	Men at risk of ischemic heart disease $n = 5,499$	Warfarin + aspirin Warfarin + placebo aspirin Placebo warfarin + aspirin Placebo + placebo	Coronary death; fatal / nonfatal MI	Medical Research Council, 1998

Note: * Denotes a partial replicate.

In January 1988 the aspirin component of the Physicians' Health Study was discontinued because evidence demonstrated convincingly that it was associated with lower rates of myocardial infarction (Steering Committee of the Physicians' Health Study Research Group, 1989). The question concerning the effect of β-carotene on cancer remains open and will be addressed by continuation of the trial. In the likely absence of an interaction between aspirin and β-carotene, the second major question of the trial will be unaffected by the closure of the aspirin component.

Another interesting example of a 2×2 factorial design is the α-tocopherol β-carotene Lung Cancer Prevention Trial, conducted in 29,133 male smokers in Finland between 1987 and 1994 (Heinonen et al., 1987; The ATBC Cancer Prevention Study Group, 1994a). In this study, lung cancer incidence is the sole outcome. It was thought possible that lung cancer incidence could be reduced by either or both interventions. When this trial was stopped in 1994, there were 876 new cases of lung cancer in the study population during the trial. Alpha-tocopherol was not associated with a reduction in the risk of cancer. Surprisingly, β-carotene was associated with a statistically significantly *increased* incidence of lung cancer (ATBC Cancer Prevention Study Group, 1994b). There was no evidence of a treatment interaction. The unexpected findings of this study have been supported by the recent results of another large trial of carotene and retinol (Thornquist et al., 1993).

The Fourth International Study of Infarct Survival (ISIS-4) was a $2 \times 2 \times 2$ factorial trial assessing the efficacy of oral captopril, oral mononitrate, and intravenous magnesium sulphate in 58,050 patients with suspected myocardial infarction (ISIS-4 Collaborative Group, 1995; Flather et al., 1994). No significant interactions among the treatments were found and each main effect comparison was based on approximately 29,000 treated versus 29,000 control patients. Among the findings was demonstration that captopril was associated with a small but statistically significant reduction in five-week mortality. The difference in mortality was 7.19% versus 7.69% (143 events out of 4319), illustrating the ability of large studies to detect potentially important treatment effects, even when they are small in relative magnitude. Mononitrate and magnesium therapy did not significantly reduce five-week mortality.

19.5 PARTIAL, FRACTIONAL, AND INCOMPLETE FACTORIALS

19.5.1 Use Partial Factorial Designs When Interactions Are Absent

Partial, or fractional, factorial designs are those that omit certain treatment groups by design. A careful analysis of the objectives of an experiment, its efficiency, and the effects it can estimate may justify not using some groups. Because many cells contribute to the estimate of any effect, a design may achieve its intended purpose without some of the cells.

In the 2×2 design all treatment groups must be present to permit estimating the interaction between A and B. However, for higher order designs, if some interactions are known biologically not to exist, certain treatment combinations can be omitted from the design and still permit estimates of other effects of interest. For example, in the $2 \times 2 \times 2$ design, if the interaction between A, B, and C is known not to exist, that treatment cell could be omitted from the design and still permit estimation of all the main effects. The efficiency would be somewhat reduced, however. Similarly the two-way interactions could still be estimated without $\overline{Y}_{ABC}$. This can be verified from the formulas above.

More generally, partial high-order designs will produce a situation termed "aliasing" in which the estimates of certain effects are algebraically identical to completely different effects. If both are biologically possible, the design will not be able to reveal which effect is being estimated. Naturally this is undesirable unless additional information is available to the investigator to indicate that some aliased effects are zero. This can be used to advantage in improving efficiency, and one must be careful in deciding which cells to exclude. See Cox (1958) or Mason and Gunst (1989) for a discussion of this topic.

The Women's Health Initiative clinical trial is a $2 \times 2 \times 2$ partial factorial design study the effects of hormone replacement, dietary fat reduction, and calcium and vitamin D on coronary disease, breast cancer, and osteoporosis (Assaf and Carleton, 1994).

The study is expected to accrue over 64,000 patients and is projected to finish in the year 2007. The dietary component of the study will randomize 48,000 women, using a 3:2 allocation ratio in favor of the control arm and nine years of follow-up. Such a large and complex trial is not without controversy (e.g., Marshall, 1993) and presents logistical difficulties, questions about adherence, and sensitivity of the intended power to assumptions that can only roughly be validated.

19.5.2   Incomplete Designs Present Special Problems

Treatment groups can be dropped out of factorial plans without yielding a fractional replication. The resulting trials have been called "incomplete factorial designs" (Byar, Herzberg, and Tan, 1993). In incomplete designs, cells are not missing by design intent but because some treatment combinations may be infeasible. For example, in a 2×2 design it may not be ethically possible to use a placebo group. In this case one would not be able to estimate the AB interaction. In other circumstances unwanted aliasing may occur, or the efficiency of the design to estimate main effects may be greatly reduced. In some cases estimators of treatment and interaction effects are biased, but there may be reasons to use a design that retains as much of the factorial structure as possible. For example, they may be the only way to estimate certain interactions.

19.6   SUMMARY

Factorial trial designs are useful in two circumstances. When two or more treatments do not interact, factorial designs can test the main effects of each using smaller sample sizes and greater precision than separate parallel groups designs. When it is essential to study treatment interactions, factorial designs are the only way to do so. The precision with which interaction effects are estimated is lower than that for main effects (in the absence of interactions), but these designs are the only ones that allow study of interactions.

When there are many treatments or factors, these designs require a relatively large number of treatment groups. In complex designs, if some interactions are known not to exist or are unimportant, it may be possible to omit some treatment groups, reduce the size and complexity of the experiment, and still estimate all of the effects of biological interest. Extra attention to the design properties is need to be certain that fractional designs will meet the intended objectives. Such fractional or partial factorial designs are of considerable use in agricultural and industrial experiments but have not been applied frequently to clinical trials.

Ethical and toxicity constraints may make it impossible to apply either a full factorial or a fractional factorial design, yielding an incomplete design. The properties of incomplete factorial designs have not been studied extensively, but they may be the best design in some circumstances.

A number of important, complex, and recent clinical trials have used factorial designs. Because of the low potential for toxicity, these designs have been more frequently applied in studies of disease prevention. Examples include the Physicians' Health Study and the Womens' Health Trial. In medical studies, the design is employed usually to achieve greater efficiency, since the treatments are unlikely to interact.

19.7   QUESTIONS FOR DISCUSSION

1. For the $2 \times 2 \times 2$ factorial design, repeat the calculation of least squares estimates from a linear model using a suitable design matrix. Do you obtain equations (19.9) through (19.11)? Explain your findings.

2. Repeat exercise 1 omitting some or all interaction terms. Explain your findings.

3. Suppose that a woman meets all of the eligibility criteria, except sex, for the Physicians' Health Study. Should she be treated with aspirin? Assume there are no contraindications and no other information about efficacy is available. Justify your answer.

4. Seasickness is a condition that is difficult to prevent and treat, although many remedies have been suggested including drugs (e.g., oral meclizine and/or scopolamine patches), wrist bands, and applying tape over the navel (with or without an aspirin tablet!). Discuss the setting and a design for a trial testing these preventives. How will you control important factors such as age, history of previous episodes, diet, environmental factors, state of mind, and the placebo effect?

20

CROSSOVER DESIGNS

20.1 INTRODUCTION

In the usual parallel or independent groups design, subjects receive a *single* therapy (or combination of therapies), and the groups are treated concurrently. An alternative design, which is useful in some circumstances, is to administer each treatment to every subject at different times in the study as a way to permit within-patient comparisons of treatment effects. Because subjects switch or crossover from one treatment to another by this strategy, these designs are called crossover trials. Stated more formally, crossover trials are those in which each patient is given more than one treatment, each at different times in the study, with the intent of estimating differences between them. Senn (1993, p. 3) defines them explicitly as trials

> in which subjects are given sequences of treatments with the object of studying differences between individual treatments (or sub-sequences of treatments).

In crossover designs the treatments are given during different time periods and all subjects receive more than one treatment, though not usually simultaneously. These studies are the only commonly encountered clinical trials in which subjects are not nested within treatments.

There is a large literature on crossover designs, reflecting their instinctive appeal, and dealing with limitations and controversies. Because the test for the principle limitation of carryover effects (discussed below) is inefficient, crossover designs have not been well received by all statisticians. However, this situation may be changing. Brief but useful methodologic discussions of these designs can be found in Hills and Armitage (1979), Brown (1980), Fleiss (1986), Matthews (1988), and Grieve (1982). A recent overview of the practical use of such designs is given by Cleophas and Tavenier (1995), and some aspects of optimal crossover designs are explored by Jones and Donev (1996). More specialized references will be mentioned in context later.

The simplest crossover trial is the two-treatment (*A* and *B*), two-period design. In this type of study there are two treatment periods and subjects are randomized to receive either *A* followed by *B* or *B* followed by *A*. In the remainder of this chapter, I will refer almost exclusively to this *AB/BA* design because it is the easiest crossover to understand, and illustrates many of the important points about these types of trials. However, more complex crossover trials may be employed in various clinical circumstances where the basic design is applicable. Extensions of this design include using more than one period for each treatment, a single treatment period for each of three or more treatments, or incomplete block designs, where not all patients receive all treatments. These more complex designs will not be covered here, but discussions of them can be found in Senn (1993) or Jones and Kenward (1989).

20.1.1   Other Ways of Giving Multiple Treatments Are Not Crossovers

There are other types of studies in which subjects receive sequences of treatments. For example, in some cancer treatment trials, patients might be randomized between two groups: one that receives $A \to B \to C$ or one that receives $A \to B$, where *A* must be given first, *B* second, and *C* third. This study design as usually implemented is actually a test of the incremental effect of treatment *C* and not a test of the differences between the components. Most of these types of studies using sequences of treatments are not crossover trials because the experimental structure (and intent) does not permit assessing differences between the individual treatments. The sequence of administration of treatments is not a feature controlled by the experimenter and most of these trials are best viewed as parallel tests of treatment combinations.

In factorial designs, some patients also receive more than one treatment. However, basic crossovers are different from these because some patients receive more than one treatment *simultaneously* in factorial trials. Even so, it is possible to construct factorial crossovers. In such a trial, for example, there might be four treatment periods and four treatments for every patient: placebo, *A*, *B*, and *A + B*. The order in which a patient would receive the treatments in this type of trial could be determined by random assignment to permuted blocks. However, most factorial trials are not crossovers, and vice versa.

20.1.2   Treatment Periods May Be Randomly Assigned

In a crossover trial subjects are not randomized to treatment in the same sense as they are in parallel group designs. This is even true of so-called randomized crossover trials because all participants receive all treatments. In these studies only the *order* of administering the treatments is randomized. Because of this the validity of the treatment comparison does not depend on the randomization as it might in parallel-groups designs. For example, randomization does not guarantee the expectation of an unbiased comparison of treatments in a crossover trial. Instead, the validity of the treatment comparison depends on additional assumptions or findings described below.

At first, the diminished role of randomization in crossover trials seems counterintuitive or wrong. However, because these studies estimate within-patient differences in treatment effects, the balance of confounding factors like that induced by randomization in parallel-groups designs is not an issue. In other words, with respect to prognostic factors, the treatment groups are identical in crossover trials, simply because the same individuals receive both treatments. Despite this design the treatment groups are *not* identical in crossover trials in other important ways. Specifically, the treatment groups differ with respect to their recent history of exposure to other potentially effective treatments.

This discussion highlights the primary source of difficulty with crossover trials: the comparability of the treatment groups (periods) is not guaranteed by the structure of the trial alone. Comparability depends also on the treatment effects being confined to the period of their administration and follow-up. Investigators may not know at the start of the trial if such an assumption is warranted. Having comparable treatment groups depends partly on the outcome of the trial. This is quite different from a randomized, concurrent control, parallel-groups design, where the estimate of the treatment difference is valid regardless of whether either, both, or neither treatment is effective, and regardless of the duration of efficacy.

20.2   ADVANTAGES AND DISADVANTAGES

Based on the discussion so far, we can already anticipate some problems with the application, analysis, or interpretation of crossover designs. However, there are circumstances where this type of design offers considerable advantages over a parallel-groups design. Investigators should fully understand the strengths and weaknesses of crossover trials so that they can be used effectively when the setting permits. Crossover designs seem to be somewhat more applicable in developmental trials than in comparative ones.

20.2.1   Crossover Designs Can Increase Precision

The primary strength of crossover trials is increased efficiency. Because each patient "serves as his or her own control," and the within-subject variability is usually less than the between-subject variability, the sample size for a crossover trial will be lower than that for a comparable parallel-groups design. The crossover design takes advantage of making treatment comparisons based on within-rather than between-subject differences. This allows the treatment difference to be estimated with greater precision, reducing the number of study subjects that are needed.

There are two effects that contribute to the greater efficiency of crossover designs. First, because each patient receives both treatments, the trial needs only half as many subjects as an independent groups design to yield the same precision in the estimated treatment difference. Second, the sample size can often be reduced further because within-subject responses to treatments are usually positively correlated. This also reduces the variance of the estimated treatment difference, further increasing efficiency.

For example, suppose that treatment effects in a AB/BA design are estimated with variance σ^2 in each patient on each treatment and the average response on treatments A and B are $\overline{Y}_A$ and $\overline{Y}_B$, respectively. Also suppose that there are no carryover or period effects. If $\widehat{\Delta}_{AB} = \overline{Y}_A - \overline{Y}_B$ is the estimated treatment difference, then

$$\text{var}(\widehat{\Delta}_{AB}) = \frac{\sigma^2}{n} + \frac{\sigma^2}{n} - 2\,\text{cov}(\overline{Y}_A, \overline{Y}_B) \qquad (20.1)$$

$$= 2\frac{\sigma^2}{n}(1 - \rho_{AB}),$$

where n is the sample size in each group and ρ_{AB} is the within-subject correlation of responses on treatments A and B (assumed to be the same for all individuals).

In parallel-groups designs, the correlation of responses between the treatment groups is zero because the groups are composed of different individuals. In crossover trials, the responses on the two treatments are correlated because they arise from the same subject. If the correlation between the responses for each individual is positive, as one might generally expect, the crossover trial estimates Δ_{AB} with a smaller variance than an independent groups design would. If the correlation is zero, the variance of $\widehat{\Delta}_{AB}$ is the same as an independent groups design, which uses $2n$ total patients. If, for some reason, the correlation between responses on A and B is large and negative (an unlikely circumstance), the crossover trial could be less efficient (i.e., have a larger variance) than an independent groups design.

20.2.2 A Crossover Design Might Improve Recruitment

Another potential advantage of crossover trials is that patient recruitment may be easier in some circumstances because all subjects receive all treatments under investigation. Sometimes patients may be unwilling to accept a no-treatment arm in a trial but be willing to delay therapy to the second treatment period. This could make it feasible to conduct a comparison that would otherwise be impractical for recruitment or be viewed as unethical.

For example, suppose that we are interested in the effect of mild exercise (versus no intervention) on patients' sense of well-being in diseases like advanced cancer or AIDS. Although of unproven benefit, the intervention will probably sound like a good idea to many patients. When they find out it is being studied, they may prefer an exercise program over usual care. In a situation like this, a parallel-groups design may have many "drop-ins" in the exercise group, particularly among patients who already feel better or are more active. In a crossover design, patients may be willing to postpone an exercise program to a second treatment period, knowing that they can derive benefit from it later if it is effective. Thus a two-period crossover design could help recruitment and compliance in both treatment periods.

Administering two or more treatments to every subject may require more time or be more inconvenient, discouraging participation. This would certainly be the case if outcomes could only be assessed after diagnostic procedures such as X rays, blood drawing, biopsies, lengthy questionnaires, or other tests. Acceptance might be greater if self-reports of symptoms are used. If the underlying disease is life threatening, diagnostic inconveniences might be unimportant to the patients. However, in chronic diseases, where crossover trials may be more useful, factors that increase patient acceptance could be important determinants of accrual.

20.2.3 Carryover Effects Are a Potential Problem

A potential problem with crossover designs is the possibility that the treatment effect from one period might continue to be present during the following period. There are a number of ways in which this "carryover" effect can happen. For example, the drug or treatment agent might physiologically persist during the second period. This might be prevented with a sufficiently long "washout" period between the treatment periods. The washout period is the interval between treatment periods, during which the previous treatment effect wears off and the patient's disease status returns to its baseline level.

Second, the first treatment could effect a permanent change or cure in the underlying condition of the patient. In this circumstance the treatment given during the second period could look artificially superior. Finally, the underlying condition of the patient could change during the second period, and the treatment effect could depend on the patient's condition. This would constitute a true treatment by period interaction. Of course, this same type of temporal trend and treatment by time interaction can occur in parallel-groups designs, where it could also disturb the estimate of treatment effect.

The possibility of carryover effects has been the focus of many concerns regarding crossover trials. If there are differences in the carryover effects in the two treatments (a likely situation unless both are zero), the design can yield biased estimates of the treatment effect, unless data from the second period are discarded (Freeman, 1989). The basic design is very efficient at estimating within-subject differences in treatment main effects, but not very efficient at estimating carryover effects. Some analytic approaches (discussed below) have suggested ways of estimating and dealing with carryover effects. However, it is probably true that the data themselves from a crossover trial are not very helpful in assessing carryover effects. A better approach is to use the design of the study, particularly a washout period, to be certain that carryover is not a problem.

A potential problem related to carryover is *treatment by period interaction*. A treatment by period interaction means that the treatment effect is not constant in the different treatment periods (i.e., over time). In AB/BA crossover trials, carryover and treatment by period interaction are not distinguishable. More generally, however, they can be separated. Treatment by time interactions are not unique to crossover studies and occur in other types of trials, such as parallel-groups designs. Carryover effects can cause treatment by period interactions, which explains why there is concern over both in crossover trials.

20.2.4 Dropouts Have Strong Effects

Two factors contribute to an increased likelihood of dropouts in crossover trials. First, the trial duration is longer than a comparable study using independent groups. This provides more opportunities for patients to drop out. Second, each participant is exposed to more drugs or treatments, increasing the chance of side effects that could contribute to dropping out.

The consequences of a dropout in a crossover trial may be more severe than in a parallel-groups design. If a participant drops out of a crossover trial in the second study period, for example, simple analyses cannot use the data from only the first period. Thus the data loss can be more significant than that from a single dropout in a parallel-groups trial. Recently, however, some authors have suggested ways of using incomplete observations in crossover trials (Feingold and Gillespie, 1996).

20.2.5 Analysis Is More Complex Than Parallel-Groups Designs

Crossover trials require more careful analysis than randomized parallel groups designs. The reason for added concern in crossover trials is the possibility of carryover effects and the need to determine that they have not confounded the estimates of treatment effect. Analytic approaches proposed for crossover trials use either (1) a staged plan, where carryover effects are studied in the first stage and, if none are found, the main effects are estimated in the second stage, or (2) baseline measurements in each period (Kenward and Jones, 1987) that can be used to test for carryover.

More recently Senn (1993) has suggested that analyses do not need to test for carryover using a staged approach. Instead, carryover should be actively controlled by the design of the trial. This and other details regarding analysis are discussed in the next section.

20.2.6   Prerequisites Are Needed to Apply Crossover Designs

Despite the potential benefits and efficiencies of the crossover design, there are serious limitations to its widespread use. The first restriction is that investigators must have some knowledge of the sign and magnitude of the within-patient correlation between responses. As indicated above, a large negative value might make a crossover design counterproductive, although this is unlikely to be the case.

Second, the underlying disease must have a constant intensity during all treatment periods. If the disease is cured by one of the treatments or can be expected to disappear in a short (relative to the treatment periods) time, the crossover design will not be applicable. Also, if the condition is improving or worsening substantially, the treatment periods will not have the same baseline, and either the first or second one administered might artificially look better.

Third, the effect of the treatment needs to be restricted to the period in which it is applied. Equivalently, the treatment periods must be separated by a sufficient length of time for the effects of the earlier treatment to subside. If the washout period is too short, the latter treatment period will be biased by carryover effects from the earlier treatment. If there is carryover, the investigator will observe the simultaneous effect of two or more treatments but attribute it only to the most recently administered one. Finally, crossover trials can create added inconvenience for the patient and can present some challenging problems during analysis.

Investigators should be quite certain of the validity of the setting in which a crossover trial is used and the expected view of such a trial by colleagues and regulators. For example, the potential for biased estimates of treatment effects from poor crossover designs led the Food and Drug Administration in 1977 to conclude that such designs are a second choice to completely randomized or randomized block designs (FDA, 1977; O'Neill, 1978). They stated:

> [the crossover design] is not the design of choice in clinical trials where unequivocal evidence of treatment effect is required ... in most cases, the completely randomized (or randomized block) design with baseline measurements will be the design of choice because it furnishes unbiased estimates of treatment effects without appeal to any modeling assumptions save those associated with the randomization procedure itself.

Although not suited to some studies of chronic disease like cancer, crossover trials are well suited to other types of trials. For example, they might be a good design for bioavailability trials. For some diseases requiring chronic medication, like arthritis, angina, asthma, hypertension, and diabetes, crossover trials might be an efficient way to compare treatments.

20.3   ANALYSIS

In the last 30 years a number of approaches have been suggested for analyzing crossover trials or modifying their designs to cope with deficiencies. This section sketches some of these proposals and illustrates strengths and weaknesses of both the underlying design and the analytic strategies. The reader interested in more depth in the statistical details can refer to Senn (1993) or Jones and Kenward (1989). Bayesian approaches to crossover trials are discussed by Grieve (1985).

The classical approach to analyzing crossover experiments was given by Grizzle (1965), who suggested conducting a preliminary test of carryover. If the carryover effect cannot be ignored, the data from the first treatment period can be analyzed as though it arose from a parallel-groups design. The data from the second period are discarded. In the absence of carryover effects, treatment effects are estimated from within-patient differences, using a linear model like the one discussed below.

An uncomplicated AB/BA design can be analyzed in a very straightforward manner. We calculate within-subject differences (treatment effects), $\delta = Y_A - Y_B$, and test for the effects of interest using appropriate averages of the δ's and their estimated standard error. The individual δ's are independent of one another. For example, a period effect implies that treatment effects are different in the two periods. In other words,

$$z = \frac{\bar{\delta}_1 - \bar{\delta}_2}{\sqrt{\text{var}\{\bar{\delta}_1\} + \text{var}\{\bar{\delta}_2\}}} \tag{20.2}$$

should differ only randomly from zero in the absence of a period effect. Therefore z has a standard normal distribution under the null hypothesis. Similarly the overall treatment effect can be estimated by averaging the estimates from each period,

$$z = \frac{\bar{\delta}_1 + \bar{\delta}_2}{2\sqrt{\text{var}\{\bar{\delta}_1\} + \text{var}\{\bar{\delta}_2\}}} = \frac{\bar{\bar{\delta}}}{\sqrt{\text{var}\{\bar{\bar{\delta}}\}}} . \tag{20.3}$$

Under the null hypothesis, this statistic also has a standard normal distribution. Although this serves as an introduction to the analysis of basic crossover trials, more general approaches are based on linear models, discussed next.

20.3.1   Analysis Can Be Based on a Cell Means Model

Assume that individual responses arise from a linear model with error terms that are normally distributed. The group means for a two-period crossover design with treatments A and B, one observation on each study subject in each period, and a possible carryover effect can be parameterized as shown in Table 20.1. This parameterization allows the estimates to be summarized in a manner similar to that used in Chapter 19 for factorial designs. In this model, β_0 is the effect of treatment A alone, β_1 is the increment in treatment effect attributable to B, β_2 is the period effect, and β_3 is the carryover effect. We cannot distinguish between carryover and treatment by period interaction because a term for the latter would appear in the model with, and not be separable from, β_3. Mean responses in each treatment-period cell are denoted by $\bar{Y}$ with appropriate subscripts.

Suppose that there is no treatment by period interaction, namely, $\beta_3 = 0$. A nonzero period effect, $\beta_2 \neq 0$, means that the individual treatment effects are different in each period. Therefore we can estimate β_2 by averaging the treatment effect differences in the two periods,

$$\widehat{\beta_2} = \tfrac{1}{2}\left(\bar{Y}_{B2} - \bar{Y}_{B1} + \bar{Y}_{A2} - \bar{Y}_{A1}\right). \tag{20.4}$$

TABLE 20.1 Cell Means Model for a Two-Period Crossover Trial

Treatments	Treatment Period	
	1	2
A then B	$\overline{Y}_{A1} = \beta_0$	$\overline{Y}_{B2} = \beta_0 + \beta_1 + \beta_2$
B then A	$\overline{Y}_{B1} = \beta_0 + \beta_1$	$\overline{Y}_{A2} = \beta_0 + \beta_2 + \beta_3$

There are two estimates of the effect of treatment B compared with A, one from each period. They can be averaged to estimate β_1,

$$\widehat{\beta_1} = \tfrac{1}{2}\left(\overline{Y}_{B2} - \overline{Y}_{A2} + \overline{Y}_{B1} - \overline{Y}_{A1}\right). \tag{20.5}$$

More generally, we must consider the possibility that $\beta_3 \neq 0$. A treatment by period interaction means that the incremental effects of treatment A (or B) in each period are not the same. We can estimate β_3 by

$$\widehat{\beta_3} = \left(\overline{Y}_{A2} + \overline{Y}_{B1} - \overline{Y}_{B2} - \overline{Y}_{A1}\right). \tag{20.6}$$

If β_3 is not zero, then we must estimate the treatment difference as

$$\widehat{\beta_1} = \left(\overline{Y}_{B1} - \overline{Y}_{A1}\right),$$

using only the data from the first period.

Suppose that each $\overline{Y}$ is estimated with variance σ^2/n and that in each treatment group the within-person correlation of responses is ρ. Then equation (20.1) can be used to calculate the variance of the difference between treatments in the same group. The formulas for $\widehat{\beta_2}$ and $\widehat{\beta_1}$ given above can be used to show that

$$\text{var}\{\widehat{\beta_2}\} = \text{var}\{\widehat{\beta_1}\} = \frac{\sigma^2}{n}(1 - \rho).$$

In comparison, the variance of the interaction effect is

$$\text{var}\{\widehat{\beta_3}\} = 4\frac{\sigma^2}{n}(1 + \rho),$$

which is at least four times larger than var$\{\widehat{\beta_2}\}$ for $\rho \geq 0$ (recall equation 20.1). Therefore any crossover trial designed to reliably detect main effects of treatment will have a less efficient test for the carryover effect. However, the carryover effect is critical to detect because its presence affects both the analysis and interpretation of the trial. In the presence of clinically important carryover effects, a crossover design is no more efficient than an independent-groups trial. In fact a crossover design is relatively inefficient in testing the assumption of no carryover effect, a conclusion supported by Brown (1980).

Some researchers feel that it is not necessary to establish definitively the absence of carryover effects. Instead, the carryover effect should be small in comparison to the treatment effect. This would seem to permit a test of the carryover effect that has high power to rule out large values of β_3 but may not reject for smaller values. A perspective on this view is given by Poloniecki and Pearce (1983).

The carryover effect can arise from more than one source. If the washout period is inadequate, the effect of the first treatment may persist into the second period. Also the first treatment may alter the patient's condition permanently, perhaps without effecting a cure. Finally the treatment effect may be proportional to the disease intensity. In a two-period crossover trial these complications are indistinguishable from one another and some or all of them must be assumed not to exist to utilize all of the data from the trial.

More General Linear Model Approach

Assume a linear model with Gaussian errors for the AB/BA crossover design of the form

$$E\{Y\} = \beta_0 + \beta_1 T + \beta_2 P + \beta_3 T \times P,$$

where Y is the response and T and P are indicator variables for treatment group and period, respectively. This model contains both a period effect and a treatment by period interaction (or carryover effect). The design matrix, $\mathbf{X}$, for such an experiment with n subjects per group has dimensions $4n \times 4$ and can be written

$$\mathbf{X} = \begin{bmatrix} 1 & \frac{1}{2} & \frac{1}{2} & \frac{1}{4} \\ 1 & -\frac{1}{2} & -\frac{1}{2} & \frac{1}{4} \\ 1 & \frac{1}{2} & \frac{1}{2} & \frac{1}{4} \\ 1 & -\frac{1}{2} & -\frac{1}{2} & \frac{1}{4} \\ \vdots & \vdots & \vdots & \vdots \\ 1 & \frac{1}{2} & -\frac{1}{2} & -\frac{1}{4} \\ 1 & -\frac{1}{2} & \frac{1}{2} & -\frac{1}{4} \\ 1 & \frac{1}{2} & -\frac{1}{2} & -\frac{1}{4} \\ 1 & -\frac{1}{2} & \frac{1}{2} & -\frac{1}{4} \end{bmatrix},$$

where each pair of rows corresponds to the same subject treated with different treatments in different periods. In the design matrix, the successive columns correspond to an intercept, the treatment group, the period, and the treatment by period interaction (calculated by multiplying the treatment and period values). The treatment and period variable coding is chosen to be symmetric around zero and have a one-unit difference. This corresponds to the cell means approach discussed above. The vector of responses corresponding to this design matrix has length $4n$ and is written

$$\mathbf{Y} = \begin{bmatrix} Y_{A1} \\ Y_{B1} \\ Y_{A2} \\ Y_{B2} \\ \vdots \end{bmatrix},$$

where the subscripts indicate the treatment period and subject number.

Ordinary least squares cannot be used to estimate the model parameters because each pair of responses is correlated, having arisen from the same individual. Instead, weighted least squares estimates must be obtained using

$$\widehat{\beta} = (\mathbf{X}'\Sigma^{-1}\mathbf{X})^{-1}\mathbf{X}'\Sigma^{-1}\mathbf{Y}, \tag{20.7}$$

where Σ is the covariance matrix of responses (Draper and Smith, 1981). If we assume that the variance of all responses is identical ($= \sigma^2$) and the within-subject covariances (or correlations) are equal ($= \gamma$), Σ has dimensions $4n \times 4n$ and

$$\Sigma = \begin{bmatrix} \sigma^2 & \gamma & 0 & 0 & \cdots \\ \gamma & \sigma^2 & 0 & 0 & \cdots \\ 0 & 0 & \sigma^2 & \gamma & \cdots \\ 0 & 0 & \gamma & \sigma^2 & \cdots \\ \vdots & \vdots & \vdots & \vdots & \ddots \end{bmatrix} = \sigma^2 \times \begin{bmatrix} 1 & \rho & 0 & 0 & \cdots \\ \rho & 1 & 0 & 0 & \cdots \\ 0 & 0 & 1 & \rho & \cdots \\ 0 & 0 & \rho & 1 & \cdots \\ \vdots & \vdots & \vdots & \vdots & \ddots \end{bmatrix}.$$

The block diagonal structure arises because pairs of responses are correlated within, but not between, individuals. In general, γ (ρ) will be positive because within-subject responses are positively correlated with one another. However, this is not absolutely required. In any case,

$$\Sigma^{-1} = \frac{1}{\sigma^4 - \gamma^2} \times \begin{bmatrix} \sigma^2 & -\gamma & 0 & 0 & \cdots \\ -\gamma & \sigma^2 & 0 & 0 & \cdots \\ 0 & 0 & \sigma^2 & -\gamma & \cdots \\ 0 & 0 & -\gamma & \sigma^2 & \cdots \\ \vdots & \vdots & \vdots & \vdots & \ddots \end{bmatrix}$$

$$= \frac{1}{\sigma^2(1-\rho^2)} \times \begin{bmatrix} 1 & -\rho & 0 & 0 & \cdots \\ -\rho & 1 & 0 & 0 & \cdots \\ 0 & 0 & 1 & -\rho & \cdots \\ 0 & 0 & -\rho & 1 & \cdots \\ \vdots & \vdots & \vdots & \vdots & \ddots \end{bmatrix}.$$

In a model with a treatment effect, period effect, and a treatment by period interaction, the matrices above yield

$$(\mathbf{X}'\Sigma^{-1}\mathbf{X})^{-1} = \frac{\sigma^2}{n} \times \begin{bmatrix} \frac{1}{4}(1+\rho) & 0 & 0 & 0 \\ 0 & 1-\rho & 0 & 0 \\ 0 & 0 & 1-\rho & 0 \\ 0 & 0 & 0 & 4(1+\rho) \end{bmatrix},$$

which is the covariance matrix of the parameter estimates. Note the independence of the estimates using this variable coding (i.e., all covariances are zero) and that the treatment by period interaction is less precisely estimated (variance $= 4(\sigma^2 + \gamma)/n$) relative to the treatment difference (variance $= (\sigma^2 - \gamma)/n$). In other words, for a fixed sample size, the power to detect a treatment by period interaction of specified magnitude will be lower than that with which other effects can be detected. Then the parameter estimates from equation (20.7) are

$$
\widehat{\beta} = \begin{bmatrix} \overline{Y}_{..} \\ \overline{Y}_{A.} - \overline{Y}_{B.} \\ \frac{1}{2}\{(\overline{Y}_{A1} - \overline{Y}_{B1}) - (\overline{Y}_{A2} - \overline{Y}_{B2})\} \\ (\overline{Y}_{A1} - \overline{Y}_{A2}) + (\overline{Y}_{B1} - \overline{Y}_{B2}) \end{bmatrix}.
$$

The exact form of the estimates depends on the model parameterization. In this case the treatment by period or carryover effect is estimated by the sum of the period differences in the treatments. These estimates are the same, apart from algebraic sign, as those developed above from the cell means approach (equations 20.4–20.6).

20.3.2   Other Issues in Analysis

There are many other issues regarding crossover trials and their analysis that cannot be covered in depth in this chapter. A few will be mentioned here to guide additional study. Baseline measurements taken at the beginning of each treatment period can be used in an analysis of covariance to improve effect estimates or the test of carryover. Here we assume that baseline measurements at the second period are not affected by carryover (i.e., there has been a sufficient washout). See Jones and Kenward (1989), Senn (1993), and Willan and Pater (1986) for discussions.

Much attention has been given to estimating carryover effects from the data, despite the poor efficiency of such tests. The assessment of carryover may depend on its magnitude relative to the treatment effect (Willan and Pater, 1986) or its direction (Lechmacher, 1991). In any case, concern about carryover effects can be reduced by using designs with more than two treatment periods (Laska, Meisner, and Kushner, 1983).

Crossover trials may be indicated in many circumstances where binary or other responses arise. An overview of some proposed approaches is given by Kenward and Jones (1987). Binary outcomes and dropouts in multiple period analyses are discussed by McKnight and Van Den Erden (1993). Poisson responses are discussed by Layard and Arvesen (1978) and nonparametric tests of the data have been suggested (Senn, 1993). Analyses can be approached from the perspective of multivariate response (Grender and Johnson, 1993).

20.3.3   Classic Case Study

A classic example of a two-period crossover trial has been given by Hills and Armitage (1979). Children with enuresis were treated with a new drug or placebo for 14 days. (The name and characteristics of the drug are not provided by the authors.) The number of dry nights out of 14 for each child is shown in Table 20.2. The study design and resulting data are not complicated, and the authors used normal theory tests to assess the outcome of the trial.

The mean differences ($\pm$ s.e.m.) are 2.82 (± 0.84) for group I and 1.24 (± 0.86) for group II. The period effect is tested by equation (20.2) with the standard error equal to $\frac{1}{2}\sqrt{0.84^2 + 0.86^2} = 0.60$. Thus the test for period effect is

$$
z = \frac{2.82 - 1.24}{2 \times 0.60} = 1.32,
$$

TABLE 20.2   A Two-Period Crossover Trial: Treatment of Enuresis

	Group I			Group II	
Patient Number	Period 1 Drug	Period 2 Placebo	Patient Number	Period 1 Placebo	Period 2 Drug
1	8	5	2	12	11
3	14	10	5	6	8
4	8	0	8	13	9
6	9	7	10	8	8
7	11	6	12	8	9
9	3	5	14	4	8
11	6	0	15	8	14
13	0	0	17	2	4
16	13	12	20	8	13
18	10	2	23	9	7
19	7	5	26	7	10
21	13	13	29	7	6
22	8	10			
24	7	7			
25	9	0			
27	10	6			
28	2	2			

Source: Data from Hills and Armitage, (1979).

which is not statistically significant. The test for overall treatment effect uses equation (20.3) or

$$z = \frac{2.82 + 1.24}{2 \times 0.60} = 3.38,$$

which is significant at the 0.01 level. Thus the drug is effective at increasing the number of dry nights in a 14-day interval.

20.4   SUMMARY

Crossover trials are those in which study participants receive all treatments under investigation, each in a different study period. Between periods, a washout period is used to allow the effects of the previous treatment to disappear. Because the treatment effect is estimated within rather than between patients, crossovers are more efficient than parallel-groups designs. Although the design is not well suited to some acute diseases or types of outcomes, it is particularly well suited to investigations such as bioavailability trials.

Crossover studies are subject to important limitations, particularly finding the proper clinical setting and actively determining that carryover and treatment by period interactions do not destroy the applicability of the design. For two-treatment, two-period crossover trials, the effects of carryover and treatment by period interactions are indistinguishable. Carryover can be assessed by preliminary studies and eliminated by design. Statistical tests for carryover after the data have been collected are inefficient and probably not helpful. Ancillary measurements, such as baseline covariates at the beginning of each treatment period, can improve the performance of crossover designs.

20.5 QUESTIONS FOR DISCUSSION

1. Instead of the design matrix given in Section 20.3.1, suppose that zero–one variable coding had been employed. What would happen to the covariance matrix of parameter estimates and the estimates themselves? Discuss the implications.

2. Describe the assumptions under the normal theory analysis of the enuresis two-period crossover trial. Are the assumptions met by the data collected? Can you suggest at least two alternative approaches to the analysis?

21

META-ANALYSES

21.1 INTRODUCTION

A meta-analysis or systematic review is a formal method for taking the findings of several or many clinical trials (or less frequently, observational studies) and combining, synthesizing, or integrating them. "Overview" is also descriptive of the techniques used, but this term is often used to describe reviews that are not quantitative. The National Library of Medicine (NLM) has endorsed the term "meta-analysis." In 1989 the NLM began a MeSH heading for meta-analysis and named it a publication type in 1993. "Collaborative re-analysis" probably better describes what is actually done, but this term is less frequently used.

Because of the ubiquitous problem of having to combine evidence from several trials, scientists have been engaging in this activity, formally or informally, for many years. Olkin (1995a, b) gives an interesting historical perspective on meta-analyses and attributes early ones to Karl Pearson, R.A. Fisher, and other statisticians. Between 1930 and 1950, agriculture was a stimulus for combining evidence (and for developing other statistical methods). Between 1950 and 1970, there was little research on meta-analysis methods (Olkin, 1995a).

An early proposal to minimize subjective analysis of similar experiments in the literature using a quantitative approach was given by Light and Smith (1971). Several methods were used and discussed in the social sciences in the 1970s but were not widely adopted. The term "meta-analysis" was first used by Glass (1976) to describe a statistical pooling method like that commonly used now. Glass defined meta-analysis as

> ... the statistical analysis of a large collection of analysis results from individual studies for the purpose of integrating the findings. It connotes a rigorous alternative to the casual, narrative discussions of research studies which typify our attempts to make sense of a large volume of research literature.

While the definitions do not necessitate using randomized studies, this is the context in which the clinically most reliable and important meta-analyses are conducted. The discussion in this chapter refers principally to meta-analyses of randomized trials.

Meta-analyses migrated into the medical sciences years ago and now are heavily used by clinical researchers (Jones, 1995). Formal efforts to develop the field in recent times spread from the social sciences (Hedges and Olkin, 1985) to medicine in the 1980s, especially oncology and epidemiology. One of the earliest statistical overviews in medicine was that of Chalmers, examining the effect of warfarin on myocardial infarction outcomes (Chalmers et al., 1977). Antman et al. (1992) wrote an influential paper pointing out discrepancies between quantitative evidence and expert opinion, supporting the utility of meta-analysis. Helpful reviews of meta-analysis are given by Dickersin and Berlin (1992), Finney (1995), and Lau et al. (1997). An excellent source for depth and diversity of opinion is the proceedings of the Potsdam International Consultation on Meta-Analysis published in the *Journal of Clinical Epidemiology* (Volume 48(1), 1995). A very nice nontechnical summary of systematic reviews, particularly with regard to history, has been given by Moynihan (2004).

One of the principal reasons why meta-analyses are helpful to clinical researchers today is the large number of trials being conducted. The number of randomized clinical trials probably approaches 10,000 per year. Synthesizing results informally from many studies can be difficult and confusing. Also researchers are frequently interested in small treatment differences (one of the main reasons for performing randomized comparisons) because they can have major public health importance when applied to common diseases. Important differences may be obscured by the variability of individual studies, which is why results in the literature often seem to disagree. However, combining evidence frequently shows consistency and permits one to estimate small but clinically significant effects.

The earliest meta-analyses required only information in the published trial reports. While this made overviews accessible to many readers, it is now clear that more than the published information is usually necessary to perform rigorous overviews. The meta-analysts need to obtain and analyze patient data from each trial to produce a credible result. Because individual patient data are not published routinely, a large additional effort is required to conduct a proper meta-analysis.

Presently the Cochrane collaboration (www.cochrane.org) is the major driving force for meta-analysis worldwide. This ambitious project attempts to synthesize all the randomized clinical trials performed for a given therapeutic question to allow the most definitive assessments possible. It is a source of both clinical evidence and improved methods for meta-analysis.

21.1.1 Meta-Analyses Formalize Synthesis and Increase Precision

The primary purpose of an meta-analysis is analytic rather than descriptive. The analyst looks for consistency in studies that may not be readily apparent from selective or casual reading. The studies selected for synthesis are those that essentially test the same hypothesis or compare the same treatments, but they need not be literally identical in design. Also the analyst looks to explain heterogeneity or differences in studies. There are several reasons why this activity is useful. First, because randomized clinical trials (RCTs) are large and expensive, we would like to gain as much information from them as possible. Second, we intuitively expect well-performed RCTs to "agree with" one another. Third, RCTs tend to be externally valid with respect to relative treatment effects, even if the cohort selected for study is atypical with respect to the diseased population.

An expert's review of the field is one alternative to a formal meta-analysis. However, the typical clinical review can be incomplete and nonquantitative. When there is uncertainty about the consistency of the findings in different studies, incomplete and nonquantitative assessments can be disadvantageous. An expert might discount findings where the effects of treatment are small, or go against the prevailing theory. Also an expert could be misled by focusing only on the statistical significance of findings (rather than the magnitude of the effects). Despite the limitations of reviews based entirely on expert opinion, there will always be circumstances in which they are useful. For example, if there are only a few messy studies, an expert opinion might be preferable to a formal analysis.

Meta-analysis attempts to counteract the deficiencies of informal reviews by (1) basing data on a comprehensive literature review, sometimes using relevant data that have not yet been published (e.g., negative findings that were not published), and (2) using formal methods for combining data and estimates of treatment effect from the studies. However, we must always ask if the question posed by a meta-analysis is well defined so that its apparent precision is useful. Subjectivity also enters the meta-analysis when deciding on which studies to include or exclude.

In addition to a firm base for inference, the statistical tools employed in the typical meta-analysis are rigorous and more reliable than those used in other types of reviews. Formal methods are used to combine various types of endpoints such as group means, odds ratios, and hazard ratios. Also some methods using random effects models attempt to account explicitly for study-to-study variability. Finally the statistical power of meta-analyses is often very high. Small treatment differences can be estimated with a high degree of precision and can yield highly statistically significant results. The high statistical power can sometimes allow reliable assessment of secondary endpoints that individual studies could not. When the disease under study is common, the public health importance of even small treatment benefits can be large.

21.2 A SKETCH OF META-ANALYSIS METHODS

21.2.1 Meta-Analysis Necessitates Prerequisites

Meta-analysis, like all formal research methods, needs to proceed from a foundation of planning and design. The basic steps required are (1) formulation of a purpose and specification of an outcome for the analysis, (2) identification of relevant studies, (3) establishing inclusion and exclusion criteria for studies, (4) data abstraction and acquisition, (5) data analysis, and (6) dissemination of results and conclusions. There is a general consensus that individual level data are preferred for meta-analysis (e.g., see Stewart and Clarke, 1995). Weaknesses in any of these steps can compromise the validity and strength of the meta-analysis.

Because meta-analyses require so much effort, they should not be undertaken casually or without clear purpose. A difficult but essential part of meta-analysis, like a clinical trial, is choosing an important feasible question to address. This chapter can be of little help in selecting a question, except to emphasize that both clinical and methodological knowledge is required to assess the importance and feasibility of the study plan.

21.2.2   Many Studies Are Potentially Relevant

Retrieval

Having been stimulated by an appropriate question, the meta-analysis investigator will use several sources to identify trials on which to base the analysis. Personal knowledge is an important source of information but will usually be incomplete. Experts in the field can be consulted but may miss, or be unaware of, some relevant clinical trials. Experts are an important and reliable way to gather information about trials that have not yet appeared in traditional media.

Usually a large number of relevant and important trials can be found using computerized searches of literature databases, such as MEDLINE, EMBASE, or the Cochrane Controlled Trials Registry. Such searches are possible in most medical libraries at minimal cost and can be accomplished quickly. In all cases well-defined terms such as Medical Subject headings (MeSH) should be used (www.nlm.nih.gov/mesh/meshhome.html). The reference lists from selected publications are an important additional source for identifying relevant trials. Additional expertise may be required to be sure that foreign language studies are identified by computerized searches, but such studies are equally valuable (Moher et al., 1996). Human recollections and computer retrievals may miss recently published trials. Therefore it may be important to search for recent publications by hand.

Limitations of Retrieval

The retrieval methods outlined have unavoidable limitations. Some sources of information may not be covered by the computerized searches discussed above. This "fugitive literature" includes government reports, book chapters, dissertations, and conference proceedings. Studies published outside the time window in which the meta-analyst searches can be easily overlooked. Finally we must consider the difficulty of dealing with nonrandomized studies. If they are excluded from the meta-analysis, important evidence may be lost. If they are included, there is a risk of introducing another source of bias.

"Publication bias" can result when the medical literature is used as a sole source for trials. The published literature is neither a complete repository nor a random sample of trials actually performed. For example, studies that do not show statistically significant differences or treatment effects, so-called negative trials, are less likely to appear in the literature than positive ones, creating publication bias. It is our scientific nature to be attracted more to "progress" than to its apparent absence. Furthermore the literature probably contains a higher proportion of results that are type I errors than statistical significance levels would suggest.

Publication bias is a serious concern for meta-analyses. Several authors have discussed the implications of publication bias and suggested methods for compensating for it (Begg and Berlin, 1988; Dear and Begg, 1992; Hedges, 1992; Ioannidis et al., 1997). Complete ascertainment of unpublished trials is an important step, as is an emphasis on the magnitude of treatment effects rather than their statistical significance.

Information retrieval for meta-analyses in the future would benefit from two improvements today. The first is prospective registration of trials when they are initiated in a comprehensive database (Dickersin and Rennie, 2003). A database of this type would facilitate meta-analyses later and would be an important source of information about clinical trials as a method of investigation. If used properly, it could also improve policy decisions by regulators and sponsors. The Cochrane Collaboration (Chalmers,

1993) comes close to this goal. A second area for improvement is complete reporting of trial results, which might be facilitated by formal guidelines (e.g., Standards of Reporting Trials Group, 1994). This last point is discussed at length in Chapter 18.

21.2.3 Select Studies

Eligibility Criteria for Study Inclusion

Meta-analysts establish specific eligibility criteria for including clinical trials in their study. The purpose of these criteria is to reduce bias in the selection of trials and to increase the reproducibility of the findings. When drafting such inclusion criteria, investigators must consider the specific treatment employed in each trial. To include a trial in the meta-analysis, we must be able to justify that its test of treatment effect is similar in design to those from other studies selected. Often this means selecting studies of only a specific agent, combination, or modality, but sometimes it permits a broader definition of "treatment." For example, the Early Breast Cancer Trialists' Collaborative Group (EBCTCG) performed a large and important meta-analysis looking at the benefit of chemotherapy in breast cancer (EBCTCG, 1990). For some of these analyses, the chemotherapy combination did not have to be identical or be given at the same dose. Nevertheless, the results showed clearly the benefit of treatment.

Many clinical trials yield more than one publication. However the principal end-point paper will usually be most relevant to the meta-analysis. Sometimes it may be reasonable to exclude trials employing a very small sample size or those with limited follow-up. Trials with low power may be methodologically weak in other ways, and therefore of questionable validity. In any case, small studies are not likely to influence the results of a meta-analysis greatly.

Abstracting Data

It may be helpful to use a data abstraction form to organize and collect the information from each clinical trial included in the meta-analysis. Using this method, investigators can help control the selection and removal of patients from the analysis, resolve questions about ancillary treatments, determine the clinical outcome for each patient, and identify and collect information on important prognostic factors. The reliability and reproducibility of these steps may be increased by a suitable data abstraction form. Also investigator bias can be reduced using masking and by separating data collection tasks in appropriate ways.

21.2.4 Plan the Statistical Analysis

The main issues in planning the statistical analysis are choosing an effect estimate, deciding on the unit of analysis (trial versus individual patient), quality scoring of studies, and selecting the specific statistical methods to use. The estimate of treatment effect can be an average effect size calculated from all the studies. Examples of effect sizes used in meta-analyses include mean differences (standardized), risk ratios, correlations, and p-values. Although any suitable measure of effect is probably appropriate, p-values cannot substitute for effect. Some quantitative issues in meta-analysis are discussed by Greenland (1987). A comparison of statistical methods is given by Brockwell and Gordon (2001).

Observed minus expected frequencies is a commonly used summary of effect (illustrated below). This method may not be optimal when effects are present (Greenland and Salvan, 1990), but it is relatively simple and illustrates the principles of analysis.

In any case, the overall effect estimate can also be adjusted for prognostic factors. Adjustment may be indicated if the prognostic factor composition of the trial cohorts are very different. Like individual clinical trials, investigators must plan for subgroup analyses and deal with treatment nonadherence and losses to follow-up. It may be useful to cumulate the results chronologically rather than show each study individually. Olkin (1995b) gives an interesting example of this.

A quality rating system for the trials included has been suggested as a way to improve the strength of evidence coming from meta-analyses. For example, the analysis could be stratified or weighted according to the perceived quality. A "sensitivity analysis," using quality scoring as an indicator to include or exclude each trial, could help validate the findings obtained in this way. There is a tendency for more recently conducted studies to be higher quality than old ones because of methodologic improvements and having more experienced and knowledgeable investigators. However, there are exceptions to this rule. Also errors or scientific misconduct by some individuals associated with a particular trial does not necessarily reduce the quality of all the data. Despite the potential utility of the quality scales used in meta-analyses, they are subjective and their reliability and validity remain unknown.

21.2.5 Summarize the Data Using Observed and Expected

A widely used statistical method for meta-analyses is based on calculating the deviation of the observed result in the treatment group within each trial from the result expected under a null hypothesis. This is "observed minus expected" or $O - E$. Observed and expected calculations can be based on statistics such as number of events, survival rates, or other appropriate clinical endpoints. The $O - E$ method avoids comparing results from one trial with those from another. The sum of the deviations is divided by an estimate of its variability to yield an overall test of the null hypothesis. If there is no treatment effect, the $O - E$ values are equally likely to be positive or negative, and their sum will differ only randomly from 0.

When the null hypothesis is not true, the $O - E$ method may not be the optimal one for assessing the risk ratio. However, it is not difficult to calculate and illustrates the important characteristics of the method.

Suppose that number of events is the primary outcome measure for studies comparing treatment versus control. In the ith study, the total number of participants is N_i, the number in the treatment group is n_i, and the total number of events observed is K_i. Define the overall event rate to be $p_i = K_i / N_i$ and the expected number of events in the treated group to be $E_i = p_i n_i$. The null hypothesis states that the treatment has no effect, meaning the treatment assignment and outcome are independent. Randomization, and independence of treatment assignment and outcome, means that we can view the n_i individuals assigned to the treatment group as a simple random sample of all N_i participants.

Under these assumptions the number of events observed in the treatment group, O_i, follows a hypergeometric distribution (Johnson and Kotz, 1969). The hypergeometric probability distribution is specified by

$$\Pr\{O_i = x\} = \frac{\binom{K_i}{x}\binom{N_i - K_i}{n_i - x}}{\binom{N_i}{n_i}} ,$$

where $\max\{0, n_i - (N_i - K_i)\} \leq x \leq \min\{K_i, n_i\}$. The expected value of this distribution is $\mathcal{E}\{O_i\} = E_i = n_i p_i$ and the variance is

$$\mathrm{var}\{O_i\} = n_i p_i (1 - p_i) \frac{N_i - n_i}{N_i - 1} = f_i K_i \left(\frac{N_i - K_i}{N_i - 1} \right) (1 - f_i),$$

where $f_i = n_i / N_i$. These formulas allow us to calculate an overall test statistic from M studies using the sum of $O - E$ and its variance (E is not a random variable),

$$Z = \frac{\sum_{i=1}^{M} (O_i - E_i)}{\sqrt{\sum_{i=1}^{M} \mathrm{var}\{O_i\}}},$$

which will have a standard normal distribution under the null hypothesis. Large negative values of Z imply that the treatment reduces the expected number of events.

Example 41 The methods sketched above are illustrated in the meta-analysis of long-term antiplatelet therapy discussed by Peto, Collins, and Gray (1995), and the Antiplatelet Trialists' Collaboration (1994). In the 11 randomized trials conducted, therapy consisted of aspirin (ASA), dipyramidole, sulphinpyrazone, or combinations and the endpoints were vascular events defined as myocardial infarction, stroke, or vascular death. The data from the trials and meta-analysis calculations are summarized in Table 21.1. The individual and overall odds ratios with 99% confidence limits are shown in Figure 21.1, which is the typical Forest plot for summarizing the results. The size of the plot symbols is proportional to the sample size in each study. The aggregate effect is an odds ratio of 0.75 ($p < 0.00001$), demonstrating the efficacy of antiplatelet drugs in reducing vascular events. The strength and consistency of these findings are greater than one would be likely to find with either an informal review or examination of only a few of the trials. Even though 8 of the 11 trials have confidence intervals that include an odds ratio of 1.0, the aggregate evidence from the meta-analysis in favor of treatment is compelling.

TABLE 21.1   Randomized Trials of Prolonged Antiplatelet Therapy

Trial	Therapy	Treatment	Control	$O - E$	Variance
Cardiff I	ASA	58/615	76/624	−8.5	29.9
Cardiff II	ASA	129/847	185/878	−25.2	64.2
PARIS I	ASA or ASA + dipyridamole	244/1620	77/406	−12.7	43.3
PARIS II	ASA + dipyridamole	154/1563	218/1565	−31.9	82.0
AMIS	ASA	395/2267	427/2257	−16.9	168.2
CDP-A	ASA	88/758	110/771	−10.2	43.1
GAMIS	ASA	39/317	49/309	−5.6	18.9
ART	Sulphinpyrazone	102/813	130/816	−13.8	49.8
ARIS	Sulphinpyrazone	38/365	57/362	−9.7	20.7
Micristin	ASA	65/672	106/668	−20.8	37.3
Rome	Dipyramidole	9/40	19/40	−5.0	4.6
All	Any	1321/9877	1685/9914	−160.1	562.0

Note: Odds ratios and 99% confidence limits are shown in Figure 21.1.
Source: From Peto, Collins, and Gray (1995).

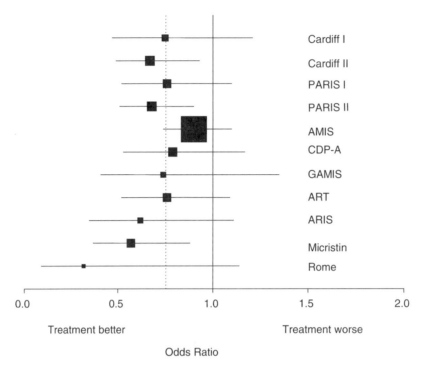

Figure 21.1   Meta-analysis of antiplatelet therapy. Observed odds ratios and 99% confidence limits from 11 randomized clinical trials. The dotted line shows the overall odds ratio (= 0.75). Data are presented in Table 21.1.

21.3   OTHER ISSUES

21.3.1   Meta-Analyses Have Practical and Theoretical Limitations

Researchers do not universally agree on the theoretical validity of meta-analysis, its practical application, or interpretation. Some of problems are illustrated by Boden (1992) and Feinstein (1995). Bailar (1995) suggests that improperly conducted meta-analyses can produce misleading results and provides some examples. Difficulties in practical applications include selecting studies of high quality and properly executing and interpreting the analysis. Greenland (1994) cautions against uncritical acceptance of some widely used meta-analytic methods, such as graphical summaries, random effect models, and quality scores. These concerns are discussed from a more favorable perspective by Olkin (1994) in his commentary. Meta-analysis methods have been increasingly used with observational studies, an application viewed dimly by Shapiro (1994). He provides examples of such studies that are of questionable validity.

We must keep in mind that meta-analyses are observational studies subject to the limitations and potential biases of such methods. They are not the same as confirmed randomized trials, but are weaker or stronger depending on the studies that comprise them, the definition of "treatment," and the methods used to synthesize the results. It is likely that more experience with methods and more attention to reporting of clinical trials and observational studies will be needed for meta-analyses to reach their full potential.

When planning or interpreting meta-analyses, investigators should keep in mind limitations of the method. As with all types of studies, there is sometimes a discrepancy between statistical and clinical significance. Often the confidence intervals placed on the overall treatment effects in meta-analyses do not account for sources of variability in the studies. For example, if the confidence intervals are calculated assuming that the individual treatment effects are "randomly" sampled from a homogeneous population of treatment effects, they may be optimistically narrow. In the same way that poor quality data undermine a clinical trial, flawed trials will invalidate a meta-analysis. Khan, Daya, and Jadad (1996) emphasize the importance of trial quality on the final meta-analysis. Finally clinical investigators should not be encouraged to perform (small) underpowered randomized trials with the hope of combining them later in a meta-analysis.

21.3.2   Meta-Analysis Has Taught Useful Lessons

Despite limitations and continued development of methods, meta-analyses have already provided researchers with some important lessons. They provide a quantitative method for synthesizing the results from more than one study that can establish a basis for policy decisions. Meta-analyses have highlighted and improved the quality of reporting clinical trials in medical journals and have suggested ways in which the methodology of trials can be improved. They are a resource for planning new investigations and can suggest if a trial is not needed because of the weight of existing evidence or if a new study may be informative. Finally meta-analyses provide a historical perspective on treatment issues and create a comprehensive quality controlled database that may be useful for other questions in the future.

Meta-analyses can also directly affect treatment decisions. When used for this purpose, they will sometimes generate controversy. One example is the assessment of mortality in patients with cardiovascular disease who are taking calcium channel blockers (Furberg, Psaty, and Meyer, 1995; Opie and Messerli, 1995; Kloner, 1995; Yusuf, 1995). The meta-analysis suggested that such drugs (nifedipine, in particular) may cause increased mortality.

Investigators also use meta-analyses to assist in the design and monitoring of clinical trials. This is a natural role for studies that synthesize the available evidence regarding a treatment. An in-depth perspective on the use of meta-analyses for this purpose was provided by the Conference on Meta-Analysis in the Design and Monitoring of Clinical Trials, published in *Statistics in Medicine* (Vol. 15(12), 1996).

21.4   SUMMARY

Meta-analysis is a formal quantitative process for combining the evidence about treatment effect from more than one clinical trial. Because the data from many similar studies can be combined using the method, meta-analyses have the potential to demonstrate treatment effects with a high degree of precision. This high precision can demonstrate small, but clinically important, treatment effect convincingly. Because of the varied content and quality of clinical trial reports, meta-analyses should be performed only on patient-level data. Careful planning, comprehensive data collection, and a formal approach to statistical methods is necessary for a high-quality meta-analysis.

Meta-analyses have important weaknesses. Meta-analyses are not experimental designs, and their validity relies on retrieving existing data and on the quality of the studies that generated the data. Incomplete retrieval of trials or data can bias the result. Because positive trial results are probably more likely to appear in the literature than negative ones (publication bias), meta-analyses can overestimate the benefits of treatment.

Meta-analyses have had a positive impact on the quality of clinical trials and their publications.

21.5 QUESTIONS FOR DISCUSSION

1. When an expert conducts a review of a particular issue, he or she may do so without the formal statistical methods of a meta-analysis. Compare and contrast this to a meta-analysis and discuss the advantages and disadvantages of each approach.

2. Discuss the similarities and differences between a cumulative meta-analysis and a sequential randomized trial.

22

MISCONDUCT AND FRAUD IN CLINICAL RESEARCH

22.1 INTRODUCTION

Falsifications and fabrications in science have a long history, even if one considers only well-documented cases. Allegations of data fabrication or falsification and other actions that would be unacceptable by today's standards have been made against Ptolemy, Galileo, Newton, Dalton, Mendel, and others. For example, Ptolemy is said to have used the astronomical observations of Hipparchus of Rhodes and claimed them as his own in formulating his theories. Newton apparently fudged calculations in his *Principia* to agree with his gravitational theory, and used his presidency of the Royal Society to discredit Leibniz's credible claim of having invented the calculus. Investigation and re-analysis of Mendel's work strongly suggests falsification of data to correspond with theory. This case is particularly interesting to statisticians (Fisher, 1936).

Manipulating data to reinforce theory may not be uncommon in the history of science. In the nineteenth century, Charles Babbage, the inventor of the calculating machine, lamented the widespread practices of data "trimming," "cooking," and "forgery" in the sciences (Babbage, 1830). For a somewhat controversial but sobering review of these and other cases of possible misconduct, including more modern ones, see Broad and Wade (1982), Kohn (1986), or Miller and Hersen (1992). Friedlander (1995) reviews a number of cases in the biological and basic sciences, particularly in the context of situations on the fringes of science. Altman and Hernon (1997) also offer a broad perspective on misconduct.

Although many of the historical instances are arguable either way because of lack of evidence, recent instances clearly involve misconduct. The circumstances surrounding the discovery of Piltdown man is an interesting and prominent example from the twentieth century, which involved outright fraud or trickery. In 1912, a human skull and the jaw of an ape were planted in a gravel pit on Piltdown Common, near Lewes, England, to suggest that the earliest human was from Great Britain. Belief in the legitimacy of the artifacts extended over many years and fooled many knowledgeable scientists. It was not until 1953 that the bones were proved to be crudely faked, although some scientists suspected and attempted to expose it much earlier (Matthews, 1981; Kohn, 1986). It was not until 1996 that solid evidence about the identity of the perpetrator was published (Gardiner and Currant, 1996; Gardiner, 2003).

There have been many well-documented instances of misconduct from the basic medical sciences. One example is the work of Spector on the protein kinase cascade leading to the phosphorylation of sodium-potassium ATPase (McKean, 1981). The findings could not be replicated by several scientists who attempted it and were discovered to be at least partly fraudulent (Broad and Wade, 1982; Kohn, 1986; Racker and Spector, 1981). A second example is the human Hodgkin's disease cell lines of Long, found to actually be from the owl monkey (Harris et al., 1981; Wade, 1981). One of the most controversial, interesting, and unfortunate cases in recent times is that of Thereza Imanishi-Kari, who was accused of misconduct when she worked with Nobel Laureate David Baltimore, but was later exonerated on appeal. The best concise summary of this complex case is given by Kevles (1996) and a more complete description is Kevles (1998). In 1996 five papers related to the genetics of leukemia were retracted after an admission of data fabrication by a graduate student. The senior author on these papers was the director of the National Center for Human Genome Research. These more recent cases indicate that historical instances of misconduct are not a consequence of lack of sophistication on the part of the scientific community.

In the clinical sciences there have been extensive and painful episodes of misconduct in recent times (Engler et al., 1987; Relman, 1983). Today there seems to be an increase in reports of fraud or misconduct in many areas of science. Discussions in the context of nursing can be found in Morrison (1990), Denham (1993), and Chop and Silva (1991). The true incidence depends on our definition of misconduct (discussed below). However, there is much more activity in science now than in the past, and most cases of possible fraud are treated with more openness and public scrutiny than they were even 25 years ago. The incidence may be no higher than in the past, but the number of events and our awareness may be greater. Heightened awareness and vigilance, as discussed below, may actually have decreased the frequency of misconduct cases today.

Clinical trials, being a mix of basic, statistical, and clinical sciences, are as vulnerable to misconduct as any other field. Recent history demonstrates the nature and serious consequences of misconduct involving trials, and public and governmental reaction to it. Such cases are thankfully uncommon despite the pressures and inherent messiness of clinical research. The Poisson case, discussed in detail in Section 22.4.2, is prototypical with regard to the different perspectives from which misconduct is seen and the potential damages, both scientific and public.

Relatively little has been written about misconduct related to statistical methods other than fabrication and falsification of data. One exception is the discussion of fraudulent statistical methods in the context of risk assessment by Bross (1990). Bross defines a fraudulent statistical method as one that gives false or misleading results and is intended to deceive. He places certain quantitative risk assessment methods in this category, a notion that might make many biostatisticians uncomfortable.

22.1.1 Integrity and Accountability Are Critically Important

In recent years public officials have increasingly questioned the ability of the scientific community to ensure integrity in the conduct of research. Clinical trials have not been free from this increasing scrutiny. The trend is both a consequence of highly visible instances of misconduct by scientists and changing public perceptions about the research process. Along with the increase in the magnitude and influence of the research enterprise in recent years, scientists have been asked for greater accountability when using public funds. Also research efforts are becoming more complex and require collaborations between institutions and investigators that may necessitate greater accountability. Finally many research efforts are commercialized rapidly or are sponsored by industry, even if they take place in academic settings, leading to demands for more careful review (Schwartz, 1991). Not only do these situations increase public awareness, but they also contribute to an evolving standard for integrity.

Scientific research is basically a creative enterprise, but one that relies on trust. It is paradoxical that this scientific process of balanced skepticism must ultimately rely on a foundation of trust. Because it does, violations of that trust through misconduct or fraud threaten science in the most fundamental way. A well-rounded discussion from this perspective is given by Fuchs and Westervelt (1995). The violation of trust may explain the disproportionate publicity and attention that misconduct cases often receive. Those engaged in science have taken the issue of misconduct seriously. Extensive studies and recommendations have been made by the Institute of Medicine (IOM, 1989) and the National Academy of Sciences (NAS, 1992a, b) in addition to the attention given by the Public Health Service (discussed below). The issues have also been considered by the Society of University Surgeons (Simmons et al., 1991).

Although not less important, the basis of trust in science is distinctly different from that in other creative activities such as art, music, or literature. Historically clinical researchers have relied on trust, honesty, objectivity, and the error-correcting nature of the scientific method to guarantee the integrity of research. Peer review, in both the funding of research and the publication of results, has been a major component of the process. Unfortunately, personal or group ambition sometimes conflicts with scientific ethic. When objectivity is lost for this or any other reason, the research process is prone to breakdown. All types of error become more likely: honest error, self-deception, sloppiness, dishonesty, and criminal behavior.

Many views of fraud, its consequences, detection, and prevention are taken from the perspective of basic science research. However, it may be more important to detect fraud in clinical trials than in basic research for several reasons. First, basic research may be more self-correcting than clinical trials by its very nature, particularly because many fundamental experimental findings are replicated independently. Second, clinical trials usually have greater or more immediate consequences for the public health and welfare than basic research. Finally, clinical trials often have a higher public profile than basic research. In any case, being able to conduct clinical trials is a privilege granted to investigators by both the patient participants in their studies and the public. This trust is violated by any acts that are disrespectful of the study participants or the research process.

22.1.2 Fraud and Misconduct Are Difficult to Define

"Fraud" is a term that is often informally applied to dishonest acts or intentional deception. However, the legal interpretation of the term "fraud" in the United States implies injury or damage to victims. Because this criterion is often not applicable to scientific research, the term "misconduct" is probably more appropriate to describe deviant scientific behavior. Unfortunately, "misconduct" can also be applied appropriately to a variety of actions by researchers, some of which are not especially consequential or relevant to the objectives of scientific research (e.g., sexual harassment). Although difficult to define satisfactorily (see below), misconduct is often easy to identify. An exception that illustrates the difficulty in defining misconduct is theft of intellectual property. There is a fine line between this and the appropriate dissemination of new ideas. Several authors have discussed this point (Judson, 1994; McCutchen, 1991; Rennie, Flanagin, and Glass, 1991; Relman, 1984; Swazey, Anderson, and Louis, 1993).

Science is unique in that the scientific method *is* the integrity of the discipline. This orientation toward process has evolved mostly since the eighteenth century but has been applied to clinical trials only in the last 45 years. The concept of ethics in science can become distorted by the idea that science is composed of a fixed set of rules. The only "rule" in science is the search for truth, a goal that is merely ancillary to most other human endeavors. In fact the scientific ethic is determined by the structure of the process rather than the outcome (Friedlander, 1995).

Questions about the definition of scientific integrity and misconduct are therefore questions about what elements of the process are integral to the scientific method. In other words, the meaning of scientific integrity depends on the meaning of science. The general norms of science are (1) intellectual integrity (honesty) and objectivity, (2) tolerance, (3) doubt of certitude, (4) recognition of error, (5) unselfish engagement, and (6) communal spirit (Bulger, 1994; Cournand, 1977; Denzin and Lincoln, 2000). These also apply to clinical investigators. In clinical trials the question is further complicated by the ethics of the practitioner–patient relationship. Concern and respect for the patient may not always be coincident with scientific goals.

One could define misconduct as significant deviations from these norms. However, misconduct is usually defined more in terms of a lack of intellectual integrity and objectivity. There have been formal definitions offered for scientific misconduct or fraud by at least four organizations: The National Academy of Sciences (NAS), The Public Health Service (PHS), the National Science Foundation (NSF), and The Royal College of Physicians (RCP).

National Academy of Sciences

The NAS Panel that studied the integrity of the research process in the United States (NAS, 1992a) distinguished between three types of inappropriate behavior by scientists: (1) misconduct in science, (2) questionable research practices, and (3) other misconduct. These are summarized in Table 22.1.

The NAS defined misconduct in science as data fabrication, data falsification, or plagiarism, but did not include errors of judgment, differences of opinion, or mistakes in the recording, selection, or analysis of data. Questionable research practice is conduct that goes against, and may be detrimental to, traditional values underlying the research process. These can include failing to maintain records, exploiting research subordinates or subjects, and using inappropriate analytic methods to enhance the significance of research findings. There are no clear standards of behavior regarding questionable research practices and no general agreement as to the seriousness of such actions. Other forms of misconduct are not unique to science but can occur in the process of any type of research. These include harassment of individuals, misuse of funds, and conflicts of interest. Conflicts of interest are common and do not necessarily constitute misconduct. However, because they can affect scientific objectivity, conflicts of interest need to be treated openly, and occasionally will disqualify the opinions of some investigators.

TABLE 22.1   Elements of Misconduct in Science as Defined by the National Academy of Sciences

Types of Inappropriate Behavior

Misconduct in Science

- Data fabrication (making up data)
- Data falsification (changing data values)
- Plagiarism

Questionable Research Practices

- Failing to retain important data for a reasonable period of time
- Inadequate research records
- Refusing to allow reasonable access to data by others
- Using inappropriate statistical methods to enhance the significance of findings
- Exploiting or inadequately supervising research subordinates
- Naming authors without regard to significant contribution to the research reported

Other Misconduct

- Sexual and other harassment
- Misuse of funds
- Gross negligence
- Vandalism
- Violation of research regulations
- Conflict of commitment or interest

Public Health Service

The Public Health Service has regulations defining misconduct in science and guiding approaches to allegations of misconduct. Government agencies such as NIH, the Centers for Disease Control and Prevention, and the FDA are covered by these regulations. In 1989 the PHS (DHHS, 1989, p. 32447) defined scientific misconduct as

> fabrication, falsification, plagiarism, or other practices that seriously deviate from those that are commonly accepted within the scientific community for proposing, conducting, or reporting research. It does not include honest error or honest differences in interpretations or judgments of data.

Unfortunately, the regulations did not explicitly define fabrication, falsification, or plagiarism. Furthermore they do not define deviations from commonly accepted practice.

The practical application of this rule resulted in a number of actions being seen as misconduct by the PHS. These include failure to perform research supported by PHS while claiming progress in reports, naming authors on publications without their knowledge, selective reporting of primary data, unauthorized use of data from another researcher, false reporting of patient status in clinical research, and use of improper and faulty statistical methods.

In 1999 DHHS announced changes in its approach to allegations of misconduct. HHS adopted the definition of research misconduct from the National Science and Technology Council after it was finalized by the Office of Science and Technology Policy (OSTP, 2000). The definition is as follows.

Research misconduct is defined as fabrication, falsification, or plagiarism in proposing, performing, or reviewing research, or in reporting research results. Fabrication is making up data or results and recording or reporting them. Falsification is manipulating research materials, equipment, or processes, or changing or omitting data or results such that the research is not accurately represented in the research record. Plagiarism is the appropriation of another person's ideas, processes, results, or words without giving appropriate credit. Research misconduct does not include honest error or differences of opinion.

A finding of research misconduct requires that: there be a significant departure from accepted practices of the relevant research community; and the misconduct be committed intentionally, or knowingly, or recklessly; and the allegation be proven by a preponderance of evidence.

Procedures for inquiries are discussed below in the section on institutional approaches to misconduct. If misconduct is found, penalties can range from a letter of reprimand to prohibiting individuals from participating in certain research activities.

National Science Foundation
The NSF historically defined misconduct as (1) fabrication, falsification, plagiarism, or other serious deviation from accepted practices, or (2) retaliation against whistle-blowers who act in good faith (NSF, 1987, 1991). Additionally the NSF extended the definition to nonresearch activities such as education and outlined procedures and actions to be taken in cases of possible misconduct. Full investigations take place only if an initial inquiry finds it necessary. Following the DHHS changes described above, the NSF policy was modified to be consistent with them (NSF, 2002).

Royal College of Physicians
In England the RCP defined "misconduct" as piracy, plagiarism, and fraud (Royal College of Physicians, 1991). They define *piracy* as taking ideas from others without acknowledgment. *Plagiarism* is the copying of ideas, data, or text without credit. *Fraud* is defined as "deliberate deception, usually the invention of data."

Overview of Definitions
It is interesting that none of the early definitions of misconduct or fraud explicitly mentioned the intent of the perpetrator. The current NSF definition corrects this. It is commonly understood that misconduct is a purposeful act, unlike error, which is inadvertent. Often the motive for misconduct can be established. The absence of a motive does not disprove misconduct, but is more typical of error. Thus we demand more stringent evidence to label an act as misconduct and little evidence to call it an error. Although there is significant overlap in these definitions, problem areas remain. Some discussed below include the definition of due process and review, the role of whistle-blowers, and how to cope with false accusations of misconduct.

Even though all of these definitions are sensible, I will use the National Academy of Sciences definition and avoid the term *fraud,* except in the informal sense of the word.

22.2   RESEARCH PRACTICES

Research practices in science are influenced by six major factors: (1) scientific norms, (2) specific disciplines, (3) individuals and teams, (4) institutions, (5) government, and (6) society. Scientists and clinical researchers share common training, values, and experiences, which help to establish scientific norms, at least within the discipline. Even so, it is not clear to what extent individual researchers adhere to the norms because of competing pressures such as research funding and competition for academic recognition and positions. A perspective on this is given in a series of short articles by Cohen, Marshall, and Taubes (1995). Specific disciplines assist in establishing good research practices through both specific guidelines and diffuse practices such as peer review. In clinical trials good research practices are founded on principles outlined in places such as the Helsinki Declaration and the Belmont Report (Section 3.3).

Individual scientists and research teams are important determinants of good research practices through the process of "mentoring" or training young scientists. Mentoring helps encourage important values such as creativity and collaboration. Similarly institutional policies are strong determinants of the environment in which research is conducted. In fact universities and other research institutions may be most important in fostering integrity in scientific research (Danforth and Schoenhoff, 1992). Some academic institutions have adopted guidelines that they expect faculty to observe when conducting research (National Academy of Sciences, 1992b). However, many institutions have no formal guidelines regarding authorship and mentoring. Because of specific and highly public instances of scientific misconduct, most institutions have adopted policies for handling such cases. They are required by the federal government to have policies for handling misconduct to be eligible for federally sponsored research grants. Additionally institutions must submit an assurance that they will follow the policies.

In contrast, government agencies have policies that regulate many areas of research. These include informed consent in clinical trials, animal research, and laboratory safety. Policies and procedures to regulate the treatment of alleged misconduct do not have wide agreement. The history of the development of the Office of Research Integrity (ORI) is an example of such differences of opinion (ORI, 1993a). Societal concerns about waste, mismanagement, and fraud in the use of government funds has translated into requirements for greater accountability from research scientists. Similarly worries about real or imagined ethical abuses, such as those surrounding the use of human subjects for radiation experiments during the early years of the Cold War, highlight the public's concern for accountability (Advisory Committee on Human Radiation Experiments, 1995).

22.2.1 Misconduct Has Been Uncommon

By most published estimates, the frequency of cases of scientific misconduct is low. The actual number of cases is probably higher than that reported (Swazey, Anderson, and Louis, 1993). Between 1980 and 1990 there were fewer than 100 recognized cases of misconduct in science (National Academy of Sciences, 1992a). Given the number of research investigators and research grants, this number appears to be small. It is likely that the number of true cases is underreported because of allegations that were investigated by other means. From detailed studies of investigated cases, it appears that plagiarism is the most common type of scientific misconduct. For example, it accounted for half of the 41 allegations reviewed by the National Science Foundation in 1990 (National Science Foundation, 1990). There have been publicized cases of misconduct in other countries as well, including Great Britain (Freestone and Mitchell, 1993; Lock, 1988; Lock, 1995; Wells, 1992), Australia (Lock, 1990), India (Jayaraman, 1991), and Germany (Eisenhut, 1990; Foesling, 1984). A noteworthy case from Australia is connected with lawsuits over Bendectin, a drug manufactured and used in the United States for nausea and vomiting during pregnancy (Humphrey, 1992; Skolnick, 1990).

In the United States, reviews of publicly reported cases have been done by Woolf (1981, 1986, 1988). In 1988 there was a NSF/AAAS survey of 392 deans of graduate studies at institutions affiliated with the Council of Graduate Schools. Forty percent of the responders indicated they were aware of *possible* cases of misconduct. Other reports and surveys concerning misconduct are given in the NAS report (National Academy of Sciences, 1992a).

Currently the frequency of misconduct can only be determined approximately. Data are available for Public Health Service Research in the United States through the Office of Research Integrity and are available for public viewing at the ORI Web site. Between 1993 and 1999, the number of misconduct cases closed by the ORI varied between 25 and 58 per year. The total case load included institutional inquiries, institutional investigations, ORI inquiries, and ORI investigations. In 2000, federal investigations became the responsibility of the Office of the Inspector General (OIG). Approximately half of all cases appear to result in findings of misconduct. The ORI does not report the number of research projects under its jurisdiction. In 1999, for example, 25 investigations were closed by ORI and/or institutions with the finding of 13 cases of misconduct. Some cases are carried over from year-to-year and not all findings are upheld on appeal.

Between 2000 and 2002, the total number of investigations performed by institutions under changing federal policy was between 25 and 32. Findings of misconduct varied between 46% and 74% of cases. These numbers are not directly comparable to the earlier decade because of changes in the way misconduct cases are investigated and managed by the OIG. In any case, the number of scientific investigations relevant to these numerators is quite large. For example, tens of thousands of clinical trials alone are conducted in the United States each year. Thus the frequency of misconduct is low.

22.2.2   Causes of Misconduct

There are several factors that can contribute to the occurrence of misconduct. These include academic pressure to publish, financial gain, professional vanity, lack of understanding of the research process, and psychiatric illness. Of these, academic pressure is widely regarded as a common contributing cause that may come from diverse sources. Appointment and promotions often depend as much on the quantity of publication as on their quality. Productivity also contributes greatly to obtaining and keeping research grants. Closely related to this type of academic pressure is professional vanity and ego. Recognition by peers is greatly prized by researchers and comes with many tangible and abstract rewards.

Financial gain may not be a common cause of misconduct, at least among academic investigators. They are usually not in a position to profit financially from a particular research project. With increasing technology transfer out of academic centers into commercial enterprises, the opportunity for investigators to realize personal financial gain is greater now than in the past. It is possible that more cases of misconduct motivated by money will arise in the future.

There are very few data regarding the role of psychiatric illness in scientific misconduct. Response to stress and self-expectations are likely to be important determinants.

The egregiousness of some of the cases of misconduct reported, however, suggests that more than just academic pressures are at work. These can easily explain an isolated incident, but many cases of misconduct illustrate long-term patterns of behavior, attitudes, and a lack of integrity that would be out of place in any professional setting. There has been little or no psychological work in this area, partly because such cases are so uncommon. General discussions are given by Miller (1992) and Thelen and DiLorenzo (1992).

Clinical research and clinical trials have characteristics that may increase pressures on investigators. There is constant pressure to accrue patients to meet the temporal and scientific objectives of the trial. Eligibility and exclusion criteria might be perceived as an obstacle in such circumstances. Similarly the specifications of the protocol may occasionally conflict with the needs of an individual patient. While there is never a need to conform to the protocol if the physician does not believe it is in the patient's best interest, a conflict can create pressure to falsify data items to give the appearance of conformity.

22.3 GENERAL APPROACH TO ALLEGATIONS OF MISCONDUCT

22.3.1 Government

The first systematic public discussion of reports of fraud in biomedical research was in the 1980 Congressional hearings of the Oversight and Investigations Subcommittee of the House Science and Technology Committee. The hearings were in response to public disclosure of instances of misconduct at four major research centers. In 1985, Congress required Public Health Service (PHS) grantees to adopt administrative procedures for reviewing reports of scientific fraud (Public Law 100-504). Guidelines for doing so were issued by the PHS in 1986. By 1989, most institutions receiving more than 100 PHS awards had adopted such procedures. However, nearly 80% of institutional grantees had no guidelines. Final regulations requiring institutional guidelines were published by PHS in 1989 (42 CFR 50). The PHS consists of the National Institutes of Health, the Centers for Disease Control and Prevention, the Food and Drug Administration, the Substance Abuse and Mental Health Services Administration, the Health Resources Services Administration, the Agency for Health Care Policy and Research, the Agency for Toxic Substances and Disease Registry, and the Indian Health Service.

Investigating allegations of scientific misconduct requires a two-stage process: an inquiry and an investigation. The inquiry is usually confidential and within the institution. The extent of the inquiry may be determined locally. Institutions may elect to perform an inquiry formally or informally using existing committees. An institutional inquiry can determine that an allegation of misconduct lacks merit, at which time the matter is closed. If the inquiry determines that an investigation is warranted, a formal process of investigation must be initiated. The inquiry itself cannot determine guilt. If an investigation is required, the institution must notify the sponsoring agency. The investigation is usually performed by a special appointed panel of objective experts. In the event that the institution lacks the resources to be thorough, fair, and objective, it can request that the PHS conduct its own investigation through the Office of Research Integrity.

A principal component of PHS handling of allegations of scientific misconduct is the Office of Research Integrity (ORI). The background of this Office is as follows: In

1989, the Office of Scientific Integrity (OSI) was formed in the NIH Director's Office and the Office of Scientific Integrity Review (OSIR) was formed in the Office of the Assistant Secretary for Health in the PHS. The OSIR was responsible for establishing PHS policies for misconduct in science, while the OSI was to implement the policies and conduct investigations. In 1992, the OSI and OSIR were merged into a single office called the ORI, under the Assistant Secretary, making these activities independent of PHS funding agencies. In June 1993, the ORI became independently established by statute within the Department of Health and Human Services (HHS) with the director reporting to the Secretary of HHS. In 1995, the ORI was placed in the Office of Public Health and Science, which is headed by the Assistant Secretary for Health.

The National Science Foundation (NSF) has a policy similar to the PHS. The responsible entity is the Office of the Inspector General (OIG) established by Congress in 1989 (Inspector General Act Amendments, 1988). The OIG is also responsible for auditing grants and contracts awarded by the NSF. The policy of the NSF is to make research institutions responsible as much as possible for dealing with all aspects of alleged misconduct in science. In response to allegations of misconduct, institutions must conduct investigations that the OIG deems fair and complete.

Role of the Office of Research Integrity

The role of the ORI in misconduct investigations has been defined in PHS regulations. Following modification to the DHHS procedures in 1999, the role of the ORI has changed slightly. A perspective on the ORI for the bench scientist is given by Pascal (2000) but is also useful for trialists. Responsibilities of the ORI include (1) developing policies, procedures and regulations related to the detection, investigation, and prevention of research misconduct; (2) reviewing and monitoring misconduct investigations; (3) recommending findings and administrative actions to the Assistant Secretary for Health, subject to appeal; (4) implementing programs to teach the responsible conduct of research, promote research integrity, prevent research misconduct, and improve the handling of allegations of research misconduct; (5) providing technical assistance to institutions that respond to allegations of research misconduct; (6) conducting policy analyses; (7) assisting the Office of the General Counsel (OGC) to present cases before the HHS Departmental Appeals Board; (8) administering programs for maintaining institutional assurances, and responding to allegations of retaliation against whistleblowers. In addition the ORI publishes a quarterly newsletter that provides details on closed cases and publishes an annual report (e.g., ORI, 1995) where additional information can be found.

Although the ORI can become involved in misconduct cases through a variety of mechanisms, their formal role is subject to certain limitations. For example, the ORI has authority only in cases involving Department of Health and Human Services funding (PHS, NIH, etc.). Furthermore the types of misconduct on which it may rule are restricted by the regulatory definition (above). Since 1992, DHHS has had a hearing mechanism (Departmental Appeal Board [DAB]) by which findings of misconduct can be reviewed formally and legally at the departmental level. When this happens, findings such as data fabrication must be demonstrated to be "material" (i.e., substantive), and the ORI must prove intent for the findings to stand. If an assessment of misconduct is affirmed, the actions that ORI can take are limited to debarring the investigator from receiving federal funds and/or serving on advisory committees for a limited period of time.

Despite the limited-sounding nature of these actions, investigation by an institution and/or the ORI is difficult for all associated with the case. The cases cast a shadow on scientific work, consume time and resources of both the accused and investigating staff, and may leave lingering questions about different but seemingly related matters. In some high-profile cases many individuals not directly involved in the study may experience a significant impact from an investigation.

In clinical trials investigation by ORI is motivated by concerns about the health of the public, public and media awareness of the issue, and the need to correct potential scientific errors. Because clinical trials are often large, expensive, and not likely to be repeated in the same form, there is concern that they are not as self-correcting as other scientific enterprises. Also the ORI cannot easily conduct an investigation without the cooperation and assistance of the trial coordinating center and local institution. For multicenter trials, the institution being investigated, the coordinating center, the trial sponsor, and ORI must work together. This consumes considerable resources. These activities affect the patients on the trials, the investigators, and the science community at large.

Aside from the responsibilities of the institution described below, the OIG (rather than ORI) conducts any fact-finding required by the government. The Assistant Secretary for Health makes the final decision regarding findings of scientific misconduct and administrative actions. The HHS Departmental Appeals Board hears appeals from those accused. The hearing panels include two scientists rather than none, a recent change.

22.3.2 Institutions

Like the PHS, most medical institutions have a structured, staged approach to examining allegations of scientific misconduct. Institutions have the primary responsibility for investigating allegations of scientific misconduct. The following sketch is based on policies and procedures at one institution, independent of, but in conformance with, recommended PHS procedures. The preliminary inquiry into good-faith allegations of misconduct is performed by an advisory committee appointed by the Dean that maintains strict confidentiality. The purpose of the preliminary inquiry is to determine if the allegations have merit. If the inquiry finds the allegations were without merit, the process is terminated. The implicated person or persons and the accuser will be notified of the findings and dismissal of the allegations no later than when the preliminary inquiry is completed. The case is then considered closed, and no notification of outside organizations is required.

If the allegations of misconduct appear to have merit based on the preliminary inquiry, a full investigation takes place by an ad hoc committee. If this committee finds that misconduct has taken place, a full report is made for adjudication by a Standing Committee on Discipline. This Committee is responsible for initiating any immediate actions outside the university that are required by the findings, such as notification of journals, sponsors, or the public. The Standing Committee reports its finding to the Advisory Board of the Medical Faculty, which will consider what appropriate disciplinary actions are needed.

Virtually every research institution has formal guidelines for dealing with allegations of misconduct. For example, a typical example is provided by the faculty policies for *Procedures for Dealing with Issues of Research Misconduct* of the Johns Hopkins University School of Medicine (Johns Hopkins, 2004). This university, like most, also has separate formal faculty policies for conflict of commitment, rules and guidelines for responsible conduct of research, and grievance procedures. These provide an overview of a typical academic approach to misconduct allegations.

22.3.3 Problem Areas

Defining and dealing with deviant behavior by scientists is a difficult problem. Doing so raises questions about defining and recognizing such behavior, double standards for scientists compared with other professionals, and personal loyalties. There seems to be no consensus among scientists, research sponsors, and the public on these issues. An interesting perspective on the lack of consensus is given by Dyson (1993). One could also raise serious questions about what role, if any, government should play in dealing with scientific misconduct (Hansen and Hansen, 1991; Klein, 1993).

Despite progress and agreement in formally defining misconduct in science, guidelines for how to approach specific allegations, and steps for prevention, there are a number of areas in which current standards can be improved. The *definition of misconduct* finds no universal agreement, especially concerning areas such as incompetence and reckless or extremely biased interpretation of results. *Requirements for evidence of misconduct* is another area for potential disagreement. Careless error and poor judgment might be considered by some, but not by others, as evidence of misconduct. It is probably true that all clinical trials contain data errors and circumstances of poor judgment. However, one must ascertain the motives and intent of the investigators before misconduct can be established. There is *no universal standard of proof* applied by research institutions or the PHS/NSF.

The appropriate nature and extent of *due process* for those accused of misconduct is another difficult area. An excellent recent discussion of these issues is given by Mello and Brennan (2003) who point out the disparities between the due process afforded individuals in civil and criminal cases compared to an accused scientist. The consequences of a finding of misconduct are potentially serious. Investigators can be prevented from receiving funding for a period of years and have their reputations damaged. There is often disagreement about the nature of an appropriate process for adjudication and appeal. The case cited at the beginning of this chapter regarding Thereza Imanishi-Kari illustrates both the consequences of allegations and the impact of lack of due process. In June 1996 a federal appeals board cleared Imanishi-Kari of charges of misconduct, an action that did little to satisfy critics on both sides of the debate (Singer, 1996; Weiss, 1996). One should also read the account of this case by Healy (1996) for a perspective on this issue. Similarly the *role of the courts* is not well defined, leaving the possibility that some findings could be challenged on legal grounds.

Not all institutions are able to provide the same quality of investigation for allegations of misconduct. This *lack of uniformity* could create injustices. The quality of ORI investigations is sometimes criticized, and it is not clear whether or not they should conduct investigations. Also we must consider the possibility that whistle-blowers may face reprisals by individuals or institutions exposed. However, the possibility exists that *false accusations* could be made to damage the reputations of good scientists and that whistle-blowing may be a convenient posture for such accusers to take.

In many cases researchers who are not a party to the case may be interested in the findings of investigations. This is sometimes true for cases in which no finding of misconduct is made. Despite the *Freedom of Information Act*, some information is held confidential by ORI. This restriction can present problems for those trying to understand more about the nature and extent of misconduct, or use the information in other appropriate ways. The role, if any, of *peer review* in detecting or controlling misconduct is controversial. Study of many cases suggests that it is an inadequate mechanism with which to detect misconduct. However, other structures, such as authorship guidelines and requirements for IRB approval, seem like sensible steps to discourage misconduct.

22.4   CHARACTERISTICS OF SOME MISCONDUCT CASES

There are many misconduct cases that one could choose that highlight important points about scientific integrity and how these problems are examined and resolved. Many detailed descriptions are provided by Broad and Wade (1982), Kohn (1986), and Miller and Hersen (1992). Two cases of misconduct are particularly interesting. The first is noteworthy because of its extensiveness, boldness, and relevance to many aspects of the problem of scientific misconduct. It is a "modern" case, amplified by the power of medical journals to disseminate information. The second is recent and more directly related to the conduct, analysis, and interpretation of clinical trials. It is also "modern" because the sensationalism of the episode was created or distorted by the media. Both cases have useful lessons for clinical investigators reflecting on the integrity of the scientific process.

Anticipating the details below, it is interesting and disturbing to see the extent to which retrospective views of these high-profile cases by review committees find fault with the supervision of the individuals in question. Frequently there are clues that falsification is taking place, particularly when seen after the fact. At the time, however, with the limited information available to any one collaborator or supervisor, the evidence may not be sufficient to question the basis of trust, which is so fundamental to the scientific process.

22.4.1   Darsee Case

An extensive case of misconduct came to light in 1981, when a 34-year-old clinical investigator, John R. Darsee, M.D., was found by colleagues at Harvard to have fabricated data in one study. Evidence subsequently revealed that Darsee had fabricated data, beginning when he was an undergraduate student at Notre Dame, continuing through his residency and fellowship at Emory, and into his fellowship at Brigham and Women's Hospital, an affiliate of Harvard. Details of this extensive case can be found in an editorial by Relman (1983) and a review by Culliton (1983).

Chronology of Events
The index act of fabrication took place on May 21, 1981, during a laboratory experiment on dogs. Dr. Darsee was seen obtaining data over a few hours but labeling them to make it appear that they had been collected over a period of two weeks. The action took place within plain sight of co-workers. An internal investigation of Dr. Darsee's research was begun by his supervisor. At first there was no evidence of more extensive fabrication and Darsee was allowed to continue participating in an NIH collaborative study. Even so, the offer of a faculty position was withdrawn and his NIH fellowship was terminated.

In November 1981, the fabrications and falsifications appeared to extend to a major NIH collaborative study. The data from Harvard were found to be strongly different from those obtained at other sites in the multi-institutional study. At this time an external committee was appointed by the Dean of Harvard to investigate. In December 1981, NIH appointed a second panel to conduct an independent audit, and after six months another group, composed of senior NIH staff, was asked to review the panel's findings. Subsequently investigations at Emory and even of undergraduate days at Notre Dame were performed. Extensive problems were found in Darsee's work from Emory (Knox, 1983).

Findings

At least 17 published papers and 53 abstracts written or coauthored by Darsee were ultimately retracted because of fraud or fabrication. In some cases, his coauthors had too little contact with the research to realize that fabrications had taken place, while in others, Darsee invented the data and published without the knowledge of the coauthors. In a few circumstances Darsee insisted that individuals be listed as coauthors over their objections because they had been helpful in the past. In retrospect, some of the reports in the medical literature were incredible.

For example, one paper, published in 1981 in the *New England Journal of Medicine*, studied taurine levels in myocardial tissue (Darsee and Heymsfield, 1981). It was based on a family of 43 members and stated that twice-yearly blood samples were drawn from each person on two consecutive days after fasting overnight. Additionally 24-hour urine samples were said to have been collected on everyone. One patient was a 2-year-old child. Claims were made regarding tissue samples which, in hindsight, were highly questionable. The paper states that myocardial tissue was obtained within four hours of death from three patients and that cardiac biopsies were done before pacemaker insertions. These claims were considered unlikely by investigators, in retrospect. For these procedures (and many other of Darsee's clinical studies) no IRB approvals were requested. In this paper Darsee acknowledged help from three individuals who probably did not exist. Thus the fabrication extended beyond falsifying data to include inventing collaborators.

Perspective

Because of the length of time involved and extensiveness of the fabrications, it is evident than little can be done to stop an unscrupulous scientist, even when he or she collaborates with knowledgeable and reasonably attentive colleagues. This, then, is the first lesson to be learned from the Darsee affair. A second and third lesson are to see the extent to which science is not self-correcting with fraudulent data and the inability of peer review to detect it. Most of Darsee's work was peer reviewed by good editorial processes and was able to pass without suspicion.

With special regard to clinical trials, a fourth lesson is the need for explicit guidelines and oversight for collection, maintenance, and analysis of data. The fifth feature of the Darsee case, and one that has contributed to widespread improvement in the last decade, is a focus on the responsibilities and contributions of coauthors. To be a coauthor requires substantial input into study design, analysis, interpretation, and writing of the research report. Finally, it is evident from this case that misconduct investigations may need to examine a researcher's entire work over many years. Patterns there may be important for verifying true misconduct.

22.4.2 Poisson (NSABP) Case

The National Surgical Adjuvant Breast and Bowel Project (NSABP) is a large multicenter collaborative oncology clinical trials group that has pioneered, among other treatments, breast conserving surgical treatment for breast cancer in the United States. The NSABP has been headquartered in Pittsburgh under the leadership of Dr. Bernard Fisher, involves as many as 5000 physicians at 484 institutions in North America, and has been funded by the National Cancer Institute (NCI) since about 1960. The data falsification by a single investigator within the NSABP produced enormous controversy from various perspectives and taught many lessons to clinical trialists. A detailed look at the chronology of events is essential to understanding the facts and consequences of the case. Additional details can be found in NCI Press Releases (1994a–f), Goldberg and Goldberg (1994), Angell and Kassirer (1994), Cohn (1994), and an exchange of letters in the *New England Journal of Medicine* (Vol. 330, pp. 1458–1462).

Chronology of Events

In June 1990, an NSABP data manager noted discrepancies in data received from one institution, L'Hôpital Saint-Luc in Montreal, Canada. The principal investigator at L'Hôpital Saint-Luc was Roger Poisson, M.D. This institution was soon audited twice, in September and December 1990, by representatives of the NSABP Coordinating Center. As a result of findings in the audits, the NSABP suspended accrual by that institution in February 1991 and notified the NCI that problems had been found. The NCI notified the Office of Scientific Integrity (soon to be ORI) and the Food and Drug Administration (FDA), after which both agencies began independent investigations of the matter.

In May 1991, Dr. Poisson was listed in the Public Health Service alert system and was disqualified for life by the FDA from receiving investigational drugs. The ORI filed a formal complaint against him in September, 1991. While the administrative proceedings were happening, NCI and NSABP officials were not allowed to comment publicly on the investigation. The NSABP re-analyzed the clinical trials under suspicion internally after deleting the questionable data, and found the new results to be nearly identical to those published originally. Both NCI and ORI were convinced that continuing to rely on the previously published results would pose no risk to the public health.

By December 1992, the ORI completed its report, released it to the NSABP and L'Hôpital Saint-Luc in February 1993, and published the findings in the *ORI Newsletter* (ORI, 1993b) and the *Federal Register* (DHHS, 1993b). A notice also appeared in the *NIH Guide for Grants and Contracts* (NIH, June 25, 1993). The case did not come to public attention until well after these events had taken place, as a consequence of newspaper articles published in the *Chicago Tribune,* beginning March 13, 1994. The media and public attention seemed surprising to many individuals knowledgeable about the case because the investigation had been closed and published for nearly a year, and because government and academic officials knew that the research findings were not sensitive to the inclusion or exclusion of data from the Montreal institution. These facts were slower to be disseminated than the anxiety over the fabrications.

Since March 1994, other instances of irregularities in NSABP studies came to light including concern over patient consents and data quality at the Memorial Cancer Research Foundation of Southern California, and audits were undertaken at institutions in Louisiana, New York, and California. On March 28, 1994, Dr. Fisher was forced to resign as head of the NSABP, and on April 13, 1994, Congressman John D. Dingell, chair of the Subcommittee on Oversight and Investigations of the House of Representatives, held investigations focused on the NIH's response to the events. Dr. Fisher was sharply criticized by Mr. Dingell. In July 1994, Fisher filed a lawsuit against the University of Pittsburgh for violating his rights to due process when he was forced to resign. On March 6, 1995, Fisher filed suit against NCI because 148 of his papers had been labeled with "scientific misconduct" warnings on computer listings. After a hearing in March 1995, the labels were removed, pending a final judgment on the suit.

Findings

The ORI was able to document 115 instances of data falsification or fabrication in data arising from 99 patients at L'Hôpital Saint-Luc. Dr. Poisson had participated in 22 NSABP clinical trials, placing about 1500 patients on study. The fabrications, some using duplicate records to avoid detection, involved 7% of the patients. Fourteen of the 22 trials had been published. Research subjects with falsified data were included in four publications from the NSABP. Study B-06 compared lumpectomy with or without radiation therapy to mastectomy for the treatment of early stage breast cancer. Study B-13 compared chemotherapy to no further treatment among women with surgically resected node negative, ER negative disease. Study B-14 compared Tamoxifen to placebo in women with surgically resected node negative, ER positive cancer. Protocol B-16 compared Tamoxifen plus chemotherapy to Tamoxifen alone among postmenopausal women with node positive breast cancer.

The fabrications and falsifications found were consequential for patient eligibility but not outcome. There was documentation of falsified estrogen receptor (ER) values in some patients, alterations of dates of surgery and biopsy to meet eligibility requirements, and accrual of ineligible patients. Consent forms could not be found for three patients who were placed on study protocols. The ORI deemed all data from L'Hôpital Saint-Luc unreliable on the basis of these findings. They recommended that all future NSABP trials be published without such data, that the group publish a re-analysis of previous studies without the questionable data, and that Dr. Poisson be prevented from receiving federal funds. At NCI's request, Dr. Poisson was replaced as Saint-Luc's principal investigator in March 1991. NCI asked Dr. Fisher to publish a manuscript re-analyzing the studies after excluding data from Saint-Luc. Since the initial findings, some irregularities in data from 11 other institutions participating in NSABP trials were found.

Perspective

The NSABP case is remarkable from a variety of perspectives. From the first perspective we see the blatant fabrication of data in important clinical trials by an experienced and respected researcher. Dr. Poisson's violation of the public and scientific trust was dealt with quickly by government and academic authorities. The penalties imposed were appropriate. From a second perspective, only eligibility criteria were fabricated, and there was no evidence of harm to individuals or to the quantitative findings of the trials. These facts make this episode of misconduct appear to have less impact than many other cases. Nevertheless, its very existence makes it consequential. Dr. Poisson seemed unable to get past this point (Poisson, 1994).

Third, the case touches on the responsibilities of clinical investigators, sponsors, regulators, and the media when they uncover evidence of misconduct. Dr. Fisher was criticized for failing to publish revised manuscripts quickly, according to requests by NCI and others. The NSABP's misjudgment of the need for this and the disproportionate public response to the case can surely be considered honest error. Many officials knew the quantitative unimportance of the data falsifications on the study results and the fact that numerous other well-performed clinical trials strongly supported the findings of the NSABP studies B-06, B-13, B-14, and B-16. They could probably have alleviated much public concern by emphasizing this.

Readers wishing to obtain a detailed perspective on the NSABP case and useful references in as compact a form as possible should read the *New England Journal of Medicine* editorial by Angell and Kassirer (1994), and replies in the correspondence section of the same issue. The case is made there that journal editors are among those who need to know when findings of misconduct are made and changes to the published literature may be required. The journal also argues that they should be notified during the investigation. This is clearly problematic because many investigations do not result in a finding of misconduct.

Finally, we must consider the forces that can transform a misconduct case into front-page news after it has been closed for a year. Even after two years of investigation, an additional year following closure by the ORI, and the fact that considerable knowledge was in hand concerning the lack of impact of the findings on the public health and results of the trial, a large amount of tension and anxiety was created among all the parties involved with the case. One has to consider the possibility of political or other forces being at work.

Impact of the NSABP Case

The NSABP case will have a lasting effect on the clinical trials community. One of the most useful perspectives on the case, particularly the consequences for the study leaders, is that from Europe (Peto, Collins, Sackett, et al., 1997). There have been formal procedural changes in the way that NCI and ORI disseminate information about misconduct cases. In particular, journals will be automatically notified of such findings (Broder, 1994; Bivens and Macfarlane, 1994). Many NIH-sponsored clinical trials coordinating centers have examined their methods for data quality control and monitoring in the light of some criticism (e.g., Cohen, 1994; Weiss et al., 1993). Interestingly NCI-sponsored studies are already among the most carefully monitored. It seems likely that monitoring of NIH-sponsored trials, particularly the role and conduct of the TEMC, will be more formal and uniform in the future (see Chapter 14). There is likely to be increased attention to quality control procedures, particularly auditing, in the future despite its dubious value in suppressing misconduct (Shapiro and Charrow, 1989). However, there is probably little that an audit can do to discourage a determined and unscrupulous investigator who maintains duplicate records to avoid detection.

Clinical trials are naturally messy, though not fraudulent because of it. It would be unfortunate to see too much effort used to thoroughly clean up data errors in clinical trials without knowing if this would improve their validity. Excessive auditing or other quality control procedures could increase the cost of trials and thereby reduce their size or number. It is probably true that all trials have a range of error in them, but provided the error rate is low, it is unlikely that it has much impact on their validity. Scientific integrity demands that investigators try to get the most reliable answers they can and not spend resources unless it will help correct a significant problem. For trial designers it is important to have only eligibility criteria that are truly necessary.

Two other issues come into sharp focus upon reviewing the NSABP case. First is the rather poor treatment that NSABP officials have received throughout the affair by some government and university officials. At times the trialists seemed to be treated like those who perpetrated the misconduct rather than uncovered it. It took many years for the ORI to close cases involving NSABP officials. Perhaps institutions and/or the ORI could promote standards for civility and collaboration when performing these investigations. Furthermore it is difficult to find trustworthy, innovative, experienced clinical researchers like most of the NSABP members. The NSABP has contributed greatly over the years to clinical science but was judged quickly and harshly by some. All trialists feel vulnerable with regard to the actions of their collaborators, making the Poisson case very sobering.

It is not obvious that re-analyses of clinical trials in question because of misconduct should discard all the data from the institution where misconduct was found. In the NSABP case the violations were in eligibility criteria and not outcomes, making it debatable as to whether or not any of the ineligible patients should be excluded. Consider the discussion in Chapter 6 concerning design validity compared with biological validity. In a survey of opinions about falsified data, many trial participants stated that they would like to see the analysis include problem records (McIntyre, Kornblith, and Coburn, 1996). For example, if the eligibility criteria had been violated through honest error, the "all patients randomized" analysis would include them and would be a valid comparison of the treatments. If the misconduct had been solely financial, for example, we would probably not consider discarding data. While in the NSABP case it seems prudent to exclude the cases, we think this way, in part, because we know that the results remain unaffected by our action. What if the study results depended on the presence or absence of those participants?

Misconduct cases such as the NSABP experience bring into sharp focus the different frameworks in which the actions of scientists will be viewed. Scientists tend to view their own actions, misconduct or not, in a purely scientific framework. However, suspicion or proof of misconduct forces us to view and manage the situation from perspectives that may be primarily nonrational. A larger relevant perspective is that of the public's health, safety, and perception. When dealing with allegations of misconduct, researchers need to appreciate the consequences of their actions when viewed publicly. This is often the most important perspective. Some actions, which are (only) scientifically correct, might be seen as arrogant by the public. Openness, conservatism in dealing with risks, efficiency, and a willingness to make changes in process are nonscientific attributes that can greatly reassure the public.

22.4.3 Two Recent Cases from Germany

It is worth examining two recent misconduct cases from Germany where the mechanisms for dealing with suspected fraud remain somewhat cumbersome. The first case began in 1997 and involved investigations of the work of Freidhelm Herrmann, a cancer researcher who worked at the universities of Freiburg, Berlin, and Ulm, and subsequently went into private practice (Abbott, 2000). An independent task force reviewed 347 of Herrmann's papers spanning 1985 to 1996. The fraud investigators identified 94 papers that contain manipulated or suspect data, 53 of which were coauthored by a colleague who admitted fabricating data. An equal number were coauthored by Herrmann's former department director who described himself as an honorary author and not knowledgeable about the experiments conducted.

An additional 132 papers were cleared of problems. However, the remaining 121 papers could not be satisfactorily evaluated because the inquiry was limited, in part, by uncooperative coauthors. Some frequent collaborators of Herrmann's provided no information to the investigation. The German research council, Deutsche Forschungsgemeinschaft (DFG) then expanded the investigation into some earlier work. However, at the end of the inquiry, the DFG removed certain conclusions of the task force from its report, angering the fraud investigators (Schiermeier, 2002). The dispute centered on one paper from *Blood* (Vol. 84, pp. 1421–1426, 1994) and one from *New England Journal of Medicine* (Vol. 333, pp. 283–287, 1995) thought by the task force to contain irregularities and improper data. So this major case of fraud, though conclusively proved, ended in a dispute among the individuals investigating it and lingering questions about process.

A year later a second case in Germany came to light, and was related directly to the conduct and reporting of a single clinical trial. The case centered around Alexander Kugler of Göttingen University Hospital and the publication of a promising clinical trial of 17 renal cancer patients treated with a novel vaccine (*Nature Medicine* Vol. 6, pp. 332–336, 2000). A picture used by the authors was apparently taken from the Internet. The university investigated the study and the publication and concluded that Kugler was guilty of misconduct because of his handling of data (Tuffs, 2002; Vogel and Bostanci, 2002). For example, 4 or 5 of the treated patients did not meet the published eligibility criteria for the trial. Some primary data were missing, and the vaccine preparation was inadequately documented. Additionally the study was not conducted under an ethics board approved protocol as indicated in the publication. Other authors were deemed not guilty of misconduct, in part, because Kugler did not involve them sufficiently in the research or report.

The Kugler case continued to generate controversy because it took so long to evolve, a partial consequence of strict patient privacy laws in Germany. The finding of no culpability by the other 14 authors of the original publication was also controversial. In 2003, a year after the investigation was concluded and three and a half years after the original publication, the paper was finally retracted (*Nature Medicine*, Vol. 9, p. 1093, 2003). The explanation of the process by the editors of *Nature Medicine* was understandable, but somewhat odd. For example, they stated that authors should voluntarily retract a publication, but yet the paper in question would have been unacceptable to the journal because of the lack of ethics approval of the study. It remains to be seen whether or not the scientific direction of the original vaccine research is worthwhile.

22.5 LESSONS

22.5.1 Recognizing Fraud or Misconduct

In most of the well-known cases of scientific misconduct, the improper actions took place over a prolonged period and came to light after a single precipitating event. Usually the published reports cannot be distinguished from legitimate science, except in retrospect and with the awareness of other questionable or improper actions by the researcher. In fact, as in the Poisson case, the published reports may still accurately convey much quantitative information. The peer review publication process is not an efficient way to recognize improper behavior.

It is unusual to question findings after they appear solely on the basis of information contained in the published report. An interesting exception in the basic sciences is the paper by Kuznetsov (1989), supposedly containing molecular biological evidence against evolution, and a critique of the work by Larhammar (1994, 1995). Kuznetsov reported finding an "antievolutionary factor" that functioned through messenger RNA. His work was published in a somewhat unusual place for its evolution-related conclusions. Larhammar became skeptical enough to study it only after seeing the paper cited by a creationist.

In reviewing Kuznetsov's work, Larhammar noted a number of discrepancies: the methods were unusual and poorly documented; the report contained unusual precision and consisted mostly of tables and numbers; biological inconsistencies were ignored; nonexistent journals were cited several times; two references by the author himself were cited but the journals could not be found, and the papers were not even on his own list of publications; an article attributed to another researcher in a known journal did not exist; and many papers cited had grammatically incorrect and illogical titles. Larhammar suggests that the paper by Kuznetsov contains fabricated data and references. If so, it is unusual in that it could have been detected by a knowledgeable and skeptical review.

Falsification Is More Likely in Some Areas

In the conduct of clinical trials falsifications are more likely for certain source data. This can be a consequence of having some data sources supervised by individuals who are pressured to keep schedules and accruals up to date and/or who may not share the clinical and scientific ethic that discourages falsification. Support staff and junior faculty members may not have sufficient ethics training to help them resolve problems caused by this kind of pressure. The first potential problem area in data is eligibility and exclusion criteria. Known disqualifying characteristics of the patient may be ignored or falsified. Age, dates of clinical landmarks, or other numerical data can be easily altered. Similarly previous treatments or events in the medical history may be suppressed.

When the protocol calls for serial measurements or documentation of diagnostic tests, reports from previous clinic visits may be propagated as current. Biopsy, blood, urine, or other specimens may be carried along or replaced with those from other patients. Entire clinical visits can be fabricated or dates changed to meet protocol specifications. Pill counts or medication diaries can be altered. This latter type of behavior may be common among patients who wish to appear compliant with the rules of the study. Unfortunately, all of these areas are also subject to honest error. Although it is difficult to distinguish between error and falsification on the basis of a few events, investigators should be aware of and suspicious about a pattern of "mistakes."

Detecting Problems

Some circumstances are so unlikely that they suggest misconduct. Although it is a matter of judgment as to how unusual circumstances are, investigators may be alerted to possible misconduct by observing the following points: Although pleasing, high accrual rates (combined with low rejection rates) may represent a problem, especially if eligibility is not documented in a timely fashion. Real patients and clinic visits are prone to stretch or violate the rigid criteria of the protocol. If this appears not to happen because the need for exceptions or variances never arises, but the data are carefully and completely recorded, there may be a problem.

Discrepancies between source documents and the database can also be a sign of trouble. There may be source documents missing, or clinic logs may be different from the data recorded. These and other similar problems may be noted by quality assurance or audit personnel. Auditing is an important tool for assuring data quality, but may not be sufficient when used without additional measures (Shapiro and Charrow, 1989; Weiss et al., 1993). Collaborative groups of investigators have no universal audit policy (Cohen, 1994). Finally, certain behavior is suspicious. This includes an individual frequently scheduling patient visits when no supervisors are present (e.g., after hours), refusing to have colleagues cover for him or her, or refusing to undergo quality assurance checks. Allegations (formal or informal) or suspicions by other staff members may be a response to suspicious behavior.

22.5.2 Misconduct Cases Yield Other Lessons

When reviewing the process and findings of many misconduct cases, some common themes emerge that seem to characterize the cases, perhaps even contributing to the findings. Not all of these occur in each case of misconduct, but they can help us to understand the roots of inappropriate behavior by researchers. One potential problem area is the use of questionable or *unverifiable data* such as those transmitted orally or in an interview. Data arising in this fashion may be more subject to error and bias of all types than objectively collected ones. For the same reason they may be more prone to falsification and should be avoided, if possible. However, there are circumstances in which it is worth collecting and using such data. It is up to the designers and administrators of the study to anticipate and minimize problems. *Unconventional research methods or design* may contribute to, or be associated with misconduct. Of course, such methods may also be associated with breakthroughs and innovation. In either case unconventional methods probably deserve more scrutiny than methods that are well known.

Plagiarism of ideas or words is explicitly defined as misconduct and is sometimes associated with a more widespread and long-term pattern of inappropriate behavior. Many times the discovered incident of misconduct is not the first, and a careful review reveals a more extensive history of repeated plagiarism. Some misconduct cases originate with, or are perpetuated by, a *lack of respect for or rapport* among colleagues. Manifestations of this might include more acrimonious, personal, extensive, emotional, and longer term disagreements than those typical of scientific debate. Evidence for a lack of respect may need to be viewed in light of its potential association with misconduct.

Many investigations of misconduct have one or more *collections of data* as a focus. Often the data are the main issue. In other cases questions about origination, authenticity, or quality of the data are central to the issue of misconduct. Thus documentation, organization, and quality control procedures for data are vital in both review of allegations and prevention of misconduct. Individuals responsible for data need proper technical and ethical training. They should not work alone or without adequate supervision, be overworked, or be placed under undue pressure to meet performance guidelines. Extra risk of inappropriate conduct is probably present if problems with interpersonal relationships occur in conjunction with such pressures.

Investigators should take particular care with *politically high-profile research* or when using *controversial drugs or devices*. Scientific misconduct is not more probable in such cases, perhaps being even less likely than elsewhere. However, any instances of questionable practice in these situations are likely to be examined in a forum that amplifies errors and discourages impartial discourse. In these settings guilt may be assumed before an adequate review of allegations can be accomplished.

Finally it is evident from the recent German cases that there continues to be short-comings in the process for investigating allegations of misconduct, and that such cases are often surrounded by inconsistent thinking by inquiry boards, colleagues, the community, and the editorial process. Scientists continue to be shocked and bewildered by misconduct cases, which is a strong rationale to consider the problem as rationally as possible in advance of an unfortunate specific case.

22.6   CLINICAL INVESTIGATORS' RESPONSIBILITIES

Clinical investigators have responsibilities for both patient care and the integrity of the research process. In specific research areas investigators may have additional responsibilities beyond those discussed here, to ensure the correctness and quality of their findings. An investigator may comply with standards as an individual or, in the case of many academic environments, the department or university may furnish some critical components as part of a general support for research. In many cases cooperative mechanisms are in place to assist or support some basic needs. In any case, the investigator must assume the final responsibility for all aspects of the research.

There are a few written guidelines outlining the investigator's responsibilities, particularly in collaborative research projects or those subject to governmental oversight. For example, when investigators hold Investigational New Drug (IND) licenses, they are subject to rules and regulations promulgated by the Food and Drug Administration (FDA). Similarly NIH and individual institutions have requirements in specific cases. NIH and institutions will hold investigators responsible for standards of research practices. Ignorance of the requirements is not an adequate defense for deficiencies.

In recent years some multicenter clinical trial protocols have been prepared by companies or other sponsoring organizations without naming a principal investigator (PI). This may make some aspects of study development more rapid or convenient. Presumably the sponsor expects to perform all of the duties of the PI. While the PI role is a difficult and time-consuming one and research sponsors actually do much of the work, failing to name a principal investigator creates many problems. Important responsibilities (see below) are diffused and could be overlooked. Because there is no identified focus for responsibility, participating institutional investigators may not be able to obtain all the information they require from a single source. Decisions that need to be made during the conduct of the study could be delayed. Failure to name an experienced clinician as PI is an unnecessary and serious shortcoming.

22.6.1   General Responsibilities

At least implicitly, clinical investigators assume, as a minimum, the following responsibilities for clinical trials under their supervision:

1. Developing a clinical hypothesis and a feasible approach to answer the study questions.
2. Establishing statistically valid endpoints for the clinical objectives of the study.
3. Defining the data to be collected.
4. Documenting the feasibility and scientific rationale for the approach.
5. Constructing record keeping forms and systems.

6. Maintaining a data system for timely, complete, and accurate recording of data elements, which are necessary to meet the objectives of the study. The data should be accessible to authorized investigators at all times for quality control purposes (but not necessarily to the clinical investigators).

7. Performing and documenting quality assurance procedures and results. This includes participation in audits based on predefined standards. Multicenter collaborations often come with these requirements.

8. Providing a mechanism for timely and objective review of safety data, pathological review of specimens, and subjectively derived data by a Data Safety and Monitoring Board or other appropriate mechanism.

9. Timely and complete detection and reporting of unexplained deaths on study and/or adverse drug reactions or other side effects of treatment.

10. Accounting for investigational drugs and devices.

11. Ensuring and documenting that the project complies with regulatory requirements and other oversight. These can come from the institution, the hospital, the study sponsor, NIH, or the FDA.

12. Certifying the clinical, research, and ethics competence of coinvestigators and other physicians or health professionals participating in the project.

13. Meeting data ownership, access, information systems, and storage of data for time period requirements of NIH or other institutions.

22.6.2 Additional Responsibilities Related to INDs

When the investigator is using an unlicensed drug for research purposes, an IND approval from the FDA is required. Thus the investigator must submit the IND application to the agency. The application and approval process requires:

1. Documentation on Form 1572 of coinvestigators and participating investigators (those treating patients on the study but not listed as coinvestigators). This requires names, addresses, location of the study, and the Institutional Review Board (IRB) that is responsible for the research.

2. The coinvestigators must report adverse experiences and deaths to the principal investigator (PI) so that they may be included in an annual report.

3. The PI must collect curriculum vitae from all participating investigators and documentation of education, training, and experience.

4. The PI must name or serve as a study monitor to oversee all aspects of the investigation.

5. During the study, adverse events must be reported to the FDA within 10 days of the event. An annual progress report to the FDA is required, and at study termination a final report is sent to the FDA and IRB.

6. For NIH studies, Form 310 (IRB review documentation) is required. Similarly the PI must maintain documentation of all collaborating IRBs and ensure that the consent forms contain necessary information about risks, benefits, and treatment alternatives.

7. The PI is responsible for timely distribution to all investigators, in writing, of all study amendments.

8. None of the federal reporting requirements substitute for local institutional requirements.

22.6.3 Sponsor Responsibilities

Some of the responsibilities sketched above may be met, in part, by the investigator's institution or sponsor. When an individual investigator submits an IND application to the FDA, he or she is the sponsor. These might include:

1. Expertise and timely consulting regarding:
   (a) Study design and performance
   (b) Biostatistical support for study design and analysis
   (c) Regulatory and policy issues
2. Support for clinical research functions too complex for single investigators, including:
   (a) Audit process and certification of study performance
   (b) Maintenance of an accurate, quality controlled, flexible, and powerful database
   (c) Expert and collegial peer review of research project development
3. Efficient research support services, including:
   (a) Patient registration/randomization
   (b) Eligibility checking
   (c) IND administration
   (d) Repository and tracking of consent documents
   (e) Files of study protocol documents
4. Data management/coordination support
5. Preparation of files for statistical analysis
6. Statistical analysis

22.7 SUMMARY

Scientific misconduct has a long history, but its origins, causes, and exact frequency are obscure. Because funding for research is scarce, public oversight of science is more attentive than it has been at many times in the past. Prominent instances of misconduct by scientists also increase public awareness and scrutiny. Unfortunately, peer review and replication of results are not sufficient safeguards against misconduct.

Workable definitions of misconduct are difficult because it is hard to define the norms of science. These include, but may not be limited to, honesty and objectivity, tolerance, doubt of certitude, recognition of error, unselfish engagement, and communal spirit. Significant departures from any of these norms would likely be seen as misconduct. In the United States a widely held view of the definition of scientific misconduct comes from the National Academy of Sciences. They define misconduct as plagiarism, data fabrication, and data falsification but also delineate questionable research practices and other forms of misconduct.

Research institutions and sponsors can approach this problem most effectively with education and prevention. Failing that, they have formal procedures for dealing with allegations of misconduct. For research funded by the U.S. Public Health Service, misconduct cases are overseen by OIG and ORI, which ensures that institutions have adequate mechanisms for dealing with allegations, oversees institutional investigations, or conducts its own investigations. The NSF has an entity with a similar role.

Review of the circumstances surrounding some misconduct cases in the clinical sciences suggests that investigators can learn some valuable lessons about preventing or detecting misconduct. First is the benefit of increased awareness of the potential for problems and maintaining a healthy skepticism about unusual findings. Individuals with little supervision and a great deal of responsibility and pressure are more likely to behave inappropriately. Subjective or unverifiable data may be more prone to error, including honest mistakes. Recurrent irregularities can also be a sign of an underlying cause. Finally, investigators should be more attentive to unconventional research methods, individuals with interpersonal difficulties or other stress, and controversies over data, methods, or treatments.

22.8   QUESTIONS FOR DISCUSSION

1. Discuss the role of "whistle-blowers" in identifying and investigating scientific misconduct. You might consult the report by Rossiter (1992). Is the role for such a person the same as or different from their role in other areas such as financial impropriety?

2. Discuss your views on due process for scientists accused of misconduct. What are the required elements of the process you favor (e.g., investigation, defense, judgment, appeal), and where should they be located?

3. Discuss changes in the peer review process that might discourage or detect data fabrication or falsification. How practical are your suggestions?

4. Discuss the role of didactic education in preventing scientific misconduct. Give examples if you can find them.

5. Read Hill (1995) and discuss ways in which it might be applicable to peer review of manuscripts.

APPENDIX A

DATA AND PROGRAMS

A.1 INTRODUCTION

Much of the data used in this book and some computer programs for analyzing them are available in electronic form. The programs are written using software packages or languages available for IBM compatible personal computers. Most use the *Statistical Analysis System* (SAS) for Windows (SAS Institute, 1999) or *MLAB* (Civilized Software, 1996). In some cases *S-PLUS* (SPLUS, 2001) programs are also provided. Each of the programs has been developed and run on a personal computer using a 360 megahertz processor, where they typically take only a few seconds to run. All the data and programs are available on the Web site that supports this book (www.cancerbiostats.onc.jhmi.edu/Piantadosi_ClinicalTrials). The same analyses described here, or ones similar to them, could be accomplished in a variety of ways. However, I have chosen uncomplicated methods that are available to many readers using common software and hardware. Readers interested in other data sets useful for practice or examples can consult Andrews and Herzberg (1985) or Hand et al. (1994).

MLAB (*M*odeling *LAB*oratory) may be less familiar to some readers and therefore requires a few words of explanation. *MLAB* was originally developed at the National Institutes of Health for interactive use in constructing and testing mathematical models. In the late 1980s it was rewritten, expanded, and improved for the microcomputer platform. It is currently available for DOS, Windows, and Mac from Civilized Software, Inc. As a development tool, *MLAB* contains some ideal features. For example, it is command driven with a natural, mathematically oriented syntax. It allows flexible function definitions by the user, including differential equation defined models, and incorporates full matrix and graphics capabilities. It is particularly useful and fast for fitting sophisticated models (e.g., systems of nonlinear differential equations) to data. It is exceptionally well suited to tasks where model development, as contrasted with parameter estimation, is the primary focus.

Clinical Trials: A Methodologic Perspective, 2E, by S. Piantadosi
Copyright © 2005 John Wiley & Sons, Inc.

A.2   DATA

Data and analysis programs for many of the examples from Chapter 16 are provided.

A.3   DESIGN PROGRAMS

The following programs are provided on the website that accompanies this book. They will run on IBM compatible microcomputers running Windows. The programs are written in the C++ language and are quite fast, but they require a math coprocessor. Source code is not available and context sensitive help is minimal. These programs are constantly being improved, corrected, and developed, but are not actively supported by the author. The user assumes all responsibility for their content, correctness, appropriate application, and consequences. In general, the relevant .zip file must be downloaded and extracted. The windows setup file must then be run to install the program.

A.3.1   Power and Sample Size Program

This program contains a number of modules for power and sample size calculation, similar to that provided in the first edition of this book. However, the programs have been rewritten, upgraded, expanded, and now run under Windows. For all calculation screens, the entry fields correspond to the statistical parameters from the original references, which should be consulted when using the modules. Following is a description of some of the modules available.

Exact Binomial Confidence Limits
The input fields are the one-sided percentile characterizing the confidence limits (e.g., 0.975 for two-sided 95% bounds), the numerator of the proportion, the denominator of the proportion, and the value of p at the confidence bound. See Chapter 11. The program assumes that confidence bounds are symmetric with respect to the tail area.

Often, the first three parameters will be specified and the program will solve for p. However, the program can solve for any parameter in terms of the other three by pressing the appropriate function key. The calculations are performed using beta distributions, so that noninteger values for numerators and denominators are permitted (and are often required to solve the equations exactly). The users is responsible for appropriate rounding. When one side of the confidence bound is specified in this way, the other side is also calculated. Additionally the program displays a sketch of the point estimate and confidence bounds.

Optimal Two-stage Designs
This module is also available as a stand-alone. Very little calculation is required in this program, except to determine some factorials and the probability of rejection (Chapter 11). The required inputs are $p0$, $p1$, α, β, a search range (upper and lower total sample sizes to limit the search), and a search method (shortcut or complete). The shortcut method searches only a subset of designs likely to contain optima. For thorough searches, the program is very time-consuming because it enumerates all possible designs within the range specified by the user.

The complete method is highly reliable, provided enough time is available and the search range actually contains an optimum. If the search range contains a global optimum, the algorithm will find it. Unfortunately, there is no simple way to determine if an optimum is local or global except to examine a sufficiently broad search range. This program could require several minutes or longer on a fast computer for some design problems (e.g., those where $p0$ and $p1$ are close). Others can be solved in a matter of seconds.

Logrank Power and Sample Size
Design parameters are input via a single screen and the solution of one parameter in terms of the remaining ones is obtained by pressing the appropriate function key. Type I and II error probabilities can be specified as quantiles of the normal distribution (the usual way) or in terms of the actual probabilities. Although these calculations are not difficult to perform or program, the user should take care to note that the formula is written in terms of the number of events (e.g., deaths or recurrences) and not the "sample size." In the presence of censoring, the sample size will be larger than the number of events.

Conditional Power
Conditional power calculations using B-values can be accomplished by the program. Limited context-sensitive help is available in this program and the notation differs slightly from that employed in Chapter 14. However, the user can consult the reference and the program should be self-explanatory. Like the interfaces for the other programs, parameters are entered on a single screen. Treatment effects under alternative hypotheses can be calculated with the assistance of a second screen.

A.3.2 Blocked Stratified Randomization

This program generates numbered, blocked, randomized treatment assignments in the form of a logbook for use by a study coordinator. Up to 10 treatment arms can be specified. The program should be run separately for each stratum with appropriate titles, assignment numbering, and number prefixes. The block size, or a mixture of block sizes, and the allocation ratio are specified by the user. The assignment list generated in this way can be stored as a computer file or printed. The pseudorandom number generator is documented in the printout, and it has a user-specified starting seed.

A.3.3 Continual Reassessment Method

The original program to assist with this method was written for *Mathematica* for *Windows* (Wolfram, 2003). The program has been rewritten as a C++ stand-alone. Several CRM methods have been described in the literature. This one is based on the proposed implementation by the New Approaches to Brain Tumor Therapy consortium (Piantadosi, Fisher, and Grossman, 1998). The program is internally documented.

A.4 MATHEMATICA CODE

Mathematica (Version 5) code supporting many of the equations in the chapters is provided on the website in a single file. This code exists to support some of the tables or other computations in the book, and to help students free themselves from extensive calculation. Mathematica can appear cryptic to a newcomer, but it is a wonderful tool for those needing quantitative investigation. The author makes no guarantees of any kind regarding use of this code.

APPENDIX B

NOTATION AND TERMINOLOGY

B.1 INTRODUCTION

The purpose of this appendix is twofold: to provide definitions of terms and statistical notations that are commonly used, and to provide additional details about some statistical concepts found in the literature related to clinical trials. The definitions provided here are not authoritative in the sense of representing a consensus of experts. However, they are internally consistent and are in accord with common usage. More details and additional references can be found in Everitt (1995), Kotz and Johnson (1988), and Meinert (1996). In the following, defined terms are in **boldface**. Comments, clarifications, and examples are in *italic face*.

B.2 NOTATION

Greek and Roman letters are commonly used in two ways in statistical writing. The first is to represent specific values, functions, or variables. The second is as a shorthand for certain mathematical calculations that would be inconvenient or tedious to write out fully. Almost all symbols are used in the first context, while only a few are typically used in the second. In the following, the explanation offered for each symbol is for the one most frequently encountered. However, usage varies and depends on the specific context.

Frequently the quantities represented by symbols are a single number or constant (scalars). However, sometimes they represent a vector or matrix. Vectors or matrices are usually denoted by boldface type, although vectors are sometimes denoted by an underscore. The quantities here are not vectors or matrices unless explicitly stated.

Clinical Trials: A Methodologic Perspective, 2E, by S. Piantadosi
Copyright © 2005 John Wiley & Sons, Inc.

B.2.1 Greek Letters

- α (alpha): type I error rate; also used as a model parameter.
- β (beta): type II error rate; equally often used to denote the coefficient or coefficients that are to be estimated from the data in a regression model.
- **B** (beta, uppercase): beta function, $B(a, b) = \Gamma(a)\Gamma(b)/\Gamma(a + b)$; or beta distribution function.
- γ (gamma): a variable or model parameter. $\gamma(\cdot)$ is also the incomplete gamma function, a generalization of $\Gamma(\cdot)$, the complete gamma function.
- Γ (gamma, uppercase): gamma or factorial function, $\Gamma(n + 1) = n!$.
- δ (delta): often denotes a difference of two quantities, $\delta = x_1 - x_2$.
- Δ (delta, uppercase): denotes a change in a measurement or a difference between two quantities. Also frequently represents a ratio of hazard rates, $\Delta = \lambda_1/\lambda_2$, where λ_1 and λ_2 are hazard rates of interest.
- ϵ (epsilon): random error term in a model; usually implies an error with a normal distribution.
- ζ (zeta): a variable.
- η (eta): a variable.
- θ (theta): a variable or a statistical parameter to be estimated.
- ι (iota): infrequently used.
- κ (kappa): a statistic measuring rater agreement commonly used for categorical outcomes.
- Λ (lambda, uppercase): cumulative hazard.
- λ (lambda): hazard or failure rate.
- μ (mu): mean of some normal distribution.
- ν (nu): degrees of freedom, e.g., in a chi-squared or F statistic.
- ξ (xi): a variable.
- o (omricron): infrequently used.
- π (pi): a proportion; also used in the more familiar context 3.14159...
- $\prod$ (pi, uppercase or larger): product sign, $\prod_{i=1}^{n} x_i = x_1 \cdot x_2 \cdot x_3 \cdot \ldots \cdot x_n$.
- ρ (rho): correlation coefficient.
- σ (sigma): standard deviation of some normal distribution. σ^2 is the variance of a normal distribution.
- Σ (sigma, uppercase): covariance matrix.
- $\sum$ (sigma, uppercase or larger): summation sign, $\sum_{i=1}^{n} x_i = x_1 + x_2 + x_3 + \cdots + x_n$.
- τ (tau): a variable related to time.
- υ (upsilon): infrequently used.
- ' density function, $\phi(x) = \frac{1}{\sqrt{2\pi}}e^{-x^2/2}$. Sometimes same as Φ.

  rcase): the cumulative normal distribution function, $\Phi(z) =$ dx.

  χ^2, the chi-squared statistic, used for goodness of fit.

  |uently used.

- **Ω** (omega, uppercase): set of all outcomes; the last possible case.
- **ω** (omega): a point in the set Ω.

B.2.2   Roman Letters

- **B**: beta function or distribution.
- **c**: arbitrary constant.
- **d**: difference or infinitesimal difference. For example, dx is an infinitesimal change in x, as in a derivative; also number of events (deaths) in event time studies.
- **e**: base of the natural logarithms, $e = 2.71828\ldots$ Also, the exponential function (anti-logarithm).
- **E**: expectation or expected value operator.
- **f**: f(x), often a probability density function; sometimes an arbitrary function.
- **F**: F(x), often a cumulative distribution function; sometimes an arbitrary function.
- **g**: g(x), arbitrary function.
- **G**: same as F.
- **h**: h(x), arbitrary function; sometimes the hazard function in survival models.
- **H**: same as F.
- **i**: counting subscript (index) to denote an arbitrary item in a list. In the context of complex numbers such as characteristic functions (not used in this book), $i = \sqrt{-1}$.
- **j**: counting subscript like i.
- **k**: same as j.
- **l** or **ℓ**: same as j.
- **L,** or **$\mathcal{L}$**: likelihood or loglikelihood function.
- **m**: similar to n.
- **n**: number of items or a count; sample size.
- **N**: same as n.
- **p**: proportion.
- **P**: probability of an event. For example, $P[X = 1]$ is the probability that a random variable, x, equals 1. This is also often written $\Pr[X = 1]$.
- **q**: proportion, often $1 - p$.
- **r**: a counter like n.
- **s**: standard deviation of a sample, often an estimate of σ.
- **t**: variable representing time. Also the t distribution.
- **T**: time, especially a fixed constant.
- **u**: arbitrary variable, especially for integrals; also a uniform random deviate.
- **U**: uniform distribution function.
- **v**: arbitrary variable.
- **w**: a weight in a weighted average.
- **x**: arbitrary quantity; independent variable in a regression.
- **X**: often a random variable.
- **y**: arbitrary quantity; dependent variable in a regression.
- **Y**: often a random variable.

- **z**: arbitrary quantity; variable that is normally distributed.
- **Z**: often a random variable.

B.2.3 Other Symbols

- $\int$: integral sign, used in two ways. $\int f(x)dx$ is a *function*, i.e., the function whose derivative is $f(x)$. In contrast, $\int_a^b f(x)dx$ is a *number* which can often be thought of as the "area under the curve $f(x)$ between a and b."
- ∂ : infinitesimal difference in a partial derivative.
- |: vertical bar, used to specify quantities given in a conditional probability. For example, $P[A \mid B]$ is the probability of the event A given that B has occurred.
- $\widehat{}$: hat overscore, denotes an estimate of the quantity covered as opposed to the actual value. For example, $\widehat{\Theta}$ is an estimate of Θ.
- $\widetilde{}$: tilde overscore, an alternative estimator to the one denoted by $\widehat{}$.
- —: bar overscore, an average. For example, $\overline{X} = \sum_{i=1}^n X_i/n$. Also used to denote the complement of the probability of an event. For example, $P[\overline{A}] = P[not\ A] = 1 - P[A]$.
- $\underline{}$: underscore, denotes a vector quantity. For example, $\underline{\Theta} = \{\Theta_1, \Theta_2, \Theta_3, \ldots, \Theta_n\}$. Often boldface type is used instead to represent a vector.
- $\sim$: distributed as or sampled from. For example, if x is a sample from a standard Gaussian distribution, we might write $x \sim N(0, 1)$.
- **exp**: exponential (anti-logarithm) function, $\exp(\log(x)) = e^{\log(x)} = x$.
- **log**: logarithmic function. *There is only one log function, meaning that with base e. All others are multiples of it. Sometimes "ln" is used to represent the natural logarithm and "log" is reserved for base 10 logarithms. This is not good notation—a better notation for bases other than e is* $\log_b$ *where b is the base.*

B.3 TERMINOLOGY AND CONCEPTS

- **adherence**: compliance with treatment assignment or specification.
- **accrual rate:** rate at which eligible patients are entered onto a clinical trial, measured as persons per unit of time.
- **active control:** a control treatment that produces a clinically significant effect. Often refers to equivalence trials or noninferiority trials.
- **adjustment**: use of any of several statistical methods to account for the effect of **prognostic factors** or baseline characteristics when estimating differences attributable to treatments or other prognostic factors. Adjustments may be non−risk adjusted (**stratification**) or risk adjusted (**multiple regression** equations).
- **alpha error**: type I error.
- **area under the curve (AUC)**: estimated area under a time−concentration curve in ‥‥macokinetic studies. The AUC may be a predictor of biological effects such ‥icacy.
- ‥‥**ean.**
- ‥**n**: an experimental design in which the same number of observa‥‥ ‥r each factor. A balanced two-armed trial would employ an equal ‥ each arm.

- **baseline:** a defined reference point or zero time at the beginning of an evaluation or treatment period, taken as the state against which a treatment effect is measured.
- **Bayes' rule**: theorem or rule of probability that relates the conditional probabilities of two events in the following way. If B_1, B_2, ..., B_n are disjoint events and A is any other event,

$$\Pr\{B_i \mid A\} = \frac{\Pr\{A \mid B_i\}\Pr\{B_i\}}{\sum_{j=1}^{n}\Pr\{A \mid B_j\}\Pr\{B_j\}}.$$

For two events, A and B,

$$\Pr\{B \mid A\} = \frac{\Pr\{A \mid B\}\Pr\{B\}}{\Pr\{A \mid B\}\Pr\{B\} + \Pr\{A \mid \overline{B}\}\Pr\{\overline{B}\}},$$

where $\overline{B}$ represents the event not-B. In the special case where B represents the event that a person has a disease and A represents a positive diagnostic test for the disease, Bayes' rule yields

$$\Pr\{disease \mid positive\ test\} =$$

$$\frac{sensitivity \times prevalence}{sensitivity \times prevalence + (1 - specificity) \times (1 - prevalence)}.$$

- **beta error**: type II error.
- **bias**: systematic (nonrandom) error in the **estimate** of a **treatment effect** or **parameter** of interest. Also a state of mind based on opinion or perception that predisposes actions or evaluations systematically (nonobjectively).
- **binary variable**: observations that can take one of only two values (e.g., yes/no, alive/dead, success/failure). Often the possible outcomes or variable values are numerically coded as 0 or 1 for analysis purposes.
- **binomial distribution**: probability distribution for binary outcome variables that characterizes the number of "successes" in a series of independent trials with a constant probability of success for each trial. If X is a random variable that counts the number of successes in n trials, the distribution is

$$\Pr[X = x] = \binom{n}{x}p^x(1 - p)^{n-x},$$

where $\binom{n}{x} = \frac{n!}{x!(n-x)!}$ is the binomial coefficient. The expected value (mean) of X is np and the variance is $np(1 - p)$.

- **block**: a group of treatment assignments or experimental units arranged to control a particular source of variability. Usually the treatment effect or difference is estimated within each block, thereby eliminating the variability arising from the blocking factor.
- **blocked randomization**: a method of constrained randomization that exactly balances the treatment assignments at the end of each block. This prevents the treatment assignments from becoming very unbalanced. When used with **stratification**, blocked randomization reduces the variability due to the stratifying variable(s).
- **case report:** published report of a detailed clinical case history. Most such histories are published because the very existence of the case, or some aspect of it, is unique or rare.

- **case series:** clinical results of an informative series of cases. The number of cases in a series may be large or small, depending on the frequency of the condition and the practice experience of the author. Series are usually defined temporally, that is, based on a practitioner's experience with a consecutive sequence of patients having a particular diagnosis. The purpose of reporting a case series is often to make treatment inferences.

- **censoring:** incompleteness in the observation of an event time (survival), that occurs when a subject has been followed for a period of time but has not yet had the event (death). Data subject to censoring require two measurements to describe them: time at risk, and an indicator that classifies what happened at the end of the time at risk, such as alive or dead. Censoring is **not** the same as lost to follow-up.

- **clinical trial:** an **experiment** in humans designed to accurately assess the effects of a treatment or treatments by reducing **random error** and **bias**.

- **confidence interval** (informal definition): a range of values calculated from the data in which the investigator believes a true parameter value will lie with some specified probability. Informally (and incorrectly) a confidence interval is a probability statement about the true parameter value.

- **confidence interval** (formal frequentist definition): a range of values calculated from the data that will contain the true parameter with a specified probability if the experiment were repeated a large number of times. Formally, a confidence interval is a probability statement about the range of values or region. *It is customary to report confidence intervals whose coverage probability is 95%; that is, such an interval would enclose the true parameter value 95% of the time if the experiment were repeated a large number of times. Approximate 95% confidence intervals can be constructed by taking the mean ±2 standard deviations. However, there are many circumstances in which intervals other than 95% are useful.*

- **confounder:** a **prognostic factor** that is associated with both treatment and outcome and can affect both. The term is most often used in an epidemiologic context.

- **correlation coefficient:** a statistic that measures the linear relationship between a pair of variables. The correlation coefficient is a **parameter** in a bivariate normal distribution and has values between -1 and 1. A frequently used estimate is the Pearson correlation coefficient

$$\widehat{\rho} = \frac{\sum_{i=1}^{n}(x_i - \overline{x})(y_i - \overline{y})}{\sqrt{\sum_{i=1}^{n}(x_i - \overline{x})^2(y_i - \overline{y})^2}} \ ,$$

where n (x, y) pairs have been sampled. *"Correlation" is often used informally to described the degree of association between two variables.*

- **crossover:** a trial in which patients receive all the study treatments during different periods separated by a "washout" period. This permits estimating the within-patient treatment difference. *Treatment crossovers are planned, whereas unplanned switching of treatments are called **drop-in**.*

- **CTE (comparative treatment efficacy):** a type of trial design that assesses the efficacy of a new treatment relative to an alternative, placebo, standard therapy, or no treatment. These studies are often called **phase III**.

- **data**: observations or measurements structured in a way amenable to inspection and/or analysis. *Sometimes the word is used informally as a synonym for "information." However, it will not be used in that fashion here.*
- **data-dependent stopping**: process of evaluating accumulating data in a clinical trial and making a decision whether the trial should be continued or stopped because the available evidence is already convincing.
- **deterministic**: without randomness.
- **DF (dose-finding)**: a type of drug development trial design that has as a primary objective identifying the optimal dose of drug to administer. Such designs are often called **phase I**.
- **dichotomous**: having only two possible values.
- **dose escalation**: see **DF**.
- **drop-in**: in a comparative trial, a study subject who takes another treatment on the trial instead of the one to which he or she was assigned and remains available for follow-up.
- **dropouts**: study subjects who stop taking the treatment to which they were assigned but remain available for follow-up.
- **effectiveness**: the effect of a treatment when widely used in practice.
- **efficacy**: true biological effect of a treatment.
- **endpoint:** a clinical or laboratory outcome that yields the definitive information about the result of treatment for an individual study subject. *For example, death or survival time is a frequently used endpoint in studies of chronic disease progression or recurrence. In theory, the investigator need not follow the subject past the endpoint (if this were even possible); however intermediate endpoints are sometimes used because of their strong association with more definitive outcomes. Examples of intermediate endpoints are CD4+ lymphocyte counts in AIDS, PSA levels in prostate cancer, and blood pressure in cardiovascular disease.*
- **ES (expanded safety)**: a type of surveillance trial design that has a primary objective of estimating the frequency of uncommon side effects from a particular treatment. In drug development, such studies are often called **phase IV**.
- **estimate** (verb): process of determining the value for an unknown **parameter**. (noun): numerical value of an estimator when it is evaluated with some specific data.
- **eligibility**: criteria that must be satisfied for each patient before entering them in a clinical trial. Examples include disease type and extent, medical history, and organ system function.
- **evaluability**: criteria that must be satisfied before concluding that the patient has had a legitimate trial on the treatment. Examples include duration of treatment or number of cycles of therapy completed. *Unfortunately, evaluability is partially an outcome of treatment and is therefore not a proper basis for excluding patients from analysis.*
- **experiment**: a study in which the investigator makes a series of careful observations under controlled or arranged conditions. In particular, the investigator controls the treatment or exposure applied to the subject(s) by design and then carefully and thoroughly records outcome measurements. *In nonexperimental studies, the investigator lacks control over exposure or treatment assignment.*

- **experimental unit**: smallest unit or entity assigned to a particular treatment in an **experiment**. For example, individuals are often the experimental unit. However, groups of individuals such as families or households may all be assigned to receive the same treatment. See **observational unit**.

- **exponential distribution**: a probability distribution that takes on continuous positive values, x, with density function

$$f(x) = \lambda e^{-\lambda x}.$$

This distribution has expected value (mean) $1/\lambda$ and variance $1/\lambda^2$. The exponential distribution is widely used in survival and reliability analysis to model failure times.

- **factor**: a categorical variable or prognostic factor. Usually a factor has only a few levels or categories. *In experimental design, a factor refers to a variable that must be controlled. An example is "treatment" where the factor levels may be different drugs or different doses of the same drug.*

- **follow-up:** to maintain contact with a study subject and gather information required by the trial. Also activities required by the study to follow subjects. Also the time period following treatment initiation until the close of the trial.

- **fully sequential**: clinical trial designs that permit the evaluation of study results after *every* patient has been accrued. Such designs can provide convincing evidence of efficacy at the earliest possible time, provided that the type I error is properly controlled. *To use such a design, endpoints must be ascertained quickly, before the next patient is accrued.*

- **group sequential: clinical trial** designs similar to **fully sequential** ones, except that treatment comparisons are made after groups of patients have been accrued. These designs also terminate early when treatment differences are large.

- **hazard**: the instantaneous risk of failure when time to failure (e.g., survival) is the **endpoint**. If T is the observed failure time, any of three functions can be used to characterize a distribution of failure times: (1) the cumulative failure time distribution

$$F(t) = \Pr[T < t],$$

which is the probability that an individual fails in less than t units of time, (2) the failure density

$$f(t) = F'(t) = \Pr[t = T],$$

which is the probability that an individual fails in the time interval $t < T < t + \Delta t$; or (3) the hazard or instantaneous failure rate

$$h(t) = \frac{f(t)}{1 - F(t)},$$

which is the probability of failure in the interval $t < T < t + \Delta t$, given that there was no failure before t. The overall hazard in a cohort can be estimated by

$$\widehat{\lambda} = \frac{d}{\sum_{i=1}^{n} t_i},$$

where d is the number of events during the observation period, n is the number of study subjects, and $\sum t_i$ is the sum of all the observation times of the patients. This is the maximum likelihood estimate of the failure rate assuming an exponential survival model. More generally, the hazard can vary with time or individual characteristics in which case this estimate may not be appropriate. *This estimate of the overall hazard is sometimes termed the "linearized rate." Even when the exponential assumption is not true, this is a useful summary.*

- **hazard function regression**: a method for analysis of a time-related event and its correlates that accounts simultaneously for the distribution of times until the event in terms of a biomathematical **model** and for **prognostic factors** in terms of one or more regression models that modulate parameters of the biomathematical model.

- **hazard ratio**: **hazard** in one group divided by the hazard in another group. A hazard ratio of 1 implies no difference in risk. *A hazard ratio of 1.8, for example, implies that the numerator group is at persistently higher risk of failure than the denominator group. The clinical significance of an elevated hazard ratio depends on other evidence including the absolute risk, the significance level, and the clinical context.*

- **inference**: drawing conclusions while accounting for **random variability**.

- **intention to treat**: idea that patients assigned to treatments in randomized clinical trials should be analyzed according to the assigned treatment group rather than according to the treatment actually received. *Clinicians sometimes disagree with this perspective, which views clinical trials as tests of treatment policy rather than tests of treatment received. However, the clinical trials literature supports the validity of the intention-to-treat principle because it yields valid tests of the null hypothesis of no treatment difference. In studies where a large fraction of patients do not receive the assigned treatment, neither the intention-to-treat (ITT) nor the treatment received (TR) analyses necessarily yield the most clinically relevant conclusions. The ITT analysis does not correspond closely to efficacy because many patients did not receive the intended treatment, and the TR analysis is potentially biased because patients may switch treatments for reasons associated with outcome.*

- **interim analysis**: an analysis of an ongoing clinical trial, particularly one where the trial might be stopped if convincing evidence of efficacy is seen. See **data-dependent stopping**.

- **likelihood** (likelihood function): probability of observing the data under an assumed model. The likelihood is an equation that combines **data** with a probability **model** and **parameters** of interest. If one views the data as fixed, the parameters of the model can be chosen to yield the maximum value (probability) of the likelihood function. This technique, called "maximum likelihood," is very general and usually yields parameter estimates with desirable statistical properties. *For example, suppose that we have a sample of n independent observations thought to arise from a normal distribution, $\mathbf{X} = \{x_1, x_2, x_3, \ldots, x_n\}$. Using the normal probability density function and the fact that independent probabilities multiply to give a joint probability, the probability of observing the data is*

$$L(\mathbf{X}, \mu, \sigma) = \prod_{i=1}^{n} \frac{1}{\sqrt{2\pi}\sigma} e^{-\frac{(x_i-\mu)^2}{2\sigma}},$$

where μ is the mean and σ the standard deviation of the distribution from which the sample came. For example, to maximize $L(\mathbf{X}, \mu, \sigma)$ with respect to μ we take logarithms (the maximum of the log likelihood will be the same) to obtain

$$\mathcal{L}(\mathbf{X}, \mu, \sigma) = \sum_{i=1}^{n} \left(-\log(\sqrt{2\pi\sigma}) - \frac{(x_i - \mu)^2}{2\sigma} \right).$$

The maximum occurs at $\widehat{\mu}$ when the derivative with respect to μ is zero,

$$0 = \frac{\partial \mathcal{L}}{\partial \mu} = \sum_{i=1}^{n} \left(-\frac{x_i - \widehat{\mu}}{\sigma} \right)$$

or when multiplying by $-\sigma$,

$$0 = \sum_{i=1}^{n} (x_i - \widehat{\mu}).$$

This yields

$$n\widehat{\mu} = \sum_{i=1}^{n} x_i$$

or

$$\widehat{\mu} = \frac{1}{n} \sum_{i=1}^{n} x_i,$$

which is the familiar estimate of the sample mean.

- **logistic regression**: a statistical **model** in which the log **odds** of the response probability is predicted from a set of **prognostic factors**,

$$\log \left\{ \frac{p}{1 - p} \right\} = \beta_0 + \beta_1 X_1 + \beta_2 X_2 + \cdots,$$

where p is the probability of response, $X_1, X_2, \ldots$ are the predictor (explanatory) variables, and $\beta_1, \beta_2, \ldots$ are **parameters** to be estimated from the data. Thus the effect of each prognostic factor is to multiply the baseline log odds. *Note the similarity of this to proportional hazards.*

- **lost to follow-up**: trial participant who cannot be traced or contacted to determine vital status or other endpoints. *Sometimes the terms "lost to follow-up" and "dropout" are used synonymously. However, patients lost to follow-up are different from dropouts in the following ways. Individuals who are lost to follow-up could remain compliant with the assigned treatment. Dropouts are noncompliant but remain available for follow-up.* Lost to follow-up is **not** the same as censoring. For example, in a survival study patients may be actively followed to the administrative end of study, at which time some remain alive and are censored (administrative censoring). Such patients are clearly not lost to follow-up. Conversely, patients who are lost to follow-up during a trial are usually treated as censored observations under the assumption that the loss process and the death process are independent.

- **LS (large-scale) trials**: studies that randomize thousands of patients, that being an order of magnitude larger than the typical CTE trial.
- **matrix**: a two-dimensional array of numbers or mathematical objects that typically represents an operation on a **vector** or one-dimensional object. *Matrix algebra is a subject of considerable importance to statistics.*
- **maximum tolerated dose**: highest dose of a drug that can be tolerated with an acceptable or manageable level of toxicity. This is an important dose to be employed in cytotoxic therapy of cancer where the treatment typically produces serious side effects.
- **mean**: expected value or first central moment of a probability distribution or random variable. Formally, the expected value of a discrete random variable is

$$E\{X\} = \sum x_i f(x_i),$$

where the probability of each x_i is denoted by $f(x_i)$. The mean of a series of n observations usually implies the ordinary arithmetic mean or average

$$\bar{x} = \frac{1}{n} \sum_{i=1}^{n} x_i.$$

There are other means such as the geometric and harmonic. A general form is

$$g(\tilde{x}) = \frac{1}{n} \sum_{i=1}^{n} g(x_i),$$

where the mean is denoted by $\tilde{x}$ and $g(\cdot)$ is a suitable monotonic function. For example, the ordinary mean is obtained when $g(z) = z$, the harmonic mean is obtained for $g(z) = 1/z$, and the geometric mean when $g(z) = \log(z)$.
- **meta-analysis**: see **overview**.
- **minimum effective dose**: lowest dose of a drug that produces the desired clinical effect.
- **model**: a logical mathematical construct, generally containing **parameters** (β_1 above) and **variables** (X_1 above), (placemarks for data values). *When data have been entered, and the values for parameters have been estimated by statistical techniques, the model, which is generic, becomes a specific equation.*
- **monitoring**: observing the conduct of an ongoing clinical trial according to a set of pre-defined guidelines.
- **multiple comparisons**: performing many hypothesis tests or comparisons in the same study. A problem arises when attempting to interpret the results of numerous tests. By chance, 5% of tests will reject the **null hypothesis** if all tests are performed at the 5% significance level. To preserve the overall **type I error** rate at 5% (assuming this is a worthwhile goal), each test might be performed at a significance level less than 5%.
- **multiple regression**: use of statistical **models** to account for the effects of several **prognostic factors** simultaneously. *Linear* regression relates the response variable to the predictor variables through a linear function of **parameters**:

$$Y = \beta_0 + \beta_1 X_1 + \beta_2 X_2 + \cdots + \epsilon,$$

where Y is the response variable, X_1, $X_2, \ldots$ are the predictor variables, $\beta_0, \beta_1, \beta_2, \ldots$ are the parameters to be estimated from the data, and ϵ is a random error term.

- **multiplicity**: a general term applied to the circumstance where multiple statistical tests are performed, giving rise to increased chance of a type I error. This can happen when tests are done on multiple outcomes, accumulating data, or multiple subsets, for example.

- **multivariable**: analyzing several predictor (explanatory) variables simultaneously. *Suppose that a univariable analysis suggests that treatment A is better than treatment B: the hazard ratio is 2.2 (95% confidence interval 1.5–2.9) and $p = 0.02$. However, when we simultaneously account for the effects of sex and treatment using multivariable analysis we estimate the adjusted hazard ratio to be 1.35 (95% confidence interval 0.75–1.95) and $p = 0.45$. We might conclude from these results that the apparent univariable effect of treatment is due to sex and that treatment A is not likely to be of real benefit.*

- **multivariate**: analyzing several outcome variables simultaneously, all as a function of one or more predictor variables. *Note that multivariate does not mean "several predictor variables," as it is often informally used, but refers to several outcome variables. When there are several predictors, the term "multiple" and "multivariable" are probably better descriptors.*

- **nesting**: a design characteristic in which a particular effect lies hierarchically entirely within some other effect. In a simple parallel group comparison, patients receive only one treatment, meaning patients are nested within treatments. In a simple **crossover design**, patients receive both treatments, meaning patients are not nested within treatments.

- **null hypothesis**: in statistical hypothesis tests, the hypothesis of no difference between the comparison groups. It is usually established as a "straw man" to be disproved.

- **observation**: process of study; a datum from an experiment.

- **observational study**: a study design in which the investigator does not control the assignment of treatment to individual study subjects.

- **observational unit**: the entity or unit in an **experiment** on which one observation is made. This may be the same or different from the **experimental unit**. *For example, households may be the experimental unit but individuals could be the observational unit. If so, a household would be randomized so that every member receives the same treatment but a single individual in each would be followed for the outcome.*

- **odds**: probability of an outcome divided by the probability of not having that outcome. If p is the probability of the outcome, the odds equals $p/(1 - p)$. *This is the ordinary betting odds.*

- **odds ratio**: **odds** in one group divided by the odds in another group. The odds ratio is commonly used for categorical data, which might be summarized as

		Outcome	
		Yes	No
Exposure	Yes	a	b
	No	c	d

where a, b, c, and d are the counts in the various categories. The odds of outcome when the condition is present is

$$\frac{\frac{a}{a+b}}{\frac{b}{a+b}} = \frac{a}{b}.$$

When the condition is absent, the odds of outcome are

$$\frac{\frac{c}{c+d}}{\frac{d}{c+d}} = \frac{c}{d}.$$

Thus the odds ratio is

$$\frac{ad}{bc}.$$

The odds ratio is related to the risk ratio or relative risk in that when the probability of the outcome is small (i.e., $a \ll b$ and $c \ll d$), the odds ratio approximately equals the risk ratio

$$\frac{\frac{a}{a+b}}{\frac{c}{c+d}}.$$

However, the odds ratio is a useful measure of difference between groups even when it does not approximate the relative risk and arises naturally in important statistical models such as logistic regression. An odds ratio of 1 implies no difference in risk between the groups. An odds ratio of 2.3, for example, implies that the numerator group is at higher risk than the denominator group. The clinical significance of an elevated odds ratio depends on other evidence including the absolute risk, the significance level, and the clinical setting.

- **optimal biological dose:** dose of a drug or biological agent that satisfies an aggregate definition of efficacy and safety appropriate to the clinical setting. For example, an optimal biological dose of an analgesic for severe pain might be one that produces side effects with a frequency of less than 20% and improves pain in 90% of patients by (say) two or three units on a standard pain rating scale. There is no guarantee that such an optimal dose actually exists, depending on the properties of the drug.

- **overview**: a comprehensive re-analysis of published and unpublished studies, usually based on obtaining individual patient data, to investigate and quantify consistency or lack of consistency among study results.

- **parallel design**: a trial design in which patients receive only one of two or more concurrently administered treatments.

- **parameter**: a constant in a **model**, or a constant that wholly or partially characterizes a function or probability distribution.

- **phase I trial**: a **clinical trial** designed to measure the distribution, metabolism, excretion, and toxicity of a new drug. See **DF**.

- **phase II trial**: a **clinical trial** designed to test the feasibility of, and level of activity of, a new agent or procedure. See **SA**.

- **phase III trial**: a **clinical trial** designed to estimate the relative efficacy of a treatment against a standard, alternative, or placebo. See **CTE**.

- **phase IV trial**: a surveillance trial designed to estimate the frequency of uncommon side effects, toxicity, or interactions. See **ES**. In drug development such trials are sometimes initiated as postmarketing studies.

- **placebo**: a treatment that appears like a comparison treatment but with no true biological effect. Placebos are used to reduce bias in a comparison where assessment of outcome could be affected by patient or investigator knowledge that no treatment was given to one group.

- **power**: chance of detecting a difference of a specified size as being statistically significant. If the **type II error** probability is β, power $= 1 - \beta$.

- **precision**: certainty with which a measurement or **estimate** is made. *A precise measurement may not be accurate because of unrecognized bias or other errors in methodology.*

- **primary outcome:** clinical event that the trial intervention is designed to treat, delay, reduce or affect.

- **probit**: a term used to describe the cumulative normal distribution curve, especially when its sigmoidal shape is used for dose response modeling, as in bioassay. The probit curve is given by

$$p = \frac{1}{\sqrt{2\pi}} \int_{-\infty}^{y} e^{-u^2/2} du,$$

where y represents dose or logarithm of dose and p is the probability of response.

- **prognostic factor:** a variable or measurement that carries information about future clinical outcomes. Baseline prognostic factors have values fixed at study onset and never change. Time-dependent prognostic factors have values that change over time and are considerably more difficult to model statistically.

- **proportional hazards:** a mathematical assumption used in survival models in which the **hazard ratio** between two groups is assumed to be constant over time although the baseline hazard can fluctuate. For example, the usual assumption is

$$\log\left\{ \frac{\lambda(t)}{\lambda_0(t)} \right\} = \beta_1 X_1 + \beta_2 X_2 + \cdots,$$

where $\lambda_0(t)$ is the baseline hazard function which can vary over time, $\lambda(t)$ is the hazard function in the test group, $X_1, X_2, \ldots$ are the predictor variables, and $\beta_1, \beta_2, \ldots$ are **parameters** to be estimated from the data. *Because the covariate terms are additive on a logarithmic scale, we refer to this model as employing multiplicative covariate effects.*

- **protocol**: logical plans for conducting an **experiment,** clinical study**,** or **clinical trial**. *Sometimes this term is used to refer to the study or trial itself. However, it is always used here to refer to either the written document or the logical plan of the study.*

- **pseudorandom**: numbers that pass important tests of randomness but are actually generated by a nonrandom (deterministic) algorithm. Most computer generated "random" numbers are pseudorandom.

- **publication bias**: tendency for studies with positive results, namely those finding significant differences, to be published in journals in preference to those with negative findings. When journals are reviewed for summarizing study results (meta-analysis), a biased impression of treatment efficacy might then result.

- **p-value**: the conditional probability that an observed effect or one larger is due to chance given that the **null hypothesis** is true. *P-values are frequently misinterpreted as the unconditional probability of an error or the probability that the observed result is due to chance. It is important to recognize that the p-value assumes the null hypothesis is true and accounts for outcomes that have not been observed.*

- **quantile**: in a probability function, the value of the variable that yields a particular probability. For example, the normal quantile that yields a probability of 0.95 is 1.645.

- **random**: result of chance alone.

- **randomization**: assignment of patients or experimental subjects to two or more treatments by chance alone.

- **rate**: a **ratio** in which the numerator and denominator are incremental differences. Often the rate has dimensions of reciprocal time. For example, an incidence rate is the number of new cases divided by an interval of time.

- **ratio**: a fraction in which the numerator and denominator have clinical or epidemiological importance. The dimensions of the numerator and denominator may be different so that the ratio has dimensions. For example, a mortality ratio might be the number of deaths per 100,000 population.

- **risk ratio**: probability of a specified outcome in one group divided by the probability of the outcome in another group. A risk ratio of 1 implies no difference in risk.

- **sample size**: number of patients or experimental subjects on a study. *For studies with event time as an outcome (e.g., survival or disease recurrence), one must distinguish the sample size from the number of events required by the design of the study. Ideally the sample size is determined as a consequence of the need for precision in the estimate of treatment difference. However, in many cases the sample size is determined by cost, time, or other mundane practical constraints.*

- **SA (safety and efficacy)**: a type of trial design with safety and efficacy estimation as a primary objective. In drug development, such studies are often called **phase II**.

- **selection bias**: a systematic error or bias that causes a sample to be unrepresentative of the population from which it came.

- **statistic**: any function of the data; a special case is one derived to estimate some parameter that is called an estimator.

- **statistical significance**: quantitative degree, as measured by the **p-value**, to which a difference is likely to be the result of chance, assuming the **null hypothesis** is true. *Conventionally p-values less than 0.05 are judged to be "significant." However, p-values are influenced by several factors, making universal criteria impossible.*

- **statistics**: study of the inferential process, especially the planning and analysis of experiments or surveys. *Statistics deals with numerical data relating to aggregates of individuals. It is the science of collecting, analyzing, and interpreting such data.*

- **stratification:** performing a statistical procedure separately in groups (strata) to reduce the effects of the group factor. *As an adjunct to randomization, stratification can only effectively be used with blocking. As an analysis method, estimates are* *made within strata and pooled, when possible, across strata to reduce the influence of the stratifying variable.*

- **surrogate outcome**: an outcome measurement in a clinical trial that substitutes for a definitive clinical outcome or disease status. Examples of possible surrogate outcomes include prostatic specific antigen (PSA) in prostate cancer, blood pressure in cardiovascular disease, and CD4 positive lymphocyte count in AIDS. *To be useful, a surrogate outcome must be strongly (even causally) related to a definitive outcome like disease progression or length of survival but manifest relatively early after treatment. Trials designed using ideal surrogate outcomes, if they exist, may be shorter, more efficient, and as equally valid as ones using definitive endpoints.*

- **Taylor series**: a method of expanding a function as a polynomial using successive derivatives. For x in the region of a, the Taylor series is

$$f(x) = f(a) + f^{(1)}(a)(x - a)$$
$$+ f^{(2)}(a)(x - a)^2/2!$$
$$+ f^{(3)}(a)(x - a)^3/3!$$
$$+ \ldots$$
$$+ f^{(n-1)}(a)(x - a)^{n-1}/(n - 1)! + R_n,$$

where $f^{(n)}$ is the nth derivative of the function f with respect to x and R_n is a remainder that can be expressed in various forms. The utility of the formula usually comes from employing only the first two terms,

$$f(x) \approx f(a) + (x - a)f^{(1)}(a),$$

which yields an approximation for $f(x)$ in the region of the point a.

- **therapeutic ratio**: ratio of efficacy response over toxicity response. High values indicate that a treatment is beneficial without causing much chance of toxicity. Low values indicate that the chance of benefit is low compared with the chance of toxicity. For drugs, the therapeutic ratio may depend on dose such that one dose may yield the best chance of benefit with the least chance of toxicity.

- **TM (treatment mechanism)**: a type of early developmental trial design that tests the mechanism of delivering treatment to the patient. Treatment mechanisms include device function, drug bioavailability, and surgical technique. **Phase I** drug trials usually have treatment mechanism objectives, but the terms are not equivalent. See **DF**.

- **trial**: a study in which the investigator controls three elements of design: (1) treatment assigned to the subject(s), (2) ascertainment of outcomes, and (3) analysis of results. *We have carefully chosen the word "assigned" rather than using "applied" because, at least in humans, if not in all trials and experiments, the investigator cannot guarantee the application of an intended treatment.*

- **type I error**: concluding that a **treatment effect** or difference exists when, in reality, it does not (false positive).

- **type II error**: concluding that a **treatment effect** or difference does not exist when, in reality, it does (false negative).

- **univariable**: analyzing one *explanatory* variable at a time.

- **univariate**: analyzing one *outcome* variable at a time.
- **variability**: unaccounted for fluctuation (random error) in the **estimate** of a **treatment effect** or measurement of a variable.
- **variable**: a measurement that can take on different values for each experimental subject or observation.
- **vector**: a one-dimensional array of numbers or mathematical objects usually representing a point or direction in space.

APPENDIX C

ABBREVIATIONS

See also Appendix B.

ABMF: Advisory Board of the Medical Faculty

ACAPS: Asymptomatic Carotid Artery Plaque Study

ACE: angiotensin converting enzyme

ACS: American Cancer Society

AIDS: Acquired Immunodeficiency Syndrome

AMA: American Medical Association

ANCOVA: analysis of covariance

ASA: acetylsalicylic acid (aspirin)

ASA: American Statistical Association

ASCO: American Society of Clinical Oncology

AUC: area under the curve

AZT: zidovudine

BCNU: 1,3-bis (2-chloroethyl)-1-nitrosourea (carmustine)

BEMAC: Biometric and Epidemiology Methods Advisory Committee

CALGB: Cancer and Leukemia Group B

CAP: cytoxan, Adriamycin (doxorubicin), and platinum

CARET: Carotene and retinol efficacy trial

CAST: Cardiac Arrhythmia Suppression Trial

CDC: Centers for Disease Control

CDP: Coronary Drug Project

CEA: carcinoembryonic antigen

CIOMS: Council for International Organizations of Medical Sciences

COMMIT: Community Intervention Trial (for smoking cessation)

COPD: chronic obstructive pulmonary disease

CRM: continual reassessment method

CTE: comparative treatment efficacy

DHHS: Department of Health and Human Services

DF: dose-finding

DLT: dose-limiting toxicity

DSMC: Data and Safety Monitoring Committee

DSMB: Data and Safety Monitoring Board

EBCTCG: Early Breast Cancer Trialists Collaborative Group

ECMO: extracorporeal membrane oxygenation

EGRET: Epidemiology, Graphics, Estimation, and Testing

EORTC: European Organization for Research on the Treatment of Cancer

EPP: extrapleural pneumonectomy

ER: estrogen receptor

ES: expanded safety

FAP: familial adenomatous polyposis

FDA: Food and Drug Administration

GAO: General Accounting Office

HIV: human immunodeficiency virus

HPB: Health Protection Branch

IARC: Internation Agency for Research on Cancer

ICH: International Conference on Harmonisation

ICMJE: Internation Committee of Medical Journal Editors

ICU: intensive care unit

IEC: Independent Ethics Committee

IND: investigational new drug

IOM: Institute of Medicine

IRB: Institutional Review Board

ITT: intention to treat

IV: intravenous

LREC: Local Research Ethics Committee

LST: large-scale trials

LVR: lung volume reduction

MACS: Multicenter AIDS Cohort Study

MLAB: Modeling LABoratory

MLE: maximum likelihood estimate

MSE: mean squared error

MTD: maximum tolerated dose

NAS: National Academy of Sciences

NCCTG: North Central Cancer Treatment Group
NETT: National Emphysema Treatment Trial
NHLBI: National Heart Lung and Blood Institute
NIH: National Institutes of Health
NLM: National Library of Medicine
NSABP: National Surgical Adjuvant Breast and Bowel Project
NSC: National Service Center
NSF: National Science Foundation
OIG: Office of the Inspector General
ORI: Office of Research Integrity
OSI: Office of Scientific Integrity
OSIR: Office of Scientific Integrity Review
PASS: Power and Sample Size
PCP: personal care principle
PDQ: Physician Data Query
PFA: prognostic factor analyses
PH: proportional hazards
PHS: Public Health Service
PI: principal investigator
PK: pharmacokinetic
PS: performance status
PSA: prostatic specific antigen
RCP: Royal College of Physicians
RPA: recursive partitioning and amalgamation
RSS: Royal Statistical Society
SAS: Statistical Analysis System
SA: safety and activity
SPRT: sequential probability ratio test
SSD: sum of squared deviations
TDC: time-dependent covariates
TM: treatment mechanism
TR: treatment received
VOD: veno-occlusive disease
WHO: World Health Organization
WMA: World Medical Association

APPENDIX   D

NUREMBERG CODE

Following is a text of the Nuremberg Code (US Government, 1949).

The voluntary consent of the human subject is absolutely essential. This means that the person involved should have legal capacity to give consent; should be so situated as to be able to exercise free power of choice, without the intervention of any element of force, fraud, deceit, duress, overreaching, or other ulterior form of constraint or coercion; and should have sufficient knowledge and comprehension of the elements of the subject matter involved as to enable him to make an understanding and enlightened decision. This latter element requires that before the acceptance of an affirmative decision by the experimental subject there should be made known to him the nature, duration, and purpose of the experiment; the method and means by which it is to be conducted; all inconveniences and hazards reasonable to be expected; and the effects upon his health or person which may possibly come from his participation in the experiment.

- The duty and responsibility for ascertaining the quality of the consent rests upon each individual who initiates, directs or engages in the experiment. It is a personal duty and responsibility which may not be delegated to another with impunity.
- The experiment should be such as to yield fruitful results for the good of society, unprocurable by other methods or means of study, and not random and unnecessary in nature.
- The experiment should be so designed and based on the results of animal experimentation and a knowledge of the natural history of the disease or other problem under study that the anticipated results will justify the performance of the experiment.
- The experiment should be so conducted as to avoid all unnecessary physical and mental suffering and injury.

- No experiment should be conducted where there is an a priori reason to believe that death or disabling injury will occur; except, perhaps, in those experiments where the experimental physicians also serve as subjects.
- The degree of risk to be taken should never exceed that determined by the humanitarian importance of the problem to be solved by the experiment.
- Proper preparations should be made and adequate facilities provided to protect the experimental subject against even remote possibilities of injury, disability, or death.
- The experiment should be conducted only by scientifically qualified persons. The highest degree of skill and care should be required through all stages of the experiment of those who conduct or engage in the experiment.
- During the course of the experiment the human subject should be at liberty to bring the experiment to an end if he has reached the physical or mental state where continuation of the experiment seems to him to be impossible.
- During the course of the experiment the scientist in charge must be prepared to terminate the experiment at any stage, if he has probable cause to believe, in the exercise of the good faith, superior skill and careful judgment required of him that a continuation of the experiment is likely to result in injury, disability, or death to the experimental subject.

APPENDIX   E

DECLARATION OF HELSINKI

Ethical principles for medical research involving human subjects adopted by the 18th WMA General Assembly, Helsinki, Finland, June 1964 and amended by the 29th WMA General Assembly, Tokyo, Japan, October 1975; 35th WMA General Assembly, Venice, Italy, October 1983; 41st WMA General Assembly, Hong Kong, September 1989; 48th WMA General Assembly, Somerset West, Republic of South Africa, October 1996; and the 52nd WMA General Assembly, Edinburgh, Scotland, October 2000.

E.1   INTRODUCTION

1. The World Medical Association has developed the Declaration of Helsinki as a statement of ethical principles to provide guidance to physicians and other participants in medical research involving human subjects. Medical research involving human subjects includes research on identifiable human material or identifiable data.

2. It is the duty of the physician to promote and safeguard the health of the people. The physician's knowledge and conscience are dedicated to the fulfillment of this duty.

3. The Declaration of Geneva of the World Medical Association binds the physician with the words, "The health of my patient will be my first consideration," and the International Code of Medical Ethics declares that, "A physician shall act only in the patient's interest when providing medical care which might have the effect of weakening the physical and mental condition of the patient."

4. Medical progress is based on research which ultimately must rest in part on experimentation involving human subjects.

Clinical Trials: A Methodologic Perspective, 2E, by S. Piantadosi
Copyright © 2005 John Wiley & Sons, Inc.

5. In medical research on human subjects, considerations related to the well-being of the human subject should take precedence over the interests of science and society.

6. The primary purpose of medical research involving human subjects is to improve prophylactic, diagnostic and therapeutic procedures and the understanding of the etiology and pathogenesis of disease. Even the best proven prophylactic, diagnostic, and therapeutic methods must continuously be challenged through research for their effectiveness, efficiency, accessibility and quality.

7. In current medical practice and in medical research, most prophylactic, diagnostic and therapeutic procedures involve risks and burdens.

8. Medical research is subject to ethical standards that promote respect for all human beings and protect their health and rights. Some research populations are vulnerable and need special protection. The particular needs of the economically and medically disadvantaged must be recognized. Special attention is also required for those who cannot give or refuse consent for themselves, for those who may be subject to giving consent under duress, for those who will not benefit personally from the research and for those for whom the research is combined with care.

9. Research Investigators should be aware of the ethical, legal and regulatory requirements for research on human subjects in their own countries as well as applicable international requirements. No national ethical, legal or regulatory requirement should be allowed to reduce or eliminate any of the protections for human subjects set forth in this Declaration.

E.2   BASIC PRINCIPLES FOR ALL MEDICAL RESEARCH

1. It is the duty of the physician in medical research to protect the life, health, privacy, and dignity of the human subject.

2. Medical research involving human subjects must conform to generally accepted scientific principles, be based on a thorough knowledge of the scientific literature, other relevant sources of information, and on adequate laboratory and, where appropriate, animal experimentation.

3. Appropriate caution must be exercised in the conduct of research which may affect the environment, and the welfare of animals used for research must be respected.

4. The design and performance of each experimental procedure involving human subjects should be clearly formulated in an experimental protocol. This protocol should be submitted for consideration, comment, guidance, and where appropriate, approval to a specially appointed ethical review committee, which must be independent of the investigator, the sponsor or any other kind of undue influence. This independent committee should be in conformity with the laws and regulations of the country in which the research experiment is performed. The committee has the right to monitor ongoing trials. The researcher has the obligation to provide monitoring information to the committee, especially any serious adverse events. The researcher should also submit to the committee, for review, information regarding funding, sponsors, institutional affiliations, other potential conflicts of interest and incentives for subjects.

5. The research protocol should always contain a statement of the ethical consider-ations involved and should indicate that there is compliance with the principles enunciated in this Declaration.

6. Medical research involving human subjects should be conducted only by scien-tifically qualified persons and under the supervision of a clinically competent medical person. The responsibility for the human subject must always rest with a medically qualified person and never rest on the subject of the research, even though the subject has given consent.

7. Every medical research project involving human subjects should be preceded by careful assessment of predictable risks and burdens in comparison with foreseeable benefits to the subject or to others. This does not preclude the participation of healthy volunteers in medical research. The design of all studies should be publicly available.

8. Physicians should abstain from engaging in research projects involving human subjects unless they are confident that the risks involved have been adequately assessed and can be satisfactorily managed. Physicians should cease any inves-tigation if the risks are found to outweigh the potential benefits or if there is conclusive proof of positive and beneficial results.

9. Medical research involving human subjects should only be conducted if the importance of the objective outweighs the inherent risks and burdens to the subject. This is especially important when the human subjects are healthy vol-unteers.

10. Medical research is only justified if there is a reasonable likelihood that the populations in which the research is carried out stand to benefit from the results of the research.

11. The subjects must be volunteers and informed participants in the research project.

12. The right of research subjects to safeguard their integrity must always be respected. Every precaution should be taken to respect the privacy of the sub-ject, the confidentiality of the patient's information and to minimize the impact of the study on the subject's physical and mental integrity and on the personality of the subject.

13. In any research on human beings, each potential subject must be adequately informed of the aims, methods, sources of funding, any possible conflicts of interest, institutional affiliations of the researcher, the anticipated benefits and potential risks of the study and the discomfort it may entail. The subject should be informed of the right to abstain from participation in the study or to withdraw consent to participate at any time without reprisal. After ensuring that the subject has understood the information, the physician should then obtain the subject's freely-given informed consent, preferably in writing. If the consent cannot be obtained in writing, the non-written consent must be formally documented and witnessed.

14. When obtaining informed consent for the research project the physician should be particularly cautious if the subject is in a dependent relationship with the physician or may consent under duress. In that case the informed consent should be obtained by a well-informed physician who is not engaged in the investiga-tion and who is completely independent of this relationship.

15. For a research subject who is legally incompetent, physically or mentally incapable of giving consent or is a legally incompetent minor, the investigator must obtain informed consent from the legally authorized representative in accordance with applicable law. These groups should not be included in research unless the research is necessary to promote the health of the population represented and this research cannot instead be performed on legally competent persons.

16. When a subject deemed legally incompetent, such as a minor child, is able to give assent to decisions about participation in research, the investigator must obtain that assent in addition to the consent of the legally authorized representative.

17. Research on individuals from whom it is not possible to obtain consent, including proxy or advance consent, should be done only if the physical/mental condition that prevents obtaining informed consent is a necessary characteristic of the research population. The specific reasons for involving research subjects with a condition that renders them unable to give informed consent should be stated in the experimental protocol for consideration and approval of the review committee. The protocol should state that consent to remain in the research should be obtained as soon as possible from the individual or a legally authorized surrogate.

18. Both authors and publishers have ethical obligations. In publication of the results of research, the investigators are obliged to preserve the accuracy of the results. Negative as well as positive results should be published or otherwise publicly available. Sources of funding, institutional affiliations and any possible conflicts of interest should be declared in the publication. Reports of experimentation not in accordance with the principles laid down in this Declaration should not be accepted for publication.

Additional principles for medical research combined with medical care

1. The physician may combine medical research with medical care, only to the extent that the research is justified by its potential prophylactic, diagnostic or therapeutic value. When medical research is combined with medical care, additional standards apply to protect the patients who are research subjects.

2. The benefits, risks, burdens and effectiveness of a new method should be tested against those of the best current prophylactic, diagnostic, and therapeutic methods. This does not exclude the use of placebo, or no treatment, in studies where no proven prophylactic, diagnostic or therapeutic method exists.

3. At the conclusion of the study, every patient entered into the study should be assured of access to the best proven prophylactic, diagnostic and therapeutic methods identified by the study.

4. The physician should fully inform the patient which aspects of the care are related to the research. The refusal of a patient to participate in a study must never interfere with the patient-physician relationship.

5. In the treatment of a patient, where proven prophylactic, diagnostic and therapeutic methods do not exist or have been ineffective, the physician, with informed consent from the patient, must be free to use unproven or new prophylactic,

diagnostic and therapeutic measures, if in the physician's judgment it offers hope of saving life, re-establishing health or alleviating suffering. Where possible, these measures should be made the object of research, designed to evaluate their safety and efficacy. In all cases, new information should be recorded and, where appropriate, published. The other relevant guidelines of this Declaration should be followed.

APPENDIX F

NCI DATA AND SAFETY MONITORING POLICY

Approved by the NCI Executive Committee 6/22/99.

F.1 INTRODUCTION

All clinical trials supported or performed by NCI require some form of monitoring. The method and degree of monitoring should be commensurate with the degree of risk involved in participation and the size and complexity of the clinical trial. Monitoring exists on a continuum from monitoring by the principal investigator/project manager or NCI program staff to a data and safety monitoring board (DSMB). These monitoring activities are distinct from the requirement for study review and approval by an Institutional Review Board (IRB).

Throughout this policy, the term "awardee" means the awardee institution. In the case of NCI intramural research, the comparable institutional unit is the NCI.

F.2 RESPONSIBILITY FOR DATA AND SAFETY MONITORING

Responsibility for data and safety monitoring depends on the phase of the study and may be conducted by NCI program staff or contractor, by the principal investigator/project manager conducting the study, or by a DSMB. Regardless of the method used, monitoring must be performed on a regular basis. Oversight of the monitoring activity is the responsibility of NCI program staff. In the case of extramurally funded research, adherence to this NCI policy and any data and safety monitoring policies of the NCI Division making the award will be made a condition of the award.

Phase I and Phase II studies may be monitored by the principal investigator/project manager, by NCI program staff or a designee, or jointly. When conducted by the principal investigator/project manager, the awardee must have written policies and procedures describing the monitoring and reporting processes in place. The awardee's policies must be consistent with any policies of the NCI Division making the award. NCI program staff from the awarding NCI division will determine the acceptability of the awardee's policies and procedures. These will be documented in the grant, cooperative agreement or contract file and become part of the award.

All Phase III randomized clinical trials supported or performed by NCI require monitoring by a DSMB. The organization, responsibilities, and operation of the DSMB are described below.

For studies co-funded with other NIH Institutes or Centers (IC), the lead NIH IC will be responsible for monitoring the study and establishing a DSMB if necessary. Oversight of the DSMB will be the collaborative responsibility of the lead NIH IC and NCI.

F.3 REQUIREMENT FOR DATA AND SAFETY MONITORING BOARDS

Data and Safety Monitoring Boards must be established to monitor all Phase III randomized clinical trials supported or performed by NCI. Funds to support the functions and operations of the DSMBs will be provided by NCI in a fashion to be determined by each NCI Division.

F.4 RESPONSIBILITIES OF THE DSMB

Familiarize themselves with the research protocol(s) and plans for data and safety monitoring.

Review interim analyses of outcome data and cumulative toxicity data summaries to determine whether the trial should continue as originally designed, should be changed, or should be terminated based on these data. The DSMB reviews trial performance information such as accrual information. The DSMB also determines whether and to whom outcome results should be released prior to the reporting of study results.

Review reports of related studies to determine whether the monitored study needs to be changed or terminated.

Review major proposed modifications to the study prior to their implementation (e.g., termination, dropping an arm based on toxicity results or other reported trial outcomes, increasing target sample size).

Following each DSMB meeting, provide the study leadership with written information concerning findings for the trial as a whole related to cumulative toxicities observed and any relevant recommendations related to continuing, changing, or terminating the trial. A copy of this information will be provided to the NCI Division Director or designee. The study leadership will provide information on cumulative toxicities and relevant recommendations to the local principal investigators to be shared with their IRBs.

F.5 MEMBERSHIP

The DSMB voting members will be appointed for a fixed term by the principal investigator/project manager or designee. Proposed DSMB members must be reviewed and approved by the awarding NCI Division Director or designee prior to their appointment.

The Chair of the DSMB will be selected from among the voting members. Voting members of the DSMB should include physicians, statisticians, other scientists, and lay representatives selected based on their experience, reputation for objectivity, absence of conflicts of interest (and the appearance of same), and knowledge of clinical trial methodology. Program and statistical staff from the NCI will be permitted to serve as non- voting ex officio members of the DSMB at the request of the NCI Program Director.

Voting members may be from within or outside the institution1, but a majority should not be affiliated with the institution. Staff affiliated with the institution who are members of the DSMB should view themselves as representing the interest of patients and not that of the institution. Voting members directly involved with the conceptual design or analysis of a particular trial must excuse themselves from all DSMB discussion of the particular trial and must not receive that portion of the DSMB report related to the particular trial.

F.6 MEETINGS

DSMB meetings will be held at least annually and more often depending on the nature and volume of the trials being monitored. Each meeting should be divided into three parts. First, an open session in which members of the clinical trial team may be present, at the request of the DSMB, to review the conduct of the trial and to answer questions from members of the DSMB. The focus in the open session may be on accrual, protocol compliance, and general toxicity issues. Outcome results must not be discussed during this session. Following this session, a closed session involving the DSMB members and the coordinating center/statistical office statistician(s) handling the trial should be held. The statistician(s) should present and discuss the outcome results with the DSMB. A final executive session involving only DSMB members should be held to allow the DSMB opportunity to discuss the general conduct of the trial and all outcome results, including toxicities and adverse events, develop recommendations, and take votes as necessary.

A written report containing the current status of each trial monitored, and when appropriate any toxicity and outcome data, should be sent to DSMB members by the coordinating center/statistical office allowing sufficient time for the DSMB members to review the report prior to the meeting. This report should address specific toxicity concerns as well as concerns about the conduct of the trial. The report may contain recommendations for consideration by the DSMB concerning whether to close the trial, report the results, or continue accrual or follow up.

F.7 RECOMMENDATIONS FROM THE DSMB

DSMB recommendations should be based on results for the trials being monitored as well as on data available to the DSMB from other studies. It is the responsibility of the coordinating center/statistical office, trial investigator(s), NCI program staff and statisticians, and individual DSMB members to ensure that the DSMB is kept apprised of non-confidential results from other related studies that become available, and of any programmatic concerns related to trials being monitored. It is the responsibility of the

DSMB to determine the extent to which this information is relevant to its decisions related to specific trials.

DSMB recommendation(s) will be given to the trial principal investigator/project manager with a copy provided to the NCI Division Director or designee. If the DSMB recommends a study change for patient safety or efficacy reasons, or that a study be closed early due to slow accrual, the trial principal investigator/project manager must act to implement the change as expeditiously as possible. In the unlikely situation that the trial principal investigator/project manager does not concur with the DSMB, then the NCI Division Director or designee must be informed of the reason for disagreement. The trial principal investigator/project manager, DSMB Chair, and the NCI Division Director or designee will be responsible for reaching a mutually acceptable decision about the study. Confidentiality must be maintained during these discussions. However, in some cases, relevant data may be shared with other selected trial investigators and NCI staff to seek advice to assist in reaching a mutually acceptable decision.

If a recommendation is made to change a trial for other than patient safety or efficacy reasons or for slow accrual, the DSMB will provide an adequate rationale for its decision. In the absence of disagreement, policies of the NCI Division that made the award under which the trial is supported should be followed in regard to amending the protocol or changing the award.

F.8 RELEASE OF OUTCOME DATA

In general, outcome data should not be made available to individuals outside of the DSMB until accrual has been completed and all patients have completed their treatment. At this time, the DSMB may approve the release of outcome data on a confidential basis to the trial principal investigator/project manager for planning the preparation of manuscripts and/or to a small number of other investigators for purposes of planning future trials. Any release of outcome data prior to the DSMB's recommendation for general dissemination of results must be reviewed and approved by the DSMB.

F.9 CONFIDENTIALITY PROCEDURES

No communication, either written or oral, of the deliberations or recommendations of the DSMB will be made outside of the DSMB except as provided for in this policy. Outcome results are strictly confidential and must not be divulged to any non-member of the DSMB, except as indicated above in the Recommendations section, until the recommendation to release the results are accepted and implemented. Each member of the DSMB, including non-voting members, must sign a statement of confidentiality.

F.10 CONFLICT OF INTEREST

DSMB members are subject to the awardee's policies regarding standards of conduct. Individuals invited to serve on the DSMB as either voting or non-voting members will disclose any potential conflicts of interest, whether real or perceived, to the trial principal investigator/project manger and the appropriate institutional officials(s), in accordance with the institution's policies. Conflict of interest can include professional interest4, proprietary interest, and miscellaneous interest as described in the NIH Grants Policy Statement, Page II-12, and 45 CFR Part 94. Potential conflicts which develop during a member's tenure on a DSMB must also be disclosed. Decisions concerning whether individuals with potential conflicts of interest or the appearance of conflicts of interest may participate in a DSMB will be made in accordance with the institution's policies.

1. "Institution" is defined for this purpose as the awardee(s) institution and any institutions collaborating scientifically in the conceptual design or analysis of the study (beyond merely referring patients), including Cooperative Group member institutions participating in the trial, or consortia member institutions participating in the trial.

2. "Trial principal investigator/project manager" means the individual primarily responsible for the project, i.e., the principal Chair of a Cooperative Group, the principal investigator listed on a U10, P01 or R01 award, or the project manager listed on a contract award.

3. This policy in no way affects any legal appeal rights of the awardee.

4. "Professional interest" is used in the sense of the trial outcome benefiting the individual professionally.

APPENDIX G

NIH DATA AND SAFETY MONITORING POLICY

Release Date: June 10, 1998.

It is the policy of the NIH that each Institute and Center (IC) should have a system for the appropriate oversight and monitoring of the conduct of clinical trials to ensure the safety of participants and the validity and integrity of the data for all NIH-supported or conducted clinical trials. The establishment of the data safety monitoring boards (DSMBs) is required for multi-site clinical trials involving interventions that entail potential risk to the participants. The data and safety monitoring functions and oversight of such activities are distinct from the requirement for study review and approval by an Institutional Review Board (IRB).

G.1 BACKGROUND

A clinical trial entails a relationship between participants and investigators, both of whom must fulfill certain obligations for the effort to succeed. Participants must be fully informed of the study requirements throughout the conduct of the trial and should comply with the rigors of the research protocol or be allowed the opportunity to withdraw from participation. The investigators must protect the health and safety of participants, inform participants of information relevant to their continued participation, and pursue the research objectives with scientific diligence.

Although there are potential benefits to be derived from participation in clinical research, the IRBs and the NIH must ensure, to the extent possible, the safety of study participants and that they do not incur undue risk and that the risks versus benefits are continually reassessed throughout the study period. With this issuance, the NIH reaffirms the 1979 policy (NIH GUIDE, Volume 8, No. 8, June 5, 1979) developed by the NIH Clinical Trials Committee. Among its recommendations was the concept that "every clinical trial should have provision for data and safety monitoring." The Committee further acknowledged that "a variety of types of monitoring may be anticipated

Clinical Trials: A Methodologic Perspective, 2E, by S. Piantadosi
Copyright © 2005 John Wiley & Sons, Inc.

depending on the nature, size, and complexity of the clinical trial. In many cases, the principal investigator would be expected to perform the monitoring function."

In 1994, the Office of Extramural Research established the Committee on Clinical Trial Monitoring to review the oversight and management practices of the ICs for phase III clinical trials. One of the outcomes of this Committee's review was a strong recommendation that "all trials, even those that pose little likelihood of harm, should consider an external monitoring body." This policy affirms the Committee's recommendations concerning DSMBs.

G.2 PRINCIPLES OF MONITORING DATA AND SAFETY

All clinical trials require monitoring—Data and safety monitoring is required for all types of clinical trials, including physiologic, toxicity, and dose-finding studies (phase I); efficacy studies (phase II); efficacy, effectiveness and comparative trials (phase III); etc. Monitoring should be commensurate with risks—The method and degree of monitoring needed is related to the degree of risk involved. A monitoring committee is usually required to determine safe and effective conduct and to recommend conclusion of the trial when significant benefits or risks have developed or the trial is unlikely to be concluded successfully. Risk associated with participation in research must be minimized to the extent practical.

Monitoring should be commensurate with size and complexity Monitoring may be conducted in various ways or by various individuals or groups, depending on the size and scope of the research effort. These exist on a continuum from monitoring by the principal investigator or NIH program staff in a small phase I study to the establishment of an independent data and safety monitoring board for a large phase III clinical trial.

G.3 PRACTICAL AND IMPLEMENTATION ISSUES: OVERSIGHT OF MONITORING

This policy provides each IC with the flexibility to implement the requirement for data and safety monitoring as appropriate for its clinical research activities. Thus, IC staff may either conduct or sponsor the monitoring of data and safety of ongoing studies or delegate such responsibilities to a grantee or contractor. Oversight of monitoring activities is distinct from the monitoring itself and should be the responsibility of the IC regardless of whether the monitoring is performed by NIH staff or is delegated. Oversight of monitoring must be done to ensure that data and safety monitoring plans are in place for all interventional trials, that the quality of these monitoring activities is appropriate to the trial(s), and that the IC has been informed of recommendations that emanate from monitoring activities.

G.4 INSTITUTES AND CENTERS RESPONSIBILITIES

Though ICs may perform a variety of roles in data and safety monitoring and its oversight, the following are the minimum responsibilities of sponsoring ICs.

1. Prepare or ensure the establishment of a plan for data and safety monitoring for all interventional trials.

2. Conduct or delegate ongoing monitoring of interventional trials.
3. Ensure that monitoring is timely and effective and that those responsible for monitoring have the appropriate expertise to accomplish its mission.
4. Oversee monitoring activities.
5. Respond to recommendations that emanate from monitoring activities.

G.5 PERFORMANCE OF DATA AND SAFETY MONITORING

The ICs will ensure the integrity of systems for monitoring trial data and participant safety, although they may delegate the actual performance to the grantee or contractor. Monitoring must be performed on a regular basis, and conclusions of the monitoring reported to the IC. Recommendations that emanate from monitoring activities should be reviewed by the responsible official in the IC and addressed. The ICs also have the responsibility of informing trial investigators concerning the data and safety monitoring policy and procedures. Considerations such as who shall perform the monitoring activities, the composition of the monitoring group (if a group is to be used), the frequency and character of monitoring meetings (e.g., open or closed, public or private), and the frequency and content of meeting reports should be a part of the monitoring plans. IRBs should be provided feedback on a regular basis, including findings from adverse-event reports, and recommendations derived from data and safety monitoring.

Monitoring activities should be conducted by experts in all scientific disciplines needed to interpret the data and ensure patient safety. Clinical trial experts, biostatisticians, bioethicists, and clinicians knowledgeable about the disease and treatment under study should be part of the monitoring group or be available if warranted.

Ideally, participants in monitoring outcomes of a trial are in no way associated with the trial. For trials that are conducted as part of a cooperative group, a majority of the individuals monitoring outcome data should be external to the group. ICs should require policies that evaluate whether the participants have conflicts of interests with or financial stakes in the research outcome; and when these conflicts exist, policies must exist to manage these in a reasonable manner.

Generally, data and safety monitoring boards meet first in open session, attended by selected trial investigators as well as NIH program staff or project officers and perhaps industry representatives, and then in closed session where they review emerging trial data. When "masked" data are presented or discussed, no one with a proprietary interest in the outcome should be allowed. Participants in the review of "masked" or confidential data and discussions regarding continuance or stoppage of the study should have no conflict of interest with or financial stake in the research outcome. However, if there is an open session, they could be present.

Confidentiality must be maintained during all phases of the trial including monitoring, preparation of interim results, review, and response to monitoring recommendations. Besides selected NIH program staff, other key NIH staff, and trial biostatisticians, usually only voting members of the DSMB should see interim analyses of outcome data. Exceptions may be made under circumstances where there are serious adverse events, or whenever the DSMB deems it appropriate.

Individuals or groups monitoring data and safety of interventional trials will perform the following activities:

1. Review the research protocol and plans for data and safety monitoring.

2. Evaluate the progress of interventional trial(s), including periodic assessments of data quality and timeliness, participant recruitment, accrual and retention, participant risk versus benefit, performance of trial sites, and other factors that can affect study outcome. Monitoring should also consider factors external to the study when interpreting the data, such as scientific or therapeutic developments that may have an impact on the safety of the participants or the ethics of the study.

3. Make recommendations to the IC, IRB, and investigators concerning continuation or conclusion of the trial(s).

4. Protect the confidentiality of the trial data and the results of monitoring.

G.6   EXAMPLES OF MONITORING OPERATIONS

The following provides examples of appropriate types of monitoring and oversight for different types of studies. These are illustrative only. The ICs must develop and implement monitoring activities and oversight of those activities appropriate to the study, population, research environment, and the degree of risk involved.

Phase I: A typical phase I trial of a new drug or agent frequently involves relatively high risk to a small number of participants. The investigator and occasionally others may have the only relevant knowledge regarding the treatment because these are the first human uses. An IC may require the study investigator to perform continuous monitoring of participant safety with frequent reporting to IC staff with oversight responsibility.

Phase II: A typical phase II trial follows phase I studies and there is more information regarding risks, benefits and monitoring procedures. However, more participants are involved and the toxicity and outcomes are confounded by disease process. An IC may require monitoring similar to that of a phase I trial or supplement that level of monitoring with individuals with expertise relevant to the study who might assist in interpreting the data to ensure patient safety.

Phase III: A phase III trial frequently compares a new treatment to a standard treatment or to no treatment, and treatment allocation may be randomly assigned and the data masked. These studies usually involve a large number of participants followed for longer periods of treatment exposure. While short-term risk is usually slight, one must consider the long term effects of a study agent or achievement of significant safety or efficacy differences between the control and study groups for a masked study. An IC may require a DSMB to perform monitoring functions. This DSMB would be composed of experts relevant to the study and would regularly assess the trial and offer recommendations to the IC concerning its continuation.

APPENDIX   H

ROYAL STATISTICAL SOCIETY CODE OF CONDUCT

H.1   INTRODUCTION

In every civilized society rules of conduct exist for the benefit of society at large and in order to give freedom for individual members to go about their legitimate business within bounds of behaviour which are accepted and observed by their fellows.

In common with professional bodies in other fields, the RSS has formulated its own rules as a Code of Conduct to define the behaviour expected of RSS Fellows practicing in everyday professional life. This code of conduct has been drawn up to reflect the standards of conduct and work expected of all practicing statisticians. It is commended of all Fellows of the Society and is mandatory on all Professionally Qualified Fellows as defined in paragraph 1(p) of the Society's Bye-Laws.

H.2   CONSTITUTIONAL AUTHORITY

The Royal Statistical Society is a professional and learned Society which, through its members, has an obligation in the public interest to provide the best possible statistical service and advice. In general, the public has no ready means of judging the quality of professional service except from the reputation of the provider. Professional membership of the Society is an assurance of ability and integrity. Thus it is essential that the highest standards are maintained by all Fellows whenever they are acting professionally and whatever their level of qualification.

The constitutional authority for the RSS Code of Conduct derives firstly from Bye-Laws 40(g) and 47 of the Society and, secondly, formal adoption by Council.

H.3   RULES OF PROFESSIONAL CONDUCT

As an aid to understanding, these rules have been grouped into the principal duties which all Fellows should endeavor to discharge in pursuing their professional lives.

Clinical Trials: A Methodologic Perspective, 2E, by S. Piantadosi
Copyright © 2005 John Wiley & Sons, Inc.

H.3.1   The Public Interest

1. Fellows shall ensure that within their chosen fields they have appropriate knowledge and understanding of relevant legislation, regulations and standards and that they comply with such requirements.

2. Fellows shall in their professional practice have regard to basic human rights and shall avoid any actions that adversely affect such rights. Enquiries involving human subjects should, as far as practicable, be based on the freely given informed consent of subjects. The identities of subjects should be kept confidential unless consent for disclosure is explicitly obtained.

H.3.2   Duty to Employers and Clients

3. Fellows shall carry out work with due care and diligence in accordance with the requirements of the employer or client and shall, if their professional judgment is overruled, indicate the likely consequences.

4. Fellows shall not disclose or authorize to be disclosed, or use for personal gain or to benefit a third party, confidential information acquired in the course of professional practice, except with prior written permission of the employer or client, or at the direction of a court of law; Fellows should seek to avoid being put in a position where they may become privy to or party to activities or information concerning activities which would conflict with their responsibilities in 1 and 2 above.

5. Fellows shall not purport to exercise independent judgment on behalf of a client on any product or service in which they knowingly have any interest, financial or otherwise.

6. Fellows should not allow any misleading summary of data to be issued in their name. In particular, a statistical analysis may need to be amplified by a description of the way the data were selected, and the way any apparently erroneous data were corrected or rejected. Explicit statements will generally be needed about the assumptions made when selecting a method of analysis. Views or opinions based on general knowledge or belief should be clearly distinguished from views or opinions derived from the statistical analyses being reported.

H.3.3   Duty to the Profession

7. Fellows shall uphold the reputation of the Profession and shall seek to improve professional standards through participation in their development, use and enforcement, and shall avoid any action which will adversely affect the good standing of Statistics and Statisticians.

8. Fellows shall seek to advance public knowledge and understanding of statistics and to counter false or misleading statements which are detrimental to the Profession.

9. Fellows shall encourage and support fellow members in their professional development and, where possible, provide opportunities for the development of new entrants to the Profession.

10. Fellows shall act with integrity towards fellow statisticians and to members of other professions with whom they are concerned in a professional capacity, and shall avoid engaging in any activity which is incompatible with their professional status.

Whilst Fellows of the Society are free to engage in controversy, no Fellow shall cast doubt on the professional competence of another without good cause.

11. Fellows shall not make any public statement in their professional capacity unless properly qualified and, where appropriate, authorized to do so, and shall have due regard to the likely consequences of any such statement on others. Fellows shall not speak in the name of the Society, its Council or Committees, without the authorization of Council.

Professional Competence and Integrity

12. Fellows shall seek to upgrade their professional knowledge and skill and shall maintain awareness of technological developments, procedures and standards which are relevant to their field, and shall encourage their subordinates to do likewise.

13. Fellows shall seek to conform to recognized good practice including quality standards which are in their judgment relevant, and shall encourage their subordinates to do likewise.

14. Fellows shall only offer to do work or provide service which is within their professional competence and shall not lay claim to any level of competence which they do not possess, and any professional opinion which they are asked to give shall be objective and reliable.

15. Fellows shall accept professional responsibility for their work and for the work of subordinates and associates under their direction.

16. The Standards of integrity required of a professional statistician should not normally conflict with the interests of a client or employer. Fellows shall aim to avoid any such conflict and shall clearly advise their client of any such potential or actual conflict. If the conflict cannot be resolved satisfactorily the public interest and professional standards must be paramount.

17. Fellows acting in private practice, or acting independently of salaried employment, have the right of disengagement in the face of a dilemma involving professional standards or conscience. They may wish to seek advice and support from the Society.

18. Fellows in salaried employment who are in serious conflict with their employer over a matter of professional standards or conscience should notify the employer in writing of the contentious circumstances. If they are unable to resolve the conflict to their satisfaction, they are advised to refer the matter to the Society, which will advise and take such action as seems appropriate.

H.3.4 Disciplinary Procedures

19. This code sets out certain basic principles that are intended to help Fellows maintain the highest standards of professional conduct. Should a case arise where a Fellow is believed to have fallen short of the standards desired, procedures are defined within the Bye-Laws of the Society which permit the convening of a Disciplinary Sub-Committee. The report of such a Committee will be submitted to the Professional Affairs Committee who will determine what action should be taken in any particular instance. Action will be appropriate to the circumstances, and in the most serious of cases, the sanctions available to the Committee shall include removal of professional status and expulsion from the Society.

BIBLIOGRAPHY

Abbott, A. (2000). German fraud inquiry casts a wider net of suspicion. *Nature* 405: 871–872.

Abby, M., Massey, M.D., Galandiuk, S., and Polk, H.C., Jr. (1994). Peer review is an effective screening process to evaluate medical manuscripts. *JAMA* 272: 105–107.

Abel, U., and Koch, A. (1997). The mythology of randomization. *Proc. Int. Conference on Nonrandomized Comparative Clinical Studies in Heidelberg*, April 10–11, 1997.

Abel, U., and Koch, A. (1999). The role of randomization in clinical studies: Myths and beliefs. *J. Clin. Epidemiol.* 52: 487–497.

Achenwall, G. (1748). Vorbereitung zur Staatswissenschaft. This became the introduction to: *Staatsverfassung der heutigen vornehmsten europäischen Reiche und Völker im Grundrisse*. Göttingen, 1749.

Ad hoc Working Group for Critical Appraisal of the Medical Literature. (1987). A proposal for more informative abstracts of clinical articles. *An. Intern. Med.* 106: 598–604.

Adams-Campbell, L.L., Ahaghotu, C., Gaskins, M., et al. (2004). Enrollment of African Americans onto clinical treatment trials: Study design barriers. *J. Clin. Oncol.* 22: 730–734.

Advani, A.S., Atkeson, B., Brown, C.L., et al. (2002). Barriers to the participation of African-American patients with cancer in clinical trials. *Cancer* 97: 1499–1506.

Advisory Committee on Human Radiation Experiments (1995). *Final Report*. Washington, DC: Government Printing Office.

Agresti, A. (1996). *An Introduction to Categorical Data Analysis*. New York: Wiley.

Ahrens, E.H. (1992). *The Crisis in Clinical Research: Overcoming Institutional Obstacles*. New York: Oxford University Press.

Albain, K.S., Green, S., LeBlanc, M., Rivkin, S., O'Sullivan, J., and Osborne, C.K. (1992). Proportional hazards and recursive partitioning and amalgamation analyses of the Southwest Oncology Group node-positive adjuvant CMFVP breast cancer data base: A pilot study. *Breast Cancer Res. Treat.* 22: 273–284.

Albin, R.L. (2002). Sham surgery controls: intracerebral grafting of fetal tissue for Parkinson's disease and proposed criteria for use of sham surgery controls. *J. Med. Ethics* 28: 322–325.

Altman, D.G., Gore, S., Gardner, M., and Pocock, S. (1983). Statistical guidelines for contributors to medical journals. *Brit. Med. J.* 286: 1489–1493.

Altman, D.G., and Bland, J.M. (1995). Absence of evidence is not evidence of absence. *BMJ* 311: 485.

Altman, E. and Hernon, P. (Eds.) (1997). *Research Misconduct: Issues, Implications, and Strategies.* Greenwich, CT: Ablex Publishing.

d'Amador, R. (1836). Mémoire sur le calcul des probabilités appliqué à la médecine. *Bull. Acad.Roy. Méd.* 1: 622–680.

Amberson, J.B., Jr., McMahon, B.T., and Pinner, M. (1931). A clinical trial of sanocrysin in pulmonary tuberculosis. *Am. Rev. Tuberc.* 24: 401–404.

American Cancer Society (ACS) (1990). Questionable cancer practices in Tijuana and other Mexican border clinics. Statement approved by the Committee on Questionable Methods of Cancer Management.

American Cancer Society (ACS) (1993). Questionable methods of cancer management: "nutritional" therapies. *CA Cancer J. Clinicians* 43(5): 309–319.

American Medical Association (2001). Principles of Medical Ethics, June 2001. Adopted by the AMA's House of Delegates June 17, 2001.

ACAPS Group. (1992). Rationale and design for the Asymptomatic Carotid Artery Plaque Study (ACAPS). *Cont. Clin. Trials* 13: 293–314.

American Statistical Association (1995). *Ethical Guidelines for Statistical Practice.*

Amsterdam, E.A., Wolfson, S., Gorlin, R. (1969). New aspects of the placebo response in angina pectoris. *Am. J. Cardiol.* 24(3):305–306.

Andersen, B. (1990). *Methodological Errors in Medical Research.* Oxford: Blackwell Scientific.

Andersen, P.K. (1987). Conditional power calculations as an aid in the decision whether to continue a clinical trial. *Cont. Clin. Trials* 8: 67–74.

Anderson, J.R., Cain, K.C., Gelber, R.D., and Gelman, R.S. (1985). Analysis and interpretation of the comparison of survival by treatment outcome variables in cancer clinical trials. *Cancer Treat. Rep.* 69: 1139–1144.

Andrews, D.F., and Herzberg, A.M. (1985). *Data: A Collection of Problems from Many Fields for the Student and Research Worker.* New York: Springer-Verlag.

Anello, C. (1999). Emerging and recurrent issues in drug development. *Stat. Med.* 18: 2301–2309.

Angell, M. (2000a). Investigators' responsibilities for human subjects in developing countries. *N. Engl. J. Med.* 342(13): 967–969.

Angell, M. (2000b). Is academic medicine for sale? *N. Engl. J. Med.* 342: 1516–1518.

Angell, M. and Kassirer, J.P. (1994). Setting the record straight in the breast cancer trials (editorial). *N. Engl. J. Med.* 330: 1448–1450.

Angell, M. and Kassirer, J.P. (1998). Alternative medicine—The risks of untested and unregulated remedies (editorial). *N. Engl. J. Med.* 339: 839–841.

Annas, G.J. and Grodin, M.A. (1992). *The Nazi Doctors and the Nuremberg Code: Human Rights in Human Experimentation.* New York: Oxford University Press.

Anscombe, F.J. (1963). Sequential medical trials. *J. Am. Stat. Assoc.* 58: 365–383.

Ansell, S.M., Rapoport, B.L., Falkson, G., Raats, J.I., and Moeken, C.M. (1993). Survival determinants in patients with advanced ovarian cancer. *Gyn. Oncol.* 50: 215–220.

Antman, E.M., Lau, J., Kupelnick, B., Mosteller, F., Chalmers, T.C. (1992). A comparison of results of meta-analyses of randomized control trials and recommendations of clinical experts. Treatments for myocardial infarction. *JAMA* 268(2): 240–248. Comment in: *JAMA* 269(2): 214 (1993).

Anyanwu, A.C. and Treasure T. (2003). Surgical research revisited: Clinical trials in the cardio-thoracic surgical literature. *Eur. J. Cardio-thorac. Surg.* 25: 299–303.

Apfel, C.C., et al. for the IMPACT Investigators. (2004). A factorial trial of six interventions for the prevention of postoperative nausea and vomiting. *N. Engl. J. Med.* 350: 2441–2451.

Appel L.J. (1997). A clinical trial of the effects of dietary patterns on blood pressure. *N. Engl. J. Med.* 336: 1117–1124.

Appelbaum, P.S., Lidz, C.W., and Meisel, A. (1987). *Informed Consent: Legal Theory and Clinical Practice.* New York: Oxford University Press.

Appleton, D.R. (1995). What do we mean by a statistical model? *Stat. Med.* 14: 185–197.

Archer, T.P., Leier, C.V. (1992). Placebo treatment in congestive heart failure. *Cardiology* 81(2-3):125–133.

ARDS Network. (2000). Ventilation with lower tidal volumes as compared with traditional tidal volumes for acute lung injury and the acute respiratory distress syndrome. *N. Engl. J. Med.* 342: 1301–1308.

Armitage, P. (1981). Importance of prognostic factors in the analysis of data from clinical trials. *Cont. Clin. Trials* 1: 347–353.

Armitage, P., and Berry, G. (1994). *Statistical Methods in Medical Research*, 3rd ed. Oxford: Blackwell.

Armitage, P. and Colton, T., eds. (1998). *Encyclopedia of Biostatistics.* Chichester: Wiley.

ASCO (1997). Critical role of phase I clinical trials in cancer treatment. *J. Clin. Oncol.* 15: 853–859.

Ashcroft, R.E., Chadwick, D.W., Clark, S.R.L., et al. (1997). Implications of socio-cultural contexts for the ethics of clinical trials. *Health Technol. Assess.* I(9).

Assaf, A.R., and Carleton, R.A. (1994). The Women's Health Initiative clinical trial and observational study: History and overview. *Rhode Island Med.* 77: 424–427.

Astin, J.A. (1998). Why patients use alternative medicine: Results of a national study. *JAMA* 279(19): 1548–1553.

ATBC Cancer Prevention Study Group (1994a). The alpha-tocopherol beta-carotene lung cancer prevention study: Design, methods, participant characteristics, and compliance. *An. Epidemiol.* 4: 1–9.

ATBC Cancer Prevention Study Group (1994b). The effect of vitamin E and beta carotene on the incidence of lung cancer and other cancers in male smokers. *N. Engl. J. Med.* 330: 1029–1034.

Atwood, K.C. (2003). The ongoing problem with the National Center for Complementary and Alternative Medicine. *Skeptical Inq.* 27: 56–61.

Aubert, B. et al. (2001). Measurement of CP-violating asymmetries in B0 decays to CP eigen-states. *Phys. Rev. Lett.* 86(12): 2515–2522.

Aulas, J.J. (1996). Alternative cancer treatments. *Sci. Am.* 275(3): 162–163.

Babbage, C. (1830, 1970). *Reflections on the Decline of Science in England.* New York: Augustus Kelley.

Baerheim, A., and Sandvik, H. (1994). Effect of ale, garlic, and soured cream on the appetite of leeches. *BMJ* 309(6970): 1689.

Bailar, J.C., III, and Mosteller, F. (1988). Guidelines for statistical reporting in articles for medical journals: Amplifications and explanations. *An. Int. Med.* 108: 266–273.

Bailar, J.C. (1995). The practice of meta-analysis. *J. Clin. Epidemiol.* 48: 149–157.

Bailar, J.C., Louis, T.A., Lavori, P.W., and Polansky, M. (1984). Studies without internal controls. *N. Engl. J. Med.* 311: 156–162.

Bailar, J.C. (2001). The Powerful Placebo and the Wizard of Oz. *N. Engl. J. Med.* 344: 1630–1616.

Bakke, O.M., Manocchia, M., de Abajo, F., Kaitin, K.I., and Lasagna, L. (1995). Drug safety discontinuations in the United Kingdom, the United States, and Spain from 1974 through 1993: A regulatory perspective. *Clin. Pharmacol. Ther.* 58(1): 108–117.

Balis F.M. (1997). The Challenge of Developing New Therapies for Childhood Cancers. *Oncologist* 2(1): I–II.

Balke, A.A. and Pearl, J. (1994). Universal formulas for treatment effects from noncompliance data. Technical Report R-199-A, Cognitive Systems Laboratory, UCLA.

Bamshad, M.J. and Olson, S.E. (2003). Does race exist? *Sci. Am.* 289: 78–85.

Barbehenn, E., Lurie, P., and Wolfe, S.M. (2002). *Letter to HHS Secretary Tommy Thompson that raises ethical concerns about the "Alzheimer's Disease Anti-inflammatory Prevention Trial"* (ADAPT) (HRG Publication #1637). www.citizen.org/publications/release.cfm?ID=7195.

Barber, R.E., Lee, J., and Hamilton, W.K. (1970). Oxygen toxicity in man: A prospective study in patients with irreversible brain damage. *N. Engl. J. Med.* 283: 1478–1484.

Barnett, V. (1982). *Comparative Statistical Inference*, 2nd ed. New York: Wiley.

Barrett, S. (1993). "Alternative" cancer treatment. In S. Barrett and W.T. Jarvis (eds.). *The Health Robbers*. Buffalo, NY: Prometheus Books, ch. 6.

Barrow, J.D. (1991). *Theories of Everything*. Oxford: Clarendon Press.

Bartlett, R.H., Roloff, D.W. Cornell, R.G., Andrews, A.F., Dillon, P.W., and Zwischenberger, J.B. (1985). Extracorporeal circulation in neonatal respiratory failure: A prospective randomized study. *Pediatrics* 76: 479–487.

Baum, M., Houghton, J., and Abrams, K. (1994). Early stopping rules—Clinical perspectives and ethical considerations. *Stat. Med.* 13: 1459–1469.

Bayes, T. (1763). An essay towards solving a problem in the doctrine of chances. *Phil. Trans. Roy. Soc. Lond.* 53: 370–418.

Bayes, T. (1958). Reprint of the 1763 paper. *Biometrika* 45: 298–315.

Beach, M.L., and Meier, P. (1989). Choosing covariates in the analysis of clinical trials. *Cont. Clin. Trials* 10: 161S–175S.

Beauchamp, T.L., and Childress, J.F. (2001). *Principles of Biomedical Ethics*. 5th ed., New York: Oxford University Press.

Becker, N., Chambliss, C., Marsh, C., and Montemayor, R. (1995). Effects of mellow and frenetic music and stimulating and relaxing scents on walking by seniors. *Percept. Mot. Skills* 80: 411–415.

Beecher, H.K. (1955). The powerful placebo. *JADA* 159: 1602–1606.

Beecher, H.K. (1961). Surgery as placebo. A quantitative study of bias. *JAMA* 176: 1102–1107.

Beecher, H.K. (1966). Ethics and clinical research. *N. Engl. J. Med.* 274: 1354–1360.

Beecher, H.K. (1970). *Research and the Individual: Human Studies*. Boston: Little, Brown.

Begg, C.B. (1990). On inferences from Wei's biased coin design for clinical trials (with discussion). *Biometrika* 77: 467–484.

Begg, C.B., and Berlin, J.A. (1988). Publication bias: A problem in interpreting medical data. *J. Roy. Statist. Soc. A* 151: 419–463.

Begg, C.B., and Iglewicz, B.A. (1980). A treatment allocation procedure for sequential clinical trials. *Biometrics* 36: 81–90.

Begg C.B., Riedel E.R., Bach P.B., et al. (2002). Variations in morbidity after radical prostatectomy. *N. Engl. J. Med.* 346: 1138–1144.

Bekelman, J.E., Li, Y., and Gross, C.P. (2003). Scope and impact of financial conflicts of interest in biomedical research: A systematic review. *JAMA* 289: 454–465.

Benson, K., and Hartz A.J. (2000). A comparison of observational studies and randomized controlled trials. *N. Engl. J. Med.* 342: 1878–1886.

Berger, R.L., Celli, B.R., Meneghetti, A.L. et al. (2001). Limitations of randomized clinical trials for evaluating emerging operations: The case of lung volume reduction surgery. *An. Thorac. Surg.* 72: 649–657.

Berman, I., Sapers, B.L., Chang, H.H., et al. (1995). Treatment of obsessive-compulsive symptoms in schizophrenic patients with clomipramine. *J. Clin. Psychopharmacol.* 15(3): 206–210.

Bernard, C. (1865). *An Introduction to the Study of Experimental Medicine*. H.C. Greene, trans. New York, 1957.

Berry, D.A. (1985). Interim analyses in clinical trials: Classical vs. Bayesian approaches. *Stat. Med.* 4: 521–526.

Berry, D.A. (1993). A case for Bayesianism in clinical trials. *Stat. Med.* 12: 1377–1393.

Berson, E.L., Rosner, B., Sandberg, M.A., Hayes, K.C., Nicholson, B.W., Weigel-DiFranco, C., and Willet, W. (1993). A randomized trial of vitamin A and vitamin E supplementation for retinitis pigmentosa. *Arch. Ophthalmol.* 111: 761–772.

Bertram, J.S., Kolonel, L.N., and Meyskens, F.L., Jr. (1987). Rationale and strategies for chemoprevention of cancer in humans. *Cancer Res.* 47: 3012–3031.

Bhasin, S., Storer, T.W., Berman, N., et al. (1996). The effects of supraphysiologic doses of testosterone on muscle size and strength in normal men. *N. Engl. J. Med.* 335: 1–7.

Birkmeyer, J.D., Siewers, A.E., Finlayson, E.V., et al. (2002). Hospital volume and surgical mortality in the United States. *N. Engl. J. Med.* 346: 1128–1137.

Biros, M.H., Lewis, R.J., Olson, C.M., et al. (1995). Informed consent in emergency research. Consensus statement from the Coalition Conference of Acute Resuscitation and Critical Care Researchers. *JAMA* 273: 1283–1287.

Bivens, L.W., and Macfarlane, D.K. (1994). Fraud in breast cancer trials. *N. Engl. J. Med.* 330: 1461.

Blackwelder, W.C. (1982). Proving the null hypothesis in clinical trials. *Cont. Clin. Trials* 3: 345–353.

Blackwelder, W.C., and Chang, M.A. (1984). Sample size graphs for "Proving the null hypothesis."*Cont. Clin. Trials* 5: 97–105.

Bloom, B.S. (2001). What is this nonsense that complementary and alternative medicine is not amenable to controlled investigation of population effects? *Acad. Med.* 76: 1221–1223.

Blume, J.D. (2002). Tutorial in biostatistics: Likelihood methods for measuring statistical evidence. *Stat. Med.* 21: 2563–2599.

Bobbio, M., Demichelis, B., and Giustetto, G. (1994). Completeness of reporting trial results: Effect on physicians' willingness to prescribe. *Lancet* 343: 1209–1211.

Bock, R.D. (1975). *Multivariate Statistical Methods in Behavioral Research*. New York: McGraw-Hill.

Boden, W.E. (1992). Meta-analysis in clinical trials reporting: Has a tool become a weapon? (editorial). *Am. J. Cardiol.* 69: 681–686.

Bogoch, S., and Bogoch, E.S. (1994). A checklist for suitability of biomarkers as surrogate endpoints in chemoprevention of breast cancer. *J. Cell. Biochem.* suppl. 19: 173–185.

Boissel, J.P., Collet, J.P., Moleur, P., and Haugh, M. (1992). Surrogate endpoints: A basis for a rational approach. *Eur. J. Clin. Pharmacol.* 43: 235–244.

Bonita, R., and Beaglehole, R. (1996). The enigma of the decline in stroke deaths in the United States: The search for an explanation. *Stroke* 27: 370–372.

Borgen, P.I. (1991). Reviewing peer review: The NCI Clinical Alert three years later. *J. LA. State Med. Soc.* 143(3): 39–41.

Box, G.E.P., and Draper, N.R. (1987). *Empirical Model Building and Response Surfaces*. New York: Wiley.

Box, J.F. (1980). R. A. Fisher and the design of experiments, 1922–1926. *Am. Statistician* 34: 1–7.

Brandt, A.M. (1978). Racism and research: The case of the Tuskegee Syphilis Study. *Hastings Center Rep.* 8: 21–29.

Brantigan, J.W. (1995). A surgeon's perspective of medical device regulation. *J. Spinal Disorders* 8: 396–412.

Breast Cancer Action Board of Directors (1996). Policy on Tamoxifen (Nolvadex).

Breitner, J.C.S., and Meinert, C.L. (2002). *Letter to Tommy Thompson, Secretary, DHHS.* www.jhucct.com/adapt/ADAPTresponse.pdf.

Brem, H., Piantadosi, S., Burger, P.C. et al. (1995). Placebo-controlled trial of safety and efficacy of intraoperative controlled delivery by biodegradable polymers of chemotherapy for recurrent gliomas. The Polymer-Brain Tumor Treatment Group. *Lancet* 345: 1008–1012.

Brennan, T.A. (1999). Proposed revisions to the Declaration of Helsinki—Will they weaken the ethical principles underlying human research? *N. Engl. J. Med.* 341(7): 527–531.

Brett, A.S. (1981). Hidden ethical issues in clinical decision analysis. *N. Engl. J. Med.* 5: 1150–1152.

Brigden, M.L. (1995). Unproven (questionable) cancer therapies. *Western J. Med.* 163(5): 463–469.

Bristol, D.R. (1999). Clinical equivalence. *J. Biopharm. Stat.* 9(4): 549–561.

Broad, W., and Wade, N. (1982). *Betrayers of the Truth*. New York: Simon and Schuster.

Brockwell, S.E., and Gordon, I.R. (2001). A comparison of statistical methods for meta-analysis. *Stat. Med.* 20(6): 825–840.

Broder, S. (1994). Fraud in breast cancer trials. *N. Engl. J. Med.* 330: 1460–1461.

Brody, B.A. (2002). Ethical issues in clinical trials in developing countries. *Stat. Med.* 21(19):2853–2858.

Bross, I.D. (1990). How to eradicate fraudulent statistical methods: Statisticians must do science. *Biometrics* 46: 1213–1225.

Brown, B.W., Jr. (1980). The cross-over experiment for clinical trials. *Biometrics* 36: 69–79.

Brown, R.D., Whisnant, J.P., Sicks, J.D., O'Fallon, W.M., and Wiebers, D.O. (1996). Stroke incidence, prevalence, and survival: Secular trends in Rochester, Minnesota, through 1989. *Stroke* 27: 373–380.

Bruning, N. (1992) Tamoxifen on trial: Congressional hearing. *Breast Cancer Action Newsletter* (15), Dec. 1992.

Bulger, R.E. (1994). Toward a statement of the principles underlying responsible conduct in biomedical research. *Acad. Med.* 69(2): 102–107.

Bull, J.P. (1959). The historical development of clinical therapeutic trials. *J. Chron. Dis.* 10: 218–248.

Burish, T.G. and Jenkins, R.A. (1992). Effectiveness of biofeedback and relaxation training in reducing the side effects of cancer chemotherapy. *Health Psychol.* 11: 17–23.

Burton, P.R. (1994). Helping doctors to draw appropriate inferences from the analysis of medical studies. *Stat. Med.* 13: 1699–1713.

Buyse, M.E., Staquet, M.J., and Sylvester, R.J. (eds.) (1984). *Cancer Clinical Trials: Methods and Practice*. Oxford: Oxford University Press.

Byar, D.P. (1978). On combining information: Historical controls, overviews, and comprehensive cohort studies. *Recent Results in Cancer Res.* 111: 95–98.

Byar, D.P. (1980). Why data bases should not replace randomized clinical trials. *Biometrics* 36: 337–342.

Byar, D.P. (1982). Analysis of survival data: Cox and Weibull models with covariates. In V. Mike and K. Stanley (eds.), *Statistics in Medical Research*. New York: Wiley, ch. 12.

Byar, D.P. (1984). Identification of prognostic factors. In M.J. Buyse, M.J. Staquet, and R.J. Sylvester (eds.), *Cancer Clinical Trials*. Oxford: Oxford University Press, ch. 24.

Byar, D.P. (1985). Assessing apparent treatment–covariate interactions in randomized clinical trials. *Stat. Med.* 4: 255–263.

Byar, D.P. (1990). Factorial and reciprocal control designs. *Stat. Med.* 9: 55–64.

Byar, D.P. (1991). Problems with using observational databases to compare treatments. *Stat. Med.* 10: 663–666.

Byar, D.P., Green, S.B., Dor, P., Williams, E.D., Colon, J., van Gilse, H.A., Mayer, M., Sylvester, R.J., and Van Glabbeke, M. (1979). A prognostic index for thyroid carcinoma: A study of the EORTC Thyroid Cancer Cooperative Group. *Eur. J. Cancer* 15: 1033–1041.

Byar, D.P., Herzberg, A.M., and Tan, W.Y. (1993). Incomplete factorial designs for randomized clinical trials. *Stat. Med.* 12: 1629–1641.

Byar, D.P., and Piantadosi, S. (1985). Factorial designs for randomized clinical trials. *Cancer Treat. Rep.* 69: 1055–1063.

Byar, D.P., Schoenfeld, D.A., Green, S.B., et al. (1990). Design considerations for AIDS trials. *N. Engl. J. Med.* 323: 1343–1348.

Byar, D.P., Simon, R.M., Friedewald, W.T., et al. (1976). Randomized clinical trials. Perspectives on some recent ideas. *N. Engl. J. Med.* 295: 74–80.

Carroll, R.J., Ruppert, D., and Stefanski, L.A. (1995). *Measurement Error in Nonlinear Models*. London: Chapman-Hall.

Campbell, M.J., and Machin, D. (1990). *Medical Statistics*. Chichester: Wiley.

Canetta, R. (2004). What went wrong in the development of matrix metalloproteinase inhibitors? Accelerating Anticancer Agent Development and Validation. Baltimore, MD, July 13–14, 2004.

Carino, T., Sheingold, S., and Tunis, S. (2004). Using clinical trials as a condition of coverage: lessons from the National Emphysema Treatment Trial (with discussion). *Clin. Trials* 1: 108–121.

Carlin, B.P., and Sargent, D.J. (1996). Robust Bayesian approaches for clinical trial monitoring. *Stat. Med.* 15: 1093–1106.

Carpenter, W.T., Jr., Sadler, J.H., Light, P.D., Hanlon, T.E., Kurland, A.A., et al. (1983). The therapeutic efficacy of hemodialysis in schizophrenia. *N. Engl. J. Med.* 308: 669–675.

Carson, E.R., Cobelli, C., and Finkelstein, L. (1983). *The Mathematical Modeling of Metabolic and Endocrine Systems*. New York: Wiley.

Carter, B. (2003). Methodological issues and complementary therapies: Researching intangibles? *Compl. Ther. Nurs. Midwifery* 9: 133–139.

Carter, R.L., Scheaffer, R.L., and Marks, R.G. (1986). The role of consulting units in statistics departments. *Am. Stat.* 40: 260–264.

Casey, A.T., Crockard, H.A., Bland, J.M., et al. (1996). Surgery on the rheumatoid cervical spine for the non-ambulant myelopathic patient—Too much, too late? *Lancet* 347: 1004–1007.

Cassileth, B.R. and Chapman, C.C. (1996). Alternative cancer medicine: A ten-year update. *Cancer Invest.* 14(4) 396–404.

CAST Investigators (1989). Preliminary report: Effect of encainide and flecainide on mortality in a randomized trial of arrythmia suppression after myocardial infarction. *N. Engl. J. Med* 321: 406–412.

Centers for Disease Control. (1997). Paralytic poliomyelitis—United States, 1980–1994. *MMWR* 46(4): 79–83.

Challis, G.B. and Stam, H.J. (1990). The spontaneous regression of cancer: A review of cases from 1900 to 1987. *Acta Oncol.* 29: 545–550.

Chalmers, I. (1990). Underreporting research is scientific misconduct. *JAMA* 263: 1405–1408.

Chalmers, I. (1993). The Cochrane collaboration: Preparing , maintaining, and disseminating systematic reviews of the effects of health care. *An. N.Y. Acad. Sci.* 703: 156–165.

Chalmers, I. (2003). Control of selection biases: comparing like with like. In: *The James Lind Library* (www.jameslindlibrary.org).

Chalmers, I., and Silverman, W.A. (1987). Professional and public double standards on clinical experimentation. *Cont. Clin. Trials* 8: 388–391.

Chalmers, T.C. (1975a). Ethical aspects of clinical trials. *Am. J. Ophthalmol.* 79: 753–758.

Chalmers, T.C. (1975b). Randomization of the first patient. *Med. Clin. North Am.* 59: 1035–1038.

Chalmers, T.C. (1989). A belated randomized control trial. *Pediatrics* 85: 366–368.

Chalmers, T.C. (1991). Problems induced by meta-analyses. *Stat. Med.* 10: 971–980.

Chalmers, T.C. (1993). Meta-analytic stimulus for changes in clinical trials. *Statist. Meth. Med. Res.* 2: 161–172.

Chalmers, T.C., Matta, R.J., Smith, H., Jr., and Kunzler, A.M. (1977). Evidence favoring the use of anticoagulants in the hospital phase of acute myocardial infarction. *N. Engl. J. Med.* 297: 1091–1096.

Chastang, C., Byar, D.P., and Piantadosi, S. (1988). A quantitative study of the bias in estimating the treatment effect caused by omitting a balanced covariate in survival models. *Stat. Med.* 7(12): 1243–1255.

Chlebowski, R.T., Bulcavage, L., Grosvenor, M., et al. (1990). Hydrazine sulfate influence on nutritional status and survival in non-small cell lung cancer. *J. Clin. Oncol.* 8: 9–15.

Cho, M.K., and Bero, L.A. (1994). Instruments for assessing the quality of drug studies published in the medical literature. *JAMA* 272: 101–104.

Chop, R.M., and Silva, M.C. (1991). Scientific fraud: definitions, policies, and implications for nursing research. *J. Prof. Nurs.* 7: 166–171.

Chow, S.C., and Ki, F.Y.C. (1994). On statistical characteristics of quality of life assessment. *J. Biopharmaceutical Stat.* 1: 1–17.

Chow, S.C., and Ki, F.Y.C. (1996). Statistical issues in quality of life assessment. *J. Biopharmaceutical Stat.* 6: 37–48.

Chowdhury, A.M., Karim, F., Rohde, J.E., Ahmed, J., and Abed, F.H. (1991). Oral rehydration therapy: A community trial comparing the acceptability of homemade sucrose and cereal-based solutions. *Bull. WHO* 69: 229–234.

Civilized Software (1996). *MLAB Users Guide*. Bethesda, MD: Civilized Software.

Clark, G.M., Hilsenbeck, S.G., Ravdin, P.M., De Laurentiis, M., and Osborne, C.K. (1994). Prognostic factors: Rationale and methods of analysis and integration. *Breast Cancer Res. Treat.* 32: 105–112.

Cleophas, T.J.M., and Tavenier, P. (1995). Clinical trials in chronic diseases. *J. Clin. Pharmacol.* 35: 594–598.

Cobb, L.A., Thomas, G.I., Dillard, D.H. et al. (1959). An evaluation of internal-mammary-artery ligation by a double-blind technic. *N. Engl. J. Med.* 260: 1115–1118.

Cochran, W.G., and Cox, G.M. (1957). *Experimental Designs*, 2nd ed. New York: Wiley.

Cochrane, A.L. (1972). *Effectiveness and Efficiency: Random Reflections on Health Services*. London: Nuffield Provincial Hospital Trust. (Reprinted 1989 in association with the BMJ and in 1989 by *Cont. Clin. Trials* 10: 428–433.)

Coffey, D.S. (1978). *General Summary Remarks (Regarding the Workshop in Genitourinary Cancer Immunology, Iowa City, 1976)*. Department of Health, Education, and Welfare. National Cancer Institute Monograph 49. Publication No. (NIH) 78-1467.

Cohen, J. (1994). Clinical trial monitoring: Hit or miss? *Science* 264: 1534–1537.

Cohen, J., Marshall, E., and Taubes, G. (1995). Special news report: Conduct in science. *Science* 268: 1705–1718. See also the introduction by J. Benditt, p. 1705.

Cohn, I. (1994). Whither NCI and NSABP? (editorial). *Arch. Surg.* 129: 1005–1009.

Cole, W.H. (1976). *Opening Address: Spontaneous Regression of Cancer and the Importance of Finding Its Cause*. NCI Monograph No. 44: 5–9.

Coleman, W. (1987). Experimental physiology and statistical inferences. In L. Kruger, G. Gigerenzer, and M.S. Morgan (eds.), *The Probabilistic Revolution: Ideas in the Sciences*, Vol. 1. Cambridge: MIT Press.

Collett, D. (1994). *Modelling Survival Data in Medical Research*. London: Chapman and Hall.

Colvin, L.B. (1969). Metabolic fate of hydrazines and hydrazides. *J. Pharm. Sci.* 58: 1433–1443.

COMMIT (1995a). Community intervention trial for smoking cessation (COMMIT): I. Cohort results from a four-year community intervention. *Am. J. Public Health* 85: 183–192.

COMMIT (1995b). Community intervention trial for smoking cessation (COMMIT): II. Changes in adult cigarette smoking prevalence. *Am. J. Public Health* 85: 193–200.

Concato, J., Shah, N., and Horwitz, R.I. (2000). Randomized, controlled trials, observational studies, and the hierarchy of research designs. *N. Engl. J. Med.* 342, 1887–1892.

Connor, E.M., Sperling, R.S., Gelber, R., et al. (1994). Reduction of maternal–infant transmission of human immunodeficiency virus type 1 with zidovudine treatment. Pediatric AIDS Clinical Trials Group Protocol 076 Study Group. *N. Engl. J. Med.* 331: 1173–1180.

Cooper, J.D. (2004). What was wrong with the trial? What do we do now? General Thoracic Surgical Club, 17th Annual Meeting, March 11–14, Litchfield Park, Arizona.

Cooper, J.E. (1991). Balancing the scales of public interest: Medical research and privacy. *Med. J. Austr.* 155: 556–560.

Cornfield, J. (1966a). Sequential trials, sequential analysis, and the likelihood principle. *Am. Statistician.* 20: 18–23.

Cornfield, J. (1966b). A Bayesian test of some classical hypotheses with applications to sequential clinical trials. *J. Am. Stat. Assoc.* 61: 577–594.

Cornfield, J. (1969). The Bayesian outlook and its application. *Biometrics* 25: 617–657.

Cornfield, J. (1978). Randomization by group: A formal analysis. *Am. J. Epidem.* 108: 100–102.

Coronary Drug Project Research Group (1980). Influence of adherence to treatment and response of cholesterol on mortality in the Coronary Drug Project. *N. Engl. J. Med.* 303: 1038–1041.

Coughlin, S.S. (ed.) (1995). *Ethics in Epidemiology and Clinical Research: Annotated Readings*. Newton, MA: Epidemiology Resources, Inc.

Council for International Organizations of Medical Sciences (CIOMS) (1993). *International Ethical Guidelines for Biomedical Research Involving Human Subjects*. Geneva.

Cournand, A. (1977). The code of the scientist and its relationship to ethics. *Science* 198: 699–705.

Cowley, A.J., McEntegart, D.J., Hampton, J.R., et al. (1994). Long-term evaluation of treatment for chronic heart failure: A 1 year comparative trial of flosequinan and captopril. *Cardiovasc. Drugs Ther.* 8(6): 829–836.

Cox, D.R. (1972). Regression models and life-tables (with discussion). *J. Roy. Stat. Soc.* (B) 34: 187–220.

Cox, D.R., and Hinkley, D.V. (1974). *Theoretical Statistics*. London: Chapman and Hall.

Cox, D.R. (1953). Some simple tests for Poisson variates. *Biometrika* 40: 354–360.

Cox, D.R. (1958). *Planning of Experiments*. New York: Wiley.

Cox, D.R., and Oakes, D. (1984). *Analysis of Survival Data*. London: Chapman and Hall.

Croll, R.P. (1984) The noncontributing author: An issue of credit and responsibility. *Perspect. Biol. Med.* 27 (3): 401–407.

Culliton, B.J. (1983). Coping with fraud: The Darsee case. *Science* 220: 31–35.

Curd, M. and Cover, J.A. (eds.) (1998). In Science and pseudoscience. *Philosophy of Science: The Central Issues*. New York: Norton, ch. 1.

Cureton, E.E. (1968). Unbiased estimation of the standard deviation. *Am. Statistician.* 22: 22.

Curran, W.J., Jr., Scott, C.B., Horton, J., Nelson, J.S., Weinstein, A.S., Fischbach, A.J., Chang, C.H., Rotman, M., Asbell, S.O., Krisch, R.E., et al. (1993). Recursive partitioning analysis of prognostic factors in three Radiation Therapy Oncology Group malignant glioma trials. *J. Nat. Cancer Inst.* 85: 704–710.

Cuschieri, A., Fayers, P., Fielding, J., et al. (1996). Postoperative morbidity and mortality after D1 and D2 resections for gastric cancer: Preliminary results of the MRC randomised controlled surgical trial. *Lancet* 347: 995–999.

Cytel Software Corporation (1992). EaSt. A Software Package for the Design and Interim Monitoring of Group Sequential Clinical Trials. Cambridge, MA: Cytel Software.

Dalal, S.R., Fowlkes, E.B., and Hoadley, B. (1989). Risk analysis of the space shuttle: Pre-Challenger prediction of failure. *J. Am. Statist. Assoc.* 84: 945–957.

DAMOCLES Study Group. (2005). A proposed charter for clinical trial data monitoring committees: helping them to do their job well. *Lancet* 365: 711–722.

Danforth, W.H., and Schoenhoff, D.M. (1992). Fostering integrity in scientific research. *Acad. Med.* 67: 351–356.

Darsee, J.R., and Heymsfield, S.B. (1981). Decreased myocardial taurine levels and hypertaurinuria in a kindred with mitral-valve prolapse and congestive cardiomyopathy. *N. Engl. J. Med.* 304: 129–135.

Davis, S., Wright, P.W., Schulman, S.F., et al. (1985). Participants in prospective, randomized clinical trials for resected non–small cell lung cancer have improved survival compared with nonparticipants in such trials. *Cancer* 56: 1710–1718.

Dawson, L. (2004). The Salk Polio Vaccine Trial of 1954: Risks, randomization and public involvement in research. *Clin. Trials* 1: 122–130.

Day, N.E. and Walter, S.D. (1984). Simplified models of screening for chronic disease: Estimation procedures from mass screening programmes. *Biometrics* 40: 1–14.

De Angelis, C., et al. (2004). Clinical trial registration: a statement from the International Committee of Medical Journal Editors. *N. Engl. J. Med.* 351: 1250–1251.

Dear, K.B.G., and Begg, C.B. (1992). An approach for assessing publication bias prior to performing a meta-analysis. *Stat. Sci.* 7: 237–245.

Deep-Brain Stimulation for Parkinson's Disease Study Group (2001). Deep-brain stimulation of the subthalamic nucleus or the pars interna of the globus pallidus in Parkinson's disease. *N. Engl. J. Med.* 345: 956–963.

DeMets, D.L. (1990). Methodological issues in AIDS clinical trials. Data monitoring and sequential analysis – an academic perspective. *J. Acq. Immune Def. Synd.* 3(suppl. 2): S124–S133.

DeMets, D.L., and Lan, K.K. (1994). Interim analysis: The alpha spending function approach. *Stat. Med.* 13: 1341–1352.

Deming, W.E. (1986). Principles of professional statistical practice. In S. Kotz, N.L. Johnson, and C.B. Read (eds.), *Encyclopedia of Statistical Sciences*. New York: Wiley.

Denham, S.A. (1993). Stemming the tide of disreputable science: Implications for nursing. *Nurs. Forum* 28: 11–18.

Denzin, N., and Lincoln, Y.(Eds.) (2000). *Handbook of Qualitative Research*. Newbury Park, CA: Sage Publications.

Department of Health and Human Services (DHHS) (1989). Responsibilities of PHS awardee and applicant institutions for dealing with and reporting possible misconduct in science: Final rule. *Fed. Reg.* 54 (August 8): 32446–32451.

Department of Health and Human Services (DHHS) (1993a). NIH guideline for the study and evaluation of gender differences in the clinical evaluation of drugs: Notice. *Fed. Reg.* 58: 39406–39416.

Department of Health and Human Services (DHHS) (1993b). Findings of scientific misconduct: Roger Poisson, M.D., St. Luc Hospital, Montreal, Canada. *Fed. Reg.* 58(117): 33831.

Department of Health and Human Services (DHHS) (1994). NIH guidelines on the inclusion of women and minorities as subjects in clinical research. *Fed. Reg.* 59: 14508–14513.

Department of Health and Human Services (DHHS) (2000). Standards for privacy of individually identifiable health information: Final rule. *Fed. Reg.* 45 CFR parts 160 and 164, 82461–82510.

Department of Health, Education, and Welfare (1973). *Final Report of the Tuskegee Syphilis Study Ad hoc Advisory Panel*. Washington, DC: Government Printing Office.

DerSimonian, R., Charette, L.J., McPeek, B., and Mosteller, F. (1982). Reporting on methods in clinical trials. *N. Engl. J. Med.* 306: 1332–1337.

Desu, M.M., and Raghavarao, D. (1990). *Sample Size Methodology*. San Diego: Academic Press.

Detsky, A.S., and Sackett, D.L. (1985). When was a "negative" clinical trial big enough? How many patients you needed depends on what you found. *Arch. Int. Med.* 145: 709–712.

DeVita, V.T. (1991). Is a mechanism such as the NCI's Clinical Alert ever an appropriate alternative to journal peer review? *Important Advances in Oncology*: 241–254.

Dickersin, K. (1990). The existence of publication bias and risk factors for its occurrence. *JAMA* 263: 1385–1389.

Dickersin, K., and Berlin, J.A. (1992). Meta-analysis: State-of-the-science. *Epidemiol. Rev.* 14: 154–176.

Dickersin, K. and Manheimer, E. (1998). The Cochrane Collaboration: Evaluation of health care and services using systematic reviews of the results of randomized controlled trials. *Clin. Obstet. Gynecol.* 41(2): 315–331.

Dickersin, K., Min, Y-I., and Meinert, C.L. (1992). Factors influencing publication of research results. *JAMA* 267: 374–378.

Dickersin, K., and Rennie, D. (2003). Registering clinical trials. *JAMA* 290(4):516–523.

Diem, K., and Lentner, C. (eds.) (1970). *Scientific Tables*, 7th ed. Basel: CIBA-Geigy.

Diggle, P.J., Liang, K.Y., and Zeger, S.L. (1994). *Analysis of Longitudinal Data*. Oxford: Oxford University Press.

Dimick, J.B., Diener-West, M., and Lipsett, P.A. (2001). Negative results of randomized clinical trials published in the surgical literature: Equivalency or error? *Arch. Surg.* 136(7): 796–800.

Dimond, E.G., Kittle, C.F., and Cockett, J.E. (1960). Comparison of internal mammary artery ligation and sham operation for angina pectoris. *Am. J. Cardiol.* 5: 483–486.

Dixon, W.J., and Mood, A.M. (1948). A method for obtaining and analyzing sensitivity data. *J. Am. Statist. Assoc.* 43: 109–126.

Djulbegovic, B., and Clarke, M. (2001). Scientific and ethical issues in equivalence trials. *JAMA* 285: 1206–1208.

Dobson, A.J. (1983). *An Introduction to Statistical Modelling*. London: Chapman and Hall.

Dodge, Y., and Afsarinejad, K. (1985). Minimal 2 connected factorial experiments. *Comput. Stat. Data Anal.* 3: 187–200.

Doll, R. (1994). The use of meta-analysis in epidemiology: Diet and cancers of the breast and colon. *Nutr. Rev.* 52: 233–237.

Donner, A. (1984). Approaches to sample size estimation in the design of clinical trials—A review. *Stat. Med.* 3: 199–214.

Donner, A., and Klar, N. (1994). Methods for comparing event rates in intervention studies when the unit of allocation is a cluster. *Am. J. Epidemiol.* 140: 279–289.

Donner, A., and Klar, N. (2000). *Design and Analysis of Cluster Randomization Trials in Health Research*. London: Arnold.

Draper, N.R., and Smith, H. (1981). *Applied Regression Analysis*, 2nd ed. New York: Wiley.

Drazen, J.M. (2003). Controlling research trials. *N. Engl. J. Med.* 348: 1377–1380.

Drutskoy A., et al. (2004). Observation of radiative B $\longrightarrow$ phi K gamma decays. *Phys. Rev. Lett.* 92(5): 051801.

Duan-Zheng, X. (1990). *Computer Analysis of Sequential Medical Trials*. New York: Ellis Horwood.

Dupont, W.D. (1985). Randomized vs. historical clinical trials: Are the benefits worth the cost? *Am. J. Epid.* 122: 940–946.

Durrleman, S., and Simon, R. (1991). When to randomize? *J. Clin. Oncol.* 9: 116–122.

Dyson, F.J. (1993). Science in trouble. *Am. Scholar* 62: 513–525.

Early Breast Cancer Trialists' Collaborative Group (1990). *Treatment of Early Breast Cancer, Vol. 1: Worldwide Evidence 1985–1990*. Oxford: Oxford University Press.

Edgington, E.S. (1987). *Randomization Tests*, 2nd ed. New York: Dekker.

Edwards, A.W.F. (1972). *Likelihood*. Cambridge: Cambridge University Press.

Efron, B. (1971). Forcing a sequential experiment to be balanced. *Biometrika* 58: 403–417.

Efron, B., and Feldman, D. (1991). Compliance as an explanatory variable in clinical trials. *J. Am. Stat. Assoc.* 86: 9–17.

Eisenberg, D.M., Kessler, R.C., Foster, C., Norlock, F.E., Calkins, D.R., and Delbanco, T.L. (1993). Unconventional medicine in the United States: Prevalence, costs, and patterns of use. *N. Engl. J. Med.* 328(4): 246–252.

Eisenberg, L. (1977). The social imperatives of medical research. *Science* 198: 1105–1110.

Eisenhut, L.P. (1990). Universität prüft Anschuldigungen gegen Professorin. *Kolner Stadtanzeiger* (October 24).

Elashoff, J.D. (1995). *nQuery Advisor User's Manual*. Los Angeles: Dixon Associates.

Elks, M.L. (1993). The right to participate in research studies. *J. Lab. Clin. Med.* 122: 130–136.

Ellenberg, S.S., Fleming, T.R., and DeMets, D.L. (2003). *Data Monitoring Committees in Clincial Trials: A Practical Perspective*. Chichester: Wiley.

Ellenberg, S.S., and Hamilton, J.M. (1989). Surrogate endpoints in clinical trials: Cancer. *Stat. Med.* 8: 405–413.

Ellison, N.M., Byar, D.P., and Newell, G.R. (1978). Special report on Laetrile: The NCI Laetrile review. *N. Engl. J. Med.* 229: 549–552.

Elting, L.S., and Bodey, G.P. (1991). Is a picture worth a thousand medical words? A randomized trial of reporting formats for medical research data. *Meth. Inf. Med.* 30: 145–150.

Emanuel, E.J. (2003). What makes clinical research ethical? 2003 ASCO/AACR Workshop, Vail, CO.

Emanuel, E.J., Wendler, D., and Grady, C. (2000). What makes clinical research ethical? *JAMA* 283: 2701–2711.

Emerson, S.S., and Banks, P.L.C. (1994). Interpretation of a leukemia trial stopped early. In Ni. Lange et al. (eds.), *Case Studies in Biometry*. New York: Wiley, ch. 14.

Endophthalmitis Vitrectomy Study Group (1995). Results of the Endophthalmitis Vitrectomy Study. A randomized trial of immediate vitrectomy and of intravenous antibiotics for the treatment of postoperative bacterial endophthalmitis. *Arch. Ophthalmol.* 113: 1479–1496.

Engler, R.L., Covell, J.W., Friedman, P.J., Kitcher, P.S., and and Peters, R.M. (1987). Misrepresentation and responsibility in medical research. *N. Engl. J. Med.* 317: 1383–1389.

Ensign, L.G., Gehan, E.A., Kamen, D.S., and Thall, P.F. (1994). An optimal three-stage design for phase II clinical trials. *Stat. Med.* 13: 1727–1736.

Epstein, B. and Sobel, M. (1953). Life testing. *J. Am. Statist. Assoc.* 48: 486–502.

Everitt, B.S. (1989). *Statistical Methods for Medical Investigations*. New York: Oxford University Press.

Everitt, B.S. (1995). *The Cambridge Dictionary of Statistics in the Medical Sciences*. Cambridge: Cambridge University Press.

Exner, D.V., Dries, D.L., Domanski, M.J., and Cohn, J.N. (2001). Lesser response to angiotensin–converting–enzyme inhibitor therapy in black as compared with white patients with left ventricular dysfunction. *N. Engl. J. Med.* 344: 1351–1357.

Faden, R.R., and Beauchamp, T.L. (1986). *A History and Theory of Informed Consent*. New York: Oxford University Press.

Fairfield, K.M., Eisenberg, D.M., Davis, R.B., et al. (1998). Patterns of use, expenditures, and perceived efficacy of complementary and alternative therapies in HIV-infected patients. *Arch. Intern. Med.* 158(20): 2257–64.

Fang, J. (1972). *Mathematicians from Antiquity to Today*. Hauppauge, NY: Paideia Press.

Faraggi, D., and Simon, R. (1995). A neural network model for survival data. *Stat. Med.* 14: 73–82.

Faries, D. (1994). Practical modifications of the continual reassessment method for phase I cancer clinical trials. *J. Biopharm. Stat.* 4: 147–164.

Farrington, C.P., and Manning, G. (1990). Test statistics and sample size formulae for comparative binomial trials with null hypothesis of non-zero risk difference or non-unity relative risk. *Stat. Med.* 9: 1447–1454.

Farrington, C.P., and Miller, E. (2001). Vaccine trials. *Molecular Biotech.* 17: 43–58.

Fayerweather, W.E., Higginson, J., and Beauchamp, T.L. (eds.) (1991). Ethics in Epidemiology. *J. Clin. Epidemiol.* (suppl.) 44: 1S–151S.

Feingold, M., and Gillespie, B.W. (1996). Cross-over trials with censored data. *Stat. Med.* 15: 953–967.

Feinstein, A.R. (1991). Intention-to-treat policy for analyzing randomized trials: Statistical distortions and neglected clinical challenges. In J.A. Cramer and B. Spilker (eds.), *Patient Compliance in Medical Practice and Clinical Trials*. New York: Raven Press, ch. 28.

Feinstein, A.R. (1995). Meta-analysis: Statistical alchemy for the 21st century. *J. Clin. Epidemiol.* 48: 71–79.

Feldman, M., Richardson, C.T., and Fordtran, J.S. (1980a). Experience with sham feeding as a test for vagotomy. *Gastroenterology* 79: 792–795.

Feldman, M., Richardson, C.T., and Fordtran, J.S. (1980b). Effect of sham feeding on gastric acid secretion in healthy subjects and duodenal ulcer patients: Evidence for increased basal vagal tone in some ulcer patients. *Gastroenterology* 79: 796–800.

Feurer, I.D., Becker, G.J., Picus, D., Ramirez, E., Darcy, M.D., and Hicks, M.E. (1994). Evaluating peer reviews: Pilot testing of a grading instrument. *JAMA* 272: 98–100.

Filipini, G., et al. (2003a). Interferons in relapsing remitting multiple sclerosis: A systematic review. *Lancet* 361: 545–552.

Filipini, G., et al. (2003b). Interferons in relapsing remitting multiple sclerosis: Authors' reply (letter). *Lancet* 361: 1824–1825.

Finkelstein, D.M., and Schoenfeld, D.A. (eds.) (1995). *AIDS Clinical Trials*. New York: Wiley-Liss.

Finney, D.J. (1995). A statistician looks at met-analysis. *J. Clin. Epidemiol.* 48: 87–103.

Fisher, B., Costantino, J., Redmond, C., et al. (1989). A randomized clinical trial evaluating tamoxifen in the treatment of patients with node-negative breast cancer who have estrogen-receptor-positive tumors. *N. Engl. J. Med.* 320: 479–484.

Fisher, C.J., Jr., Agosti, J.M., Opal, S.M., et al., (1996). Treatment of septic shock with the tumor necrosis factor receptor: Fc fusion protein. The Soluble TNF Receptor Study Group. *N. Engl. J. Med.* 334: 1697–1702.

Fisher, L.D., Dixon, D.O., Herson, J., Frankowski, R.K., Hearron, M.S., and Peace, K.E. (1990). Intention-to-treat in clinical trials. In K.E. Peace (ed.), *Statistical Issues in Drug Research and Development*. New York: Marcel Dekker.

Fisher, R.A. (1918). The causes of human variability. *Eugen. Rev.* 10: 213–220.

Fisher, R.A. (1925). *Statistical Methods for Research Workers*. Edinburgh: Oliver and Boyd.

Fisher, R.A. (1935). *The Design of Experiments*. Edinburgh: Oliver and Boyd.

Fisher, R.A. (1936). Has Mendel's work been rediscovered? *An. Sci.* 1: 115–137.

Fisher, R.A. (1960). *The Design of Experiments*, 8th ed. New York: Hafner.

FitzGerald, G.A. (2004). Coxibs and cardiovascular disease. *N. Engl. J. Med.* 351: 1709–1711.

Flather, M., Pipilis, A., Collins, R., et al. (1994). Randomized controlled trial of oral captopril, of oral isosorbide mononitrate and of intravenous magnesium sulphate started early in acute myocardial infarction: Safety and haemodynamic effects. *Eur. Heart J.* 15: 608–619.

Fleiss, J.L. (1986a). *The Design and Analysis of Clinical Experiments*. New York: Wiley.

Fleiss, J.L. (1986b). Analysis of data from multiclinic trials. *Cont. Clin. Trials* 7: 267–275.

Fleiss, J.L. (1989). A critique of recent research on the two-treatment cross-over design. *Cont. Clin. Trials* 10: 237–243.

Fleming, T.R. (1994). Surrogate markers in AIDS and cancer trials. *Stat. Med.* 13: 1423–1435.

Fleming, T.R. (2000). Design and interpretation of equivalence trials. *Am. Heart J.* 139: S171–S176.

Fleming, T.R., Green, S.J., and Harrington D.P. (1984). Considerations for monitoring and evaluating treatment effects in clinical trials. *Cont. Clin. Trials* 5: 55–66.

Fleming, T.R. and DeMets, D.L. (1993). Monitoring of clinical trials: Issues and recommendations. *Cont. Clin. Trials* 14: 183–197.

Fleming, T.R., and DeMets, D.L. (1996). Surrogate end points in clinical trials: Are we being misled? *An. Intern. Med.* 125: 605–613.

Fleming, T.R., Prentice, R.L., Pepe, M.S., and Glidden, D. (1994). Surrogate and auxiliary endpoints in clinical trials with potential applications in cancer and AIDS research. *Stat. Med.* 13: 955–968.

Foesling, A. (1984). *Der Mogelfaktor: die Wissenschaftler und die Wahrheit. Hamburg: Rasch und Röhring* Hamburg, 20–21.

Food and Drug Administration. (1977). A report on the two-period crossover design and its applicability in trials of clinical effectiveness. Minutes of the Biometric and Epidemiology Methodology Advisory Committee (BEMAC) meeting.

Ford, I., and Norrie, J. (2002). The role of covariates in estimating treatment effects and risk in long-term clinical trials. *Stat. Med.* 21(19): 2899–2908.

Forrow, L., Taylor, W.C., and Arnold, R.M. (1992). Absolutely relative: How research results are summarized can affect treatment decisions. *Am. J. Med.* 92: 121–124.

Fox, R.C., and Swazey, J.P. (1992). *Spare Parts: Organ Replacement in American Society*. New York: Oxford University Press.

Frangakis, C.E., and Rubin, D.B. (1999). Addressing complications of intention-to-treat analysis in the combined presence of all-or-none treatment-noncompliance and subsequent missing outcomes. *Biometrika* 86: 365–379.

Frangakis, C.E., and Rubin, D.B. (2002). Principal stratification in causal inference. *Biometrics* 58(1): 21–29.

Frank, E. (1995). The Women Physicians' Health Study: Background, objectives, and methods. *J. Am. Med. Womens Assoc.* 50: 64–66.

Frazier, K. (2002). Study finds NCCAM grants have produced no useful results. *Skeptical Inq.* 26: 6–7.

Freed C.R., Breeze R.E., Fahn S. (2000). Placebo surgery in trials of therapy for Parkinson's disease. *N. Engl. J. Med.*: 342(5):353–354; Discussion 354-355.

Freed C.R., Greene P.E., Breeze R.E., et al. (2001). Transplantation of embryonic dopamine neurons for severe Parkinson's disease. *N. Engl. J. Med.* 344(10): 710–719.

Freedman, B. (1987). Equipoise and the ethics of clinical research. *N. Engl. J. Med.* 317: 141–145.

Freedman, B. (1992). A response to a purported ethical difficulty with randomized clinical trials involving cancer patients. *J. Clin. Ethics* 3(3): 231–234.

Freedman, L.S. (1982). Tables of the number of patients required in clinical trials using the logrank test. *Stat. Med.* 1: 121–129.

Freedman, L.S. and Spiegelhalter, D.J. (1989). Comparison of Bayesian with group sequential methods for monitoring clinical trials. *Cont. Clin. Trials* 10: 357–367.

Freedman, L.S., Spiegelhalter, D.J., and Parmar, M.K. (1994). The what, why and how of Bayesian clinical trials monitoring. *Stat. Med.* 13: 1371–1383.

Freedman, L., Sylvester, R., and Byar, D.P. (1989). Using permutation tests and bootstrap confidence limits to analyze repeated events data from clinical trials. *Cont. Clin. Trials* 10: 129–141.

Freedman, L.S., and Green, S.B. (1990). Statistical designs for investigating several interventions in the same study: Methods for cancer prevention trials. *JNCI* 82(11): 910–914.

Freedman, L.S. (1989). The size of clinical trials in cancer research: What are the current needs? *Br. J. Cancer* 59: 396–400.

Freedman, L.S., and Schatzkin, A. (1992). Sample size for studying intermediate endpoints within intervention trials or observational studies. *Am. J. Pub. Health* 136: 1148–1159.

Freeman, L. (2004). *Mosby's Complementary and Alternative Medicine: A Research-Based Approach.* St. Louis: Mosby.

Freeman, P.R. (1989). The performance of the two-stage analysis of two-treatment, two-period crossover trials. *Stat. Med.* 8: 1421–1432.

Freeman, T.B., Vawter, D.E., Leaverton, P.E., et al. (1999). Use of placebo surgery in controlled trials of a cellular-based therapy for Parkinson's disease. *N. Engl. J. Med.* 341: 988–992.

Freestone, D.S., and Mitchell, H. (1993). Inappropriate publication of trial results and potential for allegations of illegal share dealing. *BMJ.* 306(6885): 1112–1114.

Frei, E. III (1982). Clinical cancer research: An embattled species. *Cancer* 50: 1979–1992.

Frei, E., III, and Freireich, E.J. (1993). The clinical cancer researcher—Still an embattled species. *J. Clin. Oncol.* 11: 1639–1651.

Frei E. (1998). Therapeutic innovation: The up-front window. *Clin. Cancer Res.* 4(11): 2573–2575.

Freidman, L.M., Furberg, C.D., and DeMets, D.L. (1998). *Fundamentals of Clinical Trials*, 3rd ed. New York: Springer-Verlag.

Freiman, J.A., Chalmers, T.C., Smith, H., Jr., and Kuebler, R.R. (1978). The importance of beta, the type II error and sample size in the design and interpretation of the randomized control trial: Survey of 71 "negative" trials. *N. Engl. J. Med.* 299: 690–694.

Freund, P.A. (ed.) (1970). *Experimentation with Human Subjects.* New York: Braziller.

Fried, C. (1974). Medical experimentation: Personal integrity and social policy. Vol. 5 In Amsterdam: North-Holland.

Friedlander, M.W. (1995). *At the Fringes of Science.* Boulder, CO: Westview Press.

Friedman, M.A. (1990). If not now, when? JNCI 82: 106–108.

Frisell, J., Eklund, G., Hellstrom, L., Glas, U., and Somell, A. (1989). The Stockholm breast cancer screening trial—5-year results and stage at discovery. *Breast Cancer Res. Treat.* 13(1): 79–87.

Fuchs, S. and Westervelt, S.D. (1995). Fraud and trust in science. *Perspect. Biol. Med.* 39: 248–269.

Fugh-Berman, A. (1992). Breast cancer prevention study: Are healthy women put at risk by federally funded research? Hearing before the Human Resources and Intergovernmental Relations Subcommittee of the Committee on Government Operations House of Representatives, 102nd Congress, October 22, 1992.

Funding First. (2000). *Exceptional Returns: The Economic Value of America's Investment in Medical Research.* New York, NY.

Furberg, C.D., Psaty, B.M., and Meyer, J.V. (1995). Nifedipine: Dose-related increase in mortality in patients with coronary heart disease. *Circulation* 92: 1326–1331.

GAO (1995). Cancer Drug Research: Contrary to Allegation, NIH Hydrazine Sulfate Studies Were Not Flawed. Washington, DC: General Accounting Office. Publication GAO/HEHS-95-141.

Gail, M.H. (1974). Power computations for designing comparative Poisson trials. *Biometrics* 30: 231–237.

Gail, M.H. (1984). Nonparametric frequentist proposals for monitoring comparative survival studies. *Handbook Stat.* 4: 791–811.

Gail, M.H. (1982). Monitoring and stopping clinical trials. In V. Miké and K. Stanley (eds.), *Statistics in Medical Research.* New York: Wiley, ch. 15.

Gail, M.H. (1985). Eligibility exclusions, losses to follow-up, removal of randomized patients, and uncounted events in cancer clinical trials. *Cancer Treat. Rep.* 69(10): 1107–1112.

Gail, M.H. (1996). Use of observational data, including surveillance studies, for evaluating AIDS therapies. *Stat. Med.* 15: 2273–2288.

Gail, M.H., Tan, W.-Y., and Piantadosi, S. (1988). The size and power of tests for no treatment effect in randomized clinical trials when needed covariates are omitted. *Biometrika* 75: 57–64.

Gail, M.H., Wieand, H.S., and Piantadosi, S. (1984). Biased estimates of treatment effect in randomized experiments with non-linear regressions and omitted covariates. *Biometrika* 71(3): 431–444.

Gail, M.H., Mark, S.D., Carroll, R.J., Green, S.B., and Pee, D. (1996). On design considerations and randomization-based inference for community intervention trials. *Stat. Med.* 15: 1069–1092.

Galton, F. (1899). Strawberry cure for gout. *Nature* 60: 125.

Gangarosa, E.J., Galazka, A.M., Wolfe, C.R., et al. (1998). Impact of anti-vaccine movements on pertussis control: The untold story. *Lancet* 351: 356–361.

Garceau, A.J., Donaldson, R.M., Jr., O'Hara, E.T., Callow, A.D., Muench, H., and Chalmers, T.C. (1964). A controlled trial of prophylactic portacaval-shunt surgery. *N. Engl. J. Med.* 270: 496–500.

Gardiner, B.G. (2003). The Piltdown forgery: A re-statement of the case against Hinton. *Zoological J. Linnean Soc.* 139: 315–335.

Gardiner, B.G., and Currant, A. (1996). The Piltdown Hoax: Who Done It? Presented at the Linnean Society of London.

Gardner, M.J., Machin, D., and Campbell, M.J. (1986). Use of checklists in assessing the statistical content of medical studies. *BMJ* 292: 810–812.

Garrett, A.D. (2003). Therapeutic equivalence: Fallacies and falsification. *Stat. Med.* 22(5): 741–762.

Gatsonis, C., and Greenhouse, J.B. (1992). Bayesian methods for phase I clinical trials. *Stat. Med.* 11: 1377–1389.

Gavarret, J. (1840). *Principes généraux de statistique médicale, ou, dévelopement des règles qui doivent présider à son emploi.* Paris: Bechet jeune et Labé.

Gehan, E.A. (1961). The determination of the number of patients required in a preliminary and follow-up trial of a new chemotherapeutic agent. *J. Chron. Dis.* 13: 346–353.

Gehan, E.A., and Lemak, N.A. (1994). *Statistics in Medical Research: Developments in Clinical Trials.* New York: Plenum.

Gehan, E.A., and Schneiderman, M.A. (1990). Historical and methodological developments in clinical trials at the National Cancer Institute. *Stat. Med.* 9: 871–880.

Gehlbach, S.H. (1988). *Interpreting the Medical Literature: Practial Epidemiology for Clinicians*, 2nd ed. New York: Macmillan.

Gelber, R.D. (1996). Gemcitabine for pancreatic cancer: How hard to look for clinical benefit? An American perspective. *An. Oncol.* 7: 335–337.

Geller, N.L. (1987). Planned interim analysis and its role in cancer clinical trials. *J. Clin. Onc.* 5: 1485–1490.

George, S.L., and Desu, M.M. (1974). Planning the size and duration of a clinical trial studying the time to some critical event. *J. Chron. Dis.* 27: 15–24.

George, S.L. (1988). Identification and assessment of prognostic factors. *Sem. Onc.* 15: 462–471.

Gesensway, D. (2001). Reasons for sex-specific and gender-specific study of health topics. *An. Int. Med.* 135: 935–938.

Giardiello, F.M., Hamilton, S.R., Krush, A.J., Piantadosi, S., Hylind, L.M., Celano, P., Booker, S.V., Robinson, C.R., and Offerhaus, G.J.A. (1993). Treatment of colonic and rectal adenomas with Sulindac in familial adenomatous polyposis. *N. Engl. J. Med.* 328: 1313–1316.

Gilbert, P.B. (2000). Some statistical issues in the design of HIV-1 vaccine and treatment trials. *Stat. Meth. Med. Res.* 9(3): 207–229.

Gilbert, J.P., McPeek, B., and Mosteller, F. (1977). Statistics and ethics in surgery and anesthesia. *Science* 198: 684–689.

Gilbert, M.R., Supko, J.G., Batchelor, T., et al. (2003). Phase I clinical and pharmacokinetic study of irinotecan in adults with recurrent malignant glioma. *Clin. Cancer Res.* 9: 2940–2949.

Gillings, D.B., and Koch, G. (1991). The application of the principle of intention-to-treat in the analysis of clinical trials. *Drug Inf. J.* 25: 411–424.

Gilpin, A.K., and Meinert, C.L. (1994). Gender bias in clinical trials? *Proceedings of the Society for Clinical Trials*, Houston, TX.

Girling, D.J., Parmar, M.K.B., Stenning, S.P., Stephens, R.J., and Stewart, L.A. (2003). *Clinical Trials in Cancer: Principles and Practice.* Oxford: Oxford University Press.

Glass, G.V. (1976). Primary, secondary, and meta-analysis of research. *Educ. Res.* 5: 3–8.

Gohagan, J.K., Prorok, P.C., Kramer, B.S., and Cornett, J.E. (1994). Prostate cancer screening in the prostate, lung, colorectal and ovarian cancer screening trial of the National Cancer Institute. *J. Urol.* 152: 1905–9. Comment in: *J. Urol.* (1994) 152: 1903–1904.

Gohagan, J.K., Prorok, P.C., Hayes, R.B., Kramer, B.S., and Prostate, Lung, Colorectal and Ovarian Cancer Screening Trial Project Team (2000). The Prostate, Lung, Colorectal and Ovarian (PLCO) Cancer Screening Trial of the National Cancer Institute: History, organization, and status. *Cont. Clin. Trials* 21: 251S–272S.

Gold, J. (1975). Use of hydrazine sulfate in terminal and preterminal cancer patients: Results of investigational new drug (IND) study in 84 evaluable patients. *Oncology* 32: 1–10.

Goldberg, K.B., and Goldberg, P. (eds.) (1994). NCI apologizes for mismanagement of NSABP, says Fisher resisted criticism. *Cancer Lett.* 20 (16): April 22.

Goldstein, R. (1989). Power and sample size via MS/PC-DOS computers. *Am. Statistician* 43: 253–260.

Good, P. (1994). *Permutation Tests*. New York: Springer- Verlag.

Goodman, S.N., Zahurak, M.L., and Piantadosi, S. (1995). Some practical improvements in the continual reassessment method for phase I studies. *Stat. Med.* 14: 1149–1161.

Gordon, N.H., and Wilson, J.K. (1992). Using toxicity grades in the design and analysis of cancer phase I clinical trials. *Stat. Med.* 11: 2063–2075.

Gore, S.M., Jones, G., and Thompson, S.G. (1992). The Lancet's statistical review process: Areas for improvement by authors. *Lancet* 340: 100–102.

Gotay, C.C. (1991). Accrual to cancer clinical trials: directions from the research literature. *Soc. Sci. Med.* 33: 569–577.

Gøtzsche, P.C. (1989). Methodology and overt and hidden bias in reports of 196 double-blind trials of nonsteroidal anti-inflammatory drugs in rheumatoid arthritis. *Cont. Clin. Trials* 10: 31–56. Correction (1989) 10: 356.

Govindarajulu, Z. (2001). *Statistical Techniques in Bioassay*, 2nd ed. Basel: Karger.

Grady D., et al. (1992). Hormone therapy to prevent disease and prolong life in postmenopausal women. *An. Intern. Med.* 117: 1016–1037.

Graham, N.M.H., Park, L.P., Piantadosi, S., Phair, J.P., Mellors, J., Fahey, J.L., and Saah, A.J. (1994). Prognostic value of combined response markers among human immunodeficiency virus infected persons: Possible aid in the decision to change zidovudine monotherapy. *Clin. Infect. Dis.* 20: 352–362.

Grant, A. (1989). Reporting controlled trials. *Br. J. Obstet. Gynaecol.* 96: 397–400.

Grant, A., et al. (2001). The Ipswich childbirth study: One year follow up of alternative methods used in perineal repair. *BJOG* 108: 34–40.

Gray, J.N., Lyons, P.M., Jr., and Melton, G.B. (1995). *Ethical and Legal Issues in AIDS Research*. Baltimore: Johns Hopkins University Press.

Greco, D., Salmaso, S., Mastrantonio, P., and Pertosse, P. (1994). The Italian pertussis vaccine trial: Ethical issues (letter). *JAMA* 272: 1898–1899.

Green, A., Battistitta, D., Hart, V., et al. (1994). The Nambour skin cancer and actinic eye disease prevention trial: Design and baseline characteristics of participants. *Cont. Clin. Trials* 15: 512–522.

Green, S. (2001). Stated goals and grants of the office of Alternative Medicine/National Center for Complementary and Alternative Medicine Policy. *Scientific Review of Alternative Medicine* 5:205–207.

Green, S.B., and Byar, D.P. (1984). Using observational data from registries to compare treatments: The fallacy of omnimetrics. *Stat. Med.* 3: 361–370.

Green, S., Benedetti, J., and Crowley, J. (2002). *Clinical Trials in Oncology*, 2nd ed. London: Chapman and Hall.

Green, S.J., Fleming, T.R., and O'Fallon, J.R. (1987). Policies for study monitoring and interim reporting of results. *J. Clin. Onc.* 5: 1477–1484.

Greenberg, E.R., Baron, J.A., Tosteson, T.D., et al. (1994). A clinical trial of antioxidant vitamins to prevent colorectal adenoma. *N. Engl. J. Med.* 331: 142–147.

Greenhouse, S.W. (1990). Some historical and methodological developments in early clinical trials at the National Institutes of Health. *Stat. Med.* 9: 893–901.

Greenland, S. (1987). Quantitative methods in the review of the epidemiologic literature. *Epidem. Rev.* 9: 1–30.

Greenland, S. (1994). A critical look at some popular meta-analytic methods (with discussion). *Am. J. Epidemiol.* 140: 290–296.

Greenland, S., and Salvan, A. (1990). Bias in the one-step method for pooling study results. *Stat. Med.* 9: 247–252.

Greenwald, P. (1985). Prevention of Cancer. In V.T. DeVita, S. Hellman, and S.A. Rosenberg (eds.) *Cancer: Principles and Practice of Oncology*, 2nd ed. Philadelphia: JB Lippincott, ch. 10.

Greenwald, P., and Cullen, J.W. (1985). The new emphasis in cancer control. *J. Nat. Cancer Inst.* 74: 543–551.

Greenwood, M. (1926). The errors of sampling of the survivorship tables. In *Reports on Public Health and Statistical Subjects*, no. 33. London: HMSO, Appendix 1.

Grender, J.M., and Johnson, W.D. (1993). Analysis of crossover designs with multivariate response. *Stat. Med.* 12: 69–89.

Grieve, A.P. (1982). The two-period changeover design in clinical trials. *Biometrics* 38(2): 517.

Grieve, A.P. (1985). A Bayesian analysis of the two period cross-over design for clinical trials. *Biometrics* 42: 979–990. Corrigenda 42: 459 (1986).

Grimes, D.A. (1993). Technology follies: The uncritical acceptance of medical innovation. *JAMA* 269: 3030–3033.

Grizzle, J.E. (1965). The two-period change-over design and its use in clinical trials. *Biometrics* 21: 467–480. Corrigenda 30: 727, (1965).

Groninger, E., De Graaf, S.S., Meeuwsen-De Boer, G.J., Sluiter, W.J., and Poppema, S. (2000). Short Report: Vincristine-induced apoptosis in vivo in peripheral blood mononuclear cells of children with acute lymphoblastic leukaemia (ALL). *Br. J. Haematol.* 111(3): 875–878.

Gross, A.J., and Clark, V.A. (1975). *Survival Distributions: Reliability Applications in the Biomedical Sciences*. New York: Wiley.

Grossman, S.A., Sheidler, V.A., Swedeen, K., Mucenski, J., and Piantadosi, S. (1991). Correlation of patient and caregiver ratings of cancer pain. *J. Pain Symptom Manage.* 6(2): 53–57.

Grouin, J.-M., Day, S., and Lewis, J. (2004). Adjustment for baseline covariates: an introductory note. *Stat. Med.* 23: 697–699.

Grunbaum A. (1986). The placebo concept in medicine and psychiatry. *Psychol. Med.* 16(1): 19–38.

Guttentag, O.E. (1953). The problem of experimentation on human beings: II The physician's point of view. *Science* 117: 207–210.

Haaland, P.D. (1989). *Experimental Design in Biotechnology*. New York: Marcel Dekker.

Hadorn, D.C., Draper, D., Rogers, W.H., Keeler, E.B., and Brook, R.H. (1992). Cross-validation performance of mortality prediction models. *Stat. Med.* 11: 475–489.

Halm, E.A., Lee, C., Chassin, M.R. (2002). Is volume related to outcome in healthcare? A systematic review and methodologic critique of the literature. *An. Intern. Med.* 137: 511–520.

Halperin, M. (1952). Maximum likelihood estimation in truncated samples. *An. Math. Stat.* 23: 226–238.

Halperin, M., DeMets, D.L., and Ware, J.H. (1990). Early methodological developments for clinical trials at the National Heart, Lung, and Blood Institute. *Stat. Med.* 9: 881–892.

Halpern, D.F., and Coren, S. (1991). Handedness and life span (letter). *N. Engl. J. Med.* 324: 998. See also letters in 325: 1041–1043.

Hamilton, M., Pickering, G.W., Roberts, J.A.F., and Sowry, G.S.C. (1954). The aetiology of essential hypertension (1): The arterial pressure in the general population. *Clin. Sci.* 13: 11–35.

Hand, D.J. (1992). Statistical methods in diagnosis. *Stat. Meth. Med. Res.* 1: 49–67.

Hand, D.J., Daly, F., Lunn, A.D., McConway, K.J., and Ostrowski, E. (eds.) (1994). *A Handbook of Small Data Sets.* London: Chapman and Hall.

Hankins, F.H. (1930). Gottfried Achenwall. In E.R.A. Seligman and A. Johnson (eds.), *Encyclopaedia of the Social Sciences.* New York: Macmillan.

Hansen, B.C., and Hansen, K.D. (1995). Academic and scientific misconduct: Issues for nursing educators. *J. Prof. Nurs.* 11: 31–39.

Hansen, K.D., and Hansen, B.C. (1991). Scientific fraud and the Public Health Service Act: A critical analysis. *News* 5: 2512–2515.

Harrell, F.E., Jr.,, Lee, K.L., and Mark, D.B. (1996). Multivariable prognostic models: Issues in developing models, evaluating assumptions and adequacy, and measuring and reducing errors. *Stat. Med.* 15: 361–387.

Harris, E.K., and Albert, A. (1991). *Survivorship Analysis for Clinical Studies.* New York: Marcel Dekker.

Harris, N.L., Gang, D.L., Quay, S.C., Poppema, S., Zamecnik, P.C., Nelson-Rees, W.A., and O'Brien, S.J. (1981). Contamination of Hodgkin's disease cell cultures. *Nature* 289: 228–230.

Hawkins, B.S. (1991). Controlled clinical trials in the 1980s: A bibliography. *Cont. Clin. Trials* 12: 5–272.

Haynes, R.B., Mulrow, C.D., Huth, E.J., Altman, D.G., and Gardner, M.J. (1990). More informative abstracts revisited. *An. Intern. Med.* 113: 69–76.

Healy, B. (1996). The dangers of trial by Dingell. *New York Times,* July 3.

Hedges, L.V. (1992). Modeling publication selection effects in meta-analysis. *Stat. Sci.* 7: 246–255.

Hedges, L.V. and Olkin, I. (1985). *Statistical Methods for Meta-Analysis.* Orlando, FL: Academic Press.

Heiberger, R.M. (1989). *Computation for the Analysis of Designed Experiments.* New York: Wiley.

Heinonen, O.P., Virtamo, J., Albanes, D., et al. (1987). Beta carotene, alpha-tocopherol lung cancer intervention trial in Finland. In *Proceedings of the XI Scientific Meeting of the International Epidemiologic Association,* Helsinki, August, 1987. Helsinki: Pharmy.

Heiss, M.M., Mempel, W., Delanoff, C., et al. (1994). Blood transfusion-modulated tumor recurrence: First results of a randomized study of autologous versus allogeneic blood transfusion in colorectal cancer surgery. *J. Clin. Oncol.* 12: 1859–1867.

Heitjan, D.F., Houts, P.S., and Harvey , H.A. (1992). A decision-theoretic evaluation of early stopping rules. *Stat. Med.* 11: 673–683.

Hellman, S. (1991). Clinical Alert: A poor idea prematurely used. *Important Advances in Oncology*: 255–257.

Hellman, S., and Hellman, D.S. (1991). Of mice but not men: Problems of the randomized clinical trial. *N. Engl. J. Med.* 324: 1585–1589.

Henderson, I.C. (1995). Using clinical trial information in the practice of medicine. *Cancer J. Sci. Am.* 1: 101.

Henderson, I.C. (1990). Point/counterpoint. NCI Clinical Alert policy. Shouldn't we see the white flag before we cry victory? *JNCI* 82: 103–109.

Hennekens, C.H., and Eberlein, K. (1985). A randomized trial of aspirin and beta-carotene among U.S. physicians. *Prev. Med.* 14: 165–168.

Herbert, V. (1977). Acquiring new information while retaining old ethics. *Science* 198: 690–693.

Herbert, V. (1994). Three stakes in hydrazine sulfate's heart, but questionable cancer remedies, like vampires, always rise again (editorial). *J. Clin. Oncol.* 12: 1107–1108.

Herdan, G. (1955). *Statistics of Therapeutic Trials.* Amsterdam: Elsevier.

Herson, J. (1984). Statistical aspects in the design and analysis of phase II clinical trials. In M.J. Buyse, M.J. Staquet, and R.J. Sylvester (eds.), *Cancer Clinical Trials.* Oxford: Oxford University Press, ch. 15.

Herson, J. (1989). The use of surrogate endpoints in clinical trials (an introduction to a series of four papers). *Stat. Med.* 8: 403–404.

Herson, J., and Wittes, J. (1993). The use of interim analysis for sample size adjustment. *Drug Info. J.* 27: 753–760.

Herxheimer, A. (1993). Clinical trials: Two neglected ethical issues. *J. Med. Ethics* 19: 211 and 218.

Hill, A.B. (1961). *Principles of Medical Statistics*, 7th ed. London: The Lancet.

Hill, A.B. (1963). Medical ethics and controlled trials. *Br. Med. J.* April 20: 1043–1049.

Hill, T.P. (1995). The significant-digit phenomenon. *Am. Math. Monthly* 102: 322–327.

Hillis, A., and Seigel, D. (1989). Surrogate endpoints in clinical trials: Ophthalmologic disorders. *Stat. Med.* 8: 427–430.

Hills, M., and Armitage, P. (1979). The two-period cross-over clinical trial. *Br. J. Clin. Pharmacol.* 8: 7–20.

Hinkelmann, K., and Kempthorne, O. (1994). *Design and Analysis of Experiments: Vol. I: Introduction to Experimental Design.* New York: Wiley.

Hintze, J.L. (2002). *PASS User's Guide II: Power Analysis and Sample Size for Windows.* Kaysville, UT: NCSS.

Hirschfeld, S., and Pazdur, R. (2002). Oncology drug development: United States Food and Drug Administration perspective. *Crit. Rev. Oncol./Hemat.* 42: 137–143.

Hoel, D.G., Sobel, M., and Weiss, G.H. (1975). A survey of adaptive sampling for clinical trials. In R.M. Elashoff (ed.), *Perspectives in Biometrics.* New York: Academic Press.

Holland, P. (1986). Statistics and causal inference (with discussion). *J. Am. Stat. Assoc.* 81: 945–970.

Holtzman, W.H. (1950). The unbiased estimate of the population variance and standard deviation. *Am. J. Psychol.* 63: 615–617.

Homans, G. (1965). Group factors in worker productivity. In H. Proshansky and B. Seidenberg (eds.). *Basic Studies in Social Psychology.* New York: Holt, Rinehart, and Winston, pp. 592–604.

Horng, S., and Miller, F.G. (2002). Is placebo surgery unethical? *N. Engl. J. Med.* 347: 137–139.

Horowitz, L.G. (2001). Polio, hepatitis B and AIDS: An integrative theory on a possible vaccine induced pandemic. *Med. Hypotheses* 56(5): 677–686.

Horstmann, E., McCabe, M.S., Grochow, L., et al. (2005). Risks and benefits of phase 1 oncology trials, 1991 through 2002. *New Eng. J. Med.* 352: 895–904.

Horton, R. (1996). Surgical research or comic opera: Questions, but few answers. *Lancet* 347: 984–985.

Hoyt, W.J., Jr., (2004). Anti-vaccination fever: The shot hurt around the world. *Skeptical Inq.* 28: 21–25.

Hubbard, S.M., Martin, N.B., and Thurn, A.L. (1995). NCI's cancer information systems— Bringing medical knowledge to clinicians. *Oncology* 9: 302–314.

Hughes, M.D. (1993). Stopping guidelines for clinical trials with multiple treatments. *Stat. Med.* 12: 901–915.

Hulstaert, F., Van Belle, S., Bleiberg, H., et al. (1994). Optimal combination therapy with tropisetron in 445 patients with incomplete control of chemotherapy-induced nausea and vomiting. *J. Clin. Oncol.* 12: 2439–2446.

Humber, J.M., and Almeder, R.F. (eds.) (1998). *Alternative Medicine and Ethics*. Totowa, NJ: Humana Press.

Humphrey, G.F. (1992). Scientific fraud: The McBride case. *Med. Sci. Law* 32: 199–203.

Hutt, P.B. (1989). A history of government regulation of adulteration and misbranding of medical devices. *Food Drug Cosmetic Law J.* 44: 99–117.

Huxley, T.H. (1880). *The Crayfish, An Introduction to the Study of Zoology*. London: Kegan Paul.

Hwang, I.K., Shih, W.J., and DeCani, J.S. (1990). Group sequential designs using a family of type I error probability spending functions. *Stat. Med.* 9: 1439–1445.

Hyman, N.H., Foster, R.S., DeMeules, J.E., and Costanza, M.C. (1985). Blood transfusions and survival after lung cancer resection. *Am. J. Surg.* 149: 502–507.

ISIS-2 Collaborative Group. (1988). Randomised trial of intravenous streptokinase, oral aspirin, both, or neither among 17,187 cases of suspected acute myocardial infarction: ISIS-2. *Lancet* 2: 349–360.

ISIS-4 Collaborative Group. (1995). ISIS-4: A randomised factorial trial assessing early captopril, oral mononitrate, and intravenous magnesium sulphate in 58,050 patients with suspected acute myocardial infarction. *Lancet* 345: 669–685.

Inspector General Act Amendments. (1988). Public Law 100-504 (102 Stat. 2515).

Institute of Medicine. (IOM) (1989). *The Responsible Conduct of Research in the Health Sciences*. Washington, DC: National Academy Press.

International Agency for Research on Cancer. (IARC) (1974). *Evaluation of Carcinogenic Risk of Chemicals to Man: Some Aromatic Amines, Hydrazine and Related Substances, N-Nitroso Compounds and Miscellaneous Alkylating Agents*. Vol. 4. Lyon: IARC.

International Chronic Granulomatous Disease Cooperative Study Group. (1991). A controlled trial of interferon gamma to prevent infection in chronic granulomatous disease. *N. Engl. J. Med.* 324: 509–516.

International Committee of Medical Journal Editors. (1997). Uniform requirements for manuscripts submitted to biomedical journals. *N. Engl. J. Med.* 336: 309–315.

International Conference on Harmonisation. (1994). *Good Clinical Practice Guideline for Essential Documents for the Conduct of a Clinical Trial*. Geneva: ICH Secretariat (c/o IFPMA).

International Conference on Harmonisation. (1996). *Structure and Content of Clinical Study Reports. Draft Consensus Guideline*. Geneva: ICH Secretariat (c/o IFPMA).

Ioannidis, J.P., Cappelleri, J.C., Sacks, H.S., and Lau, J. (1997). The relationship between study design, results, and reporting of randomized clinical trials of HIV infection. *Cont. Clin. Trials* 18(5): 431–444.

Ivy, A.C. (1948). The history and ethics of the use of human subjects in medical experiments. *Science* 108: 1–5.

Iwane, M., Palensky, J., Plante, K. (1997). A user's review of commercial sample size software for design of biomedical studies using survival data. *Cont. Clin. Trials* 18: 65–83.

Iyengar, S., and Greenhouse, J.B. (1988). Selection models and the file drawer problem (with discussion). *Stat. Sci.* 3: 109–135.

Jabs, D., Enger, C., and Bartlett, J.G. (1989). Cytomegalovirus retinitis and acquired immunodeficiency syndrome. *Arch. Ophthal.* 107: 75–80.

Jacobs, L.D., et al. (1996). Intramuscular interferon beta-1a for disease progression in relapsing multiple sclerosis. *An. Neurol* 39: 285–294.

Jarvill-Taylor, K.J., Anderson, R.A., and Graves, D.J. (2001). A hydroxychalcone derived from cinnamon functions as a mimetic for insulin in 3T3-L1 adipocytes *J. Am. Coll. Nutr.* 20(4): 327–336.

Jayaraman, K.S. (1991). Gupta faces suspension. *Nature* 349: 645.

Jaynes, E.T. (1975). Confidence Intervals vs Bayesian Intervals. In W.L. Harper and C.A. Hooker (eds.), *Foundations of Probability Theory, Statistical Inference, and Statistical Theories of Science. Vol. II: Foundations and Philosophy of Statistical Inference.* Dordrecht: Reidel, pp. 175–257.

Jenks, S., and Volkers, N. (1992). Razors and refrigerators and reindeer—Oh my! *J. Nat. Cancer Inst.* 84: 1863.

Johns Hopkins University School of Medicine. (2004). *Procedures for Dealing with Issues of Research Misconduct, in Faculty Policies.* Baltimore: Johns Hopkins University School of Medicine, pp. 15–23.

Johnson, J.R., Williams, G., and Pazdur, R. (2003). End points and United States Food and Drug Administration approval of oncology drugs. *J. Clin. Oncol.* 21: 1404–1411.

Johnson, N.L., and Kotz, S. (1969). *Distributions in Statistics, Discrete Distributions.* New York: Wiley.

Johnson, N.L., and Kotz, S. (1970). *Continuous Univariate Distributions.* Boston: Houghton Mifflin.

Jones, B., and Donev, A.N. (1996). Modelling and design of cross-over trials. *Stat. Med.* 15: 1435–1446.

Jones, B., and Kenward, M.G. (1989). *Design and Analysis of Cross-over Trials.* London: Chapman and Hall.

Jones, D.R. (1995). Meta-analysis: Weighing the evidence. *Stat. Med.* 14: 137–149.

Jones, J.H. (1981). *Bad Blood: The Tuskegee Syphilis Experiment.* New York: Free Press.

Judicial Council of the American Medical Association. (1946). Supplementary report. *J. Am. Medical Assoc.* 132: 1090.

Judson, H.F. (1994). Structural transformations of the sciences and the end of peer review. *JAMA* 272: 92–95.

Jüni, P., Altman, D.G., and Egger, M. (2001). Assessing the quality of controlled clinical trials. *BMJ* 323: 42–46.

Kadane, J.B. (ed.) (1996). *Bayesian Methods and Ethics in a Clinical Trial Design.* New York: Wiley.

Kahn J. (2004). How a drug becomes "ethnic": law, commerce, and the production of racial categories in medicine. *Yale J. Health Policy Law Ethics* 4(1):1–46.

Kalbfleisch, J.D. and Prentice, R.L. (2002). *The Statistical Analysis of Failure Time Data*, 2nd ed. New York: Wiley.

Kalish, L.A., and Begg, C.B. (1985). Treatment allocation methods in clinical trials: A review. *Stat. Med.* 4: 129–144.

Kamen, J. (1993). Medical Genocide, Part 26: Hope, Heartbreak, and Horror. *Penthouse*, April.

Kamen, J. (1994). Stonewalled in the U.S.A. *Penthouse*, July.

Kaplan, E.L., and Meier, P. (1958). Nonparametric estimation from incomplete observations. *Am. Stat. Assoc. J.* 53: 457–480.

Kaptchuk, T.J. (1998). Intentional ignorance: a history of blind assessment and placebo controls in medicine. *Bull. Hist. Med.* 72: 389–433.

Kaptchuk, T.J. (2002). Control of observer biases: Masking (blinding) and placebos. In Chalmers I., Milne I., Tröhler U. (eds). *The James Lind Library* (www.jameslindlibrary.org).

Karlstrom, P.O., Bergh, T., and Lundkvist, O. (1993). A prospective randomized trial of artificial insemination versus intercourse in cycles stimulated with human menopausal gonadotropin or clomiphene citrate. *Fertil. Steril.* 59: 554–559.

Kaslow, R.A., Ostrow, D.G., Detels, R., et al. (1987). The Multicenter AIDS Cohort Study: Rationale, organization, and selected characteristics of the participants. *Am. J. Epidemiol.* 126: 310–318.

Kassaye, M., Larson, C., and Carlson, D. (1994). A randomized community trial of prepackaged and homemade oral rehydration therapies. *Arch. Ped. Adolesc. Med.* 148: 1288–1292.

Kassirer, J.P. and Campion, E.W. (1994). Peer review: crude and understudied, but indispensable. *JAMA* 272: 96–97.

Katz, J. (1972). *Experimentation with Human Beings: The Authority of Investigator, Subject, Professions, and State in the Human Experimentation Process.* New York: Russell Sage Foundation.

Keller, S.M., Groshen, S., Martini, N., and Kaiser, L.R. (1988). Blood transfusion and lung cancer recurrence. *Cancer* 62(3): 606–610.

Kelloff, G.J., Boone, C.W., Crowell, J.A., Steele, V.E., Lubet, R., and Doody, L.A. (1994). Surrogate endpoint biomarkers for phase II cancer chemoprevention trials. *J. Cell. Biochem., Suppl.* 19: 1–9.

Kelloff, G.J., Johnson, J.R., Crowell, J.A., et al. (1995). Approaches to the development and marketing approval of drugs that prevent cancer. *Cancer Epidemiol. Biomark. Prevent.* 4: 1–10.

Kemper, K.J. (1991). Pride and prejudice in peer review. *J. Clin. Epidemiol.* 44: 343–345.

Kendall, M.G. (1960). Where shall the history of statistics begin? *Biometrika* 47: 447–449.

Kennedy, S.S., Mercer, J.M., Mohr, W., and Huffine, C.W. (2002). Snake oil, ethics, and the first amendment: What's a profession to do? *Am. J. Orthopsychiat.* 72: 5–15.

Kenward, M.G., and Jones, B. (1987). A log linear model for binary cross-over data. *Appl. Statist.* 36: 192–204.

Kernan, W.N., Viscoli, C.M., Makuch, R.W., Brass, L.M., and Horwitz, R.I. (1999). Stratified randomization for clinical trials. *J. Clin. Epidemiol.* 52(1): 19–26.

Kessler, D.A., Rose, J.L., Temple, R.J., Schapiro, R., and Griffin, J.P. (1994). Therapeutic-class wars—Drug promotion in a competitive marketplace. *N. Engl. J. Med.* 331: 1350–1353.

Kevles, D.J. (1996). The assault on David Baltimore. *New Yorker*, May 27, 1996.

Kevles, D.J. (1998). *The Baltimore Case: A Trial of Politics, Science, and Character.* New York: Norton.

Khan, K.S., Daya, S., and Jadad, A.R. (1996). The importance of quality of primary studies in producing unbiased systematic reviews. *Arch. Intern. Med.* 156: 661–666.

Kitchell, J.R., Glover, R.P., and Kyle, R.H. (1958). Bilateral internal mammary artery ligation for angina pectoris; preliminary clinical considerations. *Am. J. Cardiol.* 1: 46–50.

Klein D.F. (1993). Should the government assure scientific integrity? *Acad. Med.* 68 (suppl.): S56–S59.

Kleinbaum, D.G. (1996). *Survival Analysis: A Self-learning Text.* New York: Springer.

Kloner, R.A. (1995). Nifedipine in ischemic heart disease. *Circulation* 92: 1074–1078.

Knox, R.A. (1983). Deeper problems for Darsee: Emory probe. *JAMA* 249: 2867, 2871–2873, 2876.

Knuth, D. (1981). *Seminumerical Algorithms*, 2nd ed, *Vol. 2: The Art of Computer Programming.* Reading, MA: Addison-Wesley.

Kohn, A. (1986). *False Prophets*. Oxford: Basil Blackwell.

Korn, E.L., Midthune, D., Chen, T.T., Rubinstein, L.V., Christian, M.C., and Simon, R.M. (1994). A comparison of two phase I trial designs. *Stat. Med.* 13: 1799–1806.

Kosty, M.P., Fleishman, S.B., Herndon II, J.E., et al. (1994). Cisplatin, vinblastine, and hydrazine sulfate in advanced, non–small cell lung cancer: A randomized placebo controlled, double blind phase III study of the Cancer and Leukemia Group B. *J. Clin. Oncol.* 12: 1113-1120.

Kosty, M.P., Herndon, J.E., Green, M.R., and McIntyre, O.R. (1995). Placebo-controlled randomized study of hydrazine sulfate in lung cancer (letter). *J. Clin. Oncol.* 13: 1529.

Kotz, S., and Johnson, N.L. (1988). *Encyclopedia of Statistical Sciences*. New York: Wiley.

Kraemer, H.C. and Thiemann, S. (1987). How Many Subjects? Newbury Park, CA: Sage Publications.

Krzyzanowska, M.K., Pintilie, M., and Tannock, I.F. (2003). Factors associated with failure to publish large randomized trials presented at an oncology meeting. *JAMA* 290(4): 495–501.

Kunz, R., and Oxman, A.D. (1998). The unpredictability paradox: review of empirical comparisons of randomised and non-randomised clinical trials. *BMJ* 317(7167): 1185–1190.

Kuznetsov, D.A. (1989). In vitro studies of interactions between frequent and unique mRNAs and cytoplasmic factors from brain tissue of several species of wild timber voles of Northern Eurasia, *Clethrionomys glareolus*, *Clethrionomys frater*, and *Clethrionomys gapperi*: A new criticism to a modern molecular-genetic concept of biological evolution. *Int. J. Neurosci.* 49: 43–59.

Lachin, J.M. (1981). Introduction to sample size determination and power analysis for clinical trials. *Cont. Clin. Trials* 2: 93–113.

Lachin, J.M. (1988a). Properties of simple randomization in clinical trials. *Cont. Clin. Trials* 9: 312–326.

Lachin, J.M. (1988b). Statistical properties of randomization in clinical trials. *Cont. Clin. Trials* 9: 289–311.

Lachin, J.M., Matts, J.P., and Wei, L.J. (1988). Randomization in clinical trials: Conclusions and recommendations. *Cont. Clin. Trials* 9: 365–374.

Lacourciere, Y., Lefebvre, J., Poirier, L., Archambault, F., and Arnott, W. (1994). Treatment of ambulatory hypertensives with nebivolol or hydrochlorthiazide alone and in combination. A randomized, double-blind, placebo-controlled, factorial-design trial. *Am. J. Hypertension* 7: 137–145.

Lad, T., Rubinstein, L., Sadeghi, A., et al. (1988). The benefit of adjuvant treatment for resected locally advanced non-small cell lung cancer. *J. Clin. Oncol.* 6: 9–17.

Lagakos, S.W., Lim, L., and Robins, J.M. (1990). Adjusting for early treatment termination in comparative clinical trials. *Stat. Med.* 9: 1417–1424.

Laird, N. (1983). Further comparative analyses of pretest posttest research designs. *Am. Statistician* 37: 329–330.

Laird, N., and Ware, J.H. (1982). Random effects models for longitudinal data. *Biometrics* 38: 963–974.

Lambert S.M., and Markel H. (2000). Making history: Thomas Francis, Jr, MD, and the 1954 Salk Poliomyelitis Vaccine Field Trial. *Arch. Pediatr. Adolesc. Med.* 154(5): 512–517.

Lan, K.K.G., and DeMets, D.L. (1983). Discrete sequential boundaries for clinical trials. *Biometrika* 70: 659–663.

Lan, K.K.G., DeMets, D.L., and Halperin, M. (1984). More flexible sequential and non-sequential designs in long-term clinical terms. *Commun. Statist.-Theor. Meth.* 13: 2339–2353.

Lan, K.K.G., Simon, R., and Halperin, M. (1982). Stochastically curtailed tests in long-term clinical trials. *Commun. Stat.-Sequential Anal.* 1: 207–219.

Lan, K.K.G., and Wittes, J. (1988). The B-value: A tool for monitoring data. *Biometrics* 44: 579–585.

Lancaster, H.O. (1994). *Quantitative Methods in Biological and Medical Sciences: A Historical Essay*. New York: Springer-Verlag.

Lang, T.A., and Secic, M. (1997). *How to Report Statistics in Medicine: Annotated Guidelines for Authors, Editors, and Reviewers*. Philadelphia: American College of Physicians.

Lantos, J. (1993). Informed consent. The whole truth for patients? *Cancer* (suppl.) 72: 2811–2815.

Lantos, J. (1994). Ethics, randomization, and technology assessment. *Cancer* (suppl.) 74: 2653–656.

Laporte, J.R., and Figueras, A. (1994). Placebo effects in psychiatry. *Lancet* 344(8931): 1206–1209.

Larhammar, D. (1994). Lack of experimental support for Kuznetsov's criticism of biological evolution (letter). *Int. J. Neurosci.* 77: 199–201.

Larhammar, D. (1995). Severe flaws in scientific study criticizing evolution. *Skeptical Inq.* 19: 30–31.

Laska, E., Meisner, M., and Kushner, H.B. (1983). Optimal crossover designs in the presence of carryover effects. *Biometrics* 39: 1087–1091.

Lau, J., Ioannidis, J.P., and Schmid, C.H. (1997). Quantitative synthesis in systematic reviews. *An. Intern. Med.* 127(9): 820–826.

Law, S.Y., and Wong, J. (1997). Esophageal cancer surgery: the value of controlled clinical trials. *Semin. Surg. Oncol.* 13(4): 281–287.

Layard, M.W.J., and Arvesen, J.N. (1978). Analysis of Poisson data in crossover trials. *Biometrics* 34: 421–428.

LeBlanc, M., and Crowley, J. (1992). Relative risk trees for censored survival data. *Biometrics* 48: 411–425.

Leary, W.E. (1994). Critics question ethics of U.S.-sponsored vaccine tests in Italy and Sweden. *New York Times* (Sunday, March 13).

Leber, P. (1991). Is there an alternative to the randomized controlled trial? *Psychopharm. Bull.* 27: 3–8.

Lechmacher, W. (1991). Analysis of the cross-over design in the presence of residual effects. *Stat. Med.* 10: 891–899.

Lee, E.T. (1992). *Statistical Methods for Surival Data Analysis*, 2nd ed. New York: Wiley.

Lee, S.M., Wise, R., Sternberg, A.L., Tonascia, J., and Piantadosi, S. (2004). Methodologic issues in terminating enrollment of a subgroup of patients in a multicenter randomized trial. *Clin. Trials* 1(3): 326–338.

Lee, Y.J., Ellenberg, J.H., Hirtz, D.G., and Nelson, K.B. (1991). Analysis of clinical trials by treatment actually received: Is it really an option? *Stat. Med.* 10: 1595–1605.

Lehmann, E.L. (1989). Testing Statistical Hypotheses, Second Edition. New York: Wiley.

Lemonick, M.D., and Goldstein, A. (2002). At your own risk. *Time*, April 22, 46–56.

Leventhal, B.G., and Wittes, R.E. (1988). *Research Methods in Clinical Oncology*. New York: Raven Press.

Levine, R.J. (1986). *Ethics and Regulation of Clinical Research*. 2nd ed. New Haven: Yale University Press.

Levine, R.J. (1992). Clinical trials and physicians as double agents. *Yale J. Biol. Med.* 65: 65–74.

Levine, R.J. (1999). The need to revise the Declaration of Helsinki. *N. Engl. J. Med.* 341(7): 531–534.

Levinsky, N.G. (1996). Social, institutional, and economic barriers to the exercise of patients' rights. *N. Engl. J. Med.* 334: 532–534.

Lewis, J.A., and Machin, D. (1993). Intention-to-treat—Who should use ITT? *Br. J. Cancer* 68: 647–650.

Lewis, R.J., and Berry, D.A. (1994). Group sequential clinical trials: A classical evaluation of Bayesian decision-theoretic designs. *J. Am. Statist. Assoc.* 89: 1528–1534.

Lexchin, J., Bero, L.A., Djulbegovic, B., and Clark, O. (2003). Pharmaceutical industry sponsorship and research outcome and quality: systematic review. *BMJ* 326(7400):1167–70.

Li, B., Taylor, P.R., Li, J.Y., et al. (1993). Linxian nutrition intervention trials. Design, methods, participant characteristics, and compliance. *An. Epidemiol.* 3: 577–585.

Li, D., German, D., Lulla, S., Thomas, R.G., and Wilson, S.R. (1995). Prospective study of hospitalization for asthma: A preliminary risk factor model. *Am. J. Respir. Crit. Care Med.* 151: 647–655.

Liang, K.Y., and Zeger, S.L. (1986). Longitudinal data analysis using generalized linear models. *Biometrika* 73: 13–22.

Light, R.J., and Smith, P.V. (1971). Accumulating evidence: Procedures for resolving contradictions among different research studies. *Harvard Educ. Rev.* 41: 429–471.

Liliencron, R. von (ed.). (1967). *Achenwall, in Allgemeine Deutsche Biographie*, 2nd ed. Berlin: Duncker and Humblot.

Lindley, D.V. (1965). *Introduction to Probability and Statistics from a Bayesian Viewpoint*. Cambridge: Cambridge University Press.

Lindley, D.V., and Scott, W.F. (1995). *New Cambridge Statistical Tables*, 2nd ed. Cambridge: Cambridge University Press.

Liu, P.Y., Dahlberg, S., and Crowley, J. (1993). Selection designs for pilot studies based on survival. *Biometrics* 49: 391–398.

Liu, G., and Piantadosi, S. (1997). Ridge estimation in generalized linear models and proportional hazards regressions. Unpublished manuscript.

Lock, S. (1988). Misconduct in medical research: Does it exist in Britain? *Br. Med. J.* 297: 1531–1535.

Lock, S. (1990). Medical misconduct: A survey in Britain. In J. Bailar et al. (eds.). *In Ethics and Policy in Scientific Publication*. Bethesda, MD: Council of Biology Editors.

Lock, S. (1995). Lessons from the Pearce affair: Handling scientific fraud. *BMJ* 310: 1547–1548.

Loprinzi, C.L., Kuross, S.A., O'Fallon, J.R., et al. (1994). Randomized placebo controlled evaluation of hydrazine sulfate in patients with advanced colorectal cancer. *J. Clin. Oncol.* 12: 1121–1125.

Loprinzi, C.L., Goldberg, R.M., Su, J.Q., et al. (1994). Placebo controlled trial of hydrazine sulfate in patients with newly diagnosed non–small cell lung cancer. *J. Clin. Oncol.* 12: 1126–1129.

Lorenzen, T.J., and Anderson, V.L. (1993). *Design of Experiments: A No-name Approach*. New York: Marcel Dekker.

Louis, P.C.A. (1835). *Recherches sur les effets de la saignée dans quelques maladies inflammatoires et sur l'action de l'émétique et des vésicatoires dans la pneumonie*. Paris.

Lubsen, J., and Pocock, S.J. (1994). Factorial trials in cardiology (editorial). *Eur. Heart. J.* 15: 585–588.

Luft, H.S., Bunker, J.P., and Enthoven, A.C. (1979). Should operations be regionalized? The empirical relation between surgical volume and mortality. *N. Engl. J. Med.* 301: 1364–1369.

Lunn, A.D., and McNeil, D.R. (1991). *Computer-Interactive Data Analysis*. Chichester: Wiley.

Lurie, P., and Wolfe, S.M. (1997). Unethical trials of interventions to reduce perinatal transmission of the human immunodeficiency virus in developing countries. *N. Engl. J. Med.* 337: 853–856.

Macdonald, J.S. (1990). Sometimes a great notion. *JNCI* 82: 102–104.

Machin, D., Campbell, M.J., Fayers, P.M., and Pinol, A.P.Y. (1997). *Statistical Tables for the Design of Clinical Trials*, 2nd ed. Oxford: Blackwell.

Macklin, R. (1999). The ethical problems with sham surgery in clinical research. *N. Engl. J. Med.* 341: 992–996.

Magner, L.N. (2002). *A History of the Life Sciences*, 3rd ed. New York: Marcel Dekker.

Mahfouz, A.A., Abdel-Moneim, M., al-Erian, R.A., and al-Amari, O.M. (1995). Impact of chlorination of water in domestic storage tanks on childhood diarrhoea: A community trial in the rural areas of Saudi Arabia. *J. Trop. Med. Hyg.* 98: 126–130.

Mahon, W.A., and Daniel, E.E. (1964). A method for the assessment of the reports of drug trials. *Can. Med. Assoc. J.* 90: 565–569.

Majeed, A.W., Troy, G., Nicholl, J.P., et al. (1996). Randomised, prospective, single-blind comparison of laparoscopic versus small-incision cholecystectomy. *Lancet* 347: 989–994.

Mann, H., and Djulbegovic, B. (2004). Why comparisons must address genuine uncertainties. *James Lind Library* (www.jameslindlibrary.org).

Manski, C.F. (1990). Nonparametric bounds on treatment effects. *Am. Econ. Rev., Papers Proc.* 80: 319–323.

Manson, J.E., Gaziano, J.M., Spelsberg, A., Ridker, P.M., Cook, N.R., Buring, J.E., Willett, W.C., and Hennekens, C.H. (1995). A secondary prevention trial of antioxidant vitamins and cardiovascular disease in women. Rationale, design, and methods. *An. Epidemiol.* 5: 261–269.

Mantel, N., and Haenszel, W. (1959). Statistical aspects of the analysis of data from retrospective studies of disease. *JNCI* 22: 719–748.

Marcus, D.M., and Grollman, A.P. (2002). Botanical medicines – the need for new regulations. *N. Engl. J. Med.* 347: 2073–2076.

Markman, M. (1992). Ethical difficulties with randomized clinical trials involving cancer patients: Examples from the field of gynecologic oncology. *J. Clin. Ethics* 3(3): 193–195.

Marshall, E. (1993). Women's Health Initiative draws flak. *Science* 262: 838.

Marshall, E. (1994). The politics of alternative medicine. *Science* 265: 2000–2002.

Martin, B.K., Meinert, C.L., and Breitner, J.C.S., for the ADAPT Research Group. (2002). Double placebo design in a prevention trial for Alzheimer's disease. *Cont. Clin. Trials* 23: 93–99.

Martz, H.F., and Waller, R.A. (1982). *Bayesian Reliability Analysis*. New York: Wiley.

Marubini, E., Mariani, L., Salvadori, B., et al. (1996). Results of a breast-cancer-surgery trial compared with observational data from routine practice. *Lancet* 347: 1000–1003.

Marubini, E., and Valsecchi, M.G. (1995). *Analysing Survival Data from Clinical Trials and Observational Studies*. Chichester: Wiley.

Marwick, C. (1994). Ethicist faults human research protection. *JAMA* 271: 1228–1229.

Mason, R.L., and Gunst, R.L (1989). *Statistical Design and Analysis of Experiments*. New York: Wiley.

Mastroianni, A.C., Faden R., and Federman, D. (eds.) (1994). *Women and Health Research: Ethical and Legal Issues of Including Women in Clinical Studies*. Washington, DC: National Academy Press.

Matthews, J.N.S. (1988). Recent developments in crossover designs. *Int. Stat. Rev.* 56: 117–127.

Matthews, J.R. (1995). *Quantification and the Quest for Medical Certainty*. Princeton: Princeton University Press.

Matthews, L.H. (1981). Piltdown man: The missing links. *New Scientist* 90: 282–282; 376; 450; 515–516; 578–579; 647–648; 710–711; 785; 861–862; and 90: 26–28.

Matts, J.P., and Lachin, J.M. (1988). Properties of permuted-block randomization in clinical trials. *Cont. Clin. Trials* 9: 327–344.

Max, M.B., Zeigler, D., Shoaf, S.E., et al. (1992). Effects of a single oral dose of desipramine on postoperative morphine analgesia. *J. Pain Symptom Manage.* 7: 454–462.

Mayr, E. (1997). *This Is Biology, The Science of the Living World*. Cambridge, MA: Belknap Press.

McAlister, F.A., Straus, S.E., Sackett, D.L., and Altman, D.G. (2003). Analysis and reporting of factorial trials: A systematic review. *JAMA* 289(19):2545–53.

McConnochie, K.M., Roghmann, K.J., and Pasternack, J. (1993). Developing prediction rules and evaluating patterns using categorical clinical markers: Two complementary procedures. *Med. Decis. Making* 13: 30–42.

McCullagh, P., and Nelder, J.A. (1989). *Generalized Linear Models*. London: Chapman and Hall.

McCulloch, P., Taylor, I., Sasako, M., Lovett, B., and Griffin, D. (2002). Randomised trials in surgery: problems and possible solutions. *BMJ* 324: 1448–1451.

McCutchen, C.W. (1991). Peer review: Treacherous servant, disastrous master. *Technol. Rev.* (October): 28–36.

McFadden, E. (1998). *Management of Data in Clinical Trials*. New York: Wiley.

McGinnis, L.S. (1990). Alternative therapies, 1990: An overview. *Cancer* 67: 1788–1792.

McIntyre, O.R., Kornblith, A.B., and Coburn, J. (1996). Pilot survey of opinions on data falsification in clinical trials. *Cancer Invest.* 14(4): 392–395.

McKean, K. (1981). A scandal in the laboratory. *Discover* (November): 18–23.

McKnight, B., and Van Den Eeden, S.K. (1993). A conditional analysis for two-treatment multiple-period crossover designs with binomial or Poisson outcomes and subjects who drop out. *Stat. Med.* 12: 825–834.

McLean, R.A., Sanders, W.L., and Stroup, W.W. (1991). A unified approach to mixed linear models. *Am. Statist.* 45: 54–64.

McNeill, P.M. (1993). *The Ethics and Politics of Human Experimentation*. Cambridge: Press Syndicate of the University of Cambridge.

McPeek, B., Mosteller, F., and McKneally, M. (1989). Randomized clinical trials in surgery. *Int. J. Technol. Assess. Health Care* 5: 317–332.

The Medical Research Council's General Practice Research Framework. (1998). Thrombosis prevention trial: randomised trial of low-intensity oral anticoagulants with warfarin and low-dose aspirin in the primary prevention of ischaemic heart disease in men at increased risk. *Lancet* 351: 233–241.

Meier, P. (1991). Comment (on a paper by Efron and Feldman). *J. Am. Stat. Assoc.* 86: 19–22.

Meinert, C.L. (1986). *Clinical Trials: Design, Conduct, and Analysis*. Oxford: Oxford University Press.

Meinert, C.L. (1996). *A Dictionary of Clinical Trials*. Baltimore: Johns Hopkins Center for Clinical Trials.

Meinert, C.L. (1989). Extracorporeal membrane oxygenation trials (commentary). Pediatrics 85: 365–366.

Meinert, C.L. (1989). Meta-analysis: Science or religion? *Cont. Clin. Trials* 10: 257S–263S.

Meldrum, M. (1998). A calculated risk: the Salk polio vaccine field trials of 1954. *BMJ* 317: 1233–1236.

Mello, M.M., and Brennan, T.A. (2003). Due process in investigations of research misconduct. *N. Engl. J. Med.* 349: 1280–1286.

Meyer, L., Job-Spira, N., Bouyer, J., Bouvet, E., and Spira, A. (1991). Prevention of sexually transmitted diseases: A randomised community trial. *J. Epidemiol. Commun. Health* 45: 152–158.

Mick, R., and Ratain, M.J. (1994). Bootstrap validation of pharmacodynamic models defined via stepwise linear regression. *Clin. Pharmacol. Ther.* 56(2): 217–222.

Miettinen, O. (1983). The need for randomization in the study of intended effects. *Stat. Med.* 2: 267–271.

Miké, V., Krauss, A.N., and Ross, G.S. (1993). Neonatal extracorporeal membrane oxygenation (ECMO): Clinical trials and the ethics of evidence. *J. Med. Ethics* 19: 212–218.

Miles, S.H. (2004). *The Hippocratic Oath and the Ethics of Medicine*. Oxford: Oxford University Press.

Miller, D.J. (1992). Personality factors in scientific fraud and misconduct. In D.J. Miller and M. Hersen (eds.), *Research Fraud in the Behavioral and Biomedical Sciences*. New York: Wiley, Ch. 7.

Miller, D.J., and Hersen, M. (eds.) (1992). *Research Fraud in the Behavioral and Biomedical Sciences*. New York: John Wiley and Sons.

Miller, T.P., Crowley, J., Mira, J., Schwartz, J.G., Hutchins, L., Baker, L., Natale, R., Chase, E.M., and Livingston, R. (1998). A randomized trial of chemotherapy and radiotherapy for stage 111 non-small cell lung cancer. *Cancer Therapeutics* 1: 229–236.

Mills, J.L. (1993). Data torturing. *N. Engl. J. Med.* 329: 1196–1199.

Minor, J.M., and Namini, H. (1996). Analysis of clinical data using neural nets. *J. Biopharmaceut. Stat.* 6: 83–104.

Mock, V., et al. (2005). Exercise manages fatigue during breast cancer treatment: a randomized controlled trial. *Psycho-Oncology*, in press (epub).

Moertel, C.G., Schutt, A.J., Hahn, R.G., and Reitemeier, R.S. (1974). Effects of patient selection on results on phase II chemotherapy trials in gastrointestinal cancer. *Cancer Chemother. Rep.* 58: 257–260.

Moertel, C.G., and Reier, R.J. (1969). *Advanced Gastrointestinal Cancer: Clinical Management and Chemotherapy*. New York: Harper and Row.

Moertel, C.G., Taylor, W.F., Roth, A., and Tyce, F.A. (1976). Who responds to sugar pills? *Mayo Clin. Proc.* 51(2): 96–100.

Moertel, C.G., and Thynne, G.S. (1982). Large bowel. In J.F. Holland and E. Frei III (eds.), *Cancer Medicine*, 2nd ed. Philadelphia: Lea and Febiger, pp.1830–1859.

Moertel, C.G., Fleming, T.R., Rubin, J., et al. (1982). A clinical trial of amygdalin (Laetrile) in the treatment of human cancer. *N. Engl. J. Med.* 306: 201–206.

Moher, D., Dulberg, C.S., and Wells, G.A. (1994). Statistical power, sample size, and their reporting in randomized controlled trials. *JAMA* 272: 122–124.

Moher, D., Fortin, P., Jada, A.R., et al. (1996). Completeness of reporting of trials published in languages other than English: Implications for conduct and reporting of systematic reviews. *Lancet* 347: 363–366.

Moher, D., Schulz, K.F., and Altman, D.G., for the CONSORT Group. (2001). The CONSORT statement: Revised recommendations for improving the quality of reports of parallel-group randomized trials. *JAMA* 285: 1987–1991.

Moore, R.D., Hidalgo, J., Sugland, B.W., and Chaisson, R.E. (1991). Zidovudine and the natural history of the acquired immunodeficiency syndrome. *N. Engl. J. Med.* 324: 1412–1416.

Moores, D.W.O., Piantadosi, S., and McKneally, M.F. (1989). Effect of perioperative blood transfusion on outcome in patients with surgically resected lung cancer. *An. Thorac. Surg.* 47: 346–351.

Morrison, R.S. (1990). Disreputable science: Definition and detection. *J. Adv. Nurs.* 15: 911–913.

Moseley, J.B., O'Malley, K., Petersen, N.J., et al. (2002). A controlled trial of arthroscopic surgery for osteoarthritis of the knee. *N. Engl. J. Med.* 347: 81–88.

Moss, A.J., Hall, W.J., Cannom, D.S., et al. (1996). Improved survival with an implanted defibrillator in patients with coronary disease at high risk for ventricular arrhythmia. *N. Engl. J. Med.* 335: 1933–1940.

Mosteller, F., Gilbert, J.P., and McPeek, B. (1980). Reporting standards and research strategies for controlled trials; agenda for the editor. *Cont. Clin. Trials* 1: 37–58.

Mountain, C.F. and Gail, M.H. (1981). Surgical adjuvant intrapleural BCG treatment for stage I non–small cell lung cancer. Preliminary report of the National Cancer Institute Lung Cancer Study Group. *J. Thorac. Cardiovasc. Surg.* 82: 649–657.

Moynihan, R. (2004). *Evaluating Health Services: A Reporter Covers the Science of Research Synthesis.* New York: Milbank Memorial Fund.

MPS Research Unit. (2000). *PEST 4: Operating Manual.* The University of Reading, UK.

Murfitt, R.R. (1990). United States government regulation of medical device software: A review. *J. Med. Eng. Tech.* 14: 111–113.

Murray, D.M. (1998). *Design and Analysis of Group-Randomized Trials.* Oxford: Oxford University Press.

Murray, G.D. (1991a). Statistical aspects of research methodology. *Br. J. Surg.* 78: 777–781.

Murray, G.D. (1991b). Statistical guidelines for the British Journal of Surgery. *Br. J. Surg.* 78: 782–784.

Murthy, V.H., Krumholz, H.M., Gross, C.P. (2004). Participation in cancer clinical trials: race-, sex-, and age-based disparities. *JAMA* 291(22): 2720–2726. Comment in: *JAMA* 292(8): 922 (2004); Author reply 922–923.

Nahin, R.L. (2002). Use of the best case series to evaluate complementary and alternative therapies for cancer: A systematic review. *Sem. Oncol.* 29: 552–562.

National Academy of Sciences (NAS) (1992a). *Responsible Science: Ensuring the Integrity of the Research Process*, Vol. 1. Washington, DC: National Academy Press.

National Academy of Sciences (NAS) (1992b). *Responsible Science: Ensuring the Integrity of the Research Process*, Vol. 2. Washington, DC: National Academy Press.

National Cancer Institute (1993). Final Report to the Food and Drug Administration: Hydrazine Sulfate, NSC 150014, IND 33233. Private communication.

NCI Press Office (1994a). Press release. March 24.

NCI Press Office (1994b). Press release. March 29.

NCI Press Office (1994c). Press release. April 12.

NCI Press Office (1994d). Press release. April 21.

NCI Press Office (1994e). Press release. April 22.

NCI Press Office (1994f). Press release. May 2.

National Cancer Institute (2001). *NCI Cooperative Group Data Monitoring Committee Policy.* http://ctep.cancer.gov/monitoring/guidelines.html.

National Cancer Institute (2003). *NCI Cancer Centers Policies And Guidelines Relating to The Cancer-Center Support Grant.* http://www3.cancer.gov/cancercenters/.

National Commission for Protection of Human Subjects of Biomedical and Behavioral Research. (1978). *The Belmont Report: Ethical Principles and Guidelines for the Protection of*

Human Subjects of Research. Washington, DC: DHEW Publication Number (OS) 78-0012. Appendix I, DHEW Publication No. (OS) 78-0013; Appendix II, DHEW Publication No. (OS) 78-0014.

National Science Foundation (NSF) (1987). Misconduct in science and engineering research: Final regulations. *Fed. Reg.* 52 (July 1): 24466–24470.

National Science Foundation (NSF) (1991). Misconduct in science and engineering research: Final rule. *Fed. Reg.* 56 (May 14): 22286–22290.

National Science Foundation (NSF) (1990). *Semiannual Report to the Congress.* No. 3. Washington, DC: Office of the Inspector General, NSF.

National Science Foundation (NSF) (2002). Research Misconduct. *Fed. Reg.* 67(52): 11936–11939.

Naylor, C.D., Chen, E., and Strauss, B. (1992). Measured enthusiasm: Does the method of reporting trial results alter perceptions of therapeutic effectiveness? *An. Intern. Med.* 117: 916–921.

Nelder, J.A., and Wedderburn, R.W.M. (1972). Generalized linear models. *JRSS A* 135: 370–384.

Newell, D.J. (1992). Intention-to-treat analysis: implications for quantitative and qualitative research. *Int. J. Epidemiol.* 21: 837–841.

NETT Research Group (1999). Rationale and design of The National Emphysema Treatment Trial: a prospective randomized trial of lung volume reduction surgery. *Chest* 116(6): 1750–1761.

NETT Research Group (2001). Patients at high risk of death after lung-volume-reduction surgery. *N. Engl. J. Med.* 345(15): 1075–1083.

NETT Research Group (2003a). A randomized trial comparing lung-volume-reduction surgery with medical therapy for severe emphysema. *N. Engl. J. Med.* 348: 2059–2073.

NETT Research Group (2003b). Cost Effectiveness of Lung-Volume-Reduction Surgery for Patients with Severe Emphysema. *N. Engl. J. Med.* 348: 2092–2102

Neyman, J., and Pearson, E.S. (1933). On the problem of the most efficient tests of statistical hypotheses. *Phil. Trans. Roy. Soc. A* 231: 289–337.

NHLBI (1994a). *NHLBI Guidelines for Data Quality Assurance in Clinical Trials and Observational Studies* (http://www.nhlbi.nih.gov/funding/policies/dataqual.htm).

NHLBI (1994b). *NHLBI Guide for Data and Safety Monitoring Boards.* www.nhlbi.nih.gov/funding/policies/dsmb_est.htm. See also www.nhlbi.nih.gov/funding/policies/dsm-12.htm

NIH (1993). Final findings of scientific misconduct: Roger Poisson, M.D., St. Luc Hospital, Montreal, Canada. NIH Guide for Grants and Contracts. Vol. 22, No. 23, June 25, 1993: 3.

Nissenson, A.R., Rapaport, M., Gordon, A., and Narins, R.G. (1979). Hemodialysis in the treatment of psoriasis: a controlled trial. *An. Intern. Med.* 91: 218–220.

Non–Small Cell Lung Cancer Collaborative Group (1995). Chemotherapy in non–small cell lung cancer: A meta-analysis using updated data on individual patients from 52 randomised clinical trials. *BMJ* 311: 899–909.

Noseworthy, J.H., Ebers, G.C., Vandervoort, M.K., et al. (1994). The impact of blinding on the results of a randomized, placebo-controlled multiple sclerosis clinical trial. *Neurology* 44: 16–20.

NSABP (1992). NSABP Protocol P-1 (Breast Cancer Prevention Trial). Pittsburg, PA: NSABP Center Headquarters.

Nuland, S.B. (2000). *The Mysteries Within.* New York: Simon and Schuster.

O'Brien, P.C., and Fleming, T.R. (1979). A multiple testing procedure for clinical trials. *Biometrics* 35: 549–556.

O'Bryan, T., and Walter, G. (1979). Mean square estimation of the prior distribution. *Sankhyā* A, 41: 95–108.

O'Fallon, J.R. (1985). Policies for interim analysis and interim reporting of results. *Cancer Treat. Rep.* 69(10): 1101–1106.

Office of Research Integrity (ORI) (1993a). Office of Research Integrity: An Introduction. Rockville: Department of Health and Human Services.

Office of Research Integrity (ORI) (1993b). Case Summary: Fabricated and falsified clinical trial data. *ORI Newslett.* 1(2): 2.

Office of Research Integrity (ORI) (1995). *Annual Report* 1995. Rockville, MD: Office of Research Integrity.

Office of Science and Technology Policy (OSTP) (2000). Federal Policy on Research Misconduct. *Fed. Reg.* 65(235): 76260–76264.

Office of Technology Assessment (1983). Factors affecting the impact of RCTs on medical practice. In *The Impact of Randomized Clinical Trials on Health Policy and Medical Practice.* Washington, DC: Government Printing Office, OTA-BP-H-22.

Office of Technology Assessment (1990). Unconventional Cancer Treatments. Washington, DC: Government Printing Office, OTA-H-405.

O'Hagan, A. (1994). *Kendall's Advanced Theory of Statistics, Vol. 2B: Bayesian Inference.* London: Edward Arnold.

Ohashi, Y. (1990). Randomization in cancer clinical trials: Permutation test and development of computer program. *Environ. Health Perspect.* 87: 13–17.

Olanow C.W., Goetz C.G., Kordower J.H., et al. (2003). A double-blind controlled trial of bilateral fetal nigral transplantation in Parkinson's disease. *An. Neurol.* 54(3): 403–414.

Oliver, S.G., van der Aart, Q.J.M., Agostoni-Carbone, M.L., et al. (1992). The complete DNA sequence of yeast chromosome III. *Nature* 357: 38–46.

Olkin, I. (1994). Invited commentary re: A critical look at some popular meta-analytic methods. *Am. J. Epidemiol.* 140: 297–299.

Olkin, I. (1995a). Statistical and theoretical considerations in meta-analysis. *J. Clin. Epidemiol.* 48: 133–146.

Olkin, I. (1995b). Meta-analysis: Reconciling the results of independent studies. *Stat. Med.* 14: 457–472.

O'Neill, R.T. (1978). Subjects-own-control designs in clinical drug trials: Overview of the issues with emphasis on the two treatment problem. Presented at the Annual NCDEU Meeting, Key Biscayne, Florida.

Opie, L.H., and Messerli, F.H. (1995). Nifedipine and mortality: Grave defects in the dossier. *Circulation* 92: 1068–1073.

O'Quigley, J. (1992). Estimating the probability of toxicity at the recommended dose following a phase I clinical trial in cancer. *Biometrics* 48: 853–862. Erratum in: *Biometrics* 50: 322 (1994).

O'Quigley, J., and Shen, L.Z. (1996). Continual reassessment method: A likelihood approach. *Biometrics* 52: 673–684.

O'Quigley, J., Pepe, M., and Fisher, L. (1990). Continual reassessment method: A practical design for phase I clinical trials in cancer. *Biometrics* 46: 33–48.

O'Quigley, J., and Chevret, S. (1991). Methods for dose finding studies in cancer clinical trials: A review and results of a Monte Carlo study. *Stat. Med.* 10: 1647–1664.

O'Regan, B. and Hirshberg, C. (1993). *Spontaneous Remission: An Annotated Bibliography.* Sausalito, CA: Institute of Noetic Sciences.

O'Rourke, P.P. (1991). ECMO: Where have we been? Where are we going? *Resp. Care* 36: 683–694.

O'Rourke, P.P., Crone, R.K., Vacanti, J.P., Ware, J.H., Lillehei, C.W., Parad, R.B., and Epstein, M.F. (1989). Extracorporeal membrane oxygenation and conventional medical therapy in neonates with persistent pulmonary hypertension of the newborn: A prospective randomized study. *Pediatrics* 84: 957–963.

Osler, W. (1910). *Man's Redemption of Man*. London: Constable.

Ostrander, G.K., Cheng, K.C., Wolf, J.C., Wolfe, M.J. (2004). Shark cartilage, cancer and the growing threat of pseudoscience. *Cancer Res.* 64(23): 8485–8491.

Packer, M., Carver, J.R., Rodeheffer, R.J., et al. (1991). Effect of oral milrinone on mortality in severe chronic heart failure. *N. Engl. J. Med.* 325: 1468–1475.

Palafox, N., Schatz, I., Lange, R., et al. (1997). Intravenous 20% mannitol versus intravenous 5% dextrose for the treatment of acute ciguatera: A randomized, placebo controlled, double masked trial. Personal communication.

Palta, M. (1985). Investigating maximum power losses in survival studies with nonstratified randomization. *Biometrics* 41: 497–504.

Paltiel, O., Avitzour, M., Peretz, T., et al. (2001). Determinants of the use of complementary therapies by patients with cancer. *J. Clin. Oncol.* 19(9): 2439–2448.

Parker, A.B., and Naylor, C.D. (2000). Subgroups, treatment effects, and baseline risks: Some lessons from major cardiovascular trials. *Am. Heart J.* 139(6): 952–961.

Parkinson Study Group (1993). Effects of tocopherol and deprenyl on the progression of disability in early Parkinson's disease. *N. Engl. J. Med.* 328: 176–183.

Parmar, M.K.B. (1992). Randomization Before Consent. In C.J. Williams (ed.), *Introducing New Treatments for Cancer: Practical, Ethical, and Legal Problems*. Chichester: Wiley, ch. 14.

Parmar, M.K.B., and Machin, D. (1995). *Survival Analysis: A Practical Approach*. New York: Wiley.

Parmar, M.K.B., Ungerleider, R.S., and Simon, R. (1996). Assessing whether to perform a confirmatory randomized clinical trial. *J. Nat. Cancer Inst.* 88(22): 1645–1651.

Pascal, C.B. (2000). Scientific misconduct and research integrity for the bench scientist. *PSEBM* 224: 220–230.

Passamani, E. (1991). Clinical trials—Are they ethical? *N. Engl. J. Med.* 324: 1589–1592.

Peduzzi, P., et al. (1991). Intention-to-treat analysis and the problem of crossovers: An example from the Veterans Administration coronary bypass surgery study. *J. Thorac. Cardiovasc. Surg.* 101: 481–487.

Peduzzi, P., et al. (1993). Analysis as-randomized and the problem of non-adherence: An example from the Veterans Affairs randomized trial of coronary artery bypass surgery. *Stat. Med.* 12: 1185–1195.

Pena, C.M., Rice, T.W., Ahmad, M., and Medendorp, S.V. (1992). Significance of perioperative blood transfusions in patients undergoing resection of stage I and II non-small cell lung cancers. *Chest* 102: 84–88.

Perez, C. A., Gardner, P., and Glasgow, G.P. (1984). Radiotherapy quality assurance in clinical trials. *Int. J. Radiat. Onc. Biol. Phy.* 10: 119–125.

Peto, J. (1984). The calculation and interpretation of survival curves. In M.J. Buyse, M.J. Staquet, and R.J. Sylvester (eds.), *Cancer Clinical Trials*. Oxford: Oxford University Press, ch. 21.

Peto, R. (1987). Why do we need systematic overviews of randomized trials? *Stat. Med.* 6: 233–240.

Peto, R., and Baigent, C. (1998). Trials: The next 50 years. Large scale randomised evidence of moderate benefits. *BMJ* 317: 1170–1171.

Peto, R., Collins, R., and Gray, R. (1995) Large-scale randomized evidence: Large simple trials and overviews of trials. *J. Clin. Epidemiol.* 48: 23–40.

Peto, R., Collins, R., Sackett, D., et al. (1997). The trials of Dr. Bernard Fisher: A European perspective on an American episode. *Cont. Clin. Trials* 18: 1–13.

Peto, R., Pike, M.C., Armitage, P., Breslow, N.E., Cox, D.R., Howard, S.V., Mantel, N., McPherson, K., Peto, J., and Smith, G. (1977a). Design and analysis of randomized clinical trials requiring prolonged observation of each patient. I: Introduction and design. *Br. J. Cancer.* 34: 585–612.

Peto, R., Pike, M.C., Armitage, P., Breslow, N.E., Cox, D.R., Howard, S.V., Mantel, N., McPherson, K., Peto, J., and Smith, G. (1977b). Design and analysis of randomized clinical trials requiring prolonged observation of each patient. II: Analysis and examples. *Br. J. Cancer.* 35: 1–39.

Petrelli, N.J. (2002). Clinical trials are mandatory for improving surgical cancer care. *JAMA* 287: 377–378.

Piantadosi, S. (1990). *Clinical Trials Design Program.* Cambridge, UK: BIOSOFT.

Piantadosi, S. (1990). Hazards of small clinical trials (editorial). *J. Clin. Oncol.* 8(1): 1–3.

Piantadosi, S. (1992). The adverse effect of blood transfusion in lung cancer (editorial). *Chest* 102: 608.

Piantadosi, S. (2005). Translational clinical trials. *Clin. Trials* 2: 182–192.

Piantadosi, S., Fisher, J.D., and Grossman, S. for the New Approaches to Brain Tumor Therapy Consortium (1998). Practical implementation of a modified continual reassessment method for dose finding trials. *Cancer Chemother. Pharmacol.* 41: 429–436.

Piantadosi, S., and Gail, M.H. (1995). Statistical issues arising in thoracic surgery clinical trials. In F.G. Pearson, J. Deslauriers, R.J. Ginsberg, et al. (eds.), *Thoracic Surgery.* New York: Churchill Livingstone, ch. 71.

Piantadosi, S., Graham, N.M.H., Park, L., Saah, A., Kaslow, R., Detels, R. Rinaldo, C., and Phair, J., for the Multicenter AIDS Cohort Study (MACS). (1993). Risk sets for time to AIDS and survival based on pre- and post-treatment prognostic markers. First National Conference on Human Retroviruses and Related Infections, *Am. Soc. Microbiol.,* December 12–16, 1993.

Piantadosi, S., and Liu, G. (1996). Improved designs for dose escalation studies using pharmacokinetic measurements. *Stat. Med.* 15: 1605–1618.

Piantadosi, S., and Patterson, B. (1987). A method for predicting accrual, cost, and paper flow in clinical trials. *Cont. Clin. Trials* 8: 202–215.

Plackett, R.L. and Burman, J.P. (1946). The design of optimum multifactorial experiments. *Biometrika* 33: 305–325.

Pocock, S.J. (1993). Statistical and ethical issues in monitoring clinical trials. *Stat. Med.* 12: 1459–1469.

Pocock, S.J., and Simon, R. (1975). Sequential treatment assignment with balancing for prognostic factors in the controlled clinical trial. *Biometrics* 31: 103–115.

Pocock, S.J. (1996). *Clinical Trials: A Practical Approach.* New York: Wiley.

Pocock, S.J., and Elbourne, D.R. (2000). Randomized trials or observational tribulations. *N. Engl. J. Med.* 342: 1907–1090.

Pocock, S.J., Hughes, M.D., and Lee, R.J. (1987). Statistical problems in the reporting of clinical trials: A survey of three medical journals. *N. Engl. J. Med.* 317: 426–432.

Polanyi, M. (1958). Personal Knowledge; towards a post-critical philosophy. Chicago: University of Chicago Press.

Poloniecki, J.D., and Pearce, A.C. (1983). Letter to the editor. *Biometrics* 39: 789.

Popper, K.R. (1959). *The Logic of Scientific Discovery.* London: Hutchinson.

Pordy, R.C. (1994). Cilazapril plus hydrochlorthiazide: improved efficacy without reduced safety in mild to moderate hypertension. A double-blind placebo-controlled multicenter study of factorial design. *Cardiology* 85: 311–322.

Porter, T.M. (1986). *The Rise of Statistical Thinking 1820–1900.* Princeton: Princeton University Press.

Porter R., and Rousseau, G.S. (1998). *Gout: The Patrician Malady*. New Haven: Yale University Press.

Potsdam International Consultation on Meta-Analysis (1995). *J. Clin. Epidemiol.* 48(1).

Prentice, R.L. (1989). Surrogate endpoints in clinical trials: Definition and operational criteria. *Stat. Med.* 8: 431–440.

Press, W.H., Teukolsky, S.A., Vetterling, W.T., and Flannery, B.P. (1992). *Numerical Recipes in C: The Art of Scientific Computing*, 2nd ed. Cambridge: Cambridge University Press.

Prinssen, M., Verhoeven, E.L.G., Buth, J., et al. (2004). A randomized trial comparing conventional and endovascular repair of abdominal aortic aneurysms. *N. Engl. J. Med.* 351: 1607–1618.

Prorok, P.C. (2001). Screening trials. In C. Redmond and T. Colton (eds.), *Biostatistics in Clinical Trials*. Chichester: Wiley.

Proschan, M.A., Follmann, D.A., and Geller, N.L. (1994). Monitoring multi-armed trials. *Stat. Med.* 13: 1441–1452.

Prostate Cancer Trialists' Collaborative Group (1995). Maximum androgen blockade in advanced prostate cancer: An overview of 22 randomised trials with 3283 deaths in 5710 patients. *Lancet* 346: 265–269.

Public Law 103-4.3 (1993). Clinical Research Equity Regarding Women and Minorities. Public Law 103-43, Subtitle B, Sec. 131.

Racker, E. and Spector, M. (1981). Warburg effect revisited: Merger of biochemistry and molecular biology. *Science* 213: 303–307.

Ramsey, P. (1975). *The Ethics of Fetal Research*. New Haven: Yale University Press.

Raso, J. (1998). *The Expanded Dictionary of Metaphysical Healthcare: Alternative Medicine, Paranormal Healing, and Related Methods*. Atlanta: Georgia Council Against Health Fraud.

Ratain, M.J., Mick, R., Schilsky, R.L., and Siegler, M. (1993). Statistical and ethical issues in the design and conduct of phase I and II clinical trials of new anticancer agents. *JNCI* 85: 1637–1643.

Ravdin, P.M. and Clark, G.M. (1992). A practical application of neural network analysis for predicting outcome of individual breast cancer patients. *Breast Cancer Res. Treat.* 22: 285–293.

Redmond, C. and Colton, T. (eds.) (2001). Biostatistics in Clinical Trials. Chichester: Wiley.

Reich, W.T. (ed.) (1995). *Encyclopedia of Bioethics*. New York: Simon and Schuster Macmillan.

Reiser, S.J. (2002). The ethics movement in the biological sciences: A new voyage of discovery. Overview. In R.E. Bulger, E. Heitman, and S.J. Reiser (eds.), *The Ethical Dimensions of the Biological and Health Sciences*. Cambridge: Cambridge University Press.

Relman, A.S. (1983). Lessons from the Darsee affair. *N. Engl. J. Med.* 308: 1415–1417.

Relman, A.S. (1984). Dealing with conflicts of interest. *N. Engl. J. Med.* 311: 405.

Rennie, D., Flanagin, A., and Glass, R.M. (1991). Conflicts of interest in the publication of science. *JAMA* 266: 266–267.

Rennie, D., Yank, V., and Emanuel, L. (1997). When authorship fails: A proposal to make contributors accountable. *J. Am. Med. Assoc.* 278: 579–585.

Resnick, R.H., Ishihara, A., Chalmers, T.C., Schimmel, E.M., and the Boston Inter-Hospital Liver Group. (1968). A controlled trial of colon bypass in chronic hepatic encephalopathy. *Gastroenterology* 54: 1057–1069.

Riggs, B.L., Hodgson, S.F., O'Fallon, W.M., et al. (1990). Effect of fluoride treatment on the fracture rate in postmenopausal women with osteoporosis. *N. Engl. J. Med.* 322: 802–809.

Robbins, H., and Monro, S. (1951). A stochastic approximation method. *An. Math. Stat.* 22: 400–407.

Roberson, N.L. (1994). Clinical trial participation. *Cancer* 74: 2687–2691.

Roberts, A.H., Kewman, D.G., Mercier, L., and Hovell, M. (1993). The power of nonspecific effects in healing: Implications for psychosocial and biological treatments. *Clin. Psychol. Rev.* 13: 375–391.

Robertson, D.M., and Ilstrup, D. (1983). Direct, indirect, and sham laser photocoagulation in the management of central serous chorioretinopathy. *Am J. Ophthalmol.* 95: 457–466.

Robins, J.M. (1989). The analysis of randomized and non-randomized AIDS treatment trials using a new approach to causal inference in longitudinal studies. In L. Sechrest, H. Freeman, and A. Mulley (eds.) *Health Service Research Methodology: A Focus on AIDS.* NCHSR, U.S. Public Health Service.

Rochon, P.A., Gurwitz, J.H., Cheung, C.M., Hayes, J.A., and Chalmers, T.C. (1994). Evaluating the quality of articles published in journal supplements compared with the quality of those published in the parent journal. *JAMA* 272: 108–113.

Roebruck, P., and Kühn, A. (1995). Comparison of tests and sample size formulae for proving therapeutic equivalence based on the difference of binomial probabilities. *Stat. Med.* 14: 1583–1594.

Rogers, M.E. (1970). *An Introduction to the Theoretical Basis of Nursing.* Philadelphia: Davis.

Rondel, R.K., Varley, S.A., and Webb, C.F. (eds.) (1993). *Clinical Data Management.* Chichester: Wiley.

Rose, G. (1992). *The Strategy of Preventive Medicine.* Oxford: Oxford University Press.

Rosell, R., Gomez-Codina, J., Camps, C., et al. (1994). A randomized trial comparing preoperative chemotherapy plus surgery with surgery alone in patients with non–small-cell lung cancer. *N. Engl. J. Med.* 330: 153–158.

Rossiter, E.J.R. (1992). Reflections of a whistle-blower. *Nature* 357: 434–436.

Rothenberg, M.L., Moore, M.J., Cripps, M.C., et al. (1996). A phase II trial of gemcitabine in patients with 5-FU-refractory pancreas cancer. *An. Oncol.* 7: 347–353.

Rothman, K.J., and Michels, K.B. (1994). The continuing unethical use of placebo controls. *N. Engl. J. Med.* 331: 394–398.

Roy, D.J. (1986). Controlled clinical trials: An ethical imperative. *J. Chron. Dis.* 39: 159–162.

Royal College of Physicians (1991). Fraud and misconduct in medical research. Causes, investigations and prevention. *J. Roy. Coll. Physicians, London* 25: 89–94.

Royal Statistical Society (1993). *The Royal Statistical Society: Code of Conduct.*

Royall, R.M. (1986). The effect of sample size on the meaning of significance tests. *Am. Statistician* 40: 313–315.

Royall, R.M., Bartlett, R.H., Cornell, R.G., et al. (1991). Ethics and statistics in randomized clinical trials. *Stat. Sci.* 6: 52–88.

Royall, R.M. (1992). Ignorance and altruism. *J. Clin. Ethics* 3(3): 229–230.

Royall, R.M. (1997). *Statistical Evidence: A Likelihood Paradigm.* London: Chapman and Hall.

Rozovsky, F.A. (1990). *Consent to Treatment: A Practical Guide.* Boston: Little, Brown.

Ruberg, S.J. (1995a). Dose response studies. I: Some design considerations. *J. Biopharmaceut. Stat.* 5: 1–14.

Ruberg, S.J. (1995b). Dose response studies. II: Analysis and interpretation. *J. Biopharmaceut. Stat.* 5: 15–42.

Rubin, D.B. (1974). Estimating causal effects of treatments in randomized and non-randomized studies. *J. Educ. Psychol.* 66: 688–701.

Rubin, D.B. (1990). Comment: Neyman (1923) and causal inference in experiments and observational studies. *Stat. Sci.* 5: 472–480.

Rubin, D.B. (1991). Practical implications of modes of statistical inference for causal effects and the critical role of the assignment mechanism. *Biometrics* 47: 1213–1234.

Rubin, D.B., and Schenker, N. (1991). Multiple imputation in healthcare databases: an overview and some applications. *Stat. Med.* 10: 585–598.

Rubinow, S.I. (1975). *Introduction to Mathematical Biology*. New York: Wiley.

Rubinstein, L.V., Gail, M.H., and Santner, T.J. (1981). Planning the duration of a comparative clinical trial with loss to follow-up and a period of continued observation. *J. Chron. Dis.* 34: 469–479.

Rubinstein, L.V., and Gail, M.H. (1982). Monitoring rules for stopping accrual in comparative survival studies. *Cont. Clin. Trials* 3: 325–343.

Rudick, R., et al. (2003). Interferons in relapsing remitting multiple sclerosis (letter). *Lancet* 361: 1824.

Ruffin, J.M., Grizzle, J.E., Hightower, N.C., McHardy, G., Shull, H., et al. (1969). A co-operative double-blind evaluation of gastric "freezing" in the treatment of duodenal ulcer. *N. Engl. J. Med.* 281: 16–19.

Rusch, V.W., Piantadosi, S., and Holmes, E.C. (1991). The role of extrapleural pneumonectomy in malignant pleural mesothelioma. *J. Thoracic. Cardiovasc. Surg.* 102: 1–9.

Rutstein, D. (1969). The ethical design of human experiments. *Daedalus* 98: 523–541.

Rutter, C.M., and Elashoff, R.M. (1994). Analysis of longitudinal data: Random coefficient regression modeling. *Stat. Med.* 13: 1211–1231.

Sackett, D.L. (1979). Bias in analytic research. *J. Chron. Dis.* 32: 51–63.

Sackett, D.L. (1983). On some prerequisites for a successful clinical trial. In S.H. Shapiro and T.A. Louis (eds.), *Clinical Trials*. New York: Marcel Dekker.

Sackett, D.L. (2000). Why randomized controlled trials fail but needn't: 1. Failure to gain "coal-face" commitment and to use the uncertainty principle. *CMAJ* 162: 1311–1314.

Sackett, D.L., and Gent, M. (1979). Controversy in counting and attributing events in clinical trials. *N. Engl. J. Med.* 301(26): 1410–1412.

Sadeghi, A., Payne, D., Rubinstein, L., and Lad, T. (1988). Combined modality treatment for resected advanced non–small cell lung cancer: Local control and local recurrence. *Int. J. Radiat. Oncol. Biol. Phys.* 15: 89–97.

Sahai, H., and Khurshid, A. (1996). Formulae and tables for the determination of sample sizes and power in clinical trials for testing differences in proportions for the two-sample design: A review. *Stat. Med.* 15: 1–21.

Salsburg, D. (1990). Hypothesis versus significance testing for controlled clinical trials: A dialogue. *Stat. Med.* 9: 201–211.

SAS Institute (1999). *SAS/STAT User's Guide, Version 8*. Cary, NC: SAS Institute.

SAS Institute (2004). *Getting Started with the SAS Power and Sample Size Application*. Cary, NC: SAS Institute.

Sateren, W., Trimble, E.L., Abrams, J., et al. (2002). How sociodemographics, presence of oncology specialists, and hospital cancer programs affect accrual to cancer treatment trials. *J. Clin. Oncol.* 20: 2109–2117.

Savage, L.J. (1954). *The Foundations of Statistics*. New York: Wiley.

Schafer, A. (1982). The ethics of the randomized clinical trial. *N. Engl. J. Med.* 307: 719–724.

Schechtman, K.B. and Gordon, M.O. (1994). The effect of poor compliance and treatment side effects on sample size requirements in randomized clinical trials. *J. Biopharmaceut. Stat.* 4: 223–232.

Scheiber, B. and Selby, C. (eds.) (2000). *Therapeutic Touch*. Amherst: Prometheus Books.

Schellhammer, P., Sharifif, R., Block, N., Soloway, M., Venner, P., Patterson, A.L., Sarosdy, M., Vogelzang, N., Jones, J., and Kiovenbag, G. (1995). A controlled trial of bicalutamide versus flutamide, each in combination with lutenizing hormone-releasing hormone analogue therapy, in patients with advanced prostate cancer. Casodex Combination Study Group. *Urology* 45(5): 745–752.

Schiermeier, Q. (2002). German task force outraged by changes to science fraud report. *Nature* 415: 3.

Schmucker, D.L., and Vesell, E.S. (1993). Underrepresentation of women in clinical drug trials. *Clin. Pharmacol. Ther.* 54: 11–15.

Schulman, K.A., Seils, D.M., Timbie, J.W., et al. (2002). A national survey of provisions in clinical-trial agreements between medical schools and industry sponsors. *N. Engl. J. Med.* 347(17): 1335–1341.

Schulz, K.F. (1995). Subverting randomization in controlled trials. *JAMA* 274: 1456–1458.

Schulz, K.F., Chalmers, I., Hayes, R.J., and Altman, D.G. (1995). Empirical evidence of bias: Dimensions of methodologic quality associated with estimates of treatment effects in controlled trials. *JAMA* 273: 408–412.

Schulz, S.C., van Kammen, D.P., Balow, J.E., Flye, M.W., and Bunney, W.E., Jr. (1981). Dialysis in schizophrenia: A double-blind evaluation. *Science* 211: 1066–1068.

Schuman, S.H., Olansky, S., Rivers, E., Smith, C.A., and Rambo, D.S. (1955). Untreated syphilis in the male negro. Background and current status of patients in the Tuskegee Study. *J. Chron. Dis.* 2: 543–558.

Schwartz, D., and Lellouch, J. (1967). Explanatory and pragmatic attitudes in therapeutic trials. *J. Chron. Dis.* 20: 637–648.

Schwartz, D., Flamant, R., and Lellouch, J. (1980). *Clinical Trials.* London: Academic Press.

Schwartz, R.S. (2001). Racial profiling in medical research. *N. Engl. J. Med.* 344: 1392–1393.

Scott, P.E. (2004). Medical device approvals: An assessment of the level of evidence. Ph.D. dissertation: Johns Hopkins University, Baltimore, MD.

Sechzer, J.A., Rabinowitz, V.C., Denmark, F.L., McGinn, M.F., Weeks, B.M., and Wilkens, C.L. (1994). Sex and gender bias in animal research and in clinical studies of cancer, cardiovascular disease, and depression. *An. N.Y. Acad. Sci.* 736: 21–48.

Segal, M.R. (1988). Regression trees for censored data. *Biometrics* 44: 35–47.

Selby, C., and Scheiber, B. (1996). Science or pseudoscience? Pentagon grant funds alternative health study. *Skeptical Inq.* 20: 15–17.

Seleznick, M.J., and Fries, J.F. (1991). Variables associated with decreased survival in systemic lupus erythematosus. *Sem. Arthritis Rheum.* 21: 73–80.

Senn, S. (1993). *Cross-over Trials in Clinical Research.* Chichester: Wiley.

Shalala, D. (2000). Protecting research subjects—What must be done. *N. Engl. J. Med.* 343: 808–810.

Shapiro, B.S., Venet, W., Strax, P., and Venet, L. (1988). *Periodic Screening for Breast Cancer: The Health Insurance Plan Project and Its Sequelae, 1963–1986.* Baltimore: Johns Hopkins University Press.

Shapiro, S. (1994). Meta-analysis/shmeta-analysis. *Am. J. Epidemiol.* 140: 771–778.

Shapiro, M.F., and Charrow, R.P. (1989). The role of data audits in detecting scientific misconduct. *JAMA* 261: 2505–2511.

Shapiro, A.K. (1964). A historic and heuristic definition of the placebo. *Psychiatry* 27:52–58.

Shapiro, A.K. and Shapiro, E. (1997). *The Powerful Placebo: From Ancient Priest to Modern Physician.* Baltimore: Johns Hopkins University Press.

Shapiro, H.T., and Meslin, E.M. (2001). Ethical issues in the design and conduct of clinical trials in developing countries. *N. Engl. J. Med.* 345: 139–142.

Sheiner, L.B., and Rubin, D.B. (1995). Intention-to-treat analysis and the goals of clinical trials. *Clin. Pharmacol. Therapeut.* 57(1): 6–15.

Shetty, N., Friedman, J.H., Kieburtz, K., Marshall, F.J., Oakes, D., and the Parkinson Study Group. (1999). The placebo response in Parkinson's disease. *Clin Neuropharmacol.* 22(4): 207–212.

Shimkin, M.B. (1953). The problem of experimentation on human beings: I The research worker's point of view. *Science* 117: 205–207.

Short, T.G., Ho, T.Y., Minto, C.F., Schnider, T.W., Shafer, S.L. (2002). Efficient trial design for eliciting a pharmacokinetic-pharmacodynamic model-based response surface describing the interaction between two intravenous anesthetic drugs. *Anesthesiology* 96(2):400–408.

Shuster, J.J. (1990). *Handbook of Sample Size Guidelines for Clinical Trials.* Boca Raton, FL: CRC Press.

Shuster, J.J., McWilliams, N.B., Castleberry, R., Nitschke, R., Smith, E.I., Altshuler, G., Kun, L., Brodeur, G., Joshi, V., Vietti, T., and Hayes, F.A. (1992). Serum lactate dehydrogenase in childhood neuroblastoma. *Am. J. Clin. Oncol.* 15: 295–303.

Sieber, J.E. (1993). Ethical considerations in planning and conducting research on human subjects. *Acad. Med.* (suppl.) 68: S9–S13.

Siegler, M. (1982). Confidentiality in medicine—A decrepit concept. *N. Engl. J. Med.* 307: 1518–1521.

Silverman, W.A. (1985). *Human Experimentation: A Guided Step into the Unknown.* Oxford: Oxford University Press.

Silverman, W.A. (1998). *Where's the Evidence? Debates in Modern Medicine.* Oxford: Oxford University Press.

Simmons, G.F. (1972). *Differential Equations with Applications and Historical Notes.* New York: McGraw-Hill.

Simmons, R.L., Polk, H.C., Jr., Williams, B., and Mavroudis, C. (1991). Misconduct and fraud in research: social and legislative issues symposium of the Society of University Surgeons. *Surgery* 110(1): 1–7.

Simon, J.L. (2004). *Resampling Stats software.* Resampling Stats Inc.

Simon, R. (1982). Patient subsets and variation in therapeutic efficacy. *Br. J. Clin. Pharmac.* 14: 473–482.

Simon, R. (1984). Use of regression models: Statistical aspects. In M.J. Buyse, M.J. Staquet, and R.J. Sylvester (eds.), *Cancer Clinical Trials.* Oxford: Oxford University Press, ch. 25.

Simon, R. (1989). Optimal two-stage designs for phase II clinical trials. *Cont. Clin. Trials* 10: 1–10.

Simon, R. (1991). A decade of progress in statistical methodology for clinical trials. *Stat. Med.* 10: 1789–1817.

Simon, R., Freidlin, B., Rubinstein, L., et al. (1997). Accelerated titration designs for phase I clinical trials in oncology. *J. Nat. Cancer Inst.* 89: 1138–1147.

Simon, R., Wittes, R.E., and Ellenberg, S.S. (1985). Randomized phase II clinical trials. *Cancer Treat. Rep.* 69: 1375–1381.

Simons, J.W., Jaffee, E.M., Weber, C., et al. (1997). Bioactivity of human GM-CSF gene transfer in autologous irradiated renal cell carcinoma vaccines. *Cancer Res.* 57(8): 1537–1546.

Singer, M. (1996). Assault on science. *Washington Post*, June 26.

Skolnick, A. (1990). Key witness against morning sickness drug faces scientific fraud charges. *JAMA* 263: 1468–1469, 1473.

Slud, E.V. (1994). Analysis of factorial survival experiments. *Biometrics* 50: 25–38.

Smith, G.C., and Pell, J.P. (2003). Parachute use to prevent death and major trauma related to gravitational challenge: Systematic review of randomised controlled trials. *BMJ* 327: 1459–1461.

Snedecor, G.W. and Cochran, W.G. (1980). *Statistical Methods*, 7th ed. Ames: Iowa State University Press.

Sokal, A.D. (1996a). Transgressing the boundaries: Toward a transformative hermeneutics of quantum gravity. *Social Text* 46/47: 217–252.

Sokal, A.D. (1996b). A physicist experiments with cultural studies. *Lingua Franca* (May/June): 62–64.

Solomon, M.J., and McLeod, R.S. (1998). Surgery and the randomised controlled trial: Past, present and future. *Med. J. Aust.* 169: 380–383.

Sommer, A. Tarwotjo, I., Djunaedi, E., West, K.P., Jr., and Loeden, A.A. (1986). Impact of vitamin A supplementation on childhood mortality: A randomized controlled community trial. *Lancet* 1: 1169–1173.

Sommer, A., and Zeger, S.L. (1991). On estimating efficacy from clinical trials. *Stat. Med.* 10: 45–52.

Souhami, R.(1992). Large-scale studies. In C.J. Williams (ed.), *Introducing New Treatments for Cancer: Practical Ethical and Legal Problems.* Chichester: Wiley, ch.13.

Souhami, R.L., and Whitehead, J. (eds.) (1994). Workshop on Early Stopping Rules in Cancer Clinical Trials, Robinson College, Cambridge, U.K., 13-15 April, 1993. *Stat. Med.* 13: 1289–1500.

Sparber, A., Bauer, L., Curt, G., et al. (2000). Use of complementary medicine by adult patients participating in cancer clinical trials. *Oncol Nurs. Forum* 27(4): 623–630.

Sparber, A., Wootton, J.C., Bauer, L., et al. (2000). Use of complementary medicine by adult patients participating in HIV/AIDS clinical trials. *J. Altern. Compl. Med.* 6(5):415–422.

Spector, S.A., McKinley, G.F., Lalezari, J.P., et al., (1996). Oral ganciclovir for the prevention of cytomegalovirus disease in persons with AIDS. Roche Cooperative Oral Ganciclovir Study Group. *N. Engl. J. Med.* 334(23): 1491–1497.

Sperduto, R.D., Hu, T.S., Milton, R.C., et al. (1993). The Linxian cataract studies. Two nutrition intervention trials. *Arch. Ophthalmol.* 111: 1246–1253.

Spiegel, D., Kraemer, H., Carlson, R.W., et al. (2001). Is the placebo powerless? *N. Engl. J. Med.* 345: 1276–1279.

Spiegelhalter, D.J., Freedman, L.S., and Parmar, M.K.B. (1994). Bayesian approaches to randomized trials. *J. Roy. Statist. Soc.* A157: 357–416.

Spilker, W.A. (1991). *Guide to Clinical Trials.* New York: Raven Press.

SPLUS 6 for Windows User's Guide (2001). Seattle, WA: Insightful Corporation.

Spodick, D.H. (1983). Randomize the first patient: scientific, ethical, and behavioral bases. *Am. J. Cardiol.* 51: 916–917.

Sposto, R., and Krailo, M.D. (1987). Use of unequal allocation in survival trials. *Stat. Med.* 6: 119–125.

Staden, H. von (1992). The discovery of the human body: Dissection and its cultural contexts in ancient Greece. *Yale J. Biol. Med.* 65: 223–241.

Stampfer, M.J., Buring, J.E., Willett, W., et al. (1985). The 2×2 factorial design: Its application to a randomized trial of aspirin and carotene in U.S. physicians. *Stat. Med.* 4: 111–116.

Stampfer, M., and Colditz, G. (1991). Estrogen replacement therapy and coronary heart disease: A quantitative assessment of the epidemiologic evidence. *Prev. Med.* 20: 47–63.

Standards of Reporting Trials Group. (1994). A proposal for structured reporting of randomized controlled trials. *JAMA* 272: 1926–1931; Correction: 273: 776.

Stansfield, S.K., Pierre-Louis, M., Lerebours, G., and Augustin, A. (1993). Vitamin A supplementation and increased prevalence of childhood diarrhoea and acute respiratory infections. *Lancet* 341: 578–582.

Starr, M., and Chalmers, I. (2004). *The Evolution of The Cochrane Library, 1988–2003*. Update Software, Oxford (www.update-software.com/history/clibhist.htm).

Statistics and Epidemiology Research Corporation (SERC). (1993). *EGRET SIZ Reference Manual*. Seattle: SERC.

Stebbing, L.S. (1961). *Philosophy and the Physicist*. Middlesex, UK: Penguin.

Steering Committee of the Physicians' Health Study Research Group (1989). Final report on the aspirin component of the ongoing Physicians' Health Study. *N. Engl. J. Med.* 321(3): 129–135.

Stein, C.M. (2002). Are herbal products dietary supplements or drugs? An important question for public safety. *Clin. Pharmacol. Ther.* 71(6): 411–413.

Steinbrook, R. (2003a) How best to ventilate? Trial design and patient safety in studies of the acute respiratory distress syndrome. *N. Engl. J. Med.* 348: 1393–1401.

Steinbrook, R. (2003b). Trial design and patient safety—The debate continues. *N. Engl. J. Med.* 349: 629–630.

Steinbrook, R. (2004a). Public registration of clinical trials. *N. Engl. J. Med.* 351: 315–317.

Steinbrook, R. (2004b). Registration of clinical trials—Voluntary or mandatory? *N. Engl. J. Med.* 351: 1820–1822.

Stewart, L.A., and Clarke, M.J. (1995). Practical methodology of meta-analyses (overviews) using updated individual patient data. *Stat. Med.* 14: 2057–2079.

Stevens, P., Jr. (2001). Magical thinking in complementary and alternative medicine. *Skeptical Inq.* 25: 32–37.

Stigler, S. (1986). *The History of Statistics*. Cambridge, MA: Belknap Press.

Storer, B.E. (1989). Design and analysis of phase I clinical trials. *Biometrics* 45: 925–937.

Strauss, M.B. (1968). *Familiar Medical Quotations*. Boston: Little, Brown, p. 492.

Stuart, A. and Ord, J.K. (1987). *Kendall's Advanced Theory of Statistics, Vol. 1: Distribution Theory*, 5th ed. New York: Oxford University Press.

Stuart, A., and Ord, J.K. (1991). *Kendall's Advanced Theory of Statistics, Vol. 2: Classical Inference and Relationship*, 5th ed. New York: Oxford University Press.

Sullivan, M. (2004) Placebos and treatment of pain. *Pain Med.* 5(3): 325–326.

Sullivan, M.D. (1993). Placebo controls and epistemic control in orthodox medicine. *J. Med. Philos.* 18(2): 213–231.

Swazey, J.P., Anderson, M.S., and Louis, K.S. (1993). Ethical problems in academic research. *Am. Sci.* 81: 542–553.

Tartter, P.I., Burrows, L., Kirschner, P. (1984). Perioperative blood transfusion adversely affects prognosis after resection of stage I (subset N0) non–oat cell lung cancer. *J. Thorac. Cardvasc. Surg.* 88: 659–662.

Task Force of the Working Group on Arrhythmias of the European Society of Cardiology (1994). The early termination of clinical trials: Causes, consequences, and control—with special reference to trials in the field of arrhythmias and sudden death. *Circulation* 89: 2892–2907.

Taylor, J.M.G., Cumberland, W.G., and Sy, J.P. (1994). A stochastic model for longitudinal AIDS data. *J. Am. Statist. Assoc.* 89: 727–736.

Taylor, K.M., Margolese, R.G., and Soskolne, C.L. (1984). Physicians' reasons for not entering eligible patients in a randomized clinical trial of surgery for breast cancer. *N. Engl. J. Med.* 310: 1363–1367.

ten Bokkel Huinink, W.W., Eisenhaur, E., and Swenerton, K. (1993). Preliminary evaluation of a multicenter, randomized comparative study of TAXOL (paclitaxel) dose and infusion length in platinum-treated ovarian cancer. *Cancer Treat. Rev.* 19 (suppl. C): 79–86.

Terrin, M.L. (1990). Efficient use of endpoints in clinical trials: A clinical perspective. *Stat. Med.* 9: 155–160.

Testa, M.A., and Simonson, D.C. (1996). Assessment of quality of life outcomes. *N. Engl. J. Med.* 334: 835–840.

Thall, P.F., and Simon, R.M. (1994). A Bayesian approach to establishing sample size and monitoring criteria for phase II clinical trials. *Cont. Clin. Trials* 15: 463–481.

Thall, P.F., Simon, R.M., and Estey, E.H. (1995). Bayesian sequential monitoring designs for single-arm clinical trials with multiple outcomes. *Stat. Med.* 14: 357–379.

Thall, P.F., Simon, R.M., and Estey, E.H. (1996). New statistical strategy for monitoring safety and efficacy in single-arm clinical trials. *J. Clin. Oncol.* 14: 296–303.

Thelen, M.H., and DiLorenzo, T.M. (1992). Academic pressures. In D.J. Miller and M. Hersen (eds.), *Research Fraud in the Behavioral and Biomedical Sciences*. New York: Wiley, ch. 9.

Thomas, L. (1977). Biostatistics in medicine. (editorial). *Science* 198(4318): 675.

Thornquist, M.D., Owenn, G.S., Goodman, G.E., et al. (1993). Statistical design and monitoring of the carotene and retinol efficacy trial (CARET). *Contr. Clin. Trials* 14: 308–324.

Tonelli, M.R., and Callahan, T.C. (2001). Why alternative medicine cannot be evidence-based. *Acad. Med.* 76: 1213–1220.

Topol, E.J. (2004). Failing the public health—Rofecoxib, Merck, and the FDA. *N. Engl. J. Med.* 351: 1707–1709.

Toronto Leukemia Study Group (1986). Results of chemotherapy for unselected patients with acute myeloblastic leukaemia: Effect of exclusions on interpretation of results. *Lancet* 1: 786–788.

Tuffs, A. (2002). Cancer specialist found guilty of misconduct. *BMJ* 325: 1193.

Tukey, J.W. (1977). Some thoughts on clinical trials, especially problems of multiplicity. *Science* 198: 679–684.

Turner, J.G. (1994). The effect of therapeutic touch on pain and infection in burn patients (N94-020A1). Grant No. MDA 905-94-Z-0080, Uniformed Services University of the Health Sciences.

Turner, J.G., Clark, A.J., Gauthier, D.K., and Williams, M. (1998). The effect of therapeutic touch on pain and anxiety in burn patients. *J. Adv. Nurs.* 28: 10–20.

Tygstrup, N., Lachin, J.M., and Juhl, E. (eds.) (1982). *The Randomized Clinical Trial and Therapeutic Decisions*. New York: Marcel Dekker.

Urbach, P. (1993). The value of randomization and control in clinical trials. *Stat. Med.* 12: 1421–1431.

U.S. Congress (1938). Federal Food, Drug, and Cosmetic Act of 1938. Public Law Number 75-717, 52 Stat. 1040 (1938), 21 USC 201.

U.S. Congress (1980). *The Forgotten Guinea Pigs; A Report on the Health Effects of Low Level Radiation Sustained as a Result of the Nuclear Weapons Testing Program Conducted by the United States Government*. Washington, DC: Government Printing Office.

U.S. Government (1949). *Trials of War Criminals before the Nuremberg Military Tribunals under Control Council Law No. 10*, Vol. 2. Washington, DC: U.S. Government Printing Office, pp. 181–182.

USA Today (1996). Nationline: Full Disclosure. Monday, December 16.

Valtonen, S., Timonen, U., Toivanen, P., et al. (1997). Interstitial chemotherapy with carmustine-loaded polymers for high-grade gliomas: A randomized, double-blind study. *Neurosurgery* 41(1): 44–48; Discussion 48–49.

Varmus, H., and Satcher, D. (1997). Ethical complexities of conducting research in developing countries. *N. Engl. J. Med.* 337: 1003–1005.

Verweij, J. (1996). The benefit of clinical benefit: A European perspective. *An. Oncol.* 7: 333–334.

Verweij, P.J.M., and Van Houwelingen, H.C. (1994). Penalized likelihood in Cox regression. *Stat. Med.* 13: 2427–2436.

Villanueva, C., Balanzo, J., Novella, M.T., et al., (1996). Nadolol plus isosorbide mononitrate compared with sclerotherapy for the prevention of variceal rebleeding. *N. Engl. J. Med.* 334: 1624–1629.

Vogel, G., and Bostanci, A. (2002). German inquiry finds flaws, not fraud. *Science* 298: 1531–1533.

Vollset, S.E. (1993). Confidence intervals for a binomial proportion. *Stat. Med.* 12: 809–824.

Wade, N. (1981). A diversion of the quest for truth. *Science* 211: 1022–1025.

Wald, A. (1947). *Sequential Analysis*. New York: Wiley.

Waldhausen, J.A., and Localio, A.R. (1996). Notes from the editors. *J. Thorac. Cardiovasc. Surg.* 112: 209–220.

Walter, S.D., and Day, N.E. (1983). Estimation of the duration of a preclinical disease state using screening data. *Am. J. Epidemiol.* 118: 865–886.

Walter, S.D. (1995). Methods of reporting statistical results from medical research studies. *Am. J. Epidemiol.* 141: 896–906.

Walters, D.E. (1985). An examination of the conservative nature of "classical" confidence for a proportion. *Biom. J.* 27: 851–861.

Walters, D.E. (1997). Confidence limits and log-odds ratios. *Statistician* 46(3): 433–438.

Ward, J. (1995). Phase I clinical trials. *Appl. Clin. Trials* 4(7): 44–47.

Ware, J.H. (1989). Investigating therapies of potentially great benefit: ECMO (with discussion). *Stat. Sci.* 4: 298–340.

Ware, J.H. and Antman, E.M. (1997). Equivalence trials. *N. Engl. J. Med.* 337: 1159–1161.

Ware, J.H., and Epstein, M.F. (1985). Extracorporeal circulation in neonatal respiratory failure: A prospective randomized study (commentary). *Pediatrics* 76: 849–851.

Warner, B., and Misra, M. (1996). Understanding neural networks as statistical tools. *Am. Statistiscian* 50(4): 284–293.

Weber, W.W. (1987). *The Acetylator Genes and Drug Response*. New York: Oxford University Press.

Weeks, J.C., Nelson, H., Gelber, S., Sargent, D., and Schroeder, G. for the Clinical Outcomes of Surgical Therapy (COST) Study Group. (2002). Short-term quality-of-life outcomes following laparoscopic-assisted colectomy vs. open colectomy for colon cancer: a randomized trial. *JAMA* 287: 321–328.

Wei, L.J. (1988). Exact two-sample permuatation tests based on the randomized play-the-winner rule. *Biometrika* 75: 603–606.

Wei, L.J., and Lachin, J.M. (1988). Properties of the urn randomization in clinical trials. *Cont. Clin. Trials* 9: 345–364.

Wei, L.J., and Durham, S. (1978). The randomized play-the-winner rule in medical trials. *J. Am. Stat. Assoc.* 73: 840–843.

Weijer, C., and Fuks, A. (1994). The duty to exclude: excluding people at undue risk from research. *Clin. Invest. Med.* 17: 115–122.

Weinmann, G.G., and Hyatt, R. (1996). Evaluation and research in lung volume reduction surgery. *Am. J. Resp. Crit. Care. Med.* 154: 1013–1018.

Weiss, R. (1996). Proposed shifts in misconduct reviews unsettle many scientists. *Washington Post*, June 30.

Weiss, R.B., Vogelzang, N.J., Peterson, B.A., et al. (1993). A successful system of scientific data audits for clinical trials: A report from the Cancer and Leukemia Group B. *JAMA* 270: 459–464.

Wells, F. (1992) Good clinical research practice. In C.J. Williams (ed.), *Introducing New Treatments for Cancer: Practical, Ethical, and Legal Problems*, New York: Wiley, ch. 19.

Westphal, M., Hilt, D.C., Bortey, E., et al. (2003). A phase 3 trial of local chemotherapy with biodegradable carmustine (BCNU) wafers (Gliadel wafers) in patients with primary malignant glioma. *Neuro-oncol.* 5(2): 79–88.

Wetherill, G.B. (1963). Sequential estimation of quantal response curves (with discussion). *JRSS* B25: 1–48.

Weymuller, E.A., and Goepfert, H. (1991). Uniformity of results reporting in head and neck cancer. *Head Neck* 13: 275–277. Reprinted in *Laryngoscope* 104: 784–785.

Whitehead, J. (1992). *The Design and Analysis of Sequential Clinical Trials*, 2nd ed. New York: Ellis Horwood.

Whitehead, J. (1993). The case for frequentism in clinical trials. *Stat. Med.* 12: 1405–1413.

Whitehead, J. (1994). Sequential methods based on the boundaries approach for the clinical comparison of survival times. *Stat. Med.* 13: 1357–1368.

Whitehead, J., Zhou, Y., Stallard, N., Todd, S., and Whitehead, A. (2001). Learning from previous responses in phase I dose-escalation studies. *Br. J. Clin. Pharmacol.* 52(1): 1–7.

Whittier, F.C., Evans, D.H., Anderson, P.C., and Nolph, K.D. (1983). Peritoneal dialysis for psoriasis: A controlled study. *An. Intern. Med.* 99: 165–168.

Whitworth, J.A., Morgan, D., Maude, G.H., Luty, A.J., and Taylor, D.W. (1992). A community trial of ivermectin for onchocerciasis in Sierra Leone: Clinical and parisitological responses to four doses given at six-monthly intervals. *Trans. Roy. Soc. Trop. Med. Hyg.* 86: 277–280.

Wigner, E. (1959). The Unreasonable Effectiveness of Mathematics in the Natural Sciences. Richard Courant Lecture in Mathematical Sciences delivered at New York University in 1959. Reprinted in *Symmetries and Reflections: Scientific Essays of Eugene P. Wigner*. Bloomington: Indiana University Press, 1967, pp. 222–237. (*Communicat on Pure and Applied Mathematics*, vol. 13, pp. 1–14 (no. 1 Feb. 1960)).

Wilde, D.J. (1964). *Optimum Seeking Methods*. Englewood Cliffs, NJ: Prentice-Hall.

Willan, A., and Pater, J. (1986). Carryover and the two-period cross-over clinical trial. *Biometrics* 42: 593–599.

Willan, A., and Pater, J. (1986). Using baseline measurements in the two-period cross-over clinical trial. *Cont. Clin. Trials* 7: 282–289.

Williams, C.J. (ed.) (1992). *Introducing New Treatments for Cancer: Practical, Ethical, and Legal Problems*. Chichester: Wiley.

Williams, D.H., and Davis, C.E. (1994). Reporting of assignment methods in clinical trials. *Cont. Clin. Trials* 15: 294–298.

Wilson, E.O. (1992). *The Diversity of Life*. New York: Norton.

Wilson, E.O. (1998). *Consilience: The Unity of Knowledge*. New York: Knopf.

Wilson, F.R. (1998). *The Hand*. New York: Pantheon.

Winer, B.J. (1962). *Statistical Principles in Experimental Design*, 2nd ed. New York: McGraw-Hill.

Winer, B.J. (1971). *Statistical Principles in Experimental Design*, 2nd ed. New York: McGraw-Hill.

Winget, M.D. (1996). Selected issues related to the conduct, reporting, and analysis of phase I trials. Ph.D. dissertation, Johns Hopkins University.

Wittes, J., Lakatos, E., and Probstfield, J. (1989). Surrogate endpoints in clinical trials: Cardiovascular diseases. *Stat. Med.* 8: 415–425.

Wittes, R.E. (1988). Of clinical alerts and peer review. *JNCI* 80: 984–985.

Wolfram, S. (2003). *The Mathematica Book*, 5th ed. Wolfram Media.

WHI Investigators Writing Group. (2002). Risks and benefits of estrogen plus progestin in healthy postmenopausal women. *JAMA* 288: 321–333.

Wong, W.K., and Lachenbruch, P.A. (1996). Tutorial in biostatistics: Designing studies for dose response. *Stat. Med.* 15: 343–359.

Wooding, W.M. (1994). *Planning Pharmaceutical Clinical Trials*. New York: Wiley.

Woodward, B. (1999). Challenges to human subject protections in US medical research. *JAMA* 282: 1947–1952.

Woolf, P.K. (1981). Fraud in science: How much, how serious? *Hastings Center Rep.* 11 (October): 9–14.

Woolf, P.K. (1986). Pressure to publish and fraud in science. *An. Intern. Med.* 104: 254–256.

Woolf, P.K. (1988). Deception in scientific research. *Jurimetrics* 29: 67–95.

Working Group on Recommendations for Reporting of Clinical Trials in the Biomedical Literature (1994). Call for comments on a proposal to improve reporting of clinical trials in the biomedical literature. *An. Intern. Med.* 121: 894–895.

World Medical Association (1964). *Declaration of Helsinki: Recommendations Guiding Medical Doctors in Biomedical Research Involving Human Subjects* (rev. 1975, 1983, and 1989). Helsinki: World Medical Association. See also *Br. Med. J.* 2: 177 (1964).

Wu, M., Fisher, M., and DeMets, D. (1980). Sample sizes for long-term medical trial with time-dependent dropout and event rates. *Cont. Clin. Trials* 1: 109–121.

Yates, F. (1935). Complex experiments (with discussion). *J. Roy. Statist. Soc. B* 2: 181–247.

Yates, F. (1984). Tests of significance for 2×2 contingency tables (with discussion). *J. Roy. Statist. Soc.* A147: 426–463.

Young, J.H. (1983). Food and Drug Administration. In Donald R. Whitnah (ed.), *Government Agencies*. Westport, CT.: Greenwood Press, pp. 251–257.

Yuan, C.-S., and Bieber, E.J. (2003). *Textbook of Complementary and Alternative Medicine*. Boca Raton: Parthenon.

Yusuf, S., Collins, R., and Peto, R. (1984). Why do we need some large, simple, randomized trials? *Stat. Med.* 3: 409–420.

Yusuf, S., Held, P., Teo, K.K., and Toretsky, E.R. (1990). Selection of patients for randomized controlled trials: Implications of wide or narrow eligibility criteria. *Stat. Med.* 9: 73–86.

Yusuf, S. (1995). Calcium antagonists in coronary artery disease and hypertension: Time for reevaluation? *Circulation* 92: 1079–1082.

Zeger, S.L., Liang, K.Y., and Albert, P. (1988). Models for longitudinal data: A generalized estimating equation approach. *Biometrics* 44: 1049–1060.

Zeger, S.L., and Liang, K.Y. (1986). Longitudinal data analysis for discrete and continuous outcomes. *Biometrics* 42: 121–130.

Zelen, M. (1969). Play the winner rule and the controlled clinical trial. *J. Am. Stat. Assoc.* 64: 131–146.

Zelen, M. (1979). A new design for randomized clinical trials. *N. Engl. J. Med.* 300: 1242–1245.

Zelen, M. (1983). Guidelines for publishing papers on cancer clinical trials: Responsibilities of editors and authors. *J. Clin. Oncol.* 1: 164–169.

Zelen, M. (1990). Randomized consent designs for clinical trials: An update. *Stat. Med.* 9: 645–656.

Zoubek A., Kajtar P., Flucher-Wolfram B., Holzinger B., Mostbeck G., Thun-Hohenstein L., Fink F.M., Urban C., Mutz I., Schuler D., et al. (1995). Response of untreated stage IV Wilms' tumor to single dose carboplatin assessed by "up front" window therapy. *Med. Pediatr. Oncol.* 25(1): 8–11.

AUTHOR INDEX

Clinical Trials: A Methodologic Perspective, 2E, by S. Piantadosi
Copyright © 2005 John Wiley & Sons, Inc.

SUBJECT INDEX